Nanofluids and Mass Transfer

Nanofluids and Mass Transfer

Edited by

Mohammad Reza Rahimpour
Department of Chemical Engineering, Shiraz University, Shiraz, Iran

Mohammad Amin Makarem
Department of Chemical Engineering, Shiraz University, Shiraz, Iran

Mohammad Reza Kiani
Fouman Faculty of Engineering, College of Engineering, University of Tehran, Fouman, Iran

Mohammad Amin Sedghamiz
Department of Chemical Engineering, Shiraz University, Shiraz, Iran

Elsevier
Radarweg 29, PO Box 211, 1000 AE Amsterdam, Netherlands
The Boulevard, Langford Lane, Kidlington, Oxford OX5 1GB, United Kingdom
50 Hampshire Street, 5th Floor, Cambridge, MA 02139, United States

Notices

Knowledge and best practice in this field are constantly changing. As new research and experience broaden our understanding, changes in research methods, professional practices, or medical treatment may become necessary.

Practitioners and researchers must always rely on their own experience and knowledge in evaluating and using any information, methods, compounds, or experiments described herein. In using such information or methods they should be mindful of their own safety and the safety of others, including parties for whom they have a professional responsibility.

To the fullest extent of the law, neither the Publisher nor the authors, contributors, or editors, assume any liability for any injury and/or damage to persons or property as a matter of products liability, negligence or otherwise, or from any use or operation of any methods, products, instructions, or ideas contained in the material herein.

British Library Cataloguing-in-Publication Data

A catalogue record for this book is available from the British Library

Library of Congress Cataloging-in-Publication Data

A catalog record for this book is available from the Library of Congress

ISBN: 978-0-12-823996-4

Publisher: Susan Dennis
Acquisitions Editor: Kostas KI Marinakis
Editorial Project Manager: Lindsay Lawrence
Production Project Manager: Debasish Ghosh
Cover Designer: Victoria Pearson

Typeset by MPS Limited, Chennai, India

Contents

List of contributors

Meisam Ansarpour
Department of Chemical Engineering, Faculty of Petroleum, Gas and Petrochemical Engineering, Persian Gulf University, Bushehr, Iran

Ali Bakhtyari
Chemical Engineering Department, Shiraz University, Shiraz, Iran

Asmaa F. Elelamy
Department of Mathematics, Faculty of Education, Ain Shams University, Cairo, Egypt

Yasaman Enjavi
Department of Chemical Engineering, Shiraz University, Shiraz, Iran

Morteze Esfandyari
Department of Chemical Engineering, University of Bojnord, Bojnord, Iran

Fateme Etebari
Department of Chemical Engineering, Shiraz University, Shiraz, Iran

Nayef Ghasem
Department of Chemical and Petroleum Engineering, UAE University, Alain, UAE

Kamran Ghasemzadeh
Urmia University of Technology, Urmia, Iran

Ali Hafizi
Department of Chemical Engineering, Shiraz University, Shiraz, Iran

Mohammad Hatami
Department of Mechanical Engineering, Esfarayen University of Technology, Esfarayen, Iran; International Research Center for Renewable Energy, Xi'an Jiaotong University, Xi'an, P.R. China

Zohreh-Sadat Hosseini
Department of Chemical Engineering, Shiraz University, Shiraz, Iran

Angel Huminic
Transilvania University of Brasov, Brasov, Romania

Gabriela Huminic
Transilvania University of Brasov, Brasov, Romania

Adolfo Iulianelli
Institute on Membrane Technology of the Italian National research Council (CNR-ITM), Rende, Italy

Dengwei Jing
International Research Center for Renewable Energy, Xi'an Jiaotong University, Xi'an, P.R. China

Baishali Kanjilal
Department of Bioengineering, University of California, Riverside, Riverside, CA, United States

Mohammad Reza Kiani
Fouman Faculty of Engineering, College of Engineering, University of Tehran, Fouman, Iran

Mohammad Amin Makarem
Department of Chemical Engineering, Shiraz University, Shiraz, Iran

Arameh Masoumi
Department of Chemical Engineering, Rowan University, Glassboro, NJ, United States

Hatice Mercan
Department of Mechatronics Engineering, Faculty of Mechanical Engineering, Yildiz Technical University (YTU), Yildiz, Besiktas, Istanbul, Turkey

Maryam Meshksar
Department of Chemical Engineering, Shiraz University, Shiraz, Iran

Alina Adriana Minea
Gheorghe Asachi Technical University Iasi, Iasi, Romania

Masoud Mofarahi
Department of Chemical Engineering, Faculty of Petroleum, Gas and Petrochemical Engineering, Persian Gulf University, Bushehr, Iran; Department of Chemical and Biomolecular Engineering, Yonsei University, Seoul, Republic of Korea

Iman Noshadi
Department of Bioengineering, University of California, Riverside, Riverside, CA, United States; Department of Chemical Engineering, Rowan University, Glassboro, NJ, United States

Mohammad Reza Rahimpour
Department of Chemical Engineering, Shiraz University, Shiraz, Iran

Colin A. Scholes
Department of Chemical Engineering, The University of Melbourne, Melbourne, VIC, Australia

Mohammad Amin Sedghamiz
Department of Chemical Engineering, Shiraz University, Shiraz, Iran

Nourouddin Sharifi
Department of Engineering Technology, Tarleton State University, Stephenville, TX, United States

Nazanin Abrishami Shirazi
School of Environment, College of Engineering, University of Tehran, Tehran, Iran

Ali Behrad Vakylabad
Department of Materials, Institute of Science and High Technology and Environmental Sciences, Graduate University of Advanced Technology, Kerman, Iran

Jiandong Zhou
International Research Center for Renewable Energy, Xi'an Jiaotong University, Xi'an, P.R. China

About the editors

Mohammad Reza Rahimpour

Prof. Mohammad Reza Rahimpour is a professor in Chemical Engineering at Shiraz University, Iran. He received his PhD in Chemical Engineering from Shiraz University joint with University of Sydney, Australia 1988. He started his independent career as Assistant Professor in September 1998 at Shiraz University. Prof. M.R. Rahimpour was a Research Associate at the University of Newcastle, Australia in 2003–2004 and at the University of California, Davis from 2012 until 2017. During his stay at the University of California, he developed different reaction networks and catalytic processes such as thermal and plasma reactors for upgrading lignin bio-oil to biofuel with the collaboration of UCDAVIS. He was Chair of the Department of Chemical Engineering at Shiraz University from 2005 to 2009 and from 2015 to 2020. Prof. M.R. Rahimpour leads a research group in fuel processing technology focused on the catalytic conversion of fossil fuels such as natural gas, and renewable fuels such as bio-oils derived from lignin to valuable energy sources. He provides young distinguished scholars with perfect educational opportunities in both experimental methods and theoretical tools in developing countries to investigate in-depth research in the various fields of chemical engineering including carbon capture, chemical looping, membrane separation, storage and utilization technologies, novel technologies for natural gas conversion, and improving the energy efficiency in the production and use of natural gas industries. Prof. M.R. Rahimpour has collaborated with researchers around the world on syngas and hydrogen production using catalytic membrane reactors.

Mohammad Amin Makarem

Dr. Mohammad Amin Makarem is a research associate at Shiraz University. His research interests are gas separation and purification, nanofluids, microfluidics, and green energy. In gas separation, his focus is on experimental and theoretical investigation and optimization of pressure swing adsorption process, and in the gas purification field, he is working on novel technologies such as microchannels. In recent years, he has investigated heat and mass transfer performances of nanofluids in different systems such as rotating disks, stretching sheets, and microchannels. Besides, he has collaborated in writing and editing various book chapters for prestigious publishers such as Elsevier, Springer, and Wiley. Recently, he has started editing scientific book projects in various subjects of chemical engineering.

Mohammad Reza Kiani

Mohammad Reza Kiani is a research associate at the University of Tehran. He started his research by writing a book chapter about aquatic water and water treatment. He is working on CO_2 absorption in microchannels and the specifications of microfluidic and nanofluidic systems. His research interests are modeling and experimental investigations about novel methods of CO_2 capture. Besides, he has collaborated with other researchers at Shiraz University in writing new book chapters and books from Elsevier.

Mohammad Amin Sedghamiz

Dr. Mohammad Amin Sedghamiz is a research associate at Shiraz University since 2018. He started his PhD at Shiraz University in 2012, with a focus on environmental aspects of energy production from waste materials. During his PhD, he emphasized novel solvents in chemical engineering applications, for example, he investigated the application and efficiency of ionic liquids to absorb acid gases at different conditions. During this project, the first idea for utilizing the next generation of ionic liquids, that is, deep eutectic solvents, erupted in his mind and he has continued this idea with the thermodynamic research group. Besides the research mentioned before, Dr. Sedghamiz was also a research associate in four different industrial and academic research projects collaborating with industrial professions and academic teams, for example, South Pars Gas company and the national Iranian oil company.

PART

Mass transfer basics of nanofluids

1

CHAPTER

Introduction to nanofluids, challenges, and opportunities

1

Hatice MERCAN

Department of Mechatronics Engineering, Faculty of Mechanical Engineering, Yildiz Technical University (YTU), Yildiz, Besiktas, Istanbul, Turkey

1.1 Mass transfer in nanofluids

Introducing nanoparticles to working fluids in energy and chemical systems has spawned a large number of studies and has been investigated by many researchers in the last two decades. It is observed that the mass and thermal transfer of the base fluid vary significantly with a small amount of nanoparticle addition. Physical properties like diffusivity, separation, thermal conductivity, density, viscosity, and heat capacity of the blend are affected by the concentration, size, and shape of the nanoparticles. In their pioneering study, Masuda et al. [1] reported an important improvement in the thermal conductivity with increasing concentration of ultrafine nanoparticles [Al_2O_3/water (1.3%–4.3% vol.), TiO_2/water (1.1%–2.3% vol.), and SiO_2/water (1.0%–4.3% vol.)] compared to water as a base fluid. In Fig. 1.1 [2] the fundamental mechanisms and their thermophysical impacts on the nanofluids are summarized schematically.

There is a strong analogy between heat and mass transfer. The rate of mass transfer is known to change as a result of the presence of nanoparticles compared to a single liquid phase. Other than the nanoparticle addition, the type of the nanoparticle, the nanoparticle concentration, the particle size, and temperature influence the rate of the mass transfer, thus these fundamental mechanisms may change the mass transfer coefficient. Unlike for the thermophysical properties, in the literature the observations from the experimental studies are not in good agreement with each other about how the mass transfer changes with these fundamental mechanisms. Therefore there is a need to investigate the mass transfer in nanofluids not only for every colloidal blend but also for each process.

The nanoparticles can be categorized as magnetic nanoparticles, metal and metal oxides nanoparticles, carbon-based nanoparticles and polymer-based nanoparticles (see Fig. 1.2) [2]. Preparing nanofluids with an improved stability is a complicated and expensive procedure. The nanofluids can be prepared with a single-step chemical method. Alternatively, a two-step method can be employed which can be chemical or physical. Mostly surfactants, ultrasound vibration, or ultrasonication techniques are used in the two-step methods to improve long-term stability of the dispersion. Nonhomogeneous dispersion and sedimentation are two important challenges that need to be addressed in nanofluid applications.

Nanofluids and Mass Transfer. DOI: https://doi.org/10.1016/B978-0-12-823996-4.00001-X

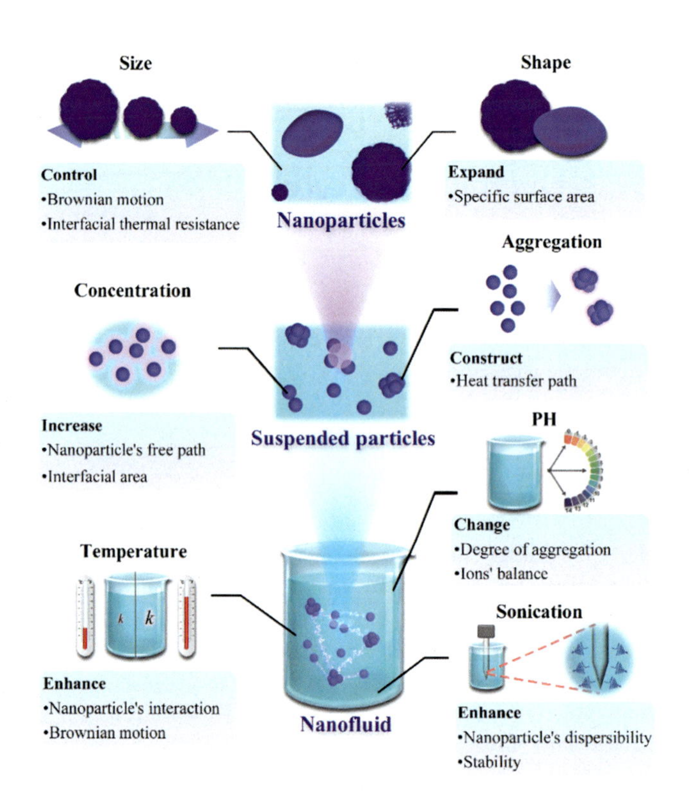

FIGURE 1.1

The fundamental mechanisms that affects the thermal conductivity of nanofluids

Adapted from Q. Lin, Z. Ning, F. Yanhui, M.E. Efstathios, Z. Gaweł, J. Dengwei, et al., A review of recent advances in thermophysical properties at the nanoscale: from solid state to colloids, Phys. Rep. 843 (2020) 1–81

The nanoparticle can be in the shape of a sphere, brick, cylinder, platelet, or blade. The experimental observations have shown that the shape of the nanoparticle affects the physical property of the nanofluid considerably. For instance, in an experimental study, Timofeeva et al. and Bijoy et al. [3,61] investigated the effects of the shape of the nanoparticles and revealed that nanofluids with platelet-shaped nanoparticles exhibit higher thermal conductivity compared to cylindrical and brick-shaped nanoparticles. Additionally, Taslimah and Daniel [4] observed that the blade-shaped nanoparticles improved the thermal conductivity more than platelet, cylindrical, brick, and spherical nanoparticles. Elias et al. [5,6] conducted a series of application-based experimental studies, where they investigated the effect of particle shape on the overall heat transfer coefficient for the heat exchangers. They reported that the cylindrical nanoparticles performed significantly better compared to other options. At this point it is not possible to derive a concrete conclusion about the effects of the nanoparticle shape on thermal conductivity of the nanofluids; it needs further consideration.

Inclusion of more than one nanoparticle to the base fluid improves the physical properties of the blend. Generally, the hybrid nanofluids demonstrate superior material properties compared to a

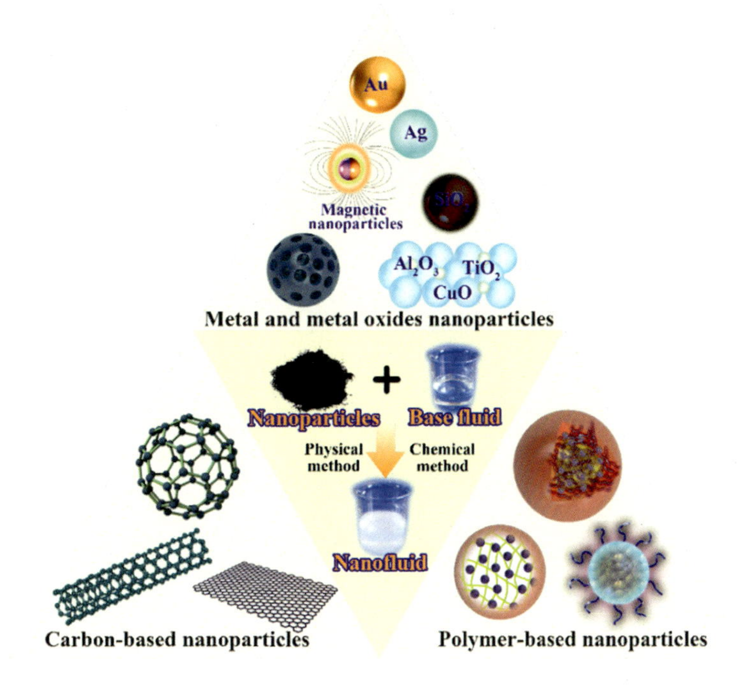

FIGURE 1.2

The classification of the nanoparticles generally used in colloidal dispersions and schematic representation of single- and two-step nanofluids preparation.

Adapted from Q. Lin, Z. Ning, F. Yanhui, M.E. Efstathios, Z. Gaweł, J. Dengwei, et al., A review of recent advances in thermophysical properties at the nanoscale: from solid state to colloids, Phys. Rep. 843 (2020) 1–81

base fluid and a subset of unitary nanofluid blend. It is important to note that the selection of the nanoparticles play an important role [7–10]. For instance, the thermal conductivity gap between the nanoparticles plays a key role for thermal conductivity of the resulting hybrid nanofluid [10]. Nabil et al. [9] reported the comparison of three different hybrid nanofluids: TiO_2-SiO_2/water-EG, Al_2O_3-G/water, and MWCBT-Fe_3O_4/water. The highest thermal conductivity ratio enhancement was observed for the TiO_2-SiO_2/water hybrid nanofluid. This was because as a hybrid nanoparticle couple TiO_2 and SiO_2 are closer to each other in their thermal conductivity compared to the other couples.

The mathematical models for suspensions do not accurately estimate the physical properties of nanofluid blends. For that reason, in the literature, there are a large number of models for the physical properties of nanofluids that procure wide variation in their estimates. Some of these models employ polynomial fits, neural network algorithms, or the Buckingham Pi theorem to relate the properties to nondimensional groups. For instance, the mathematical models to predict the nanofluid viscosity have evolved over the years, resulting in a wide range of different models with different capabilities and limitations. Initially the Einstein model was modified for nanofluids, where the Newtonian colloid viscosity is a function of concentration only [11,12]. Saito et al. [13],

Brinkman et al. [14], and Frankel and Acrivos [15] proposed modified models with improvements in concentration range and particle size range with distribution constraints, where the expression for the viscosity is still a function of the concentration [13–15]. Jang and Choi [16] reported that for smaller particle size, the nanoparticle Brownian motion becomes more prominent, thus convection becomes dominant at nanosize [16]. The nanoparticle size and density with Brownian motion features are included simultaneously by the model of Masoumin et al. [12]. Cluster size effect is taken into account by the model of Hosseini [17] and the nanolayer thicknesses with average particle radius are also a part of this formulation [17]. One common observation for these more recent models is that the mathematical expressions contain empirical coefficients that need to be determined experimentally. The proposed models cannot be directly applied in every situation; they are unique models with strict constraints. For each nanoblend with a specific concentration, temperature, and pH interval there is a specific expression in the mathematical model for a desired material property. This highlights the urgent need for the development of a generalized model to obtain the material properties of the nanofluids. This could only be possible with a clear understanding of the mass transfer mechanisms of nanofluids. These mechanisms include the diffusion and convection phenomena, where the former is due to gradient of the concentration and the latter is due to the fluid motion.

1.1.1 Mass diffusion of nanofluids

The results for diffusive characteristics of nanofluids reported in the literature can show dissimilar trends for the same nanoparticle/base fluid blend. For example, an improvement in mass diffusivity is reported for increasing concentration of the Al_3O_2/water nanofluid [18,19]. For the same colloidal mixture, a reduction in diffusivity was reported [20,21]. Additionally, Ozturk et al. [22] reported no significant change in diffusivity with concentration. Fang et al. [23], reported 26 times more diffusive Cu/water nanofluid compared to the base fluid, especially at lower concentrations [23]. These conflicting observations reveal the need for further investigations.

In the literature the mass diffusion enhancement in nanofluids is mostly linked to the random dynamic motion of the nanoparticles, which is also known as Brownian motion. The local convection due to Brownian motion of the nanoparticles enhances the thermal conductivity in nanofluids more than the classical conduction theories can foresee. A significant mass diffusion enhancement in nanofluids compared to base fluids is reported by several researchers [18,23–27]. The mass transfer and the heat transfer are two analogous processes. The mass diffusivity augmentation in nanofluids is the reason for the enhanced thermal conductivity of the nanofluid [18]. Terhani and Kelishami [27,28] investigated the mass transfer of different nanoparticle mass fractions through a supported nanoliquid membrane [27]. They constructed their model by assuming incompressible flow without chemical reactions and viscous dissipation for a nanoparticle mass fraction much less than 1, which makes the blend a dilute colloidal mixture. Using the temperature and concentration analogy they formed the following nondimensional groups: the Prandtl number and the Nusselt number for the heat transfer; and the Schmidt number and Sherwood number for the mass transfer of a nanofluid (see Table 1.1). The constants in the effective diffusivity expression, D_{nf}, provided in Table 1.1 are $c = 146$, $m1 = 0.523$, $m2 = 0.011$, and $m3 = -1.83$ [28]. They also concluded that Brownian motion enhanced the mass transfer and nanoparticles improved the convection mass transfer; however, further concentration augmentation reversed the improvement effects [27]. In a

Table 1.1 The nondimensional groups for heat and mass transfer.

Heat transfer		Mass transfer	
Prandtl number	$Pr_{nf} = \frac{\nu_{nf}}{\alpha_{nf}}$	Schmidt number	$Sc_{nf} = \frac{\nu_{nf}}{D_{nf}}$
Nusselt number	$Nu_{nf} = 2 + c\phi^{m1} Re_{nf}{}^{m2} Pr_{nf}{}^{m3}$	Sherwood number	$Sh_{nf} = 2 + c\phi^{m1} Re_{nf}{}^{m2} Sc_{nf}{}^{m3}$
Thermal conductivity	$k_{nf} = k_f(1 + c\phi^{m1} Re_{nf}{}^{m2} Pr_{nf}{}^{m3})$	Diffusivity	$D_{nf} = D_f(1 + c\phi^{m1} Re_{nf}{}^{m2} Sc_{nf}{}^{m3})$

recent study, Machrafi [29] introduced a new mechanism of nanoparticle dispersion by applying the porous flow principles and the chemical potential formula which provided a good agreement with the experimental data. This study showed that the classical free energy mixing formulation is not appropriate to explain the molecular dynamics of nanoparticle dispersions. By defining a chemical potential base for both the density and the porous medium description, where the base fluid diffuses across the particles, a better agreement with the experimental data is obtained [29].

In addition to the micromotion of the fluid, the concentration plays an important role in the mass diffusion, where the gradient of the concentration difference is the driving force for the mass transfer. Gerardi et al. [20] experimentally investigated alumina water nanofluids. Their results are in conflict with the microconvection theory presented in the previous studies, where an increase in both mass and heat transfer is expected with an increasing concentration. They reported a significant decay in self-diffusion coefficient of the nanoparticles for increasing alumina concentration [20]. Turanov and Tolmachev [30] experimentally investigated the silica–water nanofluids for volumetric concentration $\varphi \leq 0.3\%$ and compared four different silica nanoparticles with increasing particle diameters [30]. They reported for all concentrations and nanosilica diameters the solvent self-diffusion coefficient showed a reduction and the enhancement in the thermal conduction was modest, which is similar to the findings of Gerardi et al.'s study [20]. Turanov and Tolmachev [30] suggested that the diffusion decrease occurred due to water coating the nanoparticles; whereas they explained the moderate conductivity enhancement with the heat transfer resistance at the interface of the nanoparticle and water molecules [30].

1.1.2 Convective mass transfer of nanofluids

Convective mass and heat transfer is observed in nature and engineering applications ranging from atmospheric and oceanic circulations, heating of buildings to cooling of electronic devices. It is dominant in absorption processes, membrane systems, and extraction systems. During multiphase reactions, such as fermentation, hydrogenation reactions, and water treatment, the mass transfer from gas phase to liquid over the interface has a major role. The interface size as well as the concentration, the particle size, and the temperature affect this mass transfer. No significant effect on heat and mass transfer rates is observed with increasing buoyancy ratio.

Olle et al. reported an enhancement in the oxygen absorption of water with magnetic nanoparticles. They reported an improvement in mass transfer coefficient up to six times compared to the base fluid and for both dynamic and quiescent conditions. The increase in gas–liquid interface by nanoparticle addition is responsible for this increase with a strong dependency on temperature. Kim et al. [31] investigated the

0%–18.7% ammonia–water solution absorption characteristics with different nanoparticle blends considering Cu, CuO, and Al_2O_3. They reported that with the highest ammonia suspension as a base fluid, 0.1 wt.% Cu showed the highest mass transfer coefficient, and all nanofluids displayed a linear increase of mass transfer with increasing concentration [31]. However, further concentration augmentation resulted in a reverse effect on mass transfer [31,32]. Ma et al. [33] reported a similar observation of the effect of 0–0.3 wt.% CNT/ammonia nanofluids on bubble absorption [33]. They concluded that Brownian motion is promoted with nanoparticle inclusion. In a series of studies, Lee et al. [34,35] compared the performance of different nanoparticles on ammonia and carbon dioxide absorption [34,35]. In the first study, Lee et al. [34] proposed a comparison of CNTs and alumina nanoparticles where 0.01–0.06 vol.% CNT/NH_3–water and Al_2O_3/NH_3–water nanofluids are investigated for ammonia absorption. They reported a maximum rise in mass transfer coefficient at 0.02% concentration for both nanofluids. Because of the spherical shape of the Al_2O_3 nanoparticles, the drag force acting on the particles is low and the Brownian motion is stronger than the CNT nanofluid, where the nanoparticles have a higher aspect ratio. They concluded that the particle shape is a more important component that affects the gas absorption compared to particle type and size. In the second experimental study, the CO_2 absorption was investigated for increasing nanoparticle size and concentration of SiO_2/methanol and Al_2O_3/methanol nanofluids. SiO_2 and Al_2O_3 showed similar performances on CO_2 absorption and the maximum enhancement of 4.5% was reached at 0.01 vol.% Al_2O_3–methanol blends and 5.6% was reached at 0.01 vol.% SiO_2–methanol blend. Further concentration rise resulted in a reduction in CO_2 absorption because of -OH bonding to particles and aggregation [35]. Falling film absorption systems are environment-friendly refrigeration systems. Improving their absorption performance is essential to make them an alternative to the conventional vapor compression air conditioners. The absorption improvement in falling film flows of the nanofluids have been investigated by several researchers. Fe_3O_4/MDEA nanofluid is reported to absorb CO_2 with a 92.8% enhanced performance and a periodic magnetic field enhanced the mass transfer rate further, as reported by Komati and Suresh [36]. Fe-CNT/H_2O-LiBr binary nanofluid was investigated by [37] and the vapor absorption was reported for increasing concentrations. The type of the nanoparticle has more effect on the mass transfer than the nanoparticle concentration, and Fe nanoparticles exhibited better absorption. Hwang et al. [38] investigated the CO_2 absorption rate of SiO_2/water nanofluid with increasing nanoparticle sizes and increasing concentration [38]. They reported an improvement in volumetric mass transfer coefficient with increasing particle size, however, beyond a certain particle size, this improvement ceases to exist. In the study of Hussein et al. [24] the convective mass transfer characteristics of nanofluids under gravitational effects are investigated for changing aspect ratios through a porous cavity [24]. They reported that the low aspect ratio values resulted from enhanced heat and mass transfer rates. An increase of Darcy's number resulted in an increased heat transfer and decreased mass transfer, while the Rayleigh number showed a similar effect.

Park et al. [39] reported a reduction in CO_2 absorption with an increase in the concentration of colloidal silica [39]. The absorption rate showed a reverse proportionality with the nanoparticle concentration, and similar observations were reported by Partk et al. [40,41]. Lu et al. [42] experimentally studied the influence of stirring speed on the CO_2 absorption ratio of Al_2O_3, active alumina, CNT, and activate carbon nanoparticles. They conducted the experiments for increasing nanoparticle concentrations and four increasing stirring speeds. The Al_2O_3 nanoparticles showed a decline in the absorption of CO_2 for increasing concentration and a nonsignificant improvement with increased stirring speed. Active alumina, on the other hand, showed a general improvement in absorption with concentration, but the stirring resulted in an inconclusive trend. CNT/water nanofluid showed an augmentation of CO_2 absorption with increasing concentration. The stirring speeds improved the absorption at low and medium speeds, but for higher speeds the absorption remained constant. The activated carbon/water nanofluid presented a clear improvement with particle concentration, however, the lowest stirring speed resulted in the highest absorption while the absorption declined with increasing stirring speeds. These different behaviors may arise from the different hydrophilic behavior of the nanoparticles [42].

1.2 Challenges

There are several challenges for employing nanofluids in real applications. These challenges are presented in the following section with an emphasis on the role of mass transfer.

1.2.1 High cost of production and application

The material type and the synthesis method affect the nanoparticles price directly. For example, the price of some nanoparticles made of noble elements, such as titanium, gold, and silver, is high, and hence economical aspects should be considered before their application. Moreover, strict investigations should be done in laboratories using advanced and expensive equipment for the characterization of nanoparticles specifications with high accuracy, which needs both time and money.

The cost of nanofluids is not limited to the cost of the production of the nanoparticles only. For the preparation of nanofluids, one needs to employ special equipment and various physical and chemical approaches to make stable nanofluids. Therefore this preparation cost should be also included in the final cost. Moreover, in real applications where the system runs for the long term it is necessary to install some equipment such as an ultrasonic bath in the flow circuit to reduce the sedimentation of nanoparticles and consequently the clogging in the system. This is just an example of the costs required for nanofluids application. For large-scale systems with increased mass transfer rates, clearly the costs of production and application increase. Fig. 1.3 [43] presents the price and efficiency of different nanofluids versus the volume fraction of nanoparticles. The results of the figure can help designers to choose the best nanofluid considering both efficiency and price.

1.2.2 Instability and sedimentation

Dispersing nanoparticles into liquids in a homogenous way with minimum sedimentation is not an easy task. The ultrasonication, agitation, and surfactant addition are the basic remedies to overcome this difficulty. In order to maintain the repeatability of the preparation of the nanofluids, it is essential to tabulate the proper amplitude and sonication time interval and the details of the instrumentation, like brand and model. The long-term stability is essential and improved dispersion performance is vital. It is observed that at 6 min, particles settled significantly, investigated the thermal conductivity and diffusion of TiO_2/water nanofluid with 10% weight concentration for increasing temperature [44]. Cacua et al. investigated the sedimentation of the nanofluid for four different sonication times: 180, 240, 300, and 1360 s. They reported an eventual increase of absorption on the day of preparation. However, after 5 days the absorbance levels are equalized and sedimentation is observed. The agitation of the nanofluid on the first, second, and third days of preparation also resulted in no significant improvement in the dispersion [44].

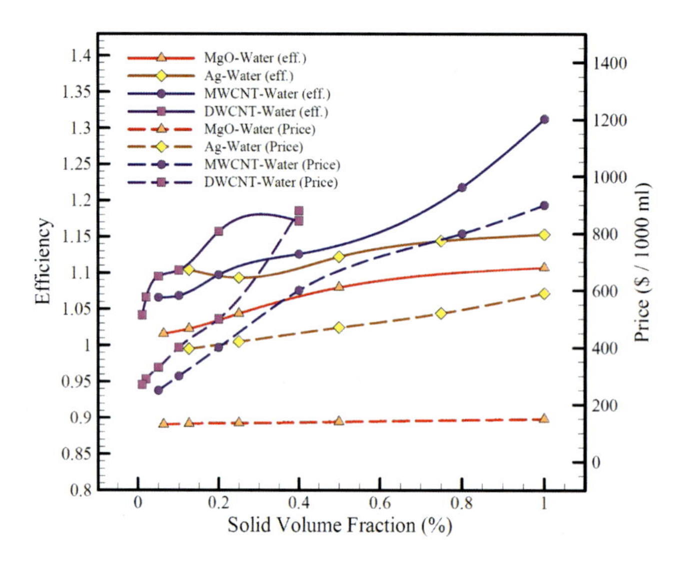

FIGURE 1.3

Efficiency and price of different nanofluid samples.

Adapted from A. Alirezaie, M.H. Hajmohammad, A. Alipour, M. Salari, Do nanofluids affect the future of heat transfer? "A benchmark study on the efficiency of nanofluids", Energy 157 (2018) 979–989

Instability has a direct relationship with sedimentation. The sedimentation of nanoparticles on the surface of thermal equipment could have negative effects on the system efficiency, since the number of dispersed nanoparticles in the working fluid reduces, and hence the ability of working fluid for heat transportation would be decreased. Sedimentation that happens due to the aggregation of nanoparticles may lead to the clogging of pipes. When this happens, the operation needs to be stopped to fix such problems, resulting in resource and time loss.

Fig. 1.4 [45] shows the sedimentation of alumina nanoparticles dispersed in an ethanol/water mixture in time. As seen, the sedimentation was very fast so that in just 1 day, most of the nanoparticles settled down on the bottom of the vessel.

Another complexity that is commonly faced in nanofluid applications is nanoparticles' affinity to agglomerate into larger groups of particles. The agglomeration decreases the surface area and the blend behaves like a microparticle fluid. The agglomeration becomes more significant with larger samples. But the majority of the studies that address the mass transfer mechanisms examine small samples, which basically neglects the potential agglomeration of the nanofluid. Therefore new experiments that consider the agglomeration effect should be undertaken.

1.2.3 Elevated pumping power

The thermal conductivity of nanofluids increases with an increase in the concentration, but meanwhile the viscosity rises as well. A higher viscosity implies a higher pressure drop and consequently requires higher pumping power. For large-scale thermal systems where a large amount of

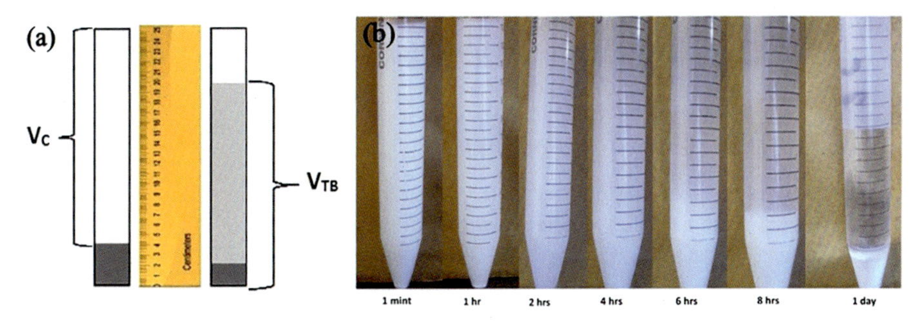

FIGURE 1.4

Sedimentation of alumina/ethanol–water with time.

Adapted from M. Abdullah, S.R. Malik, M.H. Iqbal, M.M. Sajid, N.A. Shad, S.Z. Hussain, et al. Sedimentation and stabilization of nano-fluids with dispersant, Colloids Surf. A: Physicochemical Eng. Asp. 554 (2018) 86–92

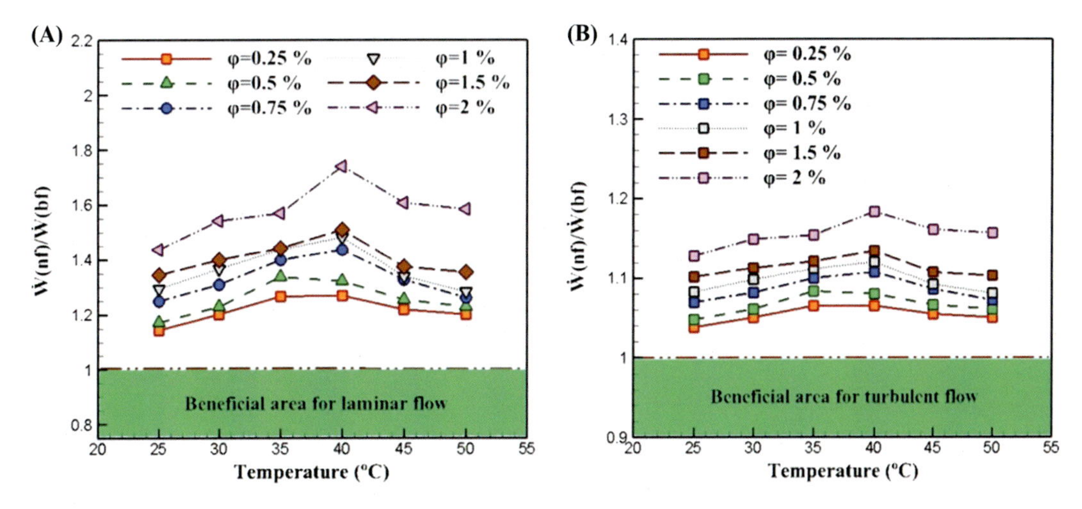

FIGURE 1.5

Pumping power variation for oil-based nanofluids: (A) laminar flow. (B) turbulent flow.

Adapted from M. Asadi, A. Asadi, S. Aberoumand, Étude expérimentale et théorique des effets de l'ajout de nanoparticules hybrides sur l'efficacité du transfert de chaleur et l'énergie de pompage d'un nanofluide à base d'huile utilisé comme liquide de refroidissement, Int. J. Refrig. 89 (2018) 83–92

nanofluid needs to be delivered from one point to another, the increased costs associated with the pumping power could be considerable. Fig. 1.5(A) and (B) [46] present the variations of pumping power for both laminar and turbulent flow. As seen, at a given volume concentration, the ratio of nanofluid pumping power to base fluid pumping power associated with laminar flow (where mass transfer rate is less than turbulent flow) is higher than that of turbulent flow. This ratio increases also with an increase in the volume fraction of the nanoparticles.

1.2.4 Erosion and corrosion of the equipment

Nanoparticles may damage the equipment surface over time. The erosion and corrosion would happen even when conventional liquids are used in the system, however, the introduction of the nanoparticles to the base fluid would increase the rate of erosion/corrosion. Mass transfer rate is certainly an important parameter that determines the amount of erosion/corrosion of surfaces of thermal equipment. Fig. 1.6 [47] indicates that the rate of materials loss increases considerably with time when the working fluid is a nanofluid. As Fig. 1.6 shows, adding nanoparticles to the base fluid resulted in destructive effects on a carbon steel pipe surface.

1.2.5 Thermal performance in turbulent flow condition

There is inconsistency in the results reported by the researchers on nanofluids heat transfer performance in the turbulent flow regime. Kim et al. [31] reported that there is no significant improvement in the thermal performance by applying the amorphous carbonic nanofluids in the turbulent flow. This implies

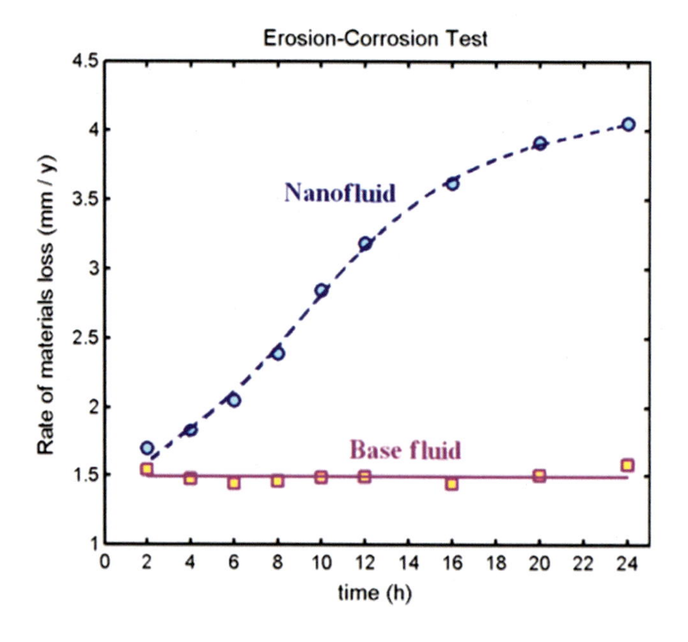

FIGURE 1.6

The comparison of material loss rate of carbon steel pipe with base fluid (sea water) and nanofluids (alumina/sea water).

Adapted from A. M. Rashidi, M. Packnezhad, M. Moshrefi-Torbati, F.C. Walsh, Erosion-corrosion synergism in an alumina/sea water nanofluid, Microfluidics Nanofluidics 17(1) (2014) 225–232

that nanofluids have a limitation in the application where turbulent flow conditions exist. This also states a contradiction over the use of different types of nanoparticles in heat transport applications.

1.2.6 Necessity of defining new mechanisms

Most of the convective heat transfer studies have been performed on oxide nanoparticles, such as Al_2O_3 and CuO. Also, most of the theoretical models and transport mechanisms have been developed by considering the spherical shape of nanoparticles. However, in recent years new types of nanoparticles with various morphologies have been introduced in heat transfer applications due to their better heat transfer capabilities compared to the oxide nanoparticles; CNTs, graphene, and metallic nanoparticles are some examples of this group of nanoparticles. But their flow behavior and thermophysical characteristics do not follow the conventional concepts and theoretical models. There are insufficient information and concepts available that can properly state the flow behavior and mechanisms of the improved thermophysical characteristics of these nanoparticles. Thus this field requires further investigations.

1.2.7 Challenges in two-phase heat transport applications

Nanorefrigerants are being considered as alternatives to the standard refrigerants due to their better thermophysical properties. In addition to their thermal conductivity, the phase change heat transfer during flow (boiling characteristics) of nanofluids should be investigated in detail to reveal their performance as refrigerants. Hernandez et al. compared the performance of R-113, R-123, and R-134a with Al_2O_3 nanoparticles with increasing concentrations and reported 1% vol. Al_2O_3/R-134a with 30 nm particle size performed with a better thermal efficiency [48]. Introducing better performing refrigerants will lower the environmental impact and nanorefrigerants are good candidates for that purpose. However, in refrigeration and heat pipe applications, nanoparticles are usually separated from the base fluid due to the phase change of the base fluid and the nanoparticles become more likely to settle down on the surfaces of the equipment. Although the use of nanofluids reduces the thermal resistance of the heat pipe and improves the heat transfer performance, nanofluids' applicability as a refrigerant is still controversial and requires further studies.

1.3 Opportunities

By overcoming the abovementioned challenges, great opportunities can be found in the development of new thermal systems using nanofluids. In the following sections some of these opportunities are discussed. Fig. 1.7 summarizes the main opportunities in the field of nanofluids.

1.3.1 Smaller equipment

Nanofluids with a higher thermal conductivity compared to common liquids can transfer heat at higher rates. Therefore for a specific rate of heat transfer, the size of thermal equipment would be smaller than that of the thermal equipment where working fluid is a common liquid, which can save space and money. A study conducted [40,49] showed that for a specific output adding Al_2O_3

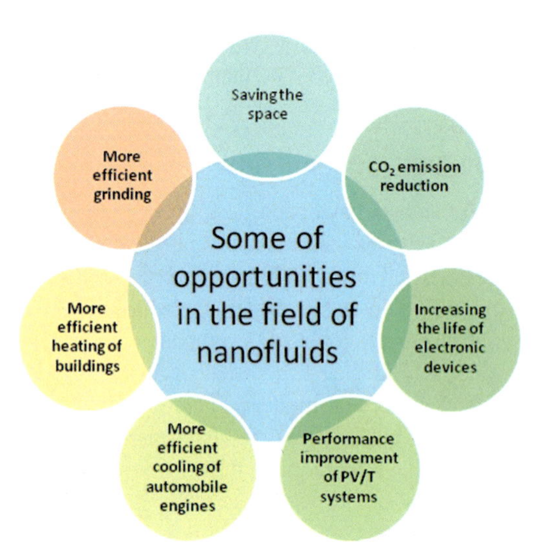

FIGURE 1.7

Main opportunities in the field of nanofluids.

to water can lead to a 24% decrease in the size of flat plate solar collectors. While the increase in nanoparticle mass transfer increases the initial costs associated with the nanofluid production, in the long term the potential of using smaller equipment can lead to savings.

1.3.2 CO_2 emission reduction

Carbon dioxide is a greenhouse gas and the rise in its emission is the main reason for global warming. Reducing CO_2 production and absorbing the emitted CO_2 are the two main remedies that can be used to avoid warming of the Earth. The CO_2 emission can be reduced by the use of nanofluids which contain nanoabsorbers. Ionic liquid and magnetic MWCNTs [50] and Al_2O_3/methanol, SiO_2/methanol [42] are used for CO_2 absorption where SiO_2–methanol was reported as a better absorbent compared to alumina. Additionally using nanofluids in the thermal systems can help to reduce CO_2 emissions in the long term through the increase in the efficiency of the system. Khullar and Tyagi [51] reported that CO_2 emission diminishes by about 2200 kg household/year if nanofluids are used in concentrating solar water heating system as the working fluid.

1.3.3 Increasing the life of electronic devices

Nanofluids can be utilized to cool electronic devices. When a nanofluid is engineered well, it can increase the cooling rate of electronic devices. Therefore the expected life span of electronic components increases with the decrease of failure risk at elevated temperatures. The microchannel flow of nanofluids is more sensitive to sedimentation and agglomeration. Additionally, phase change during electronic device cooling increases cooling performance. The pumps in the conventional heat exchangers are replaced with liquid blocks in electronic device cooling. Further investigation is necessary to have a commercially successful cooling alternative operating with nanofluids for electronic devices [52].

1.3.4 Performance improvement of PV/T systems

Electricity generation using PV/T systems is developing rapidly. Therefore the increase of efficiency of such systems is vital. The literature review shows that employing nanofluids can enhance both thermal and electrical efficiencies of a PV/T system. Mohammad and Mohammad [53] investigated the effects of three different nanoparticles, including TiO_2, ZnO, and Al_2O_3, dispersed in water on the performance of a PV/T system. It was concluded that using high mass fractions of metal oxides-based nanofluids augments the thermal performance of the PV/T significantly.

1.3.5 More efficient cooling of automobile engines

Several studies have been done on the potential use of nanofluids as coolants of automobile engines. With the use of the nanofluids in automobile applications, considerable improvement in fuel economy saving is expected, which consequently will lead to reductions in CO_2 emissions [54,55].

1.3.6 More efficient heating of buildings

Nanofluids can also be used as working fluids for heating systems in buildings. Similar to nanofluids used for cooling electronics, nanofluids used in building systems need to be transferred in high mass fractions without sedimentation. This would enable the safe use of nanofluids in building heating systems without clogging the pipes and with increased efficiency. Due to their considerably better thermophysical properties, building heating systems operating with nanofluids have smaller size, which reduces investment and operational costs. In addition, nanofluid-based heating systems offer a more environment-friendly alternative to conventional heating systems [56].

1.3.7 More efficient grinding

The grinding process causes high temperatures at the wheel–workpiece interface which develops residual stresses and undesired microdamages. The heat dissipation may be controlled in a desired fashion by employing nanofluids. Vasu and Kumar [57] reported a better surface finish of EN-31 steel block after grinding it by using emulsifier with Al_2O_3 nanoparticles. Al_2O_3, CuO, MWCNT, graphene, MoS_2/CNT, TiO_2, diamond, and graphite are common nanoparticles used to improve conventional lubricants [58]. In a series of studies Mao et al. [59,60] investigated the grinding of AISI 52100 steel using Al_2O_3/water nanofluids as a lubricant and showed the reduction of the friction coefficient during the process. An increase in concentration of nanoparticles in Al_2O_3/water nanofluid lubricant reduced the friction and grinding temperature during the grinding process. However, increased particle size worsened the final surface quality [61]. The nanolubricants are also used to minimize the lubricant quantity with improved features, mostly used in aerospace applications. Grinding can be more efficient at higher mass fractions of nanofluids since it ameliorates the cooling rate.

1.3.8 Drug delivery

In drug delivery the nanofluids are mostly used in imaging and diagnosis. Safe nanofluids should be stable with a long shelf life, biodegradable, and biocompatible [62]. The dispersion stability is essential. The surface charge of biological particles in blood is negative, therefore to avoid aggregation the nanoparticles must also be charged negatively [63]. In order to reduce the side effects and prevent the damage of healthy tissue, the targeted cancer medication is possible with magnetic delivery, magnetic fluid hyperthermia, and nanocryosurgery. The magnetic nanoparticles in the drug enable higher doses, which is essential for some types of cancers [64]. Using magnetic nanoparticles, the cancer tissue is heated and thus its growth is stopped [65]. More research is required to make full use of the potential of the use of the nanoparticles in health applications.

1.4 Conclusion

Among all the challenging applications of different types of nanofluids, determining the alteration of properties precisely, depending on different preparation techniques, concentrations, operating conditions, morphological structures, and particle distributions, is the most essential problem to be solved. The understanding of the mass transfer mechanism in quiescent and flow conditions is essential for this purpose. The stability before and during the operation of the colloidal mixture plays an important role for different application areas—from heat exchangers to drug delivery. The wide range of the application area of nanofluids will make it indisputably one of the most important working fluids in major industrial, pharmaceutical, and everyday life applications, including tribology, coating, process and extraction, pharmaceutical and medical applications, refrigeration, solar energy, and microcooling of electronic devices.

References

[1] M. Hidetoshi, E. Akira, T. Kazunari, H. Nobuo, Alteration of thermal conductivity and viscosity of liquid by dispersing ultra-fine particles. Dispersion of Al_2O_3, SiO_2 and TiO_2 ultra-fine particles, Netsu Bussei (1993) 227–233. Available from: https://doi.org/10.2963/jjtp.7.227.

[2] Q. Lin, Z. Ning, F. Yanhui, M.E. E., Ż. Gaweł, J. Dengwei, et al., A review of recent advances in thermophysical properties at the nanoscale: from solid state to colloids, Phys. Rep. (2020) 1–81. Available from: https://doi.org/10.1016/j.physrep.2019.12.001.

[3] E.V. Timofeeva, J.L. Routbort, D. Singh, Particle shape effects on thermophysical properties of alumina nanofluids, J. Appl. Phys. 106 (1) (2009). Available from: https://doi.org/10.1063/1.3155999.

[4] O.A. Taslimah, M.O. Daniel, Analysis of blasius flow of hybrid nanofluids over a convectively heated surface, Defect. Diffus. Forum (2017) 29–41. Available from: https://doi.org/10.4028/www.scientific.net/DDF.377.29.

[5] M.M. Elias, M. Miqdad, I.M. Mahbubul, R. Saidur, M. Kamalisarvestani, M.R. Sohel, et al., Effect of nanoparticle shape on the heat transfer and thermodynamic performance of a shell and tube heat exchanger, Int. Commun. Heat. Mass. Transf. 44 (2013) 93–99. Available from: https://doi.org/10.1016/j.icheatmasstransfer.2013.03.014.

[6] M.M. Elias, I.M. Shahrul, I.M. Mahbubul, R. Saidur, N.A. Rahim, Effect of different nanoparticle shapes on shell and tube heat exchanger using different baffle angles and operated with nanofluid, Int. J. Heat. Mass. Transf. 70 (2014) 289–297. Available from: https://doi.org/10.1016/j.ijheatmasstransfer.2013.11.018.
[7] N. Ahammed, L.G. Asirvatham, S. Wongwises, Entropy generation analysis of graphene–alumina hybrid nanofluid in multiport minichannel heat exchanger coupled with thermoelectric cooler, Int. J. Heat. Mass. Transf. 103 (2016) 1084–1097. Available from: https://doi.org/10.1016/j.ijheatmasstransfer.2016.07.070.
[8] S. Jana, A. Salehi-Khojin, W.H. Zhong, Enhancement of fluid thermal conductivity by the addition of single and hybrid nano-additives, Thermochim. Acta 462 (1–2) (2007) 45–55. Available from: https://doi.org/10.1016/j.tca.2007.06.009.
[9] M.F. Nabil, W.H. Azmi, K. Abdul Hamid, R. Mamat, F.Y. Hagos, An experimental study on the thermal conductivity and dynamic viscosity of TiO_2-SiO_2 nanofluids in water: ethylene glycol mixture, Int. Commun. Heat. Mass. Transf. 86 (2017) 181–189. Available from: https://doi.org/10.1016/j.icheatmasstransfer.2017.05.024.
[10] H.W. Xian, N.A.C. Sidik, R. Saidur, Impact of different surfactants and ultrasonication time on the stability and thermophysical properties of hybrid nanofluids, Int. Commun. Heat. Mass. Transf. (2020) 110. Available from: https://doi.org/10.1016/j.icheatmasstransfer.2019.104389.
[11] A. Einstein, Eine neue Bestimmung der Moleküldimensionen, Annalen der Phys. 324 (2) (1906) 289–306. Available from: https://doi.org/10.1002/andp.19063240204.
[12] M. N, S. N, B. A, A new model for calculating the effective viscosity of nanofluids, J. Phys. D: Appl. Phys. (2009) 055501. Available from: https://doi.org/10.1088/0022-3727/42/5/055501.
[13] N. Saitô, Concentration dependence of the viscosity of high polymer solutions. I, J. Phys. Soc. Jpn. 5 (1) (1950) 4–8. Available from: https://doi.org/10.1143/JPSJ.5.4.
[14] H.C. Brinkman, The viscosity of concentrated suspensions and solutions, J. Chem. Phys. 20 (4) (1952) 571. Available from: https://doi.org/10.1063/1.1700493.
[15] N.A. Frankel, A. Acrivos, On the viscosity of a concentrated suspension of solid spheres, Chem. Eng. Sci. 22 (6) (1967) 847–853. Available from: https://doi.org/10.1016/0009-2509(67)80149-0.
[16] S.P. Jang, S.U.S. Choi, Effects of various parameters on nanofluid thermal conductivity, J. Heat. Transf. 129 (5) (2007) 617–623. Available from: https://doi.org/10.1115/1.2712475.
[17] S.M. Hosseini, A.R. Moghadassi, D.E. Henneke, A new dimensionless group model for determining the viscosity of nanofluids, J. Therm. Anal. Calorim. 100 (3) (2010) 873–877. Available from: https://doi.org/10.1007/s10973-010-0721-0.
[18] S. Krishnamurthy, P. Bhattacharya, P.E. Phelan, R.S. Prasher, Enhanced mass transport in nanofluids, Nano Lett. 6 (3) (2006) 419–423. Available from: https://doi.org/10.1021/nl0522532.
[19] J. Veilleux, S. Coulombe, A total internal reflection fluorescence microscopy study of mass diffusion enhancement in water-based alumina nanofluids, J. Appl. Phys. 108 (10) (2010).
[20] C. Gerardi, D. Cory, J. Buongiorno, L.W. Hu, T. McKrell, Nuclear magnetic resonance-based study of ordered layering on the surface of alumina nanoparticles in water, Appl. Phys. Lett. 95 (25) (2009). Available from: https://doi.org/10.1063/1.3276551.
[21] V. Subba-Rao, P.M. Hoffmann, A. Mukhopadhyay, Tracer diffusion in nanofluids measured by fluorescence correlation spectroscopy, J. Nanopart. Res. 13 (12) (2011) 6313–6319. Available from: https://doi.org/10.1007/s11051-011-0607-5.
[22] S. Ozturk, Y.A. Hassan, V.M. Ugaz, Interfacial complexation explains anomalous diffusion in nanofluids, Nano Lett. 10 (2) (2010) 665–671. Available from: https://doi.org/10.1021/nl903814r.
[23] X. Fang, Y. Xuan, Q. Li, Experimental investigation on enhanced mass transfer in nanofluids, Appl. Phys. Lett. 95 (20) (2009). Available from: https://doi.org/10.1063/1.3263731.

[24] S. Hussain, H.F. Öztop, M.A. Qureshi, N. Abu-Hamdeh, Double diffusive buoyancy induced convection in stepwise open porous cavities filled nanofluid, Int. Commun. Heat. Mass. Transf. 119 (2020). Available from: https://doi.org/10.1016/j.icheatmasstransfer.2020.104949.

[25] J. Li, D. Liang, K. Guo, R. Wang, S. Fan, Formation and dissociation of HFC134a gas hydrate in nano-copper suspension, Energy Convers. Manag. 47 (2) (2006) 201–210. Available from: https://doi.org/10.1016/j.enconman.2005.03.018.

[26] B. Olle, S. Bucak, T.C. Holmes, L. Bromberg, T.A. Hatton, D.I.C. Wang, Enhancement of oxygen mass transfer using functionalized magnetic nanoparticles, Ind. Eng. Chem. Res. 45 (12) (2006) 4355–4363. Available from: https://doi.org/10.1021/ie051348b.

[27] B.M. Tehrani, A. Rahbar-Kelishami, Influence of enhanced mass transfer induced by Brownian motion on supported nanoliquids membrane: experimental correlation and numerical modelling, Int. J. Heat. Mass. Transf. 148 (2020). Available from: https://doi.org/10.1016/j.ijheatmasstransfer.2019.119034.

[28] B.M. Tehrani, A. Rahbar-Kelishami, Intensification of gadolinium(III) separation by effective utilization of nanoliquids in supported liquid membrane using Aliquat 336 as carrier, Chem. Pap. 72 (12) (2018) 3085–3092. Available from: https://doi.org/10.1007/s11696-018-0509-4.

[29] H. Machrafi, On the chemical potential of nanoparticle dispersion, Phys. Lett. A (2020).

[30] T.A. N., T.Y. V., Heat- and mass-transport in aqueous silica nanofluids, Heat. Mass. Transf. (2009) 1583–1588. Available from: https://doi.org/10.1007/s00231-009-0533-6.

[31] J. Kim, Y.T. Kang, C.K. Choi, Soret and Dufour effects on convective instabilities in binary nanofluids for absorption application, Int. J. Refrig. 30 (2) (2007) 323–328. Available from: https://doi.org/10.1016/j.ijrefrig.2006.04.005.

[32] Y.T. Kang, J.K. Kim, Comparisons of mechanical and chemical treatments and nano technologies for absorption applications, HVAC R. Res. 12 (2006) 807–819. Available from: https://doi.org/10.1080/10789669.2006.10391209.

[33] X. Ma, F. Su, J. Chen, Y. Zhang, Heat and mass transfer enhancement of the bubble absorption for a binary nanofluid, J. Mech. Sci. Technol. 21 (1813) (2007). Available from: https://doi.org/10.1007/BF03177437.

[34] J.K. Lee, J. Koo, H. Hong, Y.T. Kang, The effects of nanoparticles on absorption heat and mass transfer performance in NH_3/H_2O binary nanofluids, Int. J. Refrig. 33 (2) (2010) 269–275. Available from: https://doi.org/10.1016/j.ijrefrig.2009.10.004.

[35] J.W. Lee, J.Y. Jung, S.G. Lee, Y.T. Kang, CO_2 bubble absorption enhancement in methanol-based nanofluids, Int. J. Refrig. 34 (8) (2011) 1727–1733. Available from: https://doi.org/10.1016/j.ijrefrig.2011.08.002.

[36] S. Komati, A.K. Suresh, CO_2 absorption into amine solutions: a novel strategy for intensification based on the addition of ferrofluids, J. Chem. Technol. Biotechnol. 83 (8) (2008) 1094–1100. Available from: https://doi.org/10.1002/jctb.1871.

[37] Y.T. Kang, H.J. Kim, K.I. Lee, Heat and mass transfer enhancement of binary nanofluids for H_2O/LiBr falling film absorption process, Int. J. Refrig. 31 (5) (2008) 850–856. Available from: https://doi.org/10.1016/j.ijrefrig.2007.10.008.

[38] B.J. Hwang, S.W. Park, D.W. Park, K.J. Oh, S.S. Kim, Absorption of carbon dioxide into aqueous colloidal silica solution with different sizes of silica particles containing monoethanolamine, Korean J. Chem. Eng. 26 (3) (2009) 775–782. Available from: https://doi.org/10.1007/s11814-009-0130-x.

[39] S.W. Park, B.S. Choi, J.W. Lee, Effect of elasticity of aqueous colloidal silica solution on chemical absorption of carbon dioxide with 2-amino-2-methyl-1-propanol, Korea Aust. Rheol J. 18 (3) (2006) 133–141. http://www.rheology.or.kr/down/18-3(3).pdf.

[40] S.W. Park, B.S. Choi, S.S. Kim, J.W. Lee, Chemical absorption of carbon dioxide into aqueous colloidal silica solution containing monoethanolamine, J. Ind. Eng. Chem. 13 (1) (2007) 133–142.

[41] S.W. Park, B.S. Choi, S.S. Kim, B.D. Lee, J.W. Lee, Absorption of carbon dioxide into aqueous colloidal silica solution with diisopropanolamine, J. Ind. Eng. Chem. 14 (2) (2008) 166–174. Available from: https://doi.org/10.1016/j.jiec.2007.09.013.

[42] S. Lu, M. Xing, Y. Sun, X. Dong, Experimental and theoretical studies of CO_2 absorption enhancement by nano-Al_2O_3 and carbon nanotube particles, Chin. J. Chem. Eng. 21 (9) (2013) 983–990. Available from: https://doi.org/10.1016/S1004-9541(13)60550-9.

[43] A. Alirezaie, M.H. Hajmohammad, A. Alipour, M. Salari, Do nanofluids affect the future of heat transfer? "A benchmark study on the efficiency of nanofluids, Energy 157 (2018) 979–989. Available from: https://doi.org/10.1016/j.energy.2018.05.060.

[44] K. Cacua, S.M.S. Murshed, E. Pabón, R. Buitrago, Dispersion and thermal conductivity of TiO_2/water nanofluid: effects of ultrasonication, agitation and temperature, J. Therm. Anal. Calorim. 140 (1) (2020) 109–114. Available from: https://doi.org/10.1007/s10973-019-08817-1.

[45] M. Abdullah, S.R. Malik, M.H. Iqbal, M.M. Sajid, N.A. Shad, S.Z. Hussain, et al., Sedimentation and stabilization of nano-fluids with dispersant, Colloids Surf. A: Physicochem Eng. Asp. 554 (2018) 86–92. Available from: https://doi.org/10.1016/j.colsurfa.2018.06.030.

[46] M. Asadi, A. Asadi, S. Aberoumand, Étude expérimentale et théorique des effets de l'ajout de nanoparticules hybrides sur l'efficacité du transfert de chaleur et l'énergie de pompage d'un nanofluide à base d'huile utilisé comme liquide de refroidissement, Int. J. Refrig. 89 (2018) 83–92. Available from: https://doi.org/10.1016/j.ijrefrig.2018.03.014.

[47] A.M. Rashidi, M. Packnezhad, M. Moshrefi-Torbati, F.C. Walsh, Erosion-corrosion synergism in an alumina/sea water nanofluid, Microfluid Nanofluidics 17 (1) (2014) 225–232. Available from: https://doi.org/10.1007/s10404-013-1282-x.

[48] D.C. Hernández, C. Nieto-Londoño, Z. Zapata-Benabithe, Análisis de nanofluidos para un sistema de refrigeración, DYNA (Colomb.) 83 (196) (2016) 176–183. Available from: https://doi.org/10.15446/dyna.v83n196.50897.

[49] F. Mohd, S. Rahman, M. Saad, F. M, Potential of size reduction of flat-plate solar collectors when applying Al_2O_3 nanofluid, Adv. Mater. Res. 832 (2013) 149–153. Available from: https://doi.org/10.4028/www.scientific.net/amr.832.149.

[50] S.M. Yousefi, F. Shemirani, Carbon nanotube-based magnetic bucky gels in developing dispersive solid-phase extraction: application in rapid speciation analysis of Cr(VI) and Cr(III) in water samples, Int. J. Environ. Anal. Chem. 97 (11) (2017) 1065–1079. Available from: https://doi.org/10.1080/03067319.2017.1381236.

[51] V. Khullar, H. Tyagi, A study on environmental impact of nanofluid-based concentrating solar water heating system, Int. J. Environ. Stud. 69 (2) (2012) 220–232. Available from: https://doi.org/10.1080/00207233.2012.663227.

[52] M. Bahiraei, S. Heshmatian, Electronics cooling with nanofluids: a critical review, Energy Convers. Manag. 172 (2018) 438–456. Available from: https://doi.org/10.1016/j.enconman.2018.07.047.

[53] S. Mohammad, P.-F. Mohammad, Experimental and numerical study of metal-oxides/water nanofluids as coolant in photovoltaic thermal systems (PVT), Sol. Energy Mater. Sol. Cell (2016) 533–542. Available from: https://doi.org/10.1016/j.solmat.2016.07.008.

[54] S.A. Ahmed, M. Ozkaymak, A. Sözen, T. Menlik, A. Fahed, Improving car radiator performance by using TiO_2-water nanofluid, Eng. Sci. Technol, An. Int. J. 21 (5) (2018) 996–1005. Available from: https://doi.org/10.1016/j.jestch.2018.07.008.

[55] Z. Said, M. El Haj Assad, A.A. Hachicha, E. Bellos, M.A. Abdelkareem, D.Z. Alazaizeh, et al., Enhancing the performance of automotive radiators using nanofluids, Renew. Sustain. Energy Rev. 112 (2019) 183–194. Available from: https://doi.org/10.1016/j.rser.2019.05.052.

[56] K.D. P., D.D. K., V.R. S., Application of nanofluids in heating buildings and reducing pollution, Appl. Energy (2009) 2566–2573. Available from: https://doi.org/10.1016/j.apenergy.2009.03.021.

[57] V. Vasu, M. Manoj Kumar, Analysis of nanofluids as cutting fluid in grinding EN-31 steel, Nanomicro Lett. 3 (4) (2011) 209–214. Available from: https://doi.org/10.3786/nml.v3i4.p209-214.

[58] N.A.C. Sidik, S. Samion, J. Ghaderian, M.N.A.W.M. Yazid, Recent progress on the application of nanofluids in minimum quantity lubrication machining: a review, Int. J. Heat. Mass. Transf. 108 (2017) 79–89. Available from: https://doi.org/10.1016/j.ijheatmasstransfer.2016.11.105.

[59] M. Cong, Z. Hongfu, H. Xiangming, Z. Jian, Z. Zhixiong, The influence of spraying parameters on grinding performance for nanofluid minimum quantity lubrication, Int. J. Adv. Manuf. Technol. (2013) 1791–1799. Available from: https://doi.org/10.1007/s00170-012-4143-y.

[60] M. Cong, H. Yong, Z. Xin, G. Hangyu, Z. Jian, Z. Zhixiong, The tribological properties of nanofluid used in minimum quantity lubrication grinding, Int. J. Adv. Manuf. Technol. (2014) 1221–1228. Available from: https://doi.org/10.1007/s00170-013-5576-7.

[61] M. Bijoy, S. Rajender, D. Santanu, B. Simul, Development of a grinding fluid delivery technique and its performance evaluation, Mater. Manuf. Process. (2012) 436–442. Available from: https://doi.org/10.1080/10426914.2011.585487.

[62] W.H. De Jong, P.J.A. Borm, Drug delivery and nanoparticles: applications and hazards, Int. J. Nanomed. 3 (2) (2008) 133–149.

[63] S. Srinivasan, P.N. Sawyer, Role of surface charge of the blood vessel wall, blood cells, and prosthetic materials in intravascular thrombosis, J. Colloid Interface Sci. 32 (3) (1970) 456–463. Available from: https://doi.org/10.1016/0021-9797(70)90131-1.

[64] Q.A. Pankhurst, J. Connolly, S.K. Jones, J. Dobson, Applications of magnetic nanoparticles in biomedicine, J. Phys. D: Appl. Phys. 36 (13) (2003) R167–R181. Available from: https://doi.org/10.1088/0022-3727/36/13/201.

[65] N. Saurín, T. Espinosa, J. Sanes, F.J. Carrión, M.D. Bermúdez, Ionic nanofluids in tribology, Lubricants 3 (4) (2015) 650–663. Available from: https://doi.org/10.3390/lubricants3040650.

CHAPTER

Preparation, stability, and characterization of nanofluids

2

Mohammad Reza Kiani[1], Maryam Meshksar[2], Mohammad Amin Makarem[2] and Mohammad Reza Rahimpour[2]

[1]Fouman Faculty of Engineering, College of Engineering, University of Tehran, Fouman, Iran [2]Department of Chemical Engineering, Shiraz University, Shiraz, Iran

2.1 Introduction

Nanofluids are colloidal suspensions in which solid nanometer-size substances (nanoparticles, droplets, nanotubes, nanorods, nanowires, nanofibers, or nanosheets) are dispersed into different base fluids. Actually, nanofluids are two-phase systems with a solid phase in the liquid phase. Nanofluids are known to be promising candidates for improving heat and mass transfer properties of fluids, including their thermal diffusivity, conductivity, convective heat transfer coefficient, and viscosity [1–3]. Based on the nanoparticles type, they are categorized into metallic and nonmetallic types. In metallic nanofluids metal nanoparticles like nickel, aluminum, and copper are dispersed into the base fluid, whereas non-metal nanoparticles like carbon nanotubes (CNTs) and graphene oxide are used in nonmetallic nanofluids. The most common base fluids for the preparation of nanofluids are ethylene glycol (EG), water, acetone, oil, and decene. To study the different properties of nanofluids, their preparation and characterization methods as well as their stability are the key primary requirements. It should be noted that having a nanofluid with appropriate properties depends on its preparation method, which affects its stability.

The purpose of this chapter is to introduce new nanofluids preparation methods and their stability mechanisms for enhancing their properties compared with the base fluid. From the general point of view, the stability of the dispersed phase in the liquid phase is one of the most important issues in the two-phase systems as applying a stable nanofluid can reduce the cost and lead to the commercialization of nanofluids. The essential information of nanofluids, such as nanoparticles shape and size, chemical bonds, distribution of nanoparticles, and their stability can be measured by the use of characterization methods. Applying specific techniques for characterizing nanofluids depends on the research type and no specific standard tests are recommended to confirm nanofluids stability and homogeneity. Standard accelerator tests should be considered to confirm the stability of dispersed nanoparticles in the base fluid which are also discussed in this chapter.

2.2 Preparation of nanofluids

The first and the most crucial step in the experimental studies is nanofluids preparation, which is not only the simple dispersion of nanoparticles into the base fluid. Due to the special properties of

Nanofluids and Mass Transfer. DOI: https://doi.org/10.1016/B978-0-12-823996-4.00002-1

nanofluids, such as stable suspensions, no chemical change of solution, and accumulation of small nanoparticles, the preparation of nanofluids plays an important role in their applications [4]. Therefore to achieve a nanofluid with homogenously dispersed nanoparticles, suitable mixing and stabilization of nanoparticles, as well as consideration of the certain environmental conditions, are required. The different nanofluids preparation methods, such as one-step, two-step, phase transfer, and posttreatment method are discussed in the following.

2.2.1 Single-step method

The single-step (one-step) method is the integration of nanoparticles synthesis and nanofluid preparation in which nanoparticles can be evaporated and condensed directly at one time in the base fluid via a liquid chemical technique or physical vapor deposition method. The shape and particle size can be controlled by changing the pH value, temperature, and each component concentration. Fig. 2.1A shows different synthesis methods of one-step nanofluid preparation and Fig. 2.1B demonstrates the procedure of single-step nanofluid preparation. As can be seen in Fig. 2.1A, a submerged arc nanoparticle synthesis approach is an example of the one-step method in which an electric arc is applied to increase the arc region temperature up to 6000°C–12,000°C, causing the

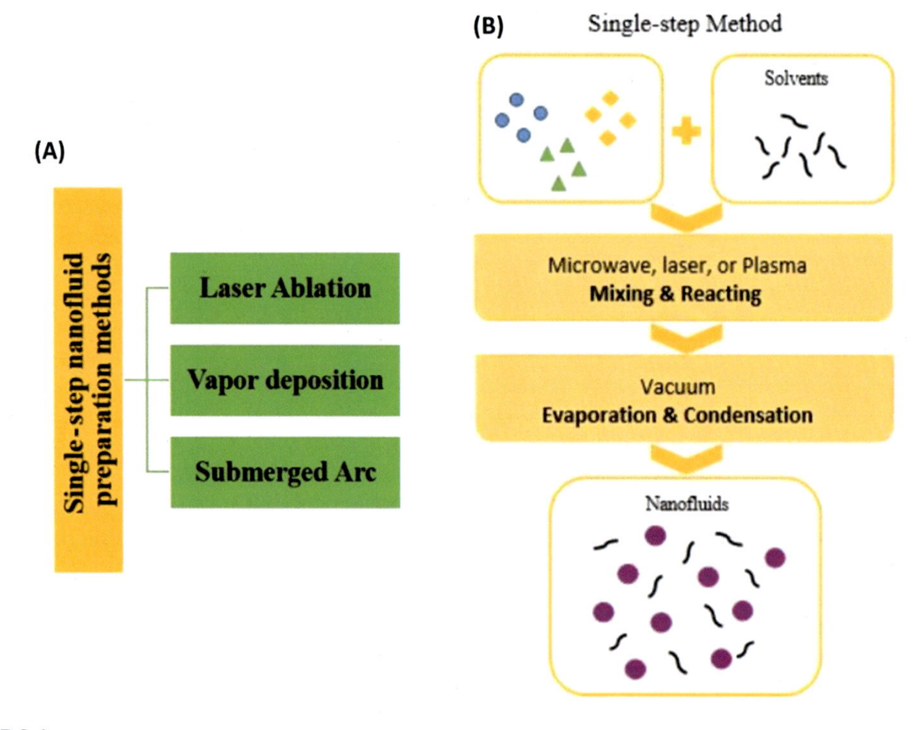

FIGURE 2.1

(A) Different single-step nanofluid preparation methods and (B) nanofluid synthesis procedure using the single-step technique.

Table 2.1 Summary of different nanofluids prepared via single-step method.

Nanoparticle type	Base fluid	Nanoparticle size (nm)	Nanoparticle loading (vol.%)	References
Cu	EG	10	0.3	[6]
Fe_3O_4	H_2O	10	4	[10]
Fe	EG	10	0.55	[11]
Cu	H_2O	75–100	0.1	[7]
Al_2O_3	H_2O	Nanorods	0.5,1, 3	[12]
Cu	H_2O	30–50	0.1–0.3	[13]

metal rod to melt and evaporate in the arc region [5]. The advantage of this method is the elimination of preparation processes such as drying, storage, transportation, and dispersion of nanoparticles, which leads to the reduction of nanoparticles agglomeration and thus enhancement of nanofluid stability. It is worth noting that only low vapor pressure fluids are compatible with this process, which limits the application of the one-step method.

Estman et al. [6] applied a one-step physical vapor condensation technique to synthesize Cu/ethylene glycol nanofluid; Cu vapor was condensed as nanoparticles in contact with flowing EG fluid. Water-based nanofluid with Cu nanoparticles was synthesized via a chemical reduction technique in Liu's [7] experiment for enhancing its thermal conductivity. A single-step method was also applied for preparing CuO nanoparticles in deionized water suspension in order to well-controll nanoparticles size and to produce uniformly distributed CuO nanofluid [5,8]. Zhu et al. [9] used a new one-step chemical technique to produce stable and nonagglomerated Cu nanofluid into the EG as a base fluid by reducing $CuSO_4 \cdot 5H_2O$ with $NaH_2PO_2 \cdot H_2O$ under microwave irradiation condition. Table 2.1 summarizes various nanofluids prepared via single-step methods.

Although stable nanofluids could be prepared via these single-step methods, the preparation process is complicated and the applied equipment needs to be of high quality. Therefore this preparation method is expensive, especially for large-scale applications. In the meantime, this method is only suitable for the production of small quantities of nanofluid.

2.2.2 Two-step method

The two-step technique is another method of nanofluids preparation in which prefabricated nanoparticles are dispersed into the base fluid. This method consists of two processing steps. In the first processing step, inert gas condensation or other suitable methods are used for synthesizing nanoparticles and then in the second step they should be dispersed in the base fluid [14]. Different two-step nanofluid preparation methods and the principles of nanofluid preparation using this method are demonstrated in Fig. 2.2A and B, respectively.

Since the nanofluid preparation step is separated from the nanoparticle synthesis, the agglomeration of nanoparticles can be taken place during both steps, particularly during processes of drying, storage, and transportation of particles. In addition to the settlement and clogging of tubes and microchannels, the agglomeration of nanoparticles can reduce thermal conductivity and mass transfer coefficient. Therefore the stability of prepared nanofluid is the most important challenge

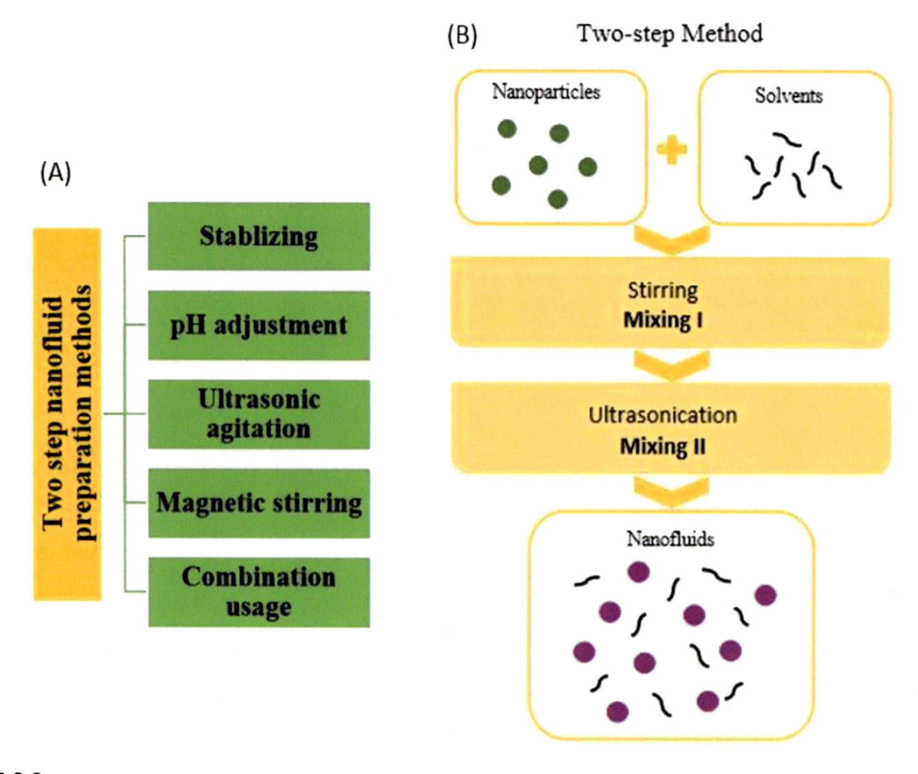

FIGURE 2.2

(A) Different two-step nanofluid preparation methods and (B) nanofluid synthesis procedure using two-step method.

of applying this technique. In order to have high stable and uniform nanoparticles, ultrasonic technology or high-shear-rate mixing devices are applied. A microfluidizer processor with over 1,000,000/s shear rate can be used for dispersing nanoparticles and reducing the cluster size in the viscoelastic fluids [15]. In addition, an appropriate surfactant such as sodium dodecylsulfate, acetic acid, polyvinylpyrrolidone (PVP), cetyltrimethylammonium bromide, oleic acid, sodium dodecylbenzenesulfonate (SDBS), and polyethylene glycol is used for stabilizing the nanoparticles in the base fluid, particularly when hydrophobic nanoparticles are applied [16]. Due to the industrial production of powder nanoparticles and scaling up the synthesis process of nanopowders by several companies, it can be found that the two-step technique is the most economical procedure. The stability of the prepared nanofluid is the most important challenge of applying this technique.

The two-step method was used to disperse Fe nanoparticles with 10 nm mean size into the EG solution in Hong et al.'s experiment [11,17]. They applied ultrasonic pulses with 700 W at 20 kHz to avoid nanoparticles agglomeration in the base fluid. Xuan et al. [5,9] prepared a copper nanofluid by dispersing Cu nanoparticles into the H_2O and oil via a two-stage method applying ultrasonic agitation and surfactant to prevent nanoparticles agglomeration. Magnetic force agitation and

Table 2.2 Summary of different nanofluids prepared via two-step method.

Nanoparticle type	Base fluid	Nanoparticle size (nm)	Nanoparticle loading (vol.%)	References
Cu	H_2O	100	7.5	[19]
SiC	H_2O	25	4.2	[20]
NCTs[a]	Engine oil	20–50	2	[21]
SiO_2	Deionized H_2O	12	4, 0.45, 1.85	[22]
TiO_2	EG/H_2O	50	0.5–1.5	[23]
TiO_2	H_2O	6	1,1.5, 2	[24]
Al_2O_3	H_2O	45,150	1,2,4, 6 wt.%	[25]
Al_2O_3	H_2O	20	40.2	[26]
ZnO	EG	50	0.005, 0.0375	[27]
Fe_2O_3	H_2O	40	0.02	[28]
Au	Ethanol	4	0.6	[29]
Au	Toluene	10–20	0.00026	[30]
NCTs	Poly oil	25 nm × 50 μm	1	[31]
NCTs	Decene	15 × 30 μm	1	[32]
CuO	Oil	Nanorods with D = 40–60 nm and L = 100–120 nm	0.8	[33]

[a] *Nanocomposite tectons.*

ultrasonic agitation are used in the preparation of Al_2O_3/H_2O and Al_2O_3/EG nanofluids by a two-step method [18]. Different nanofluids prepared via the two-step method are summarized in Table 2.2.

2.2.3 Phase transfer method

This novel nanoparticle synthesis method can be used for preparing organosols of noble metals in which nanoparticles are formed by changing the polar synthesis environment to a nonpolar one, or vice versa, as illustrated in Fig. 2.3 [34]. Therefore the poor solubility of metal ion precursors being a limitation of direct synthesis methods can be solved by the use of the phase transfer method. In addition, as the phase transfer method has the ability of creating close-packed, nanoparticle thin films and ordered arrays, it can be applied for functionalizing metallic nanoparticles with suitable ligands [35].

As an example, Hirai and Aizawa [36] synthesized stable colloidal Au particles in n-hexane and cyclohexane using phase transfer method. Au dispersed nanoparticles into the water were transferred to hexane with magnesium chloride and sodium oleate additives. This method is also applied for preparing stable kerosene-based Fe_3O_4 nanofluid in which oleic acid is grafted onto the Fe_3O_4 surface for increasing the compatibility of Fe_3O_4 nanoparticles with kerosene [37].

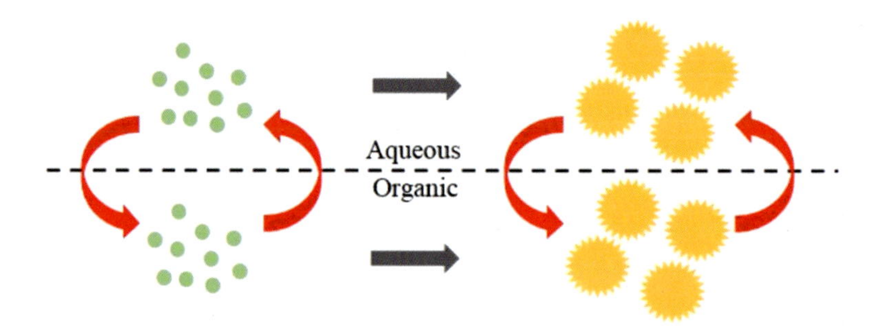

FIGURE 2.3

Phase transfer nanofluid synthesis procedure.

2.2.4 Posttreatment method

This method is suitable for systems in which agglomerated nanoparticles are formed due to the poor dispersed raw fluids [38]. It was proven that a temperature increase in posttreatment has a positive influence on the content and size of the fcc-Fe_3O_4 nanoparticles [39]. In addition, it was concluded that a high-pressure homogenizer has the best performance compared with other dispersion methods, like ultrasonic bath, stirrer, and ultrasonic disrupter, for achieving more stable nanofluids [14].

2.3 The stability of nanofluids

The stability of nanofluids is crucial in the industrial field. Since the nanofluids system is multidispersion with high surface energy, it is thermodynamically unstable, and nanoparticles dispersion can deteriorate as a result of van der Waals forces. It should be noted that as no chemical reaction occurs between dispersed nanoparticles and the base fluid, their stability depends on sedimentation and aggregation phenomena which can be offset by nanoparticles mobility due to the strong Brownian motions of dispersed nanoparticles. To prepare the stable nanofluid, the interaction between nanoparticle–nanoparticle and nanoparticle–liquid should be considered. In the following, different methods for enhancing the nanofluid stability are introduced.

2.3.1 Surfactant addition to the nanofluid

The addition of a surfactant to the nanofluid, a so-called dispersant, is an easy as well as economical method for enhancing nanofluid stability. The surfactant affects the surface characteristics of the nanofluid system in a small quantity as it consists of a hydrophobic tail portion, usually a long-chain hydrocarbon, and a hydrophilic polar head group. Actually, surfactants are employed for increasing the contact between two materials, sometimes known as wettability, by locating at the interface of the two phases. Based on the composition of the head, surfactants are divided into four classes: nonionic surfactants without charge groups on their head, like polyethylene oxide and alcohols; anionic surfactants with negatively charged head groups, such as long-chain fatty acids, sulfonates, sulfosuccinates, phosphates, and alkyl sulfates; cationic surfactants with positively charged head groups, like protonated long-chain amines and long-chain quaternary ammonium compounds; and amphoteric surfactants with zwitterionic head groups whose charge depends on the pH, such as betaines and certain lecithins [1].

Selection of an appropriate surfactant is the key issue. Generally, for polar base fluid of a nanofluid water-soluble or oil-soluble surfactants should be selected, and for nonionic surfactants we can evaluate the solubility through the term hydrophilic/lipophilic balance (HLB) value, which can be obtained easily via many handbooks: the lower the HLB number, the more oil-soluble is the surfactant, and in turn, the higher the HLB number, the more water-soluble is the surfactant. Although the addition of surfactant is an effective way for enhancing the nanoparticles dispersion, it may cause several challenges. As an example, surfactants may contaminate the heat transfer media, produce foams when heated in the heat exchange systems, and increase the thermal resistance between the nanoparticles and the base fluid, which may limit the effective thermal conductivity enhancement [40].

2.3.2 Surface modification methods: surfactant-free method

A promising method for achieving nanofluid long-term stability is using functionalized nanoparticles. Liu and Yang [41] synthesized functionalized SiO_2 nanoparticles by grafting silanes to their surface directly in an original nanoparticle solution. Chemical modification to functionalize the surface of CNTs is a common method for enhancing their stability in solvents. Hwang et al. [42] introduced hydrophilic functional groups on the nanotubes surface via a mechanochemical reaction which showed good fluidity, high stability, high thermal conductivity, and low viscosity applicable in advanced thermal systems as coolant. In another work with CNTs, wet mechanochemical reaction was applied to prepare surfactant-free nanofluids containing CNTs. Results from the infrared spectrum and zeta potential measurements revealed that the hydroxyl groups had been introduced onto the treated CNT surfaces [43]. Plasma treatment was also used for modifying the surface characteristics of diamond nanoparticles. Through this process, various polar groups were imparted on the diamond nanoparticles surface enhancing their dispersion property in water [44]. A stable dispersion of titania nanoparticles in an organic solvent of diethylene glycol dimethylether (diglyme) was successfully prepared using a ball milling process and then modified with silane coupling agents, (3-acryl-oxypropyl)trimethoxysilane and trimethoxypropylsilane, during the centrifugal bead mill process to enhance their dispersion stability in the solution [45].

2.4 Characterization techniques of nanofluids

Various methods have been used for evaluating the stability as well as properties of nanofluids. These are discussed in the following sections.

2.4.1 Measurement of zeta potential

The potential difference between the stationary layer of the solution which is attached to the nanoparticle and the dispersion medium is called Zeta potential. The repulsive interactions between the

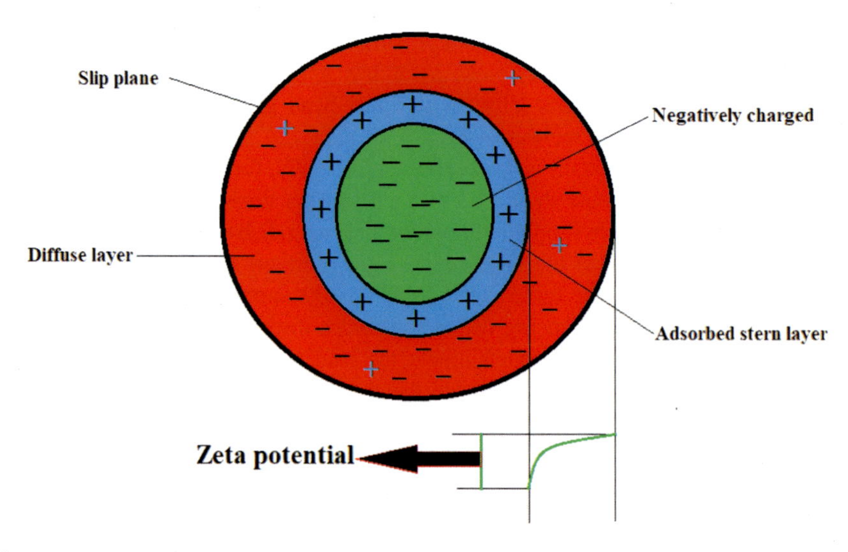

FIGURE 2.4

Zeta potential of nanofluids.

dispersant and nanoparticle can be determined by this quantitative analysis (Fig. 2.2). The unit of Zeta potential is mV and it is denoted by ζ. The nanofluid with higher Zeta potential value indicates more stability in nanosuspension. When the Zeta potential of colloidal suspension is 0 to ± 5 mV, the stability of nanofluid is poor while the agglomeration rate is high. But the nanosuspension Zeta potential of more than ± 30 mV shows that the nanofluid is stable for a long time. Furthermore, nanofluids with more than ± 60 mV Zeta potential have excellent stability [46]. Surfactant addition, surface coating, changing the suspension pH, and functionalization of nanoparticles can change the values of Zeta potential. Several researchers have investigated the Zeta potential test to evaluate the stability of nanofluids. Kim et al. [47] employed Zeta potential analysis to evaluate the stability of Au nanofluid and found the standing stability of Au particles. Allouni et al. [48] investigated the aggregation effects of TiO_2 nanoparticles in cell culture medium by the use of Zeta potential analysis experimentally. Zhu et al. [21] evaluated the Zeta potential of alumina–H_2O nanofluid under different concentrations of SDBS and different pH values (Fig. 2.4).

2.4.2 Dynamic light scattering

Dynamic light scattering (DLS) can be used to measure the size distribution of colloidal suspension nanoparticles. Fig. 2.5 indicates different parts of the DLS measurement setup. A laser is employed for illuminating the suspended nanoparticles in the base fluid. The scattered light fluctuation which arises from the Brownian motion of nanoparticles is monitored by a photon detector. By the use of light scattering theory, the diffusion coefficient can be measured from the intensity fluctuation. When the laser light passes through the colloid, the light will be scattered uniformly by nanoparticles that are less than 250 nm. By measuring the time of scattered light intensity, the nanoparticles

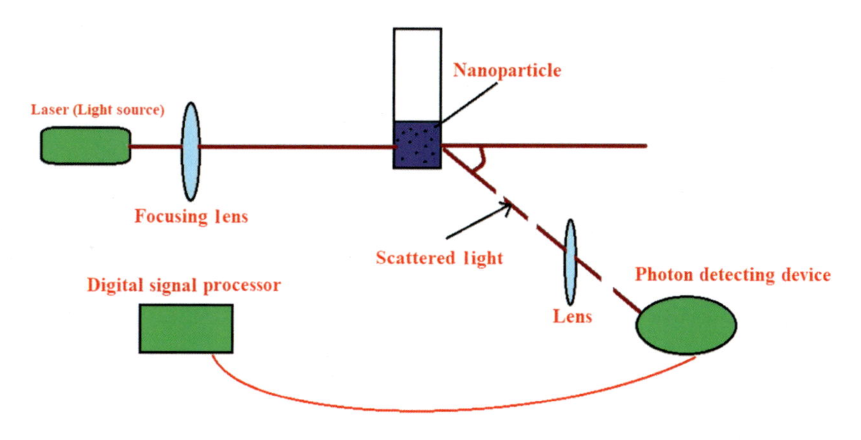

FIGURE 2.5

Schematic of dynamic light scattering measurement technique.

size in the colloid can be determined using the measured diffusion coefficient and the Stokes–Einstein equation, as below [49]:

$$R_H = \frac{K_B T}{6\pi\mu D} \tag{2.1}$$

where R_H is hydrodynamic radius, K_B is Boltzmann constant, T is absolute temperature, μ is viscosity, and D is translational diffusion coefficient. The nanoparticles size measurement at different time intervals over a long period of time illustrates the tendency for nanoparticles to agglomerate. The instability of the nanofluid can form the clusters that lead to sedimentation. Thus DLS analysis can indicate the enhancement of nanoparticles size with time due to the particles agglomeration and is known as an indicator of nanofluid instability. Kole et al. [50] compared the CuO clusters size with primary nanoparticles size and found that the CuO cluster sizes were increased up to nine times in gear oil because of the instability of the nanofluid. Sadeghi et al. [51] observed that the cluster size in Al_2O_3-H_2O nanofluid increased gradually with time which led to the aggregation of nanoparticles. Measuring very polydisperse systems is one of the most important drawbacks of DLS technique. The scattered light intensity is proportional to the nanoparticles diameter. Thus the measured sizes are correct, but nanoparticles distribution in these sizes should be weighted based on their diameter.

2.4.3 3ω technique

By considering the change in thermal conductivity of nanofluid because of agglomeration and sedimentation of nanoparticles, the 3ω technique can measure instability of nanofluids [52]. This technique can be employed in wide range of volume fraction of nanofluids. Oh et al. [53] found that the thermal conductivity of Al_2O_3-H_2O and Al_2O_3-EG nanofluids increased with time, due to the agglomeration and clustering of nanoparticles. The formation of nanoparticle clusters has an optimal degree where further agglomeration leads to a decrease in thermal conductivity [26]. The effect of sedimentation and agglomeration of nanoparticles on thermal conductivity of CuO-H_2O nanofluid has shown that the thermal conductivity reduces up to 22.5% during a 20 min time interval. The decrease in nanofluid thermal conductivity is due to the formation of clusters of nanoparticles which leads to a decrease in the level of thermal interaction and the surface-to-volume ratio.

2.4.4 Sedimentation and centrifugation

Sedimentation is the most traditional method for the evaluation of nanofluids stability. This method can evaluate the stability of nanofluid by measuring the formed sediments at the bottom of the liquid column; the schematic can be seen in Fig. 2.6. Based on this technique, the longer time for the formation of sediments indicates the superior nanofluid stability. The sedimentation technique has been employed by several researchers to evaluate the stability of nanofluids [54–56]. The centrifugation method is suggested as an alternative technique to the sedimentation method due to the lower time for the evaluation of nanofluids stability. The advantage of this technique is its much stronger centrifugal force than the gravitational force which leads to the acceleration of the sedimentation process. Singh et al. [57] applied a centrifugation procedure to enhance the stability of Ag nanoparticles into the ethanol base fluid with PVP as stabilizer. During a 10-h time interval, no sign of sedimentation was observed, whereas after a 1-month time interval sedimentation signs were observed.

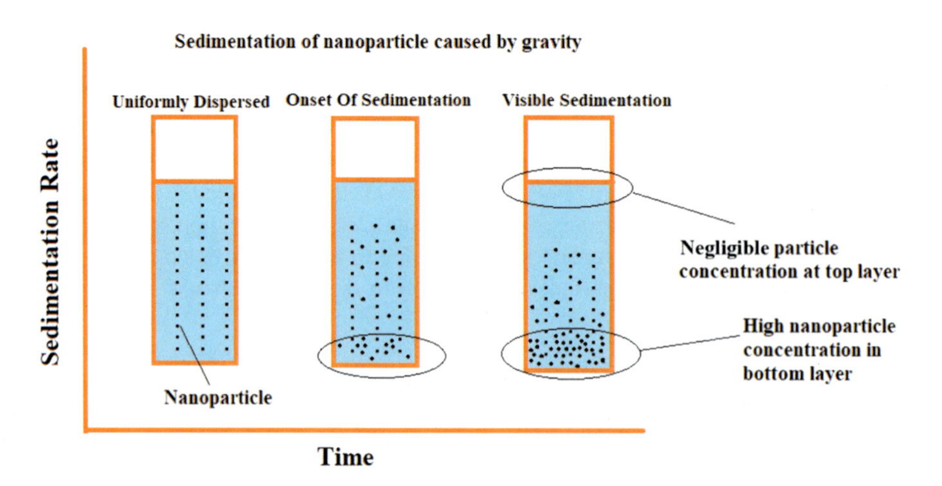

FIGURE 2.6

Schematic of sedimentation measurement method.

By applying Stokes law regime ($Re_p < 1$), the terminal settling velocity (V_t) for the smooth-spherical shape nanoparticles in the centrifuge can be calculated, via Eq. 2.2 [58]:

$$V_t = \frac{D_P^2\left(\rho_p - \rho_0\right)\omega^2 X}{18\mu} \tag{2.2}$$

where ρ_p and ρ_0 are density of nanoparticle and base fluid, D_p is the diameter of nanoparticle, μ is viscosity of base fluid, ω is centrifuge angular velocity, and X is distance between axis of rotation and specific location of the tube inside the centrifuge. From Eq. 2.2, it can be found that lower nanoparticle size, higher base fluid viscosity, and higher nominal difference between base fluid and nanoparticle density are key issues for achieving an improved stable nanofluid. It is worth noting that reducing the particle size of nanoparticles can increase the surface energy that leads to clustering of nanoparticles. Higher settling velocity leads to faster nanoparticles settling. In the general point of view, the centrifugation and sedimentation methods can evaluate the stability of nanofluid quantitatively by measuring the sediment layer height with time.

2.4.5 Measurement of transmittance and spectral absorbance

Similar to the sedimentation and centrifugation technique, this method can measure the stability of nanofluids quantitively. The nanoparticles that are suspended in the base fluid tend to absorb the light in the range of ultraviolet and visible frequency. Generally, this method can be used if the peak of adsorption of suspended nanoparticles in the base liquid is between 190 and 1100 nm [32]. In this method, the stability of nanofluid can be measured by measuring the transmittance within various time intervals. Fig. 2.7 schematically indicates the experimental setup to measure the absorbance and transmittance. In this arrangement, an He-Ni laser with 632-nm wavelength is employed to provide light source. An UV-spectrometer or Charged-coupled device (CCD) camera has been used for capturing the image at a specific wavelength. Increasing the transmittance with time indicates the instability of the nanofluid.

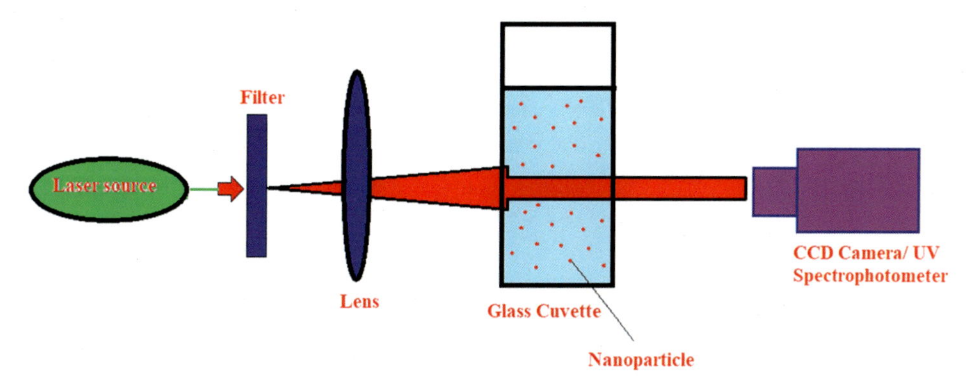

FIGURE 2.7

Schematic of transmittance/absorbance measurement technique.

Absorbance (A_λ) is proportional to nanoparticles concentration in the base fluid which is described as follows:

$$A_\lambda = \log_{10}(Io/I) = \alpha \times l \times c \tag{2.3}$$

where Io and I are, respectively, incident laser light intensity and intensity of laser light beam when it has passed through the suspended fluid; α is absorptivity, l and c are light path length and nanoparticles concentration, respectively. Decreasing the concentration of particles with time due to the sedimentation results in decreasing the absorbance value. However, measuring the stability of nanofluids with this method is difficult for dark colored nanofluids as well as nanofluids with high concentrations of nanoparticles [33,34]. Also, transmittance measurement can be employed to evaluate the stability of the nanofluid. The relation between transmittance (T_λ) and absorbance (A_λ) is as shown in Eq. 2.4:

$$T_\lambda = I/I_0 \tag{2.4}$$

2.4.6 Transmission electron microscopy

Transmission electron microscopy (TEM) can be used to observe the agglomeration of nanoparticles. Also, it is an alternative method to measure the distribution of nanoparticles size for evaluating the stability of nanofluids. This method provides a two-dimensional image of the nanoparticles dispersed into the base fluid. The electron beam can be applied in order to observe the nanosuspension features on the nanometer scale. It should be noted that the most important aspects of using TEM technique is that the samples must be dried in the suspension system to attach to the carbon matrix and therefore the agglomeration of nanoparticles can occur within the drying process as a result of the existence of nanoparticles in a different state with nanofluid. Also, the TEM technique can be integrated with the DLS method to obtain the exact sizing in nanofluid form. This procedure has been used by several researchers to evaluate the shape of various nanofluids, nanoparticles size, and nanoparticles agglomeration [52,55,56,59]. Chakraborty et al. [56,59] experimentally studied the effect of the addition of polymer and surfactant on the dispersion of Al-Cu-Zn and TiO_2 nanoparticles in aqueous solution and compared TEM stability results with sedimentation measurement and zeta potential. They concluded that TEM measurement results are consistent with the aforementioned methods.

The scanning electron microscope (SEM) method is also applicable for studying the morphology and microstructures of nanoparticles. The basic SEM method is similar to the TEM but the resolution of SEM is much lower than TEM.

2.4.7 Neuron activation analysis

One of the main problems in the characterization of nanofluids is the concentration and contamination of nanoparticles. Neuron activation analysis (NAA) is a nuclear method for determining the concentration of nanoparticles in the nanofluid. In this method, radioactive isopropes are formed after bombing the sample with neutrons and therefore the gamma decay emissons of the nanofluid are analyzed. Irradiation time depends on used nanoparticles and the gamma decay modes half-life.

Around 60 elements can be used in the NAA method. Due to the long half-life of zirconium, this method is applicable for the characterization of zirconium nanofluid.

2.4.8 Thermogravimetry analysis

The thermogravimetry analysis (TGA) method can analyze various components of nanofluid by weight. The TGA technique is suitable when the amount of polymer added to improve the nanofluid stability is known or the exact weight percentage of nanoparticles in the base liquid is evaluated. It is also applicable for determining the thermal stability of nanofluids for heat exchange processes.

2.4.9 Inductively coupled plasma

The inductively coupled plasma (ICP) method can be used to determine the composition as well as the charge of nanoparticles. This method is more time-consuming but cost-effective. By the use of the ICP method the contamination and concentration characterization of nanofluid can be determined more effectively. This method is based on the use of the light spectrum that is released from injecting substances into the high temperature plasma. Around 60 elements can be detected by the ICP test.

2.4.10 X-ray powder diffraction

X-ray powder diffraction (XRD) is a rapid analytical technique primarily used for phase identification of a crystalline structure of nanoparticles and can provide information on unit cell dimensions. The analyzed material is finely ground, homogenized, and the average bulk composition is determined. The X-rays are generated by a cathode ray tube, filtered to produce monochromatic radiation, collimated to concentrate them, and then directed toward the sample. The interaction of the incident rays with the sample produces constructive interference (and a diffracted ray) when conditions satisfy Bragg's Law ($n\lambda = 2d \sin \theta$). This law relates the wavelength of electromagnetic radiation to the diffraction angle and the lattice spacing in a crystalline sample. These diffracted X-rays are then detected, processed, and counted. By scanning the sample through a range of 2θ angles, all possible diffraction directions of the lattice should be attained because of the random orientation of the powdered material.

2.4.11 Fourier-transform infrared spectroscopy

Fourier-transform infrared spectroscopy (FT-IR) can be employed for studying the surface chemistry of solid nanoparticles. An FT-IR spectrometer collects high-resolution spectral data over a wide spectral range simultaneously.

From reviewing the characterization studies, it can be found that crucial information like nanoparticles shape, size, distribution, chemical bonds, and stability are obtained by characterization methods. But different researchers have employed various sets of procedures that are not recommended as the standard tests for confirmation of stable and homogenous nanofluid. It needs standard accelerated tests to confirm the long-term stability of prepared nanofluid. Table 2.3 summarize different methods that are applied for characterization of nanofluids by various techniques.

Table 2.3 Summary of different methods for characterizing nanofluids.

Nanoparticle	Nanoparticle concentration (vol.%)	Preparation method	Base fluid	Characterization technique	References
GNPs[a]	0.1–0.5 wt.%	Two-step	H_2O + EG	SEM-TEM-DLS	[60]
Graphite flakes	1–4	Two-step	Red wine	XRD-FT IR-Zeta potential-FE-SEM	[61]
GE	0.05, 0.1, 0.15	Two-step	DW	SEM-Zeta potential	[62]
Ag/HEG[b]	0.01–0.07	Two-step	Deionized water, EG	XRD, FT-IR, FE-SEM, TEM	[63]
ZnO	0.2–5	Two-step	EG	SEM	[64]
TiO_2	5	Two-step	Deionized water	TEM	[65]
TiO_2	0.15, 0.013, 1	Two-step	Deionized water	SEM, DLS	[66]
SiO_2 Al_2O_3	0.005–0.5	Two-step	Pure methanol	SEM, DLS	[67]
CuO/HEG	0.01–0.07	Two-step	Deionized water, EG	XRD, FT-IR, FE-SEM, TEM	[68]
CuO	0.5–2	Two-step	Oleic acid	TEM, XRD	[69]
Al_2O_3	0.1–1	Two-step	ATF	TEM	[70]

DLS, dynamic light scattering; *FT-IR*, Fourier-transform infrared spectroscopy; *SEM*, scanning electron microscope; *TEM*, transmission electron microscopy; *XRD*, X-ray powder diffraction.
[a]*Graphite Nanoplatelets.*
[b]*Graphene.*

2.5 Conclusion

This chapter has provided a brief review on different preparation techniques of nanofluids and discusses the stability and various characterization methods of nanofluids. From a general point of view, preparation methods are divided into single-step, two-step, posttreatment, and phase change methods. The advantage of the one-step method is the elimination of such preparation processes as drying, storage, transportation, and dispersion of nanoparticles into the base fluid, which can reduce the nanoparticles agglomeration, and thus enhances their stability in the nanofluids. However, the single-step method is not a cost-effective technique, especially for large-scale applications, thus leading to the application of other methods for industries. The preparation of nanofluids via the two-step technique is simpler than the single-step method but in this method the nanoparticles' stability is lower and aggregation of the nanoparticles can take place during different preparation processes. The stability of nanofluids is the main issue in the preparation process. The addition of surfactant to the nanofluid is an easy and economic method for enhancing nanofluid stability by increasing the contact between nanoparticles and base fluid. However, surfactants may contaminate the heat transfer media, produce foams, and enlarge the thermal resistance between the nanoparticles and the base fluid which may limit the effective thermal conductivity enhancement. Therefore functionalizing the nanoparticles surface is a promising method for achieving nanofluid long-term stability. To evaluate the stability as well as the properties of nanofluids, different methods are suggested, such as zeta potential, DLS, 3ω technique, centrifugation, and sedimentation. Each technique has a specific feature that must be selected to suit the intended application. Future research will investigate in detail the new methods of preparation, stability, and properties, and introduce a specific method that can be used to prepare metallic and nonmetallic nanofluids.

Abbreviations

CCD	charged-coupled device
CNT	carbon nanotubes
CTAB	cetyltrimethylammonium bromide
DLS	dynamic light scattering
EG	ethylene glycol
FT-IR	fourier-transform infrared spectroscopy
GNPs	graphite nanoplatelets
GO	graphene oxide
HLB	hydrophilic/lipophilic balance
HEG	graphene
ICP	inductively coupled plasma
NAA	neuron activation analysis
NCTs	nanocomposite tectons
PEG	poly ethylene glycol
PVP	polyvinylpyrrolidone
SDBS	sodium dodecylbenzenesulfonate
SDS	sodium dodecylsulfate
SEM	scanning electron microscope
TEM	transmission electron microscopy
TGA	thermogravimetry analysis
XRD	X-ray powder diffraction

References

[1] W. Yu, H. Xie, A review on nanofluids: preparation, stability mechanisms, and applications, J. Nanomaterials 2012 (2012).

[2] M.A. Makarem, A. Bakhtyari, M.R. Rahimpour, A numerical investigation on the heat and fluid flow of various nanofluids on a stretching sheet, Heat. Transfer: Asian Res. 47 (2) (2018) 347–365.

[3] M. Mofarahi, et al., Numerical modeling of fluid flow and thermal behavior of different nanofluids on a rotating disk, Heat. Transfer: Asian Res. 45 (4) (2016) 358–378.

[4] S. Lee, et al., Measuring thermal conductivity of fluids containing oxide nanoparticles, J. Heat Transfer 121 (1999) 280–289.

[5] C.-H. Lo, et al., Fabrication of copper oxide nanofluid using submerged arc nanoparticle synthesis system (SANSS), J. Nanopart. Res. 7 (2–3) (2005) 313–320.

[6] J.A. Eastman, et al., Anomalously increased effective thermal conductivities of ethylene glycol-based nanofluids containing copper nanoparticles, Appl. Phys. Lett. 78 (6) (2001) 718–720.

[7] M.-S. Liu, et al., Enhancement of thermal conductivity with Cu for nanofluids using chemical reduction method, Int. J. Heat. Mass. Transf. 49 (17–18) (2006) 3028–3033.
[8] C.-H. Lo, T.-T. Tsung, L.-C. Chen, Shape-controlled synthesis of Cu-based nanofluid using submerged arc nanoparticle synthesis system (SANSS), J. Cryst. Growth 277 (1–4) (2005) 636–642.
[9] H.-t Zhu, Y.-s Lin, Y.-s Yin, A novel one-step chemical method for preparation of copper nanofluids, J. Colloid Interface Sci. 277 (1) (2004) 100–103.
[10] H. Zhu, et al., Effects of nanoparticle clustering and alignment on thermal conductivities of Fe_3O_4 aqueous nanofluids, Appl. Phys. Lett. 89 (2) (2006) 023123.
[11] T.-K. Hong, H.-S. Yang, C. Choi, Study of the enhanced thermal conductivity of Fe nanofluids, J. Appl. Phys. 97 (6) (2005) 064311.
[12] Y.-H. Hung, T.-P. Teng, B.-G. Lin, Evaluation of the thermal performance of a heat pipe using alumina nanofluids, Exp. Therm. Fluid Sci. 44 (2013) 504–511.
[13] M. Khoshvaght-Aliabadi, S. Pazdar, O. Sartipzadeh, Experimental investigation of water based nanofluid containing copper nanoparticles across helical microtubes, Int. Commun. Heat Mass. Transf. 70 (2016) 84–92.
[14] Y. Hwang, et al., Production and dispersion stability of nanoparticles in nanofluids, Powder Technol. 186 (2) (2008) 145–153.
[15] Y. Wu, et al., Rheological characteristics of viscoelastic surfactant fluid mixed with silica nanoparticles, in: ASME 2013 Fourth International Conference on Micro/Nanoscale Heat and MASS Transfer, American Society of Mechanical Engineers Digital Collection, (2013).
[16] N. Nakayama, T. Hayashi, Preparation of TiO_2 nanoparticles surface-modified by both carboxylic acid and amine: dispersibility and stabilization in organic solvents, Colloids Surf. A: Physicochem. Eng. Asp. 317 (1–3) (2008) 543–550.
[17] K. Hong, T.-K. Hong, H.-S. Yang, Thermal conductivity of Fe nanofluids depending on the cluster size of nanoparticles, Appl. Phys. Lett. 88 (3) (2006) 031901.
[18] H. Xie, et al., Thermal conductivity enhancement of suspensions containing nanosized alumina particles, J. Appl. Phys. 91 (7) (2002) 4568–4572.
[19] Y. Xuan, Q. Li, Heat transfer enhancement of nanofluids, Int. J. Heat Fluid Flow. 21 (1) (2000) 58–64.
[20] X. Huaqing, et al., Study on the thermal conductivity of SiC nanofluids, J. Chin. Ceram. Soc. 29 (4) (2001) 361–364.
[21] M.-S. Liu, et al., Enhancement of thermal conductivity with carbon nanotube for nanofluids, Int. Commun. Heat Mass. Transf. 32 (9) (2005) 1202–1210.
[22] I. Tavman, et al., Experimental investigation of viscosity and thermal conductivity of suspensions containing nanosized ceramic particles, Arch. Mater. Sci. 100 (100) (2008).
[23] K.A. Hamid, et al., Effect of titanium oxide nanofluid concentration on pressure drop, ARPN J. Eng. Appl. Sci. 10 (17) (2015) 7815–7820.
[24] T. Kavitha, A. Rajendran, A. Durairajan, Synthesis, characterization of TiO2 nano powder and water based nanofluids using two step method, Eur. J. Appl. Eng. Sci. Res. 1 (2012) 235–240.
[25] J. Qu, H.-y Wu, P. Cheng, Thermal performance of an oscillating heat pipe with Al_2O_3–water nanofluids, Int. Commun. Heat Mass. Transf. 37 (2) (2010) 111–115.
[26] P.E. Gharagozloo, K.E. Goodson, Temperature-dependent aggregation and diffusion in nanofluids, Int. J. Heat Mass. Transf. 54 (4) (2011) 797–806.
[27] M. Kole, T. Dey, Thermophysical and pool boiling characteristics of ZnO-ethylene glycol nanofluids, Int. J. Therm. Sci. 62 (2012) 61–70.
[28] Y. Vermahmoudi, et al., Experimental investigation on heat transfer performance of Fe2O3/water nanofluid in an air-finned heat exchanger, Eur. J. Mech. B Fluids 44 (2014) 32–41.
[29] S.A. Putnam, et al., Thermal conductivity of nanoparticle suspensions, J. Appl. Phys. 99 (8) (2006) 084308.

[30] H.E. Patel, et al., Thermal conductivities of naked and monolayer protected metal nanoparticle based nanofluids: manifestation of anomalous enhancement and chemical effects, Appl. Phys. Lett. 83 (14) (2003) 2931–2933.

[31] S. Choi, et al., Anomalous thermal conductivity enhancement in nanotube suspensions, Appl. Phys. Lett. 79 (14) (2001) 2252–2254.

[32] H. Xie, et al., Nanofluids containing multiwalled carbon nanotubes and their enhanced thermal conductivities, J. Appl. Phys. 94 (8) (2003) 4967–4971.

[33] N. Li, et al., Nanofluids containing stearic acid-modified CuO nanorods and their thermal conductivity enhancements, Nanosci. Nanotechnol. Lett. 7 (4) (2015) 314–317.

[34] Z. Zhang, et al., Progress in enhancement of CO2 absorption by nanofluids: a mini review of mechanisms and current status, Renew. Energy 118 (2018) 527–535.

[35] J. Yang, J.Y. Lee, J.Y. Ying, Phase transfer and its applications in nanotechnology, Chem. Soc. Rev. 40 (3) (2011) 1672–1696.

[36] H. Hirai, H. Aizawa, Preparation of Stable Dispersions of Colloidal Gold in Hexanes by Phase Transfer, Elsevier, 1993.

[37] W. Yu, et al., Enhancement of thermal conductivity of kerosene-based Fe_3O_4 nanofluids prepared via phase-transfer method, Colloids Surf. A: Physicochem. Eng. Asp. 355 (1–3) (2010) 109–113.

[38] S. Oh, et al., Effect of ZnO nanoparticle morphology and post-treatment with zinc acetate on buffer layer in inverted organic photovoltaic cells, Sol. Energy 114 (2015) 32–38.

[39] L. Fang, et al., Experimental study on enhancement of bubble absorption of gaseous CO_2 with nanofluids in ammonia, J. Harbin Inst. Technol. 24 (2) (2017) 80–86.

[40] L. Chen, et al., Nanofluids containing carbon nanotubes treated by mechanochemical reaction, Thermochim. Acta 477 (1–2) (2008) 21–24.

[41] X. Yang, Z.-h Liu, A kind of nanofluid consisting of surface-functionalized nanoparticles, Nanoscale Res. Lett. 5 (8) (2010) 1324–1328.

[42] Y.-j Hwang, et al., Stability and thermal conductivity characteristics of nanofluids, Thermochim. Acta 455 (1–2) (2007) 70–74.

[43] L. Chen, H. Xie, Surfactant-free nanofluids containing double-and single-walled carbon nanotubes functionalized by a wet-mechanochemical reaction, Thermochim. Acta 497 (1-2) (2010) 67–71.

[44] K.A. Wepasnick, et al., Chemical and structural characterization of carbon nanotube surfaces, Anal. Bioanal. Chem. 396 (3) (2010) 1003–1014.

[45] I.M. Joni, et al., Dispersion stability enhancement of titania nanoparticles in organic solvent using a bead mill process, Ind. Eng. Chem. Res. 48 (15) (2009) 6916–6922.

[46] Z. Haddad, et al., A review on how the researchers prepare their nanofluids, Int. J. Therm. Sci. 76 (2014) 168–189.

[47] H.J. Kim, I.C. Bang, J. Onoe, Characteristic stability of bare Au-water nanofluids fabricated by pulsed laser ablation in liquids, Opt. Lasers Eng. 47 (5) (2009) 532–538.

[48] Z.E. Allouni, et al., Agglomeration and sedimentation of TiO_2 nanoparticles in cell culture medium, Colloids Surf. B: Biointerfaces 68 (1) (2009) 83–87.

[49] S. Bhattacharjee, DLS and zeta potential—what they are and what they are not? J. Control. Rel. 235 (2016) 337–351.

[50] M. Kole, T. Dey, Effect of aggregation on the viscosity of copper oxide–gear oil nanofluids, Int. J. Therm. Sci. 50 (9) (2011) 1741–1747.

[51] M.S. Sadeghi, et al., Analysis of hydrothermal characteristics of magnetic Al_2O_3-H_2O nanofluid within a novel wavy enclosure during natural convection process considering internal heat generation, Math. Methods Appl. Sci. (2020).

[52] N. Ali, J.A. Teixeira, A. Addali, A review on nanofluids: fabrication, stability, and thermophysical properties, J. Nanomater. 2018 (2018).
[53] D.-W. Oh, et al., Thermal conductivity measurement and sedimentation detection of aluminum oxide nanofluids by using the 3ω method, Int. J. Heat. Fluid Flow. 29 (5) (2008) 1456–1461.
[54] S. Chakraborty, et al., Synthesis of Cu-Al LDH nanofluid and its application in spray cooling heat transfer of a hot steel plate, Powder Technol. 335 (2018) 285–300.
[55] S. Chakraborty, et al., Thermo-physical properties of Cu-Zn-Al LDH nanofluid and its application in spray cooling, Appl. Therm. Eng. 141 (2018) 339–351.
[56] S. Chakraborty, et al., Experimental investigation on the effect of dispersant addition on thermal and rheological characteristics of TiO_2 nanofluid, Powder Technol. 307 (2017) 10–24.
[57] A.K. Singh, V.S. Raykar, Microwave synthesis of silver nanofluids with polyvinylpyrrolidone (PVP) and their transport properties, Colloid Polym. Sci. 286 (14) (2008) 1667–1673.
[58] R.G. Harrison, et al., Bioseparations Science and Engineering, Oxford University Press, New York, 2015.
[59] S. Chakraborty, et al., Effect of surfactant on thermo-physical properties and spray cooling heat transfer performance of Cu-Zn-Al LDH nanofluid, Appl. Clay Sci. 168 (2019) 43–55.
[60] R. Sadri, et al., A novel, eco-friendly technique for covalent functionalization of graphene nanoplatelets and the potential of their nanofluids for heat transfer applications, Chem. Phys. Lett. 675 (2017) 92–97.
[61] M. Mehrali, et al., An ecofriendly graphene-based nanofluid for heat transfer applications, J. Clean. Prod. 137 (2016) 555–566.
[62] N. Ahammed, L.G. Asirvatham, S. Wongwises, Effect of volume concentration and temperature on viscosity and surface tension of graphene–water nanofluid for heat transfer applications, J. Therm. Anal. Calorim. 123 (2) (2016) 1399–1409.
[63] T.T. Baby, S. Ramaprabhu, Synthesis and nanofluid application of silver nanoparticles decorated graphene, J. Mater. Chem. 21 (26) (2011) 9702–9709.
[64] W. Yu, et al., Investigation of thermal conductivity and viscosity of ethylene glycol based ZnO nanofluid, Thermochim. Acta 491 (1–2) (2009) 92–96.
[65] S. Murshed, K. Leong, C. Yang, Enhanced thermal conductivity of TiO_2—water based nanofluids, Int. J. Therm. Sci. 44 (4) (2005) 367–373.
[66] S. Bobbo, et al., Viscosity of water based SWCNH and TiO_2 nanofluids, Exp. Therm. Fluid Sci. 36 (2012) 65–71.
[67] C. Pang, et al., Thermal conductivity measurement of methanol-based nanofluids with Al2O3 and SiO2 nanoparticles, Int. J. Heat. Mass. Transf. 55 (21–22) (2012) 5597–5602.
[68] T.T. Baby, R. Sundara, Synthesis and transport properties of metal oxide decorated graphene dispersed nanofluids, J. Phys. Chem. C. 115 (17) (2011) 8527–8533.
[69] S. Harikrishnan, S. Kalaiselvam, Preparation and thermal characteristics of CuO–oleic acid nanofluids as a phase change material, Thermochim. Acta 533 (2012) 46–55.
[70] S. Sonawane, et al., An experimental investigation of thermo-physical properties and heat transfer performance of Al_2O_3-aviation turbine fuel nanofluids, Appl. Therm. Eng. 31 (14–15) (2011) 2841–2849.

CHAPTER

Thermophysical properties of nanofluids

3

Ali Bakhtyari[1] and Masoud Mofarahi[2,3]

[1]*Chemical Engineering Department, Shiraz University, Shiraz, Iran* [2]*Department of Chemical Engineering, Faculty of Petroleum, Gas and Petrochemical Engineering, Persian Gulf University, Bushehr, Iran*
[3]*Department of Chemical and Biomolecular Engineering, Yonsei University, Seoul, Republic of Korea*

3.1 Introduction

The increasing energy consumption, as well as declining sources, makes the industrial sector seek optimal strategies for energy utilization. In this regard, heat transfer facilities including cooling/heating equipment are of the main concern. The heat transfer facilities of the industrial sector require a dramatic improvement in the high-energy-efficiency scenario. Given the impact of working fluids upon heat transfer applications in various sectors of industries, choosing the optimal one is vital. The conventional working fluids such as water, oils, kerosene, and glycols have limited powers in terms of their thermal properties. This is the first obstacle to optimizing the performance of heat transfer facilities and then minimizing energy consumption [1−3]. Improving the thermal efficiency in terms of heat transfer capacity of the working fluid would be an effective route in this regard.

A novel strategy to improve the thermal efficiency of heat transfer facilities is to add solid particles to the working fluid, thorough which the transport properties of the fluid are enhanced. Such a strategy is not currently welcome for the liquids consisting of solid particles of millimeter-to-micrometer sizes. This is owing to their unstable state, which would result in further fast precipitation, pipeline blockage, increased erosion in pipeline and equipment, and pressure drop. Ultrafine nanosized particles are more stable than the larger ones and hence provide more contact with the fluid. Therefore employing nanosized particles is a promising means for diminishing the undesired after effects [3,4]. At the nanoscale the relative surface area (surface area-to-volume ratio) and quantum effects increase dramatically. As particle size decreases down toward the nanoscale, a major portion of atoms appear to be at the surface. This alters the properties significantly [4].

A revolution in heat transfer applications was first made by the introduction of nanoparticles (NPs) by Maxwell [5]. Masuda et al. [6] then proposed the alteration in the thermophysical properties of fluids that consisted of ultrafine particles. In 1995 a new class of fluids, known as nanofluids (NFs), was introduced by Choi and Eastman [7]. NFs are made up of a base fluid and a low mass percent of nanosized solid particles (1−100 nm) with high thermal conductivity. Thanks to their hitherto enhanced characteristics, NFs have found applications in a variety of systems including heat exchanging [2], refrigeration [8−10], automotive [11,12], solar collectors [1,13,14], nuclear reactors [15], electronics [16−18], and drilling facilities [19]. NFs opened up new vistas of

Nanofluids and Mass Transfer. DOI: https://doi.org/10.1016/B978-0-12-823996-4.00003-3

cooling/heating applications. Hence, the utilization of NFs can enhance the heat transfer rate as well as the stability of the working fluid. This would reduce costs, pipeline clogging or blockage, the size of the system, and the friction [20,21]. Extensive discussions upon the applications of NFs can be found elsewhere [3].

The enhanced thermal conductivity, heat transfer rate, and critical heat flux, as well as extreme stability, introduce NFs as promising working fluids. More discussions about the stability of NFs can be found elsewhere [4]. Employing NFs is then an efficient method to improve the heat transfer capacity [4]. Provided that NPs possess higher thermal conductivities, the heat transfer capacities of the NFs are considerably larger than those of the base fluids [15]. A variety of NPs can be employed to prepare NFs. Typical NPs include Cu, Al, K, Si, Fe, Ti, Zn, Ag, Zr, Mg, and Ce in either metallic or oxide states, as well as nonmetallic compounds such as carbons, carbides, and graphene oxide [22]. NPs can be available in different forms, such as nanotubes, nanofibers, nanoribbons, nanodots, and nanowires. Typical working fluids such as water, ethylene glycol (EG), propylene glycol (PG), polyethylene glycol (PEG), and alcohols could be utilized as the base fluid.

A variety of methods are employed to prepare NFs. These methods are classified into one-step and two-step routes, the first of which includes simultaneous NPs generation and dispersion in the base fluids [4,23]. Through the one-step routes, the intermediate steps including filtration, dehydration, drying, storage, and dispersion of NPs are eliminated, which results in the minimization of particle agglomeration and enhanced NFs stability [4,24]. The main drawback of the one-step routes is the presence of the impurities, that is, the unreacted residuals in the final NFs. The major one-step routes include direct evaporation [25–27], physical vapor condensation [28], chemical reduction [29–32], submerged arc [33,34], laser ablation [35–37], microwave irradiation [38–40], polyol process [41,42], and plasma discharging [43]. The two-step routes include the primary step, which includes synthesis, filtration, and drying of NPs, and a further step of dispersing NPs in the base fluids. The synthesis of NPs can be implemented through the reduction of precursors (transition metal salts) [44], ligand reduction and displacement from organometallics [45], microwave synthesis [46], and coprecipitation [47] methods. The dispersion step can be implemented by magnetic force agitation, ultrasonication, high-shear mixing, homogenizing, or ball milling. The two-step routes are cost-effective and thus are more suitable for large-scale operation. Hence, the two-step routes are more conventional. More details of the preparation methods can be found elsewhere [4,24,48]. The detailed procedures for the preparation of various NFs were reviewed by Devendiran and Amirtham [49].

A new class of NFs, called hybrid NFs, has been recently proposed. Hybrid NFs are prepared by dispersing either different individual NPs or nanocomposites in the base fluid. Hybrid NFs are then homogeneous mixtures of more than one nanospecies with new physical and chemical bonds. Hybrid NPs may possess better thermal and rheological characteristics compared to single-particle NFs due to synergistic effects. By a proper combination of NPs or composites through hybridizing, the desired properties and then heat transfer effects would be adjusted and optimized, even at low NPs concentrations. Hybridizing, on the other hand, makes NFs exhibit superior thermophysical properties and rheological behavior in tandem with the enhanced heat transfer characteristics [50–52]. Devendiran and Amirtham [49] reviewed the details of hybrid NFs preparation. A list of review studies that provide extensive discussions upon preparation, stability, properties, mechanism of heat transfer, applications, and challenges of NFs are introduced by Gupta et al. [3].

A major concern during the preparation of NFs is stability. Particle–particle and fluid–particle interactions determine the stability of NFs. NPs tend to agglomerate due to attractive van der Waals forces. A balance of the interparticle attractive van der Waals forces and double-layer electrical repulsive, hydration, and steric forces, on the one hand, account for the stability condition of the NFs [4,48,53]. In the case of dominant repulsive forces, no aggregates form, and thus the NFs remain stable. The dispersed NPs can be stabilized by three different mechanisms, that is, steric, electrostatic, or electrosteric stabilization. To utilize NFs in large-scale applications, long-term stability must be guaranteed. To do so, both chemical and physical methods would be applied. The chemical methods include the addition of surfactants, the adjustment of solution pH, and the modification of NPs surface, while the physical methods include ultrasonication, homogenization, and ball milling [48]. Moreover, due to a large difference between NPs and base fluids, the particles tend to sediment. Accordingly, the stability of NFs strongly depends on the characteristics of NPs and base fluids. Based on Stokes law, the terminal velocity of sedimentation of the particles is proportional to NPs size, NPs and fluid densities difference, and the inverse of NFs dynamic viscosity. Accordingly, increasing NPs size, increasing density difference, and decreasing NFs viscosity increase the sedimentation velocity, which must be avoided. Light scattering, zeta potential measurements, sedimentation and centrifugation, and spectral analysis are efficient techniques to evaluate the stability of NFs [4]. More details of stability conditions and strategies can be found elsewhere [4,48,50,53].

To establish the performance parameter of a heat transfer system (heat transfer coefficient, pressure drop, and energy efficiency), the thermophysical properties of NFs should be estimated, precisely. From a general point of view, the thermophysical properties including density, viscosity, and thermal conductivity affect the heat transfer behavior of NFs, among which thermal conductivity (k) is the most important one. Thermal conductivity is linked with the heat transfer coefficient. The density (ρ) and viscosity (μ) of NFs are also important in determining the heat transfer coefficient. Density determines the stability of NFs. It is also required to assess the weight and volume of the system that operates with NFs. The density of NFs affects the solubility power when utilized in the separation applications [54–57]. Viscosity is of immense importance in all transport phenomena in the chemical process [57,58]. Heat transfer coefficient, diffusion coefficient, and fluid flow are directly related to viscosity [57,59]. The most important impact of viscosity and density is in pipeline designing, by which the pressure drop and pumping power are determined [57]. The addition of NPs to the base fluid alters the aforementioned properties. These properties are determined by NPs characteristics (type, concentration, shape, and size), the properties of the base fluid, and temperature. A general view of the impact of various parameters on the thermophysical properties of NFs is depicted in Table 3.1. The relationship between increasing different parameters and various properties of NFs are different. In some cases, contradictory results are observed in previous studies [57]. Recent developments in the application of NFs highlight the necessity for the precise estimation of the thermophysical properties by employing empirical and/or theoretical methods [49].

Following the importance of the precise estimation of the properties, the present chapter is dedicated to giving a broader investigation of the thermophysical properties of NFs. Hence, the aforementioned properties are analyzed in the individual section of the present chapter. Furthermore, the impacts of the introduced influential parameters are assessed in each section. The methods of property estimation including the empirical and theoretical models are then introduced and discussed.

Table 3.1 The relationship between increasing different parameters and the thermophysical properties of nanofluids.

	Parameter		
Property	**NPs concentration**	**Particle size**	**Temperature**
Density (ρ)	Direct	—[a]	Inverse
Viscosity (μ)	Direct	—[a]	Inverse
Thermal Conductivity (k)	Direct	—[a]	Direct

[a] *Unknown or contradictory results.*
Adapted from I.M. Mahbubul, 4—Thermophysical properties of nanofluids, in: I.M. Mahbubul (Ed.), Preparation, Characterization, Properties and Application of Nanofluid, William Andrew Publishing, 2019. pp. 113–196 [57].

The details of the experimental apparatuses fall beyond the scope of the present contribution. It is worth noting that determining the thermophysical properties of NFs provides information about the transport phenomena. However, it does not clarify the mechanisms of these phenomena. The mechanisms of transport phenomena can be studied through molecular dynamics simulation studies, which fall beyond the scope of the present contribution. The outcome of the present contribution would shed light on the prospect of property estimation in the NFs systems. It would also help to select proper NFs for the considered processes.

3.2 Thermophysical properties

As discussed earlier, the addition of NPs into the base fluid results in considerable changes in the thermophysical properties. NPs type and characteristics change the properties to a different extent. Concentration, shape, and size of the NPs are the major effective factors, which alter the properties, significantly. Moreover, the temperature has a great impact on the properties of NFs. This section is devoted to the investigation of the thermophysical properties of NFs as well as the factors affecting these properties. To do so, density (ρ), viscosity (μ), and thermal conductivity (k) are investigated. The measurements, as well as the theoretical methods, will be then discussed.

3.2.1 Density (ρ)

Density is a measure of mass per unit volume of the system and is denoted by ρ (g/cm^3 or kg/m^3). Density affects the transport phenomena because it is related to Reynolds number (Re), Prandtl number (Pr), Schmidt number (Sc), Nusselt number (Nu), Sherwood number (Sh), friction factor (f), and pressure loss (ΔP). Hence, density estimation and determining the impact of various parameters are of great importance.

Mahbubul [57] summarized the apparatus and/or techniques employed for NFs density measurements. These techniques include weighing a known volume of the NFs, weighing NFs sample in a volumetric flask, measuring by digital density meters, measuring by hydrometers, measuring by vibrating tube densimeters, measuring by density/specific gravity meters, and calculated based on the principle of the mixture rule. A typical procedure of density measurement, as well as the accuracies and uncertainties, was also discussed by Mahbubul [57]. Density is strongly dependent on the NPs type, concentration, and temperature. However, the previous studies [57,60,61] reveal that NPs characteristics, such as size and shape, as well as NFs additives, such as stabilizers, do not affect the density of NFs, significantly. Increasing temperature results in a reduction of NFs density [62]. This is due to the effect of temperature on the density of the base fluid [63]. Solid particles possess larger densities than liquids. Typical densities are 2220 kg/m^3, 3600–3880 kg/m^3, 4175 kg/m^3, 6500 kg/m^3, and 5606 kg/m^3 for SiO_2, Al_2O_3, TiO_2, CuO, and ZnO NPs, respectively [63–65]. Therefore the density of the fluid would be increased by the addition of NPs. Hence, increasing NPs concentration increases the density of NFs. It is worth noting that there have been a few attempts to determine the density of NFs, experimentally. A major portion of the researchers, on the other hand, tend to utilize the principle of the mixture rule to determine the density of NFs, especially in the low-concentration range [19,66–77]. The first attempt to measure the density of NFs was made by Pak and Cho [64]. They measured the densities of γ-Al_2O_3/water and TiO_2/water NFs. They reported that increasing NPs concentration increased the densities of NFs. Yang et al. [78] measured the densities of the NFs with 2.0–2.5 wt.% graphite NPs in the different commercial automatic transmission fluids and synthetic base oils at 35°C and 43°C. When compared with the corresponding base fluids, the density of NFs did not show much change. This was due to the low concentration of NPs. Vajjha et al. [79] reported the experimental densities of the NFs that were composed of Al_2O_3, ZnO, and Sb_2O_5-SnO_2 NPs and a mixture of EG and water (60 wt.% EG, 40 wt.% water). The effect of NPs concentration on the density of Sb_2O_5-SnO_2/EG + water NFs is depicted in Fig. 3.1. A considerable density increase by increasing NPs concentration is observed in the whole temperature range. Fig. 3.2 also proves that increasing NPs concentration increases the

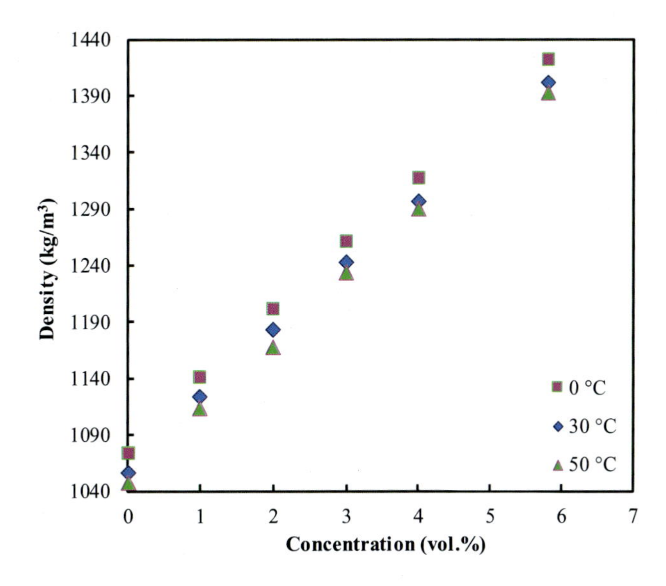

FIGURE 3.1

Effect of concentration on the density of Sb_2O_5-SnO_2/EG(60 wt.%) + water(40 wt.%) nanofluids [79].

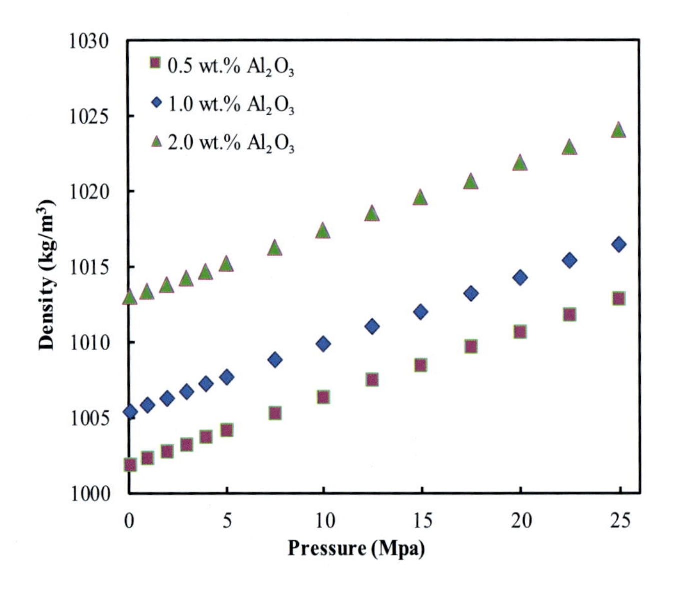

FIGURE 3.2

Effect of pressure on the density of Al_2O_3/water nanofluids at 25°C [61].

density of Al_2O_3/water NFs at various pressures. These data were collected at 25°C by Pastoriza-Gallego et al. [61]. Similar trends were also observed at 10°C and 60°C.

The densities of NFs made up of multiwalled carbon nanotube (MWCNT) particles and a mixture of water (50 vol.%) and EG (50 vol.%) were measured by Ganeshkumar et al. [80]. An ascending trend of MFs density with NPs concentration was observed. The densities of Al_2O_3/water NFs were measured in the wide ranges of NPs concentration (0.5–7.0 wt.%), temperature (10°C–40°C), and pressure (0.1–25.0 Mpa) by Pastoriza-Gallego et al. [61]. They observed that density increased with the concentration of NPs. Similar trends were reported for Al_2O_3/water [81–83], TiO_2/water [83], Al_2O_3/EG + water [62], Al_2O_3/R141b [8], CuO/lubricant [84], SnO_2/EG [85], and CuO/water [86,87] NFs. Eggers and Kabelac [88] measured the density of TiN/water and Ag/water NFs in the low-concentration range (0.0012–0.21 vol.%). Pastoriza-Gallego et al. [61] also reported that the impact of NPs size on the density was small. Sommers and Yerkes [89] determined the density of Al_2O_3/propanol NFs (0–4 wt.%) at the ambient temperature utilizing two different methods, that is, hydrometer and weighing a known volume of NFs. The employed methods were found to be 98.2%. Ho et al. [82] measured the density of Al_2O_3/water NFs at a temperature range of 10°C–40°C and an Al_2O_3 concentration range of 0–4 vol.%. They reported that the density of Al_2O_3/water NFs is strongly affected by temperature due to the sensitivity of the base fluid (water) to temperature.

The effect of temperature on the density of NFs made up of Al_2O_3 NPs and a mixture of EG and water (60 wt.% EG, 40 wt.% water) is depicted in Fig. 3.3A. Increasing temperature reduces NFs density in the whole NPs concentration range. Such a trend is observed in all of the

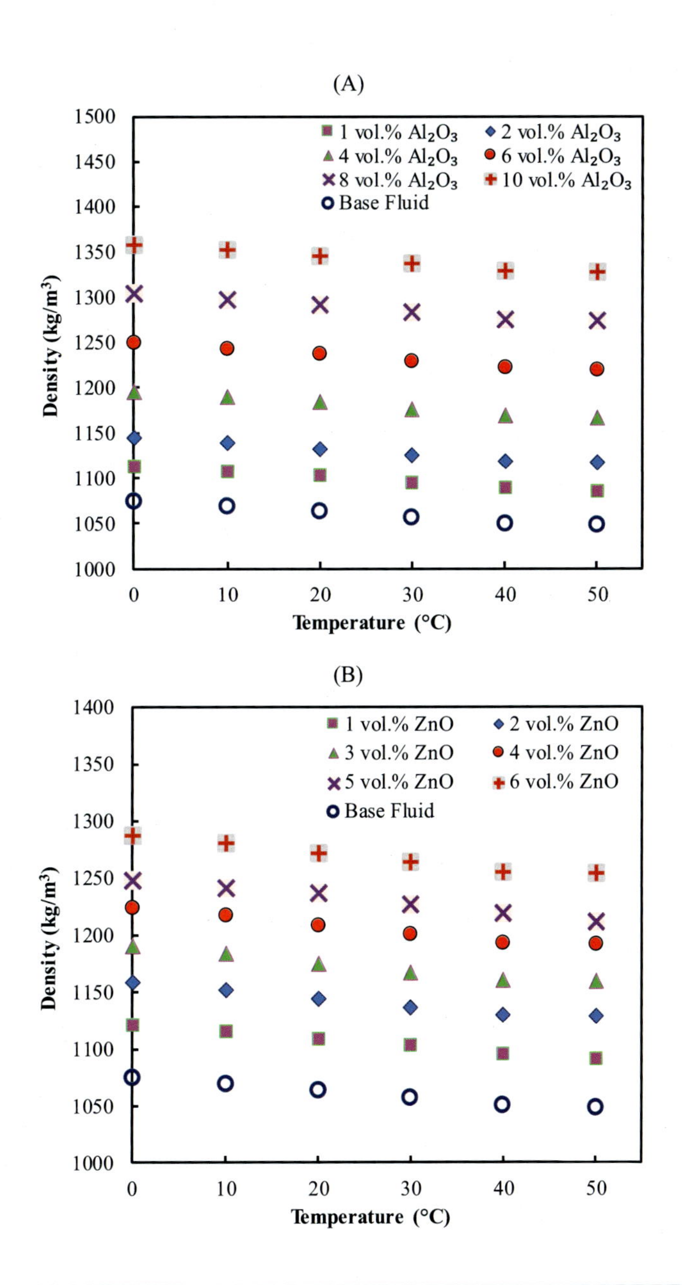

FIGURE 3.3

Effect of temperature on the densities of (A) Al_2O_3/EG(60 wt.%) + water(40 wt.%) nanofluids and (B) ZnO/EG (60 wt.%) + water(40 wt.%) nanofluids [79].

combinations of NPs and base fluids. The densities of ZnO/EG + water NFs (Fig. 3.3B) also demonstrate descending trends with increasing temperature. As per Fig. 3.3, the densities of NFs are larger than the base fluid. Moreover, increasing NPs concentration enhances density. This is due to the larger density of the solid NPs compared to the liquid base fluids. Sharifpur et al. [90] reported the experimental densities of SiO_2/water, MgO/glycerol, CuO/glycerol, and SiO_x/EG(60 wt.%) + water(40 wt.%) NFs in a 1–6 vol.% concentration range and 10°C–40°C temperature range. They utilized a digital density meter to collect the data. Increasing NPs concentration increased NFs density in the whole investigated cases. Pastoriza-Gallego et al. [91] observed that the relative density, that is, the ratio of NFs density to base fluid density in ZnO/EG NFs was temperature independent. Zafarani-Moattar and Majdan-Cegincara [65] investigated the impact of temperature (20°C–45°C) on the densities of NFs containing ZnO NPs and PEG and water as the base fluids. Density was observed to increase with NPs and PEG concentrations while increasing temperature reduced the density. Decreasing temperature and increasing TiO_2 NPs increased the densities of TiO_2/EG NFs [92]. The density of NFs made up of water and carboxylic MWCNT and carboxylic nanodiamond were measured by Alrashed et al. [93]. Their experiments were conducted at 20°C–50°C and 0.0–0.2 vol.%. Their findings reveal that the concentration of MWCNT and nanodiamond NPs has an insignificant impact on the density in the assessed concentration range. Compared to the base fluid, the density enhancements with increasing concentration were up to 0.34% and 0.73% for MWCNT and nanodiamond NPs, respectively. However, the impact of temperature on the density of NFs was significant in both cases. Mirsaeidi and Yousefi [94] reported an experimental investigation on the density of NFs made up of water, EG, and EG(40 vol.%) + water(60 vol.%) and carbon quantum dots (CQDs) NPs at various temperatures (10°C–80°C) and NPs concentrations (0.2–1.0 vol.%).

Little has been done on the impact of pressure on the NFs density. Mariano et al. [85] measured the densities of SnO_2/EG NFs at 0.1 and 45.0 Mpa pressures. They reported that increasing pressures increases NFs density. This is due to increasing the density of the base fluid. Cabaleiro et al. [92] measured the densities of TiO_2/EG NFs in a wide range of pressure (0.1–45 Mpa) at various temperatures and NPs concentrations. Pastoriza-Gallego et al. [61,91] investigated the impact of pressure on the density of Al_2O_3/water [61] and ZnO/EG [91] NFs. In both cases, increasing pressure increased NFs density. They also reported that the relative density, that is, the ratio of NFs density to base fluid density in ZnO/EG NFs was pressure independent [91]. The impact of pressure on the density of Al_2O_3/water [61] is depicted in Fig. 3.2.

Both NPs type and base fluid type affect the density of NFs. The impact of NPs type on the density of NFs that are made up of EG(60 wt.%) + water(40 wt.%) is addressed in Fig. 3.4. These data were collected by Vajjha et al. [79]. A comparison reveals that NPs type affects the density of the obtained NFs, strongly. The NFs that are composed of Sb_2O_5-SnO_2 NPs possess larger densities when compared with the Al_2O_3 and ZnO NFs.

Based on the observation of Mahbubul et al. [95], the preparation procedure also affects the density of NFs. They investigated the impact of ultrasonication time on the density of Al_2O_3/water NFs. They observed that the density increased with increasing ultrasonication time due to the good dispersion of NPs. When longer sonication is applied, NPs are properly dispersed into the base fluid. Hence, the density of NFs was found to increase as a result of the highly dispersed high-density NPs in the suspension. A linear trend was observed for the effect of sonication time on the NFs density [57,95].

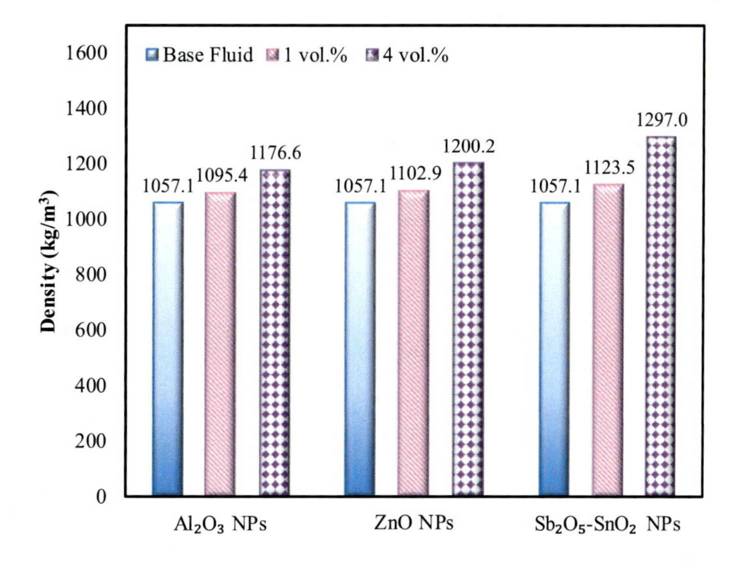

FIGURE 3.4

Effect of nanoparticles type on the density of EG(60 wt.%) + water(40 wt.%) nanofluids [79].

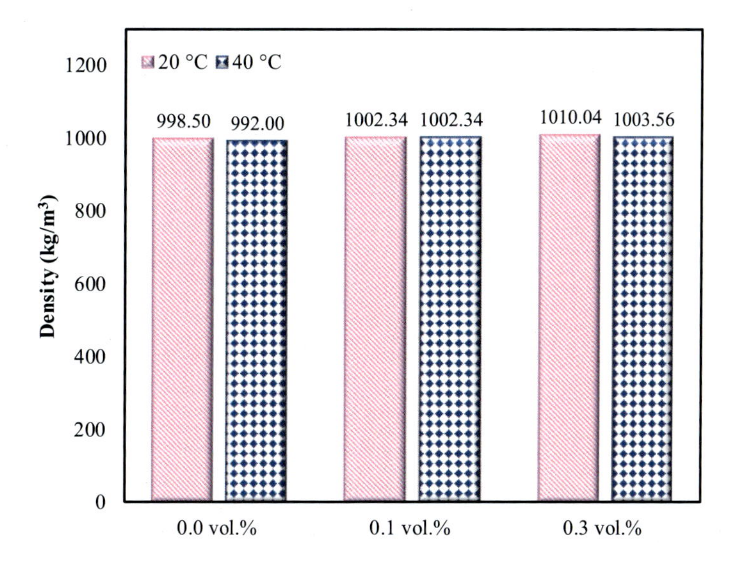

FIGURE 3.5

Density of multiwalled carbon nanotube-Fe_3O_4/water hybrid nanofluids [96].

The density data of the hybrid NFs are scarce in the literature. The density of MWCNT-Fe_3O_4/water NFs was reported by Sundar et al. [96] based on the principle of the mixture rule. The results are depicted in Fig. 3.5. Increasing temperature and decreasing NPs concentration decrease the density of hybrid NFs as well. Kannaiyan et al. [97] measured the density data of Al_2O_3-SiO_2/water hybrid NFs at 20°C–60°C and 0.05–0.20 vol.%. A literature survey reveals that the experimentally collected data of the density of NFs are rare. However, simple calculations have been developed to estimate the densities of various NFs.

3.2.1.1 Density models

Based on the principle of the mixture rule, the density of NFs can be calculated by the following equations:

$$\rho_{NF} = \sum_{i=1}^{NNPs} \varphi_{NP,i}\rho_{NP,i} + (1-\varphi_t)\rho_{BF} \tag{3.1}$$

$$\varphi_t = \sum_{i=1}^{NNPs} \varphi_{NP,i} \tag{3.2}$$

where ρ_{NF}, φ_{NP}, ρ_{NP}, and ρ_{BF} denote the density of NFs, volume fraction of the individual NPs, the density of NPs, and the density of the base fluid. φ_t is the total volume fraction of the individual NPs suspended in the NFs. These equations are applicable for the NFs with a single NPs and hybrid NFs [3,50]. Pak and Cho [64] compared the calculations of these equations against the experimental data of γ-Al_2O_3/water and TiO_2/water NFs and demonstrated good agreements. They reported a maximum deviation of 0.6% at $\varphi_{NP} = 31.6\%$. However, the capability of the principle of the mixture rule in the high concentrations of NPs is doubtful. Some experiments concluded that this model is limited to low concentrations of NPs, that is, small quantities of φ_{NP} [79,98]. Vajjha et al. [79] made a comparison of the experimentally measured NFs densities with those of the principle of the mixture rule calculations. They reported up to −8.1% deviations in the NFs made up of ZnO (up to 7 vol.%) and a mixture of EG (60 wt.%) and water (40 wt.%). However, very good agreements were observed in the NFs of Al_2O_3 (up to 10 vol.%) and Sb_2O_5-SnO_2 (up to 5.8 vol.%) NPs. The largest observed deviation in these cases was 1.2%. Qualitative comparisons are made in Fig. 3.6. Hence, it is concluded that the accuracy of this principle depends on the NPs type and concentration. A similar conclusion was reported by Pastoriza-Gallego et al. [61], Teng and Hung [81], and Ganeshkumar et al. [80] for Al_2O_3/water, γ-Al_2O_3/water, and MWCNT/EG + water NFs, respectively. They observed that the deviations increased with NPs loading. Eggers and Kabelac [88] evaluated the capability of this model for a wide range of as-published data and reported an overall deviation of ± 1%.

There is a limited number of correlations and theoretical models to calculate the density of NFs. Therefore developing accurate models to calculate the density of NFs, especially in the high-concentration range, is highly expected. Khanafer and Vafai [99] developed a simple correlation to calculate the density of Al_2O_3/water NFs as a function of temperature [T(°C)] and NPs volume fraction ($\varphi_{Al_2O_3}$). They utilized the experimental data of Ho et al. [82] to obtain Eq. 3.3 with the maximum relative error of 0.22% [99].

$$\rho = 1001.064 + 2738.6191\varphi_{Al_2O_3} - 0.2095T; 0.00 \le \varphi_{Al_2O_3} \le 0.04 \text{ and } 5 \le T(°C) \le 40 \tag{3.3}$$

Sharifpur et al. [90] observed that employing the principle of the mixture rule offers overestimation for the densities of SiO_2/water, MgO/glycerol, CuO/glycerol, and SiO_x/EG(60 wt.%) + water (40 wt.%) NFs in a 1–6 vol.% concentration range and 10°C–40°C temperature range. Larger

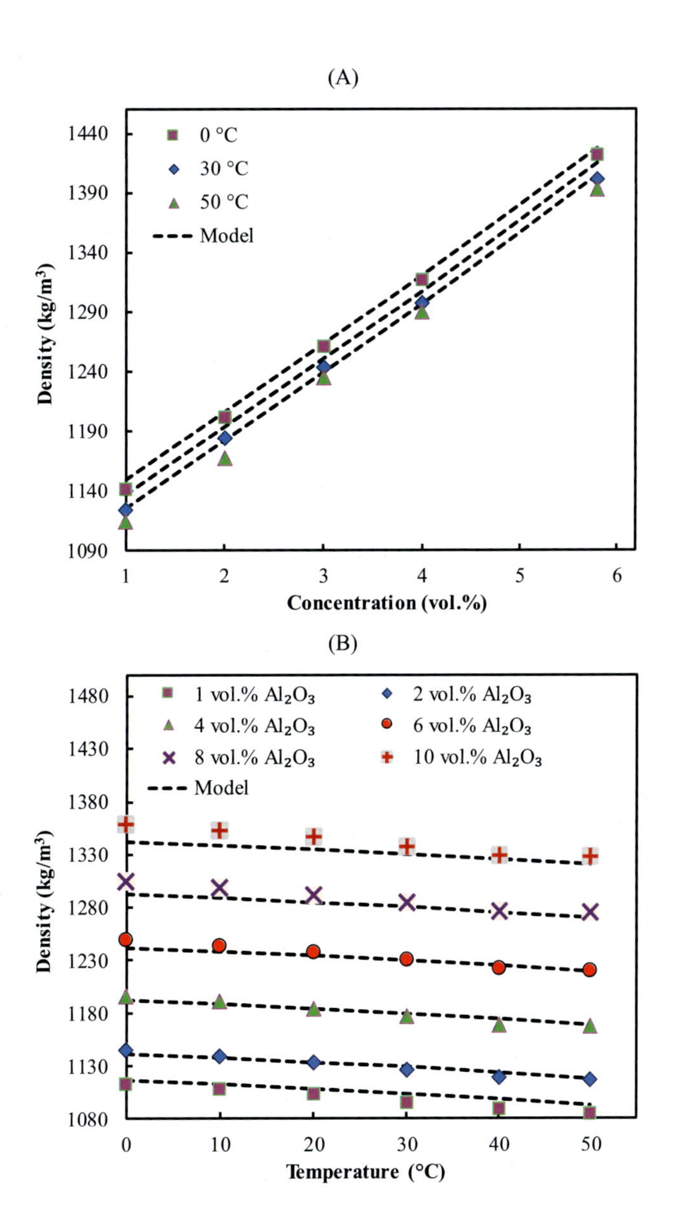

FIGURE 3.6

Comparing the calculations of the principle of the mixture rule with the experimental data of (A) Sb_2O_5-SnO_2/EG(60 wt.%) + water(40 wt.%) nanofluids [79]; and (B) Al_2O_3/EG(60 wt.%) + water(40 wt.%) nanofluids [79].

deviations were observed in the higher concentrations. Accordingly, they suggested a new model based on the interference of the nanolayer, that is, the interfacial layer between NPs and the base fluid. Accordingly, they proposed the following equations:

$$\rho'_{NF} = \frac{\rho_{NF}}{(1-\varphi_t) + \varphi_t \frac{(r_p+t_v)^3}{r_p^3}} \tag{3.4}$$

$$t_v = -0.0002833 r_p^2 + 0.0475 r_p - 0.1417 \tag{3.5}$$

Based on Eq. (3.4), the density of the NFs (ρ'_{NF}) considering the nanolayer thickness (t_v) and average NPs radius (r_p) modifies the calculated density of the principle of the mixture rule (ρ_{NF}). Hence, the new model offers good agreement compared with the principle of the mixture rule in the overall NPs concentration range. Moreover, this model considers the impact of NPs size on the density of NFs [90].

Mirsaeidi and Yousefi [94] developed a new simple correlation (Eq. 3.6) to represent the densities of CQDs-made NFs in water, EG, and EG(40 vol.%) + water(60 vol.%) base fluids at various volume fractions and temperatures.

$$\frac{\rho_{NF}}{\rho_{BF}} = A - B\exp\left(-CT{\varphi_t}^2\right) + D\exp\left(-C\varphi_t\right) \tag{3.6}$$

A, *B*, *C*, and *D* are the NFs-specific constants, which are tabulated in Table 3.2 for different NFs. This model considers the impact of temperature, NPs volume fraction, and the density of the base fluid. A major limitation of this model is the necessity of the density data of the base fluid. The R^2 of the correlations prove the accuracy of this simple model in the representation of data in the evaluated systems.

In addition to the empirical models, artificial neural networks (ANN) can also be employed to accurately estimate the physical properties of NFs, such as density. ANN models are widely utilized in fitting the data [100,101]. The ANN models are the nonlinear mathematical methods that have attracted great interest owing to the simplicity, flexibility, various training algorithms, and large modeling capacity [102]. In this regard, Karimi and Yousefi [101] developed an ANN model to determine the densities of Al_2O_3/EG(60 wt.%) + water(40 wt.%), Sb_2O_5-SnO_2/EG(60 wt.%) + water(40 wt.%), ZnO/EG(60 wt.%) + water(40 wt.%), and CuO/water NFs as a function of temperature, NPs concentration (volume fraction), the density of NPs, and the density of the fluid. They utilized 90 data points to train and 38 data points to test a genetic-algorithm-improved backpropagation network. Then, a good agreement between the predicted and the experimental data with less than 0.13% errors was managed [101]. In a further study, Yousefi and Amoozandeh [102] developed an ANN model plus principal component analysis to estimate the densities of Al_2O_3/EG

Table 3.2 The parameters of Eq. (3.6) for various nanofluids (NFs).

	Constants				
NFs	***A***	***B*** $\times 10^4$	***C***	***D*** $\times 10^2$	R^2
CQDs/EG	1.005	−63.6400	2.53900	−0.4921	0.9633
CQDs/water	1.199	56.3300	0.09417	−19.2800	0.9812
CQDs/EG(40 vol.%) + water(60 vol.%)	1.007	0.5177	0.68550	−0.7769	0.9822

Adapted from A. Mirsaeidi, F. Yousefi, Viscosity, thermal conductivity and density of carbon quantum dots nanofluids: an experimental investigation and development of new correlation function and ANN modeling. J. Therm. Anal. Calorim. (2019) 1–11 [94].

(60 wt.%) + water(40 wt.%), Sb_2O_5-SnO_2/EG(60 wt.%) + water(40 wt.%), ZnO/EG(60 wt.%) + water(40 wt.%), ZnO/PEG + water, ZnO/PEG, and TiO_2/EG NFs with 0.48% error. Alrashed et al. [93] employed an ANN model against their experimental data of MWCNT/water and nanodiamond/water NFs and found this model accurate in the estimation of these data. Kannaiyan [97] also employed an ANN model to interpret their experimental data of Al_2O_3-SiO_2/water hybrid NFs. Mirsaeidi and Yousefi [94] utilized an ANN model to correlate their experimental data of CQDs-made NFs in water, EG, and EG(40 vol.%) + water(60 vol.%) base fluids and compared the results with Eq. (3.6). They utilized temperature, volume fraction, as well as the molecular weight, acentric factor, critical temperature, and critical pressure of the base fluid, as the input variables. A 0.01% error of calculations was obtained by employing their ANN model.

3.2.2 Viscosity (μ)

Viscosity, a measure of internal resistance of the fluid to flow, is an important property in the whole applications of the fluid [39]. The frictional force between the molecules of fluid, on the one hand, creates a resistance toward fluid flow. The molecules of fluid are bound together tightly through the attractive intermolecular van der Waals forces. Since the individual molecules are tightly bound to the neighbors, they move difficultly. This characteristic appears in the viscosity of the fluid. Viscosity, also known as dynamic viscosity, is often denoted by μ (pa.s or N.s/m^2). The ratio of dynamic viscosity to density is called kinematic viscosity (υ, m^2/s). Dynamic and kinematic viscosities are related to the transport phenomena through Re, Pr, Sc, Nu, Sh, f, and ΔP parameters. Hence, viscosity estimation and determining the impact of various variables are of paramount importance.

The procedures of NFs viscosity measurement are summarized in the recent study of Mahbubul [57]. A common instrument for viscosity data collection, as well as the data acquisition procedure and uncertainties, was also discussed by Mahbubul [57]. The conventional instruments include cone and plate/parallel disks/coaxial cylinders (Couette) type viscometers [57,103].

Compared to the density, there are extensive studies on the viscosity of NFs. A summary is presented in Table 3.3. Utilizing NFs with a high concentration of NPs or with large NPs increases the probability of collision between the particles. This may result in the condensation and agglomeration of the NPs and increased flow resistance, that is, increased fluid viscosity. Viscosity strongly depends on the base fluid type, NPs characteristics (type, shape, and size), NPs concentration, temperature, and preparation method [57,129]. The viscosity of NFs decreases with increasing temperature. This could be due to the reduction of the viscosity of the base fluid. Increasing temperature, on the one hand, results in the greater kinetic and/or thermal energy of the molecules of fluid. This would make the molecules more mobile. The mobility is intensified by the Brownian motion of the NPs [130]. Hence, the attractive binding energy between the molecules is weakened, the molecules overcome the binding energy, and then the fluid viscosity decreases [131]. Increasing NPs concentration often increases the viscosity of NFs. Such an effect is in fact due to the direct effect on the NFs internal shear stress [58]. However, there are exceptions, especially in the low concentration range [132].

The effect of temperature on the viscosity of various NFs has been extensively studied in the literature. Yang et al. [78] measured the kinematic viscosities of the NFs made up of different commercial automatic transmission fluids and synthetic base oils and 2.0–2.5 wt.% graphite NPs.

Table 3.3 Summary of information on the viscosity of NFs.

NFs	NPs size (nm)	NPs shape	NPs concentration (vol.%)	T (°C)	$\mu \times 10^6$ (pa.s)	References
Al_2O_3/water	30	–[a]	0.000–0.300	20–40	990.98–630.73	[104]
Al_2O_3/water	47	Spherical or near-spherical	0.010–7.000	21–70	423.30–2016.98	[58,105]
Al_2O_3/water	36	–	1.000–7.000	22–70	429.36–2286.65	[58]
Al_2O_3/water	12,30	Spherical	0.500–1.500	20–50	651.69–1744.67	[106]
Al_2O_3/methanol	50	Cylindrical	0.050–0.250	5–25	708.00–942.00	[107]
Al_2O_3/EG	43	Near-spherical	0.500–6.600	10–50	7510.00–81510.00	[108]
Al_2O_3/EG	8	Near-spherical	0.500–3.100	10–50	8100.00–75190.00	[108]
Al_2O_3/PG + water	<50	–	0.100–1.500	10–50	10000.00–85000.00	[109]
SiO_2/water	12	Spherical	0.450–4.000	20–50	615.14–4621.45	[106]
SiO_2/methanol	190	Spherical	0.005–0.150	5–25	618.84–818.65	[110]
SiO_2/EG	>28	–	0.000–4.800	25–50	5116.48–24417.46	[111]
SiO_2/EG + water	20–100	Spherical	0.000–10.000	−35–50	1000.00–175000.00	[112]
TiO_2/water	121–180	Spherical	0.000–0.258	10–50	340.00–1400.00	[113]
TiO_2/water	20–40	Spherical	0.050–0.500	25–80	608.39–1580.42	[114]
TiO_2/water	21	–	1.000–8.000	15–60	567.91–2062.97	[115]
TiO_2/water	76	Spherical	0.240–11.200	10–70	412.00–4058.00	[116]
TiO_2/water	21	Spherical	0.200–2.000	15–35	800.00–1200.00	[117]
TiO_2/water	21	Spherical	0.200–3.000	13–55	500.00–2900.00	[118]
ZnO/water	35–40	Spherical	0.250–1.500	10–35	843.49–1732.75	[119]
ZnO/water	20	–	0.000–1.000	10–60	508.47–4182.20	[120]
ZnO/EG	40–100	Quasi-spherical	0.000–3.400	10–50	7210.00–45400.00	[91]
ZnO/EG	>4.6	Quasi-spherical	0.000–2.100	10–50	7210.00–44300.00	[91]
Ag/water	60	–	0.300–0.600	50–90	748.98–1417.35	[121]
Ag/EG	40	Near-spherical	0.125–3.000	26–55	5437.06–26066.43	[122]
Ag/EG	60	Near-spherical	0.125–3.000	26–55	5366.30–24706.96	[122]
Ag/EG	100	Near-spherical	0.125–3.000	26–55	5215.05–24139.78	[122]
Fe/EG	40	Near-spherical	0.125–3.000	24–55	5437.06–26066.43	[122]
Fe/EG	60	Near-spherical	0.125–3.000	24–55	5366.30–24706.96	[122]
Fe/EG	100	Near-spherical	0.125–3.000	24–55	5215.05–24139.78	[122]
FMWNT[b]/water	>11	–	0.000–0.250[c]	20–33	765.00–1040.00	[123]
Bentonite /pentadecane	~16	–	0.000–4.000	20–85	1100.00–193000.00	[19]

Table 3.3 Summary of information on the viscosity of NFs. ***Continued***

NFs	NPs size (nm)	NPs shape	NPs concentration (vol.%)	T (°C)	$\mu \times 10^6$ (pa.s)	References
CuO/water	11–33	Quasi-spherical	0.160–1.700	10–50	546.00–226100.00	[124]
CuO/water	29	–	0.000–6.120	−35–50	1000.00–420000.00	[125]
MgO/EG	20	–	0.000–5.000	5–60	5084.75–52542.37	[126]
MgO/PG	30–40	Spherical	0.000–1.750	28–60	8647.06–38941.18	[127]
MgO/water	20	–	0.000–1.250	25–60	441.77–1500.00	[128]

[a] *Not mentioned.*
[b] *Functionalized MWNT.*
[c] *wt.%.*

The measurements were conducted at 35°C–70°C. Increasing temperature reduces the kinematic viscosity of the whole NFs. Since both dynamic viscosity and density decrease with temperature, a fair conclusion on the impact of temperature on the dynamic viscosities of the investigated NFs cannot be made. Chen et al. [132] measured the viscosity of MWCNT/water NFs at 5°C–65°C and 0.2–1.0 vol.% and reported their measurements based on the relative viscosity, that is, the ratio of the NFs to the base fluid viscosities. Up to 55°C, the relative viscosity was almost constant in each NPs concentration, while at the temperature beyond 55°C, the relative viscosities appeared to rise considerably. The trend was unusual and no further explanation was proposed to clarify their observations. Since the absolute viscosity was not published, the range of viscosity is not clear.

The viscosities of Al_2O_3/water NFs at 10°C–60°C 0–10 wt.% were reported by Pastoriza-Gallego et al. [61]. The results are depicted in Fig. 3.7. Increasing temperature reduced the viscosity of NFs at the whole concentration range. Nguyen et al. [58] investigated the impact of temperature, as well as particle size, on the viscosities of Al_2O_3/water and CuO/water NFs at a temperature range of 20°C–75°C. The measurements were conducted at 1.0–9.4 vol.% NPs concentrations. The viscosities of CuO/water NFs were larger than those of Al_2O_3/water NFs. The differences became pronounced at the larger NPs concentrations (>5 vol.%). The viscosities of NFs decreased with temperature in the whole cases. However, viscosities tended to become almost constant at higher temperatures. In the larger NPs concentrations, the relative viscosity showed an increasing trend in the higher temperatures. In the low concentration range (<4 vol.%), all the examined NFs exhibited almost temperature-independent relative viscosities. Anoop et al. [133] measured the viscosities of CuO/EG, Al_2O_3/EG, and Al_2O_3/water NFs. Their measurements were conducted at 20°C–50°C and 0.5–6.0 vol.%. These NFs depicted Newtonian behavior. It was observed that the viscosities of these NFs decreased with an increase in temperature. Comparing the results reveals that the impact of temperature on the viscosity reduction is more severe for the EG-based NFs than for the water-based ones. The viscosities of TiO_2/water NFs were measured by Duangthongsuk and Wongwises [117]. Their measurements were conducted at NPs concentrations of 0.2–2.0 vol.% and temperatures of 15°C–35°C. Their measurements indicated that increasing NPs concentration results in larger viscosities, while temperature reduces the viscosity. Their results revealed that the relative viscosity is a strong function of temperature and NPs concentration. Both temperature and

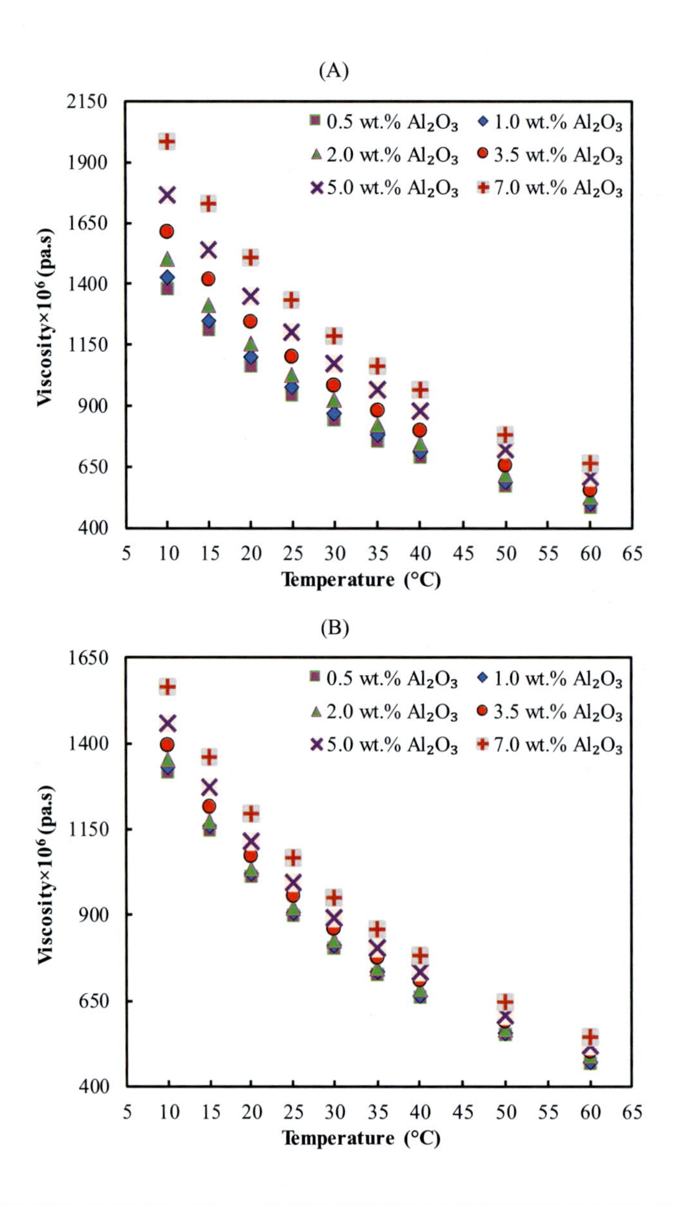

FIGURE 3.7

Effect of temperature on the viscosities of Al_2O_3/water nanofluids with nanoparticles size of (A) 8 nm and (B) 43 nm [61].

NPs concentration increase the relative viscosity of TiO_2/water NFs. Similar observations were reported by Turgut et al. [118] for TiO_2/water NFs in the ranges of 0.2–3.0 vol.% and 13°C–55°C. Temperature affects the rheological behavior of the NFs. Kole and Dey [109] measured the viscosity of NFs including Al_2O_3 NPs and a commercial car engine coolant [PG

(50%) + water(50%)] at 10°C–50°C and 0.1–1.5 vol.%. They observed that with the addition of even small quantities of NPs (0.1 vol.%), the Newtonian behavior of the fluid turns into a non-Newtonian one. In the low-concentration range (0.1–0.4 vol.%), Newtonian behavior was observed at temperatures beyond 40°C, while in the larger concentrations ($>$0.4 vol.%), the non-Newtonian behavior was evident in the whole measured range. They also illustrated that the viscosity of the Al_2O_3/PG(50%) + water(50%) NFs decreases exponentially with temperature owing to the weakened adhesion forces. Pastoriza-Gallego et al. [124] reported the viscosity data of CuO/water NFs at 10°C–50°C and 1–10 wt.% (0.16–1.70 vol.%). The results are depicted in Fig. 3.8. As per the figure, it is clear that viscosity reduces with increasing temperature. Interestingly, at 50°C, the viscosity of the NFs containing 10 wt.% of CuO NPs becomes almost equal to that of base fluid at 15°C (Fig. 3.8A). The measurements of Lee et al. [134] illustrate that increasing temperature results in the reduced viscosity of SiC/water NFs (28°C–72°C, 0.001–3.000 vol.%.). Namburu and colleagues [112,125] investigated the viscosities of CuO/EG(60 wt.%) + water(40 wt.%) (−35°C–50°C, 0.00–6.12 vol.%.) and SiO_2/EG(60 wt.%) + water(40 wt.%) (−35°C–50°C, 0–10 vol.%.). Viscosity reduction with increasing temperature was observed in the whole investigated cases. In the case of CuO/EG + water NFs, the relative viscosity reduces with temperature considerably in the larger NP loadings, while in the lower concentrations of NPs, the relative viscosity remains almost constant with increasing temperature. It was also found that CuO/EG + water NFs behave like Newtonian fluids in the entire measurement range [125]. However, the SiO_2/EG + water NFs behave like a non-Newtonian fluid at the low-temperature range (subzero condition) [112]. Increasing temperature reduces the viscosities of the NFs composed of EG + water base fluid and CuO, SiO_2, and Al_2O_3 NPs [135]. A similar conclusion is made for the CuO/PG(60 vol.%) + water (40 vol.%) [136] and Al_2O_3/water [137]. However, the observations of Prasher et al. [129] and Chen et al. [138,139] state that the NFs viscosity is temperature independent in some cases.

From the hitherto published works on the viscosity of NFs, it is inferred that increasing temperature decreases the absolute viscosity. However, the impact of temperature on the relative viscosity and the rheological behavior of the NFs could be different in each case. The rheological behavior of the NFs, that is, either Newtonian- or non-Newtonian-like behavior depends on the shear stress, base fluid type, NPs type, and NPs concentration.

NPs characteristics (type, shape, and size), as well as NPs concentration, affect the viscosity of NFs. The impact of NPs type is well discussed above. Few studies have assessed the impacts of NPs size and shape on the viscosity of NFs. However, the impact of NPs concentration has been extensively assessed. NPs concentration affects the quantity of viscosity as well as the rheological behavior of the NFs.

Das et al. [140] and Putra et al. [141] illustrated that the viscosity of Al_2O_3/water NFs increases with NPs concentration (1–4 vol.%.). Moreover, the NFs behave as the Newtonian fluids in the examined range. Chandrasekar et al. [142] also made the same conclusion. Similar conclusions were made for the NFs made up of Al_2O_3 NPs and PG as the base fluid [129]. The observations of Chevalier et al. [143] for the SiO_2/ethanol NFs with various NPs sizes (35, 94, and 190 nm) and a concentration range of 1.4–7.0 vol.% revealed the Newtonian behavior as well as the enhanced viscosities with NPs concentration. Theoretical and experimental investigations of Lu and Fan [144] illustrated that the viscosities of Al_2O_3/water and Al_2O_3/EG NFs increase with NPs concentration. The rheological behavior of the NFs composed of TiO_2 nanotubes (0.5–8.0 wt.%) and EG was investigated by Chen et al. [145]. It was found that at the larger NPs concentrations ($>$2 wt.%),

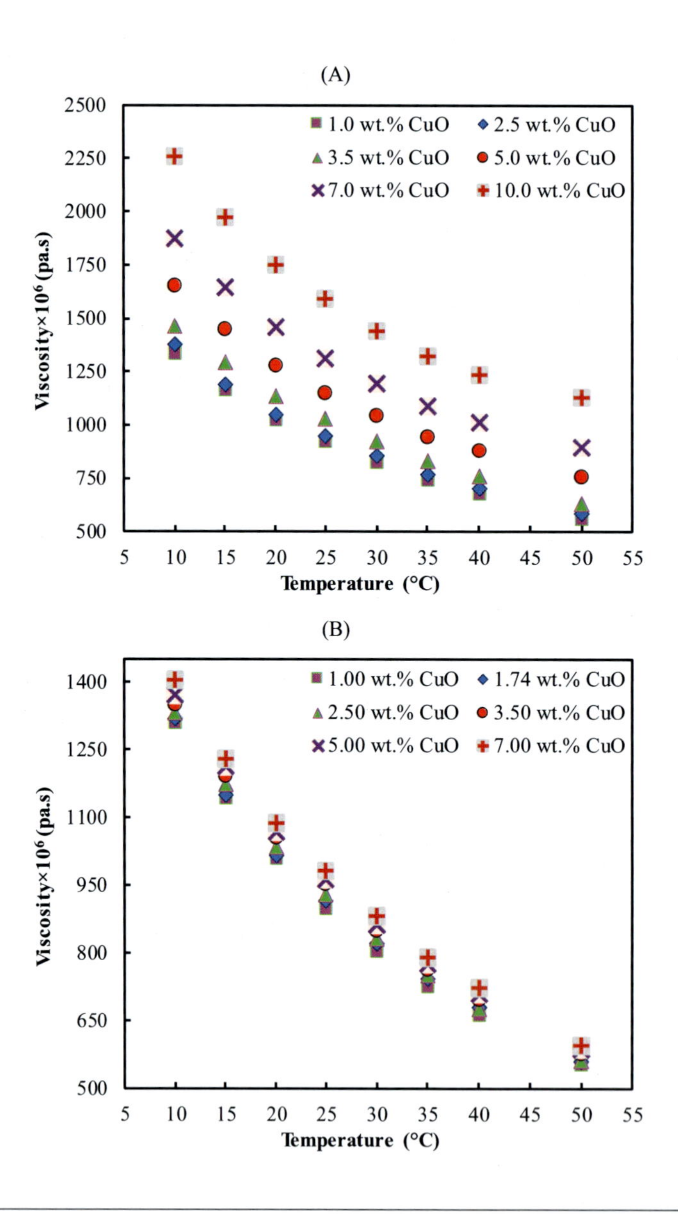

FIGURE 3.8

Effect of temperature on the viscosities of CuO/water nanoparticles with nanoparticles size of (A) 11 nm and (B) 33 nm [124].

the NFs behave as non-Newtonian (shear-thinning). NPs concentration was also observed to increase the viscosity. The measurements of Duangthongsuk and Wongwises [117] for TiO_2/water NFs (0.2–2.0 vol.%) indicated 4%–15% viscosity enhancements. Phuoc and Massoudi [146] investigated Fe_2O_3/water NFs. NPs concentration (1–4 vol.%) affects the viscosity enhancement as well as the rheological behavior of the NFs. Their results demonstrate that these NFs are the non-Newtonian (shear-thinning) ones. However, some samples with polyvinylpyrrolidone, as a dispersant, behaved as Newtonian at the NPs concentration of less than 2 vol.%. They also concluded that the dispersant type affects the rheological behavior of the NFs. Yu et al. [147] conducted an experimental study on the rheological behavior of AlN NPs in EG and PG in the wide ranges of temperature (10°C–60°C) and concentration (1–10 vol.%). It was found that the NFs with NPs concentration of larger than 5 vol.% showed non-Newtonian (i.e., shear thinning) behavior, while the addition of NPs less than 5 vol.% results in a Newtonian behavior. They also observed up to 150% viscosity enhancements with increasing NPs concentration at a specific temperature and shear rate. Most of the investigated NFs obey either Newtonian or shear-thinning patterns. However, a pseudoplastic flow behavior was also observed in some cases [148]. Viscosity enhancements with NPs concentration were also reported in a variety of studies [134,148,149]. Recently, Esmaeilzadeh et al. [19] reported the experimental viscosities of bentonite/n-pentadecane in a wide range of temperatures (20°C–85°C) and NPs concentrations (0–4 vol.%.). The results are depicted in Fig. 3.9, in terms of absolute and relative viscosities. Increasing NPs concentration increases absolute viscosity, considerably. Such an increase is larger in the lower temperatures.

Chen et al. [132] measured the relative viscosity of MWCNT/water NFs 0.2–1 vol.%. In a very low NPs concentration (i.e., <0.4 vol.%), NFs have a lower viscosity than the base fluid. This would be owing to the lubricative effect of NPs [132]. In larger concentration (i.e., >0.4 vol.%), the NFs viscosity increased with NPs concentration. Nguyen et al. [58] investigated the viscosities of Al_2O_3/water and CuO/water NFs. To draw the impact of NPs size and concentration, the relative viscosity of Al_2O_3/water NFs (NPs size of 47 nm) increased 1.12–5.3-fold when 1–12 vol.%. NPs was added. When 2.1–12.2 vol.% of the NPs with 36 nm average size was added to the fluid, up to 3.1 relative viscosity was observed. Interestingly, the viscosities of NFs with 47 nm Al_2O_3 NPs were larger than the ones with 36 nm. Larger differences were observed in the case of larger NPs concentrations (>5 vol.%). In a further study [149], they stated that the NPs size effect is more significant in the high concentration range. Anoop et al. [133] concluded that the relative viscosities of CuO/EG, Al_2O_3/EG, and Al_2O_3/water NFs increase with NPs concentration. Viscosity enhancement with increasing NPs loading was reported by Namburu and colleagues [112,125] for the CuO/EG(60 wt.%) + water(40 wt.%) (0.00–6.12 vol.%.) and SiO_2/EG (60 wt.%) + water(40 wt.%) (0–10 vol.%.) NFs. Based on the observations of Duangthongsuk and Wongwises [117], the viscosities of TiO_2/water NFs at NPs concentrations of 0.2–2.0 vol.% are 4%–15% larger than the base fluid. Similar trends were reported by Turgut et al. [118] for TiO_2/water NFs including 0.2–3.0 vol.% NPs. The measurements of Kole and Dey [109] for the NFs of Al_2O_3 NPs and a commercial car engine coolant [PG(50%) + water(50%)] at 0.1–1.5 vol.% also demonstrated the increased NFs viscosity with NPs loading. The same trend was observed by Lee et al. [134] for the SiC/water NFs (0.001–3.000 vol.%.). As per Fig. 3.7, the viscosity of Al_2O_3/water NFs at 1–7 wt.% increases with NPs concentration [61]. Fig. 3.8 also illustrates the viscosity of CuO/water NFs at 1–7 wt.% increase with NPs concentration [124]. NFs including the smaller NPs (Figs. 3.7A and 3.8A) offer considerably larger viscosities in a specific temperature and NPs concentration because of the effects imposed by the instance of the electric double-layer repulsion [61,124,150]. A summary of the viscosity enhancement by NPs concentration for various NFs is tabulated by Mahbubul [57]. Up to 430% viscosity enhancement was observed in some cases.

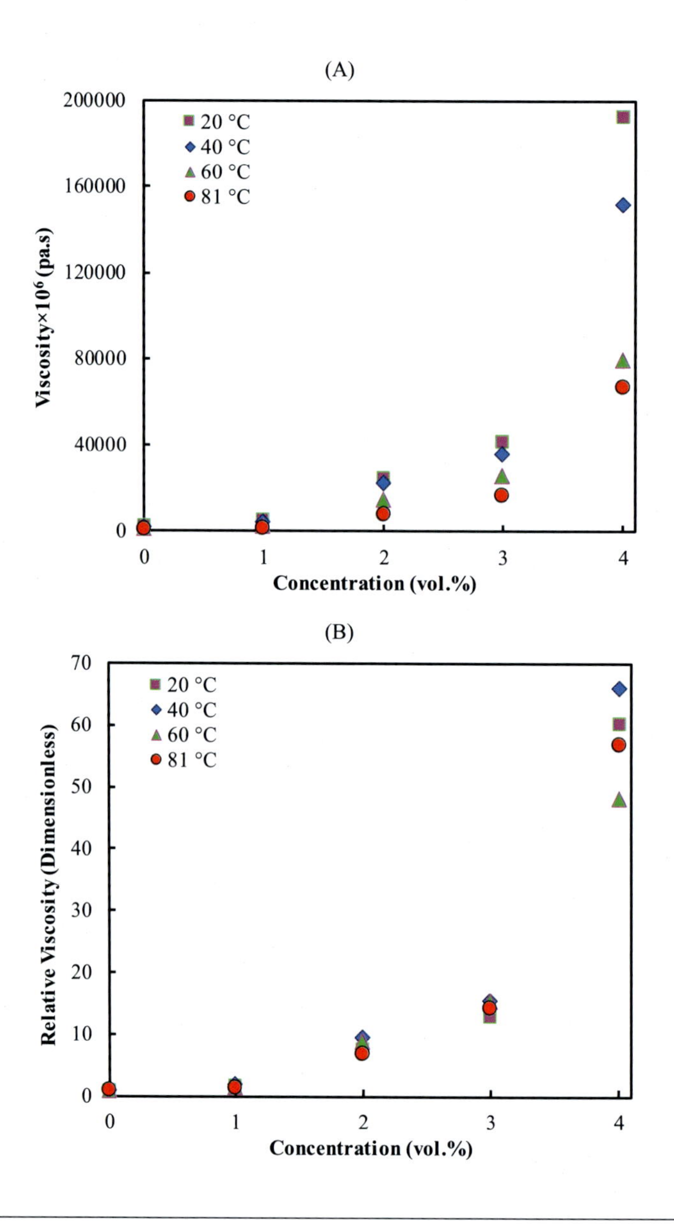

FIGURE 3.9

Effect of nanoparticles concentration on the viscosities of bentonite/n-pentadecane nanofluids in terms of (A) absolute viscosity and (B) relative viscosity [19].

The impact of NPs size on the viscosity of Ns is contradictory. It depends on the NPs type and concentration. He et al. [151] evaluated the NPs size (95, 145, and 210 nm) on the viscosity of TiO_2/water NFs. They observed that the viscosity of TiO_2/water NFs increases with NPs size. When the NPs size in a sample including 0.6 vol.% TiO_2 NPs increased (95–210 nm), more than

7.25% viscosity increase was observed. The theoretical study of Lu and Fan [144], which was based on a simplified molecular dynamic simulation, offered different trends. Based on their numerical results, the viscosities of Al_2O_3/water and Al_2O_3/EG NFs decrease with NPs size. They also stated that for NPs sizes smaller than 30 nm, the impact of NPs size on the viscosity is larger. Anoop et al. [152] also stated that the addition of smaller NPs results in a larger viscosity, especially at the larger concentrations. Based on the measurements of Pastoriza-Gallego et al. [61,124], the viscosity of CuO/water NFs, as well as Al_2O_3/water NFs, decreases with NPs size (Fig. 3.10). The differences are larger in the larger NPs concentrations.

The measurements of Chevalier et al. [143] for SiO_2/ethanol NFs also showed that NPs size (35, 94, and 190 nm) reduces the viscosity. The differences were larger at the larger NPs concentrations. Viscosity measurements of Namburu et al. [112] for SiO_2/EG(60 wt.%) + water(40 wt.%) NFs (20, 50, and 100 nm) in a wide range of temperature revealed that NPs size reduces the viscosity. The differences are larger in the low-temperature range (subzero condition).

According to the previous studies, NPs size strongly affects the viscosity of NFs. However, the trend (either ascending or descending) depends on the NPs type and concentration. The contradiction between the observations of the previous studies could be due to multiple parameters that affect the viscosity of NFs. A spread layer of fluid at the interface of the NPs−liquid is formed as a result of the electrostatic interaction between the NPs and the base fluid, that is, the first electroviscous effect [153,154]. This effect results in the increased effective particle size and then the effective volumetric concentration. Moreover, the electrostatic interaction between the NPs, that is, the second electroviscous effect has a great impact on the NPs agglomeration and thus the degrees of freedom of movement in the NFs. As NPs size reduces, the total surface area of the interface of the NPs−liquid, as well as the quantity of the NPs at the same volume concentration increases. Hence, both electroviscous effects become significant for the smaller NPs sizes. This would result in the viscosity increase [153,154]. The theoretical studies of Rubio-Hernández and colleagues [155,156] revealed that the first electroviscous effect is not solely capable of explaining the empirical observations of viscosity enhancement with NPs size. Therefore the impact of the interactions between the NPs themselves. That is, the second electroviscous effect, should be considered. In addition to this, Timofeeva et al. [154,157] illustrated that the pH of the suspension affects the viscosity of NFs, considerably. Such an effect is related to the surface charges at the interface of the NPs−liquid and interactions between the NPs. Hence, the impact of the NPs size on the viscosity must be assessed in a controlled pH of the suspension [154]. However, no pH control was applied in a majority of the studies that investigated the impact of NPs size. Timofeeva et al. [154] evaluated the impact of NPs size (16−90 nm) on the viscosity of SiC/water. It is inferred that the addition of NPs with larger sizes reduces the viscosity in the specific temperature (15°C−55°C), concentration (4.1 vol.%), and solution pH (9.4). Since the controlled pH condition was applied and the second electroviscous effect is limited, such a result could be due to a change in the area of the interface of the NPs−liquid and the reduced effective volume of the NPs, that is, the first electroviscous effect. The effect of size also depends on NPs type and concentration. It is inferred from Fig. 3.11, which compared the viscosities of the water-made NFs with CuO and Al_2O_3 NPs.

In addition to the NPs size and type, their shape affects the viscosity of NFs. However, few studies have been devoted to assessing the impact of NPs shape. Timofeeva et al. [157] investigated the effect of NPs shape on the viscosity of Al_2O_3/EG(50 vol.%) + water(50 vol.%) NFs. They assessed different shapes of NPs including platelets, blades, cylinders, and bricks experimentally

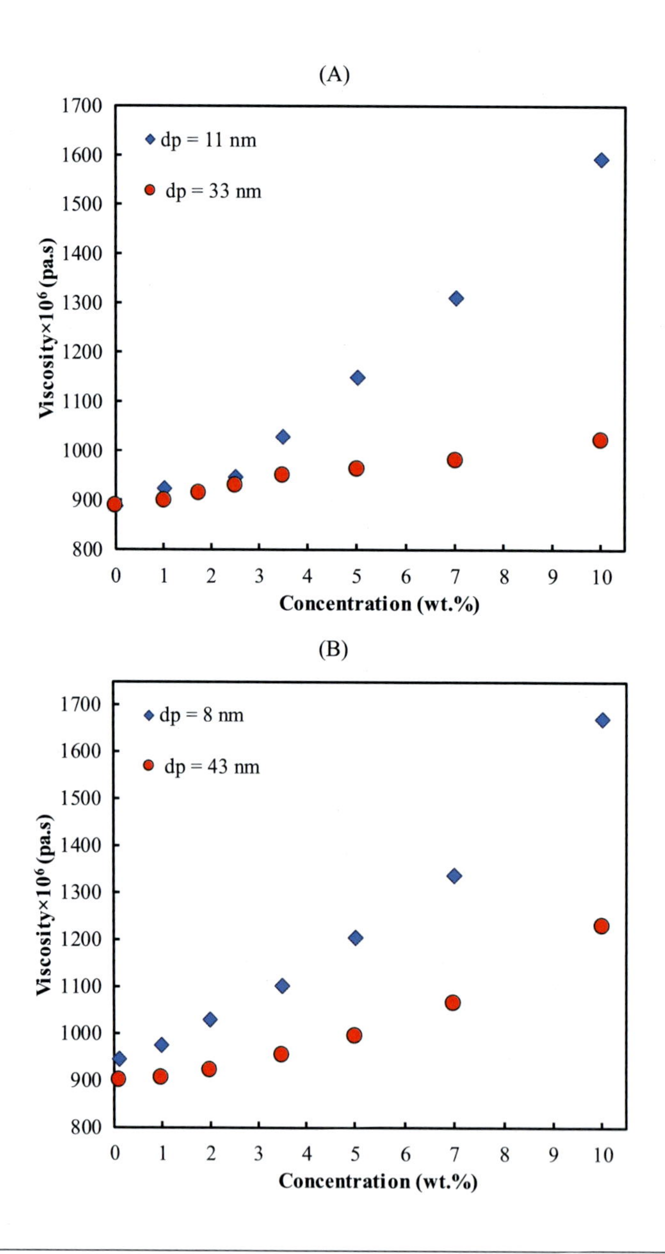

FIGURE 3.10

Effect of nanoparticles size on the viscosities of (A) CuO/water nanofluids (NFs) [124] and (B) Al_2O_3/water NFs [61].

and compared them with the theoretical results of noninteracting spheres. Their observations indicated the strong interactions between the Al_2O_3 NPs. The addition of platelet- and cylinder-shaped NPs resulted in the larger viscosities. NFs including blades showed the lowest viscosity. It was also

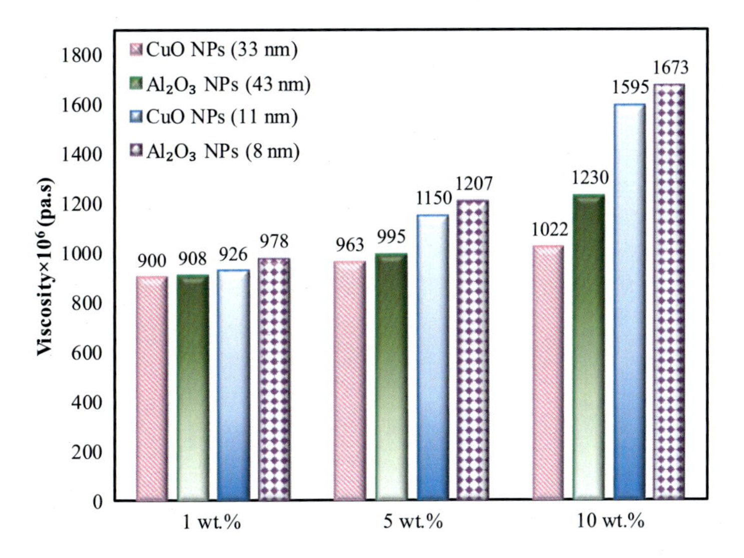

FIGURE 3.11

Effect of nanoparticles type on the viscosity of water nanofluids with different nanoparticles [61,124].

reported that the NFs including platelet- and blade-shaped NPs showed Newtonian behavior, as their viscosities were independent of the shear rate. The NFs with brick- and cylinder-shaped NPs showed non-Newtonian behavior, as their viscosities reduced with shear rate (i.e., shear thinning). Accordingly, the NPs shape and aspect ratio strongly affect the viscosity of NFs. More discussions on the impact of NPs shape and their mechanisms could be found elsewhere [60,157].

The NFs preparation procedure, such as sonication, affects the viscosity. However, there are contradictory results. Viscosity reduction with sonication time was reported in the studies of Yang et al. [158] for MWCNT/oil NFs and Mahbubul et al. for Al_2O_3/water [95] NFs. Yang et al. [158] stated that increased sonication time increases the dispersion energy, which reduces the viscosity. Moreover, the sonication time has an impact on the effective size and aspect ratio of the NPs, which would then reduce the viscosity. Wang and Guo [159] stated that the viscosity for Al_2O_3-ZrO_2/water hybrid NFs increased with increasing sonication period. Garg et al. [160] evaluated the impact of sonication time on the viscosity of MWCNT/water NFs (1 wt.%). They observed an initial viscosity enhancement in the first 40 min of sonication followed by a further reduction up to 80 min sonication. Similar trends were reported by Ruan and Jacobi [161] for MWCNT/EG NFs (0.5 wt.%) NFs. The impact of sonication time could be explained by the degree of NPs homogeneously in the base fluid and agglomeration. More discussions on the effect of preparation time on the viscosity of NFs can be found elsewhere [57,137,158,160–163]. A review of the viscosity and rheological behavior of the mono and hybrid NFs can also be found elsewhere [164].

3.2.2.1 Viscosity models

A variety of theoretical and empirical models have been hitherto proposed to estimate the viscosity of NFs. The simplest model for the viscous fluids including very small noninteracting spherical particles at low concentrations ($\varphi \geq 0.02$) is presented in Eq. (3.7) [165]. μ_{NF} and μ_{BF} are the viscosities of NFs and base fluid, respectively.

$$\mu_{NF} = \mu_{BF}(1 + 2.5\varphi) \tag{3.7}$$

A variety of parameters have been hitherto introduced as the limitations of this model, for example, NPs concentrations, the interactions of the NPs, NPs characteristics (shape, size, etc.), the necessity of base fluid data at the given condition, and the interaction of NPs with the base fluid. Therefore researchers have consistently made efforts to either modify this model and/or to develop new rigorous models. To this end, Brinkman [166] presented Eq. (3.8), which can be utilized for the larger NPs concentration (i.e., up to 4 vol.%).

$$\mu_{NF} = \frac{\mu_{BF}}{(1-\varphi)^{2.5}} \tag{3.8}$$

Ward presented Eq. (3.9), which offered higher accuracies [167].

$$\mu_{NF} = \mu_{BF}\left(1+2.5\varphi+(2.5\varphi)^2+(2.5\varphi)^3+(2.5\varphi)^4\right) \tag{3.9}$$

Lundgren [168] proposed a model for the viscosity of NFs, as follows (Eq. (3.10)):

$$\mu_{NF} = \mu_{BF}\left(1 + 2.5\varphi + \frac{25}{4}\varphi^2 + 0\varphi^3\right) \tag{3.10}$$

Eq. (3.7) was modified by Batchelor [169] through the introduction of the effect of Brownian motion and then Eq. (3.11) was obtained.

$$\mu_{NF} = \mu_{BF}(1 + 2.5\varphi + 6.2\varphi^2) \tag{3.11}$$

By introducing the impact of packing fractions, Chen et al. [138] proposed Eq. (3.12) to estimate the viscosity of NFs.

$$\mu_{NF} = \mu_{BF}(1 + 10.6\varphi + (10.6\varphi)^2) \tag{3.12}$$

A simple model was also proposed by Abedian and Kachanov [170] for the estimation of the viscosity of NFs (Eq. 3.13):

$$\mu_{NF} = \frac{\mu_{BF}}{(1 - 2.5\varphi)} \tag{3.13}$$

Although these models are simple, they require the viscosity data of the base fluid. The accuracy of these models in the estimation of the viscosities of a wide range of NFs is assessed in Table 3.4. In general, these models are often successful in the low-viscosity range. Recently, Bardool and colleagues developed two theoretical models, that is, friction theory [171] and free-volume theory [172] for the correlation of the viscosities of NFs. They utilized the friction theory in combination with two different equations of state, that is, Peng–Robinson equation of state (PR EoS) and Esmaeilzadeh–Roshanfekr equation of state (ER EoS). They obtained the best-fit parameters for a wide range of NFs. A comparison of the performance of these models in the representation of viscosity data is made in Table 3.5. Clearly, the friction models, as well as the modified version of the free-volume models, represent the viscosity of various NFs with a high degree of accuracy. Fig. 3.12 depicts the comparison of the friction models with the viscosities of Al_2O_3/water and SiO_2/EG NFs. A comparison of free-volume models and the viscosities of Al_2O_3/water and SiO_2/water NFs is also made in Fig. 3.13.

Table 3.4 A comparison of the NFs viscosity models.

NFs	NPs size (nm)	AARD% [a]						
		Einstein [b] [165]	Brinkman [c] [166]	Ward [d] [167]	Lundgren [e] [168]	Batchelor [f] [169]	Chen et al. [g] [138]	Abedian and Kachanov [h] [170]
Al_2O_3/water	30	2.32	2.32	2.32	2.32	2.32	1.58	2.32
Al_2O_3/water	47	19.54	19.15	18.94	19.03	19.03	7.82	18.94
Al_2O_3/water	36	32.30	31.71	31.40	31.53	31.54	5.79	31.40
Al_2O_3/water	12–30	30.45	30.42	30.40	30.40	30.40	24.66	30.40
Al_2O_3/methanol	50	19.43	19.43	19.42	19.42	19.42	18.44	19.42
Al_2O_3/EG	43	22.66	22.35	22.20	22.25	22.26	2.00	22.20
Al_2O_3/EG	8	26.87	26.78	26.74	26.74	26.75	16.15	26.74
SiO_2/water	12	47.15	47.07	47.03	47.04	47.04	39.64	47.03
SiO_2/methanol	190	7.97	7.97	7.97	7.97	7.97	7.50	7.97
SiO_2/EG	>28	11.45	11.13	10.97	11.02	11.03	17.86	10.97
TiO_2/water	121–180	10.54	10.54	10.54	10.54	10.54	10.14	10.54
TiO_2/water	20–40	47.80	47.80	47.80	47.80	47.80	46.94	47.80
TiO_2/water	21	25.10	24.41	24.05	24.22	24.22	16.59	24.04
TiO_2/water	76	31.36	30.62	30.20	30.45	30.45	7.53	30.18
ZnO/water	35–40	21.63	21.60	21.58	21.58	21.58	15.19	21.58
ZnO /water	20	35.06	35.05	35.05	35.05	35.05	33.05	35.05
ZnO/EG	40–100	6.24	6.11	6.05	6.07	6.07	6.88	6.05
ZnO/EG	>4.6	10.80	10.74	10.71	10.72	10.72	2.96	10.71
Ag/water	60	67.53	67.53	67.52	67.53	67.53	66.31	67.52
Fe/EG	40	13.48	13.45	13.43	13.44	13.44	10.89	13.43
Fe/EG	60	12.79	12.76	12.74	12.74	12.74	12.48	12.74
Fe/EG	100	12.25	12.21	12.20	12.20	12.20	14.10	12.20
FMWNT/water	>11	6.03	6.03	6.03	6.03	6.03	6.03	6.03

(Continued)

Table 3.4 A comparison of the NFs viscosity models. ***Continued***

NFs	NPs size (nm)	AARD%[a]						
		Einstein[b] [165]	Brinkman[c] [166]	Ward[d] [167]	Lundgren[e] [168]	Batchelor[f] [169]	Chen et al.[g] [138]	Abedian and Kachanov[h] [170]
MgO/EG	20	14.02	14.08	14.12	14.11	14.11	41.04	14.12
MgO/PG	30–40	18.75	18.82	18.85	18.84	18.84	29.55	18.85
MgO/water	20	24.15	24.13	24.13	24.13	24.13	21.11	24.13
Overall		19.43	19.28	19.21	19.23	19.24	14.90	19.20

[a] $AARD\% = \frac{100}{NDP}\sum_{i=1}^{NDP}\left|\frac{\mu_{i,exp} - \mu_{i,cal}}{\mu_{i,exp}}\right|$.
[b] *Eq. (3.7).*
[c] *Eq. (3.8).*
[d] *Eq. (3.9)*
[e] *Eq. (3.10).*
[f] *Eq. (3.11).*
[g] *Eq. (3.12).*
[h] *Eq. (3.13).*
Adapted from R. Bardool, A. Bakhtyari, F. Esmaeilzadeh, X. Wang, Nanofluid viscosity modeling based on the friction theory, J. Mol. Liq., 286 (2019) 110923 [171]; R. Bardool, A. Bakhtyari, F. Esmaeilzadeh, X. Wang, Developing free-volume models for nanofluid viscosity modeling. J. Therm. Anal. Calorim., 1 (2020) 1–14 [172].

Table 3.5 A comparison of friction and free-volume theories for the estimation of the viscosity of NFs.

NFs	NPs size (nm)	AARD%[a]			
		Friction + PR EoS [171]	Friction + ER EoS [171]	Free-volume [172]	Modified free-volume [172]
Al_2O_3/water	30	0.56	0.63	2.42	0.81
Al_2O_3/water	47	2.83	3.47	20.47	4.29
Al_2O_3/water	36	4.11	3.71	16.63	3.30
Al_2O_3/water	12–30	1.16	1.00	1.48	1.06
Al_2O_3/methanol	50	0.88	0.82	0.91	0.70
Al_2O_3/EG	43	2.94	2.47	36.40	3.34
Al_2O_3/EG	8	2.07	2.09	31.97	3.31
SiO_2/water	12	4.01	5.64	8.41	1.06
SiO_2/methanol	190	0.92	0.71	0.85	0.49
SiO_2/EG	>28	0.96	1.21	31.01	0.95
TiO_2/water	121–180	3.09	2.74	5.76	3.92
TiO_2/water	20–40	4.28	3.85	5.44	1.61
TiO_2/water	21	2.03	2.36	22.37	1.77
TiO_2/water	76	7.56	8.09	23.41	4.65
ZnO/water	35–40	1.03	0.89	13.52	1.24
ZnO/water	20	5.79	6.18	9.91	6.47
ZnO/EG	40–100	1.27	1.27	38.67	3.83
ZnO/EG	>4.6	1.15	1.18	34.34	2.31
Ag/water	60	–	–	4.18	0.51
Fe/EG	40	–	–	40.63	5.02
Fe/EG	60	–	–	41.73	4.84
Fe/EG	100	–	–	41.94	5.11
FMWNT/water	>11	–	–	1.4	0.48
MgO/EG	20	–	–	40.41	8.25
MgO/PG	30–40	–	–	34.83	3.55
MgO/water	20	–	–	7.61	3.49
Overall		2.38	2.46	19.84	2.99

[a] $\mathrm{AARD\%} = \frac{100}{\mathrm{NDP}} \sum_{i=1}^{\mathrm{NDP}} \left| \frac{\mu_{i,\mathrm{exp}} - \mu_{i,\mathrm{cal}}}{\mu_{i,\mathrm{exp}}} \right|$

Adapted from R. Bardool, A. Bakhtyari, F. Esmaeilzadeh, X. Wang, Nanofluid viscosity modeling based on the friction theory, J. Mol. Liq., 286 (2019) 110923 [171]; R. Bardool, A. Bakhtyari, F. Esmaeilzadeh, X. Wang, Developing free-volume models for nanofluid viscosity modeling. J. Therm. Anal. Calorim., 1 (2020) 1–14 [172].

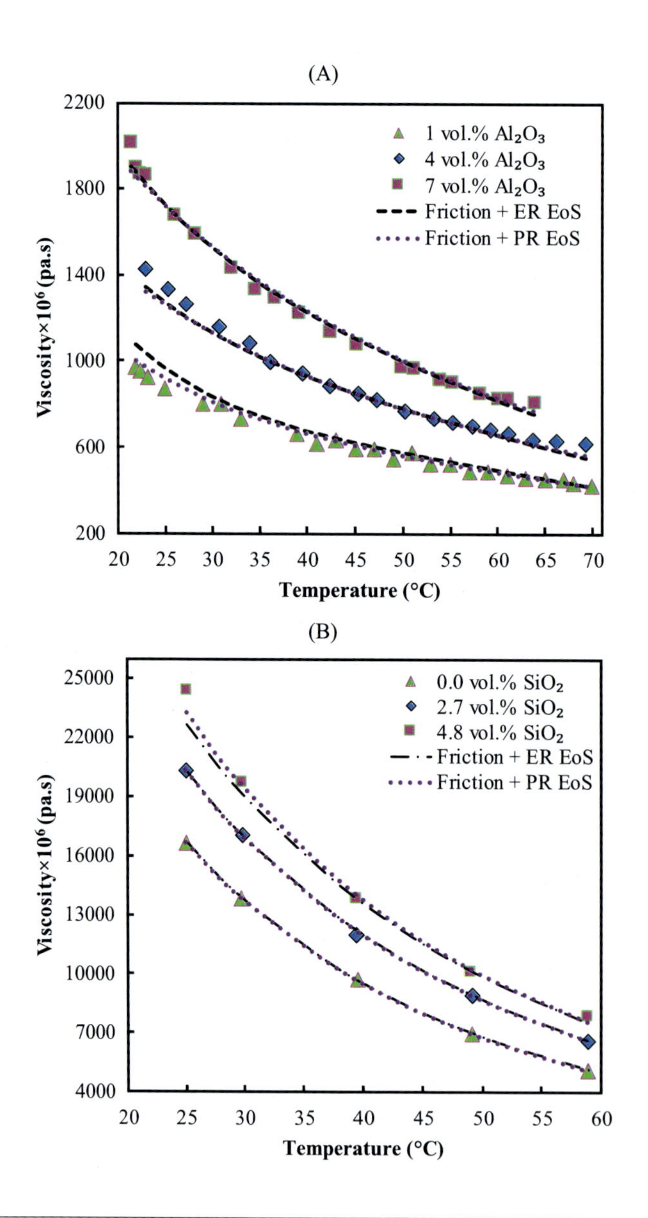

FIGURE 3.12

Comparing the calculations of the friction theory with the experimental data of (A) Al_2O_3/water (47 nm NPs) nanofluids [58] and (B) SiO_2/EG (28.3 nm nanoparticles) [111].

Adapted from R. Bardool, A. Bakhtyari, F. Esmaeilzadeh, X. Wang, Nanofluid viscosity modeling based on the friction theory, J. Mol. Liq., 286 (2019) 110923 [171].

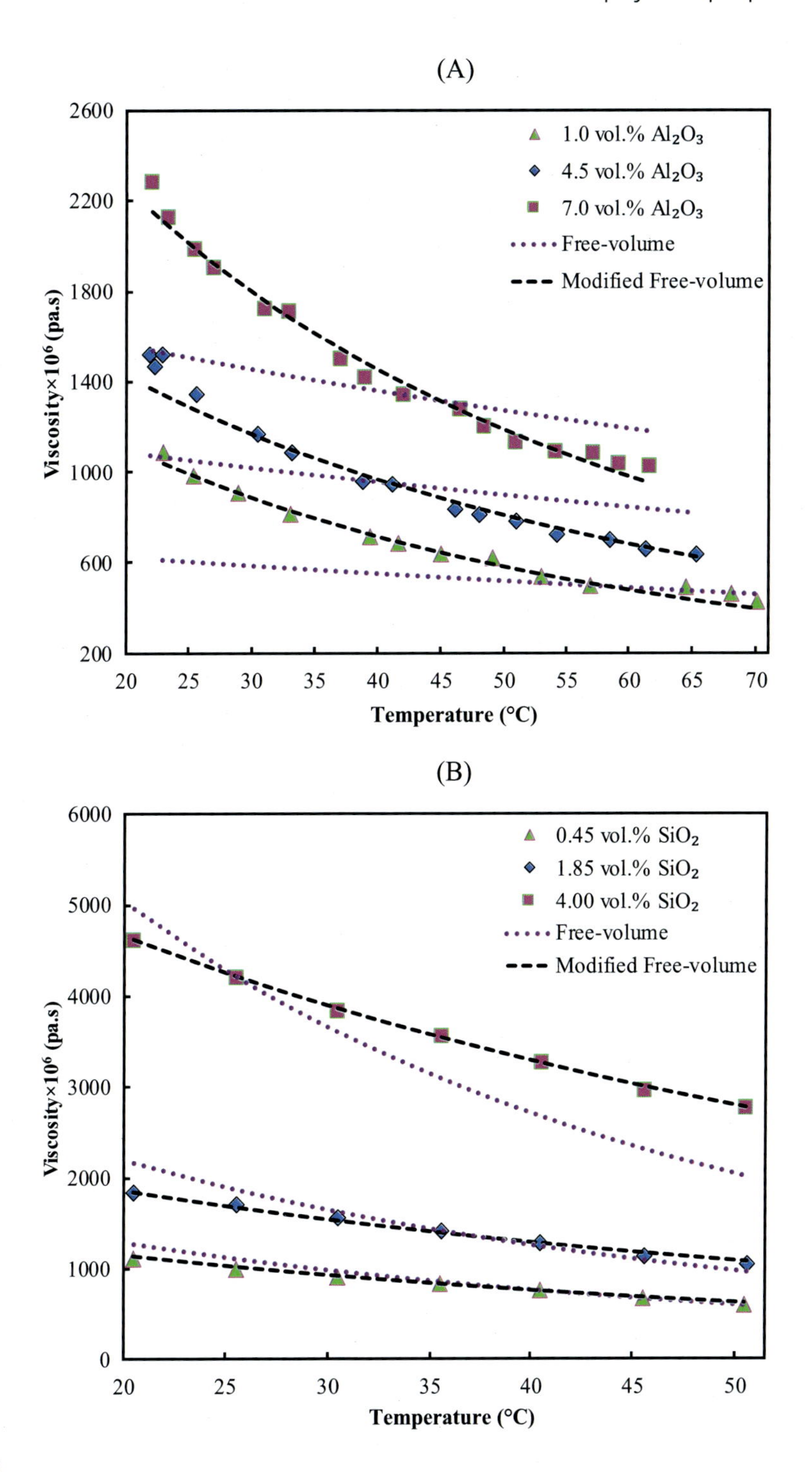

FIGURE 3.13

Comparing the calculations of the free-volume theory with the experimental data of (A) Al_2O_3/water [47 nm nanoparticles (NPs)] nanofluids [58] and (B) SiO_2/water (12 nm NPs) [106].

Adapted from R. Bardool, A. Bakhtyari, F. Esmaeilzadeh, X. Wang, Developing free-volume models for nanofluid viscosity modeling. J. Therm. Anal. Calorim., 1 (2020) 1–14 [172].

A recent review study of the theoretical, empirical, and numerical models of the viscosities of NFs was conducted by Meyer et al. [173]. They stated that a small number of the theoretical models have considered the characteristics of NFs, such as NPs size, concentration (i.e., hydrodynamic vol.%), NPs density, aggregate diameter, and the thickness of capping layer. However, the empirically developed models characterize the viscosity of NFs at various concentrations, NPs sizes, temperatures, and shear rates. The empirical models, on the one hand, often consider a single factor, mostly either concentration and/or temperature. However, the viscosity of NFs is a function of parameters including concentration, NPs size, NPs shape, shearing stress, shearing time, NPs agglomeration, base fluid type, and nanolayer formation. They also reviewed the advanced methods, including molecular dynamics simulation and artificial intelligence, and stated that these models are gradually gaining acceptance because of their capability to model the nonlinear nature of viscosity data [173]. The artificial intelligence-based models reduce the error of calculations. Compared with the conventional empirical models and correlations, these models take a wide range of affecting parameters into consideration. ANN models have been extensively employed to estimate the viscosities of NFs. In this regard, Khodadadi et al. [174], Yousefi et al. [175], Afrand et al. [176,177], Longo et al. [178], Heidari et al. [179], Esfe et al. [180,181], Atashrouz et al. [182], Karimipour et al. [183], and Bahiraei et al. [184] have developed different schemes of ANN models. Ramezanizadeh et al. [185] reviewed the utilization of machine learning models in the representation of the viscosity of NFs. Based on their investigation, the ingredients of the model structure are the heart of the models. Accordingly, the number of neurons and layers in the ANN models, the activation function, and the developed algorithm possess the greatest share in the accuracy of the models. They also compared the capabilities of ANN models against the mathematical correlations, the first of which are more accurate and confident for the estimation of the viscosity of NFs.

3.2.3 Thermal conductivity (k)

Thermal conductivity, a measure of the capability of the substance to conduct heat, is an important property in the thermal applications of the fluid. The quantity of the heat that is conducted or transferred through the unit length and unit temperature of the substance is then identified as thermal conductivity. Thermal conductivity is denoted by either k or λ (W/m.°C). Because the solids often possess larger thermal conductivities than fluids, NFs are expected to exhibit considerably larger thermal conductivity than that of the base fluid. The thermal conductivities of Al, Al_2O_3, Cu, CuO, ZnO, Pd, Au, Ag, CNT, and MWCNT NPs are 204, 27–36, 383, 18, 13, 71.8, 317, 429, 2000, and 3000 W/m.°C, respectively [186,187]. The thermal conductivities of the conventional heat transfer fluids such as water, EG, and transformer oil are 0.580–0.654 W/m.°C (at 10°C–60°C), 0.249–0.259 W/m.°C (at 20°C–60°C), and 0.111–0.108 W/m.°C (at 20°C–50°C), respectively [186]. Accordingly, the addition of NPs increases the thermal conductivity of the fluids. Such a quality was first introduced by Eastman et al. [28] when the thermal conductivity of Cu/EG NFs was reported. Various theories have been developed to explain the thermal conductivity enhancement of NFs. These theories include effective medium theory (EMT), liquid layering around NPs surface, ballistic phonon transport, Brownian motion-induced microconvection, effective conduction through percolating NPs paths, and the aggregation of NPs [22,188–193]. However, these models are not capable of representing the wide range of thermal conductivities of various NFs.

The thermal conductivity of NFs (also called effective thermal conductivity) often depends on temperature, the NPs characteristics (type, concentrations, size, shape, and thermal conductivity), and the base fluid thermal conductivity [57,194]. However, the NFs preparation method, as well as additives and pH, may affect the effective thermal conductivity [22,57,195]. A review of the recent studies reveals that increasing temperature and NPs concentration often increase the thermal conductivity, while there are still contradictions about the impact of the NPs size and shape [57,196].

The experimental determination of the thermal conductivity of NFs is still challenging. From a general point of view, Fourier's heat conduction law is developed for the experimental determination of the thermal conductivity. In this regard, Mahbubul [57], as well as Angayarkanni and Philips [4], discussed the experimental procedures of the thermal conductivity of NFs. The well-known developed techniques include basic transient hot-wire (or transient line heat source), thermal constants analyzer, steady-state parallel plate, temperature oscillation, and 3ω techniques [4,57]. There is extensive literature on the thermal conductivity of NFs. A summary is presented in Table 3.6. Discussions on the impacts of various parameters on the effective thermal conductivity will be presented in the following sections.

The effect of temperature on the thermal conductivity enhancement of various NFs has been extensively studied in the literature. Thermal conductivity enhancement is often presented in terms of relative thermal conductivity, that is, the ratio of NFs thermal conductivity to that of the base fluid (k/k_{BF}). Philip and Shima [22] presented rigorous discussions on the impact of temperature by considering the nature of NPs type. Mahbubul et al. [57] also discussed the impact of temperature on the thermal conductivity of NFs. A large number of previous studies reported enhanced relative thermal conductivity with an increase in temperature [140,160,186,187,200,201,211,212,217–236]. The least impact of temperature on the relative thermal conductivity was also reported in a variety of studies [132,192,202,204,216,237–245]. There are also reported reduced relative thermal conductivity at the highest temperatures [117,246].

A large number of studies have reported the enhancement of relative thermal conductivity with temperature. NPs type affects such enhancement. When NPs are utilized in their either metallic or oxide state, the relative thermal conductivity, as well as the impact of temperature, might differ. Patel et al. [186] found that the NFs with metallic NPs give higher enhancements than those with oxides. They collected the thermal conductivities of the NFs made up of water, EG, and transformer oil fluids and Cu, CuO, Al, and Al_2O_3 NPs. Their results reveal that increasing temperature increases the relative thermal conductivity in the whole cases. Quantitative comparisons of various NFs are made in Figs. 3.14 and 3.15. As per Fig. 3.14, increasing temperature increases the relative thermal conductivities of Al_2O_3 NPs in water (Fig. 3.14A) and EG (Fig. 3.14B) base fluids. Such an increase is observed in the whole concentration range (0.5–3.0 vol.%). Similar trends are observed in the case of CuO/EG (Fig. 3.15A) and CuO/transformer oil (Fig. 3.15B) NFs.

The research on the thermal conductivity of NFs with metallic NPs is scarce. The results of Patel et al. [186] for Cu/water and Cu/EG NFs, as well as Al/water and Al/transformer oil NFs, found enhanced relative thermal conductivities with increasing temperature. A comparison of water-based NFs that are made up of Cu and CuO NPs is made in Fig. 3.16. In the same concentration and temperature, NFs including metallic NPs (i.e., Cu) possess larger thermal conductivities. This would be due to the larger thermal conductivity of metals. Metals offer larger thermal conductivities as a result of surface vacancies, which provide easy electron transfer.

Table 3.6 Summary of information on the thermal conductivity of NFs.

NFs	NPs size (nm)	NPs concentration (vol.%)	T (°C)	Relative thermal conductivity	References
Al_2O_3/water	38.4	1.000–4.000	21–51	1.09–1.24	[197]
Al_2O_3/water	28.0	2.500–5.500	24	Up to 1.16	[198]
Al_2O_3/water	38.4	4.300	25	1.11	[199]
Al_2O_3/water	11.0	1.000	21–71	1.15	[200]
Al_2O_3/water	47.0	1.000–4.000	21–71	1.10–1.29	[200]
Al_2O_3/water	150.0	1.000	21–71	1.05	[200]
Al_2O_3/water	36.0–47.0	0.500–6.000	25–40	Up to 1.28	[201]
Al_2O_3/water	20.0	5.000	5–40	1.15	[202]
Al_2O_3/water	120.0	1.000–8.000	15–60	1.05–1.30	[203]
Al_2O_3/water	11.0–40.0	5.000	10–60	1.07–1.10	[204]
Al_2O_3/water	42.0	1.000–3.000	20–60	1.01–1.08	[205]
Al_2O_3/water	30.0–80.0	12.980	20–60	1.09–1.30	[205]
Al_2O_3/water	60.4	0.050	25	1.06	[206]
Al_2O_3/water	13.0	0.500	10–50	1.08–1.10	[95]
Al_2O_3/EG	28.0	2.500–9.000	24	Up to 1.40	[198]
Al_2O_3/EG	38.4	5.000	25	1.19	[199]
Al_2O_3/EG	60.4	0.050	25	1.14	[206]
Al_2O_3/EG + water	–[a]	1.000–10.000	25–90	1.11–1.40	[187]
Al_2O_3/EG + water	20.0	1.000–4.000	15–60	1.05–1.20	[203]
Al_2O_3/EG + water	13.0	0.200–1.000	10–50	1.01–1.08	[62]
Al_2O_3/pump oil	60.4	0.050	25	1.22	[206]
Al_2O_3/engine oil	28.0	2.500–7.500	24	Up to 1.30	[198]
Al_2O_3/R141b	13.0	0.500–2.000	5–20	1.21–1.63	[207]
TiO_2/water	30.0–80.0	12.700	20–60	1.11–1.29	[205]
TiO_2/water	21.0	1.000–8.000	15–60	1.02–1.20	[203]
TiO_2/water	21.0	3.000	13–55	1.07	[118]
TiO_2/water	15.0	2.000–5.000	25	1.24–1.30	[208]
TiO_2/water	27.0	2.000	27	1.05	[64]
TiO_2/water	25.0	0.100–1.000	25	1.10–1.15	[209]
TiO_2/water	27.0	1.000–4.30	27	1.02–1.11	[6]
TiO_2/water	34.0	0.290–0.680	25	1.02–1.06	[210]
TiO_2/water	40.0	0.60–2.600	5–40	1.01–1.06	[202]
TiO_2/water	21.0	0.240–1.920	22	1.02–1.09	[151]
TiO_2/EG + water	21.0	4.000	15–60	1.15	[203]
Ag/water	60.0–80.0	0.001	30–60	1.03–1.05	[211]
Au/toluene	3.0–4.0	0.011	30–60	1.07–1.24	[211]
SiO_2/water	30.0	31.100	20–60	1.01–1.28	[205]
SnO_2/water	4.0–5.0	0.024[b]	20–80	Up to 1.09	[212]
CuO/water	28.6	1.000–4.000	21–51	1.07–1.36	[197]

Table 3.6 Summary of information on the thermal conductivity of NFs. ***Continued***

NFs	NPs size (nm)	NPs concentration (vol.%)	T (°C)	Relative thermal conductivity	References
CuO/water	23.0	4.500–9.500	24	Up to 1.34	[198]
CuO/EG	23.0	6.000–16.000	24	Up to 1.52	[198]
Cu/EG + acid	10.0	0.300	25	1.40	[28]
Graphene/EG	0.7–1.3	5.000	10–60	1.86	[213]
Graphene Oxide/ EG	<50.0	0.050[b]	10–70	>1.40	[214]
Graphene/silicone oil	1.1–2.3	0.070[b]	20–60	Up to 1.19	[215]
Graphene oxide/ paraffin	–[a]	5.000	10–60	1.30–1.77	[216]

[a]*Not mentioned.*
[b]*wt.%.*

The observations of Yu et al. [217] reveal that Cu/EG NFs (0.5 vol.%) exhibit the relative thermal conductivities of 1.08–1.46 when the temperature rises 10°C–50°C. They also introduced that the Brownian motion of Cu NPs would be a key factor in determining the impact of the temperature on the thermal conductivity of Cu/EG NFs. Murshed et al. [222] reported that the relative thermal conductivities of Al/engine oil NFs are 1.20 and 1.37 for the NPs concentrations of 1 and 3 vol.%, respectively. The thermal conductivities of Ag/water and Au/toluene NFs were measured by Patel et al. [211] at low NPs concentrations. They observed the relative thermal conductivities of ≈1.03 and ≈1.05 for the Ag/water NFs (0.001 vol.%) at 30°C and 60°C, respectively. In the Au/toluene NFs (0.011 vol.%), the relative thermal conductivities of 1.07 and 1.14 were observed at 30°C and 60°C, respectively.

Extensive research on the impact of temperature on the NFs made up of oxide NPs is available. The observations of Murshed et al. [222] reveal that in the temperature range of 20°C–60°C, the relative thermal conductivity of Al_2O_3/EG NFs increases up to 1.09 and 1.12 in the NPs concentrations of 0.5 and 1.0 vol.%, respectively. Teng et al. [231] examined the impact of temperature (10°C–50°C) on the thermal conductivities of Al_2O_3/water NFs at various NPs sizes (20–100 nm) and concentrations (0.5–2.0 wt.%). At 10°C, 30°C, and 50°C, the relative thermal conductivities of 1.004–1.065, 1.007–1.128, and 1.012–1.147 were observed, respectively. This reveals that increasing temperature enhances the thermal conductivities of Al_2O_3/water NFs. Similar trends were reported by Das et al. [140], Chon et al. [200], Li and Peterson [201,218,221], Wong and Bhshkar [236], Sundar and Sharma [219,220], Murshed et al. [222], and Kim et al. [234] for the Al_2O_3/water NFs with different NPs sizes, concentrations, and at different temperature ranges. Kim et al. [234] also reported the thermal conductivity enhancement of water-based NFs of amorphous carbonic NPs with increasing temperature (22°C – 52°C). The results of Patel et al. [186] exhibit that temperature increases the relative thermal conductivities of Al_2O_3/EG and CuO/transformer oil NFs as well as Al_2O_3/water and CuO/water NFs. Ho and Gao [233] reported up to 17% enhancement in thermal conductivity of an n-octadecane emulsion with 10 wt.% Al_2O_3 NPs when the temperature increases up to 60°C. They also inferred the enhancement to be due to the Brownian

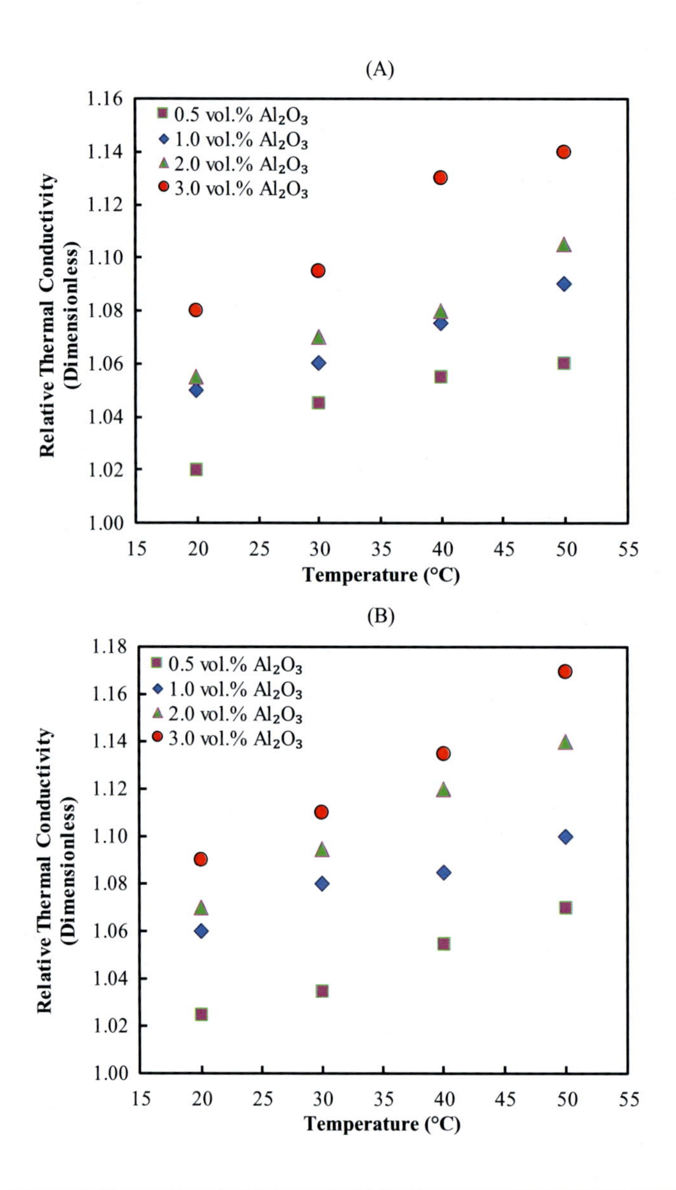

FIGURE 3.14

Effect of temperature on the relative thermal conductivities of (A) Al_2O_3/water and (B) Al_2O_3/EG nanofluids [186].

motion of NPs. Mintsa et al. [232] also observed the enhanced thermal conductivities of Al_2O_3/water and CuO/water NFs with increasing temperature. Vajjha and Das [187] measured the thermal conductivities of NFs made up of Al_2O_3, ZnO, and CuO NPs in EG(60 wt.%) + water(40 wt.%). Increasing temperature (25°C – 90°C) enhanced the thermal conductivity of NFs in all cases. As an

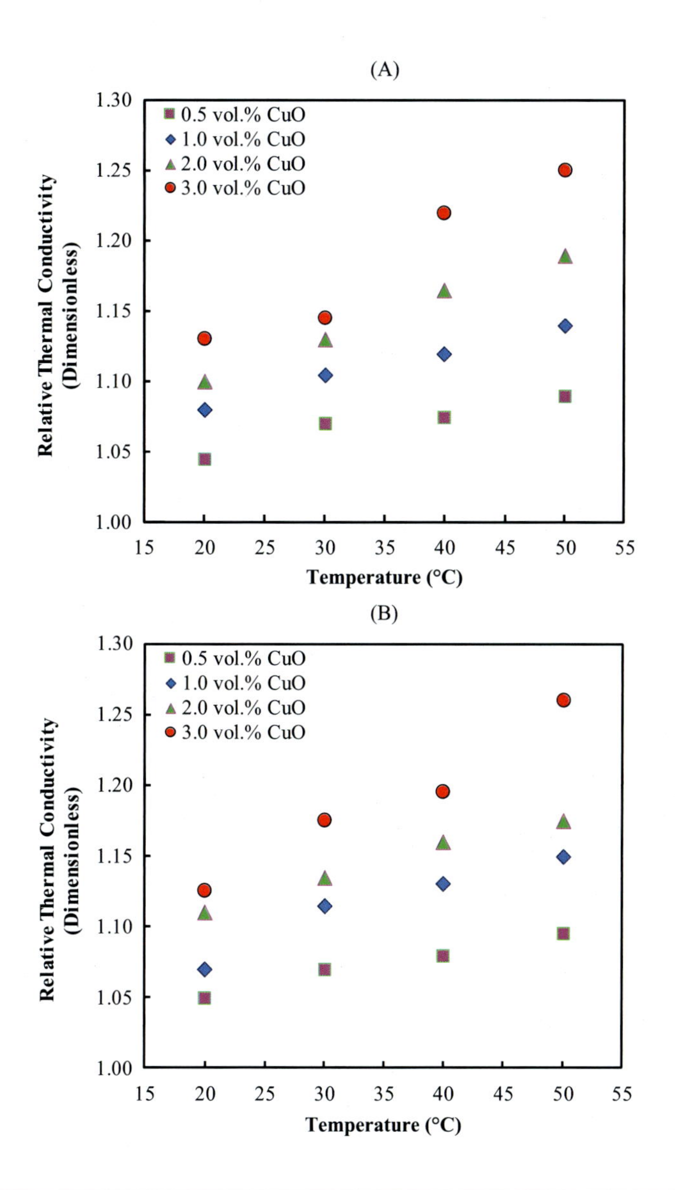

FIGURE 3.15

Effect of temperature on the relative thermal conductivities of (A) CuO/EG and (B) CuO/transformer oil nanofluids [186].

instance, when the temperature increases (25°C – 90°C) in the NFs of Al_2O_3 NPs (6 vol.%), the relative thermal conductivities of 1.224 – 1.478, that is, an almost 21.0% enhancement is observed. In the case of 10 vol.% Al_2O_3 NPs, compared to the base fluid, up to 69.0% enhancement in the relative thermal conductivity is observed at 90°C. In the case of adding ZnO NPs (7 vol.%), increasing

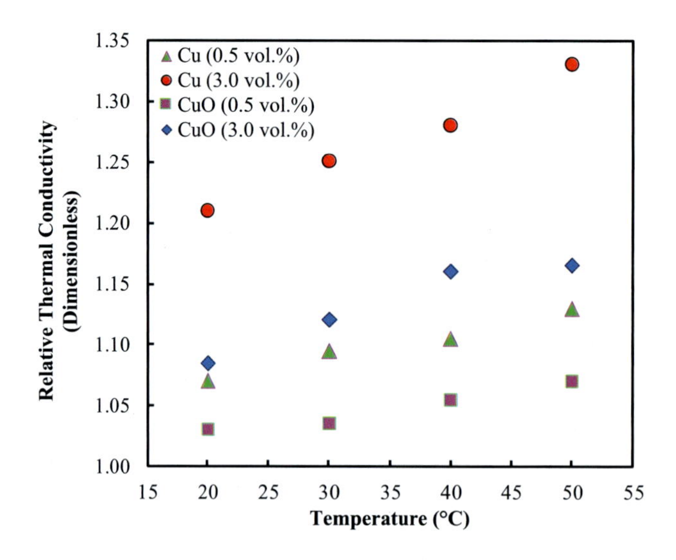

FIGURE 3.16

Effect of temperature on the relative thermal conductivities of Cu/water and CuO/water nanofluids [186].

temperature in the same range leads to an 18.0% enhancement. At 90°C, up to 48.5% enhancement in the relative thermal conductivity is observed when compared with the base fluid. A typical value of the relative thermal conductivity enhancement in the NFs of CuO NPs (6 vol.%) is 21.4% at 25°C − 90°C range. At the same concentration and 90°C, the relative thermal conductivity increases by 60.0% when compared to the base fluid [187]. Jwo et al. [235] determined the temperature dependence of relative thermal conductivity when Al_2O_3 NPs are added to a lubricant (in an R-134a refrigeration system). They observed 1.5% − 4.6% enhancement in the relative thermal conductivity of a 1.5 wt.% sample when the temperature rose 20°C–40°C. Sundar and Sharma [219,220] reported 6.5%–24.0% enhancement of relative thermal conductivity when 0.2 vol.% of CuO NPs was added to water at 30°C − 60°C. For CuO/water NFs, the relative thermal conductivity was reported also by Li and Peterson [218] to enhance with temperature (27.5°C − 34.7°C). Habibzadeh et al. [212] investigated the impact of temperature on the relative conductivity of SnO_2/water NFs at 20°C − 80°C. They observed up to approximately 8.7% enhancement at 80°C. It is inferred that the thermal conductivity of different SnO_2/water NFs increases noticeably with temperature, which would be due to the intensified Brownian motion of the NPs, as well as the decreased viscosity of the base fluid. The thermal conductivity of TiO_2/water NFs was also reported to enhance with temperature [223].

In addition to the metallic and metal oxide NPs, other classes of material could be added to the base heat transfer fluid to enhance thermal efficiency. Different nanodiamonds, as well as carbon material, are potential candidates. Yeganeh et al. [228] determined the thermal conductivity enhancements of nanodiamond/water (0.8–3.0 vol.%) at 30°C − 50°C. The increasing temperature increased the relative thermal conductivities in the whole concentration range. The largest enhancement (7.2%) was observed at 30°C for a sample with 3 vol.% NPs. The enhancement of relative thermal conductivity with temperature was up to 9.8%. A similar trend was observed by Yu et al.

[229] for the plasma-treated diamond NPs in water. The thermal conductivity of CNT/EG NFs, which was investigated by Xie and Chen [225], was found to enhance with an increase in temperature in the whole concentrations and NFs preparing durations. Amrollahi et al. [227] reported the enhanced relative thermal conductivity of CNT/EG NFs with temperature (30°C – 50°C). Similar trends were also reported for MWCNT/water NFs at 15°C – 35°C [160] and 20°C – 30°C [224] and for CNT/water NFs at 10°C – 50°C [247]. The thermal conductivities of the NFs made up of metal-dispersed MWCNTs and water and EG were investigated by Jha and Ramaprabhu [226]. In this regard, they measured the thermal conductivities of water and EG NFs with Pd-MWNTs, Au-MWNTs, and Ag-MWNTs NPs. Their results exhibited the enhancement in the relative thermal conductivities with temperature. They stated that the enhancement of thermal conductivity with temperature is caused by the increased thermal energy of the added NPs and their Brownian motion. They also observed that the sequence of thermal conductivity was Ag-MWCNT > Au-MWCNTs > Pd-MWCNTs in both water and EG NFs. This was attributed to the thermal conductivities of the NPs. Han et al. [230] assessed the impact of temperature on the hybrid NFs including polyalphaolefin lubricant as the base fluid and 0.2 vol.% of CNT–alumina–iron oxide NPs and observed an approximately 3.1% enhancement of thermal conductivity in the 10°C – 90°C temperature range.

Although the enhanced thermal conductivity of NFs with increasing temperature has been reported in a large number of studies, there exist results in the literature that indicate either temperature-independent or reduced relative thermal conductivities with temperature in some NFs. These results include the NFs with both metallic and metal oxide NPs. The results of Au/water [237], Al_2O_3/water [245], Al_2O_3/EG [245], Al_2O_3/EG + water [245], Fe_3O_4/kerosene [241], graphene oxide/water [216], graphene oxide/PG [216], graphene oxide/paraffin [216], Fe_2O_3/EG [248], and Fe_3O_4/EG [248] NFs illustrate that the absolute thermal conductivity is nearly temperature independent in these cases. Interestingly, there are some cases with temperature-enhanced absolute thermal conductivities and temperature-independent relative thermal conductivities, for example, Fe_3O_4/kerosene [240], ZnO/EG [244], Al_2O_3/hexadecane [192], Al_2O_3/water (5 vol.%, 40 nm) [204], Al_2O_3/EG (5 vol.%, 40 nm) [204], SiC/water [238], graphene/EG [213], Al_2O_3/EG [108], Fe_3O_4/kerosene [22], Fe_3O_4/hexadecane [22], Fe_3O_4/water [22], MWCNT/EG [132], and MWCNT/water (at low nanotube loading) [242] NFs. In such cases, the enhancements of the absolute thermal conductivities with temperature are due to the increase in the thermal conductivities of the base fluids.

There are rare cases that include reduced thermal conductivities at the higher temperatures. The relative thermal conductivities were found to decrease with an increase in temperature in TiO_2/water (15°C – 35°C) [117], Bi_2Te_3 nanorods/perfluoron hexane [246], and Bi_2Te_3 nanorods/hexadecane oil [246] NFs. Esmaeilzadeh et al. [19] measured the thermal conductivity of bentonite/n-pentadecane NFs, which reduced with temperature. More discussions about the impact of temperature on the thermal conductivities of NFs could be found elsewhere [22].

A higher temperature of NFs intensifies the Brownian motion of NPs, which would then increase the contribution of microconvection in heat transport phenomena. NPs move within the fluid, randomly. This random motion is Brownian motion. The Brownian motion of NPs could be represented in terms of Brownian diffusion coefficient (D_B) of the Stokes–Einstein model [249] (Eq. 3.14).

$$D_B = \frac{K_B T}{3\pi\mu d_P} \tag{3.14}$$

As per Eq. 3.14, the Brownian motion of the NPs is proportional to temperature. It is inferred that at the higher temperatures, more effective collisions occur among the NPs, which would then result in enhanced thermal conductivity [57]. The importance of Brownian motion in the thermal conductivity enhancement of the NFs can be assessed when the timescale of the movement of NPs is compared with the heat diffusion in the fluid. To this end, the required time for the NPs to displace a length equal to its size and the required time for the heat transport in the fluid by the same length are calculated by Eqs. 3.15 and 3.16, respectively [4].

$$\tau_D = \frac{3\pi\mu d_P^3}{6K_B T} \tag{3.15}$$

$$\tau_H = \frac{\rho_{BF} C_{p,BF} d_P^2}{6k_{BF}} \tag{3.16}$$

As an instance, for Cu/water NFs with NPs of 10 nm at room temperature, the ratio of τ_D to τ_H is approximately 500, which would be approximately 25 when the NPs size equals the atomic size. Therefore the Brownian motion of the NPs is slow to transport considerable quantities of heat through the NF. It is then inferred that the Brownian motion of NPs could result in particle clustering, which affects the enhancement of thermal conductivity [4]. A group of researchers believes that the Brownian motion has a small impact on the enhancement of thermal conductivity of the fluids as a result of NPs addition and thus it is not responsible for larger thermal conductivities of NFs [250,251]. However, the vital role of Brownian motion at higher fluid temperatures is often undeniable [57,252].

As discussed earlier, owing to the larger thermal conductivities of the solid NPs than the fluids, the effective thermal conductivity of NFs is larger when compared with the base fluid. Accordingly, when larger quantities of the NPs are added to the base fluid, that is, higher concentration of NPs, the enhancement of the effective thermal conductivity becomes larger. The impact of NPs concentration on the thermal conductivities of Al/EG and Al/water NFs is depicted in Fig. 3.17. Such an increase is due to the increased collisions between the NPs and species of the base fluid, which lead to enhanced thermal conductivity. It occurs in the whole experimented temperature range. A literature survey reveals that the addition of even a small quantity of NPs into the base fluid could effectively enhance the thermal conductivity. Moreover, the thermal conductivity of the NFs increases linearly with the increase in the NPs concentration, that is, vol.% [57,206,222].

When nanosized solid particles are added into the base fluid, a solid-like layer, also known as a nanolayer, is formed. This layer is formed by the fluid molecules that are close to the NPs surface. The thickness of the nanolayer is in the range of nanometers. Nanolayers work as the connections between the solid NPs and the base fluid, through which the thermal conductivity of the suspension is magnified [4,253–255]. However, a group of studies concluded that the nanolayer is not a dominating parameter for the anomalous enhancement of the thermal conductivity of NFs [4,28,256]. Cu/EG NFs are of these kinds [28]. NPs clustering and the nature of heat transport across NPs are also the determining parameters in the anomalous enhancement of the thermal conductivity of NFs, which are well discussed elsewhere [4,251].

The effect of NPs concentration on the thermal conductivity enhancement has been extensively assessed. When 4.3 vol.% of Al_2O_3 NPs were added to water, 32.4% enhancement was managed [6]. In the CuO/EG NFs, 20% enhancement was observed at 4 vol.% of CuO NPs [199]. The

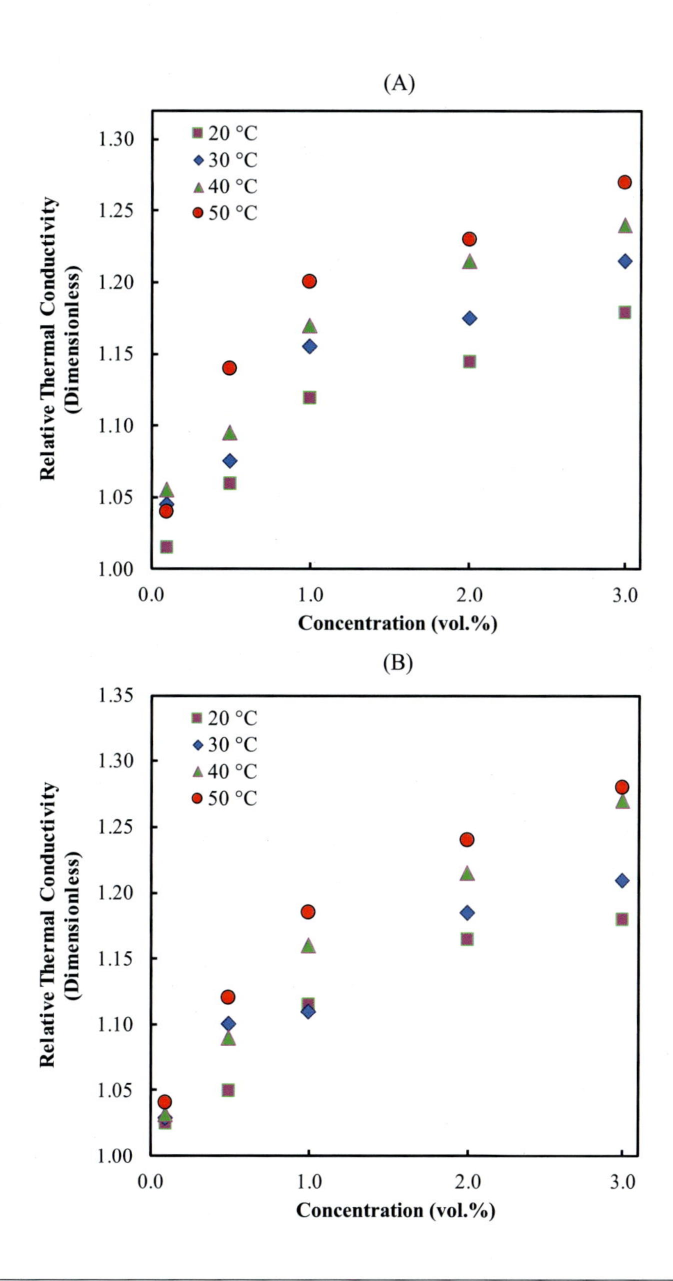

FIGURE 3.17

Effect of nanoparticles concentration on the relative thermal conductivities of (A) Al/EG and (B) Al/water nanofluids [186].

relative thermal conductivities of Cu/transformer oil were 1.2 − 1.8 at a 2.5 − 8.0 vol.% NPs concentration [257]. For Cu/water NFs, the relative thermal conductivity was 1.1 − 1.6 at 1 − 5 vol.% of NPs [257]. The observations of Shin and Lee [258] demonstrated up to 13% enhancement in the thermal conductivities of NFs made up of polyethylene and polypropylene (10 vol.%) in a blend of silicon oil and kerosene. Choi et al. [259] reported 150% thermal conductivity enhancement when 1 vol.% of MWCNT was added to polyalphaolefin oil. Liu et al. [260] observed a thermal conductivity enhancement of 12% for the CNT/EG (1 vol.%) NFs and 30% enhancement for the CNT/engine oil (2 vol.%) NFs. Up to 18% increase in the thermal conductivity of EG was observed when very small quantities (up to 10^4 ppm) of Ag NPs were added [261]. The thermal conductivity enhancement of the MWCNT/water NFs was found to be 11% at a concentration of 1 vol.% [262]. Up to 74% increase in the thermal conductivity of Cu/water (0.3 vol.%) NFs was reported by Jana et al. [263]. 16.5% increase in the thermal conductivity was observed in the Fe/EG (0.3 vol.%) NFs [209]. For MWCNT/EG (1 vol.%) NFs, a thermal conductivity enhancement of 17% was obtained [132]. A 20% enhancement in the thermal conductivity of CNT/EG NFs was managed at 2.5 vol.% [132]. In the Al_2O_3/methanol, SiO_2/methanol, and TiO_2/methanol NFs, increases in the concentration of NPs resulted in considerable enhancements in the thermal conductivities [264]. Khedkar et al. [265] reported up to 30% enhancement in Fe_3O_4/water NFs as the NPs concentration increases (0.00 − 0.04). Zhu et al. [266] also reported up to 38% enhancement in thermal conductivity in Fe_3O_4/water at a concentration of 4 vol.%. Previous studies have illustrated that thermal conductivity increases when NPs concentration increases in the Cu/EG [267], graphene oxide/water [216], graphene oxide/PG [216], graphene oxide/paraffin [216], Al_2O_3/water [268], TiO_2/water [268], and SiO_2/water [268]. Extended literature surveys of the impact of NPs concentration on the thermal conductivities of NFs can be found elsewhere [3,4,22,57,269].

In addition to temperature and concentration, NPs size affects the thermal behavior of NFs. Conflicting observations have been reported in this regard. A group of researchers has reported enhancements in the relative thermal conductivities with a decrease in the NPs size [151,152,186,187,200,201,231,232,270−274], while decreases in the relative thermal conductivity with a decrease in the NPs size were also observed in some cases [154,204,237,245,275−277]. The reviews of the experimental observations on the impact of the NPs size on the thermal conductivity of various NFs can be found elsewhere [3,4,22,57,269]. As an instance, the effect of NPs size on the relative thermal conductivity of Al_2O_3/water NFs is demonstrated in Fig. 3.18. Considerable increases are observed when smaller NPs (11 nm) are added to the base fluid.

3.2.3.1 Thermal conductivity models

A variety of theoretical and empirical models has been hitherto proposed to estimate the thermal conductivity of NFs. The first attempt was made by Maxwell [278], through which the thermal conductivity of a composite mixture was obtained by utilizing the EMT. The model includes simplifying assumptions. The NFs are considered to be two-component systems, which constitute a continuous matrix of particles (discontinuous component). The mixtures are then assumed to be under the static conditions and a low volume fraction of NPs within the matrix. Moreover, the clustering of the NPs is not considered [279]. Accordingly, the effective thermal conductivity of the NFs would be obtained as a function of the thermal conductivities of base fluid (k_{BF}), the thermal conductivity of NPs (k_{NP}), and the volume concentration of NPs (φ). This model is suitable at the low concentrations of noninteractive spherical NPs (Eq. 3.17 of Table 3.7). It is worth noting that

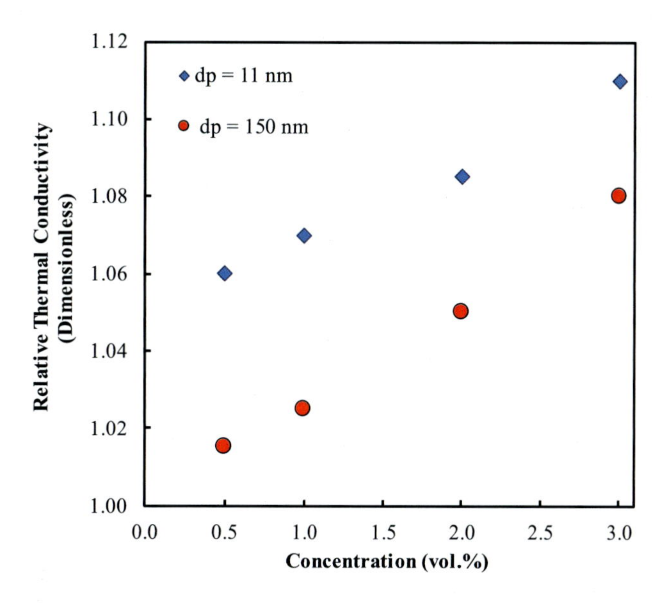

FIGURE 3.18

Effect of nanoparticles size on the relative thermal conductivities of Al_2O_3/water nanofluids [186].

at higher concentrations, the interactions between the NPs become important, and thus this model fails to predict the thermal conductivities [4]. By taking into account the role of interparticle interaction at the higher concentrations of NPs, Bruggeman [280] extended Maxwell's model to obtain Equation 18 of Table 3.7. The Hamilton–Crosser model (Equation 19 of Table 3.7), which considers a mixture of two components with different shapes and sizes, is extensively utilized in numerical studies. The performance of this model against the experimental data of Patel et al. [186] is depicted in Fig. 3.19. Clearly, the model underestimates the thermal conductivity enhancements. This is, on the one hand, because the model does not consider the effects of NPs interactions and size. A summary of the well-known models is presented in Table 3.7.

Esmaeilzadeh et al. [19] developed the following model for the thermal conductivity of bentonite/n-pentadecane NFs. The thermal conductivity was estimated based on the geometric average of the thermal conductivity of the base fluid (n-pentadecane) and NPs (bentonite). More details of the model could be found elsewhere [19]

$$\frac{k}{k_{BF}} = \left(\frac{k_{NP}}{k_{BF}}\right)^{\varphi} \tag{3.17}$$

A comparison between the experimental results and the model prediction is depicted in Fig. 3.20. Angayarkanni and Philip [4] and Gupta et al. [3] extensively reviewed the theoretical and empirical models of thermal conductivity, as well as their assumptions and applications.

In addition to the theoretical and empirical model, ANN models have been also developed to estimate the thermal conductivities of NFs. To this end, NFs made up of ferromagnetics [285], CuO [286], ZnO [287], and oxide [288] NPs have been investigated by utilizing ANN models.

Table 3.7 Summary of thermal conductivity models.

Eq.	Model	Comment	References
(17)	$\frac{k}{k_{BF}} = \frac{1 + 2\frac{k_{NP} - k_{BF}}{k_{NP} + 2k_{BF}}\varphi}{1 - \frac{k_{NP} - k_{BF}}{k_{NP} + 2k_{BF}}\varphi}$	Low concentrations of spherical NPs	[278]
(18)	$\frac{k}{k_{BF}} = 0.25\left[(3\varphi - 1)\frac{k_{NP}}{k_{BF}} + (2 - 3\varphi) + \sqrt{\Delta}\right] \Delta = (3\varphi - 1)^2\left(\frac{k_{NP}}{k_{BF}}\right)^2 + (2 - 3\varphi)^2 + 2\left(2 + 9\varphi - 9\varphi^2\right)\frac{k_{NP}}{k_{BF}}$	Considering the interparticle interactions	[280]
(19)	$\frac{k}{k_{BF}} = \frac{k_{NP} + (n-1)k_{BF} - (n-1)(k_{BF} - k_{NP})\varphi}{k_{NP} + (n-1)k_{BF} + (k_{BF} - k_{NP})\varphi}$	Hamilton–Crosser model, considering a mixture of two components with different shapes and sizes	[281]
(20)	$\frac{k}{k_{BF}} = \left[\frac{1}{1 - n\varphi A}\right]k_1 A = \frac{1}{3}\sum_{j=a,b,c}\frac{k_{NP,j} - k_{BF}}{k_{NP,j} + (n-1)k_{BF}}$	Based on the modification of the Hamilton–Crosser model (Equation 19), considering the particle–fluid interfacial layer	[254]
(21)	$\frac{k}{k_{BF}} = \frac{k_{NP} + (n-1)k_{BF} - \alpha_e(n-1)(k_{BF} - k_{NP})}{k_{NP} + (n-1)k_{BF} + \alpha_e(k_{BF} - k_{NP})}$	Considered liquid layer thickness to correct the concentration	[282]
(22)	$\frac{k}{k_{BF}} = \frac{k_{NP} + 2k_{BF} - 2(k_{BF} - k_{NP})\varphi}{k_{NP} + 2k_{BF} + (k_{BF} - k_{NP})\varphi} + \frac{\rho_{NF}\varphi C_p}{2k_{BF}}\sqrt{\frac{k_b T}{3\pi r_c \mu}}$	Considered Brownian motion, temperature, and interfacial interaction, k_b is Boltzmann constant and r_c is the apparent radius of clusters	[283]
(23)	$\frac{k}{k_{BF}} = 1 + c\frac{2k_b T}{\pi v d_P^2}\frac{\varphi r_b}{(1 - \varphi)r_b}$	Considered Brownian motion, temperature, dynamic viscosity (v), base fluid, and particle diameter (d_P)	[284]

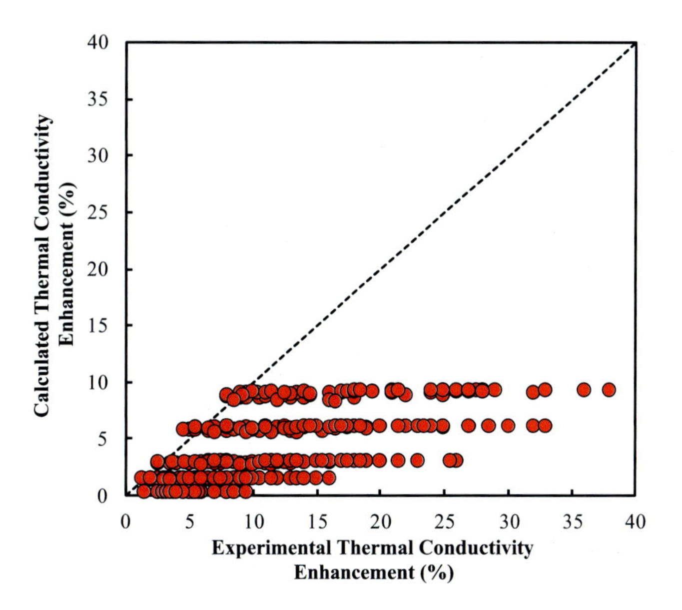

FIGURE 3.19

Comparing the calculations of the Hamilton–Crosser model with the experimental data of [186].

Furthermore, the thermal conductivities of MgO/EG [289], Cu-TiO_2/water + EG [290], Al_2O_3/water [291], Al_2O_3/carboxymethyl cellulose [292], ZnO-TiO_2/EG [293], TiO_2/carboxymethyl cellulose [292], CuO/carboxymethyl cellulose [292], Fe_3O_4/water [294], and CuO/paraffin [183] NFs have been estimated by various schemes of ANN models. Ahmadloo and Azizi assessed the ANN model for various NFs [295] and found the model to be accurate in the representation of the experimental data with errors of 1.26% and 1.44% for the collected training and test data sets, respectively.

3.3 Conclusion and outlook

Prompted by the necessity of developing NFs, as the new efficient heat transfer working fluids, we investigated the physical properties of these fluids. To this end, density, viscosity, and thermal conductivity were reviewed. The impacts of various variables such as temperature, concentration, and NPs size were assessed. A review of the theoretical estimation methods was also implemented. Although various NFs are potential working fluids for heat transfer applications, their successful development on the commercial scale relies on their performance under practical conditions. However, the assessment of physical properties could shed light on this prospect. The capability of NFs for heat transfer in the cooling/heating sections of the typical industrial sections should be evaluated, which falls beyond the scope of this study. The applications of NFs with a focus on the solar collectors and automotive heat exchangers were investigated by Gupta et al. [3]. Different application fields of NFs including transportation section, electronic cooling, energy storage, and mechanical applications are well discussed by Devendiran and Amirtham [49]. Kumar et al. [2]

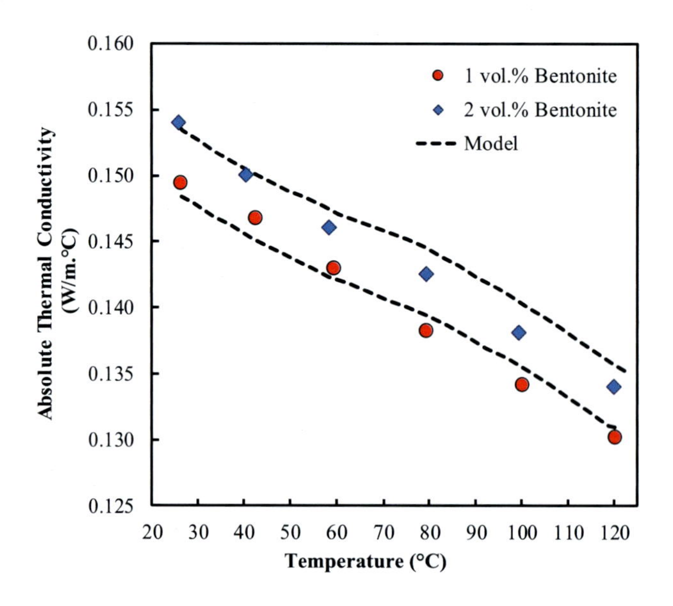

FIGURE 3.20

Comparing the calculations and experimental data of bentonite/n-pentadecane nanofluids.

F. Esmaeilzadeh, A.S. Teja, A. Bakhtyari, The thermal conductivity, viscosity, and cloud points of bentonite nanofluids with n-pentadecane as the base fluid, J. Mol. Liq., 300 (2020) 112307.

made an attempt to consider the function and performance of plate heat exchangers that employ NFs. Kasaeian et al. [1] reviewed the applications of NFs on solar collectors, photovoltaic systems, and solar thermoelectrics with a focus on the energy conversion, energy storage system, and manufacturing costs. The investigations of Mahian et al. [296] have also shed light on this path. Despite the difficulties and challenges, the development of NFs on the commercial scale is currently crucial. There is further vital scope to work theoretically and experimentally on the performance and the utilization of NFs. Investigating the environmental aspects of developing NFs is currently of potential importance. Such investigations along with the investigations of the present contribution would be beneficial in the successful development of NFs and shed light on the prospect of utilizing high-efficient heat transfer working fluids.

References

[1] A. Kasaeian, A.T. Eshghi, M. Sameti, A review on the applications of nanofluids in solar energy systems, Renew. Sustain. Energy Rev. 43 (2015) 584–598.

[2] V. Kumar, A.K. Tiwari, S.K. Ghosh, Application of nanofluids in plate heat exchanger: a review, Energy Convers. Manag. 105 (2015) 1017–1036.

[3] M. Gupta, V. Singh, R. Kumar, Z. Said, A review on thermophysical properties of nanofluids and heat transfer applications, Renew. Sustain. Energy Rev. 74 (2017) 638–670.

[4] S. Angayarkanni, J. Philip, Review on thermal properties of nanofluids: recent developments, Adv. Colloid Interface Sci. 225 (2015) 146–176.

[5] J.C. Maxwell, A Treatise on Electricity and Magnetism, Clarendon Press, 1873.

[6] H. Masuda, A. Ebata, K. Teramae, Alteration of thermal conductivity and viscosity of liquid by dispersing ultra-fine particles. Dispersion of Al_2O_3, SiO_2 and TiO_2 ultra-fine particles, Netsu Bussei 7 (1993) 227–233.

[7] Choi S.U., Eastman J.A. Enhancing Thermal Conductivity of Fluids with Nanoparticles. Argonne National Lab, Lemont, IL, 1995.

[8] I. Mahbubul, R. Saidur, M. Amalina, Thermal conductivity, viscosity and density of R141b refrigerant based nanofluid, Proc Eng. 56 (2013) 310–315.

[9] R. Saidur, S. Kazi, M. Hossain, M. Rahman, H. Mohammed, A review on the performance of nanoparticles suspended with refrigerants and lubricating oils in refrigeration systems, Renew. Sustain. Energy Rev. 15 (1) (2011) 310–323.

[10] S. Bi, K. Guo, Z. Liu, J. Wu, Performance of a domestic refrigerator using TiO_2-R600a nano-refrigerant as working fluid, Energy Convers. Manag. 52 (1) (2011) 733–737.

[11] A.M. Hussein, R. Bakar, K. Kadirgama, Study of forced convection nanofluid heat transfer in the automotive cooling system, Case Stud. Therm. Eng. 2 (2014) 50–61.

[12] K. Leong, R. Saidur, S. Kazi, A. Mamun, Performance investigation of an automotive car radiator operated with nanofluid-based coolants (nanofluid as a coolant in a radiator), Appl. Therm. Eng. 30 (17-18) (2010) 2685–2692.

[13] Z. Said, R. Saidur, M. Sabiha, A. Hepbasli, N. Rahim, Energy and exergy efficiency of a flat plate solar collector using pH treated Al_2O_3 nanofluid, J. Clean. Prod. 112 (2016) 3915–3926.

[14] R. Saidur, T. Meng, Z. Said, M. Hasanuzzaman, A. Kamyar, Evaluation of the effect of nanofluid-based absorbers on direct solar collector, Int. J. Heat Mass. Transf. 55 (21-22) (2012) 5899–5907.

[15] T.R. Barrett, S. Robinson, K. Flinders, A. Sergis, Y. Hardalupas, Investigating the use of nanofluids to improve high heat flux cooling systems, Fusion. Eng. Des. 88 (9-10) (2013) 2594–2597.

[16] A. Turgut, E. Elbasan (Eds.), Nanofluids for electronics cooling, in: 2014 IEEE 20th International Symposium for Design and Technology in Electronic Packaging (SIITME), IEEE, 2014.

[17] S.-S. Hsieh, H.-Y. Leu, H.-H. Liu, Spray cooling characteristics of nanofluids for electronic power devices, Nanoscale Res. Lett. 10 (1) (2015) 139.

[18] A. Ijam, R. Saidur, Nanofluid as a coolant for electronic devices (cooling of electronic devices), Appl. Therm. Eng. 32 (2012) 76–82.

[19] F. Esmaeilzadeh, A.S. Teja, A. Bakhtyari, The thermal conductivity, viscosity, and cloud points of bentonite nanofluids with n-pentadecane as the base fluid, J. Mol. Liq. 300 (2020) 112307.

[20] S. Masoud Hosseini, A. Moghadassi, D. Henneke, A new dimensionless group model for determining the viscosity of nanofluids, J. Therm. Anal. Calorim. 100 (3) (2010) 873–877.

[21] M. Behi, S.A. Mirmohammadi, Investigation on thermal conductivity, viscosity and stability of nanofluids. Master of Science Thesis EGI-2012, Royal Institute of Technology (KTH), School of Industrial Engineering and Management, Department of Energy Technology, Division of Applied Thermodynamics and Refrigeration, Stockholm, Sweden, 2012.

[22] J. Philip, P.D. Shima, Thermal properties of nanofluids, Adv. Colloid Interface Sci. 183 (2012) 30–45.

[23] J. Salehi, M. Heyhat, A. Rajabpour, Enhancement of thermal conductivity of silver nanofluid synthesized by a one-step method with the effect of polyvinylpyrrolidone on thermal behavior, Appl. Phys. Lett. 102 (23) (2013) 231907.

[24] Y. Li, S. Tung, E. Schneider, S. Xi, A review on development of nanofluid preparation and characterization, Powder Technol. 196 (2) (2009) 89–101.

[25] H. Akoh, Y. Tsukasaki, S. Yatsuya, A. Tasaki, Magnetic properties of ferromagnetic ultrafine particles prepared by vacuum evaporation on running oil substrate, J. Cryst. Growth 45 (1978) 495–500.

[26] S.U. Choi, J.A. Eastman, Enhanced heat transfer using nanofluids. Google Patents, US6221275B1, 2001.
[27] V. Sridhara, L.N. Satapathy, Al_2O_3-based nanofluids: a review, Nanoscale Res. Lett. 6 (1) (2011) 456.
[28] J.A. Eastman, S. Choi, S. Li, W. Yu, L. Thompson, Anomalously increased effective thermal conductivities of ethylene glycol-based nanofluids containing copper nanoparticles, Appl. Phys. Lett. 78 (6) (2001) 718–720.
[29] J. Zhu, D. Li, H. Chen, X. Yang, L. Lu, X. Wang, Highly dispersed CuO nanoparticles prepared by a novel quick-precipitation method, Mater. Lett. 58 (26) (2004) 3324–3327.
[30] S.A. Kumar, K.S. Meenakshi, B. Narashimhan, S. Srikanth, G. Arthanareeswaran, Synthesis and characterization of copper nanofluid by a novel one-step method, Mater. Chem. Phys. 113 (1) (2009) 57–62.
[31] S.U. Sandhya, S.A. Nityananda, A facile one step solution route to synthesize cuprous oxide nanofluid, Nanomater. Nanotechnol. 3 (2013) 5.
[32] A.G. Kanaras, F.S. Kamounah, K. Schaumburg, C.J. Kiely, M. Brust, Thioalkylated tetraethylene glycol: a new ligand for water soluble monolayer protected gold clusters, Chem. Commun. 20 (2002) 2294–2295.
[33] C.-H. Lo, T.-T. Tsung, L.-C. Chen, Shape-controlled synthesis of Cu-based nanofluid using submerged arc nanoparticle synthesis system (SANSS), J. Cryst. Growth 277 (1-4) (2005) 636–642.
[34] C.-H. Lo, T.-T. Tsung, H.-M. Lin, Preparation of silver nanofluid by the submerged arc nanoparticle synthesis system (SANSS), J. Alloy. Compd. 434 (2007) 659–662.
[35] H.J. Kim, I.C. Bang, J. Onoe, Characteristic stability of bare Au-water nanofluids fabricated by pulsed laser ablation in liquids, Opt. Lasers Eng. 47 (5) (2009) 532–538.
[36] S.W. Lee, S.D. Park, I.C. Bang, Critical heat flux for CuO nanofluid fabricated by pulsed laser ablation differentiating deposition characteristics, Int. J. Heat mass. Transf. 55 (23-24) (2012) 6908–6915.
[37] T.X. Phuoc, Y. Soong, M.K. Chyu, Synthesis of Ag-deionized water nanofluids using multi-beam laser ablation in liquids, Opt. Lasers Eng. 45 (12) (2007) 1099–1106.
[38] H. Wang, J.-Z. Xu, J.-J. Zhu, H.-Y. Chen, Preparation of CuO nanoparticles by microwave irradiation, J. Cryst. Growth 244 (1) (2002) 88–94.
[39] H.-t. Zhu, Y.-s. Lin, Y.-s. Yin, A novel one-step chemical method for preparation of copper nanofluids, J. Colloid Interface Sci. 277 (1) (2004) 100–103.
[40] A.K. Singh, V.S. Raykar, Microwave synthesis of silver nanofluids with polyvinylpyrrolidone (PVP) and their transport properties, Colloid Polym. Sci. 286 (14-15) (2008) 1667–1673.
[41] Y. Sun, B. Mayers, T. Herricks, Y. Xia, Polyol synthesis of uniform silver nanowires: a plausible growth mechanism and the supporting evidence, Nano Lett. 3 (7) (2003) 955–960.
[42] F. Bonet, K. Tekaia-Elhsissen, K.V. Sarathy, Study of interaction of ethylene glycol/PVP phase on noble metal powders prepared by polyol process, Bull. Mater. Sci. 23 (3) (2000) 165–168.
[43] Y.L. Hsin, K.C. Hwang, F.R. Chen, J.J. Kai, Production and in-situ metal filling of carbon nanotubes in water, Adv. Mater. 13 (11) (2001) 830–833.
[44] D.L. Van Hyning, W.G. Klemperer, C.F. Zukoski, Silver nanoparticle formation: predictions and verification of the aggregative growth model, Langmuir 17 (11) (2001) 3128–3135.
[45] Y. Chen, X. Luo, G.-H. Yue, X. Luo, D.-L. Peng, Synthesis of iron–nickel nanoparticles via a nonaqueous organometallic route, Mater. Chem. Phys. 113 (1) (2009) 412–416.
[46] S. Pati, V. Mahendran, J. Philip, A simple approach to produce stable ferrofluids without surfactants and with high temperature stability, J. Nanofluids 2 (2) (2013) 94–103.
[47] G. Gnanaprakash, S. Mahadevan, T. Jayakumar, P. Kalyanasundaram, J. Philip, B. Raj, Effect of initial pH and temperature of iron salt solutions on formation of magnetite nanoparticles, Mater. Chem. Phys. 103 (1) (2007) 168–175.
[48] N. Sezer, M.A. Atieh, M. Koç, A comprehensive review on synthesis, stability, thermophysical properties, and characterization of nanofluids, Powder Technol. 344 (2019) 404–431.

[49] D.K. Devendiran, V.A. Amirtham, A review on preparation, characterization, properties and applications of nanofluids, Renew. Sustain. Energy Rev. 60 (2016) 21–40.

[50] J.R. Babu, K.K. Kumar, S.S. Rao, State-of-art review on hybrid nanofluids, Renew. Sustain. Energy Rev. 77 (2017) 551–565.

[51] L.S. Sundar, K. Sharma, M.K. Singh, A. Sousa, Hybrid nanofluids preparation, thermal properties, heat transfer and friction factor–a review, Renew. Sustain. Energy Rev. 68 (2017) 185–198.

[52] K. Leong, K.K. Ahmad, H.C. Ong, M. Ghazali, A. Baharum, Synthesis and thermal conductivity characteristic of hybrid nanofluids–a review, Renew. Sustain. Energy Rev. 75 (2017) 868–878.

[53] I. Popa, G. Gillies, G. Papastavrou, M. Borkovec, Attractive and repulsive electrostatic forces between positively charged latex particles in the presence of anionic linear polyelectrolytes, J. Phys. Chem. B 114 (9) (2010) 3170–3177.

[54] C. Pang, J.W. Lee, Y.T. Kang, Review on combined heat and mass transfer characteristics in nanofluids, Int. J. Therm. Sci. 87 (2015) 49–67.

[55] S.-S. Ashrafmansouri, M.N. Esfahany, Mass transfer in nanofluids: a review, Int. J. Therm. Sci. 82 (2014) 84–99.

[56] Z. Zhang, J. Cai, F. Chen, H. Li, W. Zhang, W. Qi, Progress in enhancement of CO_2 absorption by nanofluids: a mini review of mechanisms and current status, Renew. Energy 118 (2018) 527–535.

[57] I.M. Mahbubul, 4—Thermophysical properties of nanofluids, in: I.M. Mahbubul (Ed.), Preparation, Characterization, Properties and Application of Nanofluid, William Andrew Publishing, 2019, pp. 113–196.

[58] C. Nguyen, F. Desgranges, G. Roy, N. Galanis, T. Maré, S. Boucher, et al., Temperature and particle-size dependent viscosity data for water-based nanofluids–hysteresis phenomenon, Int. J. Heat Fluid Flow. 28 (6) (2007) 1492–1506.

[59] G. Smith, W. Wilding, J. Oscarson, R. Rowley (Eds.), Correlation of liquid viscosity at the normal boiling point, in: Proceedings of the 15th Symposium on Thermophysical Properties, Boulder, CO, 2003.

[60] E.V. Timofeeva, W. Yu, D.M. France, D. Singh, J.L. Routbort, Nanofluids for heat transfer: an engineering approach, Nanoscale Res. Lett. 6 (1) (2011) 182.

[61] M. Pastoriza-Gallego, C. Casanova, R. Páramo, B. Barbés, J. Legido, M. Piñeiro, A study on stability and thermophysical properties (density and viscosity) of Al_2O_3 in water nanofluid, J. Appl. Phys. 106 (6) (2009) 064301.

[62] M. Elias, I. Mahbubul, R. Saidur, M. Sohel, I. Shahrul, S. Khaleduzzaman, et al., Experimental investigation on the thermo-physical properties of Al_2O_3 nanoparticles suspended in car radiator coolant, Int. Commun. Heat Mass. Transf. 54 (2014) 48–53.

[63] R.S. Vajjha, D.K. Das, A review and analysis on influence of temperature and concentration of nanofluids on thermophysical properties, heat transfer and pumping power, Int. J. Heat Mass. Transf. 55 (15-16) (2012) 4063–4078.

[64] B.C. Pak, Y.I. Cho, Hydrodynamic and heat transfer study of dispersed fluids with submicron metallic oxide particles, Exp. Heat Transf. An. Int. J. 11 (2) (1998) 151–170.

[65] M.T. Zafarani-Moattar, R. Majdan-Cegincara, Effect of temperature on volumetric and transport properties of nanofluids containing ZnO nanoparticles poly (ethylene glycol) and water, J. Chem. Thermodyn. 54 (2012) 55–67.

[66] M. Kayhani, H. Soltanzadeh, M. Heyhat, M. Nazari, F. Kowsary, Experimental study of convective heat transfer and pressure drop of TiO_2/water nanofluid, Int. Commun. Heat Mass. Transf. 39 (3) (2012) 456–462.

[67] I. Behroyan, S.M. Vanaki, P. Ganesan, R. Saidur, A comprehensive comparison of various CFD models for convective heat transfer of Al_2O_3 nanofluid inside a heated tube, Int. Commun. Heat Mass. Transf. 70 (2016) 27–37.

[68] F. Selimefendigil, H.F. Öztop, N. Abu-Hamdeh, Mixed convection due to rotating cylinder in an internally heated and flexible walled cavity filled with SiO_2−water nanofluids: effect of nanoparticle shape, Int. Commun. Heat Mass. Transf. 71 (2016) 9−19.
[69] B. Sun, W. Lei, D. Yang, Flow and convective heat transfer characteristics of Fe_2O_3−water nanofluids inside copper tubes, Int. Commun. Heat Mass. Transf. 64 (2015) 21−28.
[70] M.A. Makarem, A. Bakhtyari, M.R. Rahimpour, A numerical investigation on the heat and fluid flow of various nanofluids on a stretching sheet, Heat Transf—Asian Res. 47 (2) (2018) 347−365.
[71] M. Ansarpour, E. Danesh, M. Mofarahi, Investigation the effect of various factors in a convective heat transfer performance by ionic liquid, ethylene glycol, and water as the base fluids for Al_2O_3 nanofluid in a horizontal tube: a numerical study, Int. Commun. Heat Mass. Transf. 113 (2020) 104556.
[72] M. Sheikholeslami, M. Gorji-Bandpay, D. Ganji, Magnetic field effects on natural convection around a horizontal circular cylinder inside a square enclosure filled with nanofluid, Int. Commun. Heat Mass. Transf. 39 (7) (2012) 978−986.
[73] A. Rashad, M. Rashidi, G. Lorenzini, S.E. Ahmed, A.M. Aly, Magnetic field and internal heat generation effects on the free convection in a rectangular cavity filled with a porous medium saturated with Cu−water nanofluid, Int. J. Heat Mass. Transf. 104 (2017) 878−889.
[74] K. Leong, R. Saidur, M. Khairulmaini, Z. Michael, A. Kamyar, Heat transfer and entropy analysis of three different types of heat exchangers operated with nanofluids, Int. Commun. Heat Mass. Transf. 39 (6) (2012) 838−843.
[75] O. Manca, P. Mesolella, S. Nardini, D. Ricci, Numerical study of a confined slot impinging jet with nanofluids, Nanoscale Res. Lett. 6 (1) (2011) 188.
[76] A.A. Minea, Uncertainties in modeling thermal conductivity of laminar forced convection heat transfer with water alumina nanofluids, Int. J. Heat Mass. Transf. 68 (2014) 78−84.
[77] M.H. Esfe, S. Saedodin, M. Mahmoodi, Experimental studies on the convective heat transfer performance and thermophysical properties of MgO−water nanofluid under turbulent flow, Exp. Therm. Fluid Sci. 52 (2014) 68−78.
[78] Y. Yang, Z.G. Zhang, E.A. Grulke, W.B. Anderson, G. Wu, Heat transfer properties of nanoparticle-in-fluid dispersions (nanofluids) in laminar flow, Int. J. Heat Mass. Transf. 48 (6) (2005) 1107−1116.
[79] R. Vajjha, D. Das, B. Mahagaonkar, Density measurement of different nanofluids and their comparison with theory, Pet. Sci. Technol. 27 (6) (2009) 612−624.
[80] J. Ganeshkumar, D. Kathirkaman, K. Raja, V. Kumaresan, R. Velraj, Experimental study on density, thermal conductivity, specific heat, and viscosity of water-ethylene glycol mixture dispersed with carbon nanotubes, Therm. Sci. 21 (1 Part A) (2017) 255−265.
[81] T.-P. Teng, Y.-H. Hung, Estimation and experimental study of the density and specific heat for alumina nanofluid, J. Exp. Nanosci. 9 (7) (2014) 707−718.
[82] C. Ho, W. Liu, Y. Chang, C. Lin, Natural convection heat transfer of alumina-water nanofluid in vertical square enclosures: an experimental study, Int. J. Therm. Sci. 49 (8) (2010) 1345−1353.
[83] Z. Said, A. Kamyar, R. Saidur (Eds.), Experimental investigation on the stability and density of TiO_2, Al_2O_3, SiO_2 and $TiSiO_4$, in: IOP Conference Series: Earth and Environmental Science, IOP Publishing, 2013.
[84] M.A. Kedzierski, Viscosity and density of CuO nanolubricant, Int. J. Refrig. 35 (7) (2012) 1997−2002.
[85] A. Mariano, M.J. Pastoriza-Gallego, L. Lugo, A. Camacho, S. Canzonieri, M.M. Piñeiro, Thermal conductivity, rheological behaviour and density of non-Newtonian ethylene glycol-based SnO_2 nanofluids, Fluid Phase Equilibria 337 (2013) 119−124.
[86] V. Singh, S. Sharma, D.D. Gangacharyulu, Variation of CuO Distilled Water Based Nanofluid Properties through Circular Pipe 3 (2015) 414−420.

[87] M. Pantzali, A. Kanaris, K. Antoniadis, A. Mouza, S. Paras, Effect of nanofluids on the performance of a miniature plate heat exchanger with modulated surface, Int. J. Heat Fluid Flow. 30 (4) (2009) 691–699.
[88] J.R. Eggers, S. Kabelac, Nanofluids revisited, Appl. Therm. Eng. 106 (2016) 1114–1126.
[89] A.D. Sommers, K.L. Yerkes, Experimental investigation into the convective heat transfer and system-level effects of Al_2O_3-propanol nanofluid, J. Nanopart. Res. 12 (3) (2010) 1003–1014.
[90] M. Sharifpur, S. Yousefi, J.P. Meyer, A new model for density of nanofluids including nanolayer, Int. Commun. Heat Mass. Transf. 78 (2016) 168–174.
[91] M. Pastoriza-Gallego, L. Lugo, D. Cabaleiro, J. Legido, M. Piñeiro, Thermophysical profile of ethylene glycol-based ZnO nanofluids, J. Chem. Thermodyn. 73 (2014) 23–30.
[92] D. Cabaleiro, M.J. Pastoriza-Gallego, C. Gracia-Fernández, M.M. Piñeiro, L. Lugo, Rheological and volumetric properties of TiO_2-ethylene glycol nanofluids, Nanoscale Res. Lett. 8 (1) (2013) 286.
[93] A.A. Alrashed, M.S. Gharibdousti, M. Goodarzi, L.R. de Oliveira, M.R. Safaei, Bandarra, et al., Effects on thermophysical properties of carbon based nanofluids: experimental data, modelling using regression, ANFIS and ANN, Int. J. Heat Mass. Transf. 125 (2018) 920–932.
[94] A. Mirsaeidi, F. Yousefi, Viscosity, thermal conductivity and density of carbon quantum dots nanofluids: an experimental investigation and development of new correlation function and ANN modeling, J. Therm. Anal. Calorim. (2019) 1–11.
[95] I. Mahbubul, I. Shahrul, S. Khaleduzzaman, R. Saidur, M. Amalina, A. Turgut, Experimental investigation on effect of ultrasonication duration on colloidal dispersion and thermophysical properties of alumina–water nanofluid, Int. J. Heat Mass. Transf. 88 (2015) 73–81.
[96] L.S. Sundar, M.K. Singh, A.C. Sousa, Enhanced heat transfer and friction factor of MWCNT–Fe_3O_4/water hybrid nanofluids, Int. Commun. Heat Mass. Transf. 52 (2014) 73–83.
[97] S. Kannaiyan, C. Boobalan, F.C. Nagarajan, S. Sivaraman, Modeling of thermal conductivity and density of alumina/silica in water hybrid nanocolloid by the application of Artificial Neural Networks, Chin. J. Chem. Eng. 27 (3) (2019) 726–736.
[98] N. Ali, J.A. Teixeira, A. Addali, A review on nanofluids: fabrication, stability, and thermophysical properties, J. Nanomater (2018) 2018.
[99] K. Khanafer, K. Vafai, A critical synthesis of thermophysical characteristics of nanofluids, Int. J. Heat Mass. Transf. 54 (19-20) (2011) 4410–4428.
[100] H. Karimi, F. Yousefi, M.R. Rahimi, Correlation of viscosity in nanofluids using genetic algorithm-neural network (GA-NN), Heat Mass. Transf. 47 (11) (2011) 1417–1425.
[101] H. Karimi, F. Yousefi, Application of artificial neural network–genetic algorithm (ANN–GA) to correlation of density in nanofluids, Fluid Phase Equilibria. 336 (2012) 79–83.
[102] F. Yousefi, Z. Amoozandeh, A new model to predict the densities of nanofluids using statistical mechanics and artificial intelligent plus principal component analysis, Chin. J. Chem. Eng. 25 (9) (2017) 1273–1281.
[103] J. Mewis, N.J. Wagner, Colloidal Suspension Rheology, Cambridge University Press, 2012.
[104] J.-H. Lee, K.S. Hwang, S.P. Jang, B.H. Lee, J.H. Kim, S.U. Choi, et al., Effective viscosities and thermal conductivities of aqueous nanofluids containing low volume concentrations of Al_2O_3 nanoparticles, Int. J. Heat Mass. Transf. 51 (11-12) (2008) 2651–2656.
[105] Y.R. Sekhar, K. Sharma, Study of viscosity and specific heat capacity characteristics of water-based Al_2O_3 nanofluids at low particle concentrations, J. Exp. Nanosci. 10 (2) (2015) 86–102.
[106] I. Tavman, A. Turgut, M. Chirtoc, H. Schuchmann, S. Tavman, Experimental investigation of viscosity and thermal conductivity of suspensions containing nanosized ceramic particles, Arch. Mater. Sci. 100 (2008) 100.
[107] R. Mostafizur, R. Saidur, A.A. Aziz, M. Bhuiyan, Thermophysical properties of methanol based Al_2O_3 nanofluids, Int. J. Heat Mass. Transf. 85 (2015) 414–419.

[108] M.J. Pastoriza-Gallego, L. Lugo, J.L. Legido, M.M. Piñeiro, Thermal conductivity and viscosity measurements of ethylene glycol-based Al_2O_3 nanofluids, Nanoscale Res. Lett. 6 (1) (2011) 221.
[109] M. Kole, T. Dey, Viscosity of alumina nanoparticles dispersed in car engine coolant, Exp. Therm. Fluid Sci. 34 (6) (2010) 677–683.
[110] R. Mostafizur, A.A. Aziz, R. Saidur, M. Bhuiyan, Investigation on stability and viscosity of SiO_2–CH_3OH (methanol) nanofluids, Int. Commun. Heat Mass. Transf. 72 (2016) 16–22.
[111] V.Y. Rudyak, S. Dimov, V. Kuznetsov, On the dependence of the viscosity coefficient of nanofluids on particle size and temperature, Technical Phys. Lett. 39 (9) (2013) 779–782.
[112] P. Namburu, D. Kulkarni, A. Dandekar, D. Das, Experimental investigation of viscosity and specific heat of silicon dioxide nanofluids, Micro. Nano. Lett. 2 (3) (2007) 67–71.
[113] S. Bobbo, L. Fedele, A. Benetti, L. Colla, M. Fabrizio, C. Pagura, et al., Viscosity of water based SWCNH and TiO_2 nanofluids, Exp. Therm. Fluid Sci. 36 (2012) 65–71.
[114] Z. Said, M. Sabiha, R. Saidur, A. Hepbasli, N. Rahim, S. Mekhilef, et al., Performance enhancement of a flat plate solar collector using titanium dioxide nanofluid and polyethylene glycol dispersant, J. Clean. Prod. 92 (2015) 343–353.
[115] T. Yiamsawas, A.S. Dalkilic, O. Mahian, S. Wongwises, Measurement and correlation of the viscosity of water-based Al_2O_3 and TiO_2 nanofluids in high temperatures and comparisons with literature reports, J. Dispers. Sci. Technol. 34 (12) (2013) 1697–1703.
[116] L. Fedele, L. Colla, S. Bobbo, Viscosity and thermal conductivity measurements of water-based nanofluids containing titanium oxide nanoparticles, Int. J. Refrig. 35 (5) (2012) 1359–1366.
[117] W. Duangthongsuk, S. Wongwises, Measurement of temperature-dependent thermal conductivity and viscosity of TiO_2-water nanofluids, Exp. Therm. Fluid Sci. 33 (4) (2009) 706–714.
[118] A. Turgut, I. Tavman, M. Chirtoc, H. Schuchmann, C. Sauter, S. Tavman, Thermal conductivity and viscosity measurements of water-based TiO_2 nanofluids, Int. J. Thermophys. 30 (4) (2009) 1213–1226.
[119] K. Suganthi, K. Rajan, Temperature induced changes in ZnO–water nanofluid: zeta potential, size distribution and viscosity profiles, Int. J. Heat Mass. Transf. 55 (25-26) (2012) 7969–7980.
[120] Y. Li, H.Q. Xie, W. Yu, J. Li (Eds.), Investigation on heat transfer performances of nanofluids in solar collector, Mater. Sci. Forum, Trans Tech Publ (2011) 33–36.
[121] L. Godson, B. Raja, D.M. Lal, S. Wongwises, Experimental investigation on the thermal conductivity and viscosity of silver-deionized water nanofluid, Exp. Heat Transf. 23 (4) (2010) 317–332.
[122] M.H. Esfe, S. Saedodin, O. Mahian, S. Wongwises, Efficiency of ferromagnetic nanoparticles suspended in ethylene glycol for applications in energy devices: effects of particle size, temperature, and concentration, Int. Commun. Heat Mass. Transf. 58 (2014) 138–146.
[123] A. Amrollahi, A. Rashidi, R. Lotfi, M.E. Meibodi, K. Kashefi, Convection heat transfer of functionalized MWNT in aqueous fluids in laminar and turbulent flow at the entrance region, Int. Commun. Heat Mass. Transf. 37 (6) (2010) 717–723.
[124] M.J. Pastoriza-Gallego, C. Casanova, J.L. Legido, M.M. Piñeiro, CuO in water nanofluid: influence of particle size and polydispersity on volumetric behaviour and viscosity, Fluid Phase Equilibria 300 (1-2) (2011) 188–196.
[125] P.K. Namburu, D.P. Kulkarni, D. Misra, D.K. Das, Viscosity of copper oxide nanoparticles dispersed in ethylene glycol and water mixture, Exp. Therm. Fluid Sci. 32 (2) (2007) 397–402.
[126] H. Xie, W. Yu, W. Chen, MgO nanofluids: higher thermal conductivity and lower viscosity among ethylene glycol-based nanofluids containing oxide nanoparticles, J. Exp. Nanosci. 5 (5) (2010) 463–472.
[127] S. Manikandan, K.S. Rajan, Rapid synthesis of MgO nanoparticles and their utilization for formulation of a propylene glycol based nanofluid with superior transport properties, RSC Adv. 4 (93) (2014) 51830–51837.

[128] H. Khodadadi, D. Toghraie, A. Karimipour, Effects of nanoparticles to present a statistical model for the viscosity of MgO-Water nanofluid, Powder Technol. 342 (2019) 166–180.
[129] R. Prasher, D. Song, J. Wang, P. Phelan, Measurements of nanofluid viscosity and its implications for thermal applications, Appl. Phys. Lett. 89 (13) (2006) 133108.
[130] S.S. Murshed, S.-H. Tan, N.-T. Nguyen, Temperature dependence of interfacial properties and viscosity of nanofluids for droplet-based microfluidics, J. Phys. D: Appl. Phys. 41 (8) (2008) 085502.
[131] B.E. Poling, J.M. Prausnitz, O.C. John Paul, R.C. Reid, The Properties of Gases and Liquids, McGraw-Hill, New York, 2001.
[132] L. Chen, H. Xie, Y. Li, W. Yu, Nanofluids containing carbon nanotubes treated by mechanochemical reaction, Thermochim. Acta 477 (1-2) (2008) 21–24.
[133] K. Anoop, S. Kabelac, T. Sundararajan, S.K. Das, Rheological and flow characteristics of nanofluids: influence of electroviscous effects and particle agglomeration, J. Appl. Phys. 106 (3) (2009) 034909.
[134] S.W. Lee, S.D. Park, S. Kang, I.C. Bang, J.H. Kim, Investigation of viscosity and thermal conductivity of SiC nanofluids for heat transfer applications, Int. J. Heat Mass. Transf. 54 (1-3) (2011) 433–438.
[135] D.P. Kulkarni, D.K. Das, R.S. Vajjha, Application of nanofluids in heating buildings and reducing pollution, Appl. Energy 86 (12) (2009) 2566–2573.
[136] M. Naik, G.R. Janardhana, K.V.K. Reddy, B.S. Reddy, Experimental investigation into rheological property of copper oxide nanoparticles suspended in propylene glycol–water based fluids, ARPN J. Eng. Appl. Sci. 5 (6) (2010) 29–34.
[137] I. Mahbubul, T.H. Chong, S. Khaleduzzaman, I. Shahrul, R. Saidur, B. Long, et al., Effect of ultrasonication duration on colloidal structure and viscosity of alumina–water nanofluid, Ind. Eng. Chem. Res. 53 (16) (2014) 6677–6684.
[138] H. Chen, Y. Ding, C. Tan, Rheological behaviour of nanofluids, N. J. Phys. 9 (10) (2007) 367.
[139] H. Chen, Y. Ding, Y. He, C. Tan, Rheological behaviour of ethylene glycol based titania nanofluids, Chem. Phys. Lett. 444 (4-6) (2007) 333–337.
[140] S.K. Das, N. Putra, W. Roetzel, Pool boiling characteristics of nano-fluids, Int. J. Heat Mass. Transf. 46 (5) (2003) 851–862.
[141] N. Putra, W. Roetzel, S.K. Das, Natural convection of nano-fluids, Heat Mass. Transf. 39 (8-9) (2003) 775–784.
[142] M. Chandrasekar, S. Suresh, A.C. Bose, Experimental investigations and theoretical determination of thermal conductivity and viscosity of Al_2O_3/water nanofluid, Exp. Therm. Fluid Sci. 34 (2) (2010) 210–216.
[143] J. Chevalier, O. Tillement, F. Ayela, Rheological properties of nanofluids flowing through microchannels, Appl. Phys. Lett. 91 (23) (2007) 233103.
[144] W.-Q. Lu, Q.-M. Fan, Study for the particle's scale effect on some thermophysical properties of nanofluids by a simplified molecular dynamics method, Eng. Anal. Bound. Elem. 32 (4) (2008) 282–289.
[145] H. Chen, Y. Ding, A. Lapkin, X. Fan, Rheological behaviour of ethylene glycol-titanate nanotube nanofluids, J. Nanopart. Res. 11 (6) (2009) 1513.
[146] T.X. Phuoc, M. Massoudi, Experimental observations of the effects of shear rates and particle concentration on the viscosity of Fe_2O_3–deionized water nanofluids, Int. J. Therm. Sci. 48 (7) (2009) 1294–1301.
[147] W. Yu, H. Xie, Y. Li, L. Chen, Experimental investigation on thermal conductivity and viscosity of aluminum nitride nanofluid, Particuology. 9 (2) (2011) 187–191.
[148] W.J. Tseng, K.-C. Lin, Rheology and colloidal structure of aqueous TiO_2 nanoparticle suspensions, Mater. Sci. Eng.: A. 355 (1-2) (2003) 186–192.
[149] C. Nguyen, F. Desgranges, N. Galanis, G. Roy, T. Maré, S. Boucher, et al., Viscosity data for Al_2O_3–water nanofluid—hysteresis: is heat transfer enhancement using nanofluids reliable? Int. J. Therm. Sci. 47 (2) (2008) 103–111.

[150] D. Lee, Packing of spheres and its effect on the viscosity of suspension, J. Paint. Technol. 42 (1970) 579–584.
[151] Y. He, Y. Jin, H. Chen, Y. Ding, D. Cang, H. Lu, Heat transfer and flow behaviour of aqueous suspensions of TiO_2 nanoparticles (nanofluids) flowing upward through a vertical pipe, Int. J. Heat Mass. Transf. 50 (11-12) (2007) 2272–2281.
[152] K. Anoop, T. Sundararajan, S.K. Das, Effect of particle size on the convective heat transfer in nanofluid in the developing region, Int. J. Heat Mass. Transf. 52 (9-10) (2009) 2189–2195.
[153] B. Conway, A. Dobry-Duclaux, Viscosity of suspensions of electrically charged particles and solutions of polymeric electrolytes, Rheology, Elsevier, 1960, pp. 83–120.
[154] E.V. Timofeeva, D.S. Smith, W. Yu, D.M. France, D. Singh, J.L. Routbort, Particle size and interfacial effects on thermo-physical and heat transfer characteristics of water-based α-SiC nanofluids, Nanotechnology 21 (21) (2010) 215703.
[155] F.-J. Rubio-Hernández, E. Ruiz-Reina, A.-I. Gómez-Merino, The influence of a dynamic stern layer on the primary electroviscous effect, J. Colloid Interface Sci. 206 (1) (1998) 334–337.
[156] F. Rubio-Hernández, M. Ayúcar-Rubio, J. Velazquez-Navarro, F. Galindo-Rosales, Intrinsic viscosity of SiO_2, Al_2O_3 and TiO_2 aqueous suspensions, J. Colloid Interface Sci. 298 (2) (2006) 967–972.
[157] E.V. Timofeeva, J.L. Routbort, D. Singh, Particle shape effects on thermophysical properties of alumina nanofluids, J. Appl. Phys. 106 (1) (2009) 014304.
[158] Y. Yang, E.A. Grulke, Z.G. Zhang, G. Wu, Thermal and rheological properties of carbon nanotube-in-oil dispersions, J. Appl. Phys. 99 (11) (2006) 114307.
[159] X. Wang, L. Guo, Effect of preparation methods on rheological properties of Al_2O_3/ZrO_2 suspensions, Colloids Surf. A: Physicochem. Eng. Asp. 281 (1-3) (2006) 171–176.
[160] P. Garg, J.L. Alvarado, C. Marsh, T.A. Carlson, D.A. Kessler, K. Annamalai, An experimental study on the effect of ultrasonication on viscosity and heat transfer performance of multi-wall carbon nanotube-based aqueous nanofluids, Int. J. Heat Mass. Transf. 52 (21-22) (2009) 5090–5101.
[161] B. Ruan, A.M. Jacobi, Ultrasonication effects on thermal and rheological properties of carbon nanotube suspensions, Nanoscale Res. Lett. 7 (1) (2012) 1–14.
[162] F.W. Starr, J.F. Douglas, S.C. Glotzer, Origin of particle clustering in a simulated polymer nanocomposite and its impact on rheology, J. Chem. Phys. 119 (3) (2003) 1777–1788.
[163] J. Yu, N. Grossiord, C.E. Koning, J. Loos, Controlling the dispersion of multi-wall carbon nanotubes in aqueous surfactant solution, Carbon 45 (3) (2007) 618–623.
[164] H. Babar, M.U. Sajid, H.M. Ali, Viscosity of hybrid nanofluids: a critical review, Therm. Sci. 23 (3 Part B) (2019) 1713–1754.
[165] A. Einstein, Eine Neue Bestimmung der Moleküldimensionen, ETH Zurich, 1905.
[166] H. Brinkman, The viscosity of concentrated suspensions and solutions, J. Chem. Phys. 20 (4) (1952) 571.
[167] S. Ward, Properties of well-defined suspensions of solids in liquids, J. Oil Colour Chem. Assoc. 38 (9) (1955).
[168] T.S. Lundgren, Slow flow through stationary random beds and suspensions of spheres, J. Fluid Mech. 51 (2) (1972) 273–299.
[169] G. Batchelor, The effect of Brownian motion on the bulk stress in a suspension of spherical particles, J. Fluid Mech. 83 (1) (1977) 97–117.
[170] B. Abedian, M. Kachanov, On the effective viscosity of suspensions, Int. J. Eng. Sci. 48 (11) (2010) 962–965.
[171] R. Bardool, A. Bakhtyari, F. Esmaeilzadeh, X. Wang, Nanofluid viscosity modeling based on the friction theory, J. Mol. Liq. 286 (2019) 110923.
[172] R. Bardool, A. Bakhtyari, F. Esmaeilzadeh, X. Wang, Developing free-volume models for nanofluid viscosity modeling, J. Therm. Anal. Calorim. (2020) 1–14.

[173] J.P. Meyer, S.A. Adio, M. Sharifpur, P.N. Nwosu, The viscosity of nanofluids: a review of the theoretical, empirical, and numerical models, Heat Transf. Eng. 37 (5) (2016) 387–421.

[174] H. Khodadadi, S. Aghakhani, H. Majd, R. Kalbasi, S. Wongwises, M. Afrand, A comprehensive review on rheological behavior of mono and hybrid nanofluids: effective parameters and predictive correlations, Int. J. Heat Mass. Transf. 127 (2018) 997–1012.

[175] F. Yousefi, H. Karimi, M.M. Papari, Modeling viscosity of nanofluids using diffusional neural networks, J. Mol. Liq. 175 (2012) 85–90.

[176] M. Afrand, K.N. Najafabadi, N. Sina, M.R. Safaei, A.S. Kherbeet, S. Wongwises, et al., Prediction of dynamic viscosity of a hybrid nano-lubricant by an optimal artificial neural network, Int. Commun. Heat Mass. Transf. 76 (2016) 209–214.

[177] M. Afrand, A.A. Nadooshan, M. Hassani, H. Yarmand, M. Dahari, Predicting the viscosity of multi-walled carbon nanotubes/water nanofluid by developing an optimal artificial neural network based on experimental data, Int. Commun. Heat Mass. Transf. 77 (2016) 49–53.

[178] G.A. Longo, C. Zilio, L. Ortombina, M. Zigliotto, Application of Artificial Neural Network (ANN) for modeling oxide-based nanofluids dynamic viscosity, Int. Commun. Heat Mass. Transf. 83 (2017) 8–14.

[179] E. Heidari, M.A. Sobati, S. Movahedirad, Accurate prediction of nanofluid viscosity using a multilayer perceptron artificial neural network (MLP-ANN), Chemom. Intell. Lab. Syst. 155 (2016) 73–85.

[180] M.H. Esfe, M. Reiszadeh, S. Esfandeh, M. Afrand, Optimization of MWCNTs (10%)–Al_2O_3 (90%)/5W50 nanofluid viscosity using experimental data and artificial neural network, Phys. A: Stat. Mech. Appl. 512 (2018) 731–744.

[181] M.H. Esfe, M.R.H. Ahangar, M. Rejvani, D. Toghraie, M.H. Hajmohammad, Designing an artificial neural network to predict dynamic viscosity of aqueous nanofluid of TiO_2 using experimental data, Int. Commun. Heat Mass. Transf. 75 (2016) 192–196.

[182] S. Atashrouz, G. Pazuki, Y. Alimoradi, Estimation of the viscosity of nine nanofluids using a hybrid GMDH-type neural network system, Fluid Phase Equilibria. 372 (2014) 43–48.

[183] A. Karimipour, S. Ghasemi, M.H.K. Darvanjooghi, A. Abdollahi, A new correlation for estimating the thermal conductivity and dynamic viscosity of CuO/liquid paraffin nanofluid using neural network method, Int. Commun. Heat Mass. Transf. 92 (2018) 90–99.

[184] M. Bahiraei, S.M. Hosseinalipour, K. Zabihi, E. Taheran, Using neural network for determination of viscosity in water-TiO_2 nanofluid, Adv. Mech. Eng. 4 (2012) 742680.

[185] M. Ramezanizadeh, M.H. Ahmadi, M.A. Nazari, M. Sadeghzadeh, L. Chen, A review on the utilized machine learning approaches for modeling the dynamic viscosity of nanofluids, Renew. Sustain. Energy Rev. 114 (2019) 109345.

[186] H.E. Patel, T. Sundararajan, S.K. Das, An experimental investigation into the thermal conductivity enhancement in oxide and metallic nanofluids, J. Nanopart. Res. 12 (3) (2010) 1015–1031.

[187] R.S. Vajjha, D.K. Das, Experimental determination of thermal conductivity of three nanofluids and development of new correlations, Int. J. Heat Mass. Transf. 52 (21-22) (2009) 4675–4682.

[188] P. Keblinski, J.A. Eastman, D.G. Cahill, Nanofluids for thermal transport, Mater. Today 8 (6) (2005) 36–44.

[189] P. Keblinski, R. Prasher, J. Eapen, Thermal conductance of nanofluids: is the controversy over? J. Nanopart. Res. 10 (7) (2008) 1089–1097.

[190] R. Prasher, P.E. Phelan, P. Bhattacharya, Effect of aggregation kinetics on the thermal conductivity of nanoscale colloidal solutions (nanofluid), Nano Lett. 6 (7) (2006) 1529–1534.

[191] J. Philip, P. Shima, B. Raj, Enhancement of thermal conductivity in magnetite based nanofluid due to chainlike structures, Appl. Phys. Lett. 91 (20) (2007) 203108.

[192] J. Gao, R. Zheng, H. Ohtani, D. Zhu, G. Chen, Experimental investigation of heat conduction mechanisms in nanofluids, Clue Clustering. Nano Lett. 9 (12) (2009) 4128–4132.

[193] J. Philip, P. Shima, B. Raj, Evidence for enhanced thermal conduction through percolating structures in nanofluids, Nanotechnology 19 (30) (2008) 305706.
[194] M. Chandrasekar, S. Suresh, T. Senthilkumar, Mechanisms proposed through experimental investigations on thermophysical properties and forced convective heat transfer characteristics of various nanofluids—a review, Renew. Sustain. Energy Rev. 16 (6) (2012) 3917—3938.
[195] M.M. Tawfik, Experimental studies of nanofluid thermal conductivity enhancement and applications: a review, Renew. Sustain. Energy Rev. 75 (2017) 1239—1253.
[196] S. Murshed, K. Leong, C. Yang, Thermophysical and electrokinetic properties of nanofluids—a critical review, Appl. Therm. Eng. 28 (17-18) (2008) 2109—2125.
[197] S.K. Das, N. Putra, P. Thiesen, W. Roetzel, Temperature dependence of thermal conductivity enhancement for nanofluids, J. Heat Transf. 125 (4) (2003) 567—574.
[198] X. Wang, X. Xu, S.U. S. Choi, Thermal conductivity of nanoparticle-fluid mixture, J. Thermophys. Heat. Transf. 13 (4) (1999) 474—480.
[199] S. Lee, S.-S. Choi, S. Li, J. Eastman, Measuring thermal conductivity of fluids containing oxide nanoparticles, J. Heat Transfer 121 (1999) 280—289.
[200] C.H. Chon, K.D. Kihm, S.P. Lee, S.U. Choi, Empirical correlation finding the role of temperature and particle size for nanofluid (Al_2O_3) thermal conductivity enhancement, Appl. Phys. Lett. 87 (15) (2005) 153107.
[201] C.H. Li, G. Peterson, The effect of particle size on the effective thermal conductivity of Al_2O_3-water nanofluids, J. Appl. Phys. 101 (4) (2007) 044312.
[202] X. Zhang, H. Gu, M. Fujii, Effective thermal conductivity and thermal diffusivity of nanofluids containing spherical and cylindrical nanoparticles, Exp. Therm. Fluid Sci. 31 (6) (2007) 593—599.
[203] T. Yiamsawasd, A.S. Dalkilic, S. Wongwises, Measurement of the thermal conductivity of titania and alumina nanofluids, Thermochim. Acta 545 (2012) 48—56.
[204] E.V. Timofeeva, A.N. Gavrilov, J.M. McCloskey, Y.V. Tolmachev, S. Sprunt, L.M. Lopatina, et al., Thermal conductivity and particle agglomeration in alumina nanofluids: experiment and theory, Phys. Rev. E 76 (6) (2007) 061203.
[205] M.H. Buschmann, Thermal conductivity and heat transfer of ceramic nanofluids, Int. J. Therm. Sci. 62 (2012) 19—28.
[206] H. Xie, J. Wang, T. Xi, Y. Liu, F. Ai, Q. Wu, Thermal conductivity enhancement of suspensions containing nanosized alumina particles, J. Appl. Phys. 91 (7) (2002) 4568—4572.
[207] I. Mahbubul, R. Saidur, M. Amalina, Influence of particle concentration and temperature on thermal conductivity and viscosity of Al_2O_3/R141b nanorefrigerant, Int. Commun. Heat Mass. Transf. 43 (2013) 100—104.
[208] S. Murshed, K. Leong, C. Yang, Enhanced thermal conductivity of TiO_2—water based nanofluids, Int. J. Therm. Sci. 44 (4) (2005) 367—373.
[209] D.-H. Yoo, K. Hong, H.-S. Yang, Study of thermal conductivity of nanofluids for the application of heat transfer fluids, Thermochim. Acta 455 (1-2) (2007) 66—69.
[210] D. Wen, Y. Ding, Natural convective heat transfer of suspensions of titanium dioxide nanoparticles (nanofluids), IEEE Trans. Nanotechnol. 5 (3) (2006) 220—227.
[211] H.E. Patel, S.K. Das, T. Sundararajan, A. Sreekumaran Nair, B. George, T. Pradeep, Thermal conductivities of naked and monolayer protected metal nanoparticle based nanofluids: manifestation of anomalous enhancement and chemical effects, Appl. Phys. Lett. 83 (14) (2003) 2931—2933.
[212] S. Habibzadeh, A. Kazemi-Beydokhti, A.A. Khodadadi, Y. Mortazavi, S. Omanovic, M. Shariat-Niassar, Stability and thermal conductivity of nanofluids of tin dioxide synthesized via microwave-induced combustion route, Chem. Eng. J. 156 (2) (2010) 471—478.
[213] W. Yu, H. Xie, X. Wang, X. Wang, Significant thermal conductivity enhancement for nanofluids containing graphene nanosheets, Phys. Lett. A 375 (10) (2011) 1323—1328.

[214] M. Kole, T. Dey, Effect of prolonged ultrasonication on the thermal conductivity of ZnO–ethylene glycol nanofluids, Thermochim. Acta 535 (2012) 58–65.
[215] W. Ma, F. Yang, J. Shi, F. Wang, Z. Zhang, S. Wang, Silicone based nanofluids containing functionalized graphene nanosheets, Colloids Surf. A: Physicochem. Eng. Asp. 431 (2013) 120–126.
[216] W. Yu, H. Xie, W. Chen, Experimental investigation on thermal conductivity of nanofluids containing graphene oxide nanosheets, J. Appl. Phys. 107 (9) (2010) 094317.
[217] W. Yu, H. Xie, L. Chen, Y. Li, Investigation on the thermal transport properties of ethylene glycol-based nanofluids containing copper nanoparticles, Powder Technol. 197 (3) (2010) 218–221.
[218] C.H. Li, G. Peterson, Experimental investigation of temperature and volume fraction variations on the effective thermal conductivity of nanoparticle suspensions (nanofluids), J. Appl. Phys. 99 (8) (2006) 084314.
[219] L. Syam Sundar, K. Sharma, Thermal conductivity enhancement of nanoparticles in distilled water, Int. J. Nanopart. 1 (1) (2008) 66–77.
[220] L.S. Sundar, K. Sharma, Experimental determination of thermal conductivity of fluid containing oxide nanoparticles, Int. J. Dyn. Fluids 4 (1) (2008) 57–69.
[221] C.H. Li, G. Peterson, Experimental studies of natural convection heat transfer of Al_2O_3/DI water nanoparticle suspensions (nanofluids), Adv. Mech. Eng. 2 (2010) 742739.
[222] S. Murshed, K. Leong, C. Yang, Investigations of thermal conductivity and viscosity of nanofluids, Int. J. Therm. Sci. 47 (5) (2008) 560–568.
[223] H. Chen, W. Yang, Y. He, Y. Ding, L. Zhang, C. Tan, et al., Heat transfer and flow behaviour of aqueous suspensions of titanate nanotubes (nanofluids), Powder Technol. 183 (1) (2008) 63–72.
[224] Y. Ding, H. Alias, D. Wen, R.A. Williams, Heat transfer of aqueous suspensions of carbon nanotubes (CNT nanofluids), Int. J. Heat Mass. Transf. 49 (1-2) (2006) 240–250.
[225] H. Xie, L. Chen, Adjustable thermal conductivity in carbon nanotube nanofluids, Phys. Lett. A 373 (21) (2009) 1861–1864.
[226] N. Jha, S. Ramaprabhu, Thermal conductivity studies of metal dispersed multiwalled carbon nanotubes in water and ethylene glycol based nanofluids, J. Appl. Phys. 106 (8) (2009) 084317.
[227] A. Amrollahi, A. Hamidi, A. Rashidi, The effects of temperature, volume fraction and vibration time on the thermo-physical properties of a carbon nanotube suspension (carbon nanofluid), Nanotechnology. 19 (31) (2008) 315701.
[228] M. Yeganeh, N. Shahtahmasebi, A. Kompany, E. Goharshadi, A. Youssefi, L. Šiller, Volume fraction and temperature variations of the effective thermal conductivity of nanodiamond fluids in deionized water, Int. J. Heat Mass. Transf. 53 (15-16) (2010) 3186–3192.
[229] Q. Yu, Y.J. Kim, H. Ma, Nanofluids with plasma treated diamond nanoparticles, Appl. Phys. Lett. 92 (10) (2008) 103111.
[230] Z. Han, B. Yang, S. Kim, M. Zachariah, Application of hybrid sphere/carbon nanotube particles in nanofluids, Nanotechnology 18 (10) (2007) 105701.
[231] T.-P. Teng, Y.-H. Hung, T.-C. Teng, H.-E. Mo, H.-G. Hsu, The effect of alumina/water nanofluid particle size on thermal conductivity, Appl. Therm. Eng. 30 (14-15) (2010) 2213–2218.
[232] H.A. Mintsa, G. Roy, C.T. Nguyen, D. Doucet, New temperature dependent thermal conductivity data for water-based nanofluids, Int. J. Therm. Sci. 48 (2) (2009) 363–371.
[233] C.J. Ho, J. Gao, Preparation and thermophysical properties of nanoparticle-in-paraffin emulsion as phase change material, Int. Commun. Heat Mass Transf 36 (5) (2009) 467–470.
[234] D. Kim, Y. Kwon, Y. Cho, C. Li, S. Cheong, Y. Hwang, et al., Convective heat transfer characteristics of nanofluids under laminar and turbulent flow conditions, Curr. Appl. Phys. 9 (2) (2009) e119–e123.
[235] C.-S. Jwo, L.-Y. Jeng, H. Chang, T.-P. Teng, Experimental study on thermal conductivity of lubricant containing nanoparticles, Rev. Adv. Mater. Sci. 18 (2008) 660–666.

[236] Transport properties of alumina nanofluids, in: K.-F. Wong, T. Bhshkar (Eds.), ASME International Mechanical Engineering Congress and Exposition, 2006.
[237] N. Shalkevich, W. Escher, T. Bürgi, B. Michel, L. Si-Ahmed, D. Poulikakos, On the thermal conductivity of gold nanoparticle colloids, Langmuir 26 (2) (2010) 663–670.
[238] D. Singh, E. Timofeeva, W. Yu, J. Routbort, D. France, D. Smith, et al., An investigation of silicon carbide-water nanofluid for heat transfer applications, J. Appl. Phys. 105 (6) (2009) 064306.
[239] M. Kole, T. Dey, Thermal conductivity and viscosity of Al_2O_3 nanofluid based on car engine coolant, J. Phys. D: Appl. Phys. 43 (31) (2010) 315501.
[240] W. Yu, H. Xie, L. Chen, Y. Li, Enhancement of thermal conductivity of kerosene-based Fe_3O_4 nanofluids prepared via phase-transfer method, Colloids Surf. A: Physicochem. Eng. Asp. 355 (1-3) (2010) 109–113.
[241] K. Parekh, H.S. Lee, Magnetic field induced enhancement in thermal conductivity of magnetite nanofluid, J. Appl. Phys. 107 (9) (2010) 09A310.
[242] J. Glory, M. Bonetti, M. Helezen, M. Mayne-L'Hermite, C. Reynaud, Thermal and electrical conductivities of water-based nanofluids prepared with long multiwalled carbon nanotubes, J. Appl. Phys. 103 (9) (2008) 094309.
[243] W. Yu, H. Xie, D. Bao, Enhanced thermal conductivities of nanofluids containing graphene oxide nanosheets, Nanotechnology. 21 (5) (2009) 055705.
[244] W. Yu, H. Xie, L. Chen, Y. Li, Investigation of thermal conductivity and viscosity of ethylene glycol based ZnO nanofluid, Thermochim. Acta 491 (1-2) (2009) 92–96.
[245] M.P. Beck, Y. Yuan, P. Warrier, A.S. Teja, The thermal conductivity of alumina nanofluids in water, ethylene glycol, and ethylene glycol + water mixtures, J. Nanopart. Res. 12 (4) (2010) 1469–1477.
[246] B. Yang, Z. Han, Temperature-dependent thermal conductivity of nanorod-based nanofluids, Appl. Phys. Lett. 89 (8) (2006) 083111.
[247] A. Nasiri, M. Shariaty-Niasar, A. Rashidi, A. Amrollahi, R. Khodafarin, Effect of dispersion method on thermal conductivity and stability of nanofluid, Exp. Therm. Fluid Sci. 35 (4) (2011) 717–723.
[248] M. Pastoriza-Gallego, L. Lugo, J. Legido, M. Piñeiro, Enhancement of thermal conductivity and volumetric behavior of FexOy nanofluids, J. Appl. Phys. 110 (1) (2011) 014309.
[249] A. Einstein, Investigations on the Theory of the Brownian Movement, Courier Corporation, 1956.
[250] W. Evans, J. Fish, P. Keblinski, Role of Brownian motion hydrodynamics on nanofluid thermal conductivity, Appl. Phys. Lett. 88 (9) (2006) 093116.
[251] P. Keblinski, S. Phillpot, S. Choi, J. Eastman, Mechanisms of heat flow in suspensions of nano-sized particles (nanofluids), Int. J. Heat Mass. Transf. 45 (4) (2002) 855–863.
[252] J. Koo, C. Kleinstreuer, A new thermal conductivity model for nanofluids, J. Nanopart. Res. 6 (6) (2004) 577–588.
[253] K. Leong, C. Yang, S. Murshed, A model for the thermal conductivity of nanofluids–the effect of interfacial layer, J. Nanopart. Res. 8 (2) (2006) 245–254.
[254] W. Yu, S. Choi, The role of interfacial layers in the enhanced thermal conductivity of nanofluids: a renovated Maxwell model, J. Nanopart. Res. 5 (1-2) (2003) 167–171.
[255] W. Yu, S. Choi, The role of interfacial layers in the enhanced thermal conductivity of nanofluids: a renovated Hamilton–Crosser model, J. Nanopart. Res. 6 (4) (2004) 355–361.
[256] C.-J. Yu, A. Richter, A. Datta, M. Durbin, P. Dutta, Molecular layering in a liquid on a solid substrate: an X-ray reflectivity study, Phys. B: Condens. Matter 283 (1-3) (2000) 27–31.
[257] Y. Xuan, Q. Li, Heat transfer enhancement of nanofluids, Int. J. Heat Fluid flow. 21 (1) (2000) 58–64.
[258] S. Shin, S.-H. Lee, Thermal conductivity of suspensions in shear flow fields, Int. J. Heat Mass. Transf. 43 (23) (2000) 4275–4284.
[259] S. Choi, Z. Zhang, W. Yu, F. Lockwood, E. Grulke, Anomalous thermal conductivity enhancement in nanotube suspensions, Appl. Phys. Lett. 79 (14) (2001) 2252–2254.

[260] M.-S. Liu, M.C.-C. Lin, I.-T. Huang, C.-C. Wang, Enhancement of thermal conductivity with carbon nanotube for nanofluids, Int. Commun. Heat Mass. Transf. 32 (9) (2005) 1202–1210.
[261] T. Cho, I. Baek, J. Lee, S. Park, Preparation of nanofluids containing suspended silver particles for enhancing fluid thermal conductivity of fludis, J. Ind. Eng. Chem. 11 (3) (2005) 400–406.
[262] Y. Hwang, Y. Ahn, H. Shin, C. Lee, G. Kim, H. Park, et al., Investigation on characteristics of thermal conductivity enhancement of nanofluids, Curr. Appl. Phys. 6 (6) (2006) 1068–1071.
[263] S. Jana, A. Salehi-Khojin, W.-H. Zhong, Enhancement of fluid thermal conductivity by the addition of single and hybrid nano-additives, Thermochim. Acta 462 (1-2) (2007) 45–55.
[264] R. Mostafizur, M. Bhuiyan, R. Saidur, A.A. Aziz, Thermal conductivity variation for methanol based nanofluids, Int. J. Heat Mass. Transf. 76 (2014) 350–356.
[265] R. Khedkar, A. Kiran, S. Sonawane, K. Wasewar, S. Umare, Thermo-physical properties measurement of water based Fe_3O_4 nanofluids, Carbon Sci Technol. 5 (2013) 187–191.
[266] H. Zhu, C. Zhang, S. Liu, Y. Tang, Y. Yin, Effects of nanoparticle clustering and alignment on thermal conductivities of Fe_3O_4 aqueous nanofluids, Appl. Phys. Lett. 89 (2) (2006) 023123.
[267] H. Jiang, H. Li, Q. Xu, L. Shi, Effective thermal conductivity of nanofluids Considering interfacial nano-shells, Mater. Chem. Phys. 148 (1-2) (2014) 195–200.
[268] S. Angayarkanni, J. Philip, Effect of nanoparticles aggregation on thermal and electrical conductivities of nanofluids, J. Nanofluids 3 (1) (2014) 17–25.
[269] V.Y. Rudyak, A.V. Minakov, Thermophysical properties of nanofluids, Eur. Phys. J. E 41 (1) (2018) 1–12.
[270] M. Chopkar, P.K. Das, I. Manna, Synthesis and characterization of nanofluid for advanced heat transfer applications, Scr. Materialia 55 (6) (2006) 549–552.
[271] M. Chopkar, S. Kumar, D. Bhandari, P.K. Das, I. Manna, Development and characterization of Al_2Cu and Ag_2Al nanoparticle dispersed water and ethylene glycol based nanofluid, Mater. Sci. Eng.: B. 139 (2-3) (2007) 141–148.
[272] S.H. Kim, S.R. Choi, D. Kim, Thermal conductivity of metal-oxide nanofluids: particle size dependence and effect of laser irradiation, J. Heat Transf 129 (3) (2007) 298–307.
[273] J. Hong, S.H. Kim, D. Kim (Eds.), Effect of laser irradiation on thermal conductivity of ZnO nanofluids, J Phys Conf Ser 59 (2007) 301–304.
[274] M. Chopkar, S. Sudarshan, P. Das, I. Manna, Effect of particle size on thermal conductivity of nanofluid, Metall. Mater. Trans. A. 39 (7) (2008) 1535–1542.
[275] P. Shima, J. Philip, B. Raj, Role of microconvection induced by Brownian motion of nanoparticles in the enhanced thermal conductivity of stable nanofluids, Appl. Phys. Lett. 94 (22) (2009) 223101.
[276] M.P. Beck, Y. Yuan, P. Warrier, A.S. Teja, The Thermal Conductivity of Aqueous Nanofluids Containing Ceria Nanoparticles, American Institute of Physics, 2010.
[277] M.P. Beck, Y. Yuan, P. Warrier, A.S. Teja, The effect of particle size on the thermal conductivity of alumina nanofluids, J. Nanopart. Res. 11 (5) (2009) 1129–1136.
[278] J.C. Maxwell, A Treatise on Electricity and Magnetism, 314, Clarendon, Oxford, 1881, p. 1873.
[279] W.M. Merrill, R.E. Diaz, M.M. LoRe, M.C. Squires, N.G. Alexopoulos, Effective medium theories for artificial materials composed of multiple sizes of spherical inclusions in a host continuum, IEEE Trans. Antennas Propag. 47 (1) (1999) 142–148.
[280] V.D. Bruggeman, Berechnung verschiedener physikalischer Konstanten von heterogenen Substanzen. I. Dielektrizitätskonstanten und Leitfähigkeiten der Mischkörper aus isotropen Substanzen, Annalen der Phys. 416 (7) (1935) 636–664.
[281] R.L. Hamilton, O. Crosser, Thermal conductivity of heterogeneous two-component systems, Ind. Eng. Chem. Fundamen. 1 (3) (1962) 187–191.
[282] J. Avsec, M. Oblak, The calculation of thermal conductivity, viscosity and thermodynamic properties for nanofluids on the basis of statistical nanomechanics, Int. J. Heat Mass. Transf. 50 (21-22) (2007) 4331–4341.

[283] Y. Xuan, Q. Li, Investigation on convective heat transfer and flow features of nanofluids, J. Heat Transf. 125 (1) (2003) 151–155.
[284] D.H. Kumar, H.E. Patel, V.R.R. Kumar, T. Sundararajan, T. Pradeep, S.K. Das, Model for heat conduction in nanofluids, Phys. Rev. Lett. 93 (14) (2004) 144301.
[285] M.H. Esfe, S. Saedodin, N. Sina, M. Afrand, S. Rostami, Designing an artificial neural network to predict thermal conductivity and dynamic viscosity of ferromagnetic nanofluid, Int. Commun. Heat Mass. Transf. 68 (2015) 50–57.
[286] A. Komeilibirjandi, A.H. Raffiee, A. Maleki, M.A. Nazari, M.S. Shadloo, Thermal conductivity prediction of nanofluids containing CuO nanoparticles by using correlation and artificial neural network, J. Therm. Anal. Calorim. 139 (4) (2020) 2679–2689.
[287] A. Maleki, M. Elahi, M.E.H. Assad, M.A. Nazari, M.S. Shadloo, N. Nabipour, Thermal conductivity modeling of nanofluids with ZnO particles by using approaches based on artificial neural network and MARS, J. Therm. Anal. Calorim. (2020) 1–12.
[288] G.A. Longo, C. Zilio, E. Ceseracciu, M. Reggiani, Application of artificial neural network (ANN) for the prediction of thermal conductivity of oxide–water nanofluids, Nano Energy 1 (2) (2012) 290–296.
[289] M.H. Esfe, S. Saedodin, M. Bahiraei, D. Toghraie, O. Mahian, S. Wongwises, Thermal conductivity modeling of MgO/EG nanofluids using experimental data and artificial neural network, J. Therm. Anal. Calorim. 118 (1) (2014) 287–294.
[290] M.H. Esfe, S. Wongwises, A. Naderi, A. Asadi, M.R. Safaei, H. Rostamian, et al., Thermal conductivity of Cu/TiO2–water/EG hybrid nanofluid: experimental data and modeling using artificial neural network and correlation, Int. Commun. Heat Mass. Transf. 66 (2015) 100–104.
[291] M.H. Esfe, M. Afrand, W.-M. Yan, M. Akbari, Applicability of artificial neural network and nonlinear regression to predict thermal conductivity modeling of Al_2O_3–water nanofluids using experimental data, Int. Commun. Heat Mass. Transf. 66 (2015) 246–249.
[292] M. Hojjat, S.G. Etemad, R. Bagheri, J. Thibault, Thermal conductivity of non-Newtonian nanofluids: experimental data and modeling using neural network, Int. J. Heat Mass. Transf. 54 (5-6) (2011) 1017–1023.
[293] M.R. Safaei, A. Hajizadeh, M. Afrand, C. Qi, H. Yarmand, N.W.B.M. Zulkifli, Evaluating the effect of temperature and concentration on the thermal conductivity of ZnO-TiO2/EG hybrid nanofluid using artificial neural network and curve fitting on experimental data, Phys. A: Stat. Mech. Appl. 519 (2019) 209–216.
[294] M. Afrand, D. Toghraie, N. Sina, Experimental study on thermal conductivity of water-based Fe_3O_4 nanofluid: development of a new correlation and modeled by artificial neural network, Int. Commun. Heat Mass. Transf. 75 (2016) 262–269.
[295] E. Ahmadloo, S. Azizi, Prediction of thermal conductivity of various nanofluids using artificial neural network, Int. Commun. Heat Mass. Transf. 74 (2016) 69–75.
[296] O. Mahian, A. Kianifar, S.A. Kalogirou, I. Pop, S. Wongwises, A review of the applications of nanofluids in solar energy, Int. J. Heat Mass. Transf. 57 (2) (2013) 582–594.

CHAPTER

Mass transfer mechanisms in nanofluids

4

Ali Behrad Vakylabad

Department of Materials, Institute of Science and High Technology and Environmental Sciences, Graduate University of Advanced Technology, Kerman, Iran

4.1 Introduction

Nanofluids have, in fact, two main pillars: fluids (<95% Vol) (low thermal conductivity 10^{-1} W/(m K)) and nanomaterials (<5% Vol) (high thermal conductivity 10^{2} W/(m K)). The first may include water, ethylene glycol, and oil. Among these, water is the most widely used because of its high thermal conductivity, abundance, cheapness, and environmentally compatible characteristics [1]. But, the second can contain a wide range of the materials in the size of nanometers, which can be divided into three main sections for better understanding: (1) metal-based, (2) carbon-based, and (3) nanocomposites materials [2]. Some metal-based nanofluids are aluminum (Al) [3], iron (Fe) [4], silver (Ag) [5], copper (Cu) [6], gold (Au) [7], zinc (Zn) [8], and metal oxides like Al_2O_3[9], TiO_2[10], CuO_2[11], SiO_2[12], Fe_2O_3[13], and Fe_3O_4[14]. Next nanofluids containing carbon-based materials are listed as follows: fullerenes, nanotubes, and graphene. And, the third category of the nanofluids contain nanocomposites, including ceramic base (Al_2O_3/MoS_2, SiO_2, CuO, TiO_2, TiC, SiC, CNT) [15,16], metal-based (CoCr, Fe-MgO, Ag-MgO, Cu-Al_2O_3 [17]), and polymer-based (TiO_2, CNT) nanoarchitectures.

Much research has been done after Maxwell [18] who presented the idea of dispersing metal particles in liquids. Furthermore, the more specific ideas have been more recently published in the field of nanofluids [19]. The reports described a marked improvement in heat transfer (from low thermal conductivity 10^{-1} W/(m K) for fluids to high amounts 10^{2} W/(m K) for nanofluids [19,20]) from the laboratory [21] to large-scale applications [22]. According to the studies, dispersed nanoparticles significantly increase the energy transfer process of the base liquids due to the interaction between nanoparticles and liquid molecules. Although the mechanism of enhanced energy transfer from the suspended nanoparticles is still unclear, the prevailing hypothesis in this regard is the irregular Brownian motion of nanoparticles, which is also known as one of the main factors in increasing energy transfer [23,24]. But in terms of mass transfer in particular, since there are basic similarities in transfer phenomena (heat and mass), the success of using nanoparticles to increase heat transfer is the motivation to investigate the potential application of nanofluids to improve mass transfer. However, the relationship between mass transfer and nanofluids is more complex

Nanofluids and Mass Transfer. DOI: https://doi.org/10.1016/B978-0-12-823996-4.00004-5

than that at increased heat transfer in that while in most cases an increase in the weight percentage of the nanomaterials (e.g., nanoalumina Al_2O_3) leads to an increase in heat transfer coefficient with an almost linear behavior, increasing the mass transfer can achieve the optimal point of the weight of these nanomaterials; however, a further increase can have the opposite effect on the mass transfer [25]. Of course, possibilities have also been reported for reduced mass transfer by increasing nanomaterials in the base fluid. Particle agglomeration and the formation of larger clusters, resulting in reduced Brownian mobility, are the most important presumptions [25]. However, the role of nanofluids in the mass transfer has not yet been extensively studied and developed. In this chapter, the research done in this regard will be reviewed and summarized to pave the ways to the future horizons.

4.2 The mechanisms of mass transfer in nanofluids

To know about mass transfer in nanofluids and its mechanisms, it is very helpful to understand the nature of transmission phenomena and their relationship. The similarities between heat transfer and mass transfer (transfer phenomena in general) are very facilitating in this regard. The transfer phenomena are kinetic processes that may occur and be studied separately or jointly. Studying them apart is simpler, but both processes are modeled by similar mathematical equations in the case of diffusion and convection (there is no mass-transfer similarity to heat radiation). It is thus more efficient to consider them jointly. However, there are examples that support them separately (mass transfer apart from heat radiation). But some processes inevitably examine the two phenomena together, such as evaporative cooling and ablation. From this point, diffusion and convection are well known as transferring phenomena. However, from a deeper perspective, four important phenomena can be considered in explaining the mechanism of mass transfer in the presence of nanofluids: (1) creating microcurrents in the fluid due to the Brownian motion of the nanoparticles; (2) grazing effect [26,27]; (3) increasing the gas phase retention in the nanofluid [28,29]; and (4) improving mass transfer at the contact surface through lowering temperatures due to enhanced heat transfer of the media which, in turn, increases the absorption potential [30]. Among these, Brownian motion provides a more accurate mathematical implication for the both transfer phenomena.

Although the use of nanofluids in heat transfer is well known, their effect on increasing mass transfer is significantly higher than their effect on improving heat transfer. Therefore this issue is another important impetus for research on understanding the mechanisms and development of the use of nanofluids in the field of mass transfer. Two main categories were introduced to the concept of nanofluids from the mass transfer viewpoint: mass diffusivity and convective mass transfer [31]. From a convective mass transfer prospect, there are two key coefficients: diffusion and mass transfer [32]. A 4800% increase in the mass transfer of carbon dioxide to water in the presence of 1% by volume (Vol %) of iron oxide nanoparticles is a promising report of the convective mass transfer in nanofluids [33]. In different sections of this chapter, the application and role of nanofluids in increasing the mass transfer rate will be carefully scrutinized so that the different results obtained from various research in this field can be comparable. In this regard, it is necessary to carefully scrutinize several items in each application area: diffusion coefficient and convective mass transfer coefficient, types of nanofluids, size and concentration range of nanomaterials, mass transfer measurement method, proposed mechanisms for mass transfer, and also the most enhancement mechanism possible. There are then two accepted mass transfer mechanisms: diffusive and convective mass transfer. Relatively little research has been done on the mass transfer diffusion coefficient. However, this improvement depends on many factors, including the test conditions, material, and particle size distribution (PSD) of the nanomaterials, and the associated fluids. For example, while 20 nm alumina-water nanofluids using fluorescein dye diffusion by an optical method show a 1400% increase in diffusive mass transfer, under almost similar conditions the fluorescent dye diffusion by microfluidic approach shows no improvement. To date, limited research has shown that the largest increase in diffusive mass transfer belongs to nanofluids containing copper nanoparticles (0.5 Vol%) in water fluid (2600% improvement) [34].

Regarding the mechanism of increasing the diffusive mass transfer, there is a consensus on Brownian or microconvective motion, which results in a turbulence field through momentum transfer and continuous change in velocity with distance [1,25,35]. Measurement of oxidation of chemical reagents, including sodium sulfite, is a more direct method for evaluating and comparing mass transfer in this category. Here, the mechanism of Brownian motion of nanomaterials determines the rate of motion of nanofluid components, thereby improving mass transfer performance. Improving gas transfer to nanofluids and increasing the dissolution rate up to 600% (oxygen and ammonia gas absorption) are the most tangible examples of increased mass transfer in the presence of nanofluids [36–38]. Although a list of different mechanisms for convective mass transfer has been proposed and studied, the Brownian effect of nanomaterials and the resulting microconductivity are the most important ones [25,34,35]. Modeling can help to better understand the mechanisms. An increase of up to fivefold in volumetric mass transfer has been concluded using the models [39–43]. The fine-distributed particles in the liquid environment (gas-slurry system) lead to gas absorption by moving towards the concentration boundary film layer (the interface of gas–liquid), and then these gas molecules are excreted into the liquid bulk through these particles, also known as the grazing effect. Particle sizes should be less than the thickness of the liquid film. By absorbing themselves onto the liquid film, these fine particles not only lead to the distribution of gases adsorbed on them, but also prevent bubble coalescence, leading to an increase in the number of bubble particles and gas holdup. By investigating the effect of fine particles on multiphasic mass transfer, it is concluded that it is specifically amplified by particles especially in nano size [44,45]. Shuttle or grazing effect may effectively describe the improved gas transfer into aqueous suspensions in a turbulent media [40,46,47]. Other mechanisms are also based on the concept of creating effective turbulence at the gas–liquid interface [48–51]. Furthermore, agglomeration of particles and increased elasticity and viscosity [52] of suspension are some obstacles to the transmission of convective mass [52–57]. It is worth noting that there are hypotheses about the decrease in mass transfer under certain conditions, among which the agglomeration of nanomaterials within the nanofluid is one of the cases that eventually leads to the creation of tortuous paths in the penetration of mass [58–60]. One of the reasons for the intensification of this aggregation is the presence of excessive nanoparticles in the solution. Therefore improving mass transfer through nanofluids requires continuous research and development and optimization, since the critical point can be determined for the amount of nanomaterial, the violation of which will have a negative effect on the process [58,61–64]. In this way, other parameters including the dispersion of nanoparticles can be defined. It is possible that functionalization of nanomaterials in nanofluids for their better performance is an attractive field of research that in turn can define new limits of the PSD and volume percentage in fluids. Determining the relationship between particle size and zeta potential, as well as the zero point of charge (ZPC) for nanomaterials used in fluid can help to choose the appropriate method for optimal dispersion. This issue can be one of the important research horizons in the field of nanofluids.

4.2.1 Surfactants and nanofluids

Using 1% (Vol%) of oleic acid-modified nanohematite, a sixfold increase in mass transfer is obtained. The absorption of ammonia in water has been increased more than three times by using copper-based nanofluids. Surface activating agents (surfactants) also play a decisive role in the efficiency of nanofluids in increasing mass transfer. For example, in the presence of 700 ppm 2-ethyl 1-hexanol (surface-activating agent) in the ammonia/water system, the mass transfer increased up to 4.8-fold (480%). In this regard, three types of surfactants have been compared: 2-octanol, n-octanol, and 2-ethyl,1-hexanol. Theoretically, the use of a surfactant increases the dispersibility of the nanoparticles, which in turn increases the specific surface area available for transfer phenomena. Also, the nanofluid containing copper only (without surfactant) increases the mass transfer in the same system by 3.21 times (321%). In turn, the simultaneous addition of the copper nanoparticles and the same surfactant increases the mass transfer up to 5.31 times (531%), which shows the positive interaction of the two parameters nanoparticles and surfactants in the practical applications of the nanofluids [36]. However, the simultaneous presence of surfactants and nanomaterials can sometimes have the opposite effect on increasing the mass transfer. Of course, this is an important area of research in the sense that surfactants are necessary to increase dispersion, on the one hand, and on the other hand, they are likely to disrupt the specific properties of the base fluid and the main surface of nanomaterials. Therefore for each type of nanofluid, the appropriate surfactant must be selected and optimized.

4.2.2 From Brownian motion to diffusion

Brownian motion can explain the transmission mechanism of transferring phenomena in the most understandable way possible. Since Brownian motion can be the basis of the mass transfer mechanism, it is mathematically described in great detail [65,66,67]:

As a first step, by using the Einstein approach, the diffusion equation is obtained as a derivate of the concept of Brownian motion (Fig. 4.1).

Fig. 4.2 shows the motion of Brownian particles along the horizontal axis in a 2D view of water. In this schematic representation, x, $f(x,t)$, Δ, Φ, and $\Phi(\Delta)$ are a generic point along the horizontal axis, the number of particles at location x at time t, displacement in x direction, the probability of displacement, and probability of particle experiencing displacement of Δ, respectively.

To obtain molecular diffusion as a basic equation in mass transfer, the statistical analysis of the concept of Brownian motion can be used. The following definitions are given:

According to the conceptual schematic in Fig. 4.2, the $f(x,t).dx$ is equal to the number of particles in the differential region. And $[f(x+\Delta,t).dx.\Phi(\Delta)]$ represents the number of particles in $x+\Delta$ times the probability of displacement. The term $f(x,t).dx$ (the number of particles in the differential region) will change as the Brownian particles move about because of bombardment with the

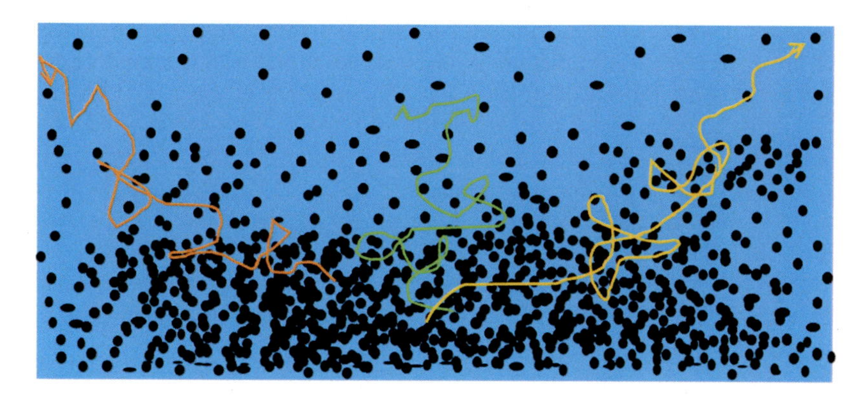

FIGURE 4.1

Brownian motion: the coarser particle colliding with smaller particles (gas or liquid molecules), and moving with different velocities in different random directions.

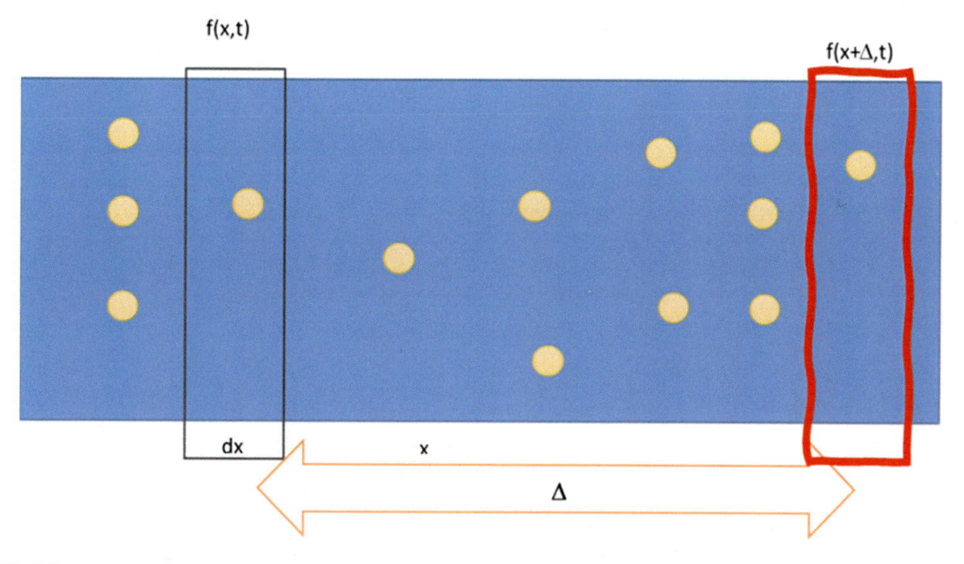

FIGURE 4.2

Simplified schematic showing diffusion of Brownian particles.

molecules of the liquid. Some particles will move out of the differential region while other particles will move in. The quite complicated movements of the particles in and out of the region is understood by considering there are 6.022×10^{23} particles (Avogadro's constant) for 1 mol (18 g) water. It means each Brownian particle is experiencing about 10^{14} collisions in a second. Einstein solved such a complicated problem by changing his perspective from a deterministic to a probabilistic view to stochastically model it. In fact, Einstein's statistical approach models the distance traveled by a particle or the displacement of a particle in the differential interval as a random variable. The question is: how many particles will end up in the area enclosed in the red block (right side of Fig. 4.2)? Assuming the distribution of particles at time t ($f(x,t)dx$), we have at a time step of τ: $f(x, t+\tau)dx$. The distance Δ from the point x represents the total number of particles in this new enclosed area. It has a value of f at the new location times $x+\Delta$. By the same logic in all directions, the entire available space will be swept differentially to obtain the probability of the total number of particles at each point.

Assuming the probability reflects net displacement in an interval, now the integration of these moments into x across the line obtains the number of particles at location $x+\Delta$ at a later time $(t+\tau)$ (Eqs. 4.1 and 4.2):

$$f(x,t+\tau)dx = dx\int_{-\infty}^{\infty} f(x+\Delta,t)\Phi(\Delta)d\Delta \tag{4.1}$$

where $\Phi(\Delta)$ is the probability of a particle experiencing displacement of Δ.

$$f(x,t+\tau) = \int_{-\infty}^{\infty} f(x+\Delta,t)\Phi(\Delta)d\Delta \tag{4.2}$$

Using the Taylor expansion (Eqs. 4.3 and 4.4):

$$f(x,t+\tau) = f(x,t) + \frac{\partial f}{\partial t}\tau \tag{4.3}$$

$$f(x+\Delta,t) = f(x,t) + \frac{\partial f}{\partial x}\Delta + \frac{1}{2}\frac{\partial^2 f}{\partial x^2}\Delta^2 \tag{4.4}$$

Substitution of the Eqs. 4.3 and 4.4 into Eq. 4.2 yields:

$$f(x,t) + \frac{\partial f}{\partial t}\tau = \int_{-\infty}^{\infty}\left(f(x,t) + \frac{\partial f}{\partial x}\Delta + \frac{1}{2}\frac{\partial^2 f}{\partial x^2}\Delta^2\right)\Phi(\Delta)d\Delta \tag{4.5}$$

Expending the terms on the right-hand side of Eq. 4.5 results in:

$$f(x,t) + \frac{\partial f}{\partial t}\tau = f(x,t)\int_{-\infty}^{\infty}\Phi(\Delta)\mathrm{d}\Delta + \frac{\partial f}{\partial x}\int_{-\infty}^{\infty}\Delta\Phi(\Delta)d\Delta + \frac{1}{2}\frac{\partial^2 f}{\partial x^2}\int_{-\infty}^{\infty}\Delta^2\Phi(\Delta)d\Delta \tag{4.6}$$

In Eq. 4.6, $f(x,t)$ is independent of Δ, so it can take out the integral.

Since the probability of displacement is symmetric from $-\infty$ to ∞ around zero, it means the negative and positive displacements are equal. Thus the second term is zero ($\int_{-\infty}^{\infty}\Delta\Phi(\Delta)d\Delta = 0$). Plus, the total probability is equal to one ($\int_{-\infty}^{\infty}\Phi(\Delta)d\Delta = 1$). Therefore the integration with the whole area must obtain 1. So, the f on both sides is canceled out and τ is shifted to the right-hand side (Eq. 4.7).

$$\frac{\partial f}{\partial t}\tau = \frac{1}{2}\frac{\partial^2 f}{\partial x^2}\int_{-\infty}^{\infty}\Delta^2\Phi(\Delta)d\Delta \tag{4.7}$$

Finally, the integral body is Eq. 4.8:

$$\frac{\partial f}{\partial t} = \frac{\partial^2 f}{\partial x^2}\frac{1}{2\tau}\int_{-\infty}^{\infty}\Delta^2\Phi(\Delta)d\Delta \tag{4.8}$$

Then, the final form of diffusion equation (Eq. 4.9) may be derived from the concept of Brownian motion:

$$\frac{\partial f}{\partial t} = D\frac{\partial^2 f}{\partial x^2} \tag{4.9}$$

The term $\frac{1}{2\tau}\int_{-\infty}^{\infty} \Delta^2\Phi(\Delta)d\Delta$ is denoted as D (diffusion coefficient) which is also known as the microscopic interpretation of the diffusion coefficient. In other words, it is representative of the average of the displacement squared. The larger the D, the faster the Brownian particles move. A very important concept of this differential equation and its solution can be obtained for the relationship between mass transfer and nanofluids.

The parameter D depends on the size of the particle, temperature, and properties of the liquid. Apparently, the smaller the particles, the faster they move at higher temperatures. To better understand, the distribution of the Brownian particles across the x axis can be derivated from the differential equation (Eq. 4.10) through finite difference approximation:

$$\frac{f(x,t+\tau)-f(x,t)}{\tau} = D\frac{f(x+\Delta,t)-2f(x,t)+f(x-\Delta,t)}{\Delta^2} \tag{4.10}$$

Rearranging Eq. 4.10 results in:

$$\frac{f(x,t+\tau)-f(x,t)}{\tau} = \frac{2D}{\Delta^2}\left(\frac{f(x+\Delta,t)+f(x-\Delta,t)}{2} - f(x,t)\right) \tag{4.11}$$

Practically, Eq. 4.11 shows the function representing the number of particles at different locations at a given time. In the same way, the left-hand side indicates the change in the number of particles at location x with respect to time and the right-hand side is showing that if the average number of particles around point x is higher than the number of particles at x, then the number of particles at x increases with time and, if the average number of particles to the left and right of x is smaller, the number of particles at x is decreasing.

From the point of view of experimental studies, D may be defined as follows (Eq. 4.12):

$$D = \frac{kTC_c}{3\pi\mu d} \tag{4.12}$$

Eq. 4.12 is the mathematical expression of the results discussed in solving the differential equations above. k is Boltzmann's constant, T denotes temperature, Cc represents the Cunningham correction factor, μ is viscosity of fluid, and d equals particle diameter. In summary, the smaller the particles ($d < 100$ nm), the larger the diffusion coefficient, D, and the longer their displacement, meaning enhanced mass transfer.

The practical methods for determining the diffusion coefficient are generally the use of a color tracer, such as fluorescein, fluorescent Rhodamine B, Rhodamine 6G. However, new innovative methods with high accuracy (pulsed field gradient nuclear magnetic resonance (PFG NMR), Cotts 13-interval pulse sequence, etc.) have also been developed that can determine the self-diffusion coefficient of the solvent [25,58,59,64,68,69].

From a conceptual point of view, Brownian motion is in fact the random jittering of particles under the bombardment of the molecules (particles) in a medium. Simply, the diffusion may be defined as the net movement as a result of Brownian motion. In nanofluids, two probable types of mass transfer may be established: (1) mass diffusion at amolecular level due to a concentration gradient [70], and (2) convective mass transfer because of the bulk flow of fluid [71]. On the first mechanism, there are promising examples: 2600% improvement in mass diffusion of fluorescent Rhodamine B in (0.5 Vol.%) nanocopper in water [34], rhodamine 6GR6G in nanofluid of 2–4 (Vol.%) nanoalumina in water [25,59]. However, it is too soon to decide on the exact effectiveness of the nanofluids from a mass transfer viewpoint [72,73]. In the second, enhanced diffusivity has been reported with the Brownian motion of the Fe_3O_4 nanoparticles [17].

4.2.3 Modeling the mass transfer mechanisms

Nanoparticle size distribution (PSD) and agglomeration as the significant parameters in nanofluids are well modeled with the Brownian Reynolds number (Eq. 4.13) to gain a full understanding of the relationship between nanoparticle properties and the mass transfer [23].

$$Re = \frac{1}{v}\sqrt{\frac{18k_bT}{\pi\rho_n d}} \tag{4.13}$$

in which v, k_b,v_n, m_n,ρ_n, d, and T are kinematic viscosity, Boltzmann constant, root-mean-square velocity of a Brownian particle, particle mass, density, diameter of the nanoparticle, and temperature, respectively.

Eq. 4.13 shows well that increasing the dimensions of nanoparticles (by aggregation or depending on the synthesis method) leads to a decrease in Reynolds number, which is equivalent to a decrease in convective displacement as the dominant mechanism of mass transfer. Also, the inverse relationship of velocity and mass is given in Eq. 4.14. For whatever reason, as the mass of the solid additive in the base fluid increases, the mass transfer will decrease.

$$v_n = \sqrt{\frac{3k_bT}{m_n}} \tag{4.14}$$

The parameters of Eq. (4.14) are defined above.

According to the literature, suspended nanoparticles due to the interaction between nanoparticles and liquid molecules, significantly increase the energy transfer process of the base liquids. Although the mechanism of enhanced energy transfer from the suspended nanoparticles is still unclear, the prevailing hypothesis in this regard is the irregular Brownian motion of nanoparticles, which is also known as one of the main factors in increasing energy transfer [23,24]. But in terms of the mass transfer in particular, since there are basic similarities in transfer phenomena (heat and mass), the success of using nanoparticles to increase heat transfer is an attempt to investigate the potential application of nanofluids to improve mass transfer. Studies show that the presence of nanodrates in fluids increases the energy of motion, which can be attributed to the Brownian displacement of particles, which results in increased heat and mass transfer.

In this regard, by testing the drip dispersion of dyes in a water-based nanofluid, the effective mass permeability of the dye in both nonionized water and in the nanofluid can be calculated and evaluated. Accordingly, suspended nanoparticles significantly increase mass transfer of the dye. Magnetic nanoparticles are also used to increase oxygen mass transfer. As a result, the aqueous solution containing these nanoparticles increases the oxygen mass transfer by several times. In addition, it has been found that the addition of CuO nanoparticles to the refrigerant water mixture improves the heat and mass transfer processes of formation and decomposition of HFC134a hydrate (CH_2FCF_3) [74]. All of the research has concluded that nanofluids clearly increase mass transfer [25].

4.2.4 Role of nanoparticles in mass transfer

The effect of nanomaterials on mass transfer can be even greater than their effect on increasing heat transfer [75]. For example, the base fluid ($LiBr/H_2O$) and iron nanoparticles and carbon nanotubes have been studied in terms of mass and heat transfer. The results show that mass transfer is significantly better than heat one. In addition, carbon nanotubes perform significantly better than iron nanoparticles in terms of transfer phenomena. Another example is oxygen uptake in the presence of normal hexadecane organic matter nanoparticles which may be increased up to 2200% [37]. Also, CO_2 absorption is intensified in the presence of silica nanoparticles [76]. In addition, more accurate measurements of diffusion coefficients have helped to understand the performance of the nanofluids. For example, the dye penetration coefficient in the pure water is equal to 5.2×10^{-10} m^2/s, while this value increases by about 21 times in the presence of nanofluids by 0.5% by weight (wt%) (1.1×10^{-8} m^2/s) [25]. Mathematical expression of these experimental results can help to better model and understand the mechanisms as well as optimization, and may be the first steps to reach a comprehensive formula to describe the relationship between nanomaterials in the base fluid and mass transfer. The first parameter is time (t_m), which considered a hidden but effective parameter in many engineering equations [25] (Eq. 4.15).

$$t_m = \frac{d^2}{2D} \tag{4.15}$$

in which d is distance traveled by diffusive molecules, t_m denotes the time required for a dye molecule to travel the distance d, and D is the diffusion coefficient of the dye in water (m^2/s).

With this simple relation, it can be understood that a molecule of a dye takes about 3.8×10^{-7} seconds (s) to travel the distance (d) of 20 nanometers in water, while this time is significantly reduced in nanofluids. Besides, in Brownian motion of a particle of size d, the time required to travel the same distance d (i.e., the diameter of the particle itself) is obtained using the Einstein–Stokes relation (Eq. 4.16) [25].

$$t_b = \frac{3\pi\eta d^3}{2k_b T} \tag{4.16}$$

where k_b, T, and η are Boltzmann constant, temperature, and the viscosity of the base fluid, respectively.

Using this relation, for nanoparticles with dimensions of 20 nm at room temperature, this time (t_b) is estimated to be 8.1×10^{-6} s. Comparing the results of the above two equations, t_m and t_b for particles of the same dimensions, it is concluded that $t_m < t_b$. Therefore mass transfer is not directly affected by the Brownian motion of the nanoparticles alone. In other words, nanoparticles are not able to push the base fluid molecules to penetrate further. However, the Brownian motion of the nanomaterials leads to the accompanying the base fluid molecules with these motions. As a result, there is more turbulence in the fluid, which can be the main reason for the increase in the mass transfer. This phenomenon is also known as the Grazing effect. Also, the following relation

(Eq. 4.17) can explain which phenomenon is more effective in the mass transfer: diffusion or convective displacement. The time (t_c) required for the transfer of molecules by the convective displacement flow is expressed by the following equation (Eq. 4.17):

$$t_c = \frac{d^2}{2v} \tag{4.17}$$

Under similar conditions of the previeous equations, the time t_c will be equal to 2.0×10^{-10} s.

Clearly, the time required to displace a molecule with a convective displacement mechanism is much shorter than that of a diffusion mechanism. Therefore increasing the mass transfer by a convective displacement mechanism due to turbulence seems to be much more probable.

Mass transfer in gas–liquid systems is one of the hottest engineering topics that can have important applications, such as carbon fixation. In this regard, TiO_2 microparticles increase the amount of CO_2 absorption in different fluids. In principle, as mentioned, in colloidal systems, the main factor in increasing the mass transfer of gas to liquid is the Grazing effect. In this case, colloidal solid particles are constantly moving between the liquid surface and the liquid bulk, as a result of which they continuously absorb the gas molecules by the surface-adsorption mechanism in this boundary layer (liquid surface and gas mass), and by moving toward the liquid mass, these gas molecules are expelled (desorbed). This phenomenon has been introduced as the main reason for the increased mass transfer [77].

The same argument can be made for the nanofluids. In fact, the increase in mass transfer can be a function of the mass of the nanomaterial (weight percentage) and the turbulence of the fluid (supplied by high agitation). In this regard, nanofluid with 1% hematite (Fe_2O_3) coated with oleic acid increases the mass transfer up to 600% in the oxygen–water system. It has also been observed that although increasing the size of nanoparticles reduces the mass transfer, increasing temperature leads to its increase. In addition, it is with good accuracy concluded that the main reason for the increase can be attributed to the increase in surface area [38]. Of course, many different factors can be involved in the relationship between nanofluids and mass transfer. One of these factors is tortuosity (τ) which is defined as shown in Eq. 4.18, in which with the increase of tortuosity, the effective diffusion coefficient in the path decreases, and as a result the mass transfer is reduced [78,79].

$$\tau = \frac{\text{Actual distance a molecule travels between two points}}{\text{Shortest distance between those two points}} \tag{4.18}$$

4.2.5 Footprint of nanofluids in the gas-liquid interaction

Mass transfer in nanofluids in the oxygen–water system and in the presence of normal hexadecane nanoparticles explains in detail the mechanisms of mass transfer. In this system, in addition to the direct effect of Brownian motions, the velocity field formed around the nanoparticles is carefully investigated. Of course, this velocity field is formed by the Brownian motion of nanoparticles (the main reason for this chapter's emphasis on Brownian motion in plain language is this). Calculations and modeling show that the distance between the nanoparticles is less than the thickness of the velocity boundary layer formed around these particles. Hence, these layers collide with each other constantly. A very important point in these studies is that in small amounts of nanomaterials the flow field around the particle increases. As a result, the momentum transfer increases due to the increase in velocity gradient around the particle. This situation means an increase in mass transfer. According to these studies, in fact, nanoparticles with two main mechanisms of random Brownian motion and displacement of liquid components as a result of this motion of the nanometer particles increase the mass transfer. Therefore we again come to the importance of Brownian motion and its effect on the transfer phenomena in the nanofluids [37].

However, in a static fluid, small particles fluctuate around the average value of the path of the motion; although the amount and direction of these motions are not precisely known, they are greater than the average value of the liquid molecules themselves. Because the diffusion coefficient of the base fluid molecules is higher than that of the nanomaterials used, the transfer motion of the liquid molecules and their diffusion rate are higher than that of the nanoparticles. As a result, in this case the Brownian motion of the nanoparticles can not have a significant effect on increasing the mass transfer rate. Here is the second important mechanism that results from the displacement of the fluid around the nanoparticles along with the motion of the nanoparticles or the creep motion (creeping flow) that decreases continuously with distance from the particle [80–83].

Mathematical modeling has shown that the distance between the nanoparticles is less than the thickness of the boundary layer around the particle. Hence, the layers constantly collide with each other. As a result, in small amounts of nanoparticles, the flow field around the particle increases. Finally, the momentum transfer increases due to the increase in the velocity gradient around the particle. Therefore the mass transfer around the nanoparticles increases. However, there is still no precise understanding of the relationship between this phenomenon (increasing velocity gradient around the particle) and increasing diffusion rate [84–86].

4.2.6 Hydrodynamics of nanofluids from mass-transfer viewpoint

Another important tool for describing and explaining the mass transfer mechanism in nanofluids is the important topic of hydrodynamics of systems containing these nanomaterials, especially the gas–nanofluid case. The addition of nanomaterials to the base fluid changes its properties and thus significantly affects its hydrodynamics (size, shape, and speed of bubble oscillation and dispersed-phase retention). Since mass transfer is a direct function of the hydrodynamics, any communication in hydrodynamics can be interpreted and translated in terms of the mass transfer. Examination of gas–liquid hydrodynamics shows that the size of the bubbles in the nanofluid is markedly smaller than that of the parent base fluid. This means increasing the specific surface area and consequently increasing the speed of the mass transfer [87,88–92].

In fact, the hydrodynamic effect shows itself in increasing the adsorption of micrometer gas bubbles on the nanoparticles within the base fluid. This increase in the presence of nanoparticles can increase up to 24%. Additionally, the stability of the nanomaterials dispersed in the base fluid has a very important effect in this case. These stabilized nanoparticles not only increase the total surface area of the gas bubbles due to their continuous contact with nanomaterials and increase their dispersibility to smaller bubbles, but also provide additional energy for dissolution through these nanomaterials.

Furthermore, the relationship between changing the amount of gas absorption and the size of the bubbles is expressed by the well-known general rules. In fact, smaller bubbles generally mean more and more bubbles. Also, according to the famous Laplace–Young relationship, the internal pressure of bubbles is inversely related to their size, that is, internal pressure increases as the bubble size decreases (Eq. 4.18).

$$\Delta P = P_{inside} - P_{outside} = \frac{2\gamma}{R} \tag{4.18}$$

In addition, increasing the solubility of bubbles by reducing their dimensions is well expressed by the Kelvin equation (Eq. 4.19). R in Eq. 4.18 denotes the average radius of the bubble curve.

$$RTln\left(\frac{P}{P_0}\right) = \frac{2\gamma V}{r} \tag{4.19}$$

where γ, ΔP, P, P_0, r, V, T, and R are surface tension, Laplace pressure the actual vapor pressure, the saturated vapor pressure, radius of the droplet, the molar volume of the liquid, temperature, and the universal gas constant, respectively. (Surface tension means the tendency of the liquid surface to shrink relative to the minimum possible surface area; and Laplace pressure is equivalent to the pressure difference between inside and outside the bubble, which creates a boundary between gas and liquid.) Furthermore, the radius of the droplet depends on its size and shape in Eq. 4.19, where if the curvature is convex, r is positive, then $P > P_0$; and if the curvature is concave, r is negative, then $P < P_0$.

Besides, provided that there is the same weight percentage of nanomaterials in the base fluid, the particle size has no significant effect on the amount of gas adsorption. This phenomenon is due to the same added energy content in all solutions, which contributes to the solubility of the bubbles. In the nanofluids with the same percentage of the nanomaterials, the number of nanoparticles is inversely related to their size. In addition, according to experimental results, in nanofluids with the same percentage of nanomaterials, the bubbles become equally small. As a result, the amount of increase in the mass transfer due to the reduction in bubble size is the same in all these cases. A typical experimental example is the presence of the nanoparticles in the ammonia and water absorption system in which the bubbles formed in the water rise to 7.5 mm, while in the presence of the nanomaterials (1% (wt%) of metallic copper nanoparticles) the bubbles rise to a height of 5.83 mm. In addition, the formation of smaller bubbles with a more regular spherical shape has been proven by the addition of the nanomaterials to the system [26,76,93].

4.3 Conclusion

This chapter, in summary, was a review of theories and experimental studies on the mechanism of mass transfer in the presence of nanofluids. There is still a long way to go to recognize this phenomenon accurately. One of the important questions that must be answered for the development of nanofluids is how to recycle and regenerate nanomaterials with the same original characteristics. In addition to the technical parameters studied in this regard, the introduction of economic parameters in future studies will prepare nanofluids to enter the large-scale and industrial stage in terms of transfer phenomena. One of these new fields can be carbon fixation, CO_2 mineralization, or CO_2 sequestration, which not only technically and economically but also socially and even politically should be given special attention. Mechanistically, the scrutinized literature in this review study have shown that Brownian motion in the nanofluids has the greatest effect on mass transfer.

Brownian motion increases both diffusion and convection in mass transfer. The presence of nanomaterials in the base fluids gives some properties, such as viscosity and surface tension, and changes in these properties can help reduce or increase the mass transfer. In fact, the presence of particles in the fluids to increase mass transfer has an optimal point because, by increasing the number of particles, problems such as agglomeration arise, which reduces the mass transfer. Simply put, the increase in heat transfer eventually leads to an increase in the energy of the nanoparticles. This added kinetic energy in the nanoparticles eventually leads to greater mobility, which is the basic concept of the mass transfer. Among the studied mechanisms, it was proven that convective displacement plays the most important role in increasing the mass transfer in the presence of nanomaterials in the base fluids, which is due to the increase of fluid turbulence or the Grazing phenomenon. Nevertheless, convective mass transfer has received a lot of attention. One of the reasons is its more convenient and facile methods of identifying and testing, some of which are stirred tank or three-phase airlift reactors, various types of absorption apparatus, gas–liquid hollow fiber membranes, solvent extraction, as well as direct measurements of mass transfer coefficient in nanofluids. Although the relationship between nanofluids and mass transfer has not been extensively investigated, studies to date have demonstrated the significant impact of nanomaterials in base fluids on mass transfer, which has heightened hopes for further in-depth research.

References

[1] H.A. Mohammed, A.A. Al-Aswadi, N.H. Shuaib, R. Saidur, Convective heat transfer and fluid flow study over a step using nanofluids: a review, Renew. Sustain. Energy Rev. 15 (2011) 2921–2939.

[2] S.K. Gupta, S. Dixit, Progress and application of nanofluids in solar collectors: an overview of recent advances, Mater. Today: Proc. (2020).

[3] M.-J. Kao, C.-C. Ting, B.-F. Lin, T.-T. Tsung, Aqueous aluminum nanofluid combustion in diesel fuel, J. Test. Evaluation 36 (2008) 186–190.

[4] Y. Gan, Y.S. Lim, Li Qiao, Combustion of nanofluid fuels with the addition of boron and ironparticles at dilute and dense concentrations, Combust. Flame 159 (2012) 1732–1740.

[5] L. Godson, B. Raja, D.M. Lal, S. Wongwises, Experimental investigation on the thermal conductivity and viscosity of silver-deionized water nanofluid, Exp. Heat. Transf. 23 (2010) 317–332.

[6] J. Garg, B. Poudel, M. Chiesa, J.B. Gordon, J.J. Ma, J.B. Wang, et al., Enhanced thermal conductivity and viscosity of copper nanoparticles in ethylene glycol nanofluid, J. Appl. Phys. 103 (2008) 074301.

[7] C.Y. Tsai, H.T. Chien, P.P. Ding, B. Chan, T.Y. Luh, P.H. Chen., Effect of structural character of gold nanoparticles in nanofluid on heat pipe thermal performance, Mater. Lett. 58 (2004) 1461–1465.

[8] B.K. Sonage, P. Mohanan, Miniaturization of automobile radiator by using zinc-water and zinc oxide-water nanofluids, J. Mech. Sci. Technol. 29 (2015) 2177–2185.

[9] J. Albadr, S. Tayal, M. Alasadi, Heat transfer through heat exchanger using Al_2O_3 nanofluid at different concentrations, Case Stud. Therm. Eng. 1 (2013) 38–44.

[10] S.M.S. Murshed, K.C. Leong, C. Yang, Enhanced thermal conductivity of TiO_2–water based nanofluids, Int. J. Therm. Sci. 44 (2005) 367–373.

[11] D.V. Guzei, A.V. Minakov, V.Y. Rudyak, A.A. Dekterev, Measuring the heat-transfer coefficient of nanofluid based on copper oxide in a cylindrical channel, Tech. Phys. Lett. 40 (2014) 203–206.

[12] P.K. Namburu, D.P. Kulkarni, A. Dandekar, D.K. Das, Experimental investigation of viscosity andspecific heat of silicon dioxide nanofluids, Micro Nano Lett. 2 (2007) 67–71.

[13] T.X. Phuoc, M. Massoudi, Experimental observations of the effects of shear rates and particle concentration on the viscosity of Fe_2O_3−deionized water nanofluids, Int. J. Therm. Sci. 48 (2009) 1294−1301.
[14] M. Abareshi, E.K. Goharshadi, S.M. Zebarjad, H.K. Fadafan, A. Youssefi, Fabrication, characterization and measurement of thermal conductivity of Fe_3O_4 nanofluids, J. Magn. Magn. Mater. 322 (2010). 901-3895.
[15] J. He, J. Sun, Y. Meng, Y. Pei, Superior lubrication performance of MoS2-Al2O3 composite nanofluid in strips hot rolling, J. Manuf. Process. 57 (2020) 312−323.
[16] S. Senthilraja, K. Vijayakumar, R. Gangadevi, A comparative study on thermal conductivity of Al_2O_3/water, CuO/water and Al_2O_3−CuO/water nanofluids, Dig. J. Nanomater. Bios. 10 (2015) 1449−1458.
[17] S. Suresh, K.P. Venkitaraj, P. Selvakumar, M. Chandrasekar, Effect of Al_2O_3−Cu/water hybrid nanofluid in heat transfer, Exp. Therm. Fluid Sci. 38 (2012) 54−60.
[18] J.C. Maxwell, A Treatise on Electricity and Magnetism, Clarendon Press, Oxford, 1873.
[19] S.U.S. Choi, J.A. Eastman, Enhancing Thermal Conductivity of Fluids with Nanoparticles, Argonne National Lab., IL, United States, 1995.
[20] J.A. Eastman, S.U.S. Choi, S. Li, W. Yu, L.J. Thompson, Anomalously increased effective thermal conductivities of ethylene glycol-based nanofluids containing copper nanoparticles, Appl. Phys. Lett. 78 (2001) 718−720.
[21] M.M. Tawfik, Experimental studies of nanofluid thermal conductivity enhancement and applications: a review, Renew. Sustain. Energy Rev. 75 (2017) 1239−1253.
[22] S.U.S. Choi, Nanofluids: from vision to reality through research, J. Heat. Transf. (2009) 131.
[23] R. Prasher, P. Bhattacharya, P.E. Phelan, Thermal conductivity of nanoscale colloidal solutions (nanofluids), Phys. Rev. Lett. 94 (2005) 025901.
[24] Y. Xuan, Q. Li, W. Hu, Aggregation structure and thermal conductivity of nanofluids, AIChEJ. 49 (2003) 1038−1043.
[25] S. Krishnamurthy, P. Bhattacharya, P.E. Phelan, R.S. Prasher, Enhanced mass transport in nanofluids, Nano Lett. 6 (2006) 419−423.
[26] J.-K. Kim, J.Y. Jung, Y.T. Kang, The effect of nano-particles on the bubble absorption performance in a binary nanofluid, Int. J. Refrig. 29 (2006) 22−29.
[27] Z. Zhang, J. Cai, F. Chen, H. Li, W. Zhang, W. Qi, Progress in enhancement of CO2 absorption by nanofluids: a mini review of mechanisms and current status, Renew. Energy 118 (2018) 527−535.
[28] L. Hendraningrat, S. Li, O. Torsæter, 'A coreflood investigation of nanofluid enhanced oil recovery', J. Pet. Sci. Eng. 111 (2013) 128−138.
[29] J. Buongiorno, L.W. Hu, G. Apostolakis, R. Hannink, T. Lucas, A. Chupin, A feasibility assessment of the use of nanofluids to enhance the in-vessel retention capability in light-water reactors, Nucl. Eng. Des. 239 (2009) 941−948.
[30] C. Pang, J.W. Lee, Y.T. Kang, Review on combined heat and mass transfer characteristics in nanofluids, Int. J. Therm. Sci. 87 (2015) 49−67.
[31] S.-S. Ashrafmansouri, M.N. Esfahany, Mass transfer in nanofluids: a review, Int. J. Therm. Sci. 82 (2014) 84−99.
[32] H. Beiki, M.N. Esfahany, N. Etesami, Laminar forced convective mass transfer of γ-Al2O3/electrolyte nanofluid in a circular tube, Int. J. Therm. Sci. 64 (2013) 251−256.
[33] S. Komati, A.K. Suresh, Anomalous enhancement of interphase transport rates by nanoparticles: effect of magnetic iron oxide on gas − liquid mass transfer, Ind. Eng. Chem. Res. 49 (2010) 390−405.
[34] X. Fang, Y. Xuan, Q. Li, Experimental investigation on enhanced mass transfer in nanofluids, Appl. Phys. Lett. 95 (2009) 203108.
[35] J. Saien, H. Bamdadi, Mass transfer from nanofluid single drops in liquid−liquid extraction process, Ind. Eng. Chem. Res. 51 (2012) 5157−5166.

[36] J.-K. Kim, J.Y. Jung, Y.T. Kang, Absorption performance enhancement by nano-particles and chemical surfactants in binary nanofluids, Int. J. Refrig. 30 (2007) 50–57.
[37] E. Nagy, T. Feczkó, B. Koroknai, Enhancement of oxygen mass transfer rate in the presence of nano-sized particles, Chem. Eng. Sci. 62 (2007) 7391–7398.
[38] B. Olle, S. Bucak, T.C. Holmes, L. Bromberg, T.A. Hatton, D.I.C. Wang, Enhancement of oxygen mass transfer using functionalized magnetic nanoparticles, Ind. Eng. Chem. Res. 45 (2006) 4355–4363.
[39] N. Frossling, Uber die verdunstung fallernder tropfen, Gerlands Beitr. Geophys. 52 (1938) 170–216.
[40] A. Kaya, A. Schumpe, Surfactant adsorption rather than "shuttle effect"? Chem. Eng. Sci. 60 (2005) 6504–6510.
[41] J.C. Lamont, D.S. Scott, An eddy cell model of mass transfer into the surface of a turbulent liquid, AIChE J. 16 (1970) 513–519.
[42] M.J. McCready, E. Vassiliadou, T.J. Hanratty, Computer simulation of turbulent mass transfer at a mobile interface, AIChE J. 32 (1986) 1108–1115.
[43] P.H. Calderbank, M.B. Moo-Young, The continuous phase heat and mass-transferproperties of dispersions, Chem. Eng. Sci. 16 (1961) 39–54.
[44] R.L. Kars, R.J. Best, The sorption of propane in slurries of active carbon in water, Chem. Eng. J. 17 (1979) 201–210.
[45] M. Zhou, F. Wang Cai, J.X. Chun, A new way of enhancing transport process–the hybrid process accompanied by ultrafine particles, Korean J. Chem. Eng. 20 (2003) 347–353.
[46] M. Kordač, V. Linek, Mechanism of enhanced gas absorption in presence of fine solid particles. Effect of molecular diffusivity on mass transfer coefficient in stirred cell, Chem. Eng. Sci. 61 (2006) 7125–7132.
[47] M. Rosu, A. Schumpe, Influence of surfactants on gas absorption into aqueous suspensions of activated carbon, Chem. Eng. Sci. 62 (2007) 5458–5463.
[48] P. Harriott, R.M. Hamilton, Solid-liquid mass transfer in turbulent pipe flow, Chem. Eng. Sci. 20 (1965) 1073–1078.
[49] R.L. Kars, R.J. Best, A.A.H. Drinkenburg, The sorption of propane in slurries of active carbon in water, Chem. Eng. J. 17 (1979) 201–210.
[50] E. Alper, B. Wichtendahl, W.-D. Deckwer, Gas absorption mechanism in catalytic slurry reactors, Chem. Eng. Sci. 35 (1980) 217–222.
[51] C.S. Lin, R.W. Moulton, G.L. Putnam, Mass transfer between solid wall and fluid streams. Mechanismand eddy distribution relationships in turbulent flow, Ind. Eng. Chem., 45, 1953, pp. 636–640.
[52] S.M.S. Murshed, K.C. Leong, C. Yang, Investigations of thermal conductivity and viscosity of nanofluids, Int. J. Therm. Sci. 47 (2008) 560–568.
[53] S. Lee, S.U.-S. Choi, S. Li, J.A. Eastman, Measuring thermal conductivity of fluids containing oxide nanoparticles, 1999.
[54] K.B. Anoop, S. Kabelac, T. Sundararajan, S.K. Das, Rheological and flow characteristics of nanofluids: influence of electroviscous effects and particle agglomeration, J. Appl. Phys. 106 (2009) 034909.
[55] N. Putra, W. Roetzel, S.K. Das, Natural convection of nano-fluids, Heat. Mass. Transf. 39 (2003) 775–784.
[56] J. Sui, L. Zheng, X. Zhang, Boundary layer heat and mass transfer with Cattaneo–Christov double-diffusion in upper-convected Maxwell nanofluid past a stretching sheet with slip velocity, Int. J. Therm. Sci. 104 (2016) 461–468.
[57] E.V. Timofeeva, A.N. Gavrilov, J.M. McCloskey, Y.V. Tolmachev, S. Sprunt, L.M. Lopatina, et al., Thermal conductivity and particle agglomeration in alumina nanofluids: experiment and theory, Phys. Rev. E 76 (2007) 061203.

[58] C. Gerardi, D. Cory, J. Buongiorno, L.-W. Hu, T. McKrell, Nuclear magnetic resonance-based study of ordered layering on the surface of alumina nanoparticles in water, Appl. Phys. Lett. 95 (2009) 253104.
[59] J. Veilleux, S. Coulombe, A total internal reflection fluorescence microscopy study of mass diffusion enhancement in water-based alumina nanofluids, J. Appl. Phys. 108 (2010) 104316.
[60] J. Veilleux, S. Coulombe, A dispersion model of enhanced mass diffusion in nanofluids, Chem. Eng. Sci. 66 (2011) 2377–2384.
[61] S.-W. Park, B.-S. Choi, J.-W. Lee, Effect of elasticity of aqueous colloidal silica solution on chemical absorption of carbon dioxide with 2-amino-2-methyl-1-propanol, Korea-Aust. Rheol. J. 18 (2006) 133–141.
[62] S.-W. Park, B.-S. Choi, S.-S. Kim, J.-W. Lee, Chemical absorption of carbon dioxide into aqueous colloidal silica solution containing monoethanolamine, J. Ind. Eng. Chem. 13 (2007) 133–142.
[63] S.-W. Park, B.-S. Choi, S.-S. Kim, B.-D. Lee, J.-W. Lee, Absorption of carbondioxide into aqueous colloidal silica solution with diisopropanolamine, J. Ind. Eng. Chem. 14 (2008) 166–174.
[64] A.N. Turanov, Y.V. Tolmachev, Heat-and mass-transport in aqueous silica nanofluids, Heat. Mass. Transf. 45 (2009) 1583–1588.
[65] D.T. Gillespie, E. Seitaridou, Simple Brownian Diffusion: An Introduction to the Standard Theoretical Models, Oxford University Press, 2013.
[66] K.Anthony. Hendrik, Brownian motion in a field of force and the diffusion model of chemical reactions, Physica 7 (1940) 284–304.
[67] F. David, Brownian Motion and Diffusion, Springer Science & Business Media, 2012.
[68] R.M. Cotts, M.J.R. Hoch, T. Sun, J.T. Markert, Pulsed field gradient stimulated echo methods for improved NMR diffusion measurements in heterogeneous systems, J. Magnetic Reson. (1969) 83 (1989) 252–266.
[69] S. Ozturk, Y.A. Hassan, V.M. Ugaz, Interfacial complexation explains anomalous diffusion in nanofluids, Nano Lett. 10 (2010) 665–671.
[70] O. Levenspiel, Chemical reaction engineering, Ind. Eng. Chem. Res. 38 (1999) 4140–4143.
[71] W.L. McCabe, J.C. Smith, P. Harriott, Unit Operations of Chemical Engineering, McGraw-hill, New York, 1967.
[72] X. Feng, D.W. Johnson, Mass transfer in SiO2 nanofluids: a case against purported nanoparticle convection effects, Int. J. Heat. Mass. Transf. 55 (2012) 3447–3453.
[73] V. Subba-Rao, P.M. Hoffmann, A. Mukhopadhyay, Tracer diffusion in nanofluids measured by fluorescence correlation spectroscopy, J. Nanopart. Res. 13 (2011) 6313–6319.
[74] J. Li, D. Liang, K. Guo, R. Wang, S. Fan, Formation and dissociation of HFC134a gas hydrate in nanocopper suspension, Energy Convers. Manag. 47 (2006) 201–210.
[75] Y.T. Kang, H.J. Kim, K.Il Lee, Heat and mass transfer enhancement of binary nanofluids for H2O/LiBr falling film absorption process, Int. J. Refrig. 31 (2008) 850–856.
[76] W.-g Kim, H.Uk Kang, K.-m Jung, S.H. Kim, Synthesis of silica nanofluid and application to CO_2 absorption, Sep. Sci. Technol. 43 (2008) 3036–3055.
[77] M.V. Dagaonkar, H.J. Heeres, A.A.C.M. Beenackers, V.G. Pangarkar, The application of fine TiO_2 particles for enhanced gas absorption, Chem. Eng. J. 92 (2003) 151–159.
[78] M. Rezakazemi, M. Darabi, E. Soroush, M. Mohammad, CO_2 absorption enhancement by water-based nanofluids of CNT and SiO_2 using hollow-fiber membrane contactor, Sep. Purif. Technol. 210 (2019) 920–926.
[79] A. Tiwari Kumar, P. Ghosh, J. Sarkar, H. Dahiya, P. Jigar, Numerical investigation of heat transfer and fluid flow in plate heat exchanger using nanofluids, Int. J. Therm. Sci. 85 (2014) 93–103.
[80] I.S. Abdelsalam, B. Muhammad Mubashir, The impact of impinging TiO_2 nanoparticles in Prandtl nanofluid along with endoscopic and variable magnetic field effects on peristaltic blood flow, Multidiscip. Model. Mater. Struct. (2018).

[81] Y. A. Cengel, Fluid mechanics (Tata McGraw-Hill Education). Malvandi, A, and DD Ganji. 2014. 'Effects of nanoparticle migration on force convection of alumina/water nanofluid in a cooled parallel-plate channel', Adv. Powder Techn., 25 (2010) 1369–1375.

[82] B. Munson Roy, T.O. Hisao, W. Wade Huebsch, P.R. Alric, Fluid Mechanics, Wiley, Singapore, 2013.

[83] P. Talebizadeh Sardari, I. Hayder Mohammed, M. Jasim Mahdi, M. Ghalambaz, M. Gillott, G.S. Walker, et al., Localized heating element distribution in composite metal foam-phase change material: Fourier's law and creeping flow effects, Int. J. Energy Res. (2021).

[84] M. Ghalambaz, A. Behseresht, J. Behseresht, Effects of nanoparticles diameter and concentration on natural convection of the Al_2O_3–water nanofluids considering variable thermal conductivity around a vertical cone in porous media, Adv. Powder Technol. 26 (2015) 224–235.

[85] A. Zaraki, M. Ghalambaz, J. AliChamkha, M. Ghalambaz, R.De Danilo, Theoretical analysis of natural convection boundary layer heat and mass transfer of nanofluids: effects of size, shape and type of nanoparticles, type of base fluid and working temperature, Adv. Powder Technol. 26 (2015) 935–946.

[86] B. Jacopo, Convective transport in nanofluids, Transact ASME 128 (2006) 240–250.

[87] K.P. Singh, P.V. Harikrishna, T. Sundararajan, K. Das Sarit, Experimental and numerical investigation into the hydrodynamics of nanofluids in microchannels, Exp. Therm. Fluid Sci. 42 (2012) 174–186.

[88] A. Arshad Waqas, H. Muhammad, Graphene nanoplatelets nanofluids thermal and hydrodynamic performance on integral fin heat sink, Int. J. Heat Mass Transfer 107 (2017) 995–1001.

[89] W. Evans, J. Fish, K. Pawel, Role of Brownian motion hydrodynamics on nanofluid thermal conductivity, Appl. Phys. Lett. 88 (2006) 093116.

[90] S. Dae Park, B.I. Cheol, Experimental study of a universal CHF enhancement mechanism in nanofluids using hydrodynamic instability, Int. J. Heat Mass Transfer 70 (2014) 844–850.

[91] T. Mustafa, Single phase nanofluids in fluid mechanics and their hydrodynamic linear stability analysis, Comput. Methods Programs Biomed. 187 (2020) 105171.

[92] A. Wakif, Z. Boulahia, S. Rachid, A semi-analytical analysis of electro-thermo-hydrodynamic stability in dielectric nanofluids using Buongiorno's mathematical model together with more realistic boundary conditions, Results Phys 9 (2018) 1438–1454.

[93] J.P. Wen, X.Q. Jia, W. Feng, Hydrodynamic and mass transfer of gas-liquid-solid three-phase internal loop airlift reactors with nanometer solid particles, Chem. Eng. Technol.: Ind. Chem.-Plant Equipment-Process Eng.-Biotechnol. 28 (2005) 53–60.

CHAPTER

Effect of nanofluids in solubility enhancement

5

Maryam Meshksar, Mohammad Amin Makarem, Zohreh-Sadat Hosseini and Mohammad Reza Rahimpour

Department of Chemical Engineering, Shiraz University, Shiraz, Iran

5.1 Introduction

Choi was the first person that introduce the term "nanofluid" to describe a novel material for heat transfer enhancement in 1995. Since then many researchers have measured or analyzed the heat transfer enhancement of nanoparticles in fluid streams, and their results were in agreement with Choi's achievements [1,2]. Actually, nanofluids comprise nanosize ($d_p = 1-100$ nm) dispersing materials like nanoparticles, nanosheets, nanotubes, nanofibers, nanowires, nanorods, and even nanosized droplets in a water/nonwater soluble liquid as a base fluid. The nanoparticles used in nanofluids could be metallic/nonmetallic, oxide, hybrid nanoparticles (mixture of different nanoparticles), ceramics, or nanoscale liquid droplets [3].

Nanofluids can be prepared via different methods like the one-step method, two-step method, phase transfer, and posttreatment method. A one-step method is used to synthesize Cu, Au, and Ag nanoparticles by evaporating and condensing nanoparticles directly at the same time in the base fluid (Fig. 5.1A) [4]. The advantage of this method is the minimal nanoparticles agglomeration and high nanoparticles stability in the base fluid. However, this procedure is difficult to scale up due to its complicated synthesis conditions [5]. A two-step method, which is extensively applied for nanofluid preparation, is to mix commercially available nanoparticles with the base fluid using physical, mechanical, or chemical mixing approaches, such as milling, sol–gel, and vapor-phase methods (Fig. 5.1B) [6]. Fig. 5.1C shows a phase transfer method in which nanoparticles are formed by changing the polar synthesis environment to a nonpolar one, or vice versa [7]. Therefore the poor solubility of metal ion precursors as a limitation of direct synthesis methods can be solved by the use of the phase transfer method [8]. The posttreatment method is suitable for systems in which agglomerated nanoparticles are formed due to the poor dispersed raw fluids [9].

It has been proven that nanofluids have enhancedthermal conductivity and diffusivity, convective heat transfer coefficient, and viscosity [10]. Also, the high specific surface area between the nanosize particles and the fluid in nanofluids causes higher thermal stability as well as higher thermal, heat, and mass efficiencies compared with base fluids [11]. Krishnamurthy et al. [12] observed faster diffusion of dye into the nanofluid than into the water.Therefore, there is no surprise that nanofluids have attracted more attention in advanced systems development [7].

Nanofluids and Mass Transfer. DOI: https://doi.org/10.1016/B978-0-12-823996-4.00007-0

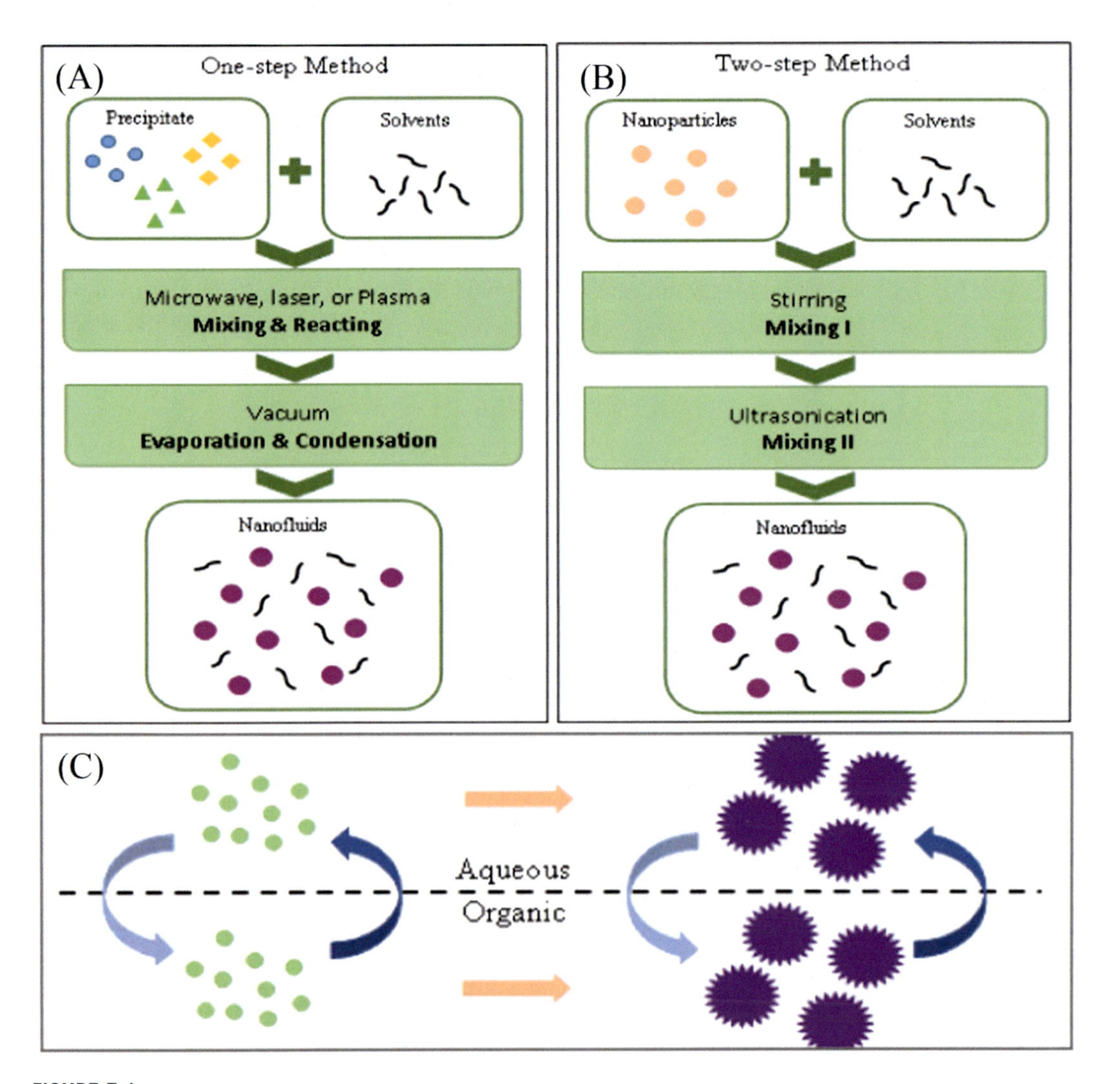

FIGURE 5.1

(A) One-step, (B) two-step, and (C) phase transfer nanofluid synthesis procedures.

As an example of nanoparticles influence on each of the nanofluid features, Eastman et al. [13] attained a 40% increase in the thermal conductivity by adding 0.3% Cu nanoparticles with 10 nm diameter into ethylene-glycol. Cho and Pak [14] tested 1–10 vol.% alumina ortitania nanoparticles in the water, causing a viscosity increase much greater than inpure water. Also, their results showed up to a 30% increase in Nusselt number by using alumina/water or titania/water systems instead of pure water which means 30% much more convective heat transfer coefficient by using nanoparticles.

Due to the fact that nanofluids have high mass and heat transfer coefficients, they are good candidates for being applied in gas absorption processes [15–17], as well as in gas hydrate formation

[18,19] and absorption refrigeration systems [20,21]. In the 1970s slurries containing fine particles were applied for gas absorption, and it was concluded that removal of gas like O_2, SO_2, etc. was improved significantly by the use of fine particles [22,23]. In the 2000s researchers started using nanoparticles for CO_2 absorption [24,25].

The nanoparticles' presence in the base fluid does not only increase the absorption rate, but also decreases the equipment sizes and enhances the selectivity in the multicomponent systems [26,27]. Since it was reported that the effect of nanoparticles on the mass transfer and solubility enhancement of the base fluid is significant, studying the effective parameters for gas solubility enhancement using nanoparticles is desirable [28].

5.2 The gas solubility enhancement mechanisms

Various mechanisms are accepted for gas transport and diffusion enhancement by the use of nanofluids. However, three theories, namely, the shuttle effect, the hydrodynamic effect, and the bubble breaking effect, are widely accepted by researchers.

5.2.1 The grazing or shuttle effect

Kars et al. [29] proposed the grazing effect in 1979 for the first time. According to this mechanism, the presence of fine particles causes the enhancement of absorption flux in three-phase systems. The particles act as shuttles meaning that they adsorb the transferred component near the gas–liquid interface and then migrate to the liquid bulk and desorb the adsorbed components. Transferring the components from the interface to the bulk, due to the particles, results in concentration reduction of reactants at the gas–liquid interface and thus increases the absorption rate. When a specific contact time is reached, the particles return to the liquid bulk, desorb the gas component, and will be regenerated [28,30]. This mechanism is shown in Fig. 5.2. Owing to this mechanism which is similar to the penetration theory, the absorption rate as well as the mass transfer coefficient will be increased by increasing the nanoparticle concentration. Also, by increasing the stirrer speed in the stirred tank reactor, the gas transportation from the gas–liquid interface into the liquid bulk will be increased which

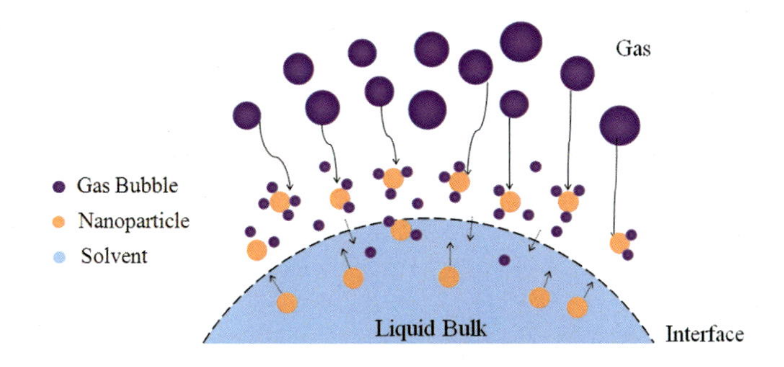

FIGURE 5.2

The grazing or shuttle effect.

causes a large mass transfer coefficient [30]. If the concentration of nanoparticles is low, the shuttle effect will be a significant event. The influence of this effect can be determined by measuring the gas diffusion coefficient into the liquid solvent phase [28].

5.2.2 The hydrodynamic or boundary mixing effect

The presence of particles in three-phase systems can also affect the hydrodynamic behavior, the mechanism of which can be seen in Fig. 5.3. The particles not only interact with the gas–liquid interface and thus break the boundary layer, but also can cause turbulence, thereupon the diffusion layer gets thinner and the mass transfer coefficient increases. As a result, nanoparticles can increase gas diffusion in the liquid film and so in the liquid bulk. This hypothesis is the result of strong microconvection movement and the Brownian motion of nanoparticles. The conclusion is that the presence of particles causes an enhancement of mass transfer [7]. It was revealed that the particles and surface interaction cause a decrease in diffusion layer thickness [30]. Kim et al. [31] found that an additional thermal eddy will be generated as a result of the nanoparticles mitigation which causes the reduction in conduction sublayer.

It should be noted that the number of nanoparticles in collision with the gas–liquid interface and the turbulence degree induced at the gas–liquid interface do not depend on the nanoparticles concentration in the liquid bulk. In this case, the more important factor is the number of nanoparticles which present at the gas–liquid interface as well as the nature of the nanoparticles' interactions with the gas–liquid interface [30].

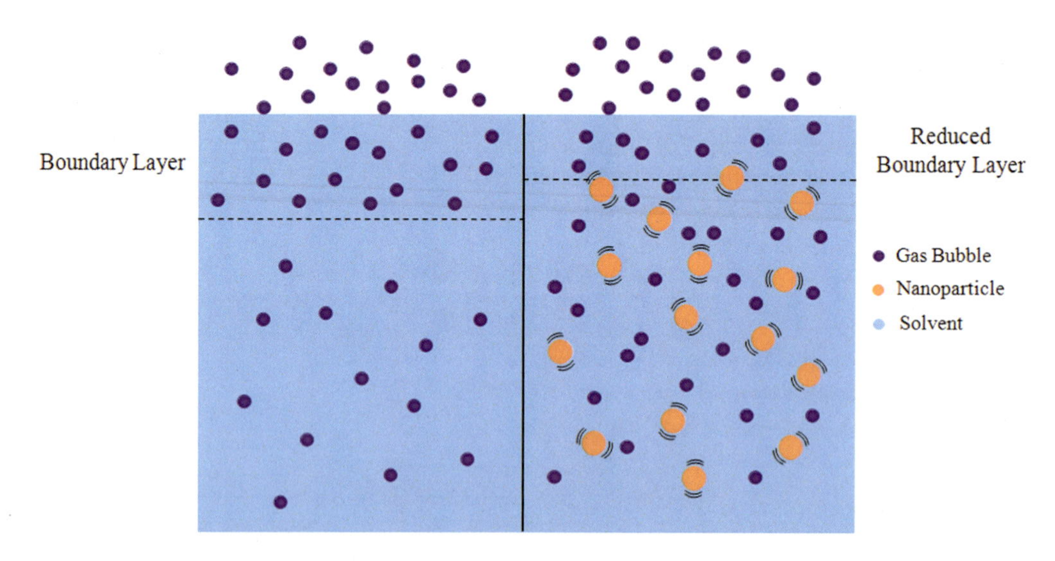

FIGURE 5.3

The hydrodynamic or boundary mixing effect.

5.2.3 The inhibition effect of bubble coalescence

In a bubbling bed reactor, the gas enters into the liquid phase via nozzles that generate fine bubbles which have a high gas–liquid interfacial area. Collisions occur when the liquid medium between two bubbles becomes thin enough. Actually, the liquid film thickness may become thin enough due to the balance between surface forces, hydrodynamic forces, and surface tension to rupture [32]. Bubble coalescence can be inhibited by adding nanoparticles. The added nanoparticles and rising bubbles collide with each other resulting in bubble coalescence or bubbles breakup. The bubble breaking effect is caused due to the fact that the specific interfacial area will be increased as a result of the nanoparticles addition. Therefore, the increased surface area affects the overall mass transfer coefficient. The mechanisms of bubble breaking effect can be observed in Fig. 5.4. Kim et al. [24] observed the formation of smaller bubble sizes in nanofluids compared with the pure absorbent. However, the bubble breaking effect was not observed in their other experimental research for absorbing CO_2 bubbles using methanol/Al_2O_3 nanofluid [33]. It also concluded that by using nanoparticle suspensions, the gas–liquid contact area was increased. Despite research into the bubble breaking effect's role in mass transfer enhancement, its role has not been clarified yet [31].

It should be noted that the gas absorption enhancement using nanofluid is done by coupling two or all three introduced mechanisms. Wang et al. [31] proposed that the bubble breaking effect has a greater role in gas absorption enhancement than other mechanisms at relatively low CO_2 flow rate. Jung et al. [34] attributed the increment in CO_2 absorption using nanofluid to both boundary mixing and bubble breaking effects.

All researchers have the same point of view that all three of these mechanisms dominate the gas, liquid, and nanoparticles interactions simultaneously. However, the contributions of each mechanism are still under discussion [4].

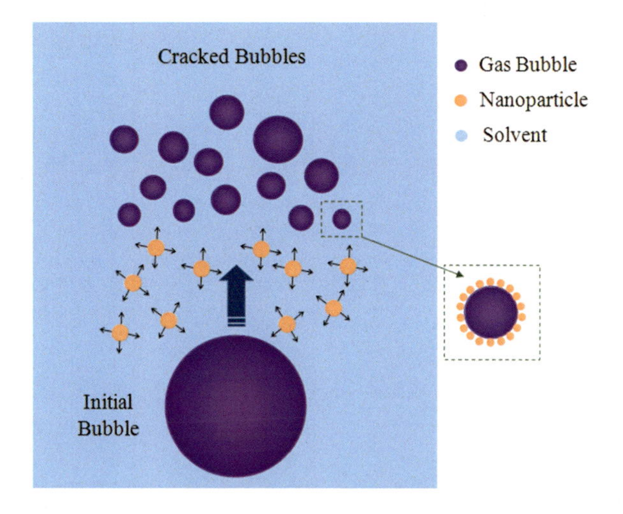

FIGURE 5.4

The bubble breaking effect.

5.3 Gas absorption enhancement by nanofluids

There are different factors, such as type of nanofluid, size and concentration of nanoparticles, temperature, pressure, and reaction gas concentration that influence the solubility enhancement of nanofluids.

5.3.1 The nanofluids type effect

As it was mentioned before, a large number of metallic and nonmetallic nanoparticles are applied in nanofluids for capturing various gases like CO_2, O_2, H_2S, etc. The enhancement results for absorbing different gases using various nanofluids are listed in Table 5.1, showing that by dispersing nanoparticles into the CO_2 gas stream, the nanofluid viscosity will be increased which is an important parameter for reactor design and heat exchangers and pumps selection.

Table 5.1 Some common nanofluids for absorption enhancement of some gases.

Nanoparticle type	Base fluid type	Gas absorption type	Enhancement (%)	Reaction condition	References
SiO_2 (0.05 wt.%)	DEA[1]	CO_2	40.0%	T = 25°C P = 0.012 MPa	[15]
SiO_2 (0.1 wt.%)	Water	CO_2	7.0%	T = 10°C P = 22 bar	[35]
SiO_2 (0.021 wt.%)	Water	CO_2	24.0%	T = 25°C P = 0.1 MPa	[24]
SiO_2 (0.05 vol.%)	Methanol	CO_2	9.7%	T = 22°C P = 0.1 MPa	[36]
SiO_2 (1.0 wt.%)	Water	SO_2	32.0%	T = 25°C P = 0.1 MPa	[37]
Al_2O_3 (0.01 vol.%)	NaCl	CO_2	12.5%	T = 20°C P = 0.1 MPa	[38]
Al_2O_3 (0.01 vol.%)	Methanol	CO_2	8.3%	T = 10°C P = 0.1 MPa	[34]
Al_2O_3 (0.01 vol.%)	Methanol	CO_2	26.0%	T = 20°C P = 0.1 MPa	[33]
Al_2O_3 (0.1 wt.%)	Water	SO_2	28.0%	T = 25°C P = 0.1 MPa	[37]
Fe_2O_3 (0.1 wt.%)	Water	SO_2	26.0%	T = 25°C P = 0.1 MPa	[37]
Fe_3O_4 (0.39 vol.%)	MDEA	CO_2	92.8%	T = 30°C P = 0.1 MPa	[39]
Fe_3O_4 (0.01 vol.%)	Na_2SO_3	O_2	60.0%	T = 37°C P = 0.1 MPa	[40]
CNT (0.05 wt.%)	Water	CO_2	32.0%	--- ---	[41]

Table 5.1 Some common nanofluids for absorption enhancement of some gases. ***Continued***

Nanoparticle type	Base fluid type	Gas absorption type	Enhancement (%)	Reaction condition	References
CNT (0.02 wt.%)	MDEA	CO_2	23.0%	T = 35°C P = 40 bar	[42]
MWCNT[2] (0.1 wt.%)	Water	CO_2	38.0%	T = 30°C P = 0.03 MPa	[43]
MWCNT (40 mg/L)	Water	CO_2	36.0%	T = 25°C P = 0.1 MPa	[44]
TiO_2 (0.8 wt.%)	MDEA	CO_2	11.5%	T = 20°C P = 0.1 MPa	[45]
TiO_2 (0.1 wt.%)	MEA	CO_2	10.0%	T = 40°C P = 0.1 MPa	[31]
Cu (0.1 wt.%)	Water	NH_3	221.0%	T = 20°C P = 0.1 MPa	[46]
Cu (2 wt.%)	DES[3]	H_2S	100.0%	T = 30°C P = 0.1 MPa	[47]
Ag (10 ppm)	Water	CH_4	13.0%	T = 11.5°C P = 5.5 MPa	[48]
Ag (10 ppm)	Water	C_2H_6	16.0%	T = 15.5°C P = 2.15 MPa	[48]
ZnO (0.1 wt.%)	Water	CO_2	14.0%	T = 5°C P = 22 bar	[35]

[1]*Diethanolamine*
[2]*Multiwalled carbon nanotubes*
[3]*Deep eutectic solvent*

As it was shown in Table 5.1, different nanofluids have been applied for CO_2 solubility enhancement. K_2CO_3/piperazine (PZ) solutions with and without silica nanoparticles showed a 12% increase in CO_2 absorption after using 0.021 wt.% synthesized SiO_2 nanoparticles. Also, due to the fact that small bubbles have large areas for mass transfer and have high solubility, the CO_2 absorption capacity coefficient of nanofluid was four times higher than H_2O without SiO_2 nanoparticles [24].

Numerous studies have been done comparing the performance of SiO_2 nanoparticles and other nanoparticles like Al_2O_3, Fe_3O_4, carbon nanotube (CNT), TiO_2, etc. in different base fluids for CO_2 absorption enhancement. Jiang et al. [49] compared the CO_2 absorption enhancement of four different nanoparticles (MgO, TiO_2, SiO_2, and Al_2O_3) into the MDEA solution. Their results showed that TiO_2-MDEA nanofluid had the best performance due to the fact that the absorption capacity of TiO_2 is the highest, which causes higher CO_2 concentration gradient and so increases the absorption enhancement. Furthermore, TiO_2-monoethanolamine (MEA) nanofluid had better performance compared with TiO_2-MDEA nanofluid because of the higher chemical reaction rate of CO_2 and MEA than of CO_2 and MDEA. The same results were reported by Fang et al. [50] and Dagaonkar et al. [51] in experiments using TiO_2 nanofluids for enhancing CO_2 absorption. The comparison between Al_2O_3, Fe_3O_4, carbon nanotube (CNT), and SiO_2 nanoparticles in the water showed that at high CO_2 concentration of 0.1 wt.

%, SiO_2 and Al_2O_3 had better absorption performance while at low CO_2 concentration of 0.02 wt.%, Fe_3O_4 and CNT were the best nanofluids. Also, CNT nanoparticles had 23% higher CO_2 adsorption capacity in MDEA than in diethanolamine (DEA) solution [42]. Lu et al. [52] figured out that CNT nanoparticles have a significantly increased effect on the CO_2 absorption capacity than Al_2O_3 nanoparticles at 25°C and 0.1 MPa. Also, CNT/H_2O nanofluid demonstrated about 16% more CO_2 absorption rate than SiO_2/H_2O nanofluid due to the higher capacity of CNT for absorbing CO_2 than silica nanoparticles [41]. Despite higher performance of SiO_2 nanoparticles than Al_2O_3 ones for CO_2 absorption [15,36], they showed lower absorption rate than ZnO nanoparticles in the water base fluid [35]. Multiwalled CNTs (MWCNTs) were also used as a nanoparticles into the Sulfinol-M fluid for comparing its performance with Fe_3O_4 nanoparticles for CO_2 absorption. The results showed a solubility increase of about 7.3% using MWCNTs compared with Fe_3O_4. Also, the absorption rate increased about 46.7% and 23.3%, in MWCNT/Sulfinol-M and Fe_3O_4/Sulfinol-M, respectively.

Nanofluids have also been applied for absorbing other gases like NH_3, SO_2, H_2S, etc. Esmaeili Faraj et al. [53] compared the H_2S absorption by the use of exfoliated graphene oxide (EGO) or silica nanoparticles into the water-based solution in a bubble column. They observed that just 0.02 wt.% EGO-water nanofluid enhanced the H_2S absorption about 40% compared with water base fluid. However, the addition of silica nanoparticles to water had a deteriorating effect on absorption of H_2S. Cu, Al_2O_3, SiO_2, and CNTs in a deep eutectic solvent (DES) composed of ETA base liquid were also used for H_2S removal. While SiO_2, Al_2O_3, and CNTs did not have an effective performance for sweetening processes at 30°C, Cu nanoparticles showed a positive effect for H_2S removal compared with ETA/DES solution without nanoparticles. In the absence of ETA, Cu/DES nanofluid had a poor desulfurization performance [47]. Al_2O_3, SiO_2, and Fe_2O_3 nanoparticles were applied in water for comparing their performance for SO_2 absorption. The optimum loading for each nanoparticle in which it had a highest absorption rate was applied for this comparison (0.1 wt.% for Al_2O_3, 0.1 wt.% for Fe_2O_3, and 1.0 wt.% for SiO_2). They showed that Fe_2O_3/water nanofluid had the highest absorption rate compared with other nanofluids [37]. Al_2O_3, Cu, and CuO nanoparticles were applied in the bubble absorber for measuring their influence on the NH_3 absorption in a NH_3/H_2O system at 20°C and 0.1 MPa. It was concluded that the addition of nanoparticles as well as the increase in their concentration increased the NH_3 absorption [46].

Nanofluids are also applicable in air conditioning and refrigeration technologies. TiO_2 and single-wall carbon nanohorns (SWCNH) were dispersed into a commercial POE lubricant for measuring their performance for solubility of tetrafluoroethane (R134a) at 30°C and 50°C. However, the results did not show a significant change in the R134a solubility after nanoparticles addition as a result of approximately equal intermolecular forces between POE oil and R134a with and without presence of nanoparticles [54].

5.3.2 The nanoparticle size effect

Several works have been done experimentally or simultaneously on the influence of particle size on the gas solubility and therefore on the absorption rate. As the surface area of nanoparticles and thus the surface area of nanofluid in contact with medium increase by decreasing the nanoparticles size, rapid dissolution will be obtained. It should be noted that the increment in solubility with the decrement in particle size is valid until the nanoparticle size reaches a particular point and after that a further decrease in solid particles has a reverse effect on solubility due to the presence of an electrical charge on the solid particles.

Kim et al. [24] found that silica nanoparticles with 30, 70, and 120 nm diameter had approximately equal CO_2 absorption capacity. They claimed that this equal CO_2 absorption capacity is due to the bubbles cracking into the same sizes at the equal nanofluid content even with different nanoparticle sizes. In another experimental study, it was concluded that by decreasing the silica nanoparticle size from 111 to 7 nm into the MEA solution, the CO_2 absorption rate was decreased in a flat-stirred vesselas a result of an increase in nanofluid viscosity by decreasing nanoparticle size [55].

5.3.3 The nanoparticle concentration effect

The nanoparticles concentration affected the nanofluid viscosity which is an important factor for the absorption process. Lee et al. [56] tested the influence of different concentrations of SiO_2 or Al_2O_3 nanoparticles in the water (0, 0.005, 0.01, 0.03, 0.05 vol.%) for CO_2 absorption and regeneration performance. They results showed the CO_2 absorption increased with increasing both SiO_2 or Al_2O_3 concentrations until 0.01 vol.%, whereas after this concentration, CO_2 absorption decreased as a result of solute concentration reduction at the interface of fluid and particles.The trend of the regeneration process was in agreement with CO_2 absorption by using SiO_2 nanoparticles, while it had an opposite trend for Al_2O_3 nanoparticles. This difference in the regeneration performance of SiO_2 and Al_2O_3 could be due to the different CO_2 reactions of these oxides and their different nanofluids characteristics. In another experimental work [38], they concluded that Al_2O_3 nanoparticles with 0.01 vol.% concentration in NaCl aqueous solution had the most effective absorption ratio and adsorption capacity as a result of low Brownian motion caused by interparticle interaction in high nanoparticle concentration. Different loadings of these two nanoparticles (0.005–0.5 wt.%) into the DEA-based fluid were also tested in Taheri et al. [15] experiment for enhancing H_2S and CO_2 absorption. The achieved results showed that by increasing both SiO_2 and Al_2O_3 nanoparticle concentrations to 0.05 wt.%, the effective CO_2 absorption ratios were first increased about 40% and 33%, respectively, and after that they both decreased. Also, the addition of SiO_2 nanoparticles deteriorated the H_2S absorption while 0.1 wt.% Al_2O_3 in DEA enhanced H_2S absorption by 14%. Karamian et al. [37] applied different mass fractions of SiO_2, Al_2O_3, and Fe_2O_3 nanoparticles (0.005–5 wt.%) into the water for SO_2 absorption. They revealed that by increasing the Al_2O_3 and Fe_2O_3 loadings to 0.1 wt.% the SO_2 absorption rate was increased about 28% and 26%, respectively, and a further increment in nanoparticles loading to 5 wt.% had a reverse effect on absorption rate. This optimum mass fraction was 1 wt.% for SiO_2 which increased the SO_2 absorption rate up to 32%. Kim et al. [24] attributed the reason for the increase in CO_2 absorption with increasing silica nanoparticles concentration from 0.01 to 0.04 wt.% in the water to smaller gas bubbles sizes in higher nanofluid concentration which cause more energy to be present in the fluid. However, it was reported in Hwang et al.'s [55] experiment that increasing silica nanoparticle concentration to 31 wt.% into the MEA causes a decrease in CO_2 absorption rate due to the solution's elasticity at 25°C and 101.3 kPa. Various concentrations of Al_2O_3 in water showed a weak CO_2 absorption enhancement in Lu et al.'s [52] experiment at 25°C and 0.1 MPa, and Jung et al. [34] concluded that among 0.005–0.1 vol.% Al_2O_3 dosage, 0.01 vol.% in the methanol solution increased the CO_2 absorption by 8.3% compared with pure methanol at 10°C and 0.1 MPa. The same optimum Al_2O_3

concentration as Jung et al.'s reports was achieved in NaCl aqueous solution which enhanced CO_2 solubility about 8.7% at equal reaction condition [38]. A comparison between Al_2O_3, TiO_2, and Fe_3O_4 nanoparticles inside a wetted-wall column showed thatCO_2 absorptionwas promoted as the concentration of Al_2O_3 nanoparticles increased. They claimed that this increment was the result of Brownian motion and grazing effect [57].

In the case of TiO_2 nanofluid, Chen et al. [58] concluded that increasing the nanoparticle concentration up to 0.1 vol.% had a negative effluence on CO_2 absorption, which was attributed to the change in nanofluid viscosity. Fang et al. [50] demonstrated that as TiO_2 nanoparticles mass fraction was increased from 1 to 5 g/L, the CO_2 absorption curve showed a maximum pick on 3 g/L TiO_2 mass fraction. This happening reasons could be summarized as by increasing the nanoparticles concentration, the agglomeration and solution viscosity will be increased which causes poor contact between gas and liquid surface. ZnO nanoparticles also have been applied for CO_2 absorption enhancement. It was concluded that with increasing ZnO nanoparticles mass content from 0.05 to 1 wt.% in water base fluid, CO_2 absorption rate was increased at 5°C. However, the difference between CO_2 solubility was decreased by increasing ZnO loading in a fixed pressure [35].

Ag nanoparticles were used for solubility enhancement of CH_4 and C_2H_6 in H_2O. It was revealed that by increasing Ag nanoparticles mass loading from 1 to 10 ppm, the solubility of both gases were increased. This increase could be explained as a result of higher surface energies of nanofluids in a higher nanoparticles loading which adsorbs more gas, thus becoming stable, as well as the Brownian motion and grazing effect of higher concentrated nanoparticles. The ethane solubility was more affected by Ag nanoparticles than methane due to the higher molecular weight of C_2H_6 than CH_4 which increases its ability for gas component adsorption on nanoparticles. It can be expressed that even low concentrations of Ag nanoparticles can promote CH_4 and C_2H_6 hydrate formation as a result of solubility increment [48].

Mohammadpour et al. [59] applied 0.0375, 0.05, 0.075, and 0.1 wt.% graphene oxide (GO) nanoparticles as well as the same percentages of sodium dodecyl sulfate (SDS) surfactant in MEA solvent separately for tested their performance in CO_2 absorption rate. Their results showed that 0.05 wt.% GO and 0.075 wt.% SDS had the highest performance compared with other concentrations which increased the CO_2 solubility up to 7.26% and 8.39%, respectively, Also, the addition of both 0.05 wt.% GO and 0.075 wt.% SDS into MEA results in 6.32% solubility enhancement.

5.3.4 The surfactant addition effect

Aggregation and sedimentation of nanoparticles are two major concerns in nanofluids usage which cause a decrease in the specific surface area of nanoparticles and thus a decrease in their activity for solubility enhancement. The repulsive force and the van der Waals attractive force—the two type of forces among nanoparticles—are the result of specific ion adsorption or surface charge. Both double layer repulsive energy (V_R) and van der Waals attractive energy (V_A) are related to the particles distances and therefore when sum of the V_R and V_A becomes zero, the particles' arrangement in the fluid solution will be at a stable distance [4].

Different parameters including ionic species, hydrodynamic conditions, nanoparticles concentration, and the nanofluid temperature can interrupt the V_R and V_A balances in the system. In order to improve the nanoparticles stability in the fluid, sonication and mechanical stirring methods are suggested [4]. Wang et al. [31] suggested the use of surfactant for long-term stability of silica nanoparticles into the MEA. 2-Ethyl-1-hexanol, 2-octanol, and n-octanol were applied as surfactants of the water base nanofluid with Al_2O_3, Cu, and CuO nanoparticles for NH_3 bubble absorption. Although the ammonia absorption enhancement was reported up to about 4.81% using surfactant in the water without nanoparticles [60], the binary solution (surfactant + nanoparticles) enhanced the NH_3 absorption about 5.32 times [61].

Depletion stabilization and electrostatic stabilization methods can also help to produce a repulsive force which stabilizes the nanoparticles. As electrostatic and steric hindrances are produced due to the double layer repulsive force, polymer chains' attachment or charge addition to the surface of the particles can stabilize the nanoparticles [5].

5.3.5 The pH effect

Another parameter that affects the gas solubility is pH of nanofluid aqueous solution. The solute charge state will be changed by changing the solution pH. If in a specific pH value of the solution molecules do not have a net electric charge, the solute will have a low solubility. As it was mentioned above, viscosity as well as the stability of nanoparticles affects their performance in mass transfer and solubility enhancement. The pH of nanofluid is an important factor for controlling their viscosity and dispersion stability for preventing agglomeration occurrence and thus reducing the nanoparticle size for having higher surface area. At a pH value near to the isometric point, the repulse forces between metal oxide particles become zero which results in the coagulation of nanoparticles and then precipitation occurs. Zhao et al. [62] reported that a stable nanofluid could not be prepared in acidic conditions as hydrogen ions were adsorbed and neutralized on the nanoparticles's surfaces, which reduced their stability.

Xian-Ju and Xin-Fang [63] tested Cu and Al_2O_3 nanoparticles in a water base liquid for measuring the effect of pH on stability and viscosity of nanofluid. Their results showed greatly higher viscosity of Al_2O_3/H_2O nanofluid than Cu/H_2O at the same nanoparticle weight fraction and pH values, which means poorer dispersion of Al_2O_3 nanoparticles than Cu nanoparticles in the water. As a result of this observation, a lower conversion of Al_2O_3/H_2O nanofluids than Cu/H_2O for absorption process can be predicted. Good alumina and Cu nanoparticles dispersion was observed at pH value of 7.5–8.9 and >7.6, respectively.Umar et al. [64] also compared CuO/H_2O and Al_2O_3/H_2O nanofluids performance in different pH values and the results showed that at pH of 8 or more, Al_2O_3 nanoparticles have a better dispersion due to their higher zeta potentials compared with CuO nanoparticles. The effect of pH on particle size and stability of silica nanoparticles in water base liquid demonstrated that by increasing pH value from 8 to 10, lower silica particles were formed in the water which have higher zeta potentials and therefore higher stability [62].

5.3.6 The temperature effect

Another effective factor that influences the solubility of gases in nanofluids is temperature which behaves in versely to solubility. Based on Le Chatelier's principle, by disturbing the system equilibrium, the system readjusts itself to counter the effect that deteriorates the equilibrium.

Therefore by increasing temperature, the solubility will be decreased as dissolution is an exothermic process. Irani et al. [65] applied GO/MDEA nanofluid for CO_2 absorption and the results showed a negative effect of temperature on CO_2 solubility. After increasing the temperature from 30°C to 60°C, the CO_2 solubility was decreased in all GO mass loading nanoparticles in MDEA. Similar results were obtained in Haghtalab et al.'s [35] experiment on SiO_2 and ZnO nanofluids for CO_2 solubility enhancement. Maleki et al. [66] compared the CO_2 solubility performance of diethylentriamine (DETA)-GO into the MDEA solution in different temperatures ranging from 30°C to 50°C. The results confirm Irani et al.'s and Haghtalab et al.'s reports in which a temperature increase had a negative effect on CO_2 solubility. The effect of temperature for absorbing H_2S using Cu/DES nanofluid showed the equal efficiency of different temperatures in the first 60 min and after that by increasing temperature from 30°C to 60°C, the sweetening efficiency was decreased due to the exothermic nature of this process [47].

5.3.7 The pressure effect

It was reported that the gas solubility in the liquid is strongly affected by pressure. After pressurizing the gas–liquid system above equilibrium condition, the gas molecules will be concentrated in the smaller area above the liquid solution media. Therefore, gas molecules' entry rate into the liquid solution will be increased until a new equilibrium point is formed. This can be also observed from Henry's law which gives a quantitative relation between gas solubility in the liquid and pressure (Eq. 5.1)

$$P = Hx \tag{5.1}$$

where P is the gas partial pressure, x is the gas mole fraction in the liquid solution, and H is the Henry's law constant. However Henry's law has some limitations including it cannot apply for high pressure systems, cannot apply for systems in which a chemical reaction occurs between solvent and solute, and can only apply for in equilibrium systems.

The effect of pressure on CO_2 solubility in the ZnO and SiO_2 nanofluids was investigated by Haghtalab et al. [35], whose results showed an increase in CO_2 content in both nanofluids solutions by increasing vessel pressure. Irani et al. [65] examined CO_2 solubility in the presence of graphene–oxide/MDEA nanofluid at different partial pressures (100–2100 kPa). Their results showed that CO_2 solubility improvement has a direct relationship with increasing pressure. Maleki et al. [66] studied CO_2 solubility in a mixture of graphene oxide functionalized by diethylenetriamine (DETA). They also reported that by increasing CO_2 partial pressure, the solubility increases. As previously mentioned, increasing temperature has a negative effect on gas solubility. However, it should be stated that this negative effect will be intensified at high pressures. Therefore it is necessary to find the optimum criteria for temperature and pressure that results in the maximum CO_2 solubility [35].

As was mentioned above, nanoparticles agglomeration is one of the important factors which decreased the gas solubility into the nanofluid. Rahmati-Abkenar and Manteghian [48] studied the effect of pressure and temperature increase on Ag nanoparticles agglomeration in the water base fluid. They revealed that by increasing both of these parameters the effect of nanoparticles for enhancing CH_4 and C_2H_6 solubility in the Ag/H_2O nanofluid was decreased as a result of nanoparticles agglomeration.

5.4 Application of nanofluids for liquid solvent solubility

The solubility of different materials in the solvents is important data in mass transfer phenomena in order to design industrial apparatus as well as to select and optimize related processes. One of the important fields of mass transfer is purifying liquid streams like separating acidic components. As nanotechnology is known as a method for improving transport phenomena, nanofluids can be used as a new transport fluid for an efficient transport rate in processes [67]. Thereforevarious types of nanofluids were applied for investigating their performance on the solubility of different components. Also, the effects of the different parameters discussed in gas solubility section using nanofluids, such as nanofluid concentration, nanoparticle size, and temperature, were checked out.

Various concentrations of Al_2O_3 and SiO_2 nanoparticles in water were used for examining their performance on salicylicacid and benzoic acid solubility at a temperature range of 20°C–60°C. The achieved results showed that alumina nanoparticles did not have a significant effect on acids solubility at temperatures below 55°C and after that by increasing temperature up to 60°C, acids had lower solubility than in water without nanoparticles. Maximum solubility reduction occurred at 0.1 vol.% alumina at 60°C for both acidic components. The same trend was observed for silica nanoparticles in which 7.2%, 9%, and 10.24% acetic acid solubility reduction were achieved at 50°C, 55°C, and 60°C, respectively, using 0.025 vol.% SiO_2/H_2O nanofluid. This optimum silica vol.% had a maximum 12.43% salicylic acid solubility reduction at 60°C. By increasing nanoparticles concentration to the optimum value (0.025 vol.% SiO_2 and 0.1 vol.% Al_2O_3) the acids solubility were first decreased and after that increased by a further increase in nanoparticle concentration. The reduction in acid solubility using silica nanoparticles was more than alumina nanofluid due to the bigger particle sizes of SiO_2 than γ-Al_2O_3, which confirms the effect of nanoparticle size on solubility. The effect of nanoparticle size on solubility can be summarized as follows: as the size of nanoparticles increases, the void spaces between solvent and solute which are filled with nanoparticles also increase. The void spaces is larger than nanoparticle volume because of its chaotic motion which would reduce solute solubility. A temperature increase also increases the chaotic motion of nanoparticles, which therefore expands void spaces and decreases solubility. The reason for solubility increasing with increasing nanoparticle concentration is probably due to the larger made particles at higher concentrations which have larger soft clusters. These soft clusters have a tendency to produce larger inert particles than void spaces between the solvent molecules. Also, they have a lower Brownian motion which decreases the void spaces and thus increases component solubility [68].

5.5 Limitations and drawbacks of nanofluids usages

In previous sections the benefits of nanofluid usage for solubility as well as mass transfer enhancement was discused. However, the addition of nanoparticles to the fluid stream always causes some drawbacks, including accumulation, sedimentation, or clogging inside columns, in addition to nanoparticles separation from fluid at outlets. There drawbacks limited the practical employment of nanoparticles in the industries [69]. Although one strategy for overcoming sedimentation and accumulation problem of nanofluid is using an appropriate surfactant as discussed in Section 5.3.4, itis difficult to overcome this issue. Furthermore, conventional nanoparticles separation techniques, such as electrophoresis, ultracentrifugation, chromatography, selective precipitation, and filtration are not continuous and need multiple steps for separation in a minimum sample volume [70]. Therefore the implementation of a low-cost and effective continuous separation process is necessary.

5.6 Conclusions and future trends

Nanotechnology is widely used in many energy systems in order to enhance operational performance. Nanofluids as a suspension of nanomaterials in the base liquid, have been used for gas absorption enhancement. This chapter reviews the different gas component absorption by the use of nanofluids. Based on previous research, the gas absorption enhancement using nanofluids depends on different factors including nanoparticle type, concentration, and size, as well as reaction temperature, pressure, and pH. The conclusions and futuresuggested directions can be summarized as follows:

- There are not uniform conclusions on the influences of nanofluids thermophysical properties (i.e., specific heat capacity, thermal conductivity, and viscosity) on the absorption enhancement.
- The hydrodynamic and grazing effects as well as the bubble coalescence inhibition effects have been described so far in detail. It can be suggested that future research can focus on the theoretical and experimental studies of connective heat transfer and mass transfer of nanofluids. Also, in order to validate and predict the achieved results, they can develop newcomprehensive models.
- The nanoparticles stability in the base fluid is an important factor for nanofluids applied in absorption applications. The achieved results showed that the modifications on the nanoparticles surface as well as an increase in their dispersions can improve their stabilities.
- The costs and capture performance of the nanofluids for the absorption application are the major issues for nanofluid selection. Also, the nanofluid regeneration for removing the absorbed gas is another issue. As an example, it was proven that CO_2 photothermal desorption by the use of MEA-based nanofluids is a promising procedure for nanofluid regeneration. Therefore researchers can also focus on the economic analysis of gas absorption and regeneration using nanofluids.

Abbreviations

CNT Carbon nanotube
CTAB Cetyltrimethylammonium bromide
DEA Diethanolamine
DES Deep eutectic solvent
DETA Diethylentriamine
EGO Exfoliated graphene oxide
GO Graphene oxide
MDEA Methyldiethanolamine
MEA Monoethanolamine
MWCNT Multiwall carbon nanotube
PEG Poly ethylene glycol
PVP Polyvinylpyrrolidone

PZ Piperazine
SDBS Sodium dodecylbenzenesulfonate
SDS Sodium dodecylsulfate
SWCNH Single-wall carbon nanohorns

References

[1] S.U. Choi, J.A. Eastman, Enhancing thermal conductivity of fluids with nanoparticles, in: Argonne National Lab., IL, (1995).

[2] J. Buongiorno, Convective transport in nanofluids, (2006).

[3] V. Fuskele, R. Sarviya, Recent developments in nanoparticles synthesis, preparation and stability of nanofluids, Mater. Today: Proc. 4 (2017) 4049–4060.

[4] W. Yu, T. Wang, A.-H.A. Park, M. Fang, Review of liquid nano-absorbents for enhanced CO_2 capture, Nanoscale 11 (2019) 17137–17156.

[5] S. Mukherjee, S. Paria, Preparation and stability of nanofluids-a review, IOSR J. Mech. Civ. Eng 9 (2013) 63–69.

[6] Y. Hwang, J.-K. Lee, J.-K. Lee, Y.-M. Jeong, S.-i Cheong, Y.-C. Ahn, et al., Production and dispersion stability of nanoparticles in nanofluids, Powder Technol. 186 (2008) 145–153.

[7] Z. Zhang, J. Cai, F. Chen, H. Li, W. Zhang, W. Qi, Progress in enhancement of CO_2 absorption by nanofluids: a mini review of mechanisms and current status, Renew. Energy 118 (2018) 527–535.

[8] J. Yang, J.Y. Lee, J.Y. Ying, Phase transfer and its applications in nanotechnology, Chem. Soc. Rev. 40 (2011) 1672–1696.

[9] S. Oh, I. Jang, S.-G. Oh, S.S. Im, Effect of ZnO nanoparticle morphology and post-treatment with zinc acetate on buffer layer in inverted organic photovoltaic cells, Sol. Energy 114 (2015) 32–38.

[10] R. Saidur, K. Leong, H.A. Mohammed, A review on applications and challenges of nanofluids, Renew. Sustain. Energy Rev. 15 (2011) 1646–1668.

[11] R. Aghehrochaboki, Y.A. Chaboki, S.A. Maleknia, V. Irani, Polyethyleneimine functionalized graphene oxide/methyldiethanolamine nanofluid: preparation, characterization, and investigation of CO_2 absorption, J. Environ. Chem. Eng. 7 (2019) 103285.

[12] S. Krishnamurthy, P. Bhattacharya, P. Phelan, R. Prasher, Enhanced mass transport in nanofluids, Nanoletters 6 (2006) 419–423.

[13] J.A. Eastman, S. Choi, S. Li, W. Yu, L. Thompson, Anomalously increased effective thermal conductivities of ethylene glycol-based nanofluids containing copper nanoparticles, Appl. Phys. Lett. 78 (2001) 718–720.

[14] B.C. Pak, Y.I. Cho, Hydrodynamic and heat transfer study of dispersed fluids with submicron metallic oxide particles, Exp. Heat. Transf. 11 (1998) 151–170.

[15] M. Taheri, A. Mohebbi, H. Hashemipour, A.M. Rashidi, Simultaneous absorption of carbon dioxide (CO_2) and hydrogen sulfide (H_2S) from $CO_2–H_2S–CH_4$ gas mixture using amine-based nanofluids in a wetted wall column, J. Nat. Gas. Sci. Eng. 28 (2016) 410–417.

[16] E. Nagy, T. Feczkó, B. Koroknai, Enhancement of oxygen mass transfer rate in the presence of nanosized particles, Chem. Eng. Sci. 62 (2007) 7391–7398.

[17] A. Golkhar, P. Keshavarz, D. Mowla, Investigation of CO_2 removal by silica and CNT nanofluids in microporous hollow fiber membrane contactors, J. Membr. Sci. 433 (2013) 17–24.

[18] S.-S. Park, S.-B. Lee, N.-J. Kim, Effect of multi-walled carbon nanotubes on methane hydrate formation, J. Ind. Eng. Chem. 16 (2010) 551–555.

[19] A. Mohammadi, M. Manteghian, A. Haghtalab, A.H. Mohammadi, M. Rahmati-Abkenar, Kinetic study of carbon dioxide hydrate formation in presence of silver nanoparticles and SDS, Chem. Eng. J. 237 (2014) 387–395.

[20] X. Ma, F. Su, J. Chen, T. Bai, Z. Han, Enhancement of bubble absorption process using a CNTs-ammonia binary nanofluid, Int. Commun. Heat. Mass. Transf. 36 (2009) 657–660.
[21] C. Amaris, M. Bourouis, M. Vallès, Passive intensification of the ammonia absorption process with $NH_3/LiNO_3$ using carbon nanotubes and advanced surfaces in a tubular bubble absorber, Energy 68 (2014) 519–528.
[22] S. Uchida, K. Koide, M. Shindo, Gas absorption with fast reaction into a slurry containing fine particles, Chem. Eng. Sci. 30 (1975) 644–646.
[23] E. Sada, B. MA, Removal of dilute sulfur dioxide by aqueous slurries of magnesium hydroxide particles, 1977.
[24] W.-G. Kim, H.U. Kang, K.-M. Jung, S.H. Kim, Synthesis of silica nanofluid and application to CO_2 absorption, Sep. Sci. Technol. 43 (2008) 3036–3055.
[25] A. Hussein, K. Kadirgama, M. Noor, Nanoparticles suspended in ethylene glycol thermal properties and applications: an overview, Renew. Sustain. Energy Rev. 69 (2017) 1324–1330.
[26] B.H. Junker, T.A. Hatton, D.I. Wang, Oxygen transfer enhancement in aqueous/perfluorocarbon fermentation systems: I. Experimental observations, Biotechnol. Bioeng. 35 (1990) 578–585.
[27] B.H. Junker, D.I. Wang, T.A. Hatton, Oxygen transfer enhancement in aqueous/perfluorocarbon fermentation systems: II. Theoretical analysis, Biotechnol. Bioeng. 35 (1990) 586–597.
[28] D.W.F. Brilman, W.P.M. van Swaaij, G. Versteeg, A one-dimensional instationary heterogeneous mass transfer model for gas absorption in multiphase systems, Chem. Eng. Processing: Process. Intensification 37 (1998) 471–488.
[29] R. Kars, R. Best, A. Drinkenburg, The sorption of propane in slurries of active carbon in water, Chem. Eng. J. 17 (1979) 201–210.
[30] J. Kluytmans, B. Van Wachem, B. Kuster, J. Schouten, Mass transfer in sparged and stirred reactors: influence of carbon particles and electrolyte, Chem. Eng. Sci. 58 (2003) 4719–4728.
[31] T. Wang, W. Yu, F. Liu, M. Fang, M. Farooq, Z. Luo, Enhanced CO_2 absorption and desorption by monoethanolamine (MEA)-based nanoparticle suspensions, Ind. Eng. Chem. Res. 55 (2016) 7830–7838.
[32] V.S. Craig, Bubble coalescence and specific-ion effects, Curr. Opin. Colloid Interface Sci. 9 (2004) 178–184.
[33] J.H. Kim, C.W. Jung, Y.T. Kang, Mass transfer enhancement during CO_2 absorption process in methanol/Al_2O_3 nanofluids, Int. J. Heat. Mass. Transf. 76 (2014) 484–491.
[34] J.-Y. Jung, J.W. Lee, Y.T. Kang, CO_2 absorption characteristics of nanoparticle suspensions in methanol, J. Mech. Sci. Technol. 26 (2012) 2285–2290.
[35] A. Haghtalab, M. Mohammadi, Z. Fakhroueian, Absorption and solubility measurement of CO_2 in water-based ZnO and SiO_2 nanofluids, Fluid phase equilibria 392 (2015) 33–42.
[36] I.T. Pineda, J.W. Lee, I. Jung, Y.T. Kang, CO_2 absorption enhancement by methanol-based Al_2O_3 and SiO_2 nanofluids in a tray column absorber, Int. J. Refrig. 35 (2012) 1402–1409.
[37] S. Karamian, D. Mowla, F. Esmaeilzadeh, The effect of various nanofluids on absorption intensification of CO_2/SO_2 in a single-bubble column, Processes 7 (2019) 393.
[38] J.W. Lee, Y.T. Kang, CO_2 absorption enhancement by Al_2O_3 nanoparticles in NaCl aqueous solution, Energy 53 (2013) 206–211.
[39] S. Komati, A.K. Suresh, CO_2 absorption into amine solutions: a novel strategy for intensification based on the addition of ferrofluids, J. Chem. Technol. Biotechnol.: Int. Res. Process Environ. Clean. Technol. 83 (2008) 1094–1100.
[40] B. Olle, S. Bucak, T.C. Holmes, L. Bromberg, T.A. Hatton, D.I. Wang, Enhancement of oxygen mass transfer using functionalized magnetic nanoparticles, Ind. Eng. Chem. Res. 45 (2006) 4355–4363.
[41] M. Darabi, M. Rahimi, A.M. Dehkordi, Gas absorption enhancement in hollow fiber membrane contactors using nanofluids: modeling and simulation, Chem. Eng. Processing: Process. Intensification 119 (2017) 7–15.

[42] B. Rahmatmand, P. Keshavarz, S. Ayatollahi, Study of absorption enhancement of CO_2 by SiO_2, Al_2O_3, CNT, and Fe_3O_4 nanoparticles in water and amine solutions, J. Chem. Eng. Data 61 (2016) 1378–1387.
[43] A. Peyravi, P. Keshavarz, D. Mowla, Experimental investigation on the absorption enhancement of CO_2 by various nanofluids in hollow fiber membrane contactors, Energy Fuels 29 (2015) 8135–8142.
[44] L. Jorge, S. Coulombe, P.L. Girard-Lauriault, Nanofluids containing MWCNTs coated with nitrogen-rich plasma polymer films for CO_2 absorption in aqueous medium, Plasma Process. Polymers 12 (2015) 1311–1321.
[45] S.H. Li, Y. Ding, X.S. Zhang, Enhancement on CO_2 bubble absorption in MDEA solution by TiO_2 nanoparticles, Advanced Materials Research, Trans Tech Publ, 2013, pp. 127–134.
[46] J.-K. Kim, J.Y. Jung, Y.T. Kang, The effect of nano-particles on the bubble absorption performance in a binary nanofluid, Int. J. Refrig. 9 (2006) 22–29.
[47] X. Liu, B. Wang, X. Lv, Q. Meng, M. Li, Enhanced removal of hydrogen sulfide using novel nanofluid system composed of deep eutectic solvent and Cu nanoparticles, J. Hazard. Materials (2020) 124271.
[48] M. Rahmati-Abkenar, M. Manteghian, Effect of silver nanoparticles on the solubility of methane and ethane in water, J. Nat. Gas. Sci. Eng. 82 (2020) 103505.
[49] J. Jiang, B. Zhao, Y. Zhuo, S. Wang, Experimental study of CO_2 absorption in aqueous MEA and MDEA solutions enhanced by nanoparticles, Int. J. Greenh. Gas. Control 29 (2014) 135–141.
[50] L. Fang, H. Liu, Y. Bian, Y. Liu, Y. Yang, Experimental study on enhancement of bubble absorption of gaseous CO_2 with nanofluids in ammonia, J. Harbin Inst. Technol. 24 (2017) 80–86.
[51] M. Dagaonkar, H. Heeres, A. Beenackers, V. Pangarkar, The application of fine TiO_2 particles for enhanced gas absorption, Chem. Eng. J. 92 (2003) 151–159.
[52] L. Sumin, X. Min, S. Yan, D. Xiangjun, Experimental and theoretical studies of CO_2 absorption enhancement by nano-Al_2O_3 and carbon nanotube particles, Chin. J. Chem. Eng. 21 (2013) 983–990.
[53] S.H. Esmaeili Faraj, M. Nasr Esfahany, M. Jafari-Asl, N. Etesami, Hydrogen sulfide bubble absorption enhancement in water-based nanofluids, Ind. Eng. Chem. Res. 53 (2014) 16851–16858.
[54] S. Bobbo, L. Fedele, M. Fabrizio, S. Barison, S. Battiston, C. Pagura, Influence of nanoparticles dispersion in POE oils on lubricity and R134a solubility, Int. J. Refrig. 33 (2010) 1180–1186.
[55] B.-J. Hwang, S.-W. Park, D.-W. Park, K.-J. Oh, S.-S. Kim, Absorption of carbon dioxide into aqueous colloidal silica solution with different sizes of silica particles containing monoethanolamine, Korean J. Chem. Eng. 26 (2009) 775–782.
[56] J.S. Lee, J.W. Lee, Y.T. Kang, CO_2 absorption/regeneration enhancement in DI water with suspended nanoparticles for energy conversion application, Appl. Energy 143 (2015) 119–129.
[57] Z. Samadi, M. Haghshenasfard, A. Moheb, CO_2 absorption using nanofluids in a wetted-wall column with external magnetic field, Chem. Eng. Technol. 37 (2014) 462–470.
[58] H. Chen, Y. Ding, C. Tan, Rheological behaviour of nanofluids, N. J. Physics 9 (2007) 367.
[59] A. Mohammadpour, M. Mirzaei, A. Azimi, S.M.T. Ghomsheh, Solubility and absorption rate of CO_2 in MEA in the presence of graphene oxide nanoparticle and sodium dodecyl sulfate, Int. J. Ind. Chem. 10 (2019) 205–212.
[60] J.-K. Kim, J.Y. Jung, J.H. Kim, M.-G. Kim, T. Kashiwagi, Y.T. Kang, The effect of chemical surfactants on the absorption performance during NH_3/H_2O bubble absorption process, Int. J. Refrig. 29 (2006) 170–177.
[61] J. Jung, J. Kim, Y. Kang, The Effect of Binary Nanofluids and Chemical Surfactants on the Absorption Performance, 2006.
[62] M. Zhao, W. Lv, Y. Li, C. Dai, H. Zhou, X. Song, et al., A study on preparation and stabilizing mechanism of hydrophobic silica nanofluids, Materials 11 (2018) 1385.
[63] W. Xian-Ju, L. Xin-Fang, Influence of pH on nanofluids' viscosity and thermal conductivity, Chin. Phys. Lett. 26 (2009) 056601.

[64] S. Umar, F. Sulaiman, N. Abdullah, et al., Investigation of the effect of pH adjustment on the stability of nanofluid, in: AIP Conference Proceedings, AIP Publishing LLC, 2018, pp. 020031.
[65] V. Irani, A. Maleki, A. Tavasoli, CO_2 absorption enhancement in graphene-oxide/MDEA nanofluid, J. Environ. Chem. Eng. 7 (2019) 102782.
[66] A. Maleki, V. Irani, A. Tavasoli, M. Vahidi, Enhancement of CO_2 solubility in a mixture of 40 wt.% aqueous N-Methyldiethanolamine solution and diethylenetriamine functionalized graphene oxide, J. Nat. Gas. Sci. Eng. 55 (2018) 219–234.
[67] H. Beiki, M.N. Esfahany, N. Etesami, Turbulent mass transfer of Al_2O_3 and TiO_2 electrolyte nanofluids in circular tube, Microfluidicsnanofluidics 15 (2013) 501–508.
[68] M.M. Fard, H. Beiki, Experimental measurement of solid solutes solubility in nanofluids, Heat. Mass. Transf. 53 (2017) 1257–1263.
[69] P. Amani, M. Amani, G. Ahmadi, O. Mahian, S. Wongwises, A critical review on the use of nanoparticles in liquid–liquid extraction, Chem. Eng. Sci. 183 (2018) 148–176.
[70] T. Salafi, K.K. Zeming, Y. Zhang, Advancements in microfluidics for nanoparticle separation, Lab. Chip 17 (2017) 11–33.

CHAPTER

6 Heat and mass transfer characteristics of magnetic nanofluids

Gabriela Huminic[1], Angel Huminic[1] and Alina Adriana Minea[2]

[1]Transilvania University of Brasov, Brasov, Romania [2]Gheorghe Asachi Technical University Iasi, Iasi, Romania

6.1 Introduction

Due to the unique features of magnetic nanoparticles (small size, high surface area, superparamagnetism, and low toxicity), these can be excellent candidates for the design and fabrication of new functional nanomaterials used in heat transfer applications. The most used magnetic nanoparticles in thermal systems are iron oxides [magnetite (Fe_3O_4) and maghemite ($\gamma - Fe_2O_3$)], owing to their ease of preparation, oxidative stability, and biocompatibility. In recent years, the ground state transition metals of iron (Fe), cobalt (Co), and nickel (Ni), as well as alloys of transition metals (FeCo, FePt, CoPt, NiPt, FeCoNi,), have been investigated.

General, a magnetic nanofluid (ferrofluid) consists of a colloidal mixture of superparamagnetic nanoparticles with sizes in the range 5–15 nm in diameter coated with a surfactant layer suspended in a nonmagnetic carrier fluid. The carrier fluid for magnetic nanofluid suitable for heat transfer applications must fulfill some requirements: high thermal conductivity, high heat capacity, and high thermal expansion coefficient [1]. In addition, to improved thermophysical properties obtained by including magnetic nanoparticles in the carrier fluid (thermal conductivity, viscosity, and heat capacity), this special class of nanofluids also exhibits improved magnetic properties. By applying an external magnetic field an intensification of the heat transfer process was noted. Compared to nonmagnetic nanofluids, the magnetic nanofluids have the following advantages: the possibility of improvement/deterioration thermophysical properties (thermal conductivity and viscosity) [2,3], the thermomagnetic convection is intense compared to the gravitational convection [3], and the energy transport process can be controlled by changing the external magnetic field and/or temperature variation inside the magnetic nanofluid [4,5].

In recent years, the magnetic nanofluids (MNFs) have attracted the attention of many researchers due to their capability to be controlled under magnetic fields. Thus the objective of this chapter is to present an overview of the literature dealing with recent investigations on MNFs used in natural convection under uniform and nonuniform magnetic field. To give the recent data, the numbers of published papers from 2010 to 2020 (up to 20th July) found in ScienceDirect under "magnetic nanofluids," "magnetic nanofluids and natural convection," "nanofluids and mass transfer," "magnetic nanofluids and mass transfer," and "natural convection and mass transfer" are summarized in Fig. 6.1. The results confirm the intensive interest and activity in the research of magnetic nanofluids

Nanofluids and Mass Transfer. DOI: https://doi.org/10.1016/B978-0-12-823996-4.00006-9

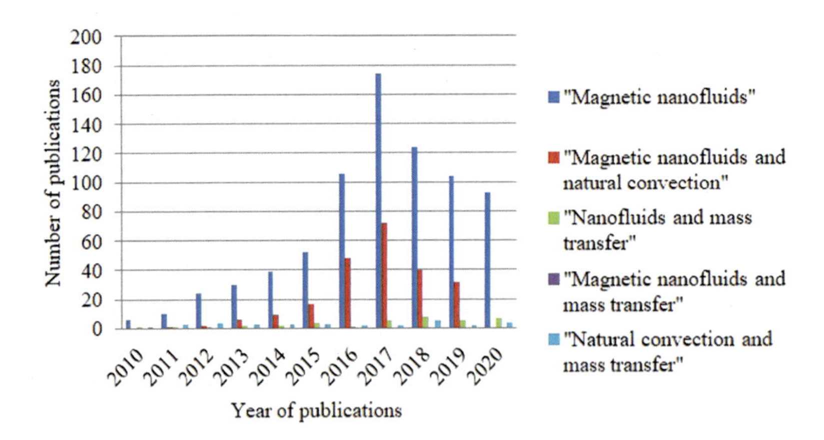

FIGURE 6.1

Magnetic nanofluids publication rate.

used in natural convection. In fact, the increase in the number of research articles dedicated to this subject shows a noticeable growth and the importance of magnetic nanofluids. As can be seen from Fig. 6.1, no article focuses on the subject of mass transfer by using magnetic nanofluids.

6.2 Heat transfer characteristics

6.2.1 Theory

Heat transfer by natural convection within an enclosure is caused by density and temperature gradients and may be used in several engineering applications. Natural convection is common in buildings (uninsulated walls, rooms), solar collectors, hot or cold liquid storage systems, heat exchangers, crystal growth, etc. The main parameters that may influence the heat transfer analysis in the enclosure are the geometry of the space (square, rectangle, triangle, cylindrical, elliptical, and spherical), as well as the physical and chemical properties of the base fluids.

Table 6.1 provides a summary of the definitions, physical interpretations, and areas of significance of the important dimensionless numbers, while Tables 6.2 and 6.3 summarize the existing correlations in literature for natural convection in enclosures with different configuration.

6.2.2 Natural convection subject to nonuniform magnetic field

The study of the influence of the magnetic field on magnetic nanofluids is a significant topic of research due to their numerous applications (engineering chemistry, physics, etc.). A special feature of magnetic nanofluids is that they can be easy to manipulate using an external magnetic field. By applying a magnetic field, the fluid motion can be controlled without having direct access to the fluid (i.e., in porous media).

Effect of spatially nonuniform magnetic field on natural convection in an enclosure filled with of Fe_3O_4–water nanofluid with a volume fraction of nanoparticles of 0.04 is investigated

Table 6.1 Main dimensionless numbers used in the analysis of the heat and mass transfer.

Dimensionless numbers	Definition equation	Physical interpretation	Area of significance
Brinkmann number	$Br = \mu U^2/(k\Delta T)$	Viscous dissipation/enthalpy change	High-speed flow
Darcy number	$Da = K/L^3$	Ratio of relative effect of the permeability of the medium to its cross-sectional area	Porous media
Eckert number	$Ec = U^2/(c_p\Delta T)$	Kinetic energy/enthalpy change	High-speed flow
Grashof number	$Gr = g\beta\Delta TL^3/\nu^2$	Buoyancy force/viscous force	Natural convection
Hartmann number	$Ha = B_0 L\sqrt{\sigma/\mu}$	Electromagnetic force/viscous forces	Natural convection
Lewis number	$Le = a/D$	Ratio between thermal and mass diffusivities	Mass transfer
Nusselt number	$Nu = hL/k$	Thermal resistance of conduction/thermal resistance of convection	Single and multiphase convection heat transfer
Prandtl number	$Pr = \nu/a$	Diffusion rate of viscous effect/diffusion rate of heat	Single and multiphase convection heat transfer
Rayleigh number	$Ra = g\beta\Delta TL^3/(\nu a)$	$Gr \bullet Pr$	Natural convection
Richardson number	$Ri = \frac{g\beta\Delta TL}{U^2}$ or Gr/Re^2	Ratio of the buoyancy term to the flow shear term	Convective heat transfer
Reynolds number	$Re = UL/\nu$	Inertial forces/viscous force	Forced convection
Schmidt number	$Sc = \nu/D$	Diffusion rate of viscous effect/diffusion rate of mass	Convective mass transfer
Sherwood numbers	$Sh = \frac{h}{D/L}$	Ratio between convective mass transfer rate and diffusion rate	Convective mass transfer

numerically by Sheikholeslami and Rashidi [21]. They considered the combined effects of ferrohydrodynamics and magnetohydrodynamics (MHD). The governing equations are solved using control volume-based finite element method (CVFEM). The influences of Magnetic number ($Mn_F = 0, 20, 60,$ and 100), Hartmann number ($Ha = 0, 5, 10,$ and 100), Rayleigh number ($Ra = 10^3, 10^4,$ and 10^5), and nanoparticle concentration on the flow and heat transfer characteristics are analyzed and discussed. They concluded that increasing Magnetic number, Rayleigh number, and nanoparticle concentration led to increasing Nusselt number, but, at the same time, the Nusselt number decreases with increasing Hartmann number. For low Rayleigh number, the heat transfer enhancement, defined as $En = \left[\left(Nu_{MNf} - Nu_{bf}\right)/Nu_{bf}\right]\bullet 100$, increases with increasing Hartmann number, but decreases with increasing Magnetic number.

For the Nusselt number the following correlation is proposed:

$$Nu_{ave} = a_{15} + a_{25}Y_3 + a_{35}Y_4 + a_{45}Y_3^2 + a_{55}Y_4^2 + a_{65}Y_4Y_3 \quad (6.1)$$

where

$$Y_3 = a_{13} + a_{23}Y_1 + a_{33}Ha^* + a_{43}Y_1^2 + a_{53}Ha^{*2} + a_{63}Y_1Ha^*$$

$$Y_4 = a_{14} + a_{24}Ha^* + a_{34}Y_2 + a_{44}Ha^{*2} + a_{54}Y_2^2 + a_{64}Y_2Ha^*$$

$$Y_1 = a_{11} + a_{21}Ra^* + a_{31}Mn_F^* + a_{41}Ra^{*2} + a_{51}Mn_F^{*2} + a_{61}Ra^*Mn_F^*$$

$$Y_2 = a_{12} + a_{22}\phi + a_{32}Ra^* + a_{42}\phi^2 + a_{52}Ra^{*2} + a_{62}\phi Ra^*$$

Table 6.2 Summary the existing correlations in literature for natural convection in enclosures with different configuration.

Geometry	Nusselt number	Comments and restrictions	Dimensionless number
Natural convection in enclosed vertical spaces [6]	$Nu_\delta = 0.42(Gr_\delta Pr)^{1/4} Pr^{0.012} \left(\frac{L}{\delta}\right)^{-0.30}$	$q_w = const$ $10^4 < Gr_\delta Pr < 10^7$ $1 < Pr < 20,000$ $10 < L/\delta < 40$	$Gr_\delta = \frac{g\beta(T_1 - T_2)\delta^3}{\nu^2}$
	$Nu_\delta = 0.46(Gr_\delta Pr)^{1/3}$	$q_w = const.$ $10^6 < Gr_\delta Pr < 10^9$ $1 < Pr < 20$ $1 < L/\delta < 40$	
Natural convection between concentric spheres [7]	$Nu_\delta \cong \frac{k_e}{k} = 0.228(Gr_\delta Pr)^{0.226}$	$0.25 \leq \delta/r_i \leq 1.5$ $1.2 \bullet 10^2 < Gr\ Pr < 1.1 \bullet 10^9$ $0.7 < Pr < 4150$	
Enclosed spaces [8]	$q = k_e A(\Delta T/\delta) Nu_\delta \cong \frac{k_e}{k} = C(Gr_\delta Pr)^n \left(\frac{L}{\delta}\right)^m$	Pure conduction for $Gr_\delta Pr < 2000$ Constants C, m, and n are given in Table 6.3	
Natural convection in closed vertical or horizontal cylindrical enclosures [9]	$Nu_f = 0.55\left(Gr_f Pr_f\right)^{1/4}$	$0.75 < L/d < 2.0$	
Natural convection in spherical and cubical enclosure [10]	$Nu_L = 0.396 Ra_L^{0.234} (L/R_i)^{0.496} Pr^{0.0162}$	Average deviation: 13.50%	$Ra_L = \frac{\rho^2 g\beta(T_1 - T_2)L^3 c_p}{\mu k}$
Natural convection in spherical, cylindrical and cubical enclosure [10]	$Nu_b = 0.585 Ra_b^{*0.236} Ra_b^* = Ra_b\left(\frac{L}{R_i}\right)$;L- gap width or hypothetical gap width, $R_0 - R_i$;	Average deviation: 14.75%	
Natural convection in vertical cylinders with internal heat generation (with adiabatic horizontal walls and isothermal vertical walls) [11]	$Nu = 0.576 Ra^{0.2024} \left(\frac{H}{D}\right)^{-0.186}$	$3 \bullet 10^{10} < Ra < 1 \bullet 10^{13}$	$Ra = Pr \bullet Gr$
	$Nu = 0.776 Ra_v^{0.246}$	–	$Ra_v = \frac{g\beta\Delta T_v D^3}{\nu a}$ – $\Delta T_v = \frac{T_t - T_b}{(H/D)}$

Table 6.3 Empirical equations for natural convection in enclosures ($Nu_\delta \cong \frac{k_e}{k} = C(Gr_\delta Pr)^n \left(\frac{L}{\delta}\right)^m$), correlation constants adjusted by Holman [8].

Fluid	Geometry	$Gr_\delta Pr$	Pr	L/δ	C	n	m	Refs.
Liquid	Vertical plate, constant heat flux or isothermal	< 2000	$k_e/k = 1.0$					
		$10^4 - 10^7$	1–20,000	10–40	0.42	1/4	−0.3	[6,12]
		$10^6 - 10^9$	1–20	1–40	0.46	1/3	0	
	Horizontal plate, isothermal, heated from below	< 1700	$k_e/k = 1.0$	–				[13–17]
		1700–6000	1–5000	–	0.012	0.6	0	
		6000–37,000	1–5000	–	0.375	0.2	0	
		$37{,}000-10^8$	1–20		0.130	0.3	0	
		$> 10^8$	1–20		0.057	1/3	0	
Gas or liquid	Verical annulus	Same as vertical plates						
	Horizontal annulus, isothermal	$6000-10^6$	1–5000	–	0.11	0.29	0	[18–20]
		$10^6 - 10^8$	1–5000	–	0.40	0.20	0	
	Spherical annulus	$120-1.1 \cdot 10^9$	0.7–4000	–	0.228	0.226	0	[7]

Coefficients a_{ij} can be found in Table 6.4.

Also, the enhancement in heat transfer can be computed as:

$$En = b_{13} + b_{23}Y_1 + b_{33}Y_2 + b_{43}Y_1^2 + b_{53}Y_2^2 + b_{63}Y_1Y_2 \tag{6.2}$$

where

$$Y_1 = b_{11} + b_{21}Ra^* + b_{31}Ha^* + b_{41}Ra^{*2} + b_{51}Ha^{*2} + b_{61}Ra^*Ha^*$$

$$Y_2 = b_{12} + b_{22}Ra^* + b_{32}Mn_F^* + b_{42}Ra^{*2} + b_{52}Mn_F^{*2} + b_{62}ReMn_F^*$$

$$Ra^* = \log(Ra), Ha^* = \frac{Ha}{10}, Mn_F^* = \frac{Mn_F}{100}$$

Coefficients b_{ij}can be found in Table 6.5.

Sheikholeslami and Shehzad [22] investigated the effect of nonuniform magnetic field on magnetic nanofluid behavior (Fe_3O_4—H_2O) in a porous enclosure using CVFEM. The numerical analyses were performed in the following conditions: Darcy number $0.01 \leq Da \leq 100$, radiation parameter $0 \leq Rd \leq 0.8$, Rayleigh number $10^3 \leq Ra \leq 10^5$, volume fraction of nanoparticles $0 \leq \phi \leq 0.04$, and Hartmann number $0 \leq Ha \leq 10$. They considered the magnetic field-dependent viscosity of nanofluid and also the shape factor effect (spherical, platelet, brick, and cylinder) on the thermal conductivity of the nanofluid.

They found that the platelet shape of nanoparticles leads to an increase in Nusselt number, but that the increase of Lorentz forces, decreases the Nusselt number. Also, the results indicated that as the Darcy number increases, the convective heat transfer convective mode becomes significant and the heat transfer rate enhances.

For the Nusselt number a new correlation is proposed:

$$\begin{aligned} Nu_{ave} = {} & 15.28 - 1.12Rd - 6.6\log(Ra) - 0.18Da^* + 0.47Ha^* + 1.07Rd\log(Ra) + 0.32RdDa^* \\ & - 0.44RdHa^* + 0.08Da^*\log(Ra) - 0.11Ha^*\log(Ra) - 0.3Da^*Ha^* - 0.9Rd^2 \\ & + 0.85(\log(Ra))^2 + 0.15(Da^*)^2 - 0.07(Ha^*)^2. \end{aligned} \tag{6.3}$$

Table 6.4 Coefficients used in Eq. (6.1) [21].

$a_{i,j}$	$i=1$	$i=2$	$i=3$	$i=4$	$i=5$	$i=6$
$j=1$	19.51647	−10.0868	−0.00804	1.525112	0.001086	0.003349
$j=2$	19.53085	−0.87952	−10.1291	−0.03518	1.525112	2.186806
$j=3$	−0.29704	1.107821	0.654435	−0.00017	0.041509	−0.24967
...1 $j=4$	−0.15202	0.647229	1.043281	0.044365	0.005999	−0.24883
...2 $j=5$	−0.07156	0.161803	0.861233	1.818237	1.846604	−3.66821

Table 6.5 Coefficients used in Eq. (6.2) [21].

b_{ij}	$i=1$	$i=2$	$i=3$	$i=4$	$i=5$	$i=6$
$j=1$	21.33634	−5.71464	0.798586	0.542492	0.064253	−0.21888
$j=2$	21.72597	−5.81042	−0.0488	0.542492	−0.01392	0.006107
$j=3$	0.061649	4.86387	−3.87977	−0.38679	0.13164	0.25619

where

$$Da^* = 0.01Da, Ha^* = 0.1Ha$$

Sheikholeslam and Shamlooei [23] studied the influence of thermal radiation on Fe_3O_4 nanofluid hydrothermal treatment in a curved cavity under the influence of external magnetic source using CVFEM. The effects of radiation parameter (Rd), Hartmann number (Ha), Rayleigh number (Ra), and volume fraction of Fe_3O_4 are investigated. The numerical analyses are performed for $Ec = 10^{-5}$ and $Pr = 6.8$, respectively. Their results demonstrated that at higher buoyancy forces, the influence of thermal radiation on convection heat transfer is more sensible. Thermal boundary layer thickness increases with increasing Rd, Ha, while it reduces with increasing Ra and ϕ. Nanofluid velocity reduced with increasing Hartmann number, but with increasing radiation parameter, an increase in velocity is noticed. Plus, they noticed that if Lorentz forces increase, the average Nusselt number decreases due to the domination of conduction heat transfer and also that by adding Fe_3O_4 nanoparticle in the base fluid, Nu_{ave} increases.

Table 6.6 shows the impact of Ha, Rd, and Ra on heat transfer improvement, $En = \left[\left(Nu_{MNf} - Nu_{bf}\right)/Nu_{bf}\right]\bullet 100$.

A correlation for Nu_{ave} is developed:

$$\begin{aligned} Nu_{ave} = {} & 6.86 - 2.5\log(Ra) + 1.5Rd + 0.05Ha + 3.34\phi + 0.69\log(Ra)Rd - 0.04\log(Ra)Ha + 2.38\log(Ra)\phi \\ & + 33\bullet 10^{-3}RdHa + 9.5Rd\phi - 0.11Ha\phi + 0.44(\log(Ra))^2 + 2.4Rd^2 + 0.5\bullet 10^{-4}Ha^2 + 102.3\phi^2 \end{aligned} \tag{6.4}$$

In another paper, Sheikholeslami and Ganji [24] numerically investigated the influence of external magnetic source on natural convection of Fe_3O_4–water nanofluid in an enclosure with a square hot cylinder. To solve the governing equations CVFEM method is used. Geometry and imposed boundary conditions are illustrated in Fig. 6.2.

They found that the temperature gradient increases with increasing inner square size r_{in}. Also, temperature increases with increasing Lorentz forces, but it reduces with both increasing volume

Table 6.6 Influence of *Ra*, *Rd*, and *Ha* on heat transfer improvement [23].

Ra	*Rd*	*Ha*	*En*
10^3	0	0	20.04866
10^3	0	20	20.06985
10^3	0.4	0	16.37888
10^3	0.4	20	16.31981
10^4	0	0	17.00093
10^4	0	20	19.85959
10^4	0.4	0	13.61591
10^4	0.4	20	16.22274
10^5	0	0	14.38823
10^5	0	20	14.64045
10^5	0.4	0	10.93439
10^5	0.4	20	11.57623

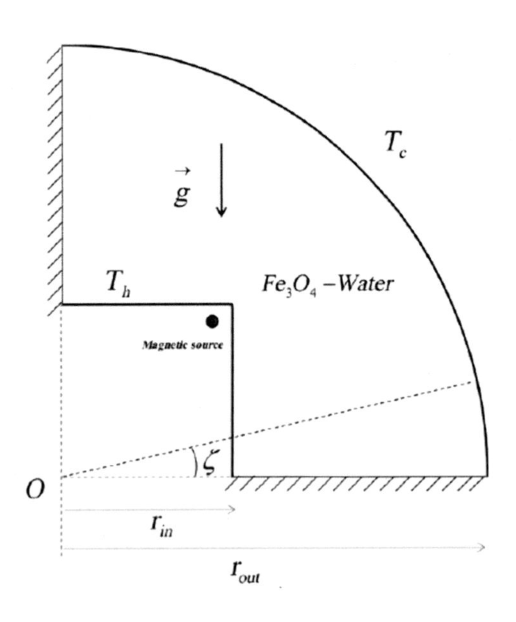

FIGURE 6.2

Schematic of the physical problem.

Reprinted from M. Sheikholeslami, D.D. Ganji, Free convection of Fe_3O_4–water nanofluid under the influence of an external magnetic source, J. Mol. Liq. 229 (2017) 530–540, Copyright 2017, with permission from Elsevier.

fraction of Fe_3O_4 and buoyancy forces. Nu_{ave} enhances with increasing r_{in} because of the decrease in distance between cold and hot walls. Increasing Lorentz forces leads to reduction of Nu_{ave} due to the domination of conduction heat transfer. For the average Nusselt number a correlation is proposed:

$$\begin{aligned} Nu_{ave} = {} & 12.12 - 2.27r_i - 7.27(\log(Ra)) - 1.29\phi + 0.16Ha + 1.5r_{in}(\log(Ra)) + 0.6r_{in}\phi \\ & - 0.04r_{in}Ha + 0.78(\log(Ra))\phi - 0.04(\log(Ra))Ha - 0.02\phi Ha \\ & + 2r_{in}^2 1.04(\log(Ra))^2 14.05\phi^2 + 1.5 \bullet 10^{-5} Ha^2 \end{aligned} \tag{6.5}$$

Sheikholeslami et al. [25] studied the effects of nonuniform magnetic field (nonuniform Lorentz force) and shape factor on Fe_3O_4–water ferrofluid flow in a porous cavity using CVFEM. The studied geometry and the imposed boundary conditions are illustrated in Fig. 6.3.

The simulations were carried out in the following conditions: radiation parameter $0 \leq Rd \leq 0.8$, Darcy number $0.01 \leq Da \leq 100$, Rayleigh number $10^3 \leq Ra \leq 10^5$, volume fractions of Fe_3O_4 nanoparticles $0 \leq \phi \leq 0.04$, Hartmann number $0 \leq Ha \leq 10$, and shape of nanoparticles: platelet, cylinder, brick and spherical. The viscosity of Fe_3O_4 ferrofluid with an external magnetic source was considered:

$$\mu_{nf} = \left(0.035\mu_0^2\overline{H}^2 + 3.1\mu_0\overline{H} - 27886.4807\phi^2 + 4263.02\phi + 316.0629\right)e^{-0.01T} \tag{6.6}$$

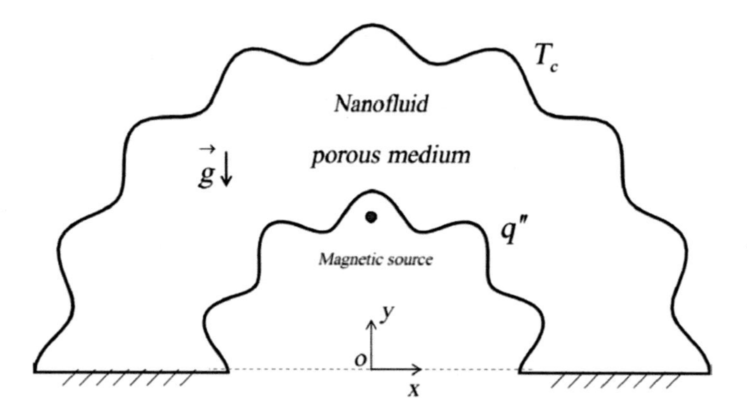

FIGURE 6.3

Schematic of the physical problem.

Reprinted from M. Sheikholeslami, D.D. Ganji, R. Moradi, Heat transfer of Fe_3O_4–water nanofluid in a permeable medium with thermal radiation in existence of constant heat flux, Chem. Eng. Sci. 174 (2017) 326–336, Copyright 2017, with permission from Elsevier.

Table 6.7 Influence of nanoparticles shape on Nusselt number for $Da = 100$, $Ra = 10^5$, $Rd = 0.8$, $\phi = 0.04$ [25].

	Ha	
Nanoparticles shape	0	10
Spherical	8.754093	6.893915
Brick	8.802073	6.930307
Cylinder	8.87772	6.987786
Platelet	8.940132	7.034996

They found that by increasing the Hartmann number, the velocity of nanofluid and heat transfer rate decreases and also that the maximum values of the Nusselt number are achieved by utilizing platelet-shaped nanoparticles (Table 6.7). Finally, a new relation for the average Nusselt number is proposed:

$$\begin{aligned} Nu_{ave} = {} & 22.3 - 3.13Rd - 10.5\log(Ra) - 0.43Da^* + 1.03Ha^* + 1.88Rd\log(Ra) \\ & + 0.36RdDa^* - 0.54RdHa^* + 0.19Da^*\log(Ra) - 0.28Ha^*\log(Ra) - 0.52Da^*Ha^* \\ & - 2.5Rd^2 + 1.37(\log(Ra))^2 + 0.29(Da^*)^2 - 0.17(Ha^*)^2 \end{aligned} \tag{6.7}$$

where $Da^* = 0.01Da$ and $Ha^* = 0.1Ha$.

The influence of variable magnetic field on nanofluid convective nonDarcy flow inside a porous cavity using CVFEM was numerical investigated by Sheikholeslami and Sadoughi [26]. Geometry and the boundary conditions are described in Fig. 6.4.

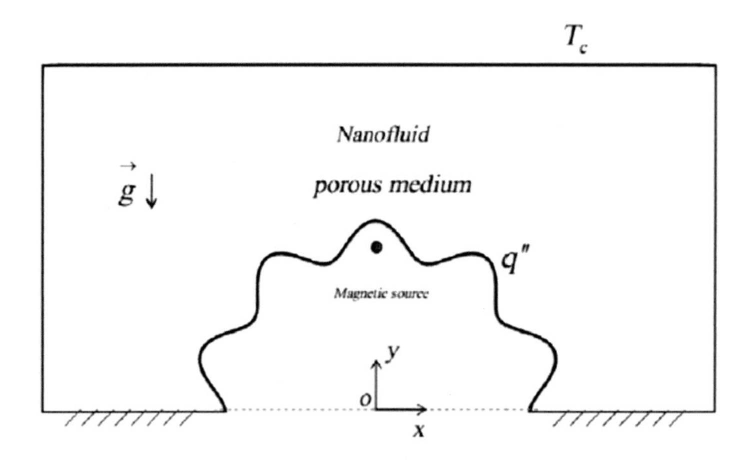

FIGURE 6.4

Schematic of the physical problem.

Reprinted from M. Sheikholeslami, M.K. Sadoughi, Numerical modeling for Fe_3O_4–water nanofluid flow in porous medium considering MFD viscosity, J. Mol. Liq. 242 (2017) 255–264, Copyright 2017, with permission from Elsevier.

Table 6.8 Influence of shape of nanoparticles on the Nusselt number for $Da = 100$, $Ra = 10^5$, $Rd = 0.8$, $\phi = 0.04$ [26].

	Ha	
Nanoparticles shape	0	10
Spherical	9.40168	7.463479
Brick	9.469653	7.504079
Cylinder	9.575843	7.568335
Platelet	9.662061	7.621197

Results revealed that the Nusselt number decreases with increasing Lorentz forces and also, that the heat transfer rate increases with increasing permeability of porous media and Rayleigh number due to an increase in convective heat transfer. And in this paper, the platelet shape of nanoparticles leads to the Nusselt number being increased (Table 6.8). The proposed relation for the average Nusselt number is given by Eq. (6.8):

$$\begin{aligned} Nu_{ave} = {} & 21.6 - 3.79Rd - 10.44\log(Ra) - 0.41Da^* + 1.02Ha^* + 2.14Rd\log(Ra) + 0.33RdDa^* \\ & - 0.44RdHa^* + 0.22Da^*\log(Ra) - 0.29Ha^*\log(Ra) - 0.72Da^*Ha^* - 2.98Rd^2 \\ & + 1.39(\log(Ra))^2 + 0.45(Da^*)^2 - 0.23(Ha^*)^2 \end{aligned} \tag{6.8}$$

where $Da^* = 0.01Da$ and $Ha^* = 0.1Ha$

Influence of external magnetic source natural convection of Fe_3O_4−water magnetic nanofluid in a porous curved cavity using CVFEM was also studied by Sheikholeslami [27]. Fig. 6.5 illustrates the geometry and the boundary conditions.

The results revealed an increase in the average Nusselt number with increasing Darcy number, due to an increase in temperature gradient near the curved wall, while a similar trend is noticed for buoyancy forces (the increase in Nusselt number with increase in Rayleigh number). Also, the increase in Lorentz forces leads to a decrease in Nusselt number due to the domination of conduction heat transfer. Effects of Da, Ha, and Ra numbers on heat transfer enhancement, defined as $En = \left[\left(Nu_{MNf} - Nu_{bf}\right)/Nu_{bf}\right]$ 100, are shown in Table 6.9.

Krakov and Nikiforov [28] analytically and numerically studied the mutual effect of gravity and the magnetic field on the onset of convection. The geometry of the cavity filled with a magnetic nanofluid is illustrated in Fig. 6.6. A uniform magnetic field $\vec{H}_\infty$ and a temperature gradient $\vec{\gamma}_\infty$ are imposed on the outer boundaries of the domain.

Their results revealed that the convective flow affects magnetic field structure in the volume of a nonconducting magnetic fluid and that the outer uniform magnetic field is turned to nonuniform magnetic field in the heated magnetic fluid.

In another paper, Sheikholeslami and Shehzad [29] numerically studied the influence of variable magnetic field on nanofluid convective nonDarcy flow using CVFEM method. Fe_3O_4−H_2O nanofluid was used in a porous cavity. Geometry and boundary conditions are illustrated in Fig. 6.7.

Their results demonstrated that the maximum values of Nusselt number are achieved for platelet-shaped nanoparticles (Table 6.10) and also, that the Darcy and Rayleigh numbers may enhance the temperature gradient. Moreover, they found that the temperature gradient decreases with increasing Hartmann number.

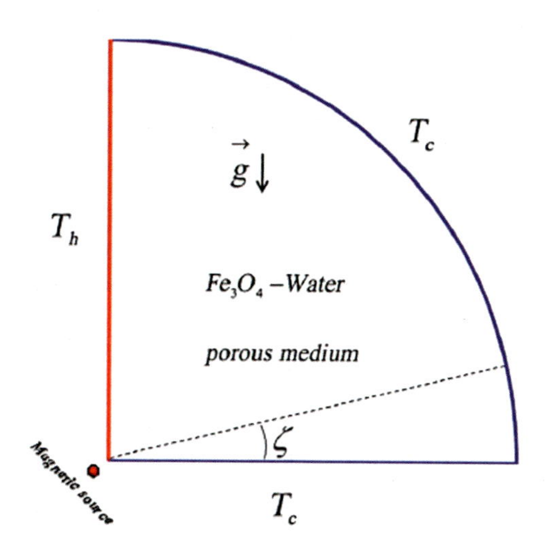

FIGURE 6.5

Schematic of the physical problem.

Reprinted from M. Sheikholeslami, Numerical simulation of magnetic nanofluid natural convection in porous media, Phys. Lett. A 381 (2017) 494–503, Copyright 2017, with permission from Elsevier.

Table 6.9 Influence of Da, Ha, and Ra on heat transfer enhancement (En) [27].

Ra	Da	Ha	En
10^3	0.01	0	9.113388
10^3	0.01	40	9.180014
10^4	0.01	0	8.528405
10^4	0.01	40	8.832767
10^5	0.01	0	9.396332
10^5	0.01	40	8.235231
10^3	100	0	8.706959
10^3	100	40	9.170092
10^4	100	0	7.511918
10^4	100	40	8.737434
10^5	100	0	9.150151
10^5	100	40	8.044214

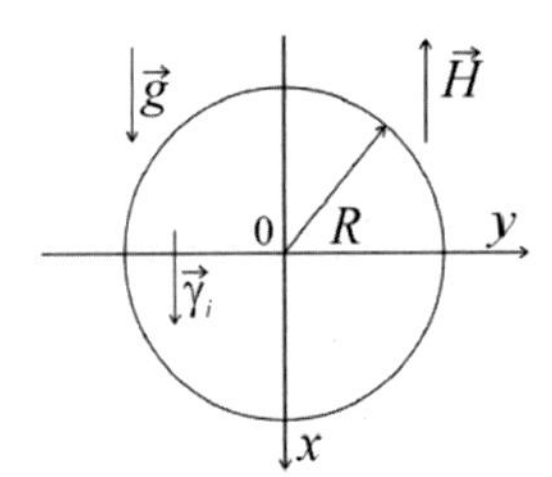

FIGURE 6.6

Schematic of the physical problem.

Reprinted from M.S. Krakov, I.V. Nikiforov, Natural convection in a horizontal cylindrical enclosure filled with a magnetic nanofluid: influence of the uniform outer magnetic field, Int. J. Therm. Sci. 133 (2018) 41–54, Copyright 2018, with permission from Elsevier.

A correlation for the average Nusselt number as a function of *Ra*, *Da*, and *Rd* is proposed:

$$\begin{aligned} Nu_{ave} = {} & 32.46 - 6.1Rd - 15.8\log(Ra) - 0.58Da^* + 1.3Ha^* + 3.2Rd\log(Ra) + 0.4RdDa^* \\ & - 0.52RdHa^* + 0.31Da^*\log(Ra) - 0.37Ha^*\log(Ra) - 0.94Da^*Ha^* - 4.8Rd^2 \\ & + 2.1(\log(Ra))^2 + 0.61(Da^*)^2 - 0.37(Ha^*)^2 \end{aligned} \tag{6.9}$$

where $Da^* = 0.01Da$, $Ha^* = 0.1Ha$

Sheikholeslami et al. [30] investigated the influence of variable external magnetic field and thermal radiation on heat transfer intensification of nanofluid in a porous curved enclosure considering the shape effect of Fe_3O_4 nanoparticles. Fig. 6.8 shows the studied geometry and imposed boundary conditions.

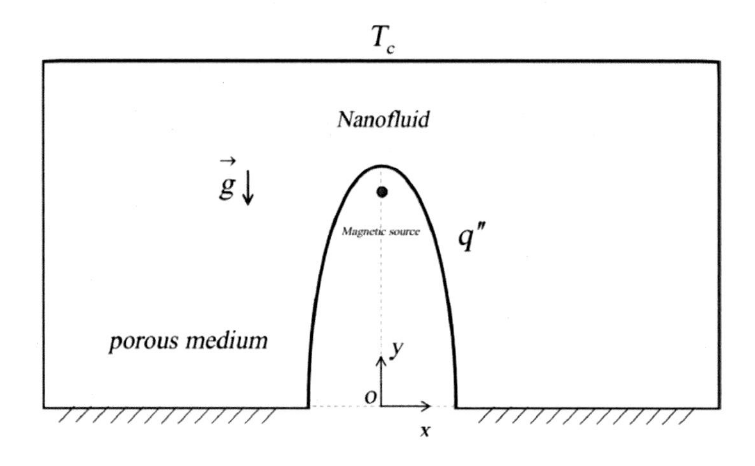

FIGURE 6.7

Schematic of the physical problem.

Reprinted from M. Sheikholeslami, S.A. Shehzad, Numerical analysis of Fe3O4–H2O nanofluid flow in permeable media under the effect of external magnetic source, Int. J. Heat Mass Transf. 118 (2018) 182–192, Copyright 2018, with permission from Elsevier.

Table 6.10 The influence of nanoparticles shape on the Nusselt number for $Da = 100$, $Ra = 10^5$, $Rd = 0.8$, and $\phi = 0.04$ [29].

	Ha	
Nanoparticles shape	0	10
Spherical	13.20747	10.67737
Brick	13.28049	10.72757
Cylinder	13.39638	10.80650
Platelet	13.4921	10.87105

The results revealed that the maximum Nusselt number is achieved for platelet-shaped nanoparticles (Table 6.11) similar to in the abovementioned papers [25,26,29]. The Nusselt number increases with increasing radiation parameter. The heat transfer rate enhances with the increasing permeability of porous media and the Rayleigh number and opposite trend for Hartmann number is observed. Plus, it can be noticed that the nanofluid velocity reduces with increasing Lorentz forces.

The proposed correlation for the average Nusselt number is given by Eq. (6.10):

$$\begin{aligned} Nu_{ave} = {} & 25.77 + 3.38Rd - 7.84\log(Ra) - 0.11Da^* + 0.23Ha^* + 1.09Rd\log(Ra) + 0.29RdDa^* \\ & - 0.22RdHa^* + 0.027Da^*\log(Ra) - 0.02Ha^*\log(Ra) - 0.114Da^*Ha^* + 2.52Rd^2 \\ & + 1.00(\log(Ra))^2 + 0.23(Da^*)^2 - 0.08(Ha^*)^2 \end{aligned} \tag{6.10}$$

where $Da^* = 0.01Da$, $Ha^* = 0.1Ha$.

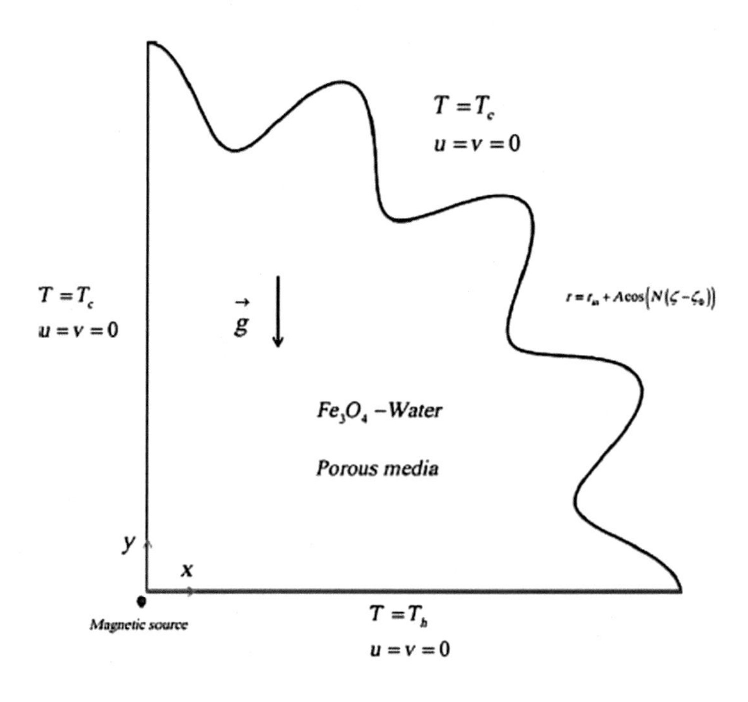

FIGURE 6.8

Schematic of the physical problem.

Reprinted from M. Sheikholeslami, M. Shamlooei, R. Moradi, Numerical simulation for heat transfer intensification of nanofluid in a porous curved enclosure considering shape effect of Fe_3O_4 nanoparticles, Chem. Eng. Processing: Process Intensif. 124 (2018) 71–82, Copyright 2018, with permission from Elsevier.

Table 6.11 The influence of nanoparticles shape on the Nusselt number for $Da = 100$, $Ra = 10^5$, $Rd = 0.8$, **and** $\phi = 0.04$ **[30].**

	Ha	
Nanoparticles shape	0	10
Spherical	20.71733	19.22733
Brick	20.91845	19.41302
Cylinder	21.23224	19.70666
Platelet	21.48711	19.94852

A half annulus cavity filled with Fe_3O_4–water nanofluid subject to the constant heat flux and the variable magnetic field was analyzed by Hatami et al. [31] using FlexPDE commercial code (Fig. 6.9).

The results revealed that at low Eckert numbers, an increase in Hartmann number leads to a decrease in the Nusselt number. This is attributed to the Lorentz force resulting from the presence of a stronger magnetic field.

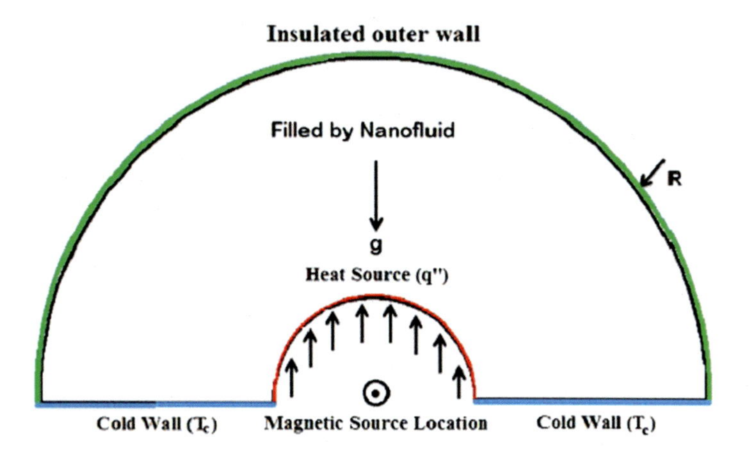

FIGURE 6.9

Schematic of the physical problem.

Reprinted from M. Hatami, J. Zhou, J. Geng, D. Jing, Variable magnetic field (VMF) effect on the heat transfer of a halfannulus cavity filled by Fe_3O_4–water nanofluid under constant heat flux, J. Magnetism Magnetic Mater. 451 (2018) 173–182, Copyright 2018, with permission from Elsevier.

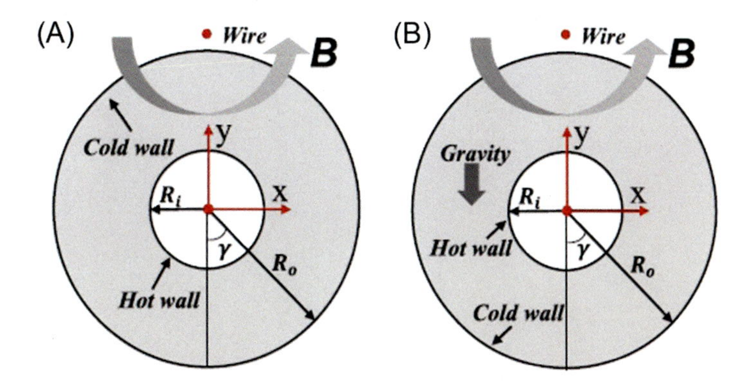

FIGURE 6.10

Schematic of the physical problem: (A) heat conduction of a stationary nanofluid in an annulus; (B) natural convection heat transfer in an annulus. Both problems apply a uniform or a nonuniform external magnetic field.

Reprinted from X.-H. Sun, M. Massoudi, N. Aubry, Z.-H. Chen, W.-T. Wu, Natural convection and anisotropic heat transfer in a ferro-nanofluid under magnetic field, Int. J. Heat Mass Transf. 133 (2019) 581–595, Copyright 2019, with permission from Elsevier.

The influence of anisotropic thermal conductivity and the Lorentz force in a magnetic nanofluid induced by nonuniform magnetic field within in a horizontal annulus enclosure was studied by Sun et al. [32]. The study is divided into two parts: in the first part, they analyzed the conduction heat transfer in an annulus enclosure, see Fig. 6.10A; and in the second part, they analyzed the same geometry in which the buoyancy force due to gravity and temperature gradient are added, see Fig. 6.10B.

They found that the isotherms become elliptic due to the anisotropic thermal conductivity, their shape changing as the direction of the magnetic field varied. As the Rayleigh number increases, it is noticed that the thermal performance is improved due to the intensification of the natural convection. As the Rayleigh number increases, it is noted that the thermal performance is improved due to the intensification of the natural convection. Also, depending on the situation, the magnetic field (Hartmann number) may increase or reduce the Nusselt number.

The study of the convective heat transfer on MWCNT–Fe_3O_4/water hybrid nanofluid in a porous medium using two variable magnetic sources was performed by Izadi et al. [33]. Configuration of the computation domain and boundary conditions are depicted in Fig. 6.11.

Results indicated that at a low Ra ($Ra = 10^4$), the average Nusselt number, Nu_{ave}, increased with both increasing magnetism number Mn_f and strength ratio γ_r values. This is due to the effects of the increase in the Kelvin forces acting on the nanofluid particles and the change in the streamline pattern acting on them. In contrast, at a high $Ra(Ra = 10^6)$, the Nu_{ave} remained unchanged with increasing Mn_f at the low values of γ_r (Fig. 6.12).

The effect of variable magnetic forces on magnetizable hybrid nanofluid (MWCNT–Fe_3O_4/H_2O) heat transfer inside a circular cavity with two circular heaters was studied by Sheikholeslami et al. [34]. Computation domain and boundary conditions are displayed in Fig. 6.13. The analyses

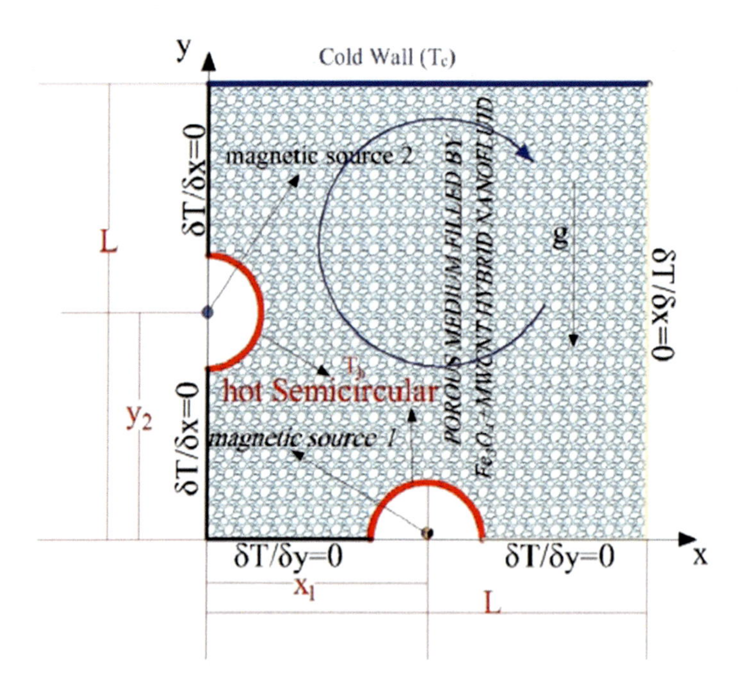

FIGURE 6.11

Schematic of the physical problem.

Reprinted from M. Izadi, R. Mohebbi, A.A. Delouei, H. Sajjadi, Natural convection of a magnetizable hybrid nanofluid inside a porous enclosure subjected to two variable magnetic fields, Int. J. Mech. Sci. 151 (2019) 154–169, Copyright 2019, with permission from Elsevier.

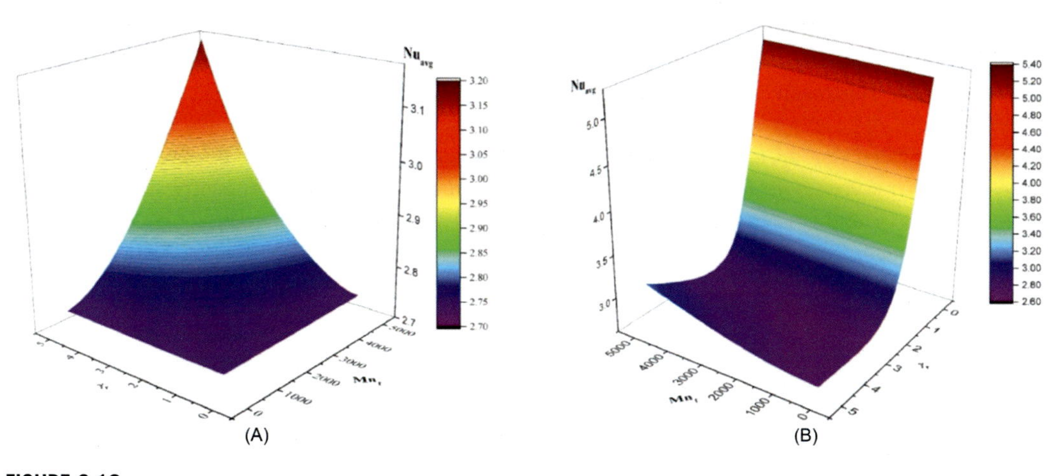

FIGURE 6.12

Average Nusselt number: (A) $\mathrm{Ra} = 10^4$, (B) $\mathrm{Ra} = 10^6$ as a function of γ_r and $\mathrm{Mn_f}$.

Reprinted from M. Izadi, R. Mohebbi, A.A. Delouei, H. Sajjadi, Natural convection of a magnetizable hybrid nanofluid inside a porous enclosure subjected to two variable magnetic fields, Int. J. Mech. Sci. 151 (2019) 154–169, Copyright 2019, with permission from Elsevier.

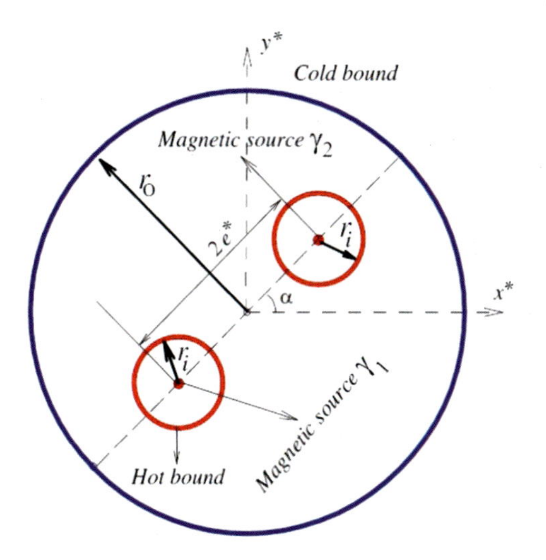

FIGURE 6.13

Schematic of the physical problem.

Reprinted from M. Sheikholeslami, S.A.M. Mehryan, A. Shafee, M.A. Sheremet, Variable magnetic forces impact on magnetizable hybrid nanofluid heat transfer through a circular cavity, J. Mol. Liq. 277 (2019) 388–396, Copyright 2019, with permission from Elsevier.

were carried out for following parameters: Rayleigh number, $10^4 \leq Ra \leq 10^6$; location angle of heaters, $0 \leq \alpha \leq \pi/2$; the magnetic strengths ratio parameter, $0.2 \leq \gamma_r \leq 5$; magnetic number $0 \leq Mn_f \leq 1000$; Hartmann number, $0 \leq Ha \leq 50$; and nanocomposite particles concentration, $0 \leq \phi \leq 0.3\%$.

Results indicated that an increase in Ha leads to increasing boundary layer thickness as a result of a reduction in convection. Plus, they found that the convection reduces with both increasing γ_r and Ha, and also that the increase in Ra decreases the heat transfer rate (Table 6.12).

6.2.3 Natural convection subject to uniform magnetic field

6.2.3.1 Natural convection in various cavity (square cavity, rectangular cavity C-shape cavity)

Shi et al. [35] studied the hybrid nanofluids, Fe_3O_4- CNT/water, in a rectangular enclosure under controllable magnetic fields. Fig. 6.14 illustrates the schematic of the experimental setup using an external magnetic field together with the rectangular simulation cavity.

The results of this study are described in Tables 6.13–6.15. As can be seen in Tables 6.13–6.15 the heat transfer is enhanced with increasing magnetic field strength.

Two correlations for the Nusselt number as a function of heating power ($\dot{Q}[W]$) and magnetic field ($H[mT]$) are proposed:

$$Nu = 0.15069\dot{Q} + 6.67555,\ R^2 = 0.97027 \tag{6.11}$$

respectively,

$$Nu = 9.121 - 1.889e^{\frac{0.875-H}{15.014}},\ R^2 = 0.983 \tag{6.12}$$

Jafarimoghaddam [36] applied a two-phase modeling framework to simulate magnetic nanofluids jets flow over a stretching/shrinking wall and found that for the momentum equation, an imposed moving wall condition together with a transverse magnetic field is enough to obtain e-jet Glauert type solution.

Table 6.12 Effect of the volume fraction for different values of Ra on Nu_{ave} in the case $\gamma_r = 0.2$, $\alpha = \pi/4$, and $Ha = 25$ [34].

Ra	ϕ	Nu_{ave}	En
10^4	0.0	2.5465	–
10^4	0.1	2.7543	8.16
10^4	0.3	2.9390	15.41
10^5	0.0	4.7435	–
10^5	0.1	4.9327	3.99
10^5	0.3	4.8280	1.78
10^6	0.0	6.1829	–
10^6	0.1	6.5280	5.58
10^6	0.3	6.7164	8.63

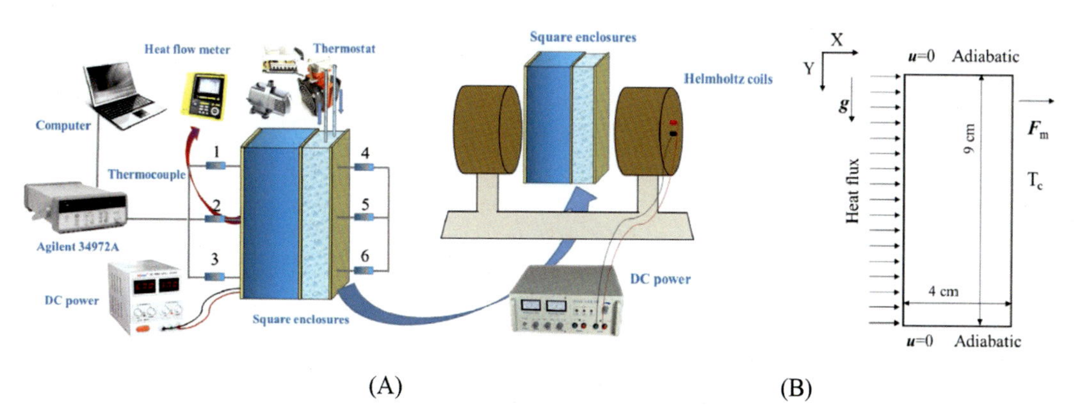

FIGURE 6.14

(A) Schematic of the experimental setup. (B) Schematic of the rectangular simulation cavity.

Reprinted from L. Shi, Y. He, Y. Hu, X. Wang, Controllable natural convection in a rectangular enclosure filled with Fe_3O_4@CNT nanofluids, Int. J. Heat Mass Transf. 140 (2019) 399–409, Copyright 2019, with permission from Elsevier.

Table 6.13 Average heat transfer coefficient, *h*, and Nusselt number, *Nu*, in the absence of magnetic field [35].

		$Fe_3O_4-CNT(4:1)$				$Fe_3O_4-CNT(2:1)$			
$\dot{Q}$[W]		0mT	5mT	10mT	20mT	0mT	5mT	10mT	20mT
11.357	$h[W/m^2K]$	147.9	147.9	152.6	155.1	149.3	155.3	163.1	167.1
	Nu	9.55	9.55	9.85	10.12	9.31	10.03	10.52	10.72
$\dot{Q}[W]$		$Fe_3O_4-CNT(1:1)$				Water			
		$0mT$	$5mT$	$10mT$	$20mT$	–			
11.357	$h[W/m^2K]$	149.6	156.4	164.8	168.4	128			
	Nu	9.32	10.10	10.57	10.78	8.33			

Influences of rotation angle and metal foam on natural convection of Fe_3O_4–water nanofluids in a rectangular cavity under an adjustable magnetic field were experimental investigated by Qi et al. [37]. Experiments were carried out in the following conditions: nanoparticle mass concentrations $0 \leq \omega \leq 0.5\%$, magnetic field directions (horizontal and vertical), magnetic field intensities $0 \leq B \leq 0.02$T, rotation angles of the cavity $0 \leq \alpha \leq 135^\circ$, and pore size (PPI) of Cu metal foam $0 \leq PPI \leq 15$. The schematic diagram of a free convection system without and with metal foam is presented in Figs. 6.15 and 6.16.

Results indicated that an increase in mass concentration of nanoparticles leads to an initial increase in Nusselt number, after which a decrease is noticed; the maximum value is obtained at a nanoparticle mass concentration $\omega = 0.3\%$.

Table 6.14 Average heat transfer, *h*, coefficient and Nusselt number, *Nu*, for various heating power [35].

$\dot{Q}[W]$		$Fe_3O_4-CNT(4:1)$	$Fe_3O_4-CNT(2:1)$	$Fe_3O_4-CNT(1:1)$	Water
3.528	$h[W/m^2K]$	116.6	119.4	120.3	113.1
	Nu	7.41	7.62	7.68	7.36
7.06	$h[W/m^2K]$	126.6	136.1	140.7	115.7
	Nu	8.13	8.43	8.48	7.58
11.357	$h[W/m^2K]$	131.6	149.3	149.6	128.3
	Nu	8.96	9.31	9.32	8.33
14.364	$h[W/m^2K]$	132.0	149.6	151.4	136.9
	Nu	9.24	9.51	9.52	8.91

Table 6.15 Average heat transfer, *h*, coefficient and Nusselt number, *Nu*, for various heating power and different magnetic fields [35].

		ΔH					
$\dot{Q}[W]$		$0mT$	$5mT$	$10mT$	$20mT$	$30mT$	$40mT$
3.528	$h[W/m^2K]$	119.4	125.5	130.7	138.6	140.2	141.8
	Nu	7.61	8.03	8.44	8.81	9.02	9.12
7.06	$h[W/m^2K]$	131.6	135.5	141.1	146.2	148.9	152.6
	Nu	8.4 0	8.63	9.08	9.31	9.36	9.82
9.246	$h[W/m^2K]$	138.3	141.2	145.6	153.2	160.1	164.4
	Nu	8.7 0	9.10	9.29	9.91	10.23	10.54
11.357	$h[W/m^2K]$	149.3	155.4	163.1	167.1	172.1	177.7
	Nu	9.38	9.97	10.29	10.44	10.74	11.11
14.364	$h[W/m^2K]$	152.2	162.4	168.1	172.4	178.0	180.3
	Nu	9.49	10.16	10.51	10.81	11.14	11.23

This study emphasized two main conclusions: applying a vertical magnetic field leads to thermal performance enhancement, while by applying a horizontal magnetic field an insignificant enhancement in thermal performance is noted; the cavity with a rotation angle of 90° having the highest thermal performance. Also, the values of Nusselt number for the cavity filled with metal foam are higher compared to those without metal foam, but an increase in PPI of metal foam can lead to a deterioration in thermal performance.

Unsteady natural convective heat transfer flow inside a square enclosure filled with magnetic nanofluids using nonhomogeneous dynamic model was studied by Al-Balushi et al. [38]. The average Nusselt number and the heat transfer enhancement for different magnetic nanoparticles $0.00 \le \phi \le 0.05$ and base fluids (water, engine oil, kerosene) are presented in Table 6.16. It is seen

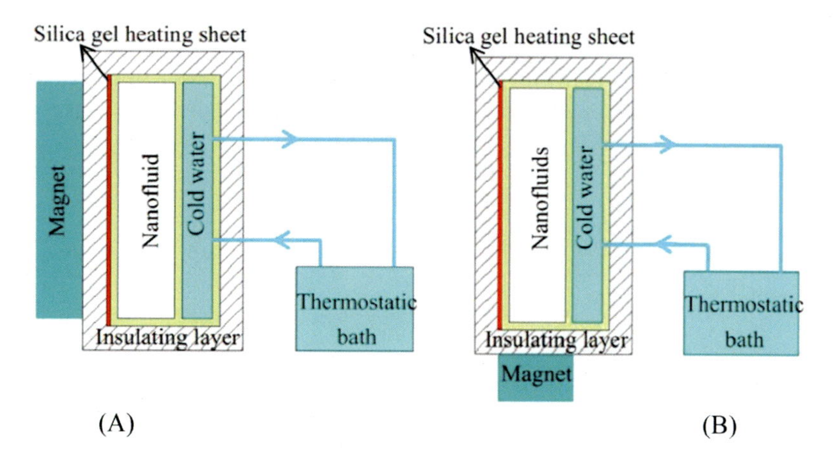

FIGURE 6.15

Schematic diagram of system without metal foam, (A) horizontal (rightward) magnetic field, (B) vertical (downward) magnetic field.

Reprinted from C. Qi, J. Tang, Z. Ding, Y. Yan, L. Guo, Y. Ma, Effects of rotation angle and metal foam on natural convection of nanofluids in a cavity under an adjustable magnetic field, Int. Commun. Heat Mass Transf. 109 (2019) 104349, Copyright 2019, with permission from Elsevier.

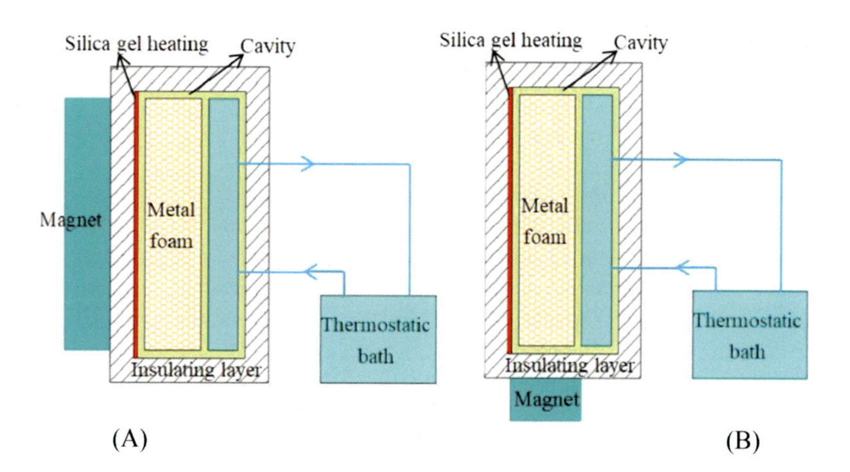

FIGURE 6.16

Schematic diagram of system with metal foam, (A) horizontal (rightward) magnetic field, (B) vertical (downward) magnetic field.

Reprinted from C. Qi, J. Tang, Z. Ding, Y. Yan, L. Guo, Y. Ma, Effects of rotation angle and metal foam on natural convection of nanofluids in a cavity under an adjustable magnetic field, Int. Commun. Heat Mass Transf. 109 (2019) 104349, Copyright 2019, with permission from Elsevier.

Table 6.16 The average Nusselt number for different volume fraction for local thermal Rayleigh number $Ra_T = 10^5$, local solutal Rayleigh number $Ra_c = 10$, the empirical shape factor of nanoparticles $n = 3$, nondimensional time $\tau = 5$, diameter of nanoparticle $d_p = 10nm$ [38].

Nanofluids	ϕ	Nu_{ave}	Increase %	Nanofluids	ϕ	Nu_{ave}	Increase %
$Fe_3O_4-H_2O$	0.00	4.0993	–	$Mn-ZnFe_2O_4-EO$	0.00	4.2194	–
	0.01	6.3325	54.5		0.01	4.7196	11.9
	0.02	8.9083	117.3		0.02	5.2354	24.1
	0.05	18.257	354.2		0.05	6.8650	62.7
$CoFe_2O_4-H_2O$	0.00	4.0993	–	Fe_3O_4-Ke	0.00	4.1809	–
	0.01	6.1409	49.8		0.01	12.2591	193.2
	0.02	8.4609	106.4		0.02	22.9944	450
	0.05	16.343	298.7		0.05	62.1741	1387.1
$Mn-ZnFe_2O_4-H_2O$	0.00	4.0993	–	$CoFe_2O_4-Ke$	0.00	4.1809	–
	0.01	6.4571	57.5		0.01	12.0820	189
	0.02	9.1872	124.1		0.02	22.4682	437.4
	0.05	18.519	351.8		0.05	66.1783	1482.9
Fe_3O_4-EO	0.00	4.2194	–	$Mn-ZnFe_2O_4-Ke$	0.00	4.1809	–
	0.01	4.7188	11.8		0.01	13.4326	221.3
	0.02	5.2360	24.1		0.02	25.8115	517.4
	0.05	6.9065	63.7		0.05	79.6022	1804
$CoFe_2O_4-EO$	0.00	4.2194	–				
	0.01	4.6706	10.7				
	0.02	5.1335	21.7				
	0.05	6.5900	56.2				

from Table 6.16 that $Mn-ZnFe_2O_4-Ke$ nanofluid has a higher average Nusselt number than other types of nanofluids.

Influence of uniform magnetic induction on heat transfer performance of aqueous ($Fe_2O_3-Al_2O_3$) hybrid ferrofluid in a rectangular cavity under magnetic induction was investigated by Giwa et al. [39]. Without considering the magnetic induction on the cavity walls, enhancement in h_{ave}, Nu_{ave}, and $\dot{Q}_{ave}$ are noted for concentrations in the range 0.05–0.2 vol.% compared to the base fluid. Maximum enhancement in heat transfer is 10.81% for a concentration of 0.10 vol.% and at $\Delta T = 35°C$. By inducing a magnetic field (118.4 G) vertically on the side wall of the cavity, the Nusselt number increases; the maximum enhancement in Nu number is 4.91% compared to the case without magnetic induction.

For the Nu_{ave} a correlation is developed:

$$Nu_{ave} = 0.721(Ra)^{0.2429}\phi^{-0.0613} \tag{6.13}$$

Joubert et al. [40] experimentally investigated the influence of volume fraction, magnetic field configuration, and magnetic field strength on Fe_2O_3 magnetic nanofluid in a differentially heated square cavity. A differentially heated square cavity with dimensions of $99 \times 96 \times 120mm$ (L x H x W) was used to investigate the behavior of the nanofluid in natural convection. Experiments are carried for

Rayleigh numbers within the range $1.7 \bullet 10^8 < Ra < 4.2 \bullet 10^8$, and volume fractions of nanoparticles between 0.05% and 0.3%. Fig. 6.17 shows the permanent magnet configuration.

They found that the maximum increase in Nusselt number (5.63%) is noted at $Ra = 4.2 \bullet 10^8$ and a volume concentration of 0.1%. Also, at $Ra = 3.18 \bullet 10^8$ and a magnetic field of 700 G [in configuration (a)] the Nusselt number is enhanced by approximately 2.80%.

Natural convection of Fe_3O_4—water nanofluid in an inclined C-shape cavity considering the effects of gravity force (downward direction) and a constant magnetic field (horizontal direction) was investigated by Rahimpour and Moraveji [41]. The geometry of cavity angle is described in Fig. 6.18. In this paper the influence of Rayleigh number (*Ra*), Hartmann number (*Ha*), inclination angle (α), nanofluids volume fraction, and cavity aspect ratio (*AR*) on the Nusselt number was examined. The results revealed that the *Ra* number and *AR* significant influence the *Nu* number values compared to other variables.

The correlation for the Nusselt number is given by:

$$Nu = (C_1 + C_2Ra + C_3Ra\ Ha)exp\left(C_4\phi AR + (C_5 + C_6Ra + C_7RaHa)AR^2\right) \tag{6.14}$$

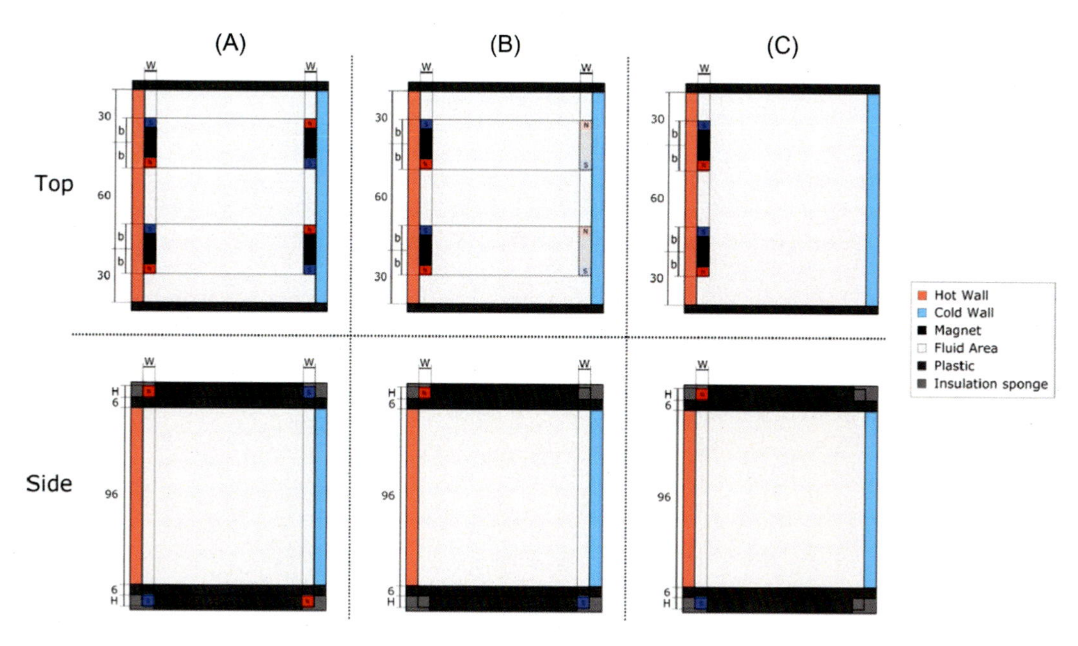

FIGURE 6.17

Permanent magnet configurations: (A) four magnets located on the top and the bottom of the cavity; (B) two magnets located on the top of the hot wall and the bottom of the cold wall; and (C) two magnets located at the top and the bottom of the hot wall.

Reprinted from J.C. Joubert, M. Sharifpur, A. Brusly Solomon, J.P. Meyer, Enhancement in heat transfer of a ferrofluid in a differentially heated square cavity through the use of permanent magnets, J. Magnetism Magnetic Mater. 443 (2017) 149–158,

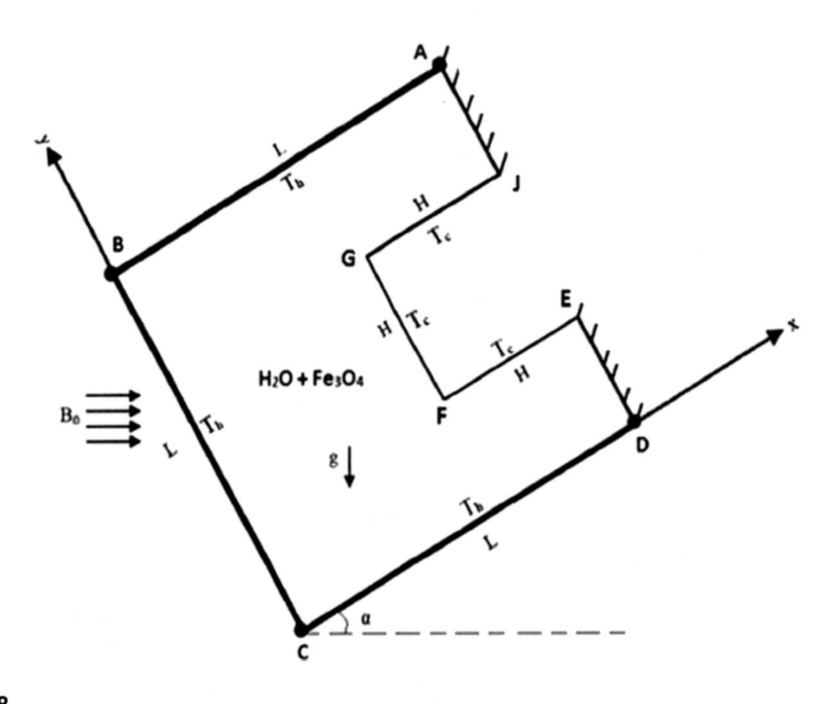

FIGURE 6.18

Schematic of the physical problem.

Reprinted from N. Rahimpour, M.K. Moraveji, Free convection of water–Fe_3O_4 nanofluid in an inclined cavity subjected to a magnetic field: CFD modeling, sensitivity analysis, Adv. Powder Technol. 28 (2017) 1573–1584, Copyright 2017, with permission from Elsevier.

Table 6.17 Constants $C_1 \ldots C_7$ values used in Eq. (6.14) [41].

Coefficient	Values
C_1	9.330E−1
C_2	2.892E−6
C_3	−2.467E−8
C_4	2.781E0
C_5	3.162E0
C_6	−1.948E−6
C_7	1.115E−8

where constants $C_1 \ldots C_7$ are given in Table 6.17.

6.2.3.2 Natural convection in annulus

Natural convection in an annulus between a triangle and rhombus enclosures filled with Fe_3O–H_2O nanofluid subject to thermal radiation using CVFEM was investigated by Dogonchi et al. [42].

Electric field and induced magnetic field characteristics are not considered. The flow configuration is shown in Fig. 6.19.

The variation of viscosity with the magnetic field is given as:

$$\mu_{nf} = \left(3.1B + 0.035B^2 + 4263.02\phi - 27886.4807\phi^2 + 316.0629\right)e^{-0.02T} \tag{6.15}$$

The effect of shape factor of nanoparticles (m), on average Nusselt number for $Ha = 20$, $Rd = 0.3$ and $AR = 0.6$ is given in Table 6.18.

A relation for the average Nusselt number is developed:

$$\begin{aligned} Nu_{ave} = {} & 1.91813 + 3.71147{\bullet}10^{-5}Ra - 0.024363Ha - 2.37661\phi + 2.97889Rd + 5.38641{\bullet}10^{-8}RaHa \\ & + 1.11560{\bullet}10^{-5}Ra\phi + 1.43417{\bullet}10^{-5}RaRd + 9.26582{\bullet}10^{-3}Ha\phi - 0.047982HaRd \\ & - 4.49466\phi Rd - 2.30170{\bullet}10^{-10}Ra^2 - 4.73437{\bullet}10^{-4}Ha^2 + 177.43905\phi^2 + 0.42772Rd^2 \end{aligned} \tag{6.16}$$

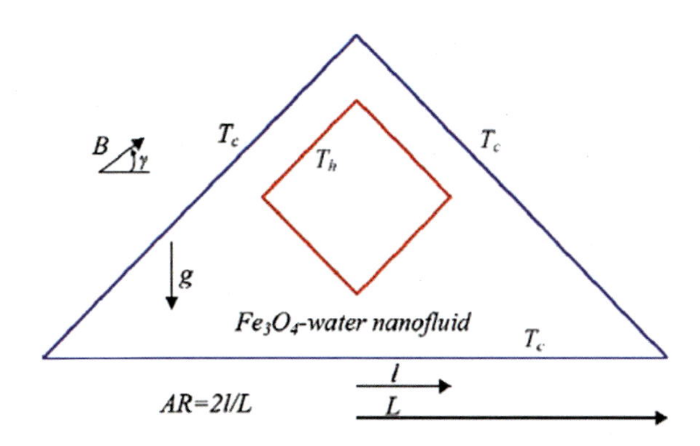

FIGURE 6.19

Schematic of the physical problem.

Reprinted from A.S. Dogonchi, M. Waqas, S.M. Seyyedi, M. Hashemi-Tilehnoee, D.D. Ganji, CVFEM analysis for Fe_3O_4–H_2O nanofluid in an annulus subject to thermal radiation, Int. J. Heat Mass Transf. 132 (2019) 473–483, Copyright 2019, with permission from Elsevier.

Table 6.18 Influence of m on average Nusselt number for $Ha = 20$, $Rd = 0.3$, and $AR = 0.6$ [42].

Ra	*m*	$Nu_{ave}(\phi = 0.02)$	$Nu_{ave}(\phi = 0.04)$
10^3	Spherical(3)	2.048989	2.112490
	Cylinder(4.8)	2.073034	2.161000
	Platelet(5.7)	2.082789	2.180628
10^4	Spherical(3)	2.436615	2.488764
	Cylinder(4.8)	2.459831	2.535751
	Platelet(5.7)	2.469254	2.554776
10^5	Spherical(3)	3.972546	4.030828
	Cylinder(4.8)	4.007588	4.100211
	Platelet(5.7)	4.021781	4.128190

In another study, Dogonchi and Hashim [43] studied thermohydrodynamics characteristics of Fe_3O_4—water nanofluid. The studied nanofluid is in an annulus between a wavy circular cylinder and a rhombus enclosure subject to a uniform magnetic field. The geometrical configuration of the physical model and boundary conditions are depicted in Fig. 6.20.

Numerical simulations for the following conditions are performed: $10^3 \le Ra \le 10^5$, $0 \le Rd \le 0.3$, $0 \le Ha \le 20, 0.4 \le AR \le 0.6$, $3 \le m \le 5.7$, $2 \le \phi \le 4$. The effects of m and ϕ on Nu_{ave} number are presented in Table 6.19 and it can be seen that the platelet nanoparticles shape achieved a higher heat transfer rate than the spherical and cylinder ones.

A correlation for the average Nusselt number is proposed:

$$Nu_{ave} = 1.36041 + 7.32027 \bullet 10^{-5} Ra - 0.034813 Ha + 4.09712 Rd - 6.01499 \bullet 10^{-7} RaHa + 3.04559 \bullet 10^{-5} RaRd - 0.032797 HaRd - 3.73236 \bullet 10^{-10} Ra^2 + 2.10637 \bullet 10^{-3} Ha^2 - 4.60920 Rd^2 \quad (6.17)$$

Recently, Dogonchi et al. [44] performed a simulation for the analysis of Fe_3O_4—H_2O nanofluid inside an annulus between two enclosures. The magnetic field and thermal radiation were considered. The configuration of the physical model and the coordinate system are described in Fig. 6.21.

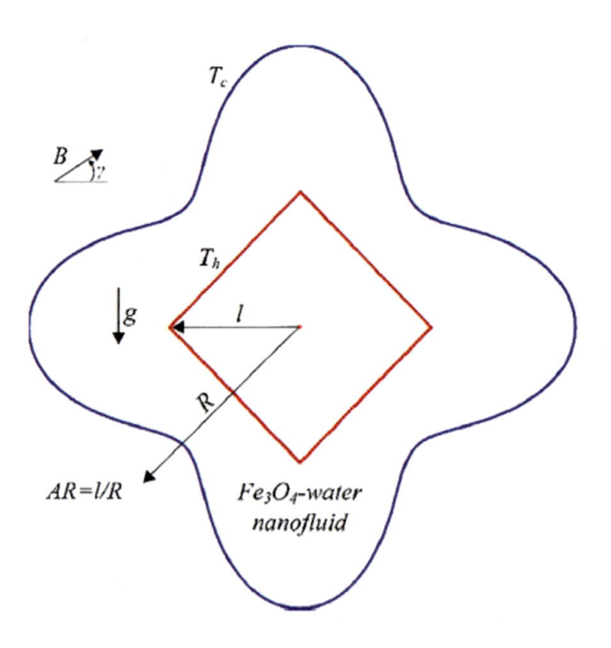

FIGURE 6.20

Schematic of the physical problem.

Reprinted from A.S. Dogonchi, Hashim, Heat transfer by natural convection of Fe_3O_4—water nanofluid in an annulus between a wavy circular cylinder and a rhombus, Int. J. Heat Mass Transf. 130 (2019) 320–332, Copyright 2019, with permission from Elsevier.

Table 6.19 Influence of m and ϕ on Nu_{ave} number for $Ha = 20$, $Rd = 0.3$, and $AR = 0.6$ [43].

Ra	m	$Nu_{ave}(\phi = 0.02)$	$Nu_{ave}(\phi = 0.04)$
10^3	Spherical(3)	1.990820	2.053496
	Cylinder(4.8)	2.014589	2.101468
	Platelet(5.7)	2.024233	2.120878
10^4	Spherical(3)	2.669636	2.694690
	Cylinder(4.8)	2.690547	2.737164
	Platelet(5.7)	2.699036	2.754361
10^5	Spherical(3)	5.104799	5.133874
	Cylinder(4.8)	5.142381	5.209005
	Platelet(5.7)	5.157682	5.239648

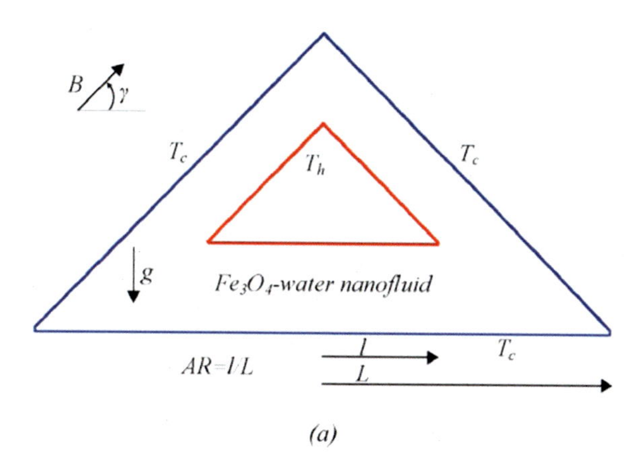

FIGURE 6.21

Schematic of the physical problem.

Reprinted from A.S. Dogonchi, Z. Asghar, M. Waqas, CVFEM simulation for Fe_3O_4–H_2O nanofluid in an annulus between two triangular enclosures subjected to magnetic field and thermal radiation, Int. Commun. Heat Mass Transf. 112 (2020) 104449, Copyright 2020, with permission from Elsevier.

The effect of m on Nu_{ave} number, together with the proposed correlation for Nu_{ave} are depicted in Table 6.20 and Eq. 6.18, respectively.

$$\begin{aligned} Nu_{ave} = {} & 1.32572 + 3.83752 \cdot 10^{-5} Ra - 0.011376 Ha + 1.46952\phi + 1.83869 Rd - 0.20959 Hs \\ & - 2.74763 \cdot 10^{-7} RaHa + 1.79527 \cdot 10^{-5} RaRd + 7.18504 \cdot 10^{-7} RaHs - 0.053442 HaRd \\ & - 0.12176 RdHs - 1.71050 \cdot 10^{-10} Ra^2 - 1.09907 \cdot 10^{-3} Ha^2 \end{aligned} \tag{6.18}$$

6.2.3.3 Natural convection in porous cavity

Natural convection of Fe_3O_4–MWCNT hybrid nanofluid within a porous enclosure by considering the magnetic effect was studied by Tlili et al. [45]. The geometrical shape of the porous enclosure

Table 6.20 Influence of m and ϕ on average Nusselt number for $Ha = 20$, $Hs = 2$ $Rd = 0.3$, and $AR = 0.4$ [44].

Ra	*m*	$Nu_{ave}(\phi = 0.02)$	$Nu_{ave}(\phi = 0.04)$
10^3	Spherical(3)	0.747387	0.800610
	Cylinder(4.8)	0.767602	0.840824
	Platelet(5.7)	0.775777	0.857002
10^4	Spherical(3)	0.942112	0.979963
	Cylinder(4.8)	0.959523	1.014968
	Platelet(5.7)	0.966578	1.029101
10^5	Spherical(3)	2.793431	2.803524
	Cylinder(4.8)	2.813491	2.842989
	Platelet(5.7)	2.821601	2.858850

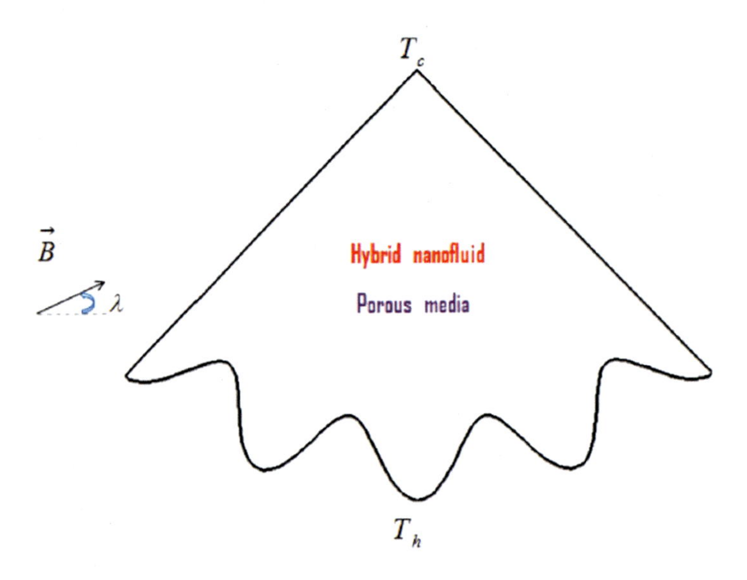

FIGURE 6.22

Geometrical shape of porous enclosure.

Reprinted from I. Tlili, M.M. Bhatti, S.M. Hamad, A.A. Barzinjy, M. Sheikholeslami, A. Shafee, Macroscopic modeling for convection of hybrid nanofluid with magnetic effects, Phys. A 534 (2019) 122136, Copyright 2019, with permission from Elsevier.

is illustrated in Fig. 6.22. The results revealed that an enhancement in radiation parameter and permeability tends to improve the average Nusselt number, Nu_{ave}, while the behavior of Ha has the opposite trend, that is it produces resistance for the Nu_{ave}.

The proposed correlation for the average number is:

$$\begin{aligned} Nu_{ave} = 3.09 + 1.23Rd + 0.064\log(Ra)Da^* + 0.02Da^* + 0.029Da^*Rd \\ - 0.09Ha^* - 0.031RdHa^* + 0.015Da^*Ha^* + 0.46\log(Ra) \end{aligned} \tag{6.19}$$

The effects of *Da*, *Ha*, and *Ra* numbers, the inclination angle of the magnetic field, the cavity aspect ratio, and the nanoparticle shape factor on the heat and flow fields in a porous enclosure filled with $Fe_3O_4-H_2O$ nanofluid under constant inclined magnetic were studied by Molana et al. [46]. The geometry of the cavity and boundary conditions are illustrated in Fig. 6.23

The influences of m and ϕ on Nu_{ave} number are described in Tables 6.21 and 6.22.

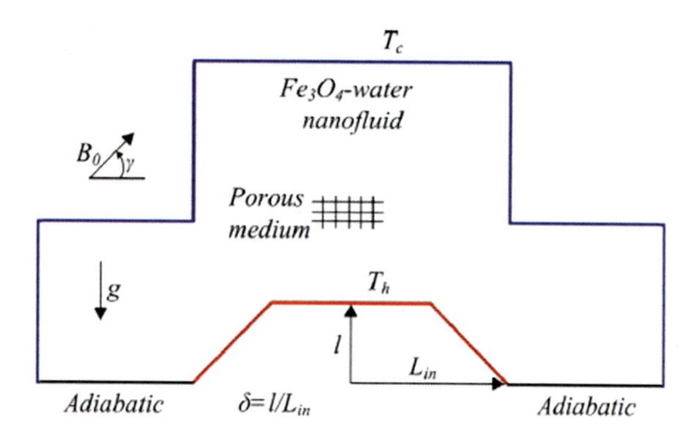

FIGURE 6.23

Schematic of the physical problem.

Reprinted from M. Molana, A.S. Dogonchi, T. Armaghani, A.J. Chamkha, D.D. Ganji, I. Tlili, Investigation of hydrothermal behavior of $Fe_3O_4-H_2O$ nanofluid natural convection in a novel shape of porous cavity subjected to magnetic field dependent (MFD) viscosity, J. Energy Storage 30 (2020) 101395, Copyright 2020, with permission from Elsevier.

Table 6.21 The effect of shape factor (m) on Nu_{ave} for $Ha = 20$, $\phi = 2\%, Da = 0.01$, $\gamma = 0°$, $\delta = 0.5$, and $Pr = 6.2$ [46].

Ra	*m*	Nu_{ave}
10^3	Spherical(3)	0.469298
	Platelet(5.7)	0.480447
	Blade(8.3)	0.487875
10^4	Spherical(3)	0.589860
	Platelet(5.7)	0.598707
	Blade(8.3)	0.605785
10^5	Spherical(3)	1.645641
	Platelet(5.7)	1.673031
	Blade(8.3)	1.687404

Table 6.22 Effect of nanoparticle volume concentration on Nu_{ave} for $Ha = 20$, $Da = 0.01$, $\gamma = 0°$, $\delta = 0.5$, and $Pr = 6.2$ [46].

Ra	ϕ	Nu_{ave}	*En*
10^4	0.00	0.448977	–
10^4	0.02	0.487875	8.66
10^4	0.04	0.527603	17.51
10^5	0.00	0.581534	–
10^5	0.02	0.605785	4.17
10^5	0.04	0.627415	7.89
10^6	0.00	1.617706	–
10^6	0.02	1.687404	4.3
10^6	0.04	1.739637	7.53

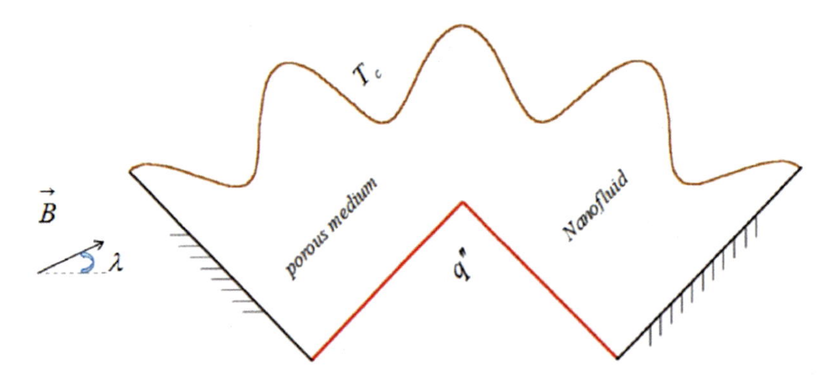

FIGURE 6.24

Geometry of porous tank and boundary conditions.

Reprinted from T.D. Manh, N.D. Nam, K. Jacob, A. Hajizadeh, H. Babazadeh, M. Mahjoub, et al., Simulation of heat transfer in 2D porous tank in appearance of magnetic nanofluid, Phys. A 550 (2020) 123937, Copyright 2020, with permission from Elsevier.

Manh et al. [47] performed an analysis of heat transfer in a 2D porous tank with a wavy cold surface filled with magnetic nanofluid. The geometry is illustrated in Fig. 6.24. They noted that the convection heat transfer is significantly enhanced with both Rayleigh and Darcy numbers and also that Nu_{ave} increases with increasing thermal radiation *Ra*, while nanoparticle shape has a negligible effect on Nu_{ave}.

The proposed correlation for the average Nusselt number is given by Eq. (6.20):

$$Nu_{ave} = 6.13 + 9.4 \cdot 10^{-3} mHa + 2.05Rd - 0.074RdHa - 0.011Da - 0.03Ha + 0.045HaRd \\ + 0.045m + 0.38\log(Ra) + 0.15Da\log(Ra) - 3.63 \cdot 10^{-3} m^2 - 0.088Ha\,Da - 0.21\log(Ra)Ha \quad (6.20)$$

Saba et al. [48] investigated the heat transfer and flow field of $CNT - Fe_3O_4$/water hybrid nanofluid inside an asymmetric long channel, whose porous walls exhibit an embracing or parting motion. They found that the heat transfer rate for $CNT - Fe_3O_4$/water hybrid nanofluid on the upper wall depicts an augmented behavior with the absolute values of deformation variable α, the porosity parameter $\mathscr{A}$, and permeation Reynolds number $\mathscr{R}$, and also that the value of Nu_{lower} (in most of the cases) occupies the higher side, increasing values of volume fraction of CNT ϕ_2, where ϕ_2 varies between $0.005 \leq \phi_2 \leq 0.06$.

6.2.3.4 Magnetohydrodynamic natural convection

MHD is defined as the hydrodynamics of any conducting fluid (nanofluid) under the influence of a magnetic field and may be expressed in terms of temperature, velocity, pressure, density, and induced magnetic fields parameters. In the MHD model, the Maxwell equations are combined with fluid flow equations that include the Lorentz force from the magnetic field. MHD flows may be found in the engineering processes, such as heat exchangers, boilers, electronics, and biomedical applications.

The heat transfer in MHD three-dimensional flow of magnetic nanofluid (Fe_3O_4−water) over a bidirectional exponentially stretching sheet under the magnetic field was numerical investigated by Ahmad et al. [49]. They considered the exponentially varying surface velocity and temperature distributions. It was demonstrated that the heat transfer rate is smaller when the larger magnetic field strength is employed, and that the heat transfer rate in the case of the bidirectional stretching surface is larger compared to the unidirectional stretching case.

The study of MHD boundary layer flow and heat transfer of Fe_3O_4/ethylene glycol nanofluid on micropolar fluid with homogeneously suspended dust particles over a stretching sheet is performed by Ghadikolaei et al. [50]. Thermal radiation and Joule heating were considered. Fig. 6.25 shows the configuration and results for the local *Nu* number. The results showed that the local *Nu* number has a direct relation with *Pr* in the *PST* case and an inverse relation in the *PHF* case in the presence of β. Also, the increase in *Ec* in the presence of *Nr*, causes the local *Nu* number to be decreased in the *PST* case and to be increased in the *PHF* case.

Analysis of MHD hybrid nanofluid natural convection heat transfer within a double-porous medium with different characteristics using a two-equation energy model was carried out by Mehryan et al. [51]. The schematic configuration of the studied problem is illustrated in Fig. 6.26.

The enhancement in heat transfer can be achieved for low values of the solid−liquid interface convection parameter, magnetic field viscosity parameter, and high thermal conductivity ratio values.

Sajjadi et al. [52] analyzed MHD natural convection in a porous media by the double multirelaxation time Lattice Boltzmann method utilizing a MWCNT−Fe_3O_4/water hybrid nanofluid. They found that by increasing the *Ra* number, the heat transfer rate increased for all studied cases, and also that with the increase in the *Ha* number the effect of *Ra* number decreases. Also, by adding the nanoparticles, an increase in the Nu_{ave} number of approximately 4.9% is noted at $Ra = 10^5$, $Da = 10^{-1}$, $Ha = 50$, and $\varepsilon = 0.9$. Results indicated also that by increasing the Darcy number the heat transfer rate is increased and the Nu_{ave} number is enhanced by porosity.

6.2.3.5 Lorentz force effect on magnetic nanofluids

The simulation of a magnetic nanofluid requires a high computational precision of the Lorentz force. The Lorentz force is the force on a charged particle due to electric and magnetic fields and can be computed as [53]

$$\vec{F} = \vec{j} \times \vec{B} \tag{6.21}$$

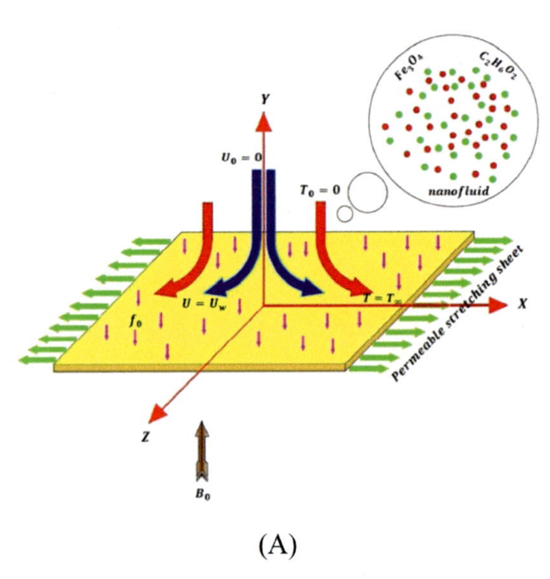

(A)

(B)

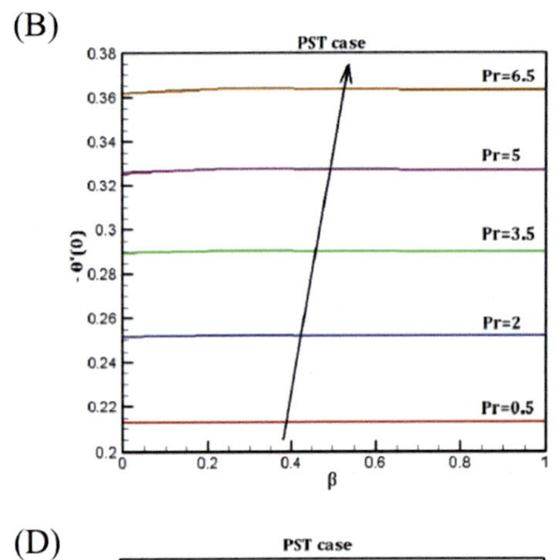

(C)

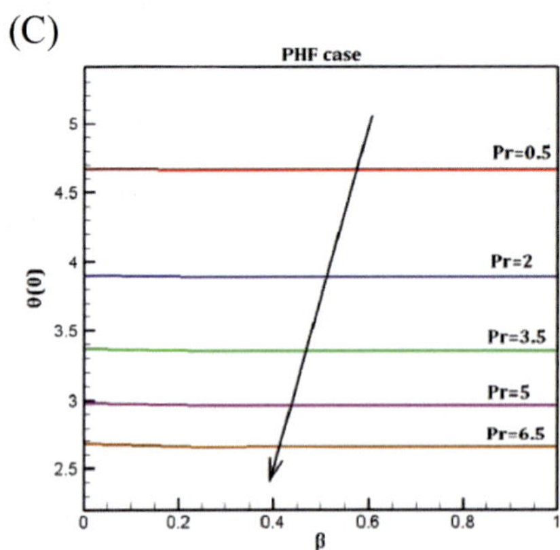

(D)

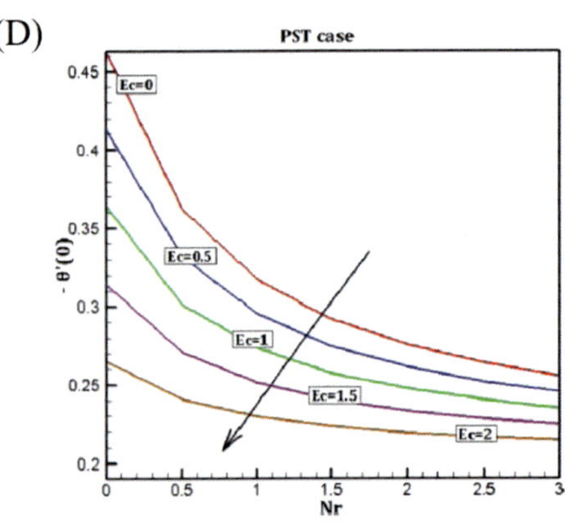

(E)

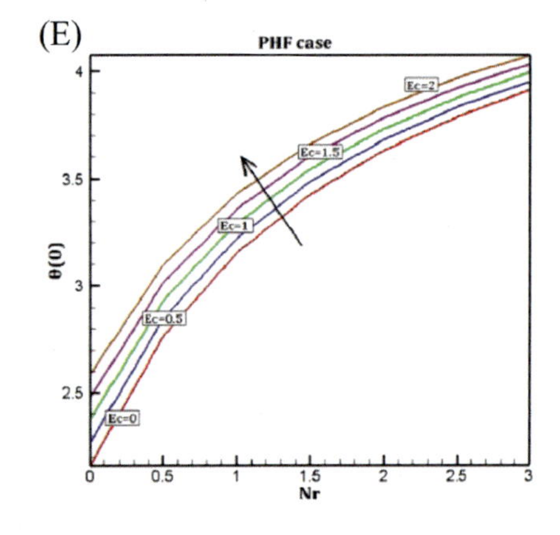

FIGURE 6.25

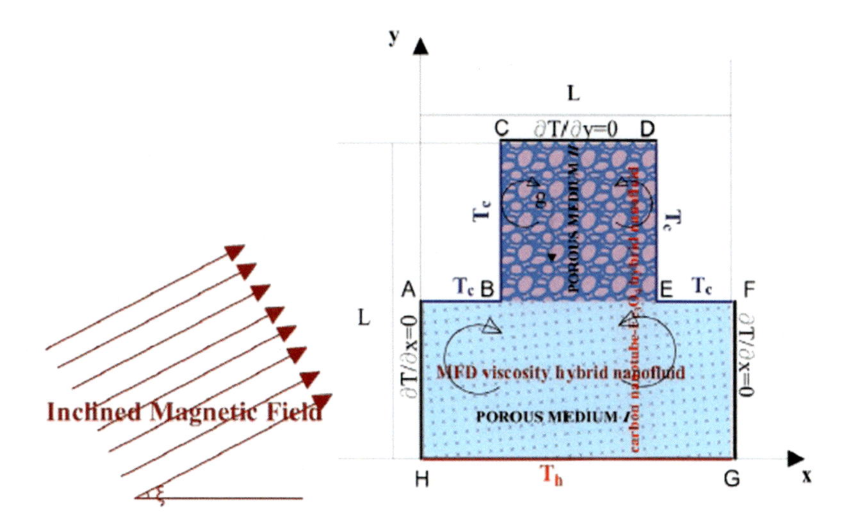

FIGURE 6.26

Schematic configuration of studied problem.

Reprinted from S.A.M. Mehryan, M.A. Sheremet, M. Soltani, M. Izadi, Natural convection of magnetic hybrid nanofluid inside a double-porous medium using two-equation energy model, J. Mol. Liq. 277 (2019) 959–970, Copyright 2019, with permission from Elsevier.

where the Ohm's law current density for fluid is

$$\vec{j} = \sigma_e\left(-\nabla\phi + \vec{U} + \vec{B}\right) \tag{6.22}$$

$\vec{F}$ is Lorentz force, N/m^3, $\vec{j}$ is electric current density, A/m^2, $\vec{B}$ is magnetic field, kg/m^2A, U is dimensionless velocity components, m/s, ϕ is electric potential, m^2kg/s^3A, and σ is electrical conductivity, s^3A^2/m^3kg.

Sheikholeslami and Shehzad [54] reported the influence of thermal radiation on ferrofluid flow by taking into account Lorentz forces using CVFEM. The variation of viscosity of Fe_3O_4 ferrofluid with the magnetic field is considered:

$$\mu_{nf} = \left(0.035B^2 + 3.1B - 27886.4807\phi^2 + 4263.02\phi + 316.0629\right)e^{-0.01T} \tag{6.23}$$

(A) Geometry of the problem, (B) influence of Prandtl number (Pr) on the local Nusselt number in the presence of fluid particle interaction parameter (β) for the prescribed surface temperature (PST); (C) influence of Pr on local Nusselt number in the presence of β for the prescribed heat flux (PHF); (D) influence of Eckert number (Ec) on local Nusselt number in the presence of radiation parameter (Nr) for the PST case; and (E) influence of Ec on local Nusselt number in the presence of Nr for the PHT case.

Reprinted from S.S. Ghadikolaei, Kh. Hosseinzadeh, D.D. Ganji, M. Hatami, Fe_3O_4–$(CH_2OH)_2$ nanofluid analysis in a porous medium under MHD radiative boundary layer and dusty fluid, J. Mol. Liq. 258 (2018) 172–185, Copyright 2018, with permission from Elsevier.

Numerical simulations were performed for various parameters: radiation parameter $0 \leq Rd \leq 0.8$, volume fraction of Fe_3O_4 nanoparticles $0 \leq \phi \leq 0.04$, Rayleigh number $10^3 \leq Ra \leq 10^5$, inclination angle $\xi = 0°$ and $90°$, and Hartmann number $0 \leq Ha \leq 40$. The results revealed that the inner wall temperature is reduced with increasing buoyancy forces but it increases with increasing *Rd* and *Ha*. Also, the *Nu* number is increased with the increase in inclination angle, *Rd* and *Ra*.

And in this study, a new correlation is developed:

$$\begin{aligned} Nu_{ave} = {} & 6.91 - 0.12\xi - 4.3Rd - 3.5\log(Ra) + 1.26Ha^* + 0.36\xi Rd + 0.2\xi\log(Ra) \\ & - 0.07\xi Ha^* + 1.61\log(Ra)Rd - 0.31\log(Ra)Ha^* - 0.45RdHa^* - 0.2\xi^2 \\ & + 0.62(\log(Ra))^2 + 1.5Rd^2 - 0.08Ha^{*2} \end{aligned} \tag{6.24}$$

where $Ha^* = 0.1Ha$.

The effect of the Lorentz force on the magnetic nanofluid in a cubic cavity together with the computation of the inductive electrical field generated by the magnetic field was investigated by Jelodari and Nikseresht [55]. Fig. 6.27 shows the studied configuration and results for the Nu_{ave} number in terms of volume fraction at different imposing magnetic field directions. Analyzing the results one can notice that there is a significant reduction in heat transfer. This is due to the sudden increase in the electrical conductivity of the base fluid. Analyzing the results one can notice that there is a significant reduction in heat transfer. This is due to the sudden increase in the electrical conductivity of the base fluid as well as the Lorentz force that reduces the fluid current field. Also, the results indicated that by imposing the magnetic field on the cavity along the Z-axis, the heat transfer is enhanced compared to that imposed on the X-direction.

A new correlation for predicting the Nu_{ave} number on the hot wall is developed:

$$Nu_m = a_1 + a_2B^* + a_3\phi + a_4B^{*2} + a_5B^*\phi + a_6\phi^2 + a_7B^{*2}\phi + a_8B^*\phi^2 + a_9\phi^3 \tag{6.25}$$

where $B^* = B/B_0$, $B_0 = 3T$ is the maximum magnitude of the magnetic field.

Coefficients $a_1 \ldots a_9$ for Eq. (6.25) are depicted in Table 6.23.

6.2.3.6 Joule heating effect on magnetic nanofluids

The Joule heating effect has a significant impact in many industrial applications, such as in geophysical streams, the petroleum industry, and nuclear engineering, and is defined as the product between the Hartmann number and the Eckert number. Joule heating enhances the heat transfer process by using electric current movement, which decreases the dynamic viscosity and increases electrical conductivity.

The natural convection of a Fe_3O_4−water nanofluid within a sinusoidal heated cavity was studied numerically by Ghaffarpasand [56]. The internal Joule heating and the external magnetic field were considered. Also they investigated the effect of internal Joule heating in order to better understand its role in fluid flow and heat transfer mechanisms. The geometry and coordinate system of physical configuration are illustrated in Fig. 6.28.

The analyses were performed in following conditions: Hartmann number $0 \leq Ha \leq 50$, Eckert number $0 \leq Ec \leq 0.075$, and volume fraction $0 \leq \phi \leq 0.06$, at various phase deviations of right vertical wall temperature distribution $0 \leq \gamma \leq \pi$. Richardson, Reynolds, and Grashof numbers were kept constant during this study ($Ri = 10, Re = 100, Gr = 10^5$). The results showed that the heat transfer rate is reduced with the increasing Hartmann or Eckert numbers. Also, applying the

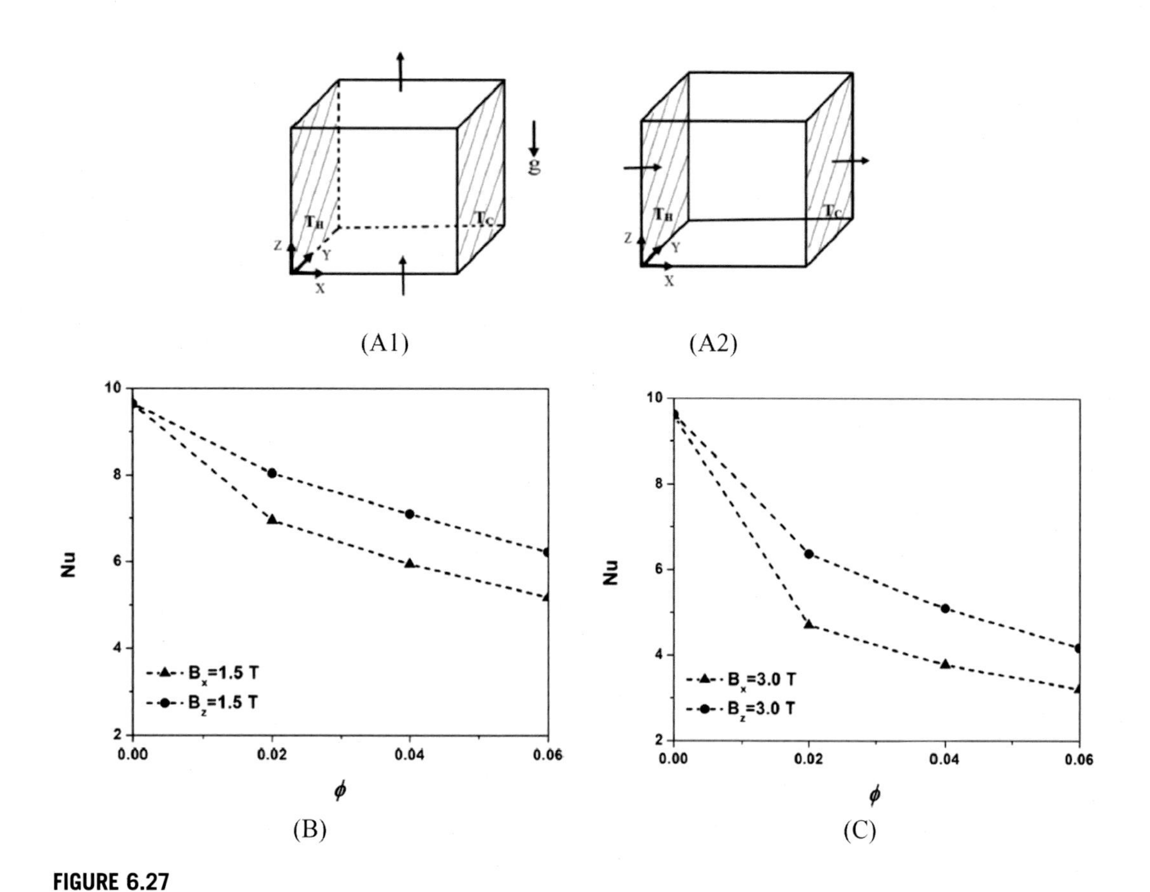

FIGURE 6.27

(A) The geometry and boundary conditions, (A1) external magnetic field in the Z-direction, (A2) external magnetic field in the X-direction. (B) The Nusselt number versus nanoparticles volume fraction in different imposing magnetic field directions with the power of 1.5 T. (C) The Nusselt number for 3T.

Reprinted from I. Jelodari, A.H. Nikseresht, Effects of Lorentz force and induced electrical field on the thermal performance of a magnetic nanofluid-filled cubic cavity, J. Mol. Liq. 252 (2018) 296–310, Copyright 2018, with permission from Elsevier.

magnetic field, and adding nanoparticles led to an improvement in the heat transfer. This improvement is accentuated by conjugating the internal Joule heating (Table 6.24).

The influence of Joule heating on MHD flow of a magnetic nanofluid in a semi porous curved channel is studied by Sajid et al. [57]. The upper and lower walls of the channel are considered porous and impermeable, respectively. Geometry of the studied problem is illustrated in Fig. 6.29.

Results showed that the *Nu* number increases with increasing dimensionless radius of curvature value (K), and decreases with increasing the values of Reynolds number (*Re*), Prandtl number (*Pr*), magnetic parameter (*M*), volume fraction of nanoparticles (ϕ), and Eckert (*Ec*) (Table 6.25).

Table 6.23 The values of the constant coefficients in Eq. (6.25), $B_0 = 3T$ [55].

	X-Direction of magnetic field $B^* = B_X/B_0$		Z-Direction of magnetic field $B^* = B_Z/B_0$	
	$Ra = 10^5$	$Ra = 10^6$	$Ra = 10^5$	$Ra = 10^6$
a_1	5.023	9.725	4.982	9.662
a_2	−0.5143	0.0992	−0.04935	0.3633
a_3	−78.49	−96.89	−66.37	−57.76
a_4	0.2658	−0.3564	−0.1522	−0.4774
a_5	−192.7	−245.9	−188	−120.7
a_6	2753	3110	2321	1316
a_7	39.34	16.38	34.59	−22.62
a_8	1952	2750	1937	1512
a_9	$-3.011 \cdot 10^4$	$-3.558 \cdot 10^4$	$-2.618 \cdot 10^4$	$-1.611 \cdot 10^4$

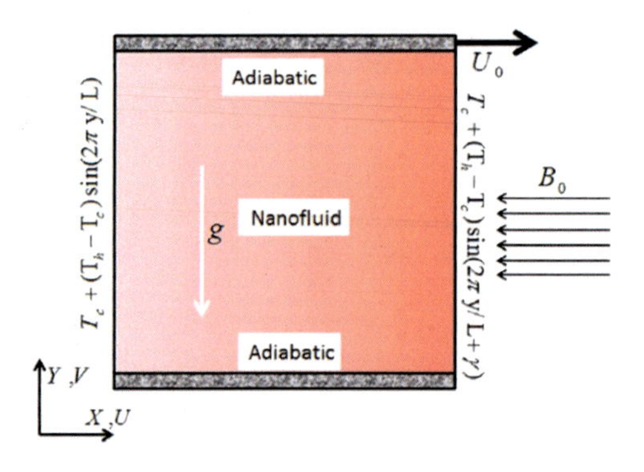

FIGURE 6.28

Geometry of the physical configuration.

Reprinted from O. Ghaffarpasand, Numerical study of MHD natural convection inside a sinusoidally heated lid-driven cavity filled with Fe_3O_4–water nanofluid in the presence of Joule heating, Appl. Math. Model. 40 (2016) 9165–9182, Copyright 2016, with permission from Elsevier.

Table 6.24 The Nu_{ave} number due to conjugate effect of Joule heating $\overline{Nu}^{\otimes}$, apply magnetic field $\overline{Nu}^{*}$, and adding nanoparticles $\overline{Nu}^{\dagger}$ [56].

Phase deviation, γ	**0**	$\frac{\pi}{4}$	$\frac{\pi}{2}$	$\frac{3\pi}{4}$	π
$\overline{Nu}^{\otimes}$	22.0%	17.0%	15.0%	15.0%	16.0%
$\overline{Nu}^{*}$	22.5%	17.0%	15.0%	14.0%	16.0%
$\overline{Nu}^{\dagger}$	3.3%	2.5%	2.0%	2.2%	1.8%

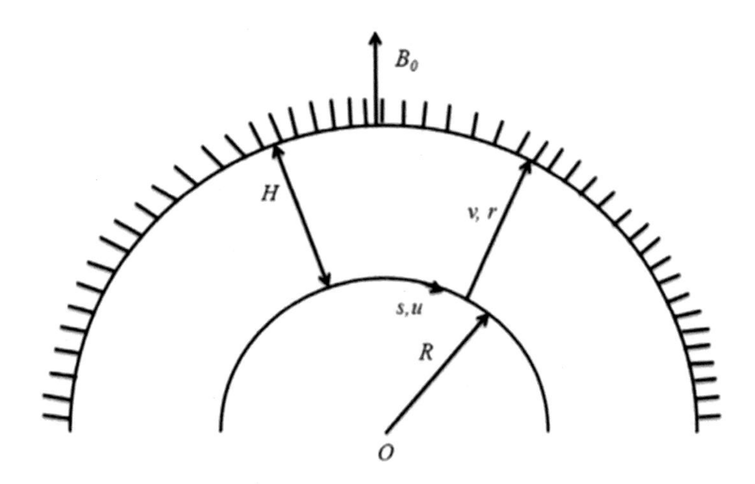

FIGURE 6.29

Geometry of the physical configuration.

Reprinted from M. Sajid, S.A. Iqbal, M. Naveed, Z. Abbas, Joule heating and magnetohydrodynamic effects on ferrofluid (Fe_3O_4) flow in a semi-porous curved channel, J. Mol. Liq. 222 (2016) 1115–1120, Copyright 2016, with permission from Elsevier.

Table 6.25 The values of the *Nu* number, $-Re_s^{-1/2}Nu_s$ for different values *K, M,* ϕ*, Re, Pr,* and *Ec* [57].

K	M	ϕ	Re	Pr	Ec	$-Re_s^{-1/2}Nu_s$
2	0.1	0.4	3	10	0.2	1.0959
4						1.3254
6						1.5433
2	0.05					1.9705
	0.10					1.0959
	0.15					0.2222
	0.10	0.1				1.3179
		0.3				1.2107
		0.6				0.4611
		0.4	2			1.5901
			4			0.5955
			5			0.0891
			3	3		2.1260
				5		1.8436
				15		0.3436
				10	0.10	1.9708
					0.15	1.5334
					0.30	0.2210

6.2.3.7 Convective instability of a magnetic nanofluids

The first studies on the convective instability of a nanofluid under a uniform magnetic field are performed by Finlayson [58] and Vaidyanathan et al. [59]. Their results revealed that by applying a magnetic field, the magnetic convection could be controlled and that the presence of a porous medium leads to an enhancement in the magnetization effect. The instability in a magnetized ferrofluid saturated porous layer was investigated by Sunil and Mahajan [60]. Their results indicated a coupling between the magnetic and buoyancy forces.

Later, the study of instability in a thin magnetic nanofluid layer saturating a porous medium was performed by Mahajan and Sharma [61]. In this paper, the effect of Brownian diffusion, thermophoresis, magnetophoresis and Darcy's law were considered. The boundary conditions imposed are the following:

1. Impermeable and conducting–Impermeable and conducting ($IMP_{LU}\&CON_{LU}$);
2. Impermeable and conducting–Impermeable with constant heat flux ($IMP_{LU}\&CON_{L}$);
3. Impermeable and conducting–Free with constant heat flux ($IMP_{L}, CON_{L}\&FRE_{U}$).

Fig. 6.30 illustrates the studied configuration and imposed boundary conditions.

As can be seen in Table 6.26, the values of critical thermal Rayleigh number (Ra_c) and critical wave number (k_c) are higher for the $IMP_{LU}\&CON_{LU}$ boundaries compared to $IMP_{LU}\&CON_{L}$ or $IMP_{L}, CON_{L}\&FRE_{U}$ boundaries. The gravity coefficient δ delays the onset of convection when $h(z_1)$ is $-z_1$or $-z_1^2$, while it advances the onset of convection when $h(z_1)$ is z_1. Thus the parameter δ depends on the gravity variation $h(z_1)$.

6.2.3.8 Homogeneous–heterogeneous reactions effect

Homogeneous–heterogeneous reactions are natural processes of chemically reacting structures, such as biochemical processes, catalysis, and combustion. The interaction between the

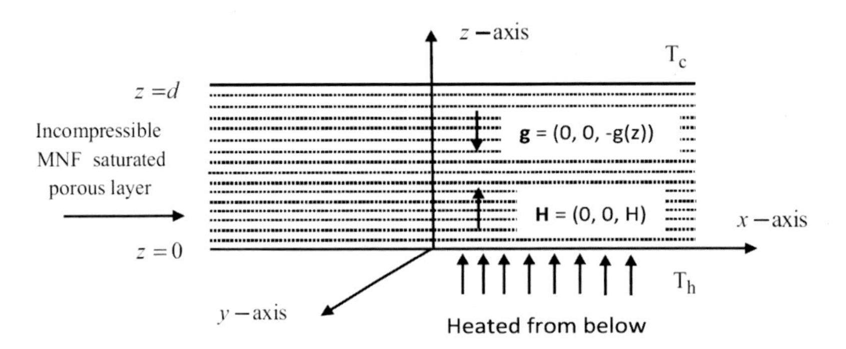

FIGURE 6.30

Geometry of the physical configuration.

Reprinted from A. Mahajan, M.K. Sharma, Convection in a magnetic nanofluid saturating a porous medium under the influence of a variable gravity field, Eng. Sci. Tech. An. Int. J. 21 (2018) 439–450, Copyright 2018, with permission from Elsevier.

Table 6.26 The values of Ra_c and k_c for water-based and ester-based magnetic nanofluids [61].

		$IMP_{LU}\&CON_{LU}$				$IMP_{LU}\&CON_{L}$				$IMP_{L}, CON_{L}\&FRE_{U}$			
		Water		Ester		Water		Ester		Water		Ester	
$h(z_1)$	δ	k_c	Ra_c	k_c	Ra_c	k_c	Ra_c	k_c	Ra_c	k_c	Ra_c	k_c	Ra_c
$-z_1$	0.0	3.03	1.649	3.05	2.024	2.10	1.028	2.12	1.206	1.35	0.415	1.41	0.479
	0.2	3.04	1.713	3.06	2.127	2.09	1.086	2.12	1.290	1.33	0.448	1.39	0.524
	0.4	3.05	1.781	3.06	2.239	2.08	1.148	2.11	1.385	1.30	0.485	1.37	0.578
	0.6	3.06	1.854	3.09	2.360	2.07	1.217	2.10	1.491	1.27	0.528	1.35	0.643
	1.0	3.09	2.012	3.11	2.635	2.06	1.374	2.08	1.750	1.17	0.634	1.26	0.819
	1.5	3.15	2.238	3.14	3.032	2.04	1.617	2.06	2.181	0.90	0.810	1.01	1.165
$-z_1^2$	0.0	3.03	1.649	3.05	2.024	2.10	1.028	2.12	1.206	1.35	0.415	1.41	0.479
	0.2	3.04	1.687	3.05	2.085	2.09	1.063	2.12	1.257	1.33	0.438	1.40	0.511
	0.4	3.04	1.726	3.06	2.149	2.08	1.100	2.11	1.312	1.31	0.464	1.39	0.547
	0.6	3.05	1.767	3.06	2.215	2.08	1.139	2.11	1.371	1.29	0.492	1.37	0.589
	1.0	3.06	1.853	3.08	2.359	2.07	1.224	2.09	1.504	1.24	0.557	1.32	0.690
	1.5	3.08	1.970	3.10	2.560	2.05	1.345	2.07	1.701	1.14	0.660	1.23	0.867
z_1	0.0	3.03	1.649	3.05	2.024	2.10	1.028	2.12	1.206	1.35	0.415	1.41	0.479
	0.2	3.03	1.588	3.04	1.930	2.10	0.976	2.13	1.131	1.37	0.387	1.43	0.440
	0.4	3.02	1.531	3.04	1.842	2.11	0.928	2.14	1.064	1.38	0.362	1.44	0.407
	0.6	3.02	1.477	3.04	1.761	2.11	0.884	2.14	1.004	1.40	0.339	1.45	0.378
	1.0	3.02	1.379	3.04	1.616	2.12	0.806	2.15	0.900	1.42	0.302	1.47	0.331
	1.5	3.01	1.270	3.04	1.462	2.14	0.724	2.16	0.796	1.44	0.264	1.48	0.286

homogeneous reactions in the bulk of the fluid and heterogeneous reactions occurring on some catalytic surfaces is generally complicated [62].

The impact of the magnetic field on Fe_3O_4 nanofluids in the presence of heterogeneous and homogeneous reactions for Blasius flow with thermal radiations was numerical studied by Sajid et al. [63]. Their results indicated that the concentration is reduced by increasing the strength of homogeneous and heterogeneous reactions, and also that the Nusselt number is reduced for large values of the radiation parameter. The effects of Prandtl number are quite opposite to those of the radiation parameter.

The unsteady 3D flow of a ferronanofluid along a vertical flat surface subject to the influence of vibrational rotations as analyzed by Kumar et al. [64]. Fig. 6.31 describes the geometry and coordinate system.

Tables 6.27 and 6.28 emphasize the relevance of volume fraction of nanoparticles (ϕ), dimensionless frequency of oscillation (ω), and Eckert number (Ec) in the control of skin friction coefficients $\left(c_{f,X}, c_{f,Y}\right)$, local Nusselt number ($Nu_X$), and reactant Sherwood numbers $\left(Sh_{\phi,f}, Sh_{\phi,s}\right)$, through the involvement of slip nanofluid (SNF) and no-slip nanofluid (NSNF) cases. As can be seen from Table 6.27, magnitudes of $\left(c_{f,X}, c_{f,Y}\right)$, (Nu_X) and $\left(Sh_{\phi,f}, Sh_{\phi,s}\right)$ are improved with Ec in the NSNF case. Similar trends also prevail in the SNF case with respect to Ec as depicted by Table 6.28.

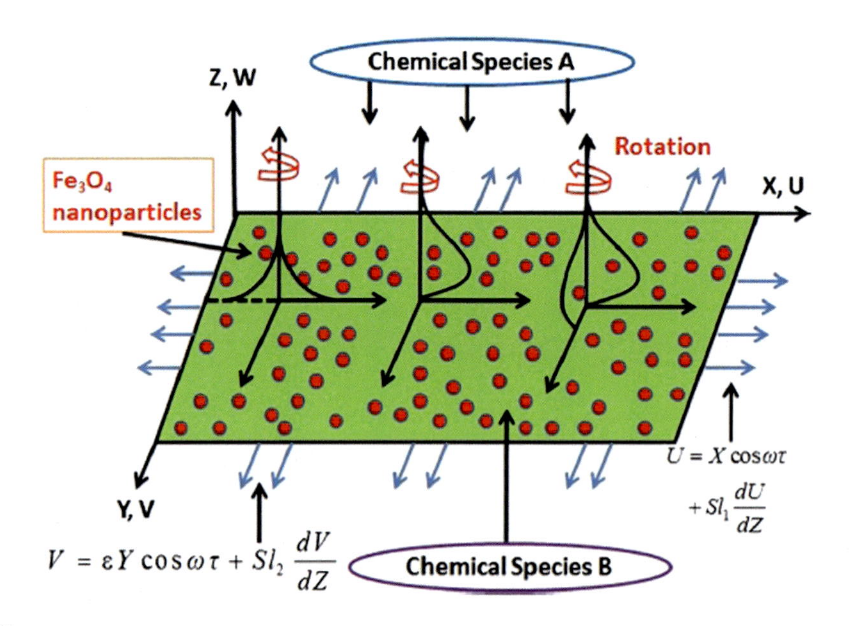

FIGURE 6.31

Geometry of the physical configuration.

Reprinted from R. Kumar, R. Kumar, S.A. Shehzad, M. Sheikholeslami, Rotating frame analysis of radiating and reacting ferro-nanofluid considering Joule heating and viscous dissipation, Int. J. Heat. Mass. Transf. 120 (2018) 540–551, Copyright 2018, with permission from Elsevier.

Tables 6.29 and 6.30 show the impacts of dimensionless magnetic field parameter (M), dimensionless rotation parameter (R), and dimensionless radiation parameter (Ra) on $c_{f,X}, c_{f,Y}$, Nu_X, and $Sh_{\phi,f}, Sh_{\phi,s}$ in both slip flow regime $Sl1 = Sl2 = Sl3 = 1$ and no-slip flow regime $Sl1 = Sl2 = Sl3 = 0$.

The effects of homogeneous–heterogeneous reactions in the flow of magnetite-nanofluid by nonlinear stretching sheet, as well as the nonlinear radiation and nonuniform heat sink/source effects were investigated by Hayat et al. [65]. Geometry and boundary conditions are described in Fig. 6.32.

The results indicated that the magnitude of heat transfer rate enhances the radiation parameter *Rd*.

6.2.3.9 Electric field effect

Sheikholeslami [66] investigated the effect of Coulomb forces on Fe_3O_4–water nanofluid hydrothermal treatment using CVFEM. The effects of supplied voltage, solid volume fraction, and Reynolds number on velocity and temperature were studied. Studied geometry and the imposed boundary conditions are depicted in Fig. 6.33. The analyses were performed for Reynolds numbers within the range $3000 \leq Re \leq 6000$, volume fractions of nanoparticles $0 \leq \phi \leq 0.04$, and supplied voltage $0 \leq \Delta\varphi \leq 10kV$. The results indicated that the Nusselt numbers in the presence of electric field with supplied voltage $\Delta\varphi = 10kV$ and $Re = 3000$, 4000, 5000, and 6000, are respectively 1.09, 2.15, 8.96, and 1.76 times higher than those in the absence of electric field (Fig. 6.34). Also, they found that an increase in Coulomb force leads to enhancement in temperature gradient along the hot wall.

Table 6.27 Values of $c_{f,X}, c_{f,Y}$, Nu_X, and $Sh_{\phi,f}, Sh_{\phi,s}$ in dimensionless velocity slip parameters, $Sl1 = Sl2 = Sl3 = 0$ case (NSNF case), for different values ϕ, ω, and Ec with thermal Grashof number, $Gr_T = 15$, Prandtl number, $Pr = 6.07$, Schmidt number, $Sc = 1$, heterogeneous species reaction parameter, $K_s = 3$, species diffusivity ratio, $\delta = 1.2$, homogeneous species reaction parameter, $K_c = 0.5$, volume fraction of nanoparticles, $\phi = 0.1$, and temperature ratio parameter, $\theta_w = 1.2$ [64].

ϕ	ω	Ec	$c_{f,X}, c_{f,Y}$	Nu_X	$Sh_{\phi,f}, Sh_{\phi,s}$
0.10	0.1	0.01	(0.2562788, −0.1230915)	−2.928238	(−0.1525883, 0.1109024)
0.15			(0.2553029, −0.1323459)	−3.249280	(−0.1388327, 0.0998251)
0.20			(0.2515604, −0.1384417)	−3.136726	(−0.1288077, 0.0918303)
0.1	0.0	0.5	(0.1009191, −0.0449148)	1.740265	(−0.2015676, 0.1488301)
	0.1		(0.2562788, −0.1230915)	−2.928238	(−0.1525883, 0.1109024)
	0.2		(−0.0354173, −0.0665581)	12.68712	(0.0150379, −0.0192684)
	0.3		(−0.1408487, 0.0389394)	5.308011	(0.3695383, −0.2965039)
	0.4		(−0.1666929, 0.1578505)	15.68491	(0.3677396, −0.2742491)
0.1	0.1	0.010	(0.2562788, −0.1230915)	−2.928238	(−0.1525883, 0.1109024)
		0.015	(0.2922708, −0.1336644)	−7.311041	(−0.1672776, 0.1231619)
		0.020	(0.3363060, −0.1443469)	−13.26837	(−0.1847091, 0.1376183)

Sheikholeslami and Ganji [67] numerical studied the influence of electric field on Fe_3O_4−water nanofluid in a permeable enclosure using CVFEM (Fig. 6.35).

The analyses were performed for Reynolds numbers within the range$3000 \leq Re \leq 6000$, volume fractions of nanoparticles $0 \leq \phi \leq 0.05$, supplied voltage $0 \leq \Delta\varphi \leq 10kV$, radiation parameter $0 \leq Rd \leq 0.8$, Darcy number $10^2 \leq Da \leq 10^5$, and platelet-, cylinder-, brick-, and spherical-shaped nanoparticles. In the presence of Coulomb force, the heat transfer rate reduces with increasing Reynolds number. The Nu_{ave} number increases with increase in $\Delta\varphi$, Rd, and Da. By increasing Da number and thermal radiation, the temperature gradient near the bottom wall is enhanced. Table 6.31 illustrates the effect of nanoparticles shape on the Nu number for $Rd = 0.8, Re = 6000, \Delta\varphi = 10$, and $\phi = 0.05$.

A correlation for the Nu_{ave} number as a function of Re, Da, Rd, and $\Delta\varphi$ is proposed:

$$\begin{aligned} Nu_{ave} = & -0.99 + 0.04\Delta\varphi - 0.18Re^* + 1.7\log(Da) + 1.04Rd - 0.038\Delta\varphi Re^* + 0.03\Delta\varphi\log(Da) \\ & + 0.12\Delta\varphi Rd - 0.07Re^*\log(Da) - 0.32Re^*Rd + 0.25\log(Da)Rd + 0.013\Delta\varphi^2 \\ & - 0.05(Re^*)^2 - 0.203(\log(Da))^2 + 0.36Rd^2 \end{aligned} \tag{6.26}$$

6.3 Mass transfer characteristics

Since studies on mass transfer in natural convection using magnetic nanofluids were not found in the literature (see Fig. 6.1), this section summarizes the research on mass transfer in magnetic nanofluids concentrated on other practical applications (gas absorption and liquid−liquid extraction).

Table 6.28 **Values of $c_{f,X}, c_{f,Y}$, Nu_X and $Sh_{\phi,f}, Sh_{\phi,s}$ in $Sl1 = Sl2 = Sl3 = 1$ case (SNF case) for different values ϕ, ω and Ec with $Gr_T = 15, Pr = 6.07, Sc = 1, K_s = 3, \delta = 1.2, K_c = 0.5, \phi = 0.1, \varepsilon = 1.2$, and $\theta_w = 1.2$ [64].**

ϕ	ω	Ec	$c_{f,X}, c_{f,Y}$	Nu_X	$Sh_{\phi,f}, Sh_{\phi,s}$
0.10	0.1	0.01	(0.0560162, −0.3792867)	−1.771486	(−0.0997377, 0.0685075)
0.15			(0.0617594, −0.3479631)	−1.935940	(−0.0809551, 0.0536214)
0.20			(0.0671695, −0.2984370)	−1.822454	(−0.0701988, 0.0452039)
0.1	0.0	0.5	(0.0255195, −0.0651031)	1.211742	(−0.1488110, 0.1064557)
	0.1		(0.0560162, −0.3792867)	−1.771486	(−0.0997377, 0.0685075)
	0.2		(0.1351163, −0.0525337)	80.89808	(0.0705516, −0.0633436)
	0.3		(−0.0289759, 0.0240781)	2.743536	(0.4115527, −0.3264074)
	0.4		(−0.0475290, 2.109572)	10.61760	(0.3602968, −0.2734647)
0.1	0.1	0.010	(0.0560162, −0.3792867)	−1.771486	(−0.0997377, 0.0685075)
		0.015	(0.0583812, −0.4828996)	−3.320996	(−0.1336381, 0.0960250)
		0.020	(0.0597260, −0.5918945)	−4.720369	(−0.1717091, 0.1267045)

Table 6.29 **Values of $c_{f,X}, c_{f,Y}$, Nu_X, and $Sh_{\phi,f}, Sh_{\phi,s}$ in $Sl1 = Sl2 = Sl3 = 0$ case (NSNF case) for different values M, R, and Ra with $Gr_T = 15, Pr = 6.07, Sc = 1, K_s = 3, \delta = 1.2, K_c = 0.5, \phi = 0.1, \varepsilon = 1.2$, and $\theta_w = 1.2$ [64].**

M	R	Ra	$c_{f,X}, c_{f,Y}$	Nu_X	$Sh_{\phi,f}, Sh_{\phi,s}$
0.2	0.6	0.2	(0.2562788, −0.1230915)	−2.928238	(−0.1525883, 0.1109024)
0.3			(0.2334637, −0.1009187)	−1.443809	(−0.1643632, 0.1210293)
0.4			(0.2175370, −0.0874276)	−0.7051078	(−0.1699836, 0.1258711)
0.2	0.3	0.2	(0.2094377, −0.0501285)	2.694426	(−0.2009968, 0.1511547)
	0.4		(0.2190444, −0.0672077)	1.765377	(−0.1909657, 0.1429109)
	0.5		(0.2323948, −0.0895975)	0.2032675	(−0.1762420, 0.1307186)
0.2	0.6	0.2	(0.2562788, −0.1230915)	−2.928238	(−0.1525883, 0.1109024)
		0.3	(0.2375168, −0.1090559)	−2.081049	(−0.1505914, 0.1091483)
		0.4	(0.2255822, −0.1008032)	−1.439185	(−0.1484432, 0.1073260)

Table 6.30 Values of $c_{f,X}, c_{f,Y}$, Nu_X and $Sh_{\phi,f}, Sh_{\phi,s}$ in $Sl1 = Sl2 = Sl3 = 1$ case (SNF case) for different values M, R, and Ra with $Gr_T = 15, Pr = 6.07, Sc = 1, K_s = 3, \delta = 1.2, K_c = 0.5, \phi = 0.1, \varepsilon = 1.2$, and $\theta_w = 1.2$ [64].

M	R	Ra	$c_{f,X}, c_{f,Y}$	Nu_X	$Sh_{\phi,f}, Sh_{\phi,s}$
0.2	0.6	0.2	(0.0560162, −0.3792867)	−1.771486	(−0.0997377, 0.0685075)
0.3			(0.0531657, −0.1825216)	−0.6763688	(−0.1347723, 0.0972164)
0.4			(0.0509297, −0.1214993)	−0.1296357	(−0.1507141, 0.1101897)
0.2	0.3	0.2	(0.0479128, −0.0470211)	3.330746	(−0.1987365, 0.1483048)
	0.4		(0.0497754, −0.0773190)	2.348849	(−0.1803139, 0.1336132)
	0.5		(0.0522783, −0.1427559)	0.8094803	(−0.1512248, 0.1102764)
0.2	0.6	0.2	(0.0560162, −0.3792867)	−1.771486	(−0.0997377, 0.0685075)
		0.3	(0.0538292, −0.2554965)	−0.9236135	(−0.0992857, 0.0680474)
		0.4	(0.0521926, −0.2042031)	−0.2559794	(−0.0978323, 0.0668422)

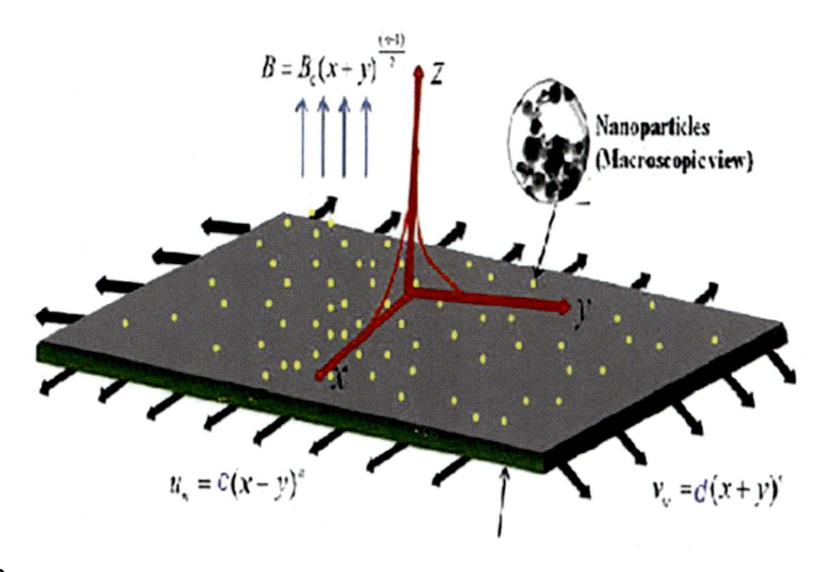

FIGURE 6.32

Flow geometry.

Reprinted from T. Hayat, M. Rashid, A. Alsaedi, Three dimensional radiative flow of magnetite-nanofluid with homogeneous–heterogeneous reactions, Results Phys. 8 (2018) 268–275, Copyright 2018, with permission from Elsevier.

Kang et al. [68] experimentally studied the heat and mass transfer in the falling film absorption process using two binary nanofluids, Fe–H_2O/LiBr and CNT–H_2O/LiBr. The considered concentrations were 0.01 and 0.1 wt.%. They found that the enhancement in the mass transfer is significantly higher than the enhancement in heat transfer. Also, their results indicated that the type of particle is more relevant than particle concentration, since they noted that for the same concentration (e.g., 0.1 wt.%), the average enhancement ratio in the mass transfer is 2.48 and 1.90 for CNT–H2O/LiBr and Fe–H2O/LiBr, respectively (Fig. 6.36).

Yang et al. [69] also performed an experimental study on the efficiency of ammonia–water falling film absorption using $ZnFe_2O_4$ and Fe_2O_3 nanoparticles. Their experimental results showed that the increase of the mass fraction of ammonia in the initial solution led to an increase in effective absorption ratio up to 70% and 50%, respectively, for Fe_2O_3 and $ZnFe_2O_4$ nanofluids. The enhancement in effective absorption ratio is attributed to the stability of the nanofluid, they showing that for a nanofluid with poor stability the microconvection and high mass transfer coefficient cannot be fully functioned.

Samadi et al. [70] proposed a new method for enhancing the mass transfer coefficient applied in the gas absorption processes using Al_2O_3/water, TiO_2/water, and Fe_3O_4/water nanofluids. In the case of Fe_3O_4/water nanofluids, the results indicated that the mass transfer rate increases with increasing volume fraction of nanoparticles and also that for all studied nanofluids the mass transfer coefficient is lower compared to water. By applying a downward magnetic field an enhancement in mass flux and mass transfer coefficient was noted, compared to those without a magnetic field.

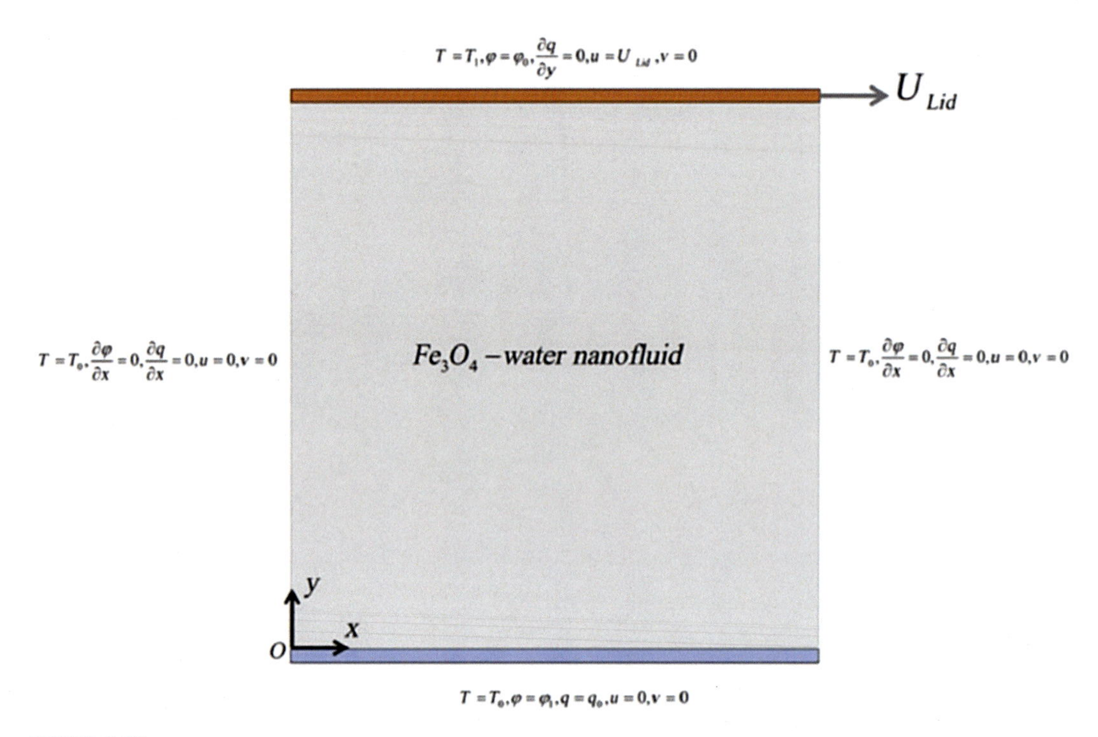

FIGURE 6.33

Geometry of the physical configuration.

Reprinted from M. Sheikholeslami, Influence of Coulomb forces on Fe_3O_4–H_2O nanofluid thermal improvement, Int. J. Hydrog. Energy 42 (2017) 821–829, Copyright 2017, with permission from Elsevier.

Olle et al. [71] carried out a study on the enhancement of oxygen mass transfer using Fe_3O_4 nanoparticles and found that the gas–liquid oxygen mass transfer was enhanced up to sixfold (600%) at nanoparticles concentrations lower than 1.0%. Also, they noted an increase in both the mass transfer coefficient and the gas–liquid interfacial area by adding nanoparticles to the base fluid.

The effect of magnetic iron oxide nanoparticles on gas–liquid mass transfer rates in a wetted wall column as well in a capillary tube was investigated by Komati and Suresh [72]. Their results showed an increase in liquid phase mass transfer coefficients. This increase depends on both the nanoparticle concentration and nanoparticle size. Also, they found that the modified Sherwood number may be a parameter that determines the magnitude of the mass transfer intensification.

The behavior of nanofluid containing magnetite and alumina nanoparticles single drops in the liquid–liquid extraction process was investigated by Saien and Bamdadi [73]. For 0.002 wt.% magnetite and alumina nanoparticles, the maximum enhancements in the mass transfer rate are 157% and 121%, respectively. The main reasons are the microconvection and particle aggregation due to the interpenetration layers.

Manikandan et al. [74] studied the volumetric mass transfer coefficient for the transfer of oxygen from air bubble to Fe_2O_3–water nanofluid in an agitated, aerated bioreactor and found an

Re = 3000
Re = 4000
Re = 5000
Re = 6000
20
16
12
8
4
Nu_{ave}
0
2
4
6
8
10
Δφ(kV)

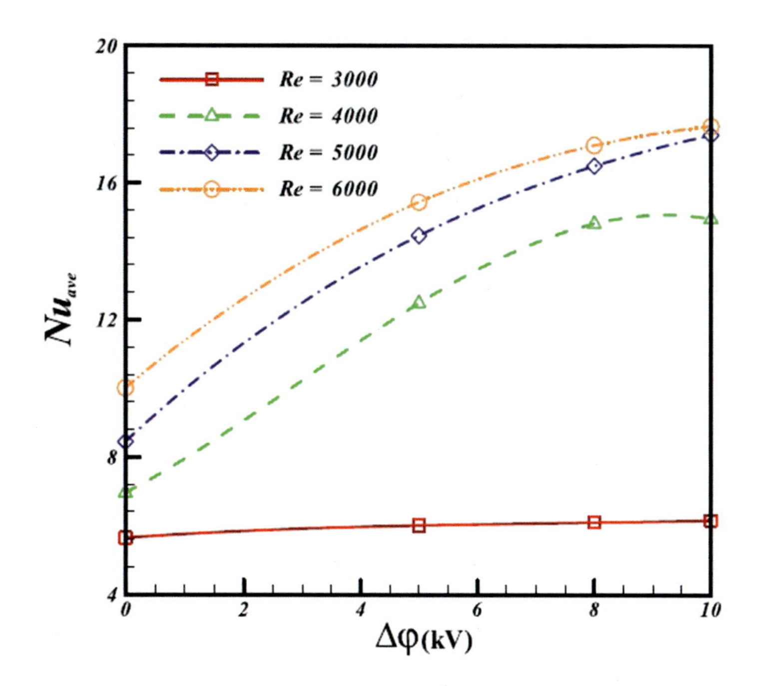

FIGURE 6.34

Effects of Reynolds number and supplied voltage on average Nusselt number.

Reprinted from M. Sheikholeslami, Influence of Coulomb forces on Fe_3O_4–H_2O nanofluid thermal improvement, Int. J. Hydrog. Energy 42 (2017) 821–829, Copyright 2017, with permission from Elsevier.

$T = T_0, \frac{\partial \varphi}{\partial n} = 0, \frac{\partial q}{\partial n} = 0, u = 0, v = 0$
$T = T_0, \varphi = \varphi_0, \frac{\partial q}{\partial y} = 0, u = 0, v = 0$
Nanofluid
Porous media
$T = T_0, \frac{\partial \varphi}{\partial n} = 0, \frac{\partial q}{\partial n} = 0, u = 0, v = 0$
y
x
O
U_{Lid}
$T = T_1, \varphi = \varphi_1, q = q_0, u = U_{Lid}, v = 0$

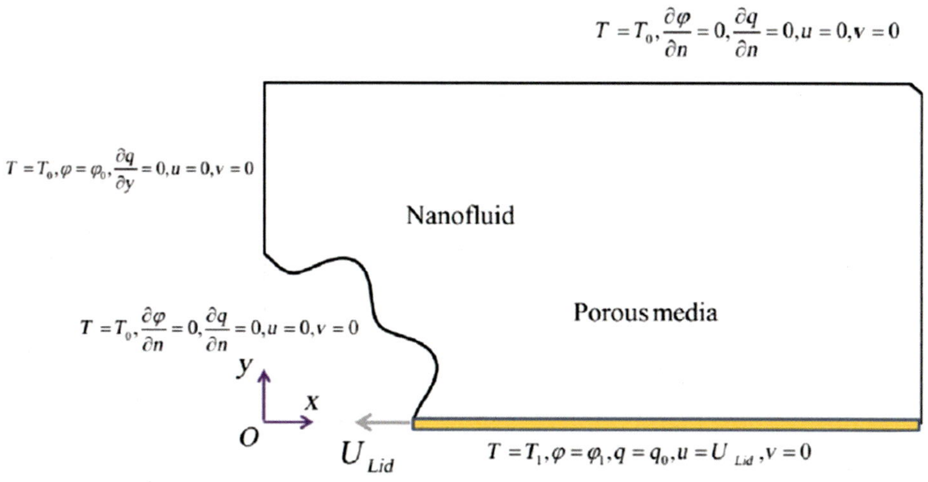

FIGURE 6.35

Geometry of the physical configuration.

Reprinted from M. Sheikholeslami, D.D. Ganji, Influence of electric field on Fe_3O_4–water nanofluid radiative and convective heat transfer in a permeable enclosure, J. Mol. Liq. 250 (2018) 404–412, Copyright 2018, with permission from Elsevier.

Table 6.31 The effect of nanoparticles shape on the *Nu* number [67].

	Da	
Nanoparticles shape	10^2	10^5
Spherical	3.225443	3.741018
Brick	3.267538	3.780537
Cylinder	3.329230	3.838521
Platelet	3.375827	3.882675

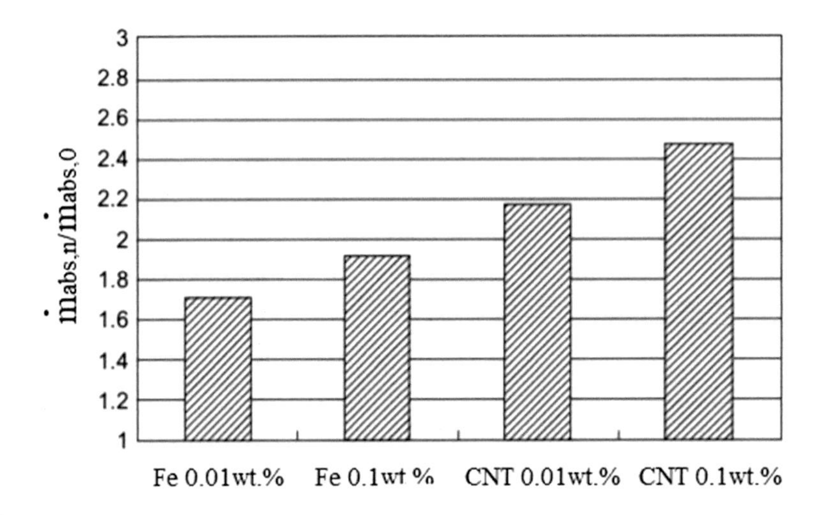

FIGURE 6.36

Enhancement in mass transfer.

Reprinted from Y.T. Kang, H.J. Kim, K.I. Lee, Heat and mass transfer enhancement of binary nanofluids for H_2O/LiBr falling film absorption process, Int. J. Refrig. 250 (2008) 850–856, Copyright 2008, with permission from Elsevier.

enhancement of 63% for a concentration of 0.065 wt.%. Also, they showed that the adding of nanoparticles contributed to enhancing the oxygen transfer through a "grazing effect." The "grazing effect" is the transfer phenomenon of a gas from the gas–liquid interface to the bulk of the liquid.

Wu et al. [75] investigated the enhancement of ammonia/water bubble absorption using Fe_3O_4 nanoferrofluid in combination with an external magnetic field. The results revealed that the combined effect of the nanofluid with the external magnetic field may lead to a significant enhancement of ammonia/water bubble absorption. Thus by increasing magnetic field intensity the magnetic field gradient is enhanced and it strengthens the squeezing action on the bubbles, while increasing nanoparticles concentration intensifies the striking action of the nanoparticles on the bubbles and increases the surface tension gradient in the solution.

Komati and Suresh [76] investigated the mass transfer coefficient in the CO_2 absorption process in a wetted wall column using magnetite/MDEA. As in the other studies, they noted an enhancement in the mass transfer coefficient with increasing concentration volume, the maximum

Table 6.32 The enhancement of mass transfer.

Nanofluids type	Experimental method	Nanoparticles volume fraction and particle size (nm)	Enhancement (ratio)	Refs.
$Fe - H_2O/LiBr$	$H_2O/LiBr$ falling film absorption	0.01/0.1 wt.%/100	1.9	[68]
$Fe_2O_3 - NH_3/H_2O$	NH_3 falling film absorption	0.2%/ < 30	1.7	[69]
$ZnFe_2O_4 - NH_3/H_2O$	NH_3 falling film absorption	0.1%/ < 30	1.5	
Fe_3O_4/water	CO_2 WWC absorption	$0 - 0.024/8 \pm 3$	No enhancement	[70]
Fe_3O_4/water	CO2 WWC absorption with magnetic field	$0 - 0.024/8 \pm 3$	1.2235	
Fe_3O_4/oleic acid	O_2 mass transfer	< 1/20	7	[71]
Fe_3O_4/water	O_2 mass transfer	0.05 − 1/13.2/15/23	2.74	[72]
Fe_3O_4/water	CO_2 mass transfer	0.05 − 0.4/13.2/15/23	1.93	
Fe_3O_4/toluene/aceticacid/ water	Liquid–liquid extraction	0.0005 − 0.005 wt.%/17	2.57	[73]
Fe_2O_4/water	O_2 absorption in H_2O-based nanofluids in an agitated, aerated bioreactor	0.022 − 0.065 wt.% /20–50[a], 120[b]	1.63	[74]
$Fe_3O_4 - NH_3/H_2O$ solution	NH_3 absorption by NH_3/H_2O-based nanofluid under external magnetic field	0.05 − 0.20 wt.%	1.0812	[75]
$Fe_3O_4 - CO_2$/MDEA solution	CO_2 absorption in nanofluid in a wetted wall column	0.02 − 0.39 vol.%/28 − 38[b]	1.928 in the absence of magnetic field; no enhancement in the presence of magnetic field is noticed	[76]

[a] *Nanoparticles size in powder state before combining with the base fluid.*
[b] *Nanoparticles size in suspension.*

enhancement being 92.8% for a nanoparticle concentration of 0.39%. Also, they concluded that applying a magnetic field does not achieve an enhancement in the mass transfer rate.

The results on the mass transfer are summarized in Table 6.32.

6.4 Conclusions and future prospects and trends

In this chapter, the studies related to heat and mass transfer of the magnetic nanofluids in natural convection have been reviewed for different boundary conditions and physical situations. The majority of the studies described in this chapter relate to the use of the magnetic nanofluids in natural convection in the presence of a uniform magnetic field. The reviewed studies emphasized the existence of two important phenomena: thermomagnetic convection and magnetoviscous effect. The thermomagnetic convection assumes that the thermophysical properties of magnetic nanofluids are constant and independent from the applied magnetic field, while the magnetoviscous effect consists of the change of the viscosity in the presence of a magnetic field. Different dimensional and nondimensional parameters based on temperature difference (ΔT) between heating and cooling surface temperature of walls are used to study the natural convective heat transfer, and the results showed that magnetic nanofluids have great potential to enhance convective heat transfer. Since the majority of the studies are focused on numerical simulations and modeling, the future investigation should be a focus on the experimental validation of the existing models. The reviewed literature indicates that the implementation of magnetic nanofluids may be an important alternative to the traditional thermal systems, if the main issues of those new working fluids (sedimentation, stability, high cost, and production difficulties) are solved. To understand the mechanisms responsible for improvement/reduction of heat and mass transfer characteristics, more theoretical and experimental investigations are required, starting from the synthesis and characterization of magnetic nanofluids up to the practical applications. In most of the research on mass transfer, only the effect of nanoparticles concentration is studied without taking into other parameters like the size, shape, or chemical nature of magnetic particles. The influence of those parameters on mass transfer as well as an extension of studies in natural convection using magnetic nanofluids may constitute future studies in this field.

Acknowledgment

This work was supported by a grant of the Romanian Ministry of Education and Research, CNCS - UEFISCDI, project number PN-III-P4-ID-PCE-2020-0353, within PNCDI III.

Nomenclature

a	thermal diffusivity, m^2/s
A	area, m^2
AR	aspect ratio
$\vec{B}$	magnetic field, kg/m^2A
B_0	applied magnetic field, G
Br	Brinkmann number
c_f	control of skin friction coefficients
c_p	specific heat, J/kgK
D	diameter, m
D	mass diffusivity, m^2/s
Da	Darcy number
Ec	Eckert number
$\vec{F}$	Lorentz force, N/m^3
g	gravity, m/s^2
Gr	Grashof number

h convective mass transfer film coefficient, m/s
H height, m
H magnetic field, mT
$\overline{H}$ external magnetic source
H_s heat source/sink variable
Ha Hartmann number
$\vec{j}$ electric current density, A/m^2
k thermal conductivity, W/mK
K permeability, m^2
k_e the effective thermal conductivity, W/mK
L characteristic length, m, or gap width or hypothetical gap width, $R_0 - R_i$, m
Le Lewis number
m nanoparticles shape factor
Mn_f magnetism Numbers
Nu Nusselt number
Pr Prandtl number
q'' heat flux, W/m^2
$\dot{Q}$ heating power, W
R_i inner body radius for the spheres, but for the cylinders and cubes it is a hypothetical radius of an inner sphere of a volume equal to the cylindrical or cubical inner body, m
R_0 outer body hypothetical radius equal to the radius of a sphere having a volume equal to the volume of the outer body, m
Ra Rayleigh number
Ra_v Rayleigh number based on the vertical temperature difference ΔT_v
Rd Radiation parameter
Ri Richardson number
Re Reynolds number
Sc Schmidt number
Sh Sherwood numbers
T temperature, K
T_b core temperature at the bottom, K
T_t core temperature at the top, K
ΔT temperature difference, K
ΔT_v vertical temperature difference, $T_t - T_b$, K
U velocity, m/s

Greek symbols

β thermal expansion coefficient, $1/K$
δ distance, m
$\Delta\varphi$ supplied voltage, kV
ε porosity
μ strength of the magnetic source
μ_r strength ratio of the two magnetic sources #1 and #2
μ dynamic viscosity, $Pa\ s$
ν kinematic viscosity, m^2/s
ρ density, kg/m^3
σ electrical conductivity $1/\Omega m$

ϕ nanoparticle volume fraction
ϕ electric potential, m^2kg/s^3A

Subscript
MNf Magnetic nanofluid
bf Base fluid

Abbreviations

CVFEM control volume-based finite element method
MHD magnetohydrodynamic

References

[1] I. Nkurikiyimfura, Y. Wang, Z. Pan, Heat transfer enhancement by magnetic nanofluids—a review, Renew. Sustain. Energy Rev. 21 (2013) 548–561.
[2] S. Odenbach, Magnetoviscous Effects in Ferrofluids, Springer-Verlang, Berlin/Heidelberg, 2002, pp. 185–201.
[3] J. Philip, P. Shima, B. Raj, Evidence for enhanced thermal conduction through percolating structures in nanofluids, Nanotechnology 19 (2008) 305706.
[4] W. Lian, Y. Xuan, Q. Li, Characterization of miniature automatic energy transport devices based on the thermomagnetic effect, Energy Convers. Manag. 50 (2009) 35–42.
[5] W. Lian, Y. Xuan, Q. Li, Design method of automatic energy transport devices based on the thermomagnetic effect of magnetic fluids, Int. J. Heat. Mass. Transf. 52 (2009) 5451–5458.
[6] R.K. MacGregor, A.P. Emery, Free convection through vertical plane layers: moderate and high Prandtl number fluids, J. Heat. Transf. 91 (1969) 391–401.
[7] N. Weber, R.E. Powe, E.H. Bishop, J.A. Scanlan, Heat transfer by natural convection between vertically eccentric spheres, ASME Transaction, J. Heat. Transf. 95 (1973) 47–52.
[8] J.P. Holman, Heat Transfer, fourth ed., McGraw-Hill, New York, 1976.
[9] L.B. Evans, N.E. Stefany, An experimental study of transient heat transfer to liquids in cylindrical enclosures, in: CEP Symposium Series, 62, 209 (1966).
[10] R.O. Warrington, R.E. Powe, The transfer of heat by natural convection between bodies and their enclosures, Int. J. Heat. Mass. Transf. 28 (1985) 319–330.
[11] M. Holzbecher, A. Steiff, Laminar and turbulent free convection vertical cylinders with internal heat generation, Int. J. Heat. Mass. Transf. 38 (1995) 2893–2903.
[12] A. Emery, N.C. Chu, Heat transfer across vertical layers, J. Heat. Transf. 87 (1965) 110–114.
[13] M. Jakob, Heat Transfer, vol. 1, John Wiley, New York, 1949.
[14] S. Globe, D. Dropkin, Natural-convection heat transfer in liquids confined by two horizontal plates and heated from below, J. Heat. Transf. 81 (1959) 24–28.
[15] E. Schmidt, Free convection in horizontal fluid spaces heated from below, in: Proc. International Heat Transfer Conference, Boulder, CO, ASME, 1961.
[16] R.J. Goldstein, T.Y. Chu, Thermal convection in a horizontal layer of air, Prog. Heat. Mass. Transf. 2 (1969) 55–75.
[17] K.G.T. Hollands, G.D. Raithby, L. Konicek, Correlation equations for free convection heat transfer in horizontal layers of air and water, Int. J. Heat. Mass. Transf. 18 (1975) 879–884.
[18] H. Krasshold, Wärmeabgabe von zylindrischen flussigkeitsschichten bei natürlichen Konvektion, Forsch. Geb. Ingenieurwes 2 (1931) 165.
[19] W. Beckmann, DieWärmeübertragung in zylindrischen Gasschichten bei natürlicher Konvektion, Forsch. Geb. Ingenieurwes 2 (1931) 186.

[20] C.Y. Liu, W.K. Mueller, F. Landis, Natural convection heat transfer in long horizontal cylindrical annuli, Int. Dev. Heat Transf. Am. Soc. Mech. Engrs 5 (1961) 976–984. New York.

[21] M. Sheikholeslami, M.M. Rashidi, Effect of space dependent magnetic field on free convection of Fe_3O_4–water nanofluid, J. Taiwan. Inst. Chem. Eng. 56 (2015) 6–15.

[22] M. Sheikholeslami, S.A. Shehzad, CVFEM for influence of external magnetic source on Fe_3O_4–H_2O nanofluid behavior in a permeable cavity considering shape effect, Int. J. Heat. Mass. Transf. 115 (2017) 180–191.

[23] M. Sheikholeslami, M. Shamlooei, Fe_3O_4–H_2O nanofluid natural convection in presence of thermal radiation, Int. J. Hydrog. Energy 42 (2017) 5708–5718.

[24] M. Sheikholeslami, D.D. Ganji, Free convection of Fe_3O_4–water nanofluid under the influence of an external magnetic source, J. Mol. Liq. 229 (2017) 530–540.

[25] M. Sheikholeslami, D.D. Ganji, R. Moradi, Heat transfer of Fe_3O_4–water nanofluid in a permeable medium with thermal radiation in existence of constant heat flux, Chem. Eng. Sci. 174 (2017) 326–336.

[26] M. Sheikholeslami, M.K. Sadoughi, Numerical modeling for Fe_3O_4–water nanofluid flow in porous medium considering MFD viscosity, J. Mol. Liq. 242 (2017) 255–264.

[27] M. Sheikholeslami, Numerical simulation of magnetic nanofluid natural convection in porous media, Phys. Lett. A 381 (2017) 494–503.

[28] M.S. Krakov, I.V. Nikiforov, Natural convection in a horizontal cylindrical enclosure filled with a magnetic nanofluid: influence of the uniform outer magnetic field, Int. J. Therm. Sci. 133 (2018) 41–54.

[29] M. Sheikholeslami, S.A. Shehzad, Numerical analysis of Fe_3O_4–H_2O nanofluid flow in permeable media under the effect of external magnetic source, Int. J. Heat. Mass. Transf. 118 (2018) 182–192.

[30] M. Sheikholeslami, M. Shamlooei, R. Moradi, Numerical simulation for heat transfer intensification of nanofluid in a porous curved enclosure considering shape effect of Fe_3O_4 nanoparticles, Chem. Eng. Processing: Process. Intensif. 124 (2018) 71–82.

[31] M. Hatami, J. Zhou, J. Geng, D. Jing, Variable magnetic field (VMF) effect on the heat transfer of a half annulus cavity filled by Fe_3O_4–water nanofluid under constant heat flux, J. Magnetism Magnetic Mater. 451 (2018) 173–182.

[32] X.-h Sun, M. Massoudi, N. Aubry, Z.-h Chen, W.-T. Wu, Natural convection and anisotropic heat transfer in a ferro-nanofluid under magnetic field, Int. J. Heat. Mass. Transf. 133 (2019) 581–595.

[33] M. Izadi, R. Mohebbi, A.A. Delouei, H. Sajjadi, Natural convection of a magnetizable hybrid nanofluid inside a porous enclosure subjected to two variable magnetic fields, Int. J. Mech. Sci. 151 (2019) 154–169.

[34] M. Sheikholeslami, S.A.M. Mehryan, A. Shafee, M.A. Sheremet, Variable magnetic forces impact on magnetizable hybrid nanofluid heat transfer through a circular cavity, J. Mol. Liq. 277 (2019) 388–396.

[35] L. Shi, Y. He, Y. Hu, X. Wang, Controllable natural convection in a rectangular enclosure filled with Fe_3O_4@CNT nanofluids, Int. J. Heat. Mass. Transf. 140 (2019) 399–409.

[36] A. Jafarimoghaddam, Two-phase modeling of magnetic nanofluids jets over a stretching/shrinking wall, Therm. Sci. Eng. Prog. 8 (2018) 375–384.

[37] C. Qi, J. Tang, Z. Ding, Y. Yan, L. Guo, Y. Ma, Effects of rotation angle and metal foam on natural convection of nanofluids in a cavity under an adjustable magnetic field, Int. Commun. Heat. Mass. Transf. 109 (2019) 104349.

[38] L.M. Al-Balushi, M.J. Uddin, M.M. Rahman, Natural convective heat transfer in a square enclosure utilizing magnetic nanoparticles, Propuls. Power Res. 8 (2019) 194–209.

[39] S.O. Giwa, M. Sharifpur, J.P. Meyer, Effects of uniform magnetic induction on heat transfer performance of aqueous hybrid ferrofluid in a rectangular cavity, Appl. Therm. Eng. 170 (2020) 115004.

[40] J.C. Joubert, M. Sharifpur, A. Brusly Solomon, J.P. Meyer, Enhancement in heat transfer of a ferrofluid in a differentially heated square cavity through the use of permanent magnets, J. Magnetism Magnetic Mater. 443 (2017) 149–158.

[41] N. Rahimpour, M.K. Moraveji, Free convection of water–Fe_3O_4 nanofluid in an inclined cavity subjected to a magnetic field: CFD modeling, sensitivity analysis, Adv. Powder Technol. 28 (2017) 1573–1584.
[42] A.S. Dogonchi, M. Waqas, S.M. Seyyedi, M. Hashemi-Tilehnoee, D.D. Ganji, CVFEM analysis for Fe_3O_4–H_2O nanofluid in an annulus subject to thermal radiation, Int. J. Heat. Mass. Transf. 132 (2019) 473–483.
[43] A.S. Dogonchi, Hashim, Heat transfer by natural convection of Fe_3O_4–water nanofluid in an annulus between a wavy circular cylinder and a rhombus, Int. J. Heat. Mass. Transf. 130 (2019) 320–332.
[44] A.S. Dogonchi, Z. Asghar, M. Waqas, CVFEM simulation for Fe_3O_4–H_2O nanofluid in an annulus between two triangular enclosures subjected to magnetic field and thermal radiation, Int. Commun. Heat. Mass. Transf. 112 (2020) 104449.
[45] I. Tlili, M.M. Bhatti, S.M. Hamad, A.A. Barzinjy, M. Sheikholeslami, A. Shafee, Macroscopic modeling for convection of hybrid nanofluid with magnetic effects, Phys. A 534 (2019) 122136.
[46] M. Molana, A.S. Dogonchi, T. Armaghani, A.J. Chamkha, D.D. Ganji, I. Tlili, Investigation of hydrothermal behavior of Fe_3O_4–H_2O nanofluid natural convection in a novel shape of porous cavity subjected to magnetic field dependent (MFD) viscosity, J. Energy Storage 30 (2020) 101395.
[47] T.D. Manh, N.D. Nam, K. Jacob, A. Hajizadeh, H. Babazadeh, M. Mahjoub, et al., Simulation of heat transfer in 2D porous tank in appearance of magnetic nanofluid, Phys. A 550 (2020) 123937.
[48] F. Saba, N. Ahmed, U. Khan, S.T. Mohyud-Din, A novel coupling of CNT–Fe_3O_4/water hybrid nanofluid for improvements in heat transfer for flow in an asymmetric channel with dilating/squeezing walls, Int. J. Heat. Mass. Transf. 136 (2019) 186–195.
[49] R. Ahmad, M. Mustafa, T. Hayat, A. Alsaedi, Numerical study of MHD nanofluid flow and heat transfer past a bidirectional exponentially stretching sheet, J. Magnetism Magnetic Mater. 407 (2016) 69–74.
[50] S.S. Ghadikolaei, Kh Hosseinzadeh, D.D. Ganji, M. Hatami, Fe_3O_4–$(CH_2OH)_2$ nanofluid analysis in a porous medium under MHD radiative boundary layer and dusty fluid, J. Mol. Liq. 258 (2018) 172–185.
[51] S.A.M. Mehryan, M.A. Sheremet, M. Soltani, M. Izadi, Natural convection of magnetic hybrid nanofluid inside a double-porous medium using two-equation energy model, J. Mol. Liq. 277 (2019) 959–970.
[52] H. Sajjadi, A. Amiri Delouei, M. Izadi, R. Mohebbi, Investigation of MHD natural convection in a porous media by double MRT lattice Boltzmann method utilizing MWCNT–Fe_3O_4/water hybrid nanofluid, Int. J. Heat. Mass. Transf. 132 (2019) 1087–1104.
[53] R.J. Singh, A.J. Chandy, Numerical investigations of the development and suppression of the natural convection flow and heat transfer in the presence of electromagnetic force, Int. J. Heat. Mass. Transf. 157 (2020) 119823.
[54] M. Sheikholeslami, S.A. Shehzad, Thermal radiation of ferrofluid in existence of Lorentz forces considering variable viscosity, Int. J. Heat. Mass. Transf. 109 (2017) 82–92.
[55] I. Jelodari, A.H. Nikseresht, Effects of Lorentz force and induced electrical field on the thermal performance of a magnetic nanofluid-filled cubic cavity, J. Mol. Liq. 252 (2018) 296–310.
[56] O. Ghaffarpasand, Numerical study of MHD natural convection inside a sinusoidally heated lid-driven cavity filled with Fe_3O_4–water nanofluid in the presence of Joule heating, Appl. Math. Model. 40 (2016) 9165–9182.
[57] M. Sajid, S.A. Iqbal, M. Naveed, Z. Abbas, Joule heating and magnetohydrodynamic effects on ferrofluid (Fe_3O_4) flow in a semi-porous curved channel, J. Mol. Liq. 222 (2016) 1115–1120.
[58] B. Finlayson, Convective instability of ferromagnetic fluids, J. Fluid Mech. 40 (1970) 753–767.
[59] G. Vaidyanathan, R. Sekar, R. Balasubramanian, Ferroconvective instability of fluids saturating a porous medium, Int. J. Eng. Sci. 29 (1991) 1259–1267.
[60] A.M. Sunil, A nonlinear stability analysis for thermoconvective magnetized ferrofluid saturating a porous medium, Transp. Porous Media 76 (2009) 327–343.
[61] A. Mahajan, M.K. Sharma, Convection in a magnetic nanofluid saturating a porous medium under the influence of a variable gravity field, Eng. Sci. Tech. An. Int. J. 21 (2018) 439–450.

[62] P.K. Kameswaran, S. Shaw, P. Sibanda, P.V.S.N. Murthy, Homogeneous–heterogeneous reactions in a nanofluid flow due to a porous stretching sheet, Int. J. Heat. Mass. Transf. 57 (2013) 465–472.

[63] M. Sajid, S.A. Iqbal, M. Naveed, Z. Abbas, Effect of homogeneous–heterogeneous reactions and magnetohydrodynamics on Fe_3O_4 nanofluid for the Blasius flow with thermal radiations, J. Mol. Liq. 233 (2017) 115–121.

[64] R. Kumar, R. Kumar, S.A. Shehzad, M. Sheikholeslami, Rotating frame analysis of radiating and reacting ferro-nanofluid considering Joule heating and viscous dissipation, Int. J. Heat. Mass. Transf. 120 (2018) 540–551.

[65] T. Hayat, M. Rashid, A. Alsaedi, Three dimensional radiative flow of magnetite-nanofluid with homogeneous–heterogeneous reactions, Results Phys. 8 (2018) 268–275.

[66] M. Sheikholeslami, Influence of Coulomb forces on Fe_3O_4–H_2O nanofluid thermal improvement, Int. J. Hydrog. Energy 42 (2017) 821–829.

[67] M. Sheikholeslami, D.D. Ganji, Influence of electric field on Fe_3O_4–water nanofluid radiative and convective heat transfer in a permeable enclosure, J. Mol. Liq. 250 (2018) 404–412.

[68] Y.T. Kang, H. Kim, K.I. Lee, Heat and mass transfer enhancement of binary nanofluids for H_2O/LiBr falling film absorption process, Int. J. Refrig. 31 (2008) 850–856.

[69] L. Yang, K. Du, X.F. Niu, B. Cheng, Y.F. Jiang, Experimental study on enhancement of ammoniaewater falling film absorption by adding nanoparticles, Int. J. Refrig. 34 (2011) 640–647.

[70] Z. Samadi, M. Haghshenasfard, A. Moheb, CO_2 absorption using nanofluids in a wetted-wall column with external magnetic field, Chem. Eng. Technol. 37 (2014) 462–470.

[71] B. Olle, S. Bucak, T.C. Holmes, L. Bromberg, T.A. Hatton, D.I.C. Wang, Enhancement of oxygen mass transfer using functionalized magnetic nanoparticles, Ind. Eng. Chem. Res. 45 (2006) 4355–4363.

[72] S. Komati, A.K. Suresh, Anomalous enhancement of interphase transport rates by nanoparticles: effect of magnetic iron oxide on gas-liquid mass transfer, Ind. Eng. Chem. Res. 49 (2010) 390–405.

[73] J. Saien, H. Bamdadi, Mass transfer from nanofluid single drops in liquid–liquid extraction process, Ind. Eng. Chem. Res. 51 (2012) 5157–5166.

[74] S. Manikandan, N. Karthikeyan, K.S. Suganthi, K.S. Rajan, Enhancement of volumetric mass transfer coefficient for oxygen transfer using Fe_2O_3–water nanofluids, Asian J. Sci. Res. 5 (2012) 271–277.

[75] W.D. Wu, G. Liu, S.-X. Chen, H. Zhang, Nanoferrofluid addition enhances ammonia/water bubble absorption in an external magnetic field, Energy Build. 57 (2013) 268–277.

[76] S. Komati, A.K. Suresh, CO_2 absorption into amine solutions: a novel strategy for intensification based on the addition of ferrofluids, J. Chem. Technol. Biotechnol. 83 (2008) 1094–1100.

CHAPTER

Conjugate heat and mass transfer in nanofluids

7

Alina Adriana Minea[1], Angel Huminic[2] and Gabriela Huminic[2]

[1]Gheorghe Asachi Technical University Iasi, Iasi, Romania [2]Transilvania University of Brasov, Brasov, Romania

7.1 Introduction

Heat and mass transfer processes are, in several industrial uses, a result of buoyancy effects affected by thermal diffusion and chemical species. Consequently, the investigation of conjugate effects of both heat and mass transmission is opportune on behalf of the enhancement of knowledge, for example, in energy, polymer/ceramics production, boosted oil recovery, food industry, the temperature and moisture distribution in several applications, and environmental studies.

Nowadays, nanofluids (NFs) are used in many industrial processes,. As their development has increased over the years, now researchers are capable of developing new and improved features of these enhanced fluids. Basically, NFs are a stable and homogenous mixture of a number of base fluids with several types of nanoparticles (NPs) with lower dimensions (<100 nm). Base fluids are acknowledged in the open literature as water, ethylene glycol, polyethylene glycol, oils, or even ionic liquids. With regard to NPs, the research extends from oxides, metals, carbon nanotubes, and graphene to chemically stable hybrid NPs.

Currently, the combined analysis of heat and mass transmission in NFs is extensively studied, especially because of the increase of thermal properties (e.g., thermal conductivity and isobaric specific heat) in a number of heat transfer routes. It has to be outlined here that the heat transfer augmentation is directly beneficial for energy consumption reduction, decreasing the processing time, and increasing the final product quality.

Nevertheless, there are two major factors involved in this theory: the base fluid and the NP type; these factors generate also the final thermal performance of the NF [1].

As a final thought, NFs have remarkable applications in solar heating and cooling processes (e.g., engine, electronics, heat exchanger equipment, and refrigeration processes), and these benefits are determined due to their increased convective heat transfer coefficient.

The intensification in NFs research is shown in Fig. 7.1, which portrays the number of studies in NFs, as well as their heat or mass transfer specific research. One can easily note that the studies dedicated to mass transfer and combined heat and mass transfer are rather scarce. A conclusion that can be drawn is that the focus of NFs research has been on their heat transfer capabilities and the thermophysical properties augmentation.

Nanofluids and Mass Transfer. DOI: https://doi.org/10.1016/B978-0-12-823996-4.00005-7

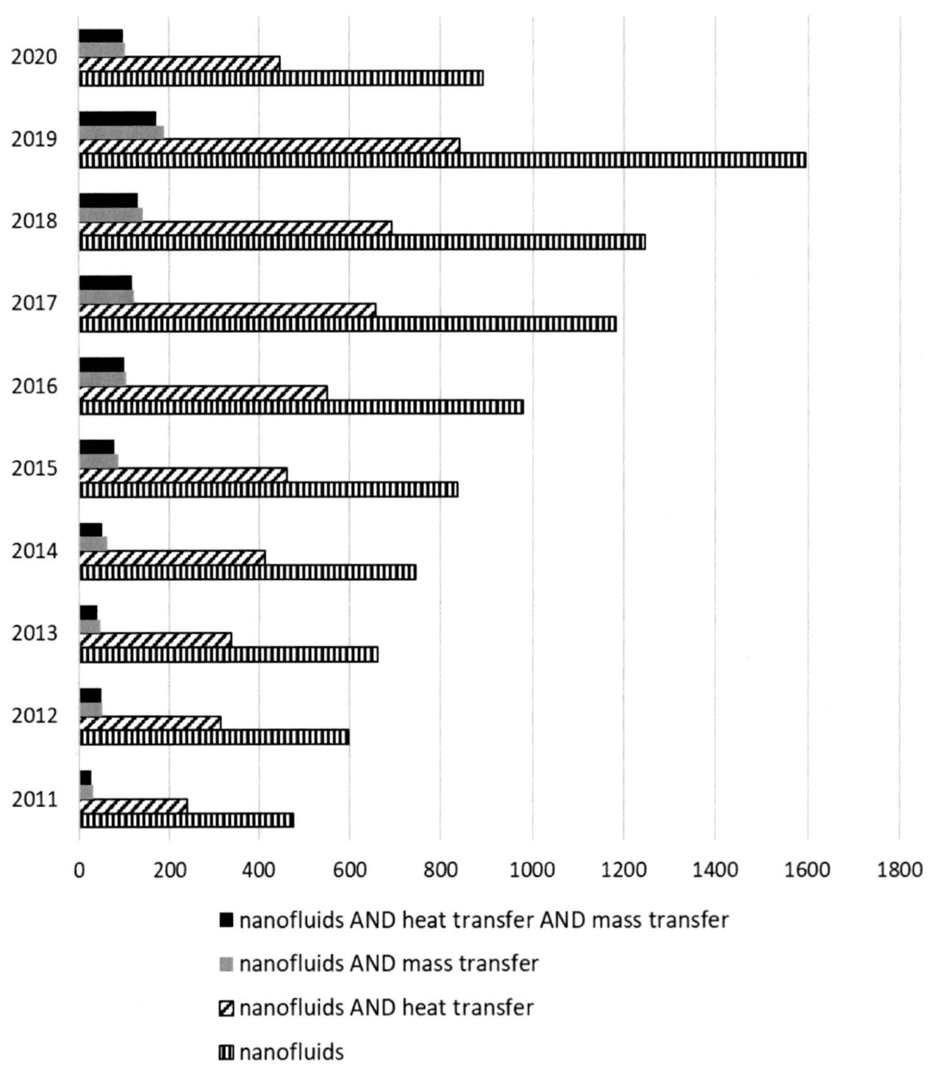

FIGURE 7.1

Nanofluids publications on different research areas, in the last 20 years.

SCOPUS.

Later in this chapter the research on combined heat and mass transfer processes will be reviewed and we will refer especially to NPs' effects on heat and mass transfer; combined heat and mass transfer mechanisms, and both liquid–liquid and liquid–gas processes (i.e., flow and pool boiling, absorption systems, etc.). Even if in the open literature there exist several parametric studies about simultaneous heat and mass transfer in NFs, these are not concrete, therefore these will not be discussed in this chapter.

7.2 Mechanisms of heat transfer in nanofluids

The mechanism of improving heat transfer was first considered in strict connection with the increase of thermal conductivity by adding solid NPs to a less conductive fluid. Keblinski et al. [2] recognized four probable reasons for this abnormal upsurge in thermal conductivity:

1. Brownian motion: NP move inside the liquid and possibly collide, thus creating an upsurge in thermal conductivity, transporting heat between NPs.
2. Molecular-level layering of the fluid at the fluid/NP interface.
3. The heat transport nature inside the NP: for instance, in crystalline solids (i.e., metallic NPs), heat is passed by phonons by transmitting lattice vibrations.
4. The NPs influence clustering: clusters may create an increase in thermal conductivity; however, large clusters are ineffective since they create sedimentation.

Pang et al. [1] performed a very good review on joint heat and mass transfer features in NFs and affirmed that, in the literature, the mechanisms for heat transfer augmentation are identified in this proportion: Brownian motion (accounted for 32.9%), interfacial layer theory (Kapitza resistance) (accounted for 22.3%), Brownian motion/aggregation and diffusion (accounted for 10.5%), and Brownian motion and thermophoresis forces (accounted for 9.4%). Plus, a number of other theories have been given: electrical double layer, near-field radiation, thermophoretic forces, flows shear thinning behavior, and thermal conductivity enrichment [1]. These latter mechanisms for heat transfer enhancement were referred to by Yu et al. [3], Chandrasekar and Suresh [4], and Wang and Fan [5].

7.3 Mechanisms of mass transfer in nanofluids

Basically, mass transfer involves two regions of different chemical concentrations in the same fluid or between two fluids. Mass transfer is actually due to a chemical species movement from a highly concentrated area to a reduced one; the driving force being, according to the first Fick law, the concentration gradient between the two zones. Mass transfer was considered analogous to conduction heat transfer and its basic laws are derived from heat transfer. The mass transfer mechanisms are:

1. Diffusion: diffusion is determined by the random molecular motion inside a fluid.
2. Convection: convection mass transfer comprises of the transport of substance between a surface (i.e., a solid or liquid one) and a moving fluid or between two relatively immiscible moving fluids.
3. Migration: migration refers to the movement of charged particles in an electric field.
4. Largely, the investigations on NFs mass transfer were focused on gas absorption (i.e., thermal absorption in two component NFs) and liquid mass diffusion. Table 7.1 summarizes a number of experimental and theoretical results on boiling heat and mass transfer in NFs.

Summarizing from the available literature for NFs mass transfer, the mechanisms are identified as [1]:

- *Shuttle or grazing effect*: NPs theoretically carry a supplementary volume of gas to the base fluid through adsorption in the gas–liquid diffusion layer and desorption in the liquid. This can increase the mass transfer coefficient [1]. Further explanations on this are provided by Alper et al. [66], Alper and Ozturk [67], and Tinge and Drinkenburg [68].

Table 7.1 Results of nanofluids (NFs) boiling heat and mass transfer enhancement.

References	Base fluid	Nanoparticles (NPs)	Method	Enhancement (ratio)	Note
Yang and Maa [6]	Water	Al_2O_3	Horizontal tube heater	Boiling heat transfer coefficient (BHTC) increases	Refers to pool boiling
Witharana [7]	Water and EG	Au	Cylindrical vessels	BHTC increases	Refers to pool boiling
	Water and EG	SiO_2	Cylindrical vessels	Reduction of BHTC	Refers to pool boiling
Das et al. [8]	Water	Al_2O_3	Cylindrical cartridge	Reduction of BHTC	Refers to pool boiling A decrease of the heat transfer performance was noted
You et al. [9]	Water	Al_2O_3	Cartridge	Critical heat flux (CHF): 2	Refers to pool boiling
Vassallo et al. [10]	Water	SiO_2	NiCr wire	CHF increases significantly	Refers to pool boiling
Bang and Chang [11]	Water	Al_2O_3	Square flat heater	Reduction of BHTC CHF increase	Refers to pool boiling
Wen and Ding [12]	Water	Al_2O_3	Flat disk heater	CHF: 1.4	Refers to pool boiling
Kim et al. [13]	Water	TiO_2	Smooth NiCr wire heater	CHF: 2	Refers to pool boiling
Kim et al. [14]	Water Water Water	Al_2O_3 ZrO_2 SiO_2	Stainless steel wire and flat plate heaters	CHF: < 1.8	Refers to pool boiling Heat transfer is weakened
Jackson [15]	Water	Au	Flat copper coupon heater	CHF: 5	Refers to pool boiling Heat transfer decreases by 20%
Prakash et al. [16]	Water	Al_2O_3	Vertical tubular heaters	–	Refers to pool boiling Rough heater surface escalate the heat transfer, while flat surface decreases it
Prakash et al. [17]	Water	Al_2O_3	Tubular heaters at various directions	–	Refers to pool boiling Horizontal and inclined heater directions exhibited a heat transfer improvement or weakening, respectively

Table 7.1 Results of nanofluids (NFs) boiling heat and mass transfer enhancement. ***Continued***

References	Base fluid	Nanoparticles (NPs)	Method	Enhancement (ratio)	Note
Chopkar et al. [18]	Water	ZrO_2	Flat surface	BHTC increase for low NP concentration	Refers to pool boiling Tetramethyl ammonium hydroxide surfactant upsurges the heat transfer
Liu and Liao [19]	Water and Alcohol (C_2H_5OH)	CuO	Flat horizontal copper surface	CHF increase	Refers to pool boiling The BHTC for both NFs are poor
Lv and Liu [20]	Water	CuO	Vertical small heated tubes	CHF and BHTC increase for surfactant-free NF	Refers to pool boiling NaDBS surfactant depreciates heat transfer
Kathiravan et al. [21]	Water	CNT	Horizontal tube	BHTC: 1.75	Refers to pool boiling
Kim et al. [22]	Water	Al_2O_3	Vertical stainless steel tube	CHF: 1.5	Refers to flow boiling
Soltani et al. [23]	Water Water	Al_2O_3 SnO_2	Vertical cylindrical glass vessel	Increase of BHTC	Refers to pool boiling
Soltani et al. [24]	cmc–water solution	Al_2O_3	Vertical cylindrical glass vessel	BHTC: 1.25	Refers to pool boiling
Henderson et al. [25]	R134a	SiO_2	Horizontal copper tube	CHF: 1.55	Refers to flow boiling
	R134a + polyolester oil	CuO	Horizontal copper tube	CHF >2	
Truong et al. [26]	Water	Diamond, ZnO, and Al_2O_3	Sandblasted and bare horizontal plate heaters	CHF: 1.35	Refers to pool boiling
Kim et al. [27]	NH_3/H_2O	Cu	Ammonia bubble absorption	Mass transfer enhancement (MTE): 3.21	Refers to bubble absorption
	NH_3/H_2O	CuO	Ammonia bubble absorption	MTE:3.11	Refers to bubble absorption
	NH_3/H_2O	Al_2O_3	Ammonia bubble absorption	MTE:3.04	Refers to bubble absorption
Kim et al. [28]	NH_3/H_2O	Cu	Ammonia bubble absorption	MTE:5.32	Refers to bubble absorption Surfactant is added

(*Continued*)

Table 7.1 Results of nanofluids (NFs) boiling heat and mass transfer enhancement. ***Continued***

References	Base fluid	Nanoparticles (NPs)	Method	Enhancement (ratio)	Note
Pang et al. [29]	NH_3/H_2O	Ag	Ammonia bubble absorption	MTE:1.55	Refers to bubble absorption
Ma et al. [30]	NH_3/H_2O	CNT	Ammonia bubble absorption	MTE:1.162	Refers to bubble absorption
Lee et al. [31]	NH_3/H_2O	Al_2O_3	Ammonia bubble absorption	MTE:1.18	Refers to bubble absorption
	NH_3/H_2O	CNTs	Ammonia bubble absorption	MTE:1.16	Refers to bubble absorption
Amaris et al. [32]	$NH_3/LiNO_3$	MWCNTs	Ammonia tubular bubble absorption	MTE:1.64	Refers to bubble absorption
Lee et al. [33]	Methanol	Al_2O_3	CO_2 bubble absorption	MTE:1.045	Refers to bubble absorption
	Methanol	SiO_2	CO_2 bubble absorption	MTE:1.056	
Torres Pineda et al. [34]	Methanol	Al_2O_3	CO_2 column tray absorption	MTE:1.094	Refers to tray absorption
	Methanol	SiO_2	CO_2 column tray absorption	MTE:1.097	The bubble breaking model was recommended
Kim et al. [35]	Methanol	Al_2O_3	CO_2 bubble absorption	MTE:1.26	Refers to bubble absorption CO2 diffusions were measured as visualization
Torres Pineda [36]	Methanol	Al_2O_3	CO_2 annular contactor absorption	MTE:1.012	Refers to absorption Viscosity was also measured
	Methanol	SiO_2	CO_2 annular contactor absorption	MTE:1.011	
	Methanol	TiO_2	CO_2 annular contactor absorption	MTE:1.046	
Kim et al. [37]	Water	SiO_2	CO_2 absorption	MTE:1.76	Refers to absorption Zeta potential was measured
	piperazine/ K_2CO_3	SiO_2	CO_2 absorption	MTE:1.12	

Table 7.1 Results of nanofluids (NFs) boiling heat and mass transfer enhancement. ***Continued***

References	Base fluid	Nanoparticles (NPs)	Method	Enhancement (ratio)	Note
Lee et al. [38]	$NaCl/H_2O$	Al_2O_3	CO_2 solubility	MTE:1.125	An analysis of the influence of the surfactant, NP size and dispersion stability of NFs was accomplished
Kang et al. [39]	$H_2O/LiBr$	Fe	$H_2O/LiBr$ falling film absorption	MTE:1.9	Refers to absorption
	$H_2O/LiBr$	CNT	$H_2O/LiBr$ falling film absorption	MTE:2.48	Refers to absorption
Kim et al. [40]	$H_2O/LiBr$	SiO_2	$H_2O/LiBr$ falling film absorption	MTE:1.18 CHF: 1.47	Refers to absorption With surfactant of GA and PVA
Lee et al. [41]	$H_2O/LiBr$	Al_2O_3	$H_2O/LiBr$ falling film absorption	MTE:1.77	Refers to absorption With surfactant (2E1H) and Arabic gum
Yang et al. [42]	NH_3/H_2O	Al_2O_3	Ammonia falling film absorption	MTE:1.3	Refers to absorption
	NH_3/H_2O	Fe_2O_3	Ammonia falling film absorption	MTE:1.7	Refers to absorption
	NH_3/H_2O	$ZnFe_2O_4$	Ammonia falling film absorption	MTE:1.5	Refers to absorption
Samadi et al. [43]	Water	γ-Al_2O_3	CO_2 WWC absorption	MTE:1.79	Refers to absorption
	Water	TiO_2	CO_2 WWC absorption	MTE:1.05	Refers to absorption
	Water	Fe_3O_4	CO_2 WWC absorption	No MTE	Refers to absorption
	Water	Fe_3O_4	CO_2 WWC absorption with magnetic field	MTE:1.2235	Refers to absorption
Golkhar et al. [44]	Water	MWCNT	CO_2 membrane absorption	MTE:1.4	Refers to absorption
	Water	SiO_2	CO_2 membrane absorption	MTE:1.2	Refers to absorption

(*Continued*)

Table 7.1 Results of nanofluids (NFs) boiling heat and mass transfer enhancement. ***Continued***

References	Base fluid	Nanoparticles (NPs)	Method	Enhancement (ratio)	Note
Li et al. [45]	Water	CuO	HFC134a hydrate formation	Enhancement	Refers to hydrate formation With surfactant (sodium dodecylbenesulfonate-6)
Park et al. [46]	Water	MWCNT	Methane hydrate formation	MTE:3	Refers to hydrate formation
Park et al. [47]	Water	MWCNT	Methane hydrate formation	MTE: ~4	Refers to hydrate formation
	Water	MWCNT	Methane hydrate formation	MTE:4.5	Refers to hydrate formation
Moraveji et al. [48]	Water	CuO	Methane hydrate formation	MTE:2.44	Refers to hydrate formation SDS surfactant was introduced
Arjang et al. [49]	Water	Ag	Methane hydrate formation	MTE:33.7	Refers to hydrate formation NFs prepared by chemical method
Mohammadi et al. [50]	Water	Ag	CO_2 hydrate formation	MTE:93.9	Refers to hydrate formation With surfactant (SDS)
Olle et al. [51]	Oleic acid	Fe_3O_4	Oxygen mass transfer	MTE:7	The amplified gas–liquid interfacial area determined a large enhancement (80% or more)
Komati et al. [52]	Water	Fe_3O_4	Oxygen mass transfer	MTE:2.74	Different kinds of aqueous ferrofluids were considered, such as polymer coated, lauric acid coated, and TMAOH coated
	Water	Fe_3O_4	CO_2 mass transfer	MTE:1.93	
Zhu et al. [53]	Water	MCM41	CO mass transfer	MTE:1.9	Improvement rest on the relation between NP and the CO molecules
Krishnamurthy et al. [54]	Water	Al_2O_3	Fluorescein dye diffusion	MTE: > 10	Velocity disturbance field was used in order to reveal the mass transfer improvement

Table 7.1 Results of nanofluids (NFs) boiling heat and mass transfer enhancement. ***Continued***

References	Base fluid	Nanoparticles (NPs)	Method	Enhancement (ratio)	Note
Fang et al. [55]	Water	Cu	Fluorescent Rhodamin B diffusion (T = 15°C)	MTE:10.71	Both Brownian motion and induced micro convection were deliberated as the possible reason for the improved mass transfer
	Water	Cu	Fluorescent Rhodamin B diffusion (T = 25°C)	MTE:26	
Veilleux et al. [56]	Water	Al_2O_3	Rhodamin 6 G (R6G) diffusion	MTE: > 10	A dispersion model that rely on the Brownian motion was proposed by the authors
Bahmanyar et al. [57]	Water	SiO_2	Liquid–liquid extraction	MTE:1.04–1.60	With 0.05 vol.% acetic acid
Saien et al. [58]	Toluene/ acetic acid/ water	Fe_3O_4	Liquid–liquid extraction	MTE:2.57	
	Toluene/ acetic acid/ water	Al_2O_3	Liquid–liquid extraction	MTE:2.21	
Kim et al. [59]	Oil (n-decane) -NH_3/H_2O	–	Ammonia absorption	MTE:1.17	With surfactant: C12E4 and Tween20
Turanov et al. [60]	Water	SiO_2	Self-diffusion of proton	No MTE	
Feng et al. [61]	Water	SiO_2	NaCl mass transfer	No MTE	
	Water	SiO_2	Oxygen mass transfer	Reduction in mass transfer	
Lu et al. [62]	Water	CNT	CO_2 stirred reactor absorption	MTE: ~ 1.9	Refers to absorption
	Water	Al_2O_3–water	CO_2 stirred reactor absorption	MTE: ~ 1.4	Refers to absorption
	Water	Activated carbon	CO_2 stirred reactor absorption	MTE: ~ 3.7	Refers to absorption
	Water	Active alumina	CO_2 stirred reactor absorption	No MTE	Refers to absorption

(Continued)

Table 7.1 Results of nanofluids (NFs) boiling heat and mass transfer enhancement. ***Continued***

References	Base fluid	Nanoparticles (NPs)	Method	Enhancement (ratio)	Note
Beiki et al. [63]	Electrolyte solution	γ-Al_2O_3	Laminar mass transfer of electrolyte	MTE:1.168	Electrolyte solution: equimolar potassium ferri-ferrocyanide (0.00985 M) in 0.5 M aqueous sodium hydroxide
Beiki et al. [64]	Electrolyte solution	γ-Al_2O_3	Turbulent mass transfer of electrolyte	MTE:1.10	
	Electrolyte solution	TiO_2	Turbulent mass transfer of electrolyte	MTE:1.18	
Keshishian et al. [65]	Electrolyte solution	SiO_2	Laminar mass transfer of electrolyte	MTE:20.94	
	Electrolyte solution	SiO_2	Turbulent mass transfer of electrolyte	No MTE	

- *Hydrodynamic effects in the gas–liquid boundary layer*: a thinner effective diffusion layer can be created due to NPs. The NPs are interacting with the gas–liquid interface and even can induce turbulence at the interface. This phenomenon was described by Yoon et al. [69] with the aid of a computational model.
- *Changes in the specific gas–liquid interface area:* this phenomenon appears due to the upsurge of gas–liquid mass transfer determined by a modification in the interface area.

More exactly, for the case of the gas–liquid mass transfer, the mass transfer improvement by the shuttle/grazing result was considered in the literature (see [1,67]) as correlated to the mass transfer mechanism, as it was defined through the penetration theory. In further detail, the refreshment of small particles adsorbing gas at the gas–liquid interface after the gas is desorbed in the liquid is analogous to the refreshment of all liquid phase features.

When talking about the hydrodynamic effects in the gas–liquid boundary layer, it can be affirmed that NPs' occurrence in the fluid phase causes the decrease of the gas bubble mounting velocity, and this phenomenon is clearly confirmed by the scattering pattern with the dependence on the NPs' underlying forces (see Yoon et al. [69] for further details and insights).

7.4 Boiling heat and mass transfer

Boiling processes consist of a phase change heating process (more exactly, it is a transfer from liquid to vapor state), wherein vapor bubbles are shaped on a heated surface or in a superheated liquid layer neighboring the heated surface. Mostly, there are two types of boiling: pool boiling and flow boiling (named also as forced convective boiling). In any case, subcooled or saturated boiling may occur, depending on the fluid temperature.

Critical heat flux (CHF) defines the thermal bound of a phenomenon where a phase change follows throughout heating (i.e., bubble formation in pool boiling), which rapidly drops the heat

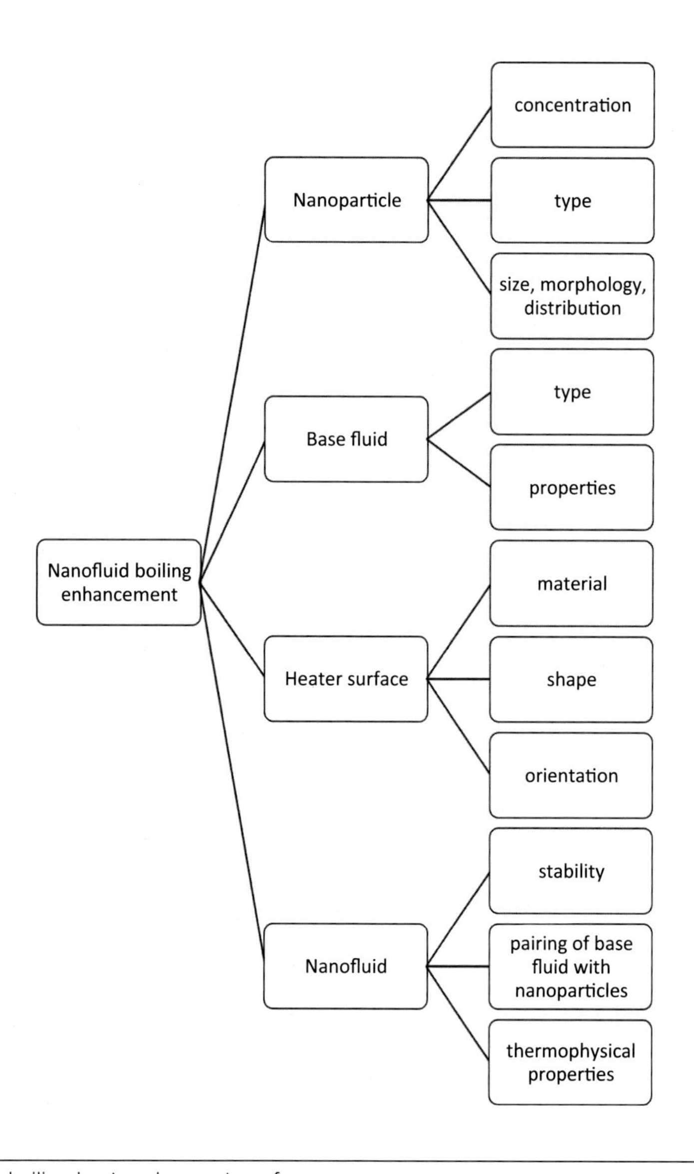

FIGURE 7.2

Factors influencing boiling heat and mass transfer.

transfer effectiveness, thus triggering local overheating of the heating surface. CHF is of tremendous relevance for all multiphase processes and for boiling, in particular.

There are extensive discussions on CHF, while a good literature review was performed by Liang and Mudawar [70] in 2018. Authors have discussed different CHF models, including CHF trigger mechanisms, and most parametric influences, referring especially to pool boiling CHF. Consequently, the augmentation of the CHF is important for every system compactness and efficiency, as well as for its operational safety. Plus, in all boiling applications, it is mandatory that the CHF is not to be surpassed.

Another important parameter when the boiling process is discussed is the boiling heat transfer coefficient (BHTC). BHTC can be evaluated only in certain defined conditions and depends on boiling heat transfer type as well on the surface geometry and dimensions. A common evaluation of BHTC can be attained with the help of the Nusselt number.

If it refers to NFs' use in boiling applications, one must mention that studies on NF boiling heat transfer actually began in 2003 and most of the available studies are in connection to pool boiling; few talk about flow boiling (i.e., even if flow boiling has lot more applications in engineering than pool boiling).

Fig. 7.2 reviews the main factors that are influencing nanofluid (NF) boiling processes. Over the years it was clearly demonstrated that the factors identified in Fig. 7.2 can act individually or in combination and influence NF boiling improvement, as will be discussed in the next sections.

Table 7.2 presents several review papers that have summarized a number of investigations on flow and pool boiling heat transfer involving NFs.

Before discussing flow and pool boiling of NFs separately, it is important to outline (see Fig. 7.3) a number of relevant applications of boiling heat transfer.

7.4.1 Pool boiling

Pool boiling requires a heated surface submerged into the liquid and any motion of the fluids is due to natural convection flow while the bubble motion is determined by buoyancy. There are several regimes of pool boiling described in the textbooks, starting with natural convection boiling, followed by nucleate boiling, transition, and finally film boiling and burnout phenomenon. All these regimes appear, in this order, when temperature increase.

When talking about CHF in pool boiling, one has to start with Kutateladze's formulation [81–83] based on a dimensional analysis that was proposed in 1948.

$$\frac{q''_{CHF}}{\rho_g h_{fg}\left[\sigma_g \frac{(\rho_f-\rho_g)}{\rho_g^2}\right]^{1/4}} = K \tag{7.1}$$

where q''_{CHF} is the *CHF*; ρ_g and h_{fg} are the vapor density and latent heat of vaporization, respectively; ρ_f is the density of the fluid; σ is the surface tension; and K is a constant. For example, a value of $K = 0.16$ was suggested by Kutateladze for pool boiling on a large horizontal flat surface.

Nevertheless, the most popular equation for pool boiling *CHF* for a flat infinite heater surface is Zuber's correlation [84]:

$$q''_{CHF} = K\rho_g^{1/2} h_{fg}\left[\sigma_g\left(\rho_f-\rho_g\right)\right]^{1/4} \tag{7.2}$$

Table 7.2 Reviews of boiling heat transfer with nanofluids (NFs).

Source	Boiling type	Discussed parameters	Comments
Fang et al. [71]	Flow boiling	Heat transfer coefficient (HTC), CHF, pressure drop, flow pattern and stability	The nanoparticles (NPs) effect on flow boiling HTC is contradictory CHF can go up to a 50% increase Pressure drop is subjective by NP sedimentation during the process
Fang et al. [72]	Pool and flow boiling	Heat transfer and CHF	NFs influences both boiling heat transfer (HT) and CHF. The influence is connected with NP type, geometry and flow pattern
Pinto and Fiorelli [73]	Pool and flow boiling	HTC	NFs boiling depends on the interaction between NP and heating surface More studies are needed
Kamel and Lezsovits [74]	Pool and flow boiling	HTC, CHF	Authors noted two main directions: NFs thermophysical properties and surface influence (sedimentation of NPs included)
Kamel et al. [75]	Pool and flow boiling	CHF	NFs are beneficial for attaining enhanced heat flux even by applying little temperature differences throughout the boiling process
Liang and Mudawar [76]	Pool boiling	HTC, CHF, boiling hysteresis	Authors noted various degree for HTC and CHF improvement
Xie et al. [77]	Pool and flow boiling	CHF	CHF is higher for pool and flow boiling
Cheng et al. [78]	Pool and flow boiling	heat transfer performance, CHF	The mechanism of boiling and two-phase flow using NFs was acknowledged as a complex phenomenon Enrichment in CHF was found to be over 50%
Moreira et al. [79]	Pool and flow boiling	HTC	HTC for pool and flow boiling can increase or decrease. This relies on surface texture and NF thermophysical properties
Kamel et al. [80]	Experimental studies of flow boiling	HTC and CHF	Authors considered that in order to apply NFs in flow boiling, it will need to improve the NF manufacturing methods

where $K = 0.138-0.157$ (see [85] for more detail and comments on this topic and Table 7.1 for several results extracted from archived literature).

Heat transfer coefficient (HTC) for NF pool boiling was commented upon in 2015 by Ciloglu and Bolukbasi [86] who affirmed that the reason behind its augmentation is the large specific surface area due to NPs, as well as the enlarged heat capacity of the fluid, Brownian motion, and interfacial liquid layering. With regard to HTC reduction, the same authors clarified that a reduction in the contact angle can go to a decrease of the active nucleation site density by forming a smoother surface due to NPs. A summary of pool boiling experimental results, identified by year, came from Fang et al. [71] and Prakash and Prasanth [87], where it was noted that the CHF enhancement goes up to 267%, depending on the NF type and experimental conditions. Also, from the literature

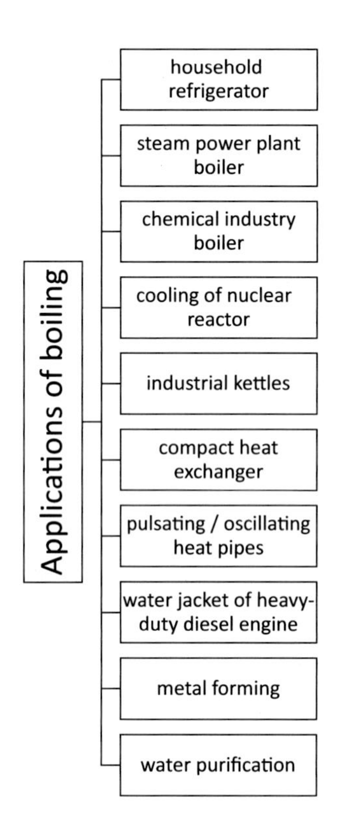

FIGURE 7.3

Boiling heat and mass transfer: applications.

survey, Fig. 7.4 plots the NPs considered for pool boiling experiments, while the base fluids are mostly water or ethylene glycol [88]. One can note from Fig. 7.4 that, as expected, the most studied NFs are those with water and Al_2O_3, TiO_2, and SiO_2, while those with CNT and graphene are less studied. Table 7.3 summarizes several results on pool boiling with NFs.

Another interesting paper was published by Wen et al. [112], who performed a comparative experimental study on two surfaces' influences on pool boiling of alumina–water NFs. They concluded that the surface geometry greatly influences the boiling heat and mass transfer and this is the main reason for scattered results in the literature. The improvement or failing of boiling heat transfer highly depends on the suspended NPs and the heating surface geometry, combined with their specific interactions. Plus, they acknowledged also that the final experimental outcomes are greatly affected by the surface modification (occurring due to NPs presence in the fluid), as well as by the number and incidence of the usage of the pool boiling surface.

In conclusion, even if a number of studies were identified in the literature in regard to the effect of NPs addition on CHF, its central physical enhancement mechanism is still uncertain. Additionally, the surfactant's influence is another relevant aspect that must be evaluated in the future, since a surfactant's presence can create additional deposition on the heater surface.

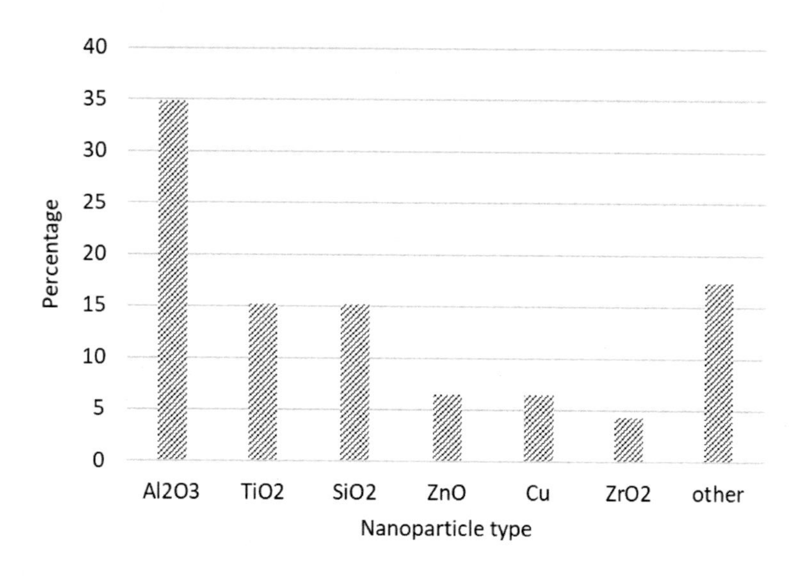

FIGURE 7.4

Nanoparticles in nanofluids experiments for pool boiling.

7.4.2 Flow boiling

Flow boiling requires a fluid flow and the fluid is forced to move inside a heated pipe or over a surface (i.e., similar to forced convection process). As for all boiling processes, CHF is the most important issue, being called also the boiling crisis [71]; thus most studies are dedicated to experiments on CHF, together with heat transfer coefficient, pressure drop, and the NF stability.

Generally, CHF is highly influenced by the NP geometry, dimensions, type, and concentration, as well as by the thermal hydraulic conditions of the studied application (i.e., mass flux, pressure, inlet subcooling temperature, etc.).

A number of studies were first depicted in Table 7.1 and the reader can easily see that a large number of NP types have been considered, as well as several base fluids (i.e., water, refrigerants, oils, and combinations). Continuing the analysis, the reader can find in Table 7.4 a summary of HTC results for flow boiling. It is worth mentioning here that a number of numerical studies are also available in the open literature; however, Table 7.4 consists of a collection of experimental results only.

From Table 7.4 one can say that even if the addition of NPs to a basic heat transfer fluid can increase the CHF, other techniques may be applied and here can be outlined the application of external magnetic field [127]. There are several experimental studies on this technique and some authors outlined also the advantage of an easier cleaning process through using magnetite based fluids (e.g., see ref. [127]). In addition, to the abovementioned outcomes, an intensification in the experimental work is needed to confirm if the CHF enhancement can be linearly enlarged by applying a magnetic field, and also what the influence of its intensity is and if a critical point appears.

Table 7.3 Outline of several results on pool boiling.

Reference	Base fluid	Nanoparticles	CHF enhancement (%)
Cieslinski [88]	Water	Al_2O_3	171
	Water	TiO_2	176
Truong [89]	Water	Al_2O_3	56
	Water	SiO_2	64
Reinho et al. [90]	Water	Al_2O_3	129
	Water	Fe_2O_3	
	Water	CNT	
Bang [91]	Water	Al_2O_3	32
You [92]	Water	Al_2O_3	200
Kim et al. [93]	Water	TiO_2	180
Coursey and Kim [94]	Water	Al_2O_3	37
Golubovic et al [95]	Water	Al_2O_3	50
	Water	Bismuth	33
Kim and Ahn [96]	Water	Al_2O_3	30
Muhammad et al. [97]	Water	TiO_2	222
Kim et al. [98]	Water	Al_2O_3	52
	Water	SiO_2	80
	Water	ZrO_2	72
Kim et al. [99]	Water	TiO_2	148–267
Sarafraz et al. [100]	Water-EG	ZrO_2	29
Sulaiman et al. [101]	Water	TiO_2	250
	Water	Al_2O_3	
	Water	SiO_2	
Padhye et al. [102]	Water	Graphene oxide	179
	Water	Graphene	84
	Water	Al_2O_3	152
Wen and Ding [12]	Water	Al_2O_3	40
Kwark et al. [103]	Water	Al_2O_3	80
	Water	CuO	
	Water	Diamond	
Kwark et al. [104]	Water	Al_2O_3	70
	Ethanol	Al_2O_3	80
Lee and Jeong [105]	Water	Al_2O_3	150–230
Suriyawong and Wongwises [106]	Water	TiO_2	30
Kole and Dey [107]	Ethylene glycol	ZnO	117
Hiswankar and Kshirsagar [108]	Water	ZnO	70–80
Milanova and Kumar [109]	Water	SiO_2	150–200
Vasallo et al. [110]	Water	SiO_2	60
Kathiravan et al. [111]	Water	Cu	25–30

Table 7.4 Results of HTC variation in flow boiling.

Reference	Base fluid	Nanoparticles (NPs)	Concentration	Results
Faukner et al. [113]	Water	Ceramic-based NP suspensions (Al2O3, Al-N)	0.25–0.5 wt.%	Both deterioration and improvement in the HTC were detected, being dependent on the operation conditions and NF type
Peng et al. [114]	R113	CuO	0–5 wt.%	HTC enhancement was up to 29.7% due to decrease of the boundary layer that was determined by the NP disturbance and the development of molecular adsorption layer on the NPs surface All the theoretical correlations are under predicting the experimental outcomes
Kim et al. [115]	Water	Al_2O_3 ZnO	0.001, 0.01, and 0.1 vol.%	HTC was amplified with mass and heat flux upsurge, for both water and NF
Boudouh et al. [116]	Water	Cu	0.00056, 0.0011, and 0.0056 vol.%	The heat transfer enhancement of Cu/water NF occurred mainly due to NF upper viscosity
Henderson et al. [117]	R134a	SiO_2	0.02, 0.04, and 0.08 vol.%	HTC decreased up to 55%, if compared to the base fluid A possible reason for this decrease can be the poor NP suspension (formation of NP clusters) and particle film resistance
	R134a/polyolester	CuO	0.02, 0.04, and 0.08 vol.%	HTC enhancement was up to 76%, larger as the NP concentration is higher
Xu and Xu [118]	Water	Al_2O_3		HTC was higher for the NF than that for pure water Reasons behind this behavior can be: (1) smaller bubble departure size for NFs and (2) thin liquid film with NP to the whole bubble shadow area is amplified
Abedini et al. [119]	Water	TiO_2	0.01%, 0.1%, 0.5%, and 2.5%	HTC decreased Authors believe that the heating surface wettability was amplified during NF boiling due to TiO_2 deposition. A solution can be the surface coating with a porous layer of NPs, that diminished the contact angle
Chehade et al. [120]	Water	Ag	0.000237 and 0.000475 vol.%	HTC enhances up to 162%, depending on Ag concentration
Rana et al. [121,122]	Water	ZnO	<0.01 vol.%	HTC amplified with NF concentration increase due to the enriched thermal conductivity and the change of heat transfer surface
Sun and Yang [123]	R141b	Cu Al Al_2O_3 CuO	0.1–0.3 wt.%	HTC increased up to 49%, percentage influenced by a growing mass fraction and mass velocity

(Continued)

Table 7.4 Results of HTC variation in flow boiling. ***Continued***

Reference	Base fluid	Nanoparticles (NPs)	Concentration	Results
Sarafraz and Hormozi [124]	Water	CuO	0.5–1.5 vol.%	HTC decreased due to NP deposition on the boiling surface, producing fouling problem
Lee et al. [125]	Water	Al_2O_3, SiC	0.01%	CHF increased up to 15%
Lee et al. [126]	Water	GO	0.01%	CHF increased up to 100%
Lee et al. [127]	Water	Magnetite	1 and 10 ppm	CHF increased up to 60%
Aminfaret al. [128]	Water	Fe_4O_3	0.01%–0.1%	CHF increased up to 56%

7.5 Techniques for enhancement of the nanofluids critical heat flux

The NF used in boiling multiphase heat transfer has a dynamic role in different research and industrial areas (i.e., nuclear reactors, fossil fuel boilers, electronic chips, power electronics cooling, spray cooling, refrigeration, and air-conditioning).

For example, even if the concept of pool boiling and the hydrodynamic theory of burnout crisis [129] were established many years ago, the maximum heat flux is restricted by the appearance of the burnout circumstances. In the moment that heat flux touches the value of CHF (i.e., CHF), the direct contact amongst the fluid and the heater dramatically reduces, causing finally the burnout of the material. Consequently, it is mandatory to find ways to augment CHF for increasing the efficiency of the specific thermal systems. On the other side, the CHF enhancement can be greatly influenced by the heater surface roughness and its wettability.

Furthermore, this section will briefly discuss the mechanisms of increasing CHF for NFs. On the other hand, the previous section clearly established and outlined that adding NPs to water or other regular heat transfer fluids is going to lead to a very good augmentation of CHF.

Thus several further mechanisms for increasing CHF of NFs are identified as:

- modifying surface characteristics
- changing flow channel structures
- integration of hybrid approaches

Modifying surface characteristics mainly refers to its roughness and the addition of possible nanocoatings. The addition of nanolayers on the heating surface is considered in the literature as a major reason for CHF improvement [129].

There are several examples in the literature, as outlined in Table 7.5, even if some studies affirm that the coating structure can determine an amplified pressure drop and weakening of morphology under high flow velocities, resulting in a decrease of CHF [138].

Other studies applied NFs to several modified geometries like macro and micro fins, surfaces with concentric circular structures, and honeycomb porous plates [77].

Table 7.5 Methods for CHF enhancement by nanocoating selection.

Reference	Nanoparticles (NPs)	Observations
Hendricks et al. [130]	ZnO	NPs are processed into carpet-like, flower-like, and distinctive nanotextured formulae
Chu et al. [131]	SiO_2	Authors manufactured a horizontally oriented surface
Chu et al. [132]	SiO_2 and CuO_2	Si NPs are deposed on a SiO_2 substrate and Cu NPs on a CuO substrate
Bang [133]	Al_2O_3	Authors coated NPs on a SS316L tube
Nazari and Saedodin [134]	Al_2O_3	Authors placed a uniform Al_2O_3 NPs layer on a plain aluminum surface
Cheedarala et al. [135]	Cu	Micro/nanoporous copper surface based on copper NPs were adopted
Yazid et al. [136]	CNT	Authors used both MWCNT and SWCNT
Ujereh et al. [137]	CNT	The increase of CNTs treatment on a silicon surface can enhance CHF and the critical heat flux is better when silicon is replaced with the copper substrate

However, studies on the optimal layer thickness are of paramount relevance to guarantee the maximum delay and to increase the CHF. Moreover, the bubble dynamics and the stability of NPs coating are other issues that need further experiments.

Changing flow channel structures is seen in the literature as another effective method to increase CHF. This method can be applied through several techniques, like changing the channel size (optimizing the flow channels dimensions), applying several treatments to augment heat dissipation, etc.

Hybrid approaches mainly refer to combined techniques, as Xie et al. acknowledged in their paper. These techniques are identified as acoustic techniques, magnetic fields, and other creative elements like hypervapotrons, notable fabrication techniques, and new materials in engineering. Nevertheless, a drawback was considered the difficulty in evaluating the single effect of each method.

In conclusion, the modification in CHF mostly relies on NPs type, material and size, the base fluid, the heated surface (in terms of shape, orientation, roughness), and eventually additives. Anyhow, it has been generally proven that the NFs concentration greatly influences both CHF and heat transfer coefficient. So, in order to augment the CHF up to a maximum level without influencing greatly the HTC, a balance of NP concentration is needed. The knowledge on CHF enhancement of NF boiling is quite deficient at this moment and further research is needed. Nevertheless, examination of the NF boiling CHF must be focused on the combined effect of NP deposition on the boiling surfaces and the correct suspension of NPs in the NFs.

7.6 Conclusion and future work and trends

Boiling heat transfer is a relevant process in many real-life applications where simultaneous heat and mass transfer occurs. The use of NFs in such a process enhances the CHF, even if the results are a bit scattered in the open literature.

A few challenging research directions are:

1. Obtaining stable NFs. Formulating stable new fluids is a very challenging aspect of NFs development, therefore it is mandatory to ensure long-term stability of these fluids, thus getting long-term stability of thermophysical and chemical properties.
2. Getting maximum benefits with lower mass concentrations of NPs. In the literature it is recommended to lower the NP concentration to keep the sedimentation and viscosity at the minimum possible. This leads to the proper identification of NPs type with maximum benefits for exploitation.
3. Identifying the most proper combination between the base fluid and NP type. This is an extremely challenging objective and more coordinated experimentation is needed.
4. Coordinated studies on the mutual effects of surface wettability and capillary wicking structure. This research may shed some light on the mechanisms of CHF augmentation.
5. Identifying proper nanocoatings (i.e., in terms of structure and width).
6. Intensifying the studies on bubble dynamics.
7. Studying also other factors that may affect the CHF, for example the pressure and its contribution to growing of dry patches.
8. Intensifying the numerical studies on boiling heat transfer (i.e., most of the studies are experimental—over 90%).
9. Elaborating different correlations and experimental equations to cover most boiling processes.

Most of the researchers that work in the area of NFs have dedicated little attention to boiling heat transfer, so the research needs to be intensified in this direction. Though the NFs applications as advanced cooling medium seem to be auspicious, the understanding of NFs' new features, as well as their development for real-life applications are extremely stimulating, mainly due to the lack of agreement between data attained from different research groups.

Nevertheless, in the renewable energy sector, NFs can be considered as a very good option for augmentation of the heat energy storage from solar collectors and also can be applied to increase the energy density. In conclusion, future exploration should also encompass the renewable energy-based uses of NFs.

Nomenclature

K constant
h_{fg} latent heat of vaporization
q''_{CHF} critical heat flux
ρ_g vapor density
ρ_f density of the fluid
σ surface tension

Abbreviations

BHTC boiling heat transfer coefficient
CHF critical heat flux

HTC heat transfer coefficient
MTE mass transfer enhancement
NF nanofluids
NP nanoparticles

References

[1] C. Pang, J.W. Lee, Y.T. Kang, Review on combined heat and mass transfer characteristics in nanofluids, Int. J. Therm. Sci. 87 (2015) 49–67.
[2] P. Keblinski, S.R. Phillpot, S.U.S. Choi, J.A. Eastman, Mechanism of heat flow in suspensions of nano-sized particles, Int. J. Heat Mass Transf. 45 (2002) 855–863.
[3] W. Yu, D.M. France, J.L. Routbort, S.U.S. Choi, Review and comparison of nanofluid thermal conductivity and heat transfer enhancements, Heat Transf. Eng. 29 (5) (2008) 432–460.
[4] M. Chandrasekar, S. Suresh, A review on the mechanisms of heat transport in nanofluids, Heat Transf. Eng. 30 (14) (2009) 1136–1150.
[5] L. Wang, J. Fan, Nanofluids research: key issues, Nanoscale Res. Lett. 5 (2010) 1241–1252.
[6] Y.M. Yang, J.R. Maa, Boiling of suspension of solid particles in water, Int. J. Heat Mass Transf. 27 (1984) 145–147.
[7] S. Witharana, Boiling of refrigerants on enhanced surfaces and boiling of nanofluids, PhD thesis, Royal Institute of Technology, 2003.
[8] S.K. Das, N. Putra, W. Roetzel, Pool boiling characterization of nano-fluids, Int. J. Heat Mass Transf. 46 (2003) 851–862.
[9] S.M. You, J.H. Kim, K.M. Kim, Effect of nanoparticles on critical heat flux of water in pool boiling of heat transfer, Appl. Phys. Lett. 83 (2003) 3374–3376.
[10] P. Vassallo, R. Kumar, S. D'Amico, Pool boiling heat transfer experiments in silica–water nano-fluids, Int. J. Heat Mass Transf. 47 (2004) 407–411.
[11] I.C. Bang, S.H. Chang, Boiling heat transfer performance and phenomena of Al_2O_3–water nano-fluids from a plain surface in a pool, Int. J. Heat Mass Transf. 48 (2005) 2407–2419.
[12] D. Wen, Y. Ding, Experimental investigation into the pool boiling heat transfer of aqueous based alumina nanofluids, J. Nanopart. Res. 7 (2005) 265–274.
[13] H. Kim, J. Kim, M. Kim, Experimental study on CHF characteristics of water-TiO_2 nano-fluids, Nucl. Eng. Technol. 39 (2006) 61–68.
[14] S.L. Kim, I.C. Bang, J. Buongiorno, L.W. Hu, Surface wettability change during pool boiling of nanofluids and its effect on critical heat flux, Int. J. Heat Mass Transf. 50 (2007) 4105–4116.
[15] J. Jackson, Investigation into the pool-boiling characteristics of gold nanofluids, MS thesis, University of Missouri-Columbia, 2007.
[16] N.G. Prakash, K.B. Anoop, S.K. Das, Mechanism of enhancement/deterioration of boiling heat transfer using stable nanoparticles suspensions over vertical tubes, J. Appl. Phys. 102 (2007). 074317-1-7.
[17] N.G. Prakash, K.B. Anoop, G. Sateesh, S.K. Das, Effect of surface orientation on pool boiling heat transfer of nanoparticle suspensions, Int. J. Multiph. Flow 34 (2008) 145–160.
[18] M. Chopkar, A.K. Das, I. Manna, P.K. Das, Pool boiling heat transfer characteristics of ZrO_2–water nanofluids from a flat surface in a pool, Heat Mass Transf. 44 (2008) 999–1004.
[19] Z.H. Liu, L. Liao, Sorption and agglutination phenomenon of nanofluids on a plain heating surface during pool boiling, Int. J. Heat Mass Transf. 51 (2008) 2593–2602.
[20] L.C. Lv, Z.H. Liu, Boiling characteristics in small vertical tubes with closed bottom for nanofluids and nanoparticles-suspensions, Heat Mass Transf. 45 (2008) 1–9.

[21] R. Kathiravan, R. Kumar, A. Gupta, R. Chandra, P.K. Jain, Pool boiling characteristics of carbon nanotube based nanofluids over a horizontal tube, J. Therm. Sci. Eng. Appl. 1 (2009). 022001-1-7.
[22] S.J. Kim, T. McKrell, J. Buongiorno, L.W. Hu, Enhancement of flow boiling critical heat flux (CHF) in alumina/water nanofluids, Adv. Sci. Lett. 2 (2009) 100−102.
[23] S. Soltani, S.G. Etemad, J. Thibault, Pool boiling heat transfer performance of Newtonian nanofluids, Heat Mass Transf. 45 (2009) 1555−1560.
[24] S. Soltani, S.G. Etemad, J. Thibault, Pool boiling heat transfer performance of Newtonian nanofluids, Int. Commun. Heat Mass Transf. 37 (2010) 29−33.
[25] K. Henderson, Y.G. Park, L. Liu, A.M. Jacobi, Flow-boiling heat transfer of R-134abased nanofluids in a horizontal tube, Int. J. Heat Mass Transf. 53 (2010) 944−951.
[26] B. Truong, L.W. Hu, J. Buongiorno, T. McKrell, Modification of sandblasted plate heaters using nanofluids to enhance pool boiling critical heat flux, Int. J. Heat Mass Transf. 53 (2010) 85−94.
[27] J. Kim, A. Akisawa, T. Kashiwagi, Y.T. Kang, Numerical design of ammonia bubble absorber applying binary nanofluids and surfactants, Int. J. Refrig. 30 (6) (2007) 1086−1096.
[28] J.K. Kim, J.Y. Jung, Y.T. Kang, Absorption performance enhancement by nanoparticles and chemical surfactants in binary nano fluids, Int. J. Refrig. 30 (1) (2007) 50−57.
[29] C. Pang, W. Wu, W. Sheng, H. Zhang, Y.T. Kang, Mass transfer enhancement by binary nanofluids (NH_3/H_2O + Ag nanoparticles) for bubble absorption process, Int. J. Refrig. 35 (8) (2012) 2240−2247.
[30] X. Ma, F. Su, J. Chen, Y. Zhang, Heat and mass transfer enhancement of the bubble absorption for a binary nanofluid, J. Mech. Sci. Technol. 21 (2007) 1813−1818.
[31] J.K. Lee, J. Koo, Y.T. Kang, The effects of nanoparticles on absorption heat and mass transfer performance in NH_3/H_2O binarynanofluids, Int. J. Refrig. 33 (2) (2010) 269−275.
[32] C. Amaris, M. Bourouis, M. Valles, Passive intensi fication of the ammonia absorption process with $NH_3/LiNO_3$ using carbon nanotubes and advanced surfaces in a tubular bubble absorber, Energy 68 (2014) 519−528.
[33] J.W. Lee, J.Y. Jung, S.G. Lee, Y.T. Kang, CO_2 bubble absorption enhancement in methanol-based nanofluids, Int. J. Refrig. 34 (8) (2011) 1727−1733.
[34] I. Torres Pineda, J.W. Lee, I. Jung, Y.T. Kang, CO_2 absorption enhancement by methanol-based Al_2O_3 and SiO_2 nanofluids in a tray column absorber, Int. J. Refrig. 35 (5) (2012) 1402−1409.
[35] J.H. Kim, C.W. Jung, Y.T. Kang, Mass transfer enhancement during CO_2 absorption process in methanol/Al_2O_3 nanofluids, Int. J. Heat Mass Transf. 76 (2014) 484−491.
[36] I. Torres Pineda, C.K. Choi, Y.T. Kang, CO_2 gas absorption by CH_3OH based nanofluids in an annular contactor at low rotational speeds, Int. J. Greenhouse Gas Control 23 (2014) 105−112.
[37] W. Kim, H.U. Kang, K. Jung, S.H. Kim, Synthesis of silica nanofluid and application to CO_2 absorption, Sep. Sci. Technol. 43 (2008) 3036−3055.
[38] J.W. Lee, Y.T. Kang, CO_2 absorption enhancement by Al_2O_3 nanoparticles in NaCl aqueous solution, Energy 53 (2013) 206−211.
[39] Y.T. Kang, H. Kim, K. Lee, Heat and mass transfer enhancement of binary nanofluids for $H_2O/LiBr$ falling film absorption process, Int. J. Refrig. 31 (5) (2008) 850−856.
[40] H. Kim, J. Jeong, Y.T. Kang, Heat and mass transfer enhancement for falling film absorption process by SiO_2 binary nanofluids, Int. J. Refrig. 35 (2012) 645−651.
[41] J.K. Lee, H. Kim, M.H. Kim, J. Koo, Y.T. Kang, The effect of additives and nanoparticles on falling film absorption performance of binary nanofluids ($H_2O/LiBr$ + nanoparticles), J. Nanosci. Nanotechnol. 9 (12) (2009) 7461−7466.
[42] L. Yang, K. Du, X.F. Niu, B. Cheng, Y.F. Jiang, Experimental study on enhancement of ammonia-water falling film absorption by adding nanoparticles, Int. J. Refrig. 34 (2011) 640−647.

[43] Z. Samadi, M. Haghshenasfard, A. Moheb, CO_2 absorption using nanofluids in a wetted-wall column with external magnetic field, Chem. Eng. Technol. 37 (3) (2014) 462–470.

[44] A. Golkhar, P. Keshavarz, D. Mowla, Investigation of CO_2 removal by silica and CNT nanofluids in microporous hollow fiber membrane contactors, J. Membr. Sci. 433 (2013) 17–24.

[45] J. Li, D. Liang, K. Guo, R. Wang, S. Fan, Formation and dissociation of HFC134a gas hydrate in nano-copper suspension, Energy Convers. Manag. 47 (2006) 201–210.

[46] S.S. Park, S.B. Lee, N.J. Kim, Effect of multi-walled carbon nanotubes on methane hydrate formation, J. Ind. Eng. Chem. 16 (2010) 551–555.

[47] S.S. Park, E.J. An, S.B. Lee, W.G. Chun, N.J. Kim, Characteristics of methane hydrate formation in carbon nanofluids, J. Ind. Eng. Chem. 18 (2012) 443–448.

[48] M.K. Moraveji, M. Golkaram, R. Davernejad, Effect of CuO nanoparticle on dissolution of methane in water, J. Mol. Liq. 180 (2013) 45–50.

[49] S. Arjang, M. Manteghian, A. Mohammaid, Effect of synthesized silver nanoparticles in promoting methane hydrate formation at 4.7 Mpa and 5.7 Mpa, Chem. Eng. Res. Des. 91 (2013) 1050–1054.

[50] M. Mohammadi, A. Manteghian, A.H. Haghtalab, M. Mohammadi, Rahmati-Abkenar, Kinetic study of carbon dioxide hydrate formation in presence of silver nanoparticles and SDS, Chem. Eng. J. 237 (2014) 387–395.

[51] B. Olle, S. Bucak, T.C. Holmes, L. Bromberg, T.A. Hatton, D.I.C. Wang, Enhancement of oxygen mass transfer using functionalized magnetic nanoparticles, Ind. Eng. Chem. Res. 45 (2006) 4355–4363.

[52] S. Komati, A.K. Suresh, Anomalous enhancement of interphase transport rates by nanoparticles: effect of magnetic iron oxide on gaseliquid mass transfer, Ind. Eng. Chem. Res. 49 (2010) 390–405.

[53] H. Zhu, B.H. Shanks, T.J. Heindel, Enhancing CO-water mass transfer by functionalized MCM41 nanoparticles, Ind. Eng. Chem. Res. 47 (2008) 7881–7887.

[54] S. Krishnamurthy, P. Bhattacharya, P.E. Phelan, Enhanced mass transport in nanofluids, Nano Lett 6 (3) (2006) 419–423.

[55] X. Fang, Y. Xuan, Q. Li, Experimental investigation on enhanced mass transfer in nanofluids, Appl. Phys. Lett. 95 (2009). 203108-1-203108-8.

[56] J. Veilleux, S. Coulombe, A total internal reflection fluorescence microscopy study of mass diffusion enhancement in water-based alumina nanofluids, J. Appl. Phys. 108 (2010). 104316-1-104316-8.

[57] A. Bahmanyar, N. Khoobi, M.R. Mozdianfard, H. Bahmanyar, The influence of nanoparticles on hydrodynamic characteristics and mass transfer performance in a pulsed liquid–liquid extraction column, Chem. Eng. Process. 50 (2011) 1198–1206.

[58] J. Saien, H. Bamdadi, Mass transfer from nanofluid single drops in liquid–liquid extraction process, Ind. Eng. Chem. Res. 51 (2012) 5157–5166.

[59] Y. Kim, J.K. Lee, Y.T. Kang, The effect of oil-droplet on bubble absorption performance in binary nanoemulsions, Int. J. Refrig. 34 (2011) 1734–1740.

[60] A.N. Turanov, Y.V. Tolmachev, Heat- and mass-transport in aqueous silica nanofluids, Heat Mass Transf 45 (2009) 1583–1588.

[61] X. Feng, D.W. Johnson, Mass transfer in SiO_2 nanofluids: a case against purported nanoparticle convection effects, Int. J. Heat Mass Transf. 55 (2012) 3447–3453.

[62] S. Lu, M. Xing, Y. Sun, X. Dong, Experimental and theoretical studies of CO_2 absorption enhancement by nano-Al_2O_3 and carbon nanotube particles, Chin. J. Chem. Eng. 21 (9) (2013) 983–990.

[63] H. Beiki, M. Nasr Esfahany, N. Etesami, Laminar forced convective mass transfer of g-Al_2O_3/electrolyte nanofluid in a circular tube, Int. J. Therm. Sci. 64 (2013) 251–256.

[64] H. Beiki, M. Nasr Esfahany, N. Etesami, Turbulent mass transfer of Al_2O_3 and TiO_2 electrolyte nanofluids in circular tube, Microfluid Nanofluid 15 (2013) 501–508.

[65] N. Keshishian, M. Nasr Esfahany, N. Etesami, Experimental investigation of mass transfer of active ions in silica nanofluids, Int. Commun. Heat Mass Transf. 46 (2013) 148–153.

[66] E. Alper, B. Wichtendahl, W.D. Deckwer, Gas absorption mechanism in catalytic slurry reactors, Chem. Eng. Sci. 35 (1e2) (1980) 217–222.
[67] E. Alper, S. Ozturk, The effect of activated carbon loading on oxygen absorption into aqueous sodium sulfide solutions in a slurry reactor, Chem. Eng. J. 32 (2) (1986) 127–130.
[68] J.T. Tinge, A.A.H. Drinkenburg, Absorption of gases into activated carbon-water slurries in a stirred cell, Chem. Eng. Sci. 47 (6) (1992) 1337–1345.
[69] S. Yoon, J.T. Chung, Y.T. Kang, The particle hydrodynamic effect on the mass transfer in a buoyant CO_2-bubble through the experimental and computational studies, Int. J. Heat Mass Transf. 73 (2014) 399–409.
[70] G. Liang, I. Mudawar, Pool boiling critical heat flux (CHF) – part 1: review of mechanisms, models, and correlations, Int. J. Heat Mass Transf. 117 (2018) 1352–1367.
[71] M.S. Kamel, F. Lezsovits, A.K. Hussein, Experimental studies of flow boiling heat transfer by using nanofluids A critical recent review, J. Therm. Anal. Calorim. 138 (2019) 4019–4043.
[72] X. Fang, R. Wang, W. Chen, H. Zhang, C. Ma, A review of flow boiling heat transfer of nanofluids, Appl. Therm. Eng 91 (2015) 1003–1017.
[73] X. Fang, Y. Chen, H. Zhang, W. Chen, A. Dong, R. Wang, Heat transfer and critical heat flux of nanofluid boiling: a comprehensive review, Renew. Sustain. Energy Rev 62 (2016) 924–940.
[74] R.V. Pinto, F.A.S. Fiorelli, Review of the mechanisms responsible for heat transfer enhancement using nanofluids, Appl. Therm. Eng 108 (2016) 720–739.
[75] M.S. Kamel, F. Lezsovits, Boiling heat transfer of nanofluids: a review of recent studies, Therm. Sci 23 (2017) 216–219.
[76] M.S. Kamel, F. Lezsovits, A.M. Hussein, O. Mahian, S. Wongwises, Latest developments in boiling critical heat flux using nanofluids: a concise review, Int. Commun. Heat Mass Transf 98 (2018) 59–66.
[77] G. Liang, I. Mudawar, Review of pool boiling enhancement by surface modification, Int. J. Heat Mass Transf 128 (2019) 892–933.
[78] S. Xie, M. Shahmohammadi Beni, J. Cai, J. Zhao, Review of critical-heat-flux enhancement methods, Int. J. Heat Mass Transf 122 (2018) 275–289.
[79] L. Cheng, G. Xia, Q. Li, J.R. Thome, Fundamental issues, technology development, and challenges of boiling heat transfer, critical heat flux, and two-phase flow phenomena with nanofluids, Heat Transf. Eng 7632 (2018) 1–36.
[80] T.A. Moreira, D.C. Moreira, G. Ribatski, Nanofluids for heat transfer applications: a review, J. Braz. Soc. Mech. Sci. Eng 40 (2018) 303–314.
[81] S.S. Kutateladze, On the transition to film boiling under natural convection, Kotloturbostroenie 3 (1948) 10–12.
[82] S.S. Kutateladze, Boiling and bubbling heat transfer under free convection of liquid, Int. J. Heat Mass Transf. 22 (1979) 281–299.
[83] S.S. Kutateladze, Boiling heat transfer, Int. J. Heat Mass Transf. 4 (1961) 31–45.
[84] N. Zuber, Hydrodynamic aspects of boiling heat transfer, Physics and Mathematics, AEC Report No. AECU-4439, 1959.
[85] S.M.S. Murshed, C.A. Nieto de Castro, M.J.V. Lourenco, M.L.M. Lopes, F.J.V. Santos, A review of boiling and convective heat transfer with nanofluids, Renew Sustain Energy Rev. 15 (2011) 2342–2354.
[86] D. Ciloglu, A. Bolukbasi, A comprehensive review on pool boiling of nanofluids, Appl. Therm. Eng. 84 (2015) 45–63.
[87] C.G. J. Prakash, R. Prasanth, Enhanced boiling heat transfer by nano structured surfaces and nanofluids, Renew Sustain Energy Rev. 82 (2018) 4028–4043.
[88] J.T. Cieslinski, Augmentation of critical heat flux in water-Al_2O_3, water-TiO_2 and water-Cu nanofluids, in: MATEC Web of Conferences, EDP Sciences, 18 (2014) 801012.

[89] B. H. Truong. Determination of pool boiling critical heat flux enhancement in nanofluids, Report MIT, June 2007.
[90] A. Reinho Neto, J.L.G. Oliveira, J.C. Passos, Heat transfer coefficient and critical heat flux during nucleate pool boiling of water in the presence of nanoparticles of alumina, maghemite and CNTs, Appl. Therm. Eng. 111 (2017) 1493–1506.
[91] I.C. Bang, Boiling heat transfer performance and phenomena of Al_2O_3-water nanofluids from a plain surface in a pool, Int. J. Heat Mass Transf. 48 (2005) 2407–2419.
[92] S.M. You, Effect of nanoparticles on critical heat flux of water in pool boiling heat transfer, Appl. Phys. Lett. 83 (2003) 3374.
[93] H.D. Kim, J. Kim, M.H. Kim, Experimental studies on CHF characteristics of nanofluids at pool boiling, Int. J. Multiph. Flow 33 (2007) 691–706.
[94] J.S. Coursey, J. Kim, Nanofluid boiling: the effect of surface wettability, Int. J. Heat Fluid Flow 29 (2008) 1577–1585.
[95] M.N. Golubovic, H.D. Madhawa, W.M. Worek, Nanofluids and critical heat flux, experimental and analytical study, Appl. Therm. Eng. 29 (2009) 1281–1288.
[96] H. Kim, H.S. Ahn, On the mechanism of pool boiling critical heat flux enhancement in nanofluids, J. Heat Transf. 132 (2010) 061501.
[97] H.M. Ali, M.M. Generous, F. Ahmad, M. Irfan, Experimental investigation of nucleate pool boiling heat transfer enhancement of TiO_2-water based nanofluids, Appl. Therm. Eng. 113 (2017) 1146–1151.
[98] S.J. Kim, I.C. Bang, J. Buongiorno, Study of pool boiling and critical heat flux enhancement in nanofluids, Bull. Pol. Acad. Sci.: Tech. Sci. 55 (2007) 211–216.
[99] H. Kim, J. Kim, M.H. Kim, Effect of nanoparticles on CHF enhancement in pool boiling of nanofluids, Int. J. Heat Mass Transf. 49 (2006) 5070–5074.
[100] M.M. Sarafraz, T. Kiani, F. Hormozi, Critical heat flux and pool boiling heat transfer analysis of synthesized zirconia aqueous nano-fluids, Int. Commun. Heat Mass Transf. 70 (2016) 75–83.
[101] M.H. Sulaiman, D. Matsuo, K. Enoki, T. Okawa, Systematic measurements of heat transfer characteristics in saturated pool boiling of water-based nanofluids, Int. J. Heat Mass Transf. 102 (2016) 264–276.
[102] R. Padhye, J. McCollum, M.L. Pantoya, Effects of nanofluids containing graphene/graphene-oxide nanosheets on critical heat flux, J. Phys. Chem. C. 119 (2015) 26547–26553.
[103] S.M. Kwark, R. Kumar, G. Moreno, Pool boiling characteristics of low concentration nanofluids, Int. J. Heat Mass Transf. 53 (2010) 972–981.
[104] S.M. Kwark, R. Kumar, G. Moreno, Nanocoating characterization in pool boiling heat transfer of pure water, Int. J. Heat Mass Transf. 53 (2010) 4579–4587.
[105] J.H. Lee, Y.H. Jeong, Experimental investigation on the CHF enhancement of pool boiling using magnetic fluid, in: Trans Korean Nuclear Society Autumn Meeting, (2010) 479–480.
[106] A. Suriyawong, S. Wongwises, Nucleate pool boiling heat transfer characteristics of TiO_2-water nanofluids at very low concentrations, Exp. Therm. Fluid Sci. 34 (2010) 992–999.
[107] M. Kole, T.K. Dey, Investigations on the pool boiling heat transfer and critical heat flux of ZnO-ethylene glycol nanofluids, Appl. Therm. Eng. 37 (2012) 112–119.
[108] S.C. Hiswankar, J.M. Kshirsagar, Determination of critical heat flux using ZnO nanofluids, IJERT 2 (2013) 2091–2095.
[109] D. Milanova, R. Kumar, Heat transfer behaviour of silica nanoparticles in pool boiling experiment, J. Heat Transf. 130 (2008) 042401.
[110] P. Vasallo, R. Kumar, S. D'Amico, Pool boiling heat transfer experiments in silica-water nano-fluids, Int. J. Heat Mass Transf. 47 (2004) 407–411.
[111] R. Kathiravan, R. Kumar, A. Gupta, Preparation and pool boiling characteristics of copper nanofluids over a flat plate heater, Int. J. Heat Mass Transf. 53 (2010) 1673–1681.

[112] D. Wen, M. Corr, X. Hu, G. Lin, Boiling heat transfer of nanofluids: the effect of heating surface modification, Int. J. Therm. Sci. 50 (2011) 480–485.

[113] D. Faukner, M. Khotan, R. Shekarriz, Practical design of a 1000 W/cm^2 cooling system, in: Proceedings of 19th IEEE SEMI-THERM Symposium, San Jose, CA, March 11–13, 2003.

[114] H. Peng, G.L. Ding, W.T. Jiang, H.T. Hu, Y.F. Gao, Heat transfer characteristics of refrigerant-based nanofluid flow boiling inside a horizontal smooth tube, Int. J. Refrig. 32 (2009) 1259–1270.

[115] S.J. Kim, T. McKrell, J. Buongiorno, L.W. Hu, Subcooled flow boiling heat transfer of dilute alumina, zinc oxide, and diamond nanofluids at atmospheric pressure, Nucl. Eng. Des. 240 (2010) 1186–1194.

[116] M. Boudouh, H.L. Gualous, M. De Labachelerie, Local convective boiling heat transfer and pressure drop of nanofluid in narrow rectangular channels, Appl. Therm. Eng. 30 (2010) 2619–2631.

[117] K. Henderson, Y.-G. Park, L.P. Liu, Flow-boiling heat transfer of R-134a-based nanofluids in a horizontal tube, Int. J. Heat Mass Transf. 53 (2010) 944–951.

[118] L. Xu, J. Xu, Nanofluid stabilizes and enhances convective boiling heat transfer in a single microchannel, Int. J. Heat Mass Transf. 55 (2012) 5673–5686.

[119] E. Abedini, A. Behzadmehr, H. Rajabnia, S.M.H. Sarvari, S.H. Mansouri, Experimental investigation and comparison of subcooled flow boiling of TiO_2 nanofluid in a vertical and horizontal tube, Proc. IMechE Part C J. Mech. Eng. Sci 227 (2013) 1742–1753.

[120] A.A. Chehade, H.L. Gualous, S. Le Masson, F. Fardoun, A. Besqet, Boiling local heat transfer enhancement in minichannels using nanofluids, Nanoscale Res. Lett. 8 (2013) 1–20.

[121] K.B. Rana, A.K. Rajvanshi, G.D. Agrawal, A visualization study of flow boiling heat transfer with nanofluids, J. Vis. 16 (2) (2013) 133–143.

[122] K.B. Rana, G.D. Agrawal, J. Mathura, U. Puli, Measurement of void fraction in flow boiling of ZnO-water nanofluids using image processing technique, Nucl. Eng. Des. 270 (2014) 217–226.

[123] B. Sun, D. Yang, Flow boiling heat transfer characteristics of nano-refrigerants in a horizontal tube, Int. J. Refrig. 38 (2014) 206–214.

[124] M.M. Sarafraz, F. Hormozi, Scale formation and subcooled flow boiling heat transfer of CuO-water nanofluid inside the vertical annulus, Exp. Therm. Fluid Sci. 52 (2014) 205–214.

[125] S.W. Lee, S.D. Park, S.R. Kang, S.M. Kim, H. Seo, D.W. Lee, I.C. Bang, Critical heat flux enhancement in flow boiling of Al_2O_3 and SiC nanofluids under low pressure and low flow conditions, Nucl. Eng. Technol. 44 (2012) 429–436.

[126] S.W. Lee, K.M. Kim, I.C. Bang, Study on flow boiling critical heat flux enhancement of graphene oxide/water nanofluid, Int. J. Heat Mass Transf. 65 (2013) 348–356.

[127] T. Lee, D.H. Kam, J.H. Lee, Y.H. Jeong, Effects of two-phase flow conditions on flow boiling CHF enhancement of magnetite-water nanofluids, Int. J. Heat Mass Transf. 74 (2014) 278–284.

[128] H. Aminfar, M. Mohammadpourfard, R. Maroofiazar, Experimental study on the effect of magnetic field on critical heat flux of ferrofluid flow boiling in a vertical annulus, Exp. Therm. Fluid Sci. 58 (2014) 156–169.

[129] R. Kamatchi, S. Venkatachalapathy, Parametric study of pool boiling heat transfer with nanofluids for the enhancement of critical heat flux: a review, Int. J. Therm. Sci. 87 (2015) 228–240.

[130] T.J. Hendricks, S. Krishnan, C. Choi, C.H. Chang, B. Paul, Enhancement of pool boiling heat transfer using nanostructured surfaces on aluminum and copper, Int. J. Heat Mass Transf. 53 (2010) 3357–3365.

[131] K.H. Chu, R. Enright, E.N. Wang, Structured surfaces for enhanced pool boiling heat transfer, Appl. Phys. Lett. 100 (2012) 241603.

[132] K.H. Chu, Y. Soo Joung, R. Enright, C.R. Buie, E.N. Wang, Hierarchically structured surfaces for boiling critical heat flux enhancement, Appl. Phys. Lett. 102 (2013) 151602.

[133] I.C. Bang, Effects of Al_2O_3 nanoparticles deposition on critical heat flux of R- 123 in flow boiling heat transfer, Nucl. Eng. Technol. 47 (2015) 398–406.

[134] A. Nazari, S. Saedodin, Critical heat flux enhancement of pool boiling using a porous nanostructured coating, Exp. Heat Transf. 30 (2017) 316–327.

[135] R.K. Cheedarala, E. Park, K. Kong, Y.B. Park, H.W. Park, Experimental study on critical heat flux of highly efficient soft hydrophilic CuO–chitosan nanofluid templates, Int. J. Heat Mass Transf. 100 (2016) 396–406.

[136] M.N.A.W.M. Yazid, N.A.C. Sidik, R. Mamat, G. Najafi, A review of the impact of preparation on stability of carbon nanotube nanofluids, Int. Commun. Heat Mass Transf. 78 (2016) 253–263.

[137] S. Ujereh, T. Fisher, I. Mudawar, Effects of carbon nanotube arrays on nucleate pool boiling, Int. J. Heat Mass Transf. 50 (2007) 4023–4038.

[138] V. Khanikar, I. Mudawar, T. Fisher, Effects of carbon nanotube coating on flow boiling in a micro-channel, Int. J. Heat Mass Transf. 52 (2009) 3805–3817.

CHAPTER

Bionanofluids and mass transfer characteristics

Baishali Kanjilal[1], Nouruddin Sharifi[2], Arameh Masoumi[3] and Iman Noshadi[1,3]

[1]*Department of Bioengineering, University of California, Riverside, Riverside, CA, United States* [2]*Department of Engineering Technology, Tarleton State University, Stephenville, TX, United States* [3]*Department of Chemical Engineering, Rowan University, Glassboro, NJ, United States*

8.1 Introduction

A nanofluid (NF), simply defined, is a colloidal dispersion of nanometer-sized particles in a fluid-based medium [1]. Examples of such NF may be metals or oxide suspensions, carbides, or even carbon nanotubes [2,3]. NFs' novel properties have been well studied and exploited in heat transfer applications, such as in fuel cells and electronics—particularly their microapplications, hybrid motorized engines, or many other industrial processes [4,5]. The most traditional dispersion medium is water due to its environment, low cost, and availability [3]. NFs also bestow advantageous applications in mass transfer [1].

Bionanofluid (BioNF) is essentially a NF system that entails bioconvection. In the process, self-propelled microorganisms form a heavier-than-water suspension, float, and come close to an upper surface, increasing their density and causing an unstable state and the suspension to fall to the bottom again [6]. The microorganisms then recontinue on this pathway [6]. Plumes of bioconvection are formed due to hydrodynamic instability, and are even used to transport cells and oxygen. This can also be extended to antiviral platforms and sensors. Many groups have examined the convection of bioorganism-filled plumes in three-dimensional chambers, wherein they were found to transport bacteria and oxygen from the upper layers. Such BioNF transport has been studied in squeezing fluid flow between parallel plates. Stagnation flow problems in BioNFs have been studied analytically by several groups with field and process variations, including chemical reactions, the effect of a magnetic field, thermal radiation, etc. [6].

In mass diffusion, the molecular phenomena entail diffusion transport of particles brought about by a potential difference due to concentration gradients and essentially entails convective and diffusive transfer, broadly [7]. Fang et al. studied the diffusion coefficient of Rhodamine B in water-based NF [8]. In general, there is a diffuse understanding of mass transport in NF structures, unlike heat transfer studies, including conflicting reports on the effect of mass diffusion brought about by adding nanoparticles to base medium.

Bioconvection was initially defined as the process of streaming patterns in dense cultures of swimming organisms. The word underwent a significant change over time to be defined as spontaneous pattern formation in microorganism suspensions. It is currently defined as the

Nanofluids and Mass Transfer. DOI: https://doi.org/10.1016/B978-0-12-823996-4.00008-2

macroscopic convection of fluid caused due to the density gradient due to swimming motile microorganisms [6]. This is caused by density stratification and enhances mixing and slows particulate settling. Khan and Makinde studied nanofluid bioconvection and showed the effect of swimming patterns on the density of the base fluid [9]. A seminal work on the modeling of such systems was laid out by Kuznetsov [10]. This was extended by Zaimi [11] in stagnation-point bioconvection flow over an oscillating stretching/shrinking surface in BioNF. Kuznetsov showed that oscillations in BioNF bioconvection could be caused by microorganism interactions or temperature changes due to heating of a NP size and weight distribution [12]. Tham showed that in BioNF, motile microorganisms improved mass transfer, mixing, and stability [13]. In recent years BioNF has been modeled in various ways, including by Ali et al. as an unsteady MHD mixed convection toward a stagnation flow and mass transfer with an induced magnetic field [14]. Effect of radiation and magnetic field BioNF flowing past a stretching sheet was modeled by Ishak [15]. Sinha and Misra studied the magnetic field's effect on stagnation flow and heat transfer over a stretching sheet [16]. Chaudhary and Merkin [9,10] modeled homogeneous–heterogeneous reactions in a stagnation-point boundary-layer flow in a BioNF [17]. This chapter attempts to traverse through brief introduction nanofluids, their application possibilities, and a review on various types of mass transfer processes with NFs followed by a general introduction to the mechanism of enhanced mass transport in NFs. The review then delves into BioNFs, their applications in mass transfer, and an elucidation of understanding their models of mass transfer. A conclusion and the future outlook is finally presented.

8.2 Present status of research in nanofluids

Currently, there is a spate of technologies that use NF properties in various industrial processes entailing heat exchangers, electronic cooling, solar energy harvesting, among others. There has been a steady increase in literature in the past decade on NF studies and applications, underscoring its rising prominence in engineering fields. Fig. 8.1 shows the various research areas under development in NF and the number of publications spanning 2010–2019 [18].

8.3 Preparation and stabilization of nanofluids

NFs are prepared by suspending particles of dimensions 1–100 nm, maintaining suitable volume fractions, in suitable fluid media [19]. NFs must be stable over the process of transport with negligible agglomeration and nil chemical change. This section briefly explores the methods of preparing NFs.

8.3.1 Preparation of nanofluids and bionanofluids

NFs can be produced by a single-step method entailing direct dispersion in the base medium fluid, wherein the particles are produced by vapor deposition or liquid chemical method [20]. It minimizes agglomeration, but reactant residues are left. The other method is to make the

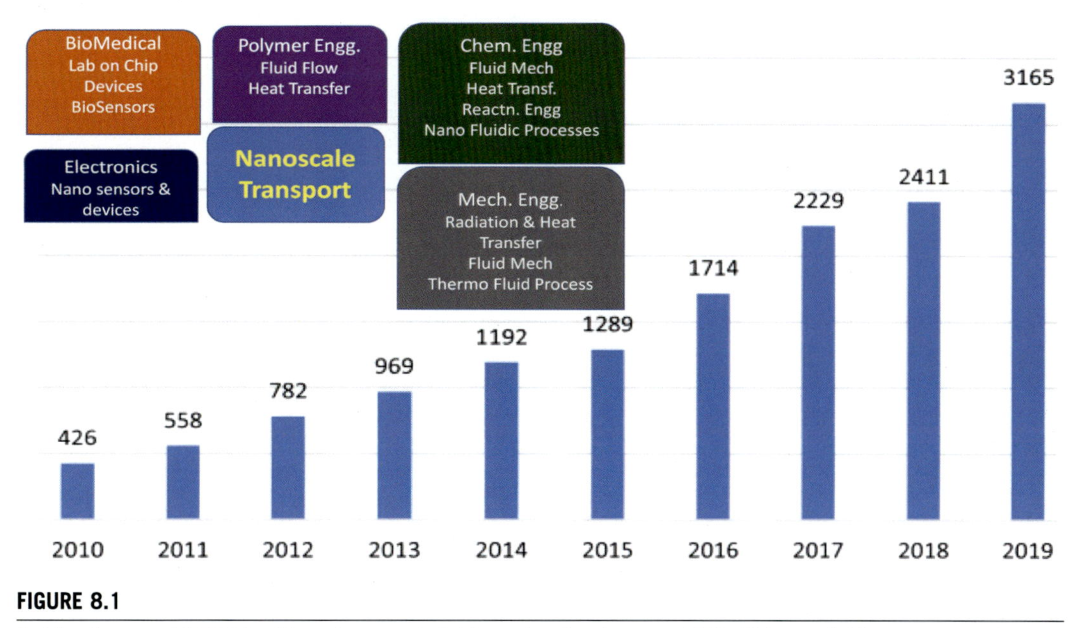

FIGURE 8.1

Publications on NF 2010–19 and research fields under development [18]. *NF*, nanofluid.

nanoparticle separately and, in a second step, disperse in the base fluid [19]. Yet another design of NF could be a micellar amphiphilic structure, dispersed in fluids or a phase inverted form of the same [2,3]. The best way to avoid agglomeration and sedimentation is to maintain a constant distance between the particles, achieved by constant concentration [4]. Table 8.1 explores data on some select NF preparation and stabilization methods. A NF is essentially considered as one in which liquid suspended NPs exist whose primary measurement is lower than 10 nm. Conventional fluids such as oils, polymeric gels, or even water could be used as the medium for doping with NPs comprising metal particles or their compounds or even organic particles which satisfy the size requirements. As with other materials with the prefix "bio" the nomenclature is open to interpretation. Hence there are works that have described BioNFs as those where NFs have been employed along with biosurfactants. Such a structure would in essence be no different from a NF and the wettability mechanism would be the only difference applied for applications such as hydrocarbon recovery. Studies abound of BioNF structures wherein the dispersant is a biosurfactant. But for the usage of bio surface active agents, such dispersions would be no different from other NFs. Thus in this chapter, an emphasis has been put on the usage of BioNF to strictly mean those NF fluids that show the effect of bioconvection, hence necessarily involving a gyrotactic microorganism, which is what makes their transport phenomenon such an interesting and dynamic topic of research and study. The bioconvection is brought about by the presence of microorganisms that are denser than the dispersing medium, most often water, capable of swimming upwards. These microorganisms accumulate at the upper surface causing the collective density to increase at that point, resulting in a disturbance in the balance of the suspension leading to a tumble and fall of the plumes of the microorganisms. This causes a macroscopic bioconvective

Table 8.1 Nanoparticle synthesis and stability.

NP/base fluid	Particle loading	Particle size	Synthetic process	Stabilization technique	References
SiO_2 in DI water	0.45%–4%	12 nm	Two-step	Ultrasonic vibration	[20]
TiO_2 in water	1%–2%	6 nm	Two-step	Ultrasonic vibration and magnetic stirring	[19]
Al_2O_3 in water	0.5%	60 nm	Two-step	Ultrasonic disrupter + homogenization by magnetic force agitation	[21]
ZnO in water	NA	NA	Two-step	150–80 acetyl acetone + sonication for 10 minutes Stability over 9 months to 1 year	[22]
CuO in distilled water	0.05 vol.%		Two-step	SDBS surfactant + sonication	[23]
Fe_3O_4 with water	NA	NA	Two-step	Sonication for 1 hour, Stability reported up to 12 hours	[24]
Cu in deionized water	NA	10 nm	Two-step	Laureate salts + ultrasonic vibration stability up to 30 hours	[25]
MWCNT in water, CNT was functionalized prior	NA	NA	Two-step	Agitation with stability up to several months	[26]
CNT in water	NA	NA	Two-step	Agitated with 0.2 wt.% chitosan, Stability up to 2 months	[27]

current and the falling microorganisms are replaced by upward swimming microorganisms. In some experimental systems, more than one kind of gyrotactic swimming microorganisms may be considered, such as the bottom-heavy alga *Chlamydomonas nivalis* and the soil bacterium species *Bacillus subtilis*. The mechanism of gyrotaxis is dependent on the microorganism species. Commonly, oxytactic bacteria and gyrotactic algae are used for developing green engineering processes which need the gravitational torque for mixing in novel applications such as nanobio-convection fuel cells. The process of gyrotaxis is also known to enhance NP suspension stability by continuous convection as well as local convection eddies entailing swirls, helical, and peristaltic flows [10,19,20]. Gyrotactic microorganisms or motile microorganisms are widely present in aquatic environments. The gyrotaxis of these organisms is responsible for many ecological phenomena essentially due to bioconvection. Microorganisms as such include algae of many types, phytoplankton algal blooms, *Chlamydomonas nivalis*, *Chlamydomonas reinhardtii*, *Dunaliella salina*, amongst other tetrahymena, ciliate, and flagellate species [10]. The inorganic suspended nanoparticles could be stabilized as per functionalization of the surface of nanoparticles or by the usage of biological/natural surfactants, such as *Sapindus mukorossi*, which are nontoxic to the microorganisms [20]. As a result of the presence of both suspended and stabilized nanoparticles as well as gyrotactic motile microorganisms, the velocity, local and global, of the fluid is more complex than can be described by mere Newtonian behavior. Essentially it becomes a second grade fluid with the velocity field having two derivatives in the stress–strain tensor. In a BioNF, the fluid medium's macroscopic movement due to a spatial density variation causes additive mobility of the motile microorganisms. A bioconvective stream in a given direction is caused by the motile microorganisms. These microorganisms are either chemotactic, oxytactic, or gyrotactic/negative gravitactic. However, the suspended NPs have zero self-motility. Their overall movement is an addendum of that due to Brownian motion, the bioconvection brought about by the motile microorganisms and some thermophoresis. The concentration of nanoparticles has to be small, in order for bioconvective motion to take place. It is essential that there be no change in the base fluid viscosity owing to the presence of the NPs [20].

8.3.2 Stabilization of nanofluids

NF stability can be visualized in several ways:

1. Surfactants can stabilize NFs by reducing fluid surface tension and prevent agglomeration. The amphiphilic structure of surfactants aids by selectively bonding to the surface of the nanoparticle and the fluid. Surfactants, however, run the risk of contamination and production of foam.
2. Alternative surface modifications have entailed joining silanes to silica NPs' surface to produce a combination of functionalized silica with NF properties that prevent surface deposition.
3. The aggregation rate is dependent on recurring impacts and the probability of coalescence during particulate collisions. An interplay of these two during the Brownian motion of particles will result in coalescence if attraction is higher than repulsion leading to suspension instability.

While Table 8.1 is not a comprehensive list of synthesis, dispersions, and treatments, it does uphold the presence of a dispersion augmenting agent as a superior method of achieving stability.

8.4 Applications of nanofluids and bionanofluids

NFs have found use in automobile and energy industries, mostly as cooling fluids in machines and devices owing to excellent heat transfer [28]. In heat exchangers, the efficiency of MWCNT and graphite NF suspensions have been offset by the obstacles of cost and particulate suspensions in the laminar flow [29]. The mass transfer characteristics of NFs find significant implications in biomedical applications [30]. NFs are used for targeted imaging and drug delivery to minimize radiation or chemocancer therapy [30]. Some NFs have antibacterial properties (ZnO) for site-specific delivery with concerted antibacterial action [30]. Micellar functionalized NF's have been studied in nanodrug delivery to supply controlled drug dosages [28]. Mass diffusion in NFs is enhanced by the Brownian motion of NPs, inducing hydrodynamic effects [31]. NF can also facilitate diffusion via reversible chemical reactions with the solvent and gas enhancing overall gas holdup [31]. Photocatalysis studies have been carried out using titanium dioxide (TiO_2) as a photoactive, inert, and inexpensive catalyst [32]. An understanding of the considerations in NF preparation processes is summarized in Fig. 8.2. The NFs' practical application will eventually be determined by the type

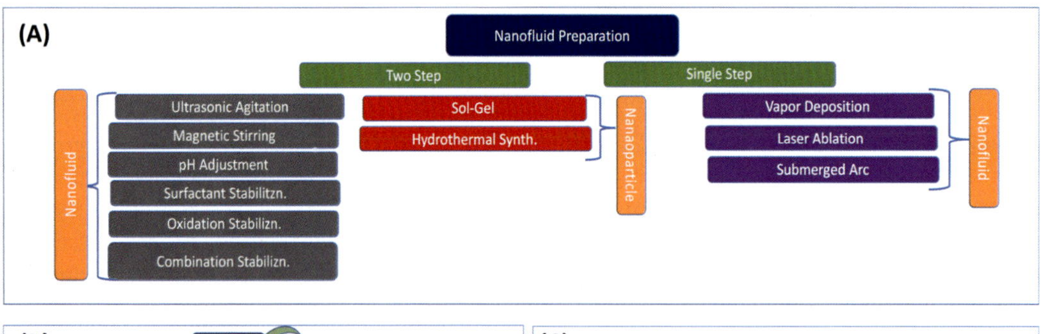

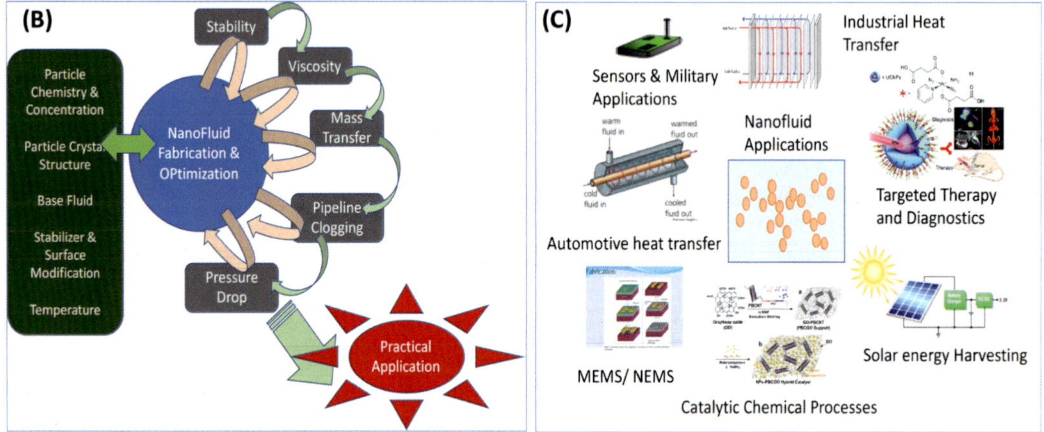

FIGURE 8.2

(A) Summary of considerations in nanofluid preparation processes. (B) Summary of considerations for the practical application of a nanofluid. (C) Schematic of the various application possibilities of nanofluids [1–4].

of NF and a gamut of considerations entailing concentration, additives and transfer requisites. This is summarized in Fig. 8.2. The use of NFs is likely to expand to the realm of biomass processing using catalytic metallic nanoparticles that are suitably dispersed. Fig. 8.2 summarizes the various application possibilities of NFs [1–4].

Applications of bionanofluids have been considered in drug delivery as well as antibacterial therapeutics. They have been considered for imaging and diagnostics such as MRI. BioNFs in medical applications need to be well characterized by way of composition, morphology, stability, biocompatibility, and agglomeration. In a drug delivery system, the structure must release the drug at the intended site in a controlled manner [6]. Drug release can be dependent on kinetics, pH, or temperature. A BioNF could be envisaged in the way of a soft micellar structure, with functionalizations on its surface that read out and seek out receptors on cellular surfaces that are expressed especially well in tumors, causing it to adhere to the surface and then allowing it to enable a transport mechanism of the drug release at the tumor site [33]. The BioNF must also be safe, noncytotoxic, and have biocompatibility. The BioNF system must also have appropriate drug loading and release, long shelf life, and must not be attacked by the body's immune system as a foreign body [33]. It should also not coagulate in response to charged entities in the body fluid. Nanocrystals are carrier-free drug NPs as a suspension of the poorly soluble drug in water with a surfactant stabilizer. They offer enhanced dissolution velocity and high bioavailability.

An example is ZnO NPs, because of their antibacterial properties and low toxicity for cancer treatment [34]. Magnetic BioNFs have also been used as drug carriers and therapeutics [35]. Their size and surface property ought to allow them to be directed using a magnetic field. Usually, here the blood acts as the carrier fluid. If the BioNF is drug-loaded, it can be injected close to the tumor and guided to reduce chemotherapy side effects [35]. Thus an understanding of the response of BioNFs to a magnetic field is extremely important [36]. An example of NPs used for this application is superparamagnetic iron oxide NPs in cancer detection and treatment. BioNFs could also be delivered using microelectromechanical biosystems through microchannels to maintain a uniform concentration of nanomedicine at channel output at an optimum temperature [33]. BioNFs have also been employed in magnetic fluid hyperthermia—the selective exposure of heat to the desired tissue using magnetic BioNF with an external magnetic field [37]. CNTs have been studied with a magnetic field to treat cancer cells. In alternative studies of controlled drug delivery systems based on a BioNF design entailing doxorubicin-loaded magnetic PLGA, poly(lactic-co-glycolic acid), microspheres have been studied for inhibition of tumor growth [33]. Magnetic bionanofluids have been used for anemia treatment entailing iron oxide stabilized with PVP (polyvinyl pyrrolidine) and 2-ethyl-6-methyl-3-hydroxy pyridine succinate. The BioNF was shown to stimulate red blood cells, hemoglobin, and hematocrit production [38]. Magnetic BioNFs have been used in MRI imaging using iron oxides, gold, and gadolinium NPs embedded in the polymeric matrix to give colloidal stability. A good contrast agent should be biologically inert and be fully metabolized without toxic effects. The BioNFS used for imaging could also be designed to have tissue-specific antibodies on the target organ [33]. TiO_2 and SiO_2-based BioNFs have also been used for antibacterial applications, like ZnO [34]. Photosensitivity has been used in harnessing light energy to produce reactive oxygen species (ROS), contributing to the antibacterial effect [33,34]. With $Mg(OH)_2$ NPs, the mechanism entails accumulating OH^- ions in cells, an increase of the pH to 10, and hence bacterial death [34]. Silver and CuO-based NFs have also been studied in bacterial inhibition and applied in wound dressings [34].

BioNFs have been specifically used for targeting papillary thyroid cancer cells, for which it was needed to be able to convert light to heat energy, be functionalized with ligands, and target and kill specific cells only [39]. MWCNTs were used because of their concentric cylindrical structure, layering, and large aspect ratio, which proffered a significant surface area for multiple bioattachments for antibodies and ligands to recognize several cell surface receptors. MWCNTs also have good localized temperature gains and remain undamaged. They can cause local hyperthermia but not bulk heating of benign tissue [33,39]. $COOH^-$ functionalized Au-decorated MWCNTs (made by plasma treatment and pulsed laser ablation techniques) were used for thiolated PEG attachment [39]. The BioNF had PEG brushes around the thin Au-coated islands and exposed COOH groups. These were conjugated to target molecule ligands—α-TSHR, thyrotropin, or thyrogen. Hybrid BioNFs include the carbon-based BioNF with further modification using an alloy, a transition metal, a semimetal, or polymer NP, all functionalized for biological targeting [39]. Fig. 8.3 illustrates the selective application of BioNFs in biomedical fields.

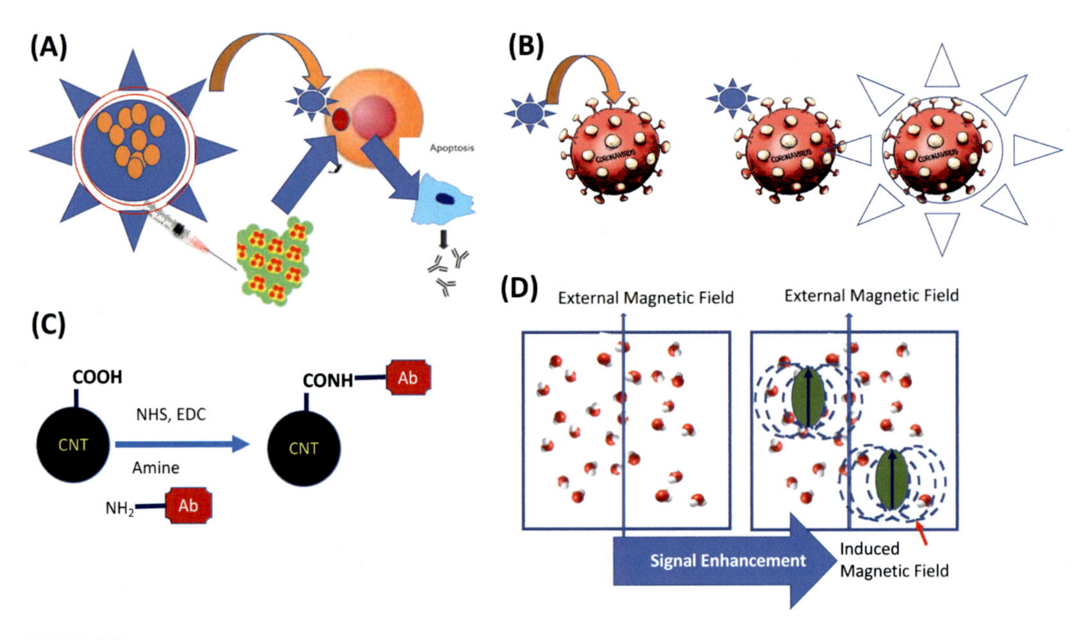

FIGURE 8.3

(A) Schematic of mechanism multiwalled carbon nanotube therapeutic bionanofluids to selectively target papillary thyroid cancer cells [39]. (B) Schematic virus specific targeting by micellar nanofluids [33]. (C) Schematic of ligand attachment to CNT nanofluids for tumor specific targeting [39]. (D) Schematic of local and global magnetic field interaction in nanofluid to obtain enhanced signals in imaging [33].

8.5 Types of mass transfer processes in nanofluids

A lot of mass transfer research in NFs have been carried out on gas absorption and liquid mass diffusion [2].

8.5.1 Bubble type absorption

Bubble type absorption is where the production of gas bubbles makes larger interface areas, hence enhancing transfer. This is used in ammonia–water systems in binary nanofluids (NH_3/H_2O) with Cu, CuO, and Al_2O_3 nanoparticles to study absorption [2]. The gas bubble breaking model was used to explain enhanced mass transfer by Ag-nanoparticles attributed to larger mass transfer interface [40]. Yet another mechanistic study, carried out by Kim et al. [41], with Al_2O_3-based NF, showed that while Al_2O_3 NP cannot break the bubble, a mushroom-like shape at the gas–liquid interface occurs during mass diffusion, caused by absorbed CO_2 in the solvent with a higher density, breaking into the solvent bulk [41].

8.5.2 Falling film absorption

This is mainly used in refrigeration and CO_2 removal [42]. In falling film absorption mass transfer resistance was dominant in the liquid flow.

8.5.3 Membrane absorption

CO_2 removal by silica and CNT NFs was studied in a shell-type microporous hollow fiber membrane [43]. Nanoparticles enhanced the CO_2 removal efficiency, and CNT was better than silica for the CO_2 removal enhancement.

8.5.4 Mass transfer with phase change

This entails hydrate formation and dissociation of pure water with nanocopper. The hydrogen bond of host molecules forms hydrate at high pressure, wherein the NPs have a lattice structure to capture the gas. [44].

8.5.5 Three-phase airlift reactor

NFs have also shown an adverse effect on mass transfer processes; for example, a reduction in O_2/H_2O mass transfer coefficient was observed in the presence of TiO_2 nanoparticles due to aggregation with increased particle concentration [44].

8.5.6 Agitated absorption reactor

Magnetic Fe_3O_4 was investigated in mass oxygen transfer in an agitated, sparged reactor with enhanced mass transfer by up to sixfold (600%) at NP volume fractions below 1% as well as gas–liquid interfacial area. The chemical absorption rate of CO_2 in amino propanol and other amines was also seen to be enhanced with colloidal silica [45] in an agitated reactor setup.

8.6 Mechanism of mass transfer enhancement in nanofluids and bionanofluids

8.6.1 Shuttle or grazing effect

Here, the NPs transport an extra volume of gas to bulk liquid. This happens via adsorption in the G-L diffusion layer. The desorption happens in the liquid bulk. This results in a large mass transfer coefficient [1].

8.6.2 Hydrodynamics in the GL layer

NPs are thought to induce turbulence at the gas–liquid interface. This thins down the diffusion layer and enhances mixing and mass transfer coefficient [1].

8.6.3 Changes in GL interface

In this, diffusion through the gas is thought to proceed proportional to solute concentration difference on either side of the gas film. Diffusion through the liquid is controlled by the solute concentration difference between the liquid–gas interface and the other side of the liquid film [1].

Fig. 8.4 schematically represents the general methods of colloid particle suspension, an example of surface treatment to achieve colloid stabilization using CNT as an example, and the mechanisms of mass transfer enhancement in NFs. Physical methods entail ultrasonication or high-pressure homogenization, which result in short-term stability [46]. Practical usage employs a combination of chemical and physical methods as shown in the picture, CNT nanofluids are prepared by the two-step process by producing dry NPs and dispersing them in an appropriate host fluid [46].

In a BioNF, in addition to the interfacial effects, there also occurs the addendum of transfer due to bioconvection owing to the spatial difference in density brought about by dense cultures of free

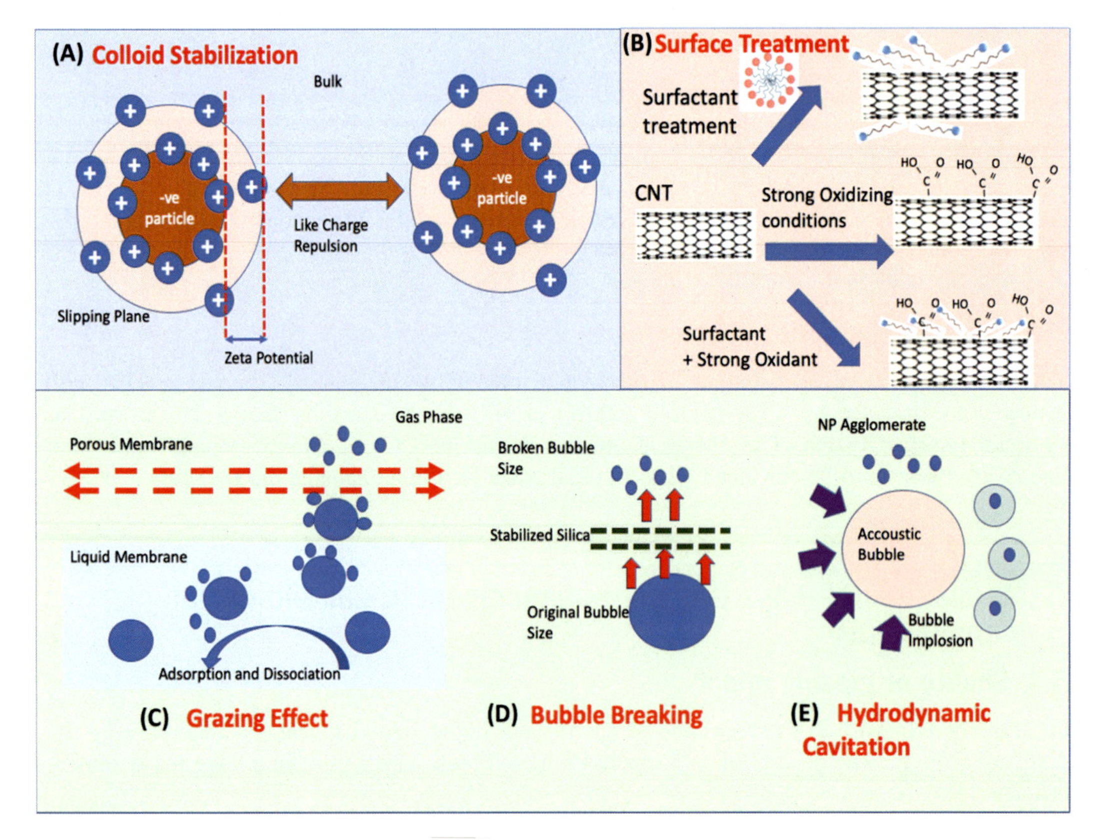

FIGURE 8.4

(A) Mechanism of colloid stabilization. (B) Surface treatment of CNT-NFs (chemical methods using acids / alkalis / surfactants) [46]. (C–E) Mechanism of mass transfer enhancement in NFs. (C) Grazing effect. (D) Bubble breaking. (E) Hydrodynamic cavitation. *NF*, nanofluid.

streaming motile microorganisms which cause streaming patterns to develop. This essentially causes local eddies and global streams which add to the thermophoretic transport of the system. Even though the microorganisms may look like Bernard cells [46] it is not due to thermal convection. Local micro-convection patterns in BioNFs also additionally develop due to a dissipative process by the microorganisms, akin to local dissipative patterns forming even in isothermal liquids [46]. From the act of heating, the gyrotactic patterns of the organisms on the BioNF are disturbed compared to isothermal suspensions. The directional swimming of the motile organisms causes a drag force due to which BioNFs behave like non-Newtonian nanofilms, visualized as dragging past each other and also contributing to the transport between them. BioNF transfer mechanisms can also be visualized as a magnetohydrodynamic second-grade fluid with dispersed NPs where flow occurs under boundary conditions with a convectively heated vertical surface [11–13,45]. At a microlevel, gyrotactic organisms along with thermal effects cause boundary layers over the suspended NPs, causing the dispersant medium to flow over the N surface in a paraboloid structure [12].

8.7 Analogy and equivalence between heat and mass transfer in nanofluids: an experimental and modeling approach

Enhanced thermal conductivity in NFs has been attributed to localized convection due to Brownian motion as one of the reasons [47]. Some theoretical models postulate that the nanoscale convection due to Brownian motion causes enhanced mixing and heat transfer beyond the realm of simple conduction. However, other studies suggest that [48] particle coupling due to interparticle potential is primarily responsible for enhanced transport. Thus some studies have chosen to experimentally elucidate microscale mass transfer of dye droplets in a water-based NF versus its mass diffusivity in the solvent [48]. It was observed in the experiment that dye diffuses faster in the presence of NF particles with peak enhancement at NF volume fraction of 0.5% [47]. Aggregation would produce fewer larger particles of much greater mass with the massive particle's consequent effect on Brownian motions. It is not just the Brownian motion but also the velocity disturbance fluid field created by NP motion responsible for enhancement. [48]. From a microscopic perspective, heat and mass transfer inside NFs are neither pure conduction, not pure molecule diffusion but a combination of microconvective heat and mass transfer. This is seen in the experiments [49] with Rhodamine B. The diffusion coefficient of the dye increases perceptibly with a change of temperature by at least 10°C. The effect is more pronounced at high NP volume fractions. While a stochastic Brownian motion enhances heat transfer, stirring induced by the NPs increases the mass transfer process inside the suspension. The energy equation of the suspensions can be written as [50]:

$$\frac{\partial T}{\partial t} + v \cdot \nabla T = a \nabla^2 T \tag{8.1}$$

where a is thermal diffusivity of the NF

The equation for mass transfer for the ith species is:

$$\frac{\partial Ci}{\partial t} + v \cdot \nabla Ci = D \nabla^2 Ci \tag{8.2}$$

where D is mass diffusivity of the NF

These two equations are the same due to the mass, and heat transfer is analogous. If the molecular diffusivity is *D*0 then the diffusive mass flux is,

$$J0 = -D0\nabla Ci \quad (8.3)$$

The random motion of the NPs induces perturbations in velocity (v′) and concentration (C_i'). The supplemental additive mass flux (*Dad*) arising from Brownian motion and microconvection is

$$Jad = <v'Ci'> = -Dad(\nabla Ci) \quad (8.4)$$

The total mass flux is

$$J = -D0(\nabla Ci) + <v'Ci'> = -(D0 + Dad)\nabla Ci \quad (8.5)$$

Making the effective mass diffusivity

$$D(nanofluid) = (D0 + Dad) \quad (8.6)$$

It means that irregular Brownian motion of the NPs causes microconvection and enhances mass transport inside the NF. Yet another experimental approach is discussed herein that attempts to understand and establish the rationale behind enhanced mass transfer in NFs. This experiment used glass capillary with fluorescein disodium dye solution on one side and alumina NF [51]. The hypothesis explores that enhancement occurs due to the diffusiophoretic motion [52] of NPs that gives rise to the fluid's convection. Diffusiophoresis is a colloidal particulate motion in the presence of a concentration gradient in a solute that interacts with the colloidal particle surface [51].

Fig. 8.5 shows the dependence of thermal transfer coefficients on the aggregation state of NFs [53]. A well-dispersed system may be thought to contain one particle per aggregate (right side). As particulate aggregation increases, thermal conductivity increases and then drops off after reaching a maximum. While effectively transport is increased with increasing NP fraction, particle agglomeration, sedimentation, and a general increase in fluid viscosity, it also controls the effective phenomenon of both mass and heat transfer and process issues including abrasion, clogging, and pressure drop. While low NP volume fractions can see rapid and intense Brownian movement, especially with an increase in temperature, NPs also agglomerate with an increase in temperature and volume fraction.

8.8 Bioconvection

Bionanofluid is essentially a NF system that entails bioconvection. In this section, we first summarize a canonical model presented by Kuznetsov on the onset of bioconvection in a horizontal layer of a BioNF and the basic solution to the model [11,12]. We then delve into its expansion into five different case studies with different conditions. Mixing in biochemical processes is not just expensive in terms of fabrication and power consumption costs, but also the heating generated due to mixing leads to damage of biological samples. This may be mitigated by bioconvection with its enhanced mass transport microvolume mixing that enhances BioNF stability, which may be particularly attractive in microfluidic devices.

The bioconvection models are essentially the following: continuum models, microscale models, and particle models [11,12]. The continuum model neglects the cell–cell interaction, chamber length scale and concentration distribution [54]. The suspension is dilute. In the microscale model [55], detailed geometry, flagellar rotation, hydrodynamic interaction are considered. In the particle

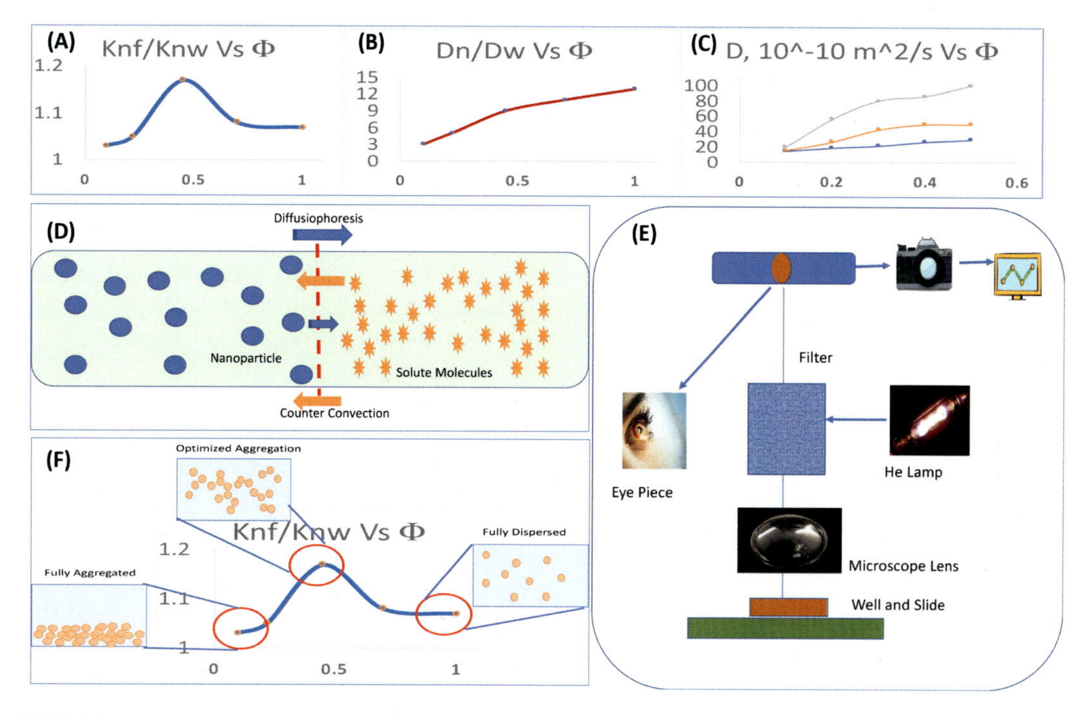

FIGURE 8.5

(A–C) Dependence of thermal transfer coefficients on aggregation state of nanofluids. (D) Mechanism of diffusiophoretic motion [52]. (E) Glass capillary fluorescein disodium dye solution and alumina experiment to understand Brownian motion in nanofluids [51]. (F) Thermal transfer coefficient and aggregation state of nanofluids [53].

model, chemotaxis is considered [56]. Microorganisms are considered individual particles in this model. The continuum model assumes that there is no cellular interaction and that the length of the chamber is much larger than the cellular dimensions, in addition to a large distribution of concentrations. Hence all variables are continuous, and the nanoparticle suspension is dilute. Despite the directional motility of the microorganism, the cellular concentration does not change and in such cases instability is incumbent on a critical value of the gyrotactic Rayleigh number. The microscale model describes microorganisms as microbes with detailing built on their geometry and movement, hydrodynamic interactions, and uptake. This model was able to establish and explain local hydrodynamic interaction between cells. In the particle model, the microorganisms are considered as mere particles without any consideration of their geometry.

8.9 A general model of bioconvection

Kuznetsov's work models BioNF bioconvection based on physical mechanisms with boundary layer slip between NPs and base fluid and also includes consideration of Brownian motion and thermophoresis. BioNFs are mostly considered water-based NFs with assumed nonagglomeration between NPs and that the NPs do not affect the direction or velocity of the swimming microorganism. Hence the model is applicable for a dilute NP suspension [11,12]. The governing equations are

$$\nabla \boldsymbol{v}^* = 0 \tag{8.7}$$

where $\boldsymbol{v}^* = (u^*, v^*, w^*)$ is the three-dimensional NF velocity

Buoyancy force and Boussinesq approximation are assumed to hold for a dilute suspension of NPs and that there is a small temperature gradient across the layer. To these assumptions, the concept of diffusion are employed (both Brownian motion and thermophoretic diffusion coefficients) as well as the volumetric heat capacities of the fluid and the NPs are introduced, and any spatial variation of the thermal conductivity is negligible to arrive at the conservation of momentum:

$$\frac{\partial \phi^*}{\partial t^*} + v^* \cdot \nabla \phi = \nabla^* \cdot \left[Db \nabla^* \phi^* \cdot \nabla^* T^* + \frac{Dt}{T^*} \nabla^* T^* \cdot \nabla^* T^* \right] \tag{8.8}$$

where ϕ^* is the nanoparticle volume fraction, *Db* and *Dt* are coefficients of Brownian and thermophoretic diffusion. This equation includes the sum of Brownian diffusion and thermophoresis. The fluid temperature variation is small. Conservation of microorganisms flux is based on bioconvection in a swimming microorganism suspension:

$$\frac{\partial n^*}{\partial t^*} = - \nabla^* \cdot j \tag{8.9}$$

where n^* is the number of microorganisms, and j is the total flux.

This includes microorganism flux due to fluid convection, self-propulsion, and diffusion. The assumption is that motion of the microorganisms can be split into components that are directional and random. The random component is controlled by diffusion, where Dm is the microorganism diffusivity. The boundary is considered to be constant for temperature and NP volumetric fraction. Even with an open to upper air surface, microorganisms will collect here in a packed layer and hence never let it be stress-free, hence justifying the use of "no-slip condition" at the upper wall. To the final solution, all physical variables are converted to dimensionless variables giving rise to the following boundary conditions:

$$\frac{\partial T}{\partial t} + v \cdot \nabla T = \nabla^2 T + \frac{Nb}{Le} \nabla \phi \cdot \nabla T + \frac{NaNb}{Le} \nabla T \cdot \nabla T \tag{8.10}$$

$$\frac{\partial \phi}{\partial t^*} + v \cdot \nabla \phi = \frac{1}{Le} \nabla^2 \phi + \frac{Na}{Le} \nabla^2 T \tag{8.11}$$

$$\frac{\partial n}{\partial t} = - \nabla \left(nv + n \frac{Q}{Lb} p - \frac{1}{Lb} \nabla n \right) \tag{8.12}$$

Le is the Lewis number for Brownian diffusion of NPs and *Lb* is the bioconvection Lewis number. *Q* is the bioconvection Péclet number. *p* is the unit vector indicating the average direction of microorganisms' swimming, and *n* is the dimensionless concentration of microorganisms. *Na* is a modified diffusivity ratio, while *Nb* is the modified particle-density increment. The basic solution for a condition which is time-independent, quiescent with varying temperature, nanoparticle volume fraction, and microorganism concentration in the z-direction only (*z* being the length in the z-direction)

$$øb = -\mathrm{Na}Tb + (1 - Na)z + Na \tag{8.13}$$

$$Tb = \frac{1 - exp\left[-\frac{(1-Na)Nb(1-z)}{Le}\right]}{1 - exp\left[-\frac{(1-Na)Nb}{Le}\right]} \tag{8.14}$$

The computation details are present by Kuznetsov [10]. The general solution is now presented in light of five case studies as follows:

8.9.1 Case study 1—three-dimensional stagnation point flow of bionanofluid with variable transport properties

We first look at the three-dimensional steady stagnation point flow of a BioNF with variable transport properties, which are dependent on concentration along with zero mass flux and thermal convective boundary conditions [6]. In the paper, the governing equations are nondimensionalized and transformed using similarity transformation generated. The controlling parameters entail viscosity, thermal conductivity, mass diffusivity, microorganism diffusivity, bioconvection Schmidt number, and Peclet number [6]. The variables are dimensionless and entail velocity, temperature, nanoparticle volume fraction, microorganisms, and microorganism rate. Bioconvection entails suspensions of self-propelled microorganisms, heavier than water, which keeps floating and sinking, affecting transporting cells and oxygen from the upper boundary layer. Stagnation studies have been carried out by many groups, including many different kinds of field effects and constant thermal physical properties. However, in BioNF, heat and mass transfer are intimately connected, and thermal conductivity and viscosity of nanofluids are a function of the local volume fraction of nanoparticles. This case study describes a 3D steady point flows of nanofluid with microorganisms with zero mass flux in boundary condition and thermal convective. All results are presented in the light of the corresponding dimensionless constants [6].

$$u = uw(x) = k1ax \tag{8.15}$$

$$v = vw(y) = k3ax \tag{8.16}$$

$$w = 0 \tag{8.17}$$

$$-k\frac{\partial T}{\partial z} = h(Tf - T) \tag{8.18}$$

$$\frac{\partial}{\partial z}[Db(\phi)\phi] + \left(\frac{Dt}{T\propto}\right)\frac{\partial T}{\partial z} = 0 \ (Passive\ control) \tag{8.19}$$

$$\phi = \phi w \ (Active\ Control) \tag{8.20}$$

The boundary conditions are $n = n0$ at $z = 0$, $u = ue(x) = k2ax$, $v = ve(x) = k2ay$, $T \rightarrow T\propto$, $\phi \rightarrow \phi\propto$, $n \rightarrow 0$, $z \rightarrow \propto$; u, v, and w are the velocities in the x, y, and z directions, respectively; $k1, k2, k3, h$ are mathematical constants, and k is the variable thermal conductivity.

T is the nanofluid temperature, and Tf is the convective surface temperature, both in kelvins. $Db(\phi)$ is the variable Brownian diffusion coefficient. Dt is the thermophoretic diffusion coefficient. $T\propto$ is the ambient fluid temperature. ϕw and $\phi\propto$ are the wall nanoparticles and ambient volume fractions, respectively. *ue* and *ve* are the dimensionless external fluid velocity in the x and y directions. If we now introduce the dimensionless variable, we get the following results:

Here c_2 is the dimensionless viscosity parameter, c_4 is the thermal conductive parameter, c_6 is the mass diffusivity parameter, c_8 is the microorganisms diffusivity parameter, η is the independent similarity variable, and $f(\eta)$, $g(\eta)$, $\theta(\eta)$, $\phi(\eta)$, and $\chi(\eta)$ are the dimensionless stream function, temperature, nanoparticle volume fraction functions, and microorganisms, respectively. The constants are B_i = Biot number, P_e = Peclet number, S_c = Schmidt number, and Sb = bioconvection Schmidt number.

The effect of variable viscosity parameter on the velocity, temperature, nanoparticles volume fraction, and microorganisms are reproduced from Alexandria Engineering Journal (2016) 55, 1983–1993 [6] as sample results in Fig. 8.6. It illustrates that the concentration boundary layer's thickness is small compared with hydrodynamic and thermal boundary layers. The detailed results are available in the journal article referenced and are briefly discussed in this case study.

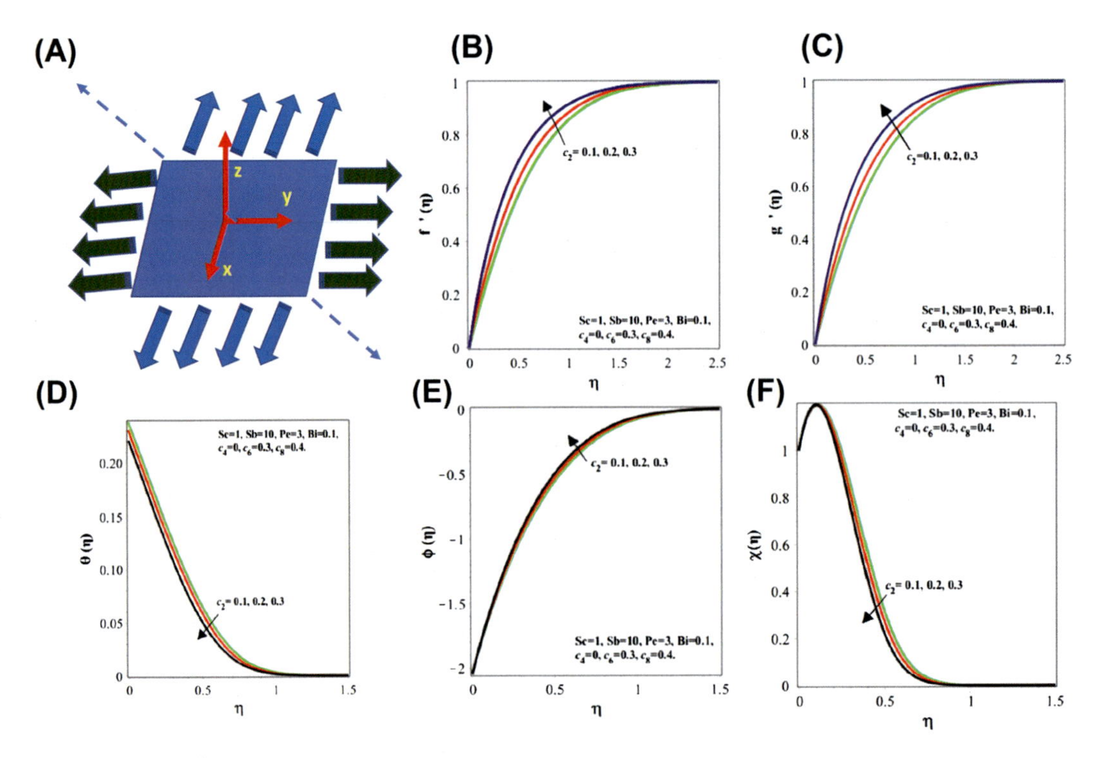

FIGURE 8.6

Three-dimensional stagnation point flow of bionanofluid with variable transport properties. (A) Physical modeling of the problem. (B–F) Effect of the dimensionless viscosity parameter on the x-component of velocity, y-component velocity, temperature, nanoparticles volume fraction, microorganism concentration.

Adapted from N.A. Amirsom, M.J. Uddin, A.I. Ismail, Three dimensional stagnation point flow of bionanofluid with variable transport properties, Alex. Eng. J. 55 (3) (2016) 1983–1993.

The dimensionless velocity components and NP volume fraction increase with viscosity, underscoring its effect on the boundary layer. Both temperature and microorganism both decrease with the increase of viscosity. It is also seen that with decreasing viscosity, the dimensionless temperature parameter decreases, indicating the growth of the thermophoresis effect and increasing the volume fraction of NP at the surface. As the mass diffusivity parameter increases, the velocity components decrease while the temperature, nanoparticle volume fraction, and microorganisms increase. The motile microorganism density at the boundary layer also decreases with increasing Schmidt numbers, as does the microorganism diffusivity relative to the porous medium's thermal diffusivity. The Schmidt number shows the relative strength of thermal diffusion to nanoparticle diffusion rates. If $S_c > 1$, thermal exceeds nanoparticle diffusivity and results in enhanced NP regime concentrations and NP concentration in the boundary layer thickness. Thus the Schmidt number increases the nanoparticle volume fraction too. The Peclet number is proportional to constant maximum cell swimming speed and inversely proportional to the diffusivity of microorganisms. When $P_e > 1$ species diffusivity will be dominated by a swimming motion, thus reducing the density of motile microorganisms. Peclet number is dependent on the type of microorganism. An increase in Lewis number causes a decrease in microorganism diffusivity relative to the porous medium's thermal diffusivity. An increase in viscosity parameter and Schmidt number causes the local mass transfer to increase.

8.9.2 Case study 2—bioconvection nanofluid slip flow past a wavy surface with applications in nanobiofuel cells

In this case study, the convective boundary layer flow of water-based bionanofluid with gyrotactic microorganisms is examined past a wavy surface using Buongiorno's nanofluid model with passive boundary conditions [57]. The wavy surface causes higher skin friction and higher local Nusselt numbers compared with a stationary surface. The physical representation of the problem and the results of its mathematical simulation are shown in Figs. 8.7 and 8.8 [57]. This case study studies free convection of BioNFs along a horizontal wavy surface with zero nanoparticle flux. This uses local similarity and nonsimilarity models and boundary conditions. For the sake of brevity and repetition, the results are reviewed directly. A moving surface irrespective of its vicissitudes shows higher Nusselt numbers with lower frictional resistance. A wavy surface simultaneously shows both high frictional resistance and Nusselt numbers. While Brownian motion shows no significant effect on skin friction, it is reduced by a buoyancy parameter owing to mass flux boundary condition, which also has thermophoresis.

On a wavy surface, the dimensionless velocity fluctuates and increases with the surface's amplitude in the stream's direction. This is because the fluid flow accelerates along the positive slope portions and decelerates along the wavy surface's negative slope portion. The variation of the dimensionless axial pressure distribution was also studied. A pressure difference drives the flow on such a surface, and hence any surface motion has a nontangible influence on the thickness of the boundary layer. Any surface motion, however, lowers surface pressure. Without a thermal slip, the surface is essentially isothermal and leads to increased dimensionless surface pressure. The surface temperature decreases with velocity as well as thermal slips.

The nanoparticle volume fraction increases with the Lewis number resulting in a decrease in the rescaled nanoparticle volume fraction boundary layer thickness. Velocity slip enhances the

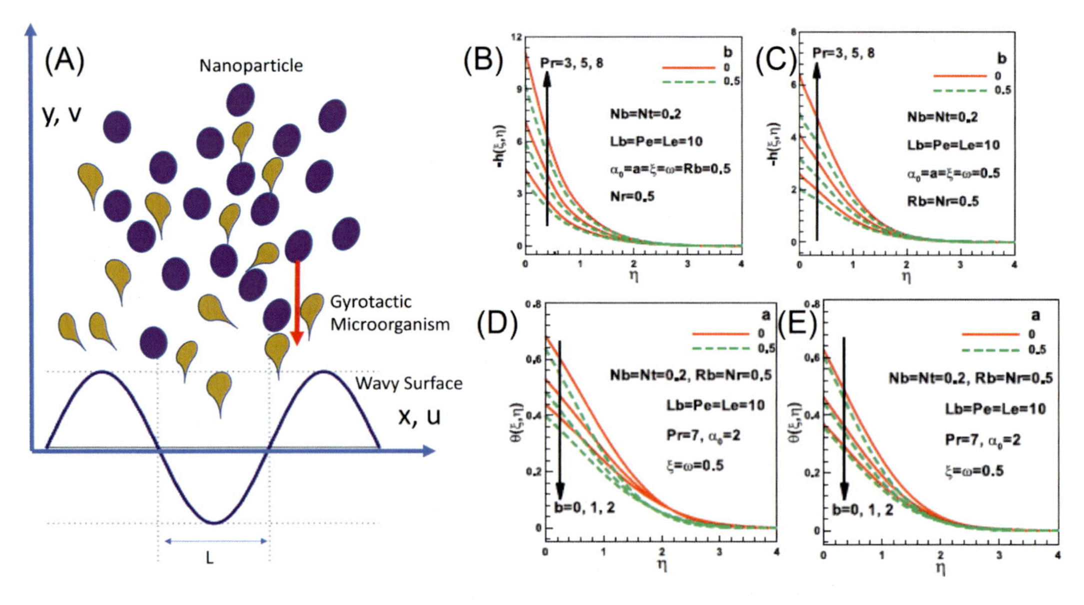

FIGURE 8.7

Model representation of the problem. (A) Biological flow diagram of wavy surface in a microbial fuel cell wall. (B, C) Dimensionless axial pressure distribution versus thermal slip and Prandtl number when (A) plate is stationary and (B) plate is moving. (D, E) Variation of dimensionless temperature with thermal and velocity slip when (D) plate is stationary and (E) plate is moving.

Adapted from M.J. Uddin, W.A. Khan, S.R. Qureshi, O.A. Bég, Bioconvection nanofluid slip flow past a wavy surface with applications in nano-biofuel cells, Chin. J. Phys. 55 (5) (2017) 2048–2063.

nanoparticle volume fraction within the boundary layer. The density of motile microorganisms is much thinner with a wavy surface. Given the microorganisms are heavier than water, upswimming causes unstable density stratification. An increase in Lewis number decreases microorganism concentration layer thickness, enhancing the viscous diffusion rate and decreasing the dimensionless velocity and hence decreasing microorganism density. The bioconvection Peclet number, as it increases, causes the speed of the microorganism to increase, and the density of the microorganisms near the surface also increases. However, if the Lewis number increases (ratio of thermal to mass diffusivity), it decreases the microorganism concentration boundary layer's thickness. An increase in Lewis number causes an increase in the viscous diffusion rate. This reduces the surface velocity and hence increases the microorganism density number.

8.9.3 Case study 3—stagnation point flow with time-dependent bionanofluid past a sheet: richardson extrapolation technique

In general, BioNF transport is envisioned as denser than water microorganisms swimming in an upward direction, accumulating toward an upper surface until there is a suspension disbalance [58].

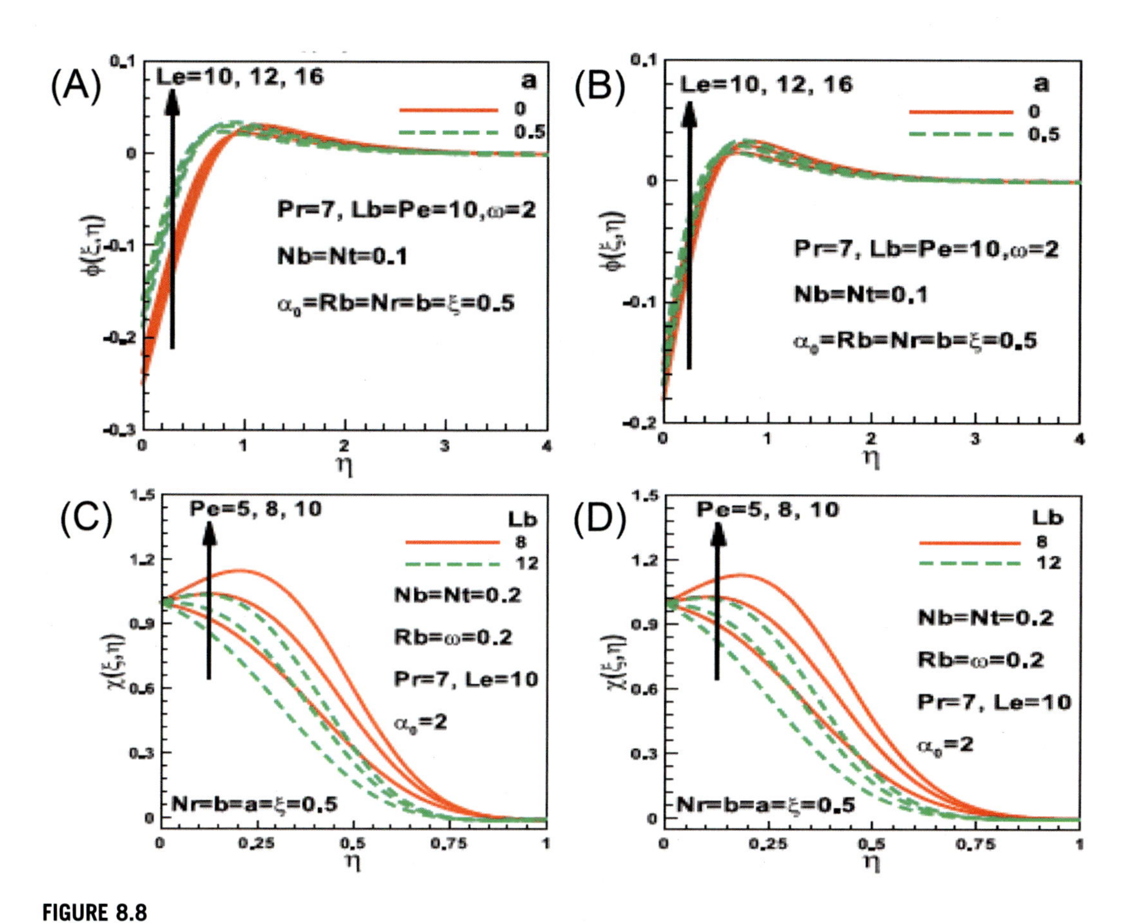

FIGURE 8.8

(A, B) Dimensionless rescaled nanoparticle volume fraction versus Lewis number and velocity slip (A) plate is stationary (B) plate is moving, (C, D) Rescaled density of motile microorganisms versus bioconvection parameters (A) plate is stationary and (B) plate is moving.

Adapted from M.J. Uddin, W.A. Khan, S.R. Qureshi, O.A. Bég, Bioconvection nanofluid slip flow past a wavy surface with applications in nano-biofuel cells, Chin. J. Phys. 55 (5) (2017) 2048–2063.

This is when the microorganisms tumble downwards. In this model case study, an unsteady two-dimensional boundary layer flow is considered along with heat transfer at the stagnation region [58]. A moving solvent surface causes this heat transfer. A bionanofluid comprises nanoparticles and motile gyrotactic microorganisms and allows the bioconvection process. This model works with water as the fluid and assumes that the nanoparticles' presence does not affect the swimming routes and velocities of the microorganisms [58].

The physical representation of the problem is schematically shown in Fig. 8.9. The computations, results of which are shown in Figs. 8.10 and 8.11, yielded several interesting results. As the

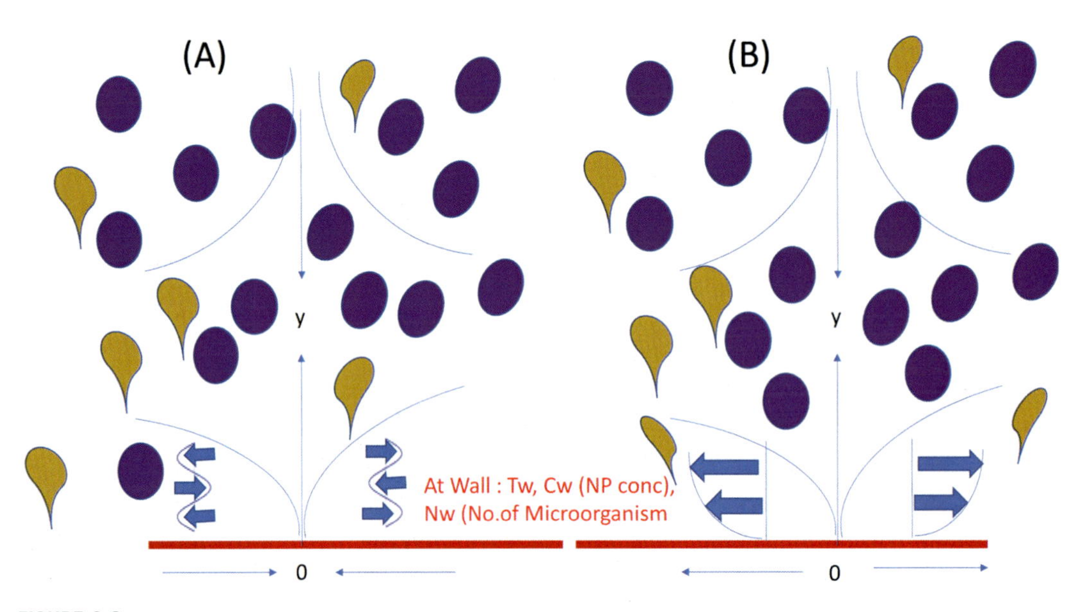

FIGURE 8.9

Stagnation point flow with time-dependent bionanofluid past a sheet: Richardson extrapolation technique. Schematic diagram of the problem. (A) Shrinking case. (B) Stretching case. [58].

Brownian motion increased (denoted by the dimensionless Brownian motion parameter) at the stretching and shrinking sheet, it led to a temperature increase. For a stretching sheet, there exists a thin thermal boundary layer and steep profiles. A shrinking sheet increases fluid temperature, reduces heat convection, and causes a shallow temperature gradient.

An increase in Brownian motion causes intermolecular collisions between the NPs and water solvent, pushes the NPs away from the sheet surface, and reduces their sheet layer density. It also causes greater macroconvection and thermal conduction. This is more pronounced with smaller NPs. Thus there is also an overall decrease in microorganism density at the wall with increasing Brownian motion. There is an increase in microorganism density for a stretching sheet, but with a shrinking sheet, the Brownian motion increases, and the organism density decreases. When the sheet shrinks in the boundary layer, it causes the microorganisms to swim away from the surface, causing a depletion in their density at the wall.

In this model, the effect of thermophoresis is studied too. As the thermophoresis parameter increases, it causes an increase in local temperature, concentration, and microorganism density. The temperature gradient causes a strong thermophoretic force displacing NPs away from the hot sheet surface to the bulk of the fluid and increases NP concentration in bulk. In the case of the shrinking sheet, the nanoparticles concentration overshoots on the shrining surface and forms a backward flow disrupting the laminar fluid flow.

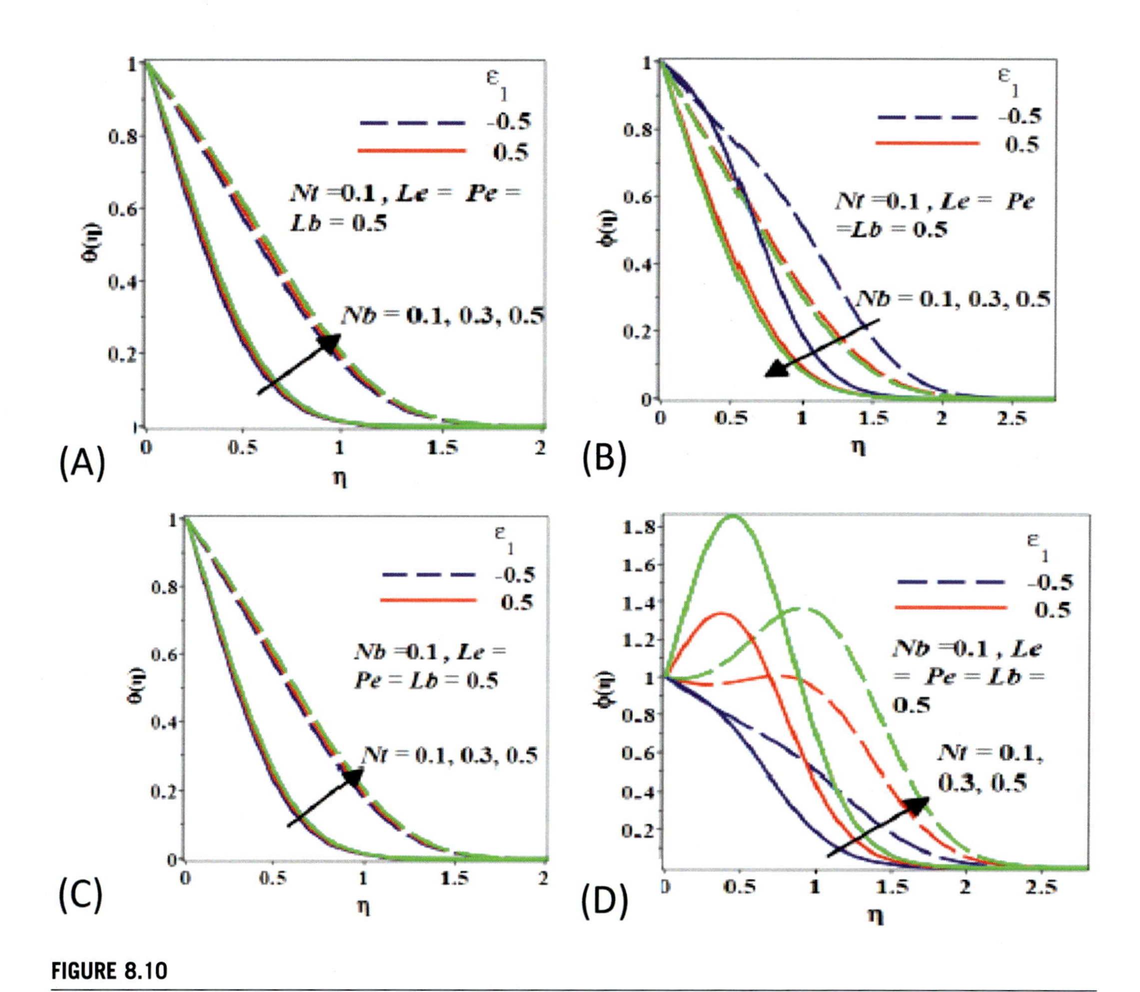

FIGURE 8.10

(A, B) Temperature and concentration of NP versus Brownian motion parameter for shrinking ($\varepsilon = -0.5$) and stretching sheets ($\varepsilon = +0.5$). (C, D) Temperature and concentration of NP with thermophoresis parameter for shrinking ($\varepsilon = -0.5$) and stretching sheets ($\varepsilon = +0.5$).

Adapted from K. Naganthran, M.F.M. Basir, S.O. Alharbi, R. Nazar, A.M. Alwatban, I. Tlili, Stagnation point flow with time-dependent bionanofluid past a sheet: Richardson extrapolation technique. Processes 7 (2019) 722.

8.9.4 Case study 4—unsteady magnetoconvective flow of bionanofluid with zero mass flux boundary condition

In this case study, an induced magnetic field stagnation point flow for an unsteady 2D laminar BioNF convection consisting of microorganisms along a vertical plate is shown. The physical configuration of the problem is shown in Fig. 8.12. This model assumed zero mass flux at the boundary layer [59]. In this case, a two-dimensional unsteady, incompressible, viscous, and conducting fluid is considered at

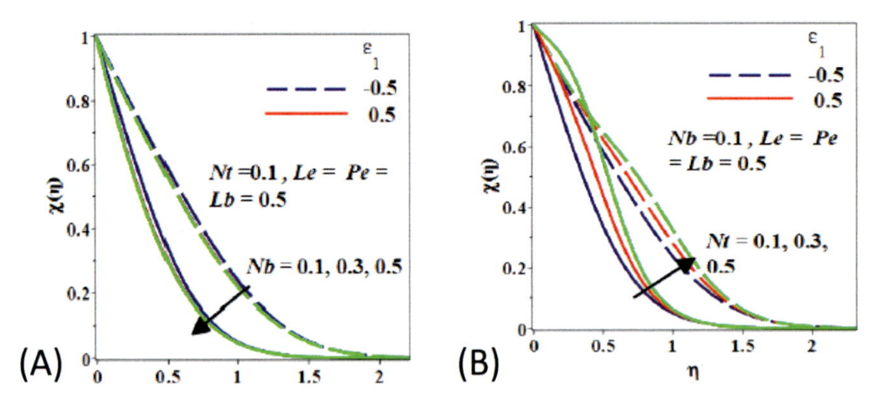

FIGURE 8.11

(A, B) Motile microorganism density versus Brownian motion parameter and thermophoresis parameter for shrinking ($\varepsilon = -0.5$) and stretching sheets ($\varepsilon = +0.5$).

Adapted from K. Naganthran, M.F.M. Basir, S.O. Alharbi, R. Nazar, A.M. Alwatban, I. Tlili, Stagnation point flow with time-dependent bionanofluid past a sheet: Richardson extrapolation technique. Processes 7 (2019) 722.

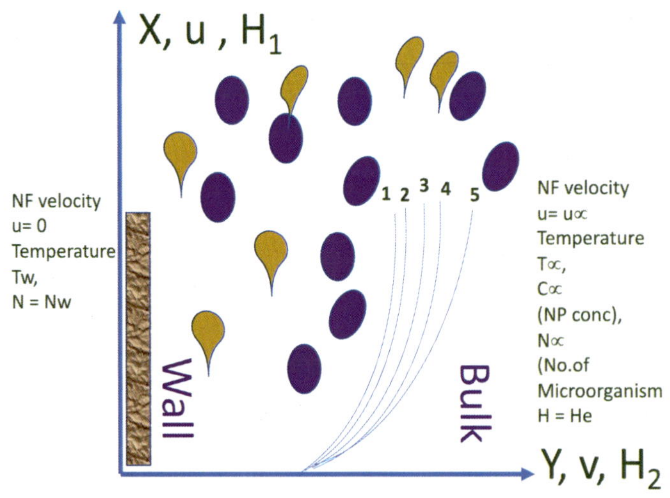

FIGURE 8.12

Unsteady magnetoconvective flow of bionanofluid with zero mass flux boundary condition—physical configuration of the problem [59].

the stagnation point. The fluid is considered hydromagnetically laminar with forced convective boundary flow of the NF with microorganisms over a solid stationary plate. While the electrical field is taken to be nil, the magnetic field is taken into account. Because the fluid is conducting, its flow will give rise to electric currents and a magnetic field, which acts normal to the surface [59].

The formulation of the problem, in addition to the conservation of velocity, also includes the perpendicular magnetic fields $H1$ and $H2$:

$$\frac{\partial H1}{\partial x} + \frac{\partial H2}{\partial y} = 0 \tag{8.21}$$

The normal component of the induced magnetic field reduces to zero when it reaches the surface. Thus the unsteadiness in fluid flow is caused by the magnetic field. In this case, as the magnetic field intensity increases, the velocity profile decreases. The presence of the magnetic pressure gradient and the modified pressure gradient increases the Lorentz force, a drag-like force, and hence causes retardation of velocity. The Lorentz drag causes resistance to reduce fluid velocity. However, their effect on the local temperature of nanoparticle volume fraction and microorganism concentration is negligible. As the induced magnetic field at the wall increases, it leads to an increase in momentum diffusivity, and this is correlated to the magnetic field diffusing into the medium. Its effect on the local temperature, NP volume fraction, and microorganism density is not as significant. Near the plate boundary, the Schmidt number increases, and the NP concentration profile increases, indicating lower solute diffusivity or high dynamic viscosity. The Schmidt number has very little effect on the velocity of NP, induced magnetic field due to movement, or local temperature and microorganisms. The bioconvection Schmidt number (i.e., that associated with the microorganism), however, strongly causes a reduction in the density of the motile microorganism as it rises.

8.9.5 Case study 5—second grade bioconvective nanofluid flow with buoyancy effect and chemical reaction

In this model, microorganisms are thought to stabilize the NF by bioconvection-generated nanoparticulate flow and buoyant force [60]. The schematic representation of the problem is drawn in Fig. 8.13.

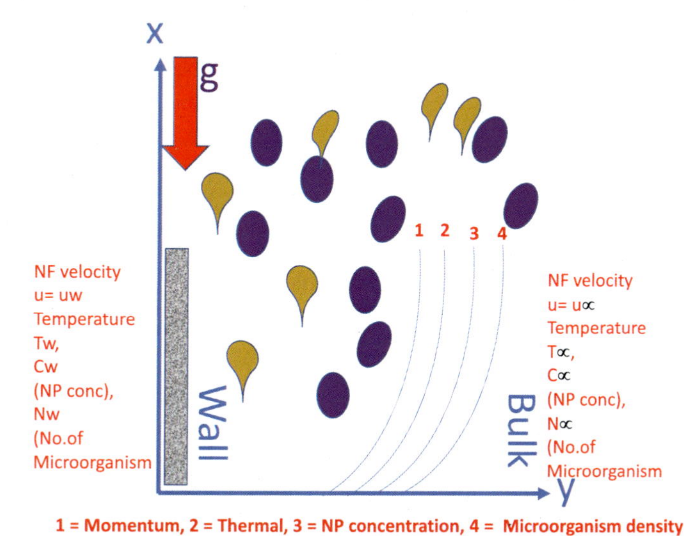

FIGURE 8.13

Second grade bioconvective nanofluid flow with buoyancy effect and chemical reaction—physical configuration of the problem [60].

In this model, a higher Rayleigh number restricts nanoparticles' upward movement by bioconvection [60]. It supports larger buoyancy opposing fluid flow in a direction. Rayleigh number with a stronger buoyant force also increases the fluid temperature.

As Brownian diffusion increases, the thermal layer is enhanced with improved heat convection and thermal distribution. Thermal distribution is also enhanced in this case with a higher Lewis number but with a decline in concentration distribution. The Brownian motion also enhances the transport of NPs between the surface and the microorganism layer. While the thermal movement of fluid is enhanced, the motile density of microorganism is also noted. The thermophoresis parameter for thermal and NP diffusion is also considered here. For intensive thermophoresis, if the plate is cold, NPs are transported from the hot to colder regions to increase NP concentration.

8.10 Conclusion and future outlook

With an ever-increasing application of BioNFs in medical and biomedical uses and environmentally benign chemical processes, the study and understanding of mass transfer are of utmost importance and of current interest with a spate of technologies which use NF properties in various industrial processes. There has been a steady increase in literature in the past decade on NF studies. The understanding of the mass transfer in BioNFs is intimately tied to heat transfer mechanisms with enhanced thermal conductivity to attribute to localized convection due to Brownian motion causing enhanced mixing. This chapter attempted to approach it via the general understanding of NF transport, then moving onto the simple canonical form of the mass transport in BioNFs, followed by five important case studies in the application.

The principal crux of BioNF transport is based on bioconvection, entailing the propelling and floating of motile microorganisms, causing a suspension density, which causes them to fall and recontinue. The plumes thus generated have been employed in the transport of cells, nutrients, and oxygen and have been applied in targeted medicine and sensors. Studies have been carried out for the past decade or more on the unique transport system of density stratification ranging from the effect of organism swimming patterns to the effect of external fields. As examples, five such case studies are presented in this chapter, as variations of the general model. In the case study involving a 3D stagnation point flow, an increase in viscosity causes enhanced thermophoretic effect and a net increase in volume fraction of NPs at the surface. With thermal diffusivity dominating over nanoparticle diffusivity, the NP concentration in the boundary layer increases and so does the local mass transfer rather than the global mass transfer. While the Peclet number is dependent on the type of the microorganism, a greater value is associated with swimming motion of microorganisms and hence a reduction in their density. In bioconvection over wavy surfaces, fluid flow was seen to increase over positive slope portion and vice versa, motile organism density was seen to be thinner on a wavy surface. An increase in Lewis number was seen to decrease microorganism layer thickness and density and enhance viscous diffusion, also attributable to an increase in thermal over mass diffusivity. With an increase in overall bioconvection and microorganism speed, their density at the surface increases. When a stretching and shrinking sheet is considered, the fluctuations in fluid temperature at the sheet surface are considered, with the effect of enhanced Brownian motion leading to intermolecular collisions, reduced sheet layer density, enhanced macroconvection, thermal conduction, and a decrease in microorganism density. In magnetoconvective flow, magnetic drag-like force retards overall velocity affecting momentum diffusivity. The presence of the magnetic pressure gradient and the modified pressure gradient increase the Lorentz force, a drag-like force, and hence cause retardation of velocity. The Lorentz drag causes resistance to reduce fluid velocity but with little additional effect on bioconvection. In the last case study, a bioconvective BioNF is considered with an inbuilt chemical process. A higher Rayleigh number restricts upward NP upward movement and increases fluid temperature leading to enhanced diffusion and heat convection. An enhanced Brownian motion increases transport of NPs between surface and microorganism layer. The basic seminal work on the general model of bioconvection was laid out by Kuznetsov. It remains the same, but can be extended to a wide number of cases entailing stagnation layers, oscillating surfaces, wavy surfaces, surfaces with differential heating, and even the effect of magnetic fields, to simulate conditions under which BioNFs may be applied in real-life circumstances for aided process enhancement.

The future outlook on BioNFs and its possible large-scale application in the mass transfer requires a greater depth in studying and understanding parameters like size, shape, surface, or morphology, and surface chemistry and its effect on mass transfer. The usage of BioNFs in industrial processes also depends upon the development of methods for recycling and separation. Thus given the diverse emergent applications, the future holds tremendous promise. Mass transfer studies and applications are, in fact, a topic of current importance, particularly in biomedical and catalytic applications, with the promise of making a significant impact in the future. Future applications of BioNFs will expand in mainly biomedical applications and take advantage of the trend of moving toward environment-friendly processes for biofuels and biorenewable chemicals.

References

[1] S. Ashrafmansour, M.N. Esfahany, Mass transfer in nanofluids: a review, Int. J. Therm. Sci. 82 (2014) 84e99.

[2] S. Ozturk, A.Y. Hassan, V.M. Ugaz, Interfacial complexation explains anomalous diffusion in nanofluids, Nano Lett. 10 (2010) 665e671.

[3] H.A. Mohammed, A.A. Al-aswadi, N.H. Shuaib, R. Saidur, Convective heat transfer and fluid flow study over a step using nanofluids: a review, Renew. Sustain. Energy Rev. 15 (2011) 2921e2939.

[4] S.U.S. Choi, Nanofluids: from vision to reality through research, ASME J. Heat Transf. 131 (3) (2009) 1e9.

[5] X.-Q. Wang, A.S. Mujumdar, Heat transfer characteristics of nanofluids: a review, Int. J. Therm. Sci. 46 (2007) 1e19.

[6] N.A. Amirsom, M.J. Uddin, A.I. Ismail, Three dimensional stagnation point flow of bionanofluid with variable transport properties, Alex. Eng. J. 55 (3) (2016) 1983–1993.

[7] W.L. McCabe, J.C. Smith, Unit Operations of Chemical Engineering, 2nd ed., McGraw-Hill, New York, 1967.

[8] W. Yu, H. Xie, L. Chen, Y. Li, Investigation of thermal conductivity and viscosity of ethylene glycol based ZnO nanofluid, Thermochim. Acta 491 (1-2) (2009) 92.

[9] W.A. Khan, O.D. Makinde, MHD nanofluid bioconvection due to gyrotactic microorganisms over a convectively heat stretching sheet, Int. J. Therm. Sci. 81 (2014) 118–124.

[10] A.V. Kuznetsov, The onset of nanofluid bioconvection in a suspension containing both nanoparticles and gyrotactic microorganisms, Int. Commun. Heat Mass Transf. 37 (10) (2010) 1421–1425.

[11] K. Zaimi, A. Ishak, I. Pop, Stagnation-point flow toward a stretching/shrinking sheet in a nanofluid containing nanoparticles and gyrotactic microorganisms, J. Heat. Transf. 136 (2014) 041705.
[12] A.V. Kuznetsov, nanofluid bioconvection in water-based suspensions containing nanoparticles and oxytactic microorganisms: oscillatory instability, Nanoscale Res. Lett. 6 (1) (2011) 100.
[13] L. Tham, R. Nazar, I. Pop, Mixed convection flow over a solid sphere embedded in a porous medium filled by a nanofluid containing gyrotactic microorganisms, Int. J. Heat. Mass. Transf. 62 (2013) 647–660.
[14] P.S. Reddy, P. Sridevi, A.J. Chamkha, MHD Natural convection boundary layer flow of nanofluid over a vertical cone with chemical reaction and suction/injection, Comput. Therm. Sci.: An. Int. J. 9 (2) (2017) 165–182.
[15] A. Ishak, Similarity solutions for flow and heat transfer over a permeable surface with convective boundary condition, Appl. Math. Comput. 217 (2010) 837–842.
[16] A. Sinha, J.C. Misra, Effect of induced magnetic field on magneto hydrodynamic stagnation point flow and heat transfer on a stretching sheet, J. Heat. Transf. 136 (11) (2014) 112701.
[17] M.A. Chaudhary, J.H. Merkin, A simple isothermal model for homogeneous-heterogeneous reactions in boundary-layer flow. I Equal diffusivities, Fluid Dyn. Res. 16 (6) (1995) 311–333.
[18] E.C. Okonkwo, I. Wole-Osho, I.W. Almanassra, Y.M. Abdullatif, T. Al-Ansari, An updated review of nanofluids various heat transfer devices, J. Therm. Anal. Calorim. (2020).
[19] T. Kavitha, A. Rajendran, A. Durairajan, Synthesis, characterization of TiO_2 nano powder and water based nanofluids using two step method, Eur. J. Appl. Eng. Scienctific Res. 1 (240) (2012) 235.
[20] I. Tavman, A. Turgut, Experimental investigation of viscosity and thermal conductivity of suspensions containing nanosized ceramic particles, Arch. Mater. Sci. Eng. 34 (2) (2008) 99.
[21] Y. Lv, Y. Zhou, C. Li, Q. Wang, B. Qi, Recent progress in nanofluids based on transformer oil: preparation and electrical insulation properties, IEEE Electr. Insulation Mag. 30 (5) (2014) 23.
[22] S. Soltani, S.G. Etemad, J. Thibault, Pool boiling heat transfer of non-Newtonian nanofluids, Int. Commun. Heat. Mass. Transf. 37 (1) (2010) 23.
[23] J.J. Michael, S. Iniyan, Performance analysis of a copper sheet laminated photovoltaic thermal collector using copper oxide—water nanofluid, Sol. Energy 119 (2015) 439.
[24] F. Asadzadeh, M.N. Esfahany, N. Etesami, Natural convective heat transfer of Fe_3O_4/ethylene glycol nanofluid in electric field, Int. J. Therm. Sci. 62 (2012) 114.
[25] Y. Xuan, Q. Li, Heat transfer enhancement of nanofluids, Int. J. Heat. Fluid Flow. 21 (2000) 58.
[26] L. Chen, H. Xie, Y. Li, W. Yu, Nanofluids containing carbon nanotubes treated by mechanochemical reaction, Thermochim. Acta. 477 (1-2) (2008) 21.
[27] T.X. Phuoc, M. Massoudi, R.-H. Chen, Viscosity and thermal conductivity of nanofluids containing multi-walled carbon nanotubes stabilized by chitosan, Int. J. Therm. Sci. 50 (1) (2011) 12.
[28] Z.L. Tyrrell, Y. Shen, M. Radosz, Fabrication of micellar nanoparticles for drug delivery through the self-assembly of block copolymers, Prog. Polym. Sci. 35 (9) (2010) 1128.
[29] J.H. Lee, K.S. Hwanga, S.P. Jang, B.H. Lee, J.H. Kim, S.U.S. Choi, et al., Effective viscosities and thermal conductivities of aqueous nanofluids containing low volume concentrations of Al_2O_3 nano particles, Int. J. Heat. Mass. Transf. 51 (11-12) (2008) 2651.
[30] D.K. Devendiran, V.A. Amirtham, A review on preparation, characterization, properties and applications of nanofluids, Renew. Sustain. Energy Rev. 60 (2016) 21.
[31] R. Taylor, S. Coulombe, T. Otanicar, P. Phelan, A. Gunawan, W. Lv, et al., Small particles, big impacts: a review of the diverse applications of nanofluids, J. Appl. Phys. 113 (2013) 011301.
[32] J. Tavares, S. Coulombe, Dual plasma synthesis and characterization of a stable copper–ethylene glycol nanofluid, Powder Technol. 210 (2) (2010) 132.
[33] M. Sheikhpour, M. Arabi, A. Kasaeian, A.R. Rabei, Z. Taherian, Role of nanofluids in drug delivery and biomedical technology: methods and applications, Nanotechnol. Sci. Appl. 13 (2020) 47–59.

[34] L. Yuan, Y. Wang, J. Wang, H. Xiao, X. Liu, Additive effect of zinc oxide nanoparticles and isoorientin on apoptosis in human hepatoma cell line, Toxicol. Lett. 225 (2) (2014). 294–30.
[35] M. Kothandapani, J. Prakash, The peristaltic transport of Carreau nanofluids under effect of a magnetic field in a tapered asymmetric channel: application of the cancer therapy, J. Mech. Med. Biol. 15 (03) (2015) 1550030.
[36] J. Estelrich, M.J. Sánchez-Martín, M.A. Busquets, Nanoparticles in magnetic resonance imaging: from simple to dual contrast agents, Int. J. Nanomed. 10 (2015) 1727–1741.
[37] Z. Hedayatnasab, F. Abnisa, W.M.A.W. Daud, Review on magnetic nanoparticles for magnetic nanofluid hyperthermia application, Mater. Des. 123 (2017) 174–196.
[38] N. Tran, T.J. Webster, Magnetic nanoparticles: biomedical applications and challenges, J. Mater. Chem. 20 (40) (2010) 8760–8767.
[39] I. Dotan, P.J.R. Roche, M. Tamilia, M. Paliouras, E.J. Mitmaker, M.A. Trifiro, Engineering multi-walled carbon nanotube therapeutic bionanofluids to selectively target papillary thyroid cancer cells, PLOS ONE 11 (6) (2016) e0158022.
[40] E. Nagy, T. Feczko, B. Koroknai, Enhancement of oxygen mass transfer rate in the presence of nanosized particles, Chem. Eng. Sci. 62 (24) (2007) 7391.
[41] C.S. Jwo, L.Y. Jeng, T.P. Teng, H. Chang, Effects of nanolubricant on performance of hydrocarbon refrigerant system, J. Vac. Sci. Technol. B 27 (3) (2009) 1473.
[42] P. Garg, J.L. Alvarado, C. Marsh, T.A. Carlson, D.A. Kessler, K. Annamalai, An experimental study on the effect of ultrasonication on viscosity and heat transfer performance of multiwall carbon nanotube-based aqueous nanofluids, Int. J. Heat Mass Transf. 52 (21-22) (2009) 5090.
[43] A. Golkhar, P. Keshavarz, D. Mowla, Investigation of CO_2 removal by silica and CNT nanofluids in microporous hollow fiber membrane contactors, J. Membr. Sci. 433 (2013) 17e24.
[44] J.P. Wen, X.Q. Jia, W. Feng, Hydrodynamic and mass transfer of gaseliquide solid three-phase internal loop airlift reactors with nanometer solid particles, Chem. Eng. Technol. 28 (2005) 53e60.
[45] J.K. Lee, J. Koo, Y.T. Kang, The effects of nanoparticles on absorption heat and mass transfer performance in NH_3/H_2O binary nanofluids, Int. J. Refrig. 33 (2) (2010) 269e275.
[46] M.M. Derakhshan, M.A. Akhavan-Behabadi, Mixed convection of MWCNT-heat transfer oil nanofluid inside inclined plain and microfin tubes under laminar assisted flow, Int. J. Therm. Sci. 99 (2016) 1.
[47] P. Bhattacharya, S.K. Saha, A. Yadav, P.E. Phelan, R.S. Prasher, Brownian dynamics simulation to determine the effective thermal conductivity of nanofluids, J. Appl. Phys. 95 (2004) 6492.
[48] P. Bhattacharya, P.E. Phelan, S. Krishnamurthy, Enhanced mass transport in nanofluids, Nano Lett. 6 (3) (2006) 419.
[49] X. Fang, Y. Xuan, Q. Li, Experimental investigation on enhanced mass transfer in nanofluids, Appl. Phys. 95 (2009) 203108.
[50] Y. Xuan, Conception for enhanced mass transport in binary nanofluids, Heat Mass Transf. 46 (2009) 277.
[51] R. Dhuriya, V. Dalia, P. Sunthar, Diffusiophoretic enhancement of mass transfer by nanofluids, Chem. Eng. Sci. 176 (2018) 632.
[52] D.C. Prieve, J.L. Anderson, J.P. Ebel, M.E. Lowell, Motion of a particle generated by chemical gradients. Part 2. Electrolytes, J. Fluid Mech. 148 (1984) 247.
[53] M.B. Bigdeli, M. Fasano, A. Cardellini, E. Chiavazzo, P. Asinari, A review on the heat and mass transfer phenomena in nanofluid coolants with special focus on automotive applications, Renew. Sustain. Energy Rev. 60 (2016) 1615.
[54] A.J. Hillesdon, T.J. Pedley, Bioconvection in suspensions of oxytactic bacteria: linear theory, J. Fluid Mech. 32410 (1996) 223–259.
[55] C. Hsu, R. Dillon, A 3D motile rod-shaped monotrichous bacterial model, Bull. Math. Biol. 71 (2009) 1228–1263.

[56] M.A. Hokins, L.J. Fauci, A computational model of the collective fluid dynamics of motile microorganisms, J. Fluid Mech. 45525 (2002) 149–174.
[57] M.J. Uddin, W.A. Khan, S.R. Qureshi, O.A. Bég, Bioconvection nanofluid slip flow past a wavy surface with applications in nano-biofuel cells, Chin. J. Phys. 55 (5) (2017) 2048–2063.
[58] K. Naganthran, M.F.M. Basir, S.O. Alharbi, R. Nazar, A.M. Alwatban, I. Tlili, Stagnation point flow with time-dependent bionanofluid past a sheet: Richardson extrapolation technique, Processes 7 (2019) 722.
[59] M.D. Faisal, M.D. Basir, M.J. Uddin, A.I.M. Ismail, Unsteady magnetoconvective flow of bionanofluid with zero mass flux boundary condition, Sains Malaysiana 46 (2) (2017) 327–333.
[60] A. Shafiq, G. Rasool, C.M. Khalique, S. Aslam, Second grade bioconvective nanofluid flow with buoyancy effect and chemical reaction, Symmetry 12 (2020) 621.

PART 2

Mass transfer modelling and simulation of nanofluids

CHAPTER

Mass transfer modeling in nanofluids: theoretical basics and model development

9

Nayef Ghasem

Department of Chemical and Petroleum Engineering, UAE University, Alain, UAE

9.1 Introduction

Nanoparticles are nanometer-sized solids and a nanofluid is a liquid containing nanoparticles. The nanoparticles are usually made of metals, carbides, metal oxides, or carbon nanotubes (CNT). Generally, fluids consist of water, oil, and ethylene glycol. The dispersion of solid nanoparticles in a base-fluid forms nanofluids with innovative thermal and flow properties. Nanofluids improve heat and mass transfer rates. The presence of nanoparticles increases the performance of the rates of mass and heat transfer. They have unique properties that make nanofluids useful for various applications.

There have been several techniques where nanofluids are used; the absorption process utilizes nanofluids in capturing CO_2 emitted to the atmosphere, and CO_2 and H_2S present in natural gas and the flue gas. Accordingly, assessing the mass transfer modeling in nanofluids is essential. Mass transfer enhancement of the CO_2 absorption in the presence of silica nanoparticles dispersed in water-based nanofluids at variable pressures (5−15 MPa) and temperatures (35°C and 45°C) was explored [1]. Results revealed that at a constant temperature, the diffusion coefficient decreased with increased pressure. The removal of carbon dioxide using nanoparticles such as silicon oxide (SiO_2), alumina (Al_2O_3), and titanium oxides (TiO_2) in water-based nanofluids as a solvent inside a polypropylene hollow fiber membrane contactor was investigated. The separation of CO_2 is influenced by the type of nanoparticles, concentration of nanoparticles, particle size, gas type, and nanofluid flow rates. A correlation was developed for the CO_2 mass transfer in nanofluid inside the hollow fiber as a Reynolds number and Schmidt number and the nanoparticle volume fraction [2].

A mathematical model was developed to predict the CO_2 absorption in nanoparticles in hollow fiber membrane contactors. The mathematical model considers the impact of Brownian motion and grazing on the enhancement mass transfer rate. The results revealed that adding a small number of silica nanoparticles enhanced the absorption of CO_2 in nanofluids [3].

The CO_2 absorption in a wetted-wall column using base fluid was experimentally investigated [4]. The study focused on silica nanoparticle (SiO_2) and water-based nanofluid on the CO_2 removal rate. Results revealed that the liquid side's mass transfer coefficient was enhanced at low absorption temperature and high liquid flow rate of nanofluid. The improvement ratio of the liquid side mass transfer coefficient in the presence of silica was over 1.37. The effect of CO_2 absorption rate

Nanofluids and Mass Transfer. DOI: https://doi.org/10.1016/B978-0-12-823996-4.00009-4

in nanofluids (CNT, Al_2O_3, Fe_3O_4, and SiO_2 solid nanoparticles in water-based fluid) used as a liquid absorbent on the mass transfer enhancement rate was inspected. The CO_2 absorption of solid nanoparticles in water-based nanofluid took place in a pilot-scale membrane system. The nanofluid flowed through the lumen side of the membrane module, and the gas mixture passed through the shell-side, concurrently. Results revealed a significant enhancement of the rate of absorption of CO_2 in nanofluids compared to pure base fluid [5]. The CO_2 absorption with nanoparticles (Al_2O_3, CNT) and micron-sized solid particles (activated carbon and Al_2O_3) in water-based nanofluid proceeding inside a well-mixed thermostatic reactor was done experimentally. Results revealed that micron-sized particles had an insignificant impact on the rate of CO_2 absorption rate [6]. The impact of nanoparticles on the CO_2 absorption in distilled water-based fluid and aqueous 4-diethylamino-2-butanol (DEAB) inside membrane contactor was mathematically modeled and analyzed. CNT and nanosized silica were used as nanoparticles. Model predictions disclosed a strong impact of nanofluid, where the base fluid is an amine solution, increasing the CO_2 absorption rate's efficiency. The separation performance of CNT nanofluid is better than silica nanofluid. Absorption performance increased with increasing amine concentration, liquid flow rate, and decreased gas flow rate [7].

The effects of TiO_2 nanoparticles in propylene carbonate-based solution on the enhancement of CO_2 absorption rate ware inspected. Results revealed that TiO_2 particle size and loading have a strong influence on the rate of CO_2 gas absorption significantly. The CO_2 removal rate first increases with solid loading and decreases with a further increase in the solid loading; there is an optimal value of solids loading [8]. CO_2 absorption from a gas mixture in nanofluids composed of silica and CNT in water-based nanofluid was modeled and simulated using the CFD technique. The model predictions were compared with experimental data available in the literature. Results disclosed that the CO_2 absorption rate improved with increasing nanofluid flow rate [9].

The mass transfer characteristics and diffusion processes of the bubbling absorption of CO_2 in Al_2O_3 nanofluids were studied. Results showed that the CO_2 absorption rate was enhanced with Al_2O_3 nanoparticles. The nanofluid viscosity increased with the addition of Al_2O_3 nanoparticles to base fluid; by contrast, surface tension remains constant [10]. They experimentally investigated TiO_2, MgO, and SiO_2 solid nanoparticles' effect on enhancing mass transfer for CO_2 absorption in MEA-based fluid. An unsteady-state mathematical model was developed for this purpose. The absorption system took place in a temperature-controlled bubble reactor. Results showed that the enhancement factor increased with increased nanoparticle concentration up to a certain point (optimal solid loading) then declined with further increases in the solid loading depending on the type of nanoparticles [11]. The effect of MDEA-based copper (Cu) and copper oxide (CuO) nanofluids on the mass transfer coefficient breakthrough time of the hydrogen sulfide removal from biogas was examined. Results revealed that the H_2S removal efficiency was higher for CuO MDEA-based nanofluids [12].

In liquid–liquid extraction, the mass transfer coefficient has a strong effect on the solute extraction phenomenon. Nanoparticles are used to enhance the coefficient of mass transfer [13]. Magnetic nanoparticles were used to enhance mass transfer under static and rotating magnetic fields. Results revealed that an increase of magnetic field induction enhanced mass transfer features [14].

Based on the heat and mass transfer analogy, a correlation was developed to predict the effective mass diffusivity of supported nanoliquid membranes [15]. The adsorption of CO_2 was improved by silica nanoparticles modified chemically. Results revealed that the adsorption rate of carbon dioxide increased by chemical modification [16]. Propylene carbonate-based TiO_2 nanofluid was used to improve the CO_2 absorption rate. The presence of TiO_2 nanoparticles enhanced the CO_2 absorption rate significantly. Results revealed that there is an optimal solid concentration beyond that enhancement factor declined [17].

Aqueous ammonia-based Fe_3O_4 nanofluid was used to improve the mass transfer of CO_2 absorption into nanofluid. Results disclosed that the overall mass transfer coefficient increased with increased solid nanoparticle concentration then decreased after reaching an optimal value. Mass transfer coefficient increased with increased ammonia concentration and decreased with increased CO_2 inlet concentration and got an optimal gas flow rate value [18].

The effect of the addition of nanoparticles on the mass transfer rate in falling film absorption was investigated. CuO nanoparticle was added to the lithium bromide solution in the falling film absorption process. Results revealed that lithium bromide solution's vapor absorption was significantly increased with CuO nanoparticles' addition [19].

The recent nanotechnology developments help to increase the gas–liquid mass transfer coefficient. The grazing effect and the Brownian motion of nanoparticles have a substantial impact on the mass transfer rate. 4-Diethylamino-2-butanol (DEAB)-based nanofluids dispersed with silica (SiO_2) nanoparticles and CNT were used to enhance the mass transfer of CO_2 absorbent taking place inside membrane contactor. Experimental investigations revealed that dispersed nanoparticles to a DEAB amine solution enhanced the CO_2 absorption and CNT perform better than silica nanoparticles [7].

A bubble-type absorber was used to absorb NH_3 from the NH_3/H_2O system using various concentrations of Al_2O_3, CuO, and Cu nanoparticles [20–22]. Results revealed that the NH_3 absorption rate increased with increased nanoparticle concentration for all types of nanoparticles. The enhancement factor was 3.21. In a comparison of nanoparticles, Cu was the most efficient. A sparged stirred tank reactor occupied with an aqueous suspension of F_3O_4 magnetic nanoparticles was experimentally investigated to enhance the gas–liquid oxygen rate of mass transfer. Results showed that the mass transfer factor's rate increased sixfold using 1% volume fraction of solid nanoparticles; magnetic nanoparticles improved the interfacial area and enhanced mass transfer coefficient [23].

Experimental observations in the airlift reactor for the oxygen–water system resulted in the reduction of mass transfer coefficient. Various concentrations of TiO_2 nanoparticles were used (1.1%, 2.2%, and 3.3 vol.%). The reduction in the mass transfer coefficient is related to the aggregation of the nanoparticles. As the nanoparticle concentration increased the mass transfer coefficient decreased due to nanoparticle amalgamation [24].

An agitated microreactor occupied with 250 nm of mesoporous silica was used to absorb CO in water-based nanofluid [6]. Observed results revealed that volumetric mass transfer significantly improved with silica nanoparticles. The enhancement factor increased from 1 to 1.55, with increased nanoparticle concentration from 0 to 0.4 wt.%. Oxygen was absorbed in water-based TiO_2 nanoparticles in an airlift reactor. The decrease of the gas holdup was related to increased viscosity and apparent density of the nanofluid [25].

Most of the research work available in the literature on nanofluids has focused on nanofluids' thermal properties and their applications in heat transfer and the enhanced heat transfer coefficient. There is a lack of information in the literature on utilizing nanofluids to enhance the mass transfer rate and the critical factor of mass transfer coefficient. This chapter investigates the use of nanofluids to enhance the mass transfer in different mass transfer systems.

9.2 Wetted-wall column

A wetted-wall vessel is used to attain mass transfer between two fluids; the gas and liquid phases. The liquid stream flows through the tube's inner wall, and the gas stream flows in correspondence with the center of the tube (Fig. 9.1); the liquid forms a thin layer that covers the vessel's inner surface. The gas stream is introduced from the bottom of the column. At the gas–liquid interface, the gas stream and the liquid stream exchanger matter in countercurrent mode.

Gas-phase stream and liquid phase stream involves absorption of carbon dioxide from a gas mixture of carbon dioxide/air (gaseous stream) into pure liquid water or nano-liquid (liquid stream). A wetted-wall column is used experimentally to measure the mass transfer coefficient values because of its simplicity to model. It is not used on an industrial scale because of its low surface area and low liquid holdup compared to packed bed columns.

Experimentally, the mass transfer coefficient in the liquid phase, k_l, is obtained for CO_2 absorption using the molar flux relation, N_A, as a function of the concentration of CO_2 solute at gas–liquid interface, $C_{A,i}$, and the CO_2 bulk concentration, $\overline{C}_{A,i}$

$$N_A = k_{l,\exp}(C_{A,i} - \overline{C}_{A,i}) \tag{9.1}$$

The concentration of CO_2 in bulk nanofluid can be considered zero. Theoretically, the liquid side mass transfer coefficient can be calculated using the penetration theory: [26]

$$k_l = 2\sqrt{\frac{D_{CO_2-H_2O}}{\pi \times \tau}} \tag{9.2}$$

where $D_{CO_2-H_2O}$ is the diffusion coefficient of CO_2 in pure water without nanoparticles [27]

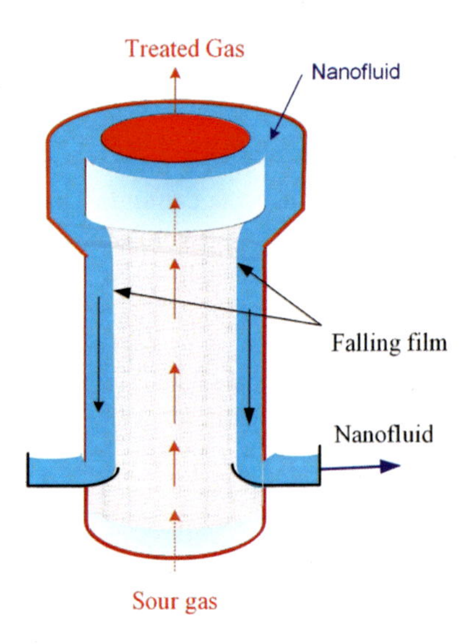

FIGURE 9.1

Schematic of a typical wetting-wall column for CO_2 absorption in nanofluid.

$$D_{CO_2-H_2O}\left(\frac{m^2}{s}\right) = 2.35 \times 10^{-6}\exp\left(-\frac{2119}{T(K)}\right) \tag{9.3}$$

where the contact time τ is in the unit of seconds and can be predicted using the following correlation: [28]

$$\tau = 3h\left(\frac{\rho_l\, g}{\mu_l}\right)^{-\frac{1}{3}}\left(\frac{3Q_l}{W}\right)^{-\frac{2}{3}} \tag{9.4}$$

where ρ_l is the liquid density (kg/m^3); μ_l is the viscosity of the liquid (kg/m/s); Q_l is the liquid volumetric flow rate; W is the cylinder's height; and h (m) is the cylinder circumference in the wetted-wall column. Experimentally, a correlation of the liquid mass transfer coefficient was attained for the CO_2 absorption in nanofluid inside a wetted-wall column [4]. The correlation is found to be a function of operating temperature, T (°C) (25°C–45°C), silica nanoparticle concentration, W (%w) (0–1%w), and flow rate of nanofluid, W (mL/min).

$$k_l = 3.06 \times 10^{-4} - 7.29 \times 10^{-6}\mathrm{T} - 6.36 \times 10^{-7}\mathrm{Q} - 3.92 \times 10^{-5}\mathrm{W} + 2.71 \times 10^{-8}\mathrm{TQ} + 2.98 \times 10^{-7}\mathrm{QW} \tag{9.5}$$

Results revealed that low temperature, high nanoparticle concentration, and high nanofluid flow rate led to a high liquid side mass transfer coefficient (Fig. 9.2). The optimal mass transfer coefficient of approximately 1.8×10^{-4} (m/s), around 40% enhanced over pure water, obtained the

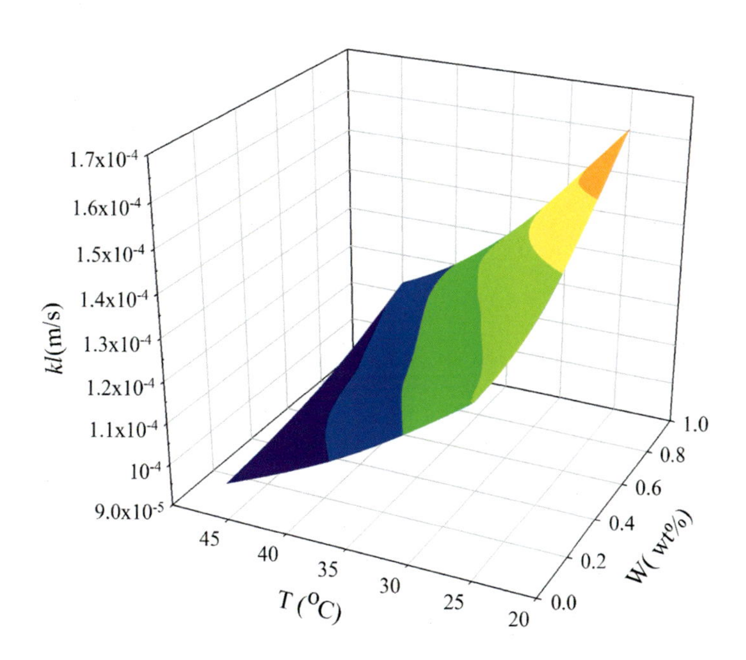

FIGURE 9.2

Effect of nanoparticle concentration and temperature on liquid side mass transfer coefficient. Nanofluid composed of SiO_2 (25 nm) dispersed in deionized water

Adapted from P. Valeh-e-Sheyda, A. Afshari, A detailed screening on the mass transfer modeling of the CO_2 absorption utilizing silica nanofluid in a wetted wall column, Process. Saf. Environ. Prot. 127 (2019) 125–132

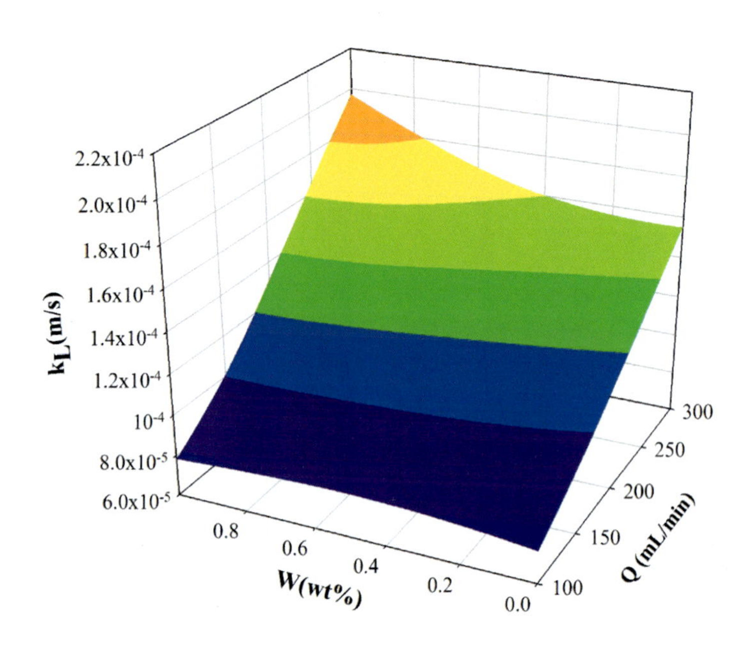

FIGURE 9.3

Effect of nanoparticle concentration and nanofluid concentration on liquid-phase mass transfer coefficient. Nanofluid composed of SiO_2 (25 nm) dispersed in deionized water

Adapted from P. Valeh-e-Sheyda, A. Afshari, A detailed screening on the mass transfer modeling of the CO_2 absorption utilizing silica nanofluid in a wetted wall column, Process. Saf. Environ. Prot. 127 (2019) 125–132

nanofluid flow rate of 280 mL/min, 1 wt.% silica concentration, and 280 L/min of nanofluid flow rate [4].

There is a significant impact of the nanofluid flow rate on the liquid side mass transfer coefficient. Fig. 9.3 reveals that as the liquid flow rate increased mass transfer coefficient increased. The effect of the flow rate of liquid is high at elevated nanofluid concentration. Table 9.1 shows particular samples of systems where mass transfer in the wetted-wall column occupied with nanofluid is improved.

9.3 Packed bed column

A packed bed column is a cylinder or a vessel filled with packing material. The purpose of a packed tower is to improve the contact between two phases (gas and liquid). A packed column is used in distillation columns, packed bed reactors, and absorbers.

The packing material such as Raschig rings can be filled randomly or filled as structured packing to increase the gas–liquid interfacial area (Fig. 9.4). Billet and Schultes [31] obtained mass transfer coefficients in packed bed absorbers. They studied 31 different binary and ternary systems, equipped with 67 different types and sizes of packing and columns of diameter from 6 cm to 1.4 m.

Table 9.1 Mass transfer in wetted-wall column occupied with nanofluid.

Absorption system	Nanoparticle	Optimum nanoparticle concentration	Enhancement factor	References
CO_2 in water	Al_2O_3	1.79 vol.%	1.79	[29]
CO_2 in water	TiO_2	0.05 vol.%	0.84	[29]
CO_2 in water and amine	Fe_3O_4	0.39 vol.%	1.928	[30]

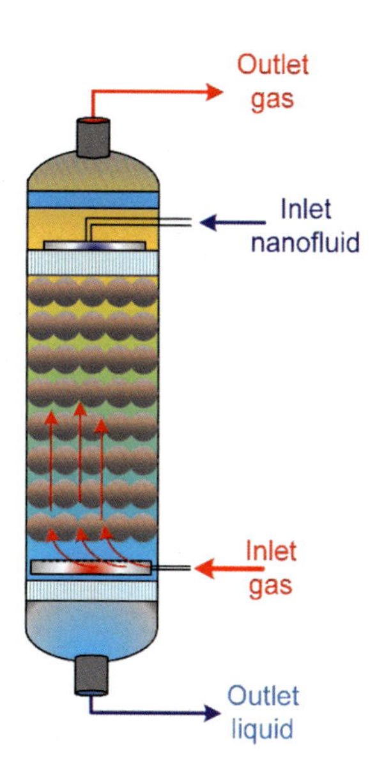

FIGURE 9.4

Schematic of a packed bed column for gas absorption with nanofluids.

Packing material should be chemically inert to the nanofluids, strong enough without extreme mass, comprise sufficient void volume for the passage of gas and liquid streams without unnecessary liquid holdup or pressure drop, and provide respectable interaction between the gas and the liquids with affordable cost.

The gas-phase mass transfer coefficients (k_G)

$$k_G = 0.1305.C_V.\left[\frac{D_G.P}{R.T}\right].\left(\frac{a}{[\varepsilon(\varepsilon-h_L)]^{0.5}}\right).\left[\frac{Re_G}{K_W}\right]^{3/4}.Sc_G^{2/3} \tag{9.6}$$

where C_V is the mass transfer factor, D_G is the gas phase diffusion coefficient, R is the ideal gas constant $\left(0.0821\,\frac{m^3.atm}{kmol.K}\right)$, ε is the void fraction, Sc_G is the Schmidt number $\left(Sc_G = \frac{\mu_G}{\rho_G.D_G}\right)$, and Re_G the Reynolds number in the gas phase $\left(Re_G = \frac{v_G.d_p.\rho_G.K_W}{(1-\varepsilon).\mu_G}\right)$ in the gas phase, where ρ_G and μ_G are the density and viscosity of the gas phase, respectively. h_L can be determined using the following equation: [31]

$$h_L = \left[12.\frac{Fr_L}{Re_L}\right]^{1/3}.\left[\frac{a_h}{a}\right]^{2/3} \tag{9.7}$$

$$K_W = 1/(1 + \frac{2}{3}.\left(\frac{1}{1-\varepsilon}\right).\frac{d_p}{D}) \tag{9.8}$$

$$d_p = 6.\left(\frac{1-\epsilon}{a}\right) \tag{9.9}$$

$$\frac{a_h}{a} = C_h.Re_L^{0.5}.Fr_L^{0.1} \quad \text{for } Re_L < 5 \tag{9.10}$$

$$\frac{a_h}{a} = 0.85.C_h.Re_L^{0.25}.Fr_L^{0.1} \quad \text{for } Re_L > 5 \tag{9.11}$$

The liquid-phase mass transfer coefficient (k_L) for base fluid without nanoparticles:

$$k_L = 0.757.C_L.\left[\frac{D_L.a.v_L}{\varepsilon.h_L}\right]^{0.5} \tag{9.12}$$

The mass transfer factor, C_L, and the mass transfer area per unit volume, $a(m^2/m^3)$ can be found in Table 9.2. More information can be found elsewhere [33].

The overall mass transfer coefficient, K_m is calculated using the following equation:

$$\frac{1}{K_m} = \frac{1}{k_G.a_h} + \frac{H.M_W}{P.k_L.a_h.\rho_L} \tag{9.13}$$

where H is Henry's constant (atm), P is the column operating pressure (atm), and a_h is the effective surface area of the packing material (m^{-1}). Mass transfer enhancement rates in nanofluids were experimentally studied [34]. A semiempirical correlation for predicting the mass transfer coefficient was developed. Experiments were performed for the absorption of CO_2 in water-based silica nanofluid. The presence of nanoparticles enhanced the mass transfer coefficient [35]. The effect of holdup of silica nanoparticles and various molarities of MEA-based fluid aqueous solution is shown in Fig. 9.5. It is evident that as the MEA aqueous solution increases, the mass transfer coefficient increases (Fig. 9.5).

Table 9.2 Mass transfer parameters [32].

Packing material, rings	a	C_L	C_V
50-mm metal Top Pak	75	1.326	0.389
50-mm Metal Hiflow	92	1.168	0.408
50-mm ceramic ball	121	1.227	0.415
25-mm metal VSP	205	1.376	0.405

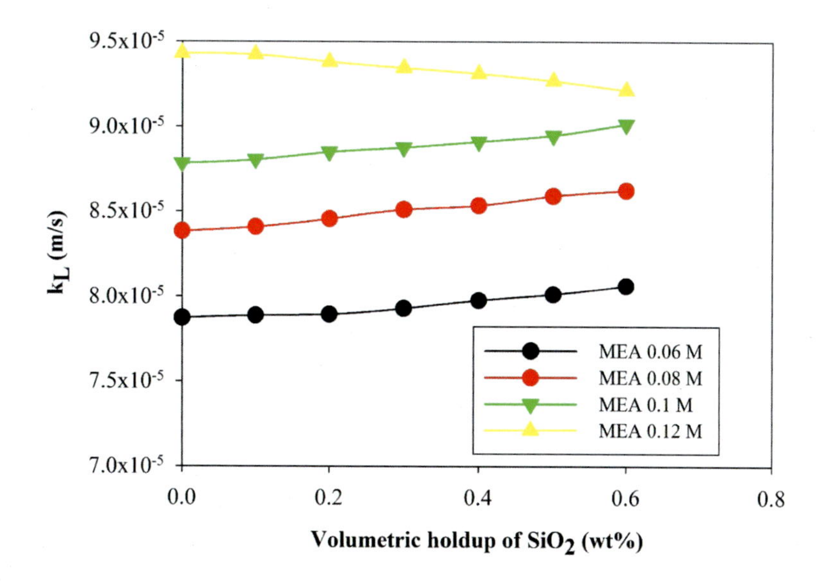

FIGURE 9.5

Mass transfer coefficient versus volumetric holdup of SiO_2 nanoparticles at various aqueous MEA-based nanofluid concentrations.

Adapted from T. Ramprasad, R. Khanolkar, A.K. Suresh, Mass-transfer rate enhancement in nanofluids: packed column studies and a design basis, Ind. Eng. Chem. Res. 58 (2019) 7670–7680

9.4 High-pressure vessel

Nanofluids are potentially useful for different applications, and they play a significant role in mass transfer enhancement. Mass transfer of CO_2 in SiO_2 water-based nanofluids increased with temperature rise and an increase in silica holdup. High-pressure stainless steel vessels are generally applied for mass transfer measurement (Fig. 9.6). In this apparatus, temperature and volume are fixed, and the gas's pressure decreases with time. The measured pressures versus time determine the mass transfer coefficient. Nanofluids improve the mass transfer coefficient [1,6,36].

The diffusion process in the high-pressure cell is described by Fick's second law [37]:

$$\frac{\partial C_A}{\partial t} = D_A \frac{\partial^2 C_A}{\partial z^2} \tag{9.14}$$

For nonideal mixtures, the diffusion coefficient is corrected as:

$$D_A = D_A^{'} \left(1 + \frac{\partial ln\varnothing}{\partial x} \right) \tag{9.15}$$

where x and $\varnothing$ are the mole fraction and the fugacity coefficient of component i, respectively.

The boundary conditions are

$$\text{At } t = 0 \quad C_A = C_{A0}$$

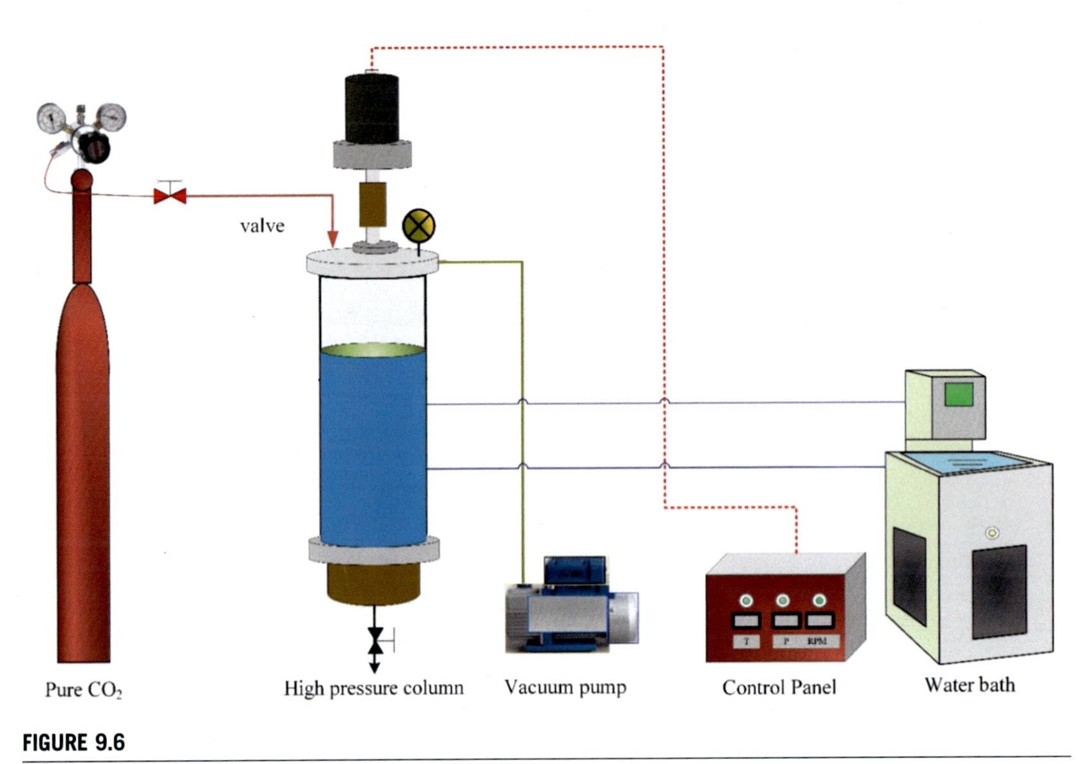

FIGURE 9.6

High-pressure vessel for gas absorption in nanofluids.

$$\text{At } z = 0 \quad \frac{\partial C_A}{\partial z} = 0$$

$$\text{At } z = L \quad C_A = C_{AL}$$

where the initial and boundary concentration of component A in the liquid phase are C_{A0} and C_{AL}, respectively. The value of the mass transfer coefficient was estimated theoretically using the correlation available in the literature. The model's predicted results were validated experimentally employing a high-pressure cell. Fick's second law was used to update the CO_2 diffusion coefficient in the higher-pressure cell without mixing. Results revealed that the mass transfer coefficient increased with increased nanoparticle concentration to a certain extent. Fig. 9.7 shows the effect of silica nanoparticle loading in the base fluid on the liquid side mass transfer coefficient at two different temperatures (35°C, 45°C). Results revealed that at specific solid nanoparticle concentration in the base fluid, mass transfer decreased with increased solvent temperature.

The effect of silica nanoparticles (weight percent) dispersed in a water-based nanofluid at different initial pressures and constant temperature of 35°C on the mass transfer coefficient is shown in Fig. 9.8. The figure shows that the solid loading has a noticeable effect on the mass transfer coefficient at low SiO_2 nanoparticle concentration, and the effect becomes insignificant at high solid concentration. There is a slight increase in the interface of the mass transfer coefficient of carbon

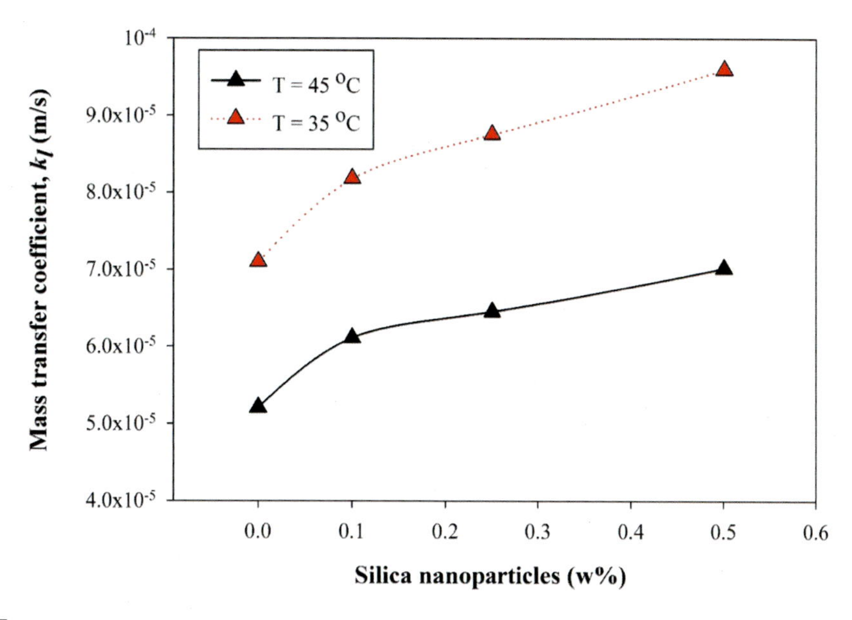

FIGURE 9.7

Effect of nanoparticle concentration at two different temperatures and constant initial pressure of 5 MPa calculated using high-pressure cell.

Adapted from S. Farzani Tolesorkhi, F. Esmaeilzadeh, M. Riazi, Experimental and theoretical investigation of CO_2 mass transfer enhancement of silica nanoparticles in water, Pet. Res. 3 (2018) 370–380

dioxide with an increase in initial vessel pressure. The mixing of nanofluid during the absorption process was found to enhance the mass transfer coefficient [6]. Fig. 9.9 shows the mixing rate's influence on two types of nanofluid (water-based alumina and CNT). The mass transfer coefficient of CNT nanofluid shows better performance compared to alumina nanoparticles.

Adapted from S. Farzani Tolesorkhi, F. Esmaeilzadeh, M. Riazi, Experimental and theoretical investigation of CO_2 mass transfer enhancement of silica nanoparticles in water, Pet. Res. 3 (2018) 370–380

Adapted from S. Lu, M. Xing, Y. Sun, X. Dong, H. Mohammaddoost, A. Azari, et al., Experimental and theoretical Studies of CO_2 absorption enhancement by nano-Al_2O_3 and carbon nanotube particles, Chin. J. Chem. Eng. 21 (2013) 983–990

9.5 Liquid–gas membrane contactor

Liquid–gas contactors are processes intended to enhance mass transfer between a liquid and a gas phase because of their contact [38]. Traditional contactors such as packed bed columns (structured or random) and tray towers require direct mixing between the liquid and gas phases; by contrast, membrane contactors do not need direct dispersion of the liquid phase into the gas phase. The microporous

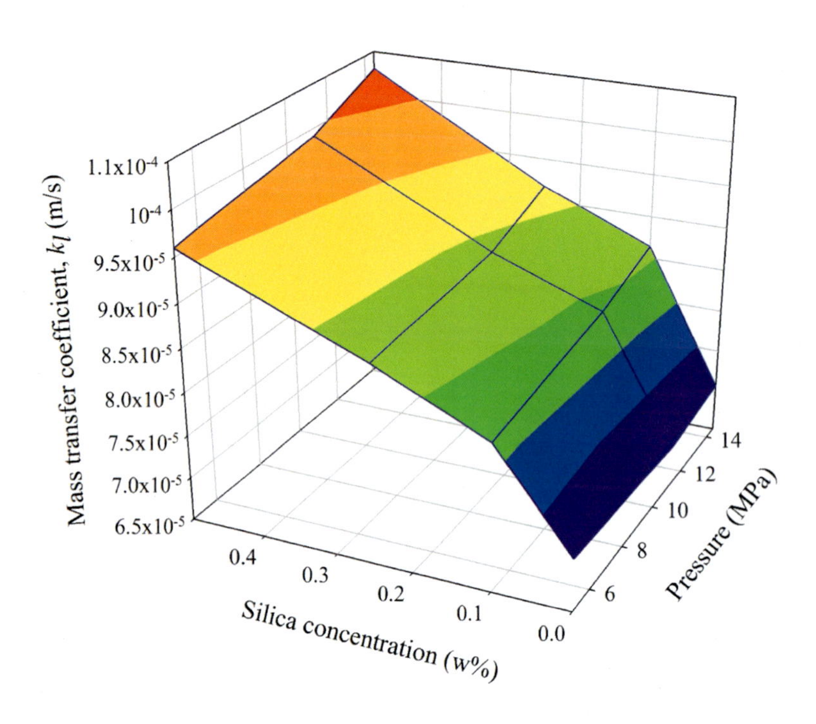

FIGURE 9.8

Effect of silica nanoparticle loading and initial pressure on the mass transfer coefficient at a constant temperature of 45°C measured in the high-pressure vessel.

membrane is a physical barrier between a liquid phase and a gas phase [39]. Optimal operating conditions are needed to attain a high mass transfer rate and high mass transfer coefficient [40]. Membrane interfacial area (a) is 10,000 m^2/m^3, which is 20 times higher than packed columns. The interfacial area is the main factor in predicting the mass transfer coefficient [41]. The gas–liquid membrane contactor constructed from a membrane module equipped with hollow fibers with a low inner diameter (50–100 μm) increases the liquid–gas interfacial area between the gas and liquid [42]. Fig. 9.10 shows a hollow fiber membrane contactor, nanofluid is flowing in the tube side of the hollow fiber membrane, and the gas phase is injected in the shell-side of the membrane module [44,45].

Adapted from N. Ghasem, Chemical absorption of CO_2 enhanced by nanoparticles using a membrane contactor: modeling and simulation, Membr. (Basel). 9 (2019)

The mass transfer coefficient, k_p, is determined by the following correlation [6]:

$$Sh = \frac{k_p d_p}{D_{nf}} = 2 \tag{9.16}$$

The diffusion coefficient in nanofluid can be determined as follows [6,9,46,47], where D_{nf} is the diffusion coefficient in the presence of solid nanoparticles.

$$D_{nf} = D_{bf}\left(1 + 1650Re^{0.039}Sc^{1.064}\varnothing^{0.203}\right) \tag{9.17}$$

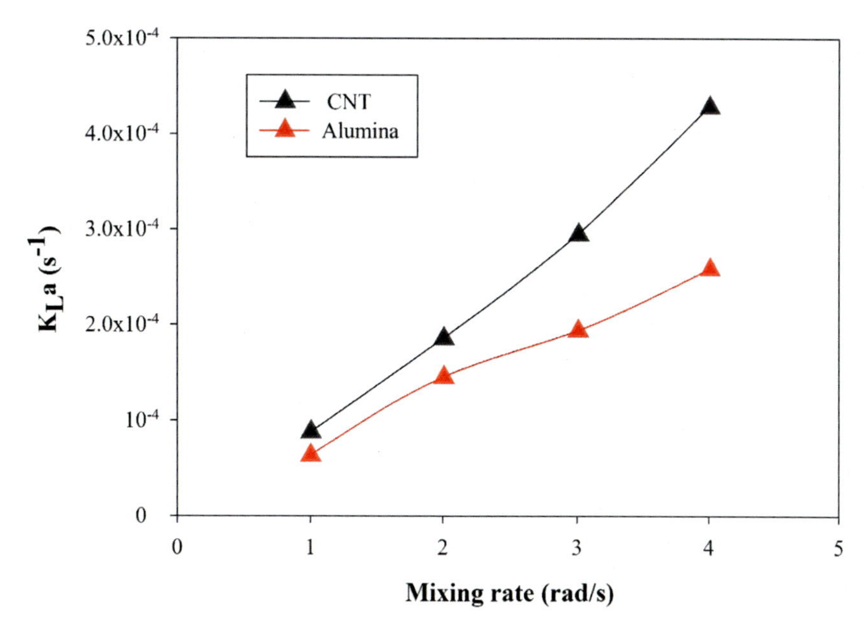

FIGURE 9.9

Effect of mixing rate and the enhancement of the mass transfer coefficient.

where D_{bf} is the diffusion coefficient of CO_2 in the base fluid, Re is the Reynolds number, and Sc is the Schmidt number determined as follows:

$$Re = \sqrt{\frac{18KT\rho^2}{\pi d_p \rho_p \mu}} \tag{9.18}$$

where ρ and μ are density and viscosity of nanofluid, respectively; ρ_p and μ_p are the density and viscosity of nanoparticles, respectively; K is the Boltzmann constant; and T the temperature. The material balance equation for a gas solution in the nanoparticles and the amount of gas CO_2 adsorbed by the nanoparticles, q:

$$\varnothing \rho_p V \frac{\partial q}{\partial z} = k_p a_p (C_{CO_2} - C_s) \tag{9.19}$$

The amount of solute gas adsorbed in the solid nanoparticles (q) and the CO_2 concentration at the liquid–solid interface (C_s) are related to the Longmuir adsorption isotherm model: [6]

$$q = q_m \frac{k_d C_s}{1 + k_d C_s} \tag{9.20}$$

The maximum amount of solute gas adsorbed by nanoparticle (q_m) and the Langmuir constant (k_d) depends on the type of the nanoparticles, 29.45 mol/kg, and 0.00049 m^3/mole for CNT, and 1.2 mol/kg and 0.008 for silica nanoparticles, respectively [16]. Values of the mass transfer coefficient for various nanoparticle concentration are shown in Table 9.3 [48].

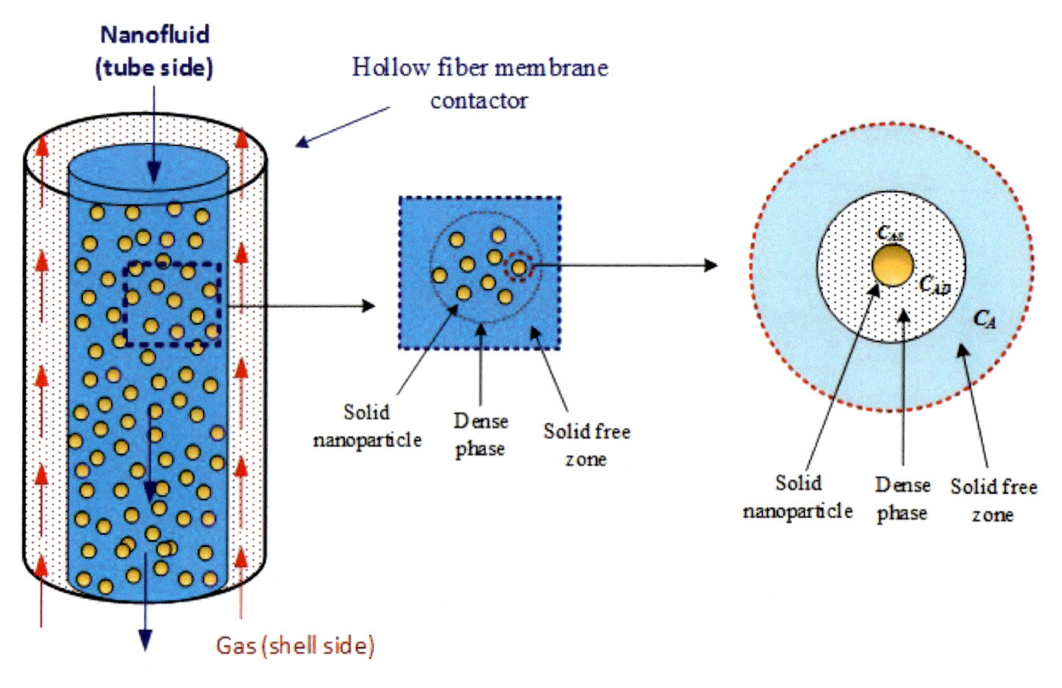

FIGURE 9.10

Schematic of gas–liquid hollow fiber membrane contactor with nanofluid absorbent [43].

Table 9.3 Mass transfer coefficient (k_l) for various nanofluids [48].

Wt.% nanoparticle	k_L (m/s)
0% (distilled water)	2.57×10^{-5}
0.25 wt.% silica	2.73×10^{-5}
0.5 wt.% silica	2.71×10^{-5}
0.25 wt.% CNT	2.73×10^{-5}
0.5 wt.% CNT	2.70×10^{-5}

9.6 Bubble column

Bubble columns are used for different purposes because the process is simple to operate, has perfect mixing, has no moving parts, and high mass transfer rates are achievable (Fig. 9.11), along with the capability to accommodate a wide range of residence times by manipulating the gas and liquid flow rates.

Gas is injected at the bottom of the column by a suitable distributor [49].

Gas bubbles move upward, leading to intensive mixing of the liquid phase. Usually, gas and liquid are continuously fed to the column in a countercurrent mode [50]. SiO_2, Al_2O_3, and Fe_2O_3

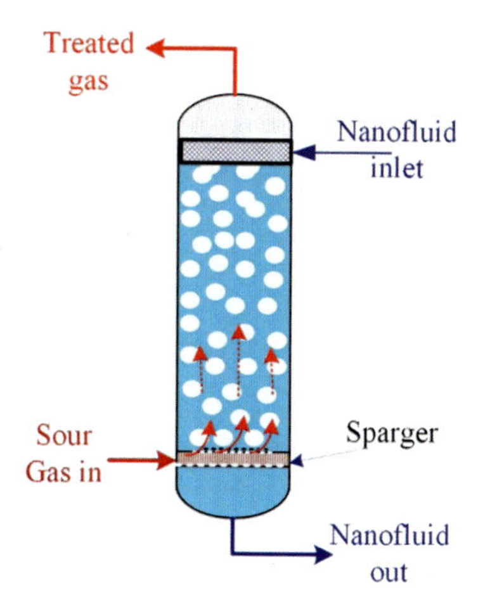

FIGURE 9.11

Bubble column for gas absorption in nanofluids.

Table 9.4 Liquid side mass transfer coefficient (k_L) in different water-based nanofluids [50].

Impurity gas	Solid nanoparticle	k_L (m/s)
CO_2	None (pure water)	1.03×10^{-4}
CO_2	SiO_2	1.935×10^{-4}
CO_2	Fe_2O_3	2.324×10^{-4}
CO_2	Al_2O_3	2.092×10^{-4}
SO_2	None (pure water)	1.450×10^{-4}
SO_2	SiO_2	2.493×10^{-4}
SO_2	Fe_2O_3	2.186×10^{-4}
SO_2	Al_2O_3	2.063×10^{-4}

water-based nanofluid are used to absorb CO_2 and SO_2 in the bubble column. The resultant mass transfer coefficients are shown in Table 9.4.

The mass transfer coefficients can be predicted from the following correlation for different gas and nanofluid absorption processes [50]:

$$\frac{Sh_{nf}}{Sh_{bf}} = 1.3643\left(\frac{Sc_{nf}}{Sc_{bf}}\right)^{0.6125} \tag{9.21}$$

The mass transfer coefficient of gas flowing in the bubble bed column is estimated with the following correlation [51].

$$Sh_{bf} = 0.6\, Re_{bf}^{1/2} Sc_{bf}^{1/3} \tag{9.22}$$

The Reynolds number for base fluid

$$Re_b = \frac{U_b d_b}{\nu_{nf}} \tag{9.23}$$

where U_b is the rising velocity of the bubble in the column (approximately 0.21 m/s) and d_p is the bubble's diameter (approximately 7 mm). The Schmidt number (Sc_{nf}) of the base fluid in the bubbling column filled with nanofluid is determined by:

$$Sc_{nf} = \frac{v_{nf}}{D_{nf}} \tag{9.24}$$

The Sherwood number of gas flowing in bubble column filled with nanofluid, Sh_{nf}:

$$Sh_{nf} = \frac{k_{L,nf}.d_p}{D_{nf}} \tag{9.25}$$

where $k_{L,nf}$ is the liquid side mass transfer coefficient of gas absorption in the bubbling bed column, d_p is the bubble diameter, and D_{nf} is the diffusion coefficient in a nanofluid. There is an insignificant change of Reynolds number between pure base fluid (Re_{bf}) and nanofluid (Re_{nf}) in a single rising bubble absorber ($v_{bf} = v_{nf}$). Accordingly, there is an insignificant effect of the Reynolds number on the Sherwood number.

$$\frac{Sh_{nf}}{Sh_{bf}} = K\left(\frac{Sc_{nf}}{Sc_{bf}}\right)^n \tag{9.26}$$

Where the viscosity of the nanofluid (μ_{nf}) as a function of base fluid viscosity (μ_{bf}) and the void fraction of nanoparticles in base fluid (φ) is [52]:

$$\mu_{nf} = \mu_{bf}(1-\varphi)^{2.5} \tag{9.27}$$

The density of the nanofluid (ρ_{nf}) as a function of base fluid density (ρ_{bf}) and volume void fraction (φ) is:

$$\rho_{nf} = \varphi\rho_{bf} + (1-\varphi)\rho_{bf} \tag{9.28}$$

The kinematic viscosity of nanofluid, v_{nf}

$$v_{nf} = \mu_{nf}/\rho_{bf} \tag{9.29}$$

The volume percent void fraction of the solid nanoparticle with the base fluid (φ) as a function of weight percent of solid nanoparticles in base fluid [$w(\%)$], nanoparticles bulk density (ρ_b) (SiO_2, Al_2O_3, Fe_2O_3, 2.196, 3.98, 5.242 g/cm^3, respectively), and ρ_{bf} is the density of base fluid:

$$\varphi(\%\text{vol}) = \frac{w(\%\text{wt})}{w(\%\text{wt}) + \left(\frac{\rho_b}{\rho_{bf}}\right)[100 - w(\%\text{wt})]} \tag{9.30}$$

Table 9.5 shows various absorption systems where the bubble column's mass transfer is engaged with various nanoparticles.

9.7 Airlift reactor

An airlift reactor is frequently used for biosystems (Fig. 9.12). Similar to the bubble column, bubbling air in the riser achieves essential mixing in the reactor.

Table 9.5 Mass transfer in bubble column occupied with nanofluid.

Absorption in nanofluid-based solvent	Nanoparticle	Optimum nanoparticle concentration	Enhancement factor	References
NH_3 in water	Al_2O_3, Cu	0.1 wt.% Cu	1.79	[20]
CO_2 in water	TiO_2, SiO_2, CNT	0.07 vol.% CNT	1.78	[53]
CO_2 in water and amine	SiO_2	0.02 wt.% SiO_2	1.24	[54]
CO_2 in methanol	Al_2O_3	0.01 vol.% Al_2O_3	1.08	[55]
CO_2 in MEA, MDEA	MgO, SiO_2, TiO_2	0.6 vol.% TiO_2	1.34	[11]

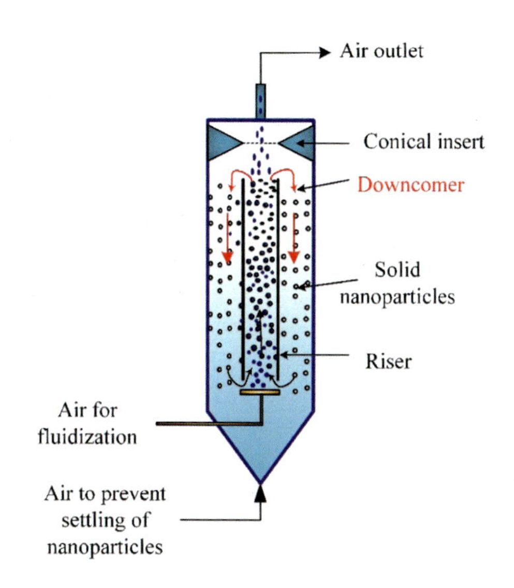

FIGURE 9.12

Schematic of airlift reactor filled with nanofluids.

The airlift reactor holds an inner flow tube that generates the drafting force essential for liquid circulation and potential enhancement by mixing. The draft tube divides the entire reactor into two halves (riser, downcomer). Air is injected in the riser and moves upwards; by contrast, the downcomer degasses the media. The density gradient between the downcomer and riser causes continuous circulation.

For example, an airlift reactor occupied with 12 nm TiO_2 nanoparticles and different volume fractions (1.1%, 2.2%, and 3.3%) was dispersed in water used as a solvent. The prediction of the mass transfer performance of the downcomer and riser in the three phases (solid nanoparticle, liquid water, gas) loop airlift reactor was mathematically modeled as follows [24,56,57]. The mass transfer coefficient (K_{Li}) of the liquid phase is found using the following correlation for riser (R) and downcomer (D): [56]

$$K_{\mathrm{Li}} = \frac{2}{\sqrt{\pi}}\sqrt{\frac{D_L}{\theta_i}} \quad \text{for } i = R \text{ or} \tag{9.31}$$

where D_L is the diffusivity. The exposure time (θ_i) is a function of the length (l_i) of the eddy and the velocity (u_i) of the fluctuation as follows:

$$\theta_i = \frac{l_i}{u_i} \tag{9.32}$$

The eddy length (l_i) is calculated from energy dissipation per unit mass (ξ_i) and the kinematic viscosity of the liquid (v_i):

$$l_i = \left(\frac{v_i^3}{\xi_i}\right)^{1/4} \tag{9.33}$$

The dissipation factor (ξ_i):

$$\xi_i = \varepsilon_{Gi} U_s g \tag{9.34}$$

The overall gas holdup (ε_G):

$$\epsilon_G = m\varepsilon_{\mathrm{GR}} + (1-m)\varepsilon_{\mathrm{GR}} \tag{9.35}$$

where the fractional cross section of the draught tube (m) is the ration of the riser area (A_R) to total area (A):

$$m = \frac{A_R}{A}$$

where ϵ_{Gi} is the gas holdup and the slipping gas velocity (U_s):

$$\frac{U_s \mu_L}{\sigma_L} = 2.25\left(\frac{\sigma^3 \rho_L}{g\mu_L^4}\right)^{-0.273}\left(\frac{\rho_L}{\rho_G}\right)^{0.03} \tag{9.36}$$

where σ_L is the liquid service tension and the solid-liquid density (ρ_L) is a function of water density (ρ_W), solid density (ρ_s), and the volume fraction of the solid nanoparticle (ε_s):

$$\rho_L = (1-\varepsilon_s)\rho_W + \epsilon_s \rho_s \tag{9.37}$$

The solid–liquid apparent viscosity (μ_L) is [58]:

$$\mu_L = \frac{\mu_W}{(1-\epsilon_s)^{2.5}} \tag{9.38}$$

Advantages of airlift reactor

- Homogeneous mixing.
- No moving parts, hence low maintenance cost.
- Energy requirements are low.

Main disadvantages

- High pressure is required for pumping air for fluidization and for preventing solid nanoparticles from settling.

Examples of mass transfer systems in a bubble column with nanofluid solvents are shown in Table 9.6.

Table 9.6 Mass transfer in bubble column occupied with nanofluid.

Absorption in nanofluid-based solvent	Nanoparticle	Optimum nanoparticle concentration	Enhancement factor	References
O_2 in H_2O	TiO_2 (12 nm)	1.1 vol.%	0.75	[56]
O_2 in H_2O	TiO_2 (10 nm)	3.3 vol.%	0.8	[25]

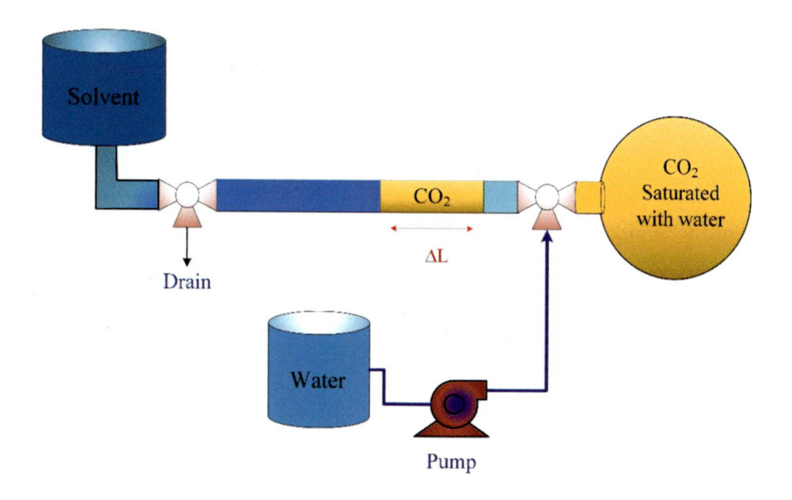

FIGURE 9.13

Schematic diagram of capillary tube apertures [34,59].

9.8 Capillary tube

The capillary tube apparatus sketched in Fig. 9.1 was used to investigate the rate of mass transfer enhancement by nanoparticles. The apparatus consists of a tube of a small inside diameter (e.g., 5 mm), a three-way valve, a balloon for gas (e.g., CO_2), reservoirs for liquids (e.g., nanofluid) [30,35]. The liquid in the chamber sucked in the direction to suck in some of the impurity gas. The position of the water slug controls the absorption process with time (Fig. 9.13).

The relation between the changes in the length of the gas compartment to the instantaneous rate of absorption is [60]:

$$C_g \frac{-dl}{dt} = R_A \tag{9.39}$$

where A is the concentration of gas in the chamber, C_g; l is the variable length of the chamber; R_A is the rate of absorption. The transient diffusion in physical absorption is:

$$R_A = C_A^* \sqrt{\frac{D}{\pi t}} = C_g - \frac{dL}{dt} \tag{9.40}$$

where C_A^* is the solubility of the gas A in the liquid, D is the diffusivity of species A. Integration of the above equation gives:

$$L_1 - L_2 = \frac{k_L C_A^* t}{C_g} = 2\frac{C_A^*}{C_g}\sqrt{\frac{D}{\pi}}\sqrt{t} \tag{9.41}$$

where L_1 and L_2 are the initial and final length of the gas slug, k_L is the mass transfer coefficient.

The average mass transfer coefficient is:

$$k_L = \frac{(L_1 - L_2)C_g}{C_A^* t} \tag{9.42}$$

9.9 Advantages and disadvantages of nanofluids

Nanofluids have many advantages among which are the following:

1. Nanoparticles have a high specific surface area and, consequently, more mass and heat transfer between fluids and nanoparticles.
2. Low clogging of nanoparticles compared to conventional slurry solutions.
3. Properties of nanoparticles are adjustable by varying the concentration of nanoparticles.
4. Dispersion stability is high when good sonication of dispersed particles in the base fluid is applied.

Despite the promising application of liquid nanofluids for heat and mass transfer improvements, the system suffers from the flowing drawbacks

1. The production cost of nanoparticles are high.
2. Nanofluid production is difficult.

9.10 Challenges for nanofluid applications in mass transfer technology

Various absorption systems were used in the literature to employ nanofluids to enhance the mass transfer rate of various undesired gas. A wetted-wall column is used to accomplish mass transfer between two streams: gas and liquid phase streams. Packed bed columns replaced traditional solvents with nanofluids to improve mass transfer coefficient and can be used as on the industrial scale. Packed material helps in increasing the contact area between nanofluids and the absorbent gas. Packed bed columns occupied with nanofluids enhanced mass transfer of CO_2 with an increase in nanoparticle holdup. Bubble columns are widely used for various purposes because the process is simple to operate, they are well-mixed, high mass transfer rates are attainable, they are competent for housing varied residence time by controlling the gas and liquid flow rates.

An airlift reactor is frequently used as a bioreactor because it is a slow process that is challenging to use in gas treatment plants with high operational capacity. The capillary tube apparatus is mainly used for the measurement of mass transfer coefficient at the lab scale or in small capacity plants.

The main challenges of nanofluids are:

1. The discrepancy between experimental data and theoretical model prediction.
2. The suspension characterization of nanoparticles in the base fluid is low.
3. The physical phenomena responsible for the abnormal behavior for nanofluids are not clear.
4. The clustering influence of nanoparticles in base fluids.

9.11 Conclusion and future work

Nanofluids are a suspension of nanoparticles, characteristically in the range of 2–100 nm. Nanofluids have many exciting properties. High interfacial area and absorption affinity allow the removal of impurity gases such as CO_2, H_2S, and SO_2. The increase in the coefficient of mass transfer in gas absorption by nanofluids results from the grazing effect and Brownian motion; both mechanisms play essential roles in enhancing the mass transfer coefficient. The diffusivity of gases into the metal oxide nanoparticles and solid loading in nanofluids are crucial parameters that influence the mass transfer coefficient. The addition of solid nanoparticles to base fluids increases the diffusion coefficient.

More attention is required in the future due to the discrepancy between experimental data and theoretical model predictions. Further modeling and experimental investigations are essential for finding the key factors inducing the performance of nanofluids. The higher viscosity of nanofluids than conventional base fluids is a drawback because of the increase in associate power required for pumping and recycling nanofluids during the regeneration process. More attention should be given to achieve the long-term stability of nanofluids. Suitable mixing techniques are required to maintain nanoparticles' suspension in base fluids and avoid settlement or clustering of the nanoparticles. Ultrasonication or engaging a suitable surfactant is essential.

Many mass transfer apparatus types have used nanofluids in gas absorption and enhanced mass transfer on an industrial and commercial scale. Using nanofluids in these technologies improves the efficiency of mass transfer processes compared to commercial solvents. This chapter explores the theoretical basics and model development of mass transfer modeling in nanofluids for several mass transfer types of equipment, such as packed bed column, high-pressure vessel, liquid–gas membrane contactor, bubble column, airlift reactor, and capillary tube. Specific units are used on an industrial scale, such as bubble columns and packed bed towers, and in the lab. Pilot-scale wetted-wall columns, membrane contactors, and airlift reactors are used at lower capacities than conventional packed bed and bubble columns. The wetted-wall column is easy to model and utilize compared to the packed bed tower, membrane contactor, and bubble column.

References

[1] S. Farzani Tolesorkhi, F. Esmaeilzadeh, M. Riazi, Experimental and theoretical investigation of CO_2 mass transfer enhancement of silica nanoparticles in water, Pet. Res. 3 (2018) 370–380. Available from: https://doi.org/10.1016/j.ptlrs.2018.09.002.

[2] H. Mohammaddoost, A. Azari, M. Ansarpour, S. Osfouri, Experimental investigation of CO_2 removal from N_2 by metal oxide nanofluids in a hollow fiber membrane contactor, Int. J. Greenhouse Gas Control 69 (2018) 60–71. Available from: https://doi.org/10.1016/j.ijggc.2017.12.012.
[3] M. Darabi, M. Rahimi, A. Molaei Dehkordi, Gas absorption enhancement in hollow fiber membrane contactors using nanofluids: modeling and simulation, Chem. Eng. Process. Process Intensif. 119 (2017) 7–15. Available from: https://doi.org/10.1016/j.cep.2017.05.007.
[4] P. Valeh-e-Sheyda, A. Afshari, A detailed screening on the mass transfer modeling of the CO_2 absorption utilizing silica nanofluid in a wetted wall column, Process. Saf. Environ. Prot. 127 (2019) 125–132. Available from: https://doi.org/10.1016/j.psep.2019.05.009.
[5] A. Peyravi, P. Keshavarz, D. Mowla, Experimental investigation on the absorption enhancement of CO_2 by various nanofluids in hollow fiber membrane contactors, Energy Fuels 29 (2015) 8135–8142. Available from: https://doi.org/10.1021/acs.energyfuels.5b01956.
[6] S. Lu, M. Xing, Y. Sun, X. Dong, H. Mohammaddoost, A. Azari, et al., Experimental and theoretical studies of CO_2 absorption enhancement by nano-Al_2O_3 and carbon nanotube particles, Chin. J. Chem. Eng. 21 (2013) 983–990. Available from: https://doi.org/10.1016/S1004-9541(13)60550-9.
[7] M. Saidi, CO_2 absorption intensification using novel DEAB amine-based nanofluids of CNT and SiO_2 in membrane contactor, Chem. Eng. Process. – Process Intensif. 149 (2020) 107848. Available from: https://doi.org/10.1016/j.cep.2020.107848.
[8] Y. Zhang, B. Zhao, J. Jiang, Y. Zhuo, S. Wang, The use of TiO_2 nanoparticles to enhance CO_2 absorption, Int. J. Greenhouse Gas Control 50 (2016) 49–56. Available from: https://doi.org/10.1016/j.ijggc.2016.04.014.
[9] N. Hajilary, M. Rezakazemi, M. Saidi, M. Darabi, M. Rahimi, A. Molaei Dehkordi, et al., C.F.D. modeling of CO_2 capture by water-based nanofluids using hollow fiber membrane contactor, Int. J. Greenh. Gas. Control. 77 (2018) 88–95. Available from: https://doi.org/10.1016/j.ijggc.2018.08.002.
[10] J.H. Kim, C.W. Jung, Y.T. Kang, Mass transfer enhancement during CO_2 absorption process in methanol/Al_2O_3 nanofluids, Int. J. Heat. Mass. Transf. 76 (2014) 484–491. Available from: https://doi.org/10.1016/j.ijheatmasstransfer.2014.04.057.
[11] J. Zong Jiang, L. Liu, B. min Sun, Model study of CO_2 absorption in aqueous amine solution enhanced by nanoparticles, Int. J. Greenhouse Gas Control 60 (2017) 51–58. Available from: https://doi.org/10.1016/j.ijggc.2017.02.006.
[12] M. Ma, C. Zou, Effect of nanoparticles on the mass transfer process of removal of hydrogen sulfide in biogas by MDEA, Int. J. Heat Mass. Transf. 127 (2018) 385–392. Available from: https://doi.org/10.1016/j.ijheatmasstransfer.2018.06.091.
[13] E.H. Moghadam, H. Bahmanyar, F. Heshmatifar, M. kasaie, H. Ziaei-Azad, The investigation of mass transfer coefficients in a pulsed regular packed column applying SiO_2 nanoparticles, Sep. Purif. Technol. 176 (2017) 15–22. Available from: https://doi.org/10.1016/j.seppur.2016.11.044.
[14] N. Azimi, M. Rahimi, Magnetic nanoparticles stimulation to enhance liquid-liquid two-phase mass transfer under static and rotating magnetic fields, J. Magn. Magn. Mater. 422 (2017) 188–196. Available from: https://doi.org/10.1016/j.jmmm.2016.08.092.
[15] B.M. Tehrani, A. Rahbar-Kelishami, Influence of enhanced mass transfer induced by Brownian motion on supported nanoliquids membrane: experimental correlation and numerical modelling, Int. J. Heat Mass. Transf. 148 (2020) 119034. Available from: https://doi.org/10.1016/j.ijheatmasstransfer.2019.119034.
[16] H.-K. Song, K. Won Cho, K.-H. Lee, Adsorption of carbon dioxide on the chemically modified silica adsorbents, J. Non. Cryst. Solids 242 (1998) 69–80. Available from: https://doi.org/10.1016/S0022-3093(98)00793-5.
[17] B. Xu, H. Gao, X. Luo, H. Liao, Z. Liang, Mass transfer performance of CO_2 absorption into aqueous DEEA in packed columns, Int. J. Greenhouse Gas Control 51 (2016) 11–17. Available from: https://doi.org/10.1016/j.ijggc.2016.05.004.

[18] J. Zong Jiang, B. Zhao, Y. Zhuo, S. Wang, B. Xu, H. Gao, et al., Mass transfer enhancement during CO_2 absorption process in methanol/Al_2O_3 nanofluids, Int. J. Greenhouse Gas Control 50 (2014) 49–56. Available from: https://doi.org/10.1016/j.ijggc.2016.05.004.

[19] H. Gao, F. Mao, Y. Song, J. Hong, Y. Yan, Effect of adding copper oxide nanoparticles on the mass/heat transfer in falling film absorption, Appl. Therm. Eng. 181 (2020) 115937. Available from: https://doi.org/10.1016/j.applthermaleng.2020.115937.

[20] J.K. Kim, J.Y. Jung, Y.T. Kang, The effect of nano-particles on the bubble absorption performance in a binary nanofluid, Int. J. Refrig. 29 (2006) 22–29. Available from: https://doi.org/10.1016/j.ijrefrig.2005.08.006.

[21] Y.T. Kang, H.J. Kim, K. Il Lee, Heat and mass transfer enhancement of binary nanofluids for H_2O/LiBr falling film absorption process, Int. J. Refrig. 31 (2008) 850–856. Available from: https://doi.org/10.1016/j.ijrefrig.2007.10.008.

[22] S.S. Ashrafmansouri, M.N. Esfahany, Mass transfer in nanofluids: a review, Int. J. Therm. Sci. 82 (2014) 84–99. Available from: https://doi.org/10.1016/j.ijthermalsci.2014.03.017.

[23] B. Olle, S. Bucak, T.C. Holmes, L.E. Bromberg, T.A. Hatton, D.I.C. Wang, Enhancement of oxygen mass transfer using functionalized magnetic nanoparticles, Ind. Eng. Chem. Res. 45 (2006) 4355–4363. Available from: https://doi.org/10.1021/ie051348b.

[24] J.P. Wen, X.Q. Jia, W. Feng, Hydrodynamic and mass transfer of gas-liquid-solid three-phase internal loop airlift reactors with nanometer solid particles, Chem. Eng. Technol. 28 (2005) 53–60. Available from: https://doi.org/10.1002/ceat.200407034.

[25] W. Feng, J. Wen, J. Fan, Q. Yuan, X. Jia, Y. Sun, Local hydrodynamics of gas-liquid-nanoparticles three-phase fluidization, Chem. Eng. Sci. 60 (2005) 6887–6898. Available from: https://doi.org/10.1016/j.ces.2005.06.006.

[26] R. Higbie, The rate of absorption of a pure gas into still liquid during short periods of exposure, Trans. AIChE. 31 (1935) 365–389.

[27] G.F. Versteeg, W.P.M. van Swaal, Solubility and diffusivity of acid gases (CO_2, N_2O) in aqueous alkanolamine solutions, J. Chem. Eng. Data 33 (1988) 29–34. Available from: https://doi.org/10.1021/je00051a011.

[28] E.N.L.R. Byron Bird, W.E. Stewart, Transport Phenomena, second ed., John Wiley & Sons, New York, 2007. https://doi.org/10.1002/aic.690070245.

[29] Z. Samadi, M. Haghshenasfard, A. Moheb, CO_2 absorption using nanofluids in a wetted-wall column with external magnetic field, Chem. Eng. Technol. 37 (2014) 462–470. Available from: https://doi.org/10.1002/ceat.201300339.

[30] Srinivas Komati, A.K. Suresh, CO_2 absorption into amine solutions: a novel strategy for intensification based on the addition of ferrofluid, J. Chem. Technol. Biotechnol. 83 (2008) 1163–1169. Available from: https://doi.org/10.1002/jctb.

[31] A. Pérez Sánchez, E.J. Pérez Sánchez, R. Segura Silva, Design of a packed-bed absorption column considering four packing types and applying matlab, Nexo Rev. Científica. 29 (2016) 83–104. Available from: https://doi.org/10.5377/nexo.v29i2.4577.

[32] J. Benitez, Principles and Modem Applications of Mass Transfer Operations, second ed., John Wiley and Sons, Hoboken, NJ, 2009.

[33] J. Salimi, M. Haghshenasfard, S.G. Etemad, CO_2 absorption in nanofluids in a randomly packed column equipped with magnetic field, Heat. Mass. Transf. Und Stoffuebertragung 51 (2015) 621–629. Available from: https://doi.org/10.1007/s00231-014-1439-5.

[34] R.U. Khanolkar, A.K. Suresh, Enhanced mass transfer rates in nanofluids: experiments and modeling, J. Heat Transf. 137 (2015) 6–11. Available from: https://doi.org/10.1115/1.4030219.

[35] T. Ramprasad, R. Khanolkar, A.K. Suresh, Mass-transfer rate enhancement in nanofluids: packed column studies and a design basis, Ind. Eng. Chem. Res. 58 (2019) 7670–7680. Available from: https://doi.org/10.1021/acs.iecr.9b00770.

[36] N.A. Policarpo, P.R. Ribeiro, Experimental measurement of gas-liquid diffusivity, Braz. J. Pet. Gas. 5 (2011) 171–188. Available from: https://doi.org/10.5419/bjpg2011-0017.

[37] M.R. Riazi, A new method for experimental measurement of diffusion coefficients in reservoir fluids, J. Pet. Sci. Eng. 14 (1996) 235–250. Available from: https://doi.org/10.1016/0920-4105(95)00035-6.

[38] F. Cao, H. GeGao, H. Gao, et al., Investigation of mass transfer coefficient of CO_2 absorption into amine solutions in hollow fiber membrane contactor, Energy Proc 114 (2017) 621–626. Available from: https://doi.org/10.1016/j.egypro.2017.03.1204.

[39] A. Marjani, A.T. Nakhjiri, A.S. Taleghani, S. Shirazian, Mass transfer modeling CO_2 absorption using nanofluids in porous polymeric membranes, J. Mol. Liq. 318 (2020) 114115. Available from: https://doi.org/10.1016/j.molliq.2020.114115.

[40] Z. Qi, E.L. Cussler, Microporous hollow fibers for gas absorption. I. Mass transfer in the liquid, J. Memb. Sci. 23 (1985) 321–332. Available from: https://doi.org/10.1016/S0376-7388(00)83149-X.

[41] K. Villeneuve, D. Albarracin Zaidiza, D. Roizard, S. Rode, E. Nagy, T. Feczkó, et al., Enhancement of oxygen mass transfer rate in the presence of nanosized particles, Chem. Eng. Sci. 62 (2007) 7391–7398. Available from: https://doi.org/10.1016/j.ces.2007.08.064.

[42] A. Gabelman, S.-T.T. Hwang, Hollow fiber membrane contactors, J. Memb. Sci. 159 (1999) 61–106. Available from: https://doi.org/10.1016/S0376-7388(99)00040-X.

[43] N. Ghasem, Chemical absorption of CO_2 enhanced by nanoparticles using a membrane contactor: modeling and simulation, Membr. (Basel) 9 (2019) 150. Available from: https://doi.org/10.3390/membranes9110150.

[44] N. Ghasem, Modeling and simulation of CO_2 absorption enhancement in hollow-fiber membrane contactors using CNT–water-based nanofluids, J. Membr. Sci. Res. 5 (2019) 295–302. Available from: https://doi.org/10.22079/jmsr.2019.100177.1239.

[45] N. Ghasem, Modeling and simulation of the simultaneous absorption/stripping of CO_2 with potassium glycinate solution in membrane contactor, Membr. (Basel) 10 (2020) 72. Available from: https://doi.org/10.3390/membranes10040072.

[46] A. Bahmanyar, N. Khoobi, M.M.A. Moharrer, H. Bahmanyar, Mass transfer from nanofluid drops in a pulsed liquid-liquid extraction column, Chem. Eng. Res. Des. 92 (2014) 2313–2323. Available from: https://doi.org/10.1016/j.cherd.2014.01.024.

[47] M. Rezakazemi, M. Darabi, E. Soroush, M. Mesbah, CO_2 absorption enhancement by water-based nanofluids of CNT and SiO_2 using hollow-fiber membrane contactor, Sep. Purif. Technol. 210 (2019) 920–926. Available from: https://doi.org/10.1016/j.seppur.2018.09.005.

[48] A. Golkhar, P. Keshavarz, D. Mowla, Investigation of CO_2 removal by silica and CNT nanofluids in microporous hollow fiber membrane contactors, J. Memb. Sci. 433 (2013) 17–24. Available from: https://doi.org/10.1016/j.memsci.2013.01.022.

[49] P.H. Calderbank, A.C. Lochiel, Mass transfer coefficients, velocities and shapes of carbon dioxide bubbles in free rise through distilled water, Chem. Eng. Sci. 19 (1964) 485–503. Available from: https://doi.org/10.1016/0009-2509(64)85075-2.

[50] S. Karamian, D. Mowla, F. Esmaeilzadeh, The effect of various nanofluids on absorption intensification of CO_2/SO_2 in a single-bubble column, Processes 7 (2019) 393. Available from: https://doi.org/10.3390/pr7070393.

[51] J.M.T. Vasconcelos, S.P. Orvalho, S.S. Alves, Gas-liquid mass transfer to single bubbles: effect of surface contamination, AIChE J. 48 (2002) 1145–1154. Available from: https://doi.org/10.1002/aic.690480603.

[52] P.C. Mishra, S. Mukherjee, S.K. Nayak, A. Panda, A brief review on viscosity of nanofluids, Int. Nano Lett. 4 (2014) 109–120. Available from: https://doi.org/10.1007/s40089-014-0126-3.

[53] S.H. Kim, W.G. Kim, H.U. Kang, K.M. Jung, Synthesis of silica nanofluid and application to CO_2 absorption, Sep. Sci. Technol. 43 (2008) 3036–3055. Available from: https://doi.org/10.1080/01496390802063804.

[54] M.H.K. Darvanjooghi, M.N. Esfahany, S.H. Esmaeili-Faraj, Investigation of the effects of nanoparticle size on CO_2 absorption by silica-water nanofluid, Sep. Purif. Technol. 195 (2018) 208–215. Available from: https://doi.org/10.1016/j.seppur.2017.12.020.

[55] J.Y. Jung, J.W. Lee, Y.T. Kang, CO_2 absorption characteristics of nanoparticle suspensions in methanol, J. Mech. Sci. Technol. 26 (2012) 2285–2290. Available from: https://doi.org/10.1007/s12206-012-0609-y.

[56] J.P. Wen, X.Q. Jia, W. Feng, J.P. Wen, J. Fan, Q. Yuan, et al., Hydrodynamic and mass transfer of gas-liquid-solid three-phase internal loop airlift reactors with nanometer solid particles, Chem. Eng. Sci. 60 (2005) 3815–3820. Available from: https://doi.org/10.1002/ceat.200407034.

[57] J. Chen, M. Liu, L. Zhang, J. Zhang, L. Jin, Application of nano TiO_2 towards polluted water treatment combined with electro-photochemical method, Water Res. 37 (2003) 3815–3820. Available from: https://doi.org/10.1016/S0043-1354(03)00332-4.

[58] S.K. Das, N. Putra, W. Roetzel, Pool boiling characteristics of nanofluids, Int. J. Heat Mass. Transf. 46 (2003) 851–862. Available from: https://doi.org/10.1016/S0017-9310(02)00348-4.

[59] S. Komati, A.K. Suresh, Anomalous enhancement of interphase transport rates by nanoparticles: effect of magnetic iron oxide on gas-liquid mass transfer, Ind. Eng. Chem. Res. 49 (2010) 390–405. Available from: https://doi.org/10.1021/ie900302z.

[60] T. Ramprasad, R. Khanolkar, A.K. Suresh, S. Komati, A.K. Suresh, CO_2 absorption into amine solutions: a novel strategy for intensification based on the addition of ferrofluid, J. Chem. Technol. Biotechnol. 83 (2008) 1163–1169. Available from: https://doi.org/10.1002/jctb.

CHAPTER

Mass transfer modeling in nanofluids: numerical approaches and challenges

10

Mohammad Hatami[1,2], Asmaa F. Elelamy[3] and Dengwei Jing[2]

[1]Department of Mechanical Engineering, Esfarayen University of Technology, Esfarayen, Iran [2]International Research Center for Renewable Energy, Xi'an Jiaotong University, Xi'an, P.R. China [3]Department of Mathematics, Faculty of Education, Ain Shams University, Cairo, Egypt

10.1 Nanofluid mass transfer

Choi [1] presented the idea of nanofluids, which alludes to creative fluid containing base fluids with uniform and stable suspension of nanosized particles. A fluid which contains small volumetric amounts of nanometer-sized particles is known as a nanofluid (smaller than 100 nm), and the particles are called nanoparticles [2]. Nanofluids can be considered as composites that include nanometer-sized solid particles dissipated in regular liquid for heat transfer, for example ethylene glycol, water, toluene, and motor oil. The nanoparticles utilized in nanofluids are regularly made of metals (such as Al, Cu), metal oxides (Al_2O_3, TiO_2, CuO, SiO_2), nonmetals (graphite, carbon nanotubes), carbides (SiC), or nitrides (AlN, SiN) [3]. Nanoparticle colloids have explicit physical characteristics that make them supportive for an extensive range of uses such as paints and coatings, earthenware, curing movement, and food organizations. Nanofluids are called supercoolant since they can hold heat more than ordinary fluids, so they can reduce the system size. By the addition of nanoparticles to the base liquid, the decrease in coefficient of heat transfer can be obtained. Most investigations show an upgrade in heat transfer. Mass dispersion is a molecular wonder that alludes to the species diffusive transport because of concentration gradients in a blend. However, convective mass exchange happens at the point that the fluid streams; which some mass is moved from one spot then onto the next by the mass smooth movement. Examination of mass exchange in nanofluids can be isolated into two principle groups [4]:

1. The principal group of studies contemplates dispersion coefficients in nanofluids, for instance, the connection between exact NaCl diffusivity (d_{NaCl}) and estimated diffusivity (d_M) affected by density-driven nanoflow should be communicated as the capacity of the Peclet number (Pe) [5].

$$d_M = d_{NaCl}\left(1 + \frac{Pe^2}{48}\right), Pe = \frac{K\Delta\rho g d_p}{2\mu d_{NaCl}}, \tag{10.1}$$

where K is the Darcy permeability K, d_p is the membrane pore diameter (0.4 lm), g refers to gravitational acceleration, and μ indicates solution dynamic viscosity.

Nanofluids and Mass Transfer. DOI: https://doi.org/10.1016/B978-0-12-823996-4.00010-0

2. The second group of studies centers on examining the convective mass exchange coefficients in nanofluids. Most examinations on mass transfer in nanofluids have endeavored on the subsequent gathering studies and there are restricted explores concentrating on mass diffusivities in nanofluid frameworks.

Mass exchange quality in a cycle is assessed by the coefficient of mass transfer. From that point of view, the upgrade factor, that is the proportion of the coefficients of mass transfer for the nanofluid and the base liquid, is utilized to show the level of improvement in mass exchange. Relative to the various frameworks and systems, an assortment of mass exchange coefficients emerge. Leaving the mass transfer segment alone, a gas stage A, and its partial pressure p_A, the mass exchange interface territory is defined as a, with mass transfer rate by χ_A, likewise for the fluid stage is signified by $A*$. The transferred mass at point per unit time is communicated as [6]:

$$\text{For Gas phase:} \quad \chi_A = k_g a(p_A - p_{A*}), \tag{10.2}$$

$$\text{For Gas phase:} \quad \chi_A = k_l a(c_A - c_{A*}), \tag{10.3}$$

where k_g and k_l indicate the gas and liquid mass transfer coefficient, p_{A*} is the partial pressure interface phase for the component $A*$, c_A, and c_{A*} are the molar concentrations of A in the liquid phase and of the components $A*$.

10.1.1 Basic equations of nanofluid mass transfer

The model of mass exchange may be created for the nanofluid by presenting the nanofluid as a two-segment blend including nanoparticles and base fluid for incompressible flow, no compound responses, debilitate mixture (nanoparticle mass portion fraction, $(\varphi \ll 1)$), negligible external force, and viscous dissipation. The impact of nanoparticles nearness in the nanofluid framework, a gathering of four equations which contains momentum and mass equations were considered. The nanofluid continuity equation can be presented by [7]:

$$\nabla \cdot \overline{u} = 0, \tag{10.4}$$

The nanofluid momentum equations are written as:

$$\rho_{nnf} \frac{D\overline{u}}{Dt} = \nabla \cdot \overline{\tau} + \overline{g}, \tag{10.5}$$

The Cauchy stress tensor $\overline{\tau}$ of Newton's fluid is

$$\overline{\tau} = -pI + \mu\left(\nabla\overline{u} + (\nabla\overline{u})^T\right), \tag{10.6}$$

where p shows the pressure, μ indicates the variable dynamic viscosity, and I is the identity tensor. The nanoparticles preservation equations with no chemical responses can be composed as:

$$\frac{\partial \phi}{\partial t} + \overline{u} \cdot \nabla \phi = -\frac{1}{\rho_P} \nabla \cdot G_P, \tag{10.7}$$

where t is the time and G_p presents the diffusion mass flux produced by the nanoparticle. Brownian motion can be written as:

$$G_p = \rho_p D_B \nabla \phi, \tag{10.8}$$

The Stokes–Einstein Brownian diffusivity D_B is represented as:

$$D_B = \frac{k_B T}{3\pi\mu d_p}, \tag{10.9}$$

where k_B indicates the Boltzmann's constant, T refers to the nanofluid temperature, and nanoparticle diameter is d. The nanofluid mass equation can be written as [8]:

$$\frac{\partial S}{\partial t} + \overline{V}\cdot\nabla S = -D\nabla^2 S + \rho_p D_B \nabla\phi \tag{10.10}$$

where S denotes the transported species concentration and D states the diffusion coefficient. There are various techniques for the mathematical numerical estimation of partial differential equations for nanofluid heat and mass transfer, among them, prominently embraced are the control volume finite element method, finite element method, finite difference method, and finite volume method. These control volume finite element method, finite element, finite difference, and finite volume methods necessitate that every halfway partial differential equation (PDE) be changed into its proportional arrangement of algebraic equations that rely upon the quantity of elements into which the physical area is isolated. In this way more the number of complicated PDEs, are the higher number of algebraic produced equations. As a rule it tends to be said that the number of equations are multiples of the number of PDEs engaged with a wonder. Any endeavor either at numerical level or during execution of the calculation to lessen the number of PDEs would decrease the computational exertion and time needed to show up at the arrangement and solution. The accompanying segments are an endeavor to present the full subtleties of these mathematical techniques that should be taken care of by the computer in order to explain and solve nanofluid heat and mass exchange [9].

10.2 Finite element method

The finite element method (FEM) includes separating the space of the issue into an assortment of subdomains, and every subdomain spoke to be a great deal of element equations to the base problem, tracked by deliberately recombining all arrangements of element equations into an overall course of action of equations for the last assessment. FEM's fame has been expanding because of the more prominent flexibility and it offers in demonstrating complex calculations. It has critical capacities in obliging general boundary conditions and adjustable properties of material with an unmistakable structure and flexibility that assists with building broadly useful programming for different applications. FEM has a strong hypothetical establishment, which includes unwavering quality and makes it conceivable to investigate and assess error in the estimated solution, numerically and mathematically [10]. Some current specialists have utilized the finite element method in numerous fields for heat and mass exchange in nanofluid fields. The impact of SiO_2 nanoparticles expansion on CO_2 partition has been actualized by Azam et al. [11]. They applied the well-known convection–diffusion model for the mass exchange, and mathematically understood it by utilizing the finite element method. The Galerkin finite element technique for the characteristic convection of Cu-Al_2O_3 crossover nanofluid in a slanted sinusoidal creased walled in area filled partially with a permeable medium was performed by Hakim et al. [12]. The Galerkin finite element procedure was utilized by Farooq et al. [13] for the heat transfer and fluid flow attributes of nanofluid in a U-shaped hole under attractive magnetic field. Additionally Mohsen et al. [14] performed the Galerkin finite element method with the expectation of complementary convection cross hybrid nanofluid flow and energy transport inside the warmed square permeable chamber differentially under the impact of an intermittent inclined magnetic field. The nanofluid concentration distribution and the management temperature profile inside biotissue were discussed by Yun et al. [15], and the finite element method was utilized for solving the nannofluid transport on the removal of harmful cells and multiphysics equations. The mass and heat exchange conduct of nanofluids inside a hexagonal alcove within a nonuniform attractive field including the ferro-hydrodynamic (FHD) and magnetohydrodynamic (MHD) effects has been presented by Ghalambaz et al. [16]. In their study, the governing equations for heat and mass exchange and the related boundary conditions were considered in the powerless structure, and afterward the FEM was utilized to integrate the equations. Also, an unstructured three-sided matrix structure was chosen to discretize the governing equations into a lot of algebraic equations. A second-order precision method was used to discretize the equations of momentum. At that point, a straight discretization plot was utilized for the progression of heat and nanoparticles transport. The equations were defined as completely coupled, where the Newton method was used to analysis the equations, iteratively. At that point, the computations were repeated until the residuals reached below 10^{-6}. The Galerkin finite element method (GFEM) approach was created by Emad et al. [17] to investigate the treatment of MHD applied to mixed convection inside an annulus fenced in area loaded up with a non-Newtonian nanofluid through the introducing governing equations and the boundary conditions. Ammar et al. [18] portrayed the nanofluid heat transfer which is displayed by utilizing a two-stage Buongiorno model. Water-based nanoliquid movement and energy transport inside a cubical cavity under the effect of the moving cold vertical surface and centrally centrally-located solid cylinder has been focused upon mathematically by Alsabery et al. [19]. They applied the finite element procedure and nondimensional crude factors; likewise the two-phase Buongiorno's nanofluid approach with thermophoresis and Brownian diffusion impacts has been utilized. Water-based nanofluid motion as well as the energy transport in a cubical enclosure was investigated numerically by Alsabery et al. [19], considering influence of the moving cold vertical plate and center-position cylinder. They applied the FEM and nondimensional primitive variables for their analysis, and they also used the two-phase Buongiorno's nanoliquid approach with thermophoresis and Brownian diffusion effects to describe the modeling.

10.2.1 Finite element method application on nanofluid heat and mass transfer

A two-dimensional convectional consistent stream is inside a pit with L as length, in which the underneath left corner of the gap is considered to be a three-sided solid divider with length of d. It is unmistakable to confirm the whole space remains as a square, as shown in Fig. 10.1. The dimensionless system of basic equations for the convectional laminar flow displayed were shown by Ammar [18]:

$$\nabla \cdot \overline{u} = 0, \tag{10.11}$$

$$\overline{u} \cdot \nabla \overline{u} = -\nabla P + \frac{\rho_f}{\rho_{nf}} \frac{\mu_{nf}}{\mu_f} \frac{1}{R_e} \nabla^2 \overline{u} + \frac{(\rho\beta)_f}{\rho_{nf}\beta_f} R_i \Theta + \frac{\rho_f}{\rho_{nf}} \frac{\sigma_{nf}}{\sigma_f} \overline{u} \times \overline{B}^*, \tag{10.12}$$

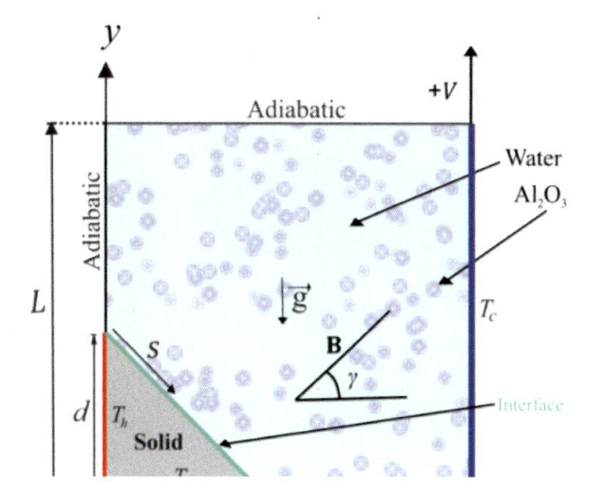

FIGURE 10.1

Coordinate system for convectional square cavity [18].

$$
\begin{aligned}
\overline{u}\cdot\nabla\Theta = & \frac{(\rho C_p)_f}{(\rho C_p)_{nf}}\frac{k_{nf}}{k_f}\frac{1}{R_eP_r}\nabla^2\Theta + \frac{(\rho C_p)_f}{(\rho C_p)_{nf}}\frac{D_B^*}{R_eP_rL_e}\nabla\Theta\cdot\nabla\chi^* \\
& + \frac{(\rho C_p)_f}{(\rho C_p)_{nf}}\frac{D_T^*}{R_eP_rL_eN_{BT}}\frac{(\nabla\Theta)^2}{1+\delta\Theta},
\end{aligned} \tag{10.13}
$$

$$
\nabla^2\Theta_w = 0, \tag{10.14}
$$

where $\overline{u}$, Θ, χ^*, and Θ_w, and T_c are dimensionless variables of velocity vector, nanofluid temperature, nanoparticles local volume fraction, the solid triangular temperature, and fixed cold temperature, respectively, and fully described in [18]. The nanofluid properties can be considered as [18,19]:

$$
\left.\begin{aligned}
\frac{\mu_{nf}}{\mu_f} &= 1/\left(1-34.87\left(d_p/d_f\right)^{-0.3}\phi^{1.03}\right) \\
\frac{k_{nf}}{k_f} &= 1/\left(1+4.4R_e^{\,0.4}P_r^{\,0.66}\left(T/T_f\right)^{10}\left(\tfrac{k_p}{k_f}\right)^{10}\phi^{0.66}\right)
\end{aligned}\right\}, \tag{10.15}
$$

$$
\left.\begin{aligned}
&(\rho C_p)_{nf} = (1-\phi)(\rho C_p)_f + \phi(\rho C_p)_p,\ \rho_{nf} = (1-\phi)\rho_f + \phi\rho_p, \\
&(\rho\beta)_{nf} = (1-\phi)(\rho\beta)_f + \phi(\rho\beta)_p,\ \alpha_{nf} = k_{nf}/(\rho C_p)_{nf} \\
&\frac{\sigma_{nf}}{\sigma_f} = 1 + \frac{3(\sigma_p/\sigma_f-1)\phi}{(\sigma_p/\sigma_f+2)-(\sigma_p/\sigma_f-1)\phi}
\end{aligned}\right\}, \tag{10.16}
$$

Where $D_B = k_bT_C/3\pi\mu_f d_p$, $D_T = 0.26\,k_bT_C/\left(2k_f+k_p\right)\frac{\mu_f}{\rho_f}\phi$, $R_e = \frac{U_0L}{f}$, $R_i = \frac{G_r}{R_e^{\,2}}$, $S_c = {}_f/D_{B_0}$, $N_{BT} = \varphi D_{B_0}T_c/D_T(T_h-T_c)$, $L_e = k_f/(\rho C_p)_f\phi D_B$, $G_r = g\varsigma_f(T_h-T_c)L^3/{}_f^2$, $H_a = BL\sqrt{\sigma_f/\mu_f}$ and $P_r = \sqrt{v_f/\alpha_f}$ are defined, where D_B and D_T are referenced Brownian diffusion and thermophoretic coefficients, respectively,R_e refers to the Reynolds number, R_i indicates the Richardson number, S_c is the Schmidt number, N_{BT} denotes the diffusivity ratio parameter, L_e is the Lewis number, G_r is the Grashof number, H_a is the Hartman number, and P_r is the Prandtl number. So, by above defined numbers, the dimensionless boundary conditions can be introduced as:

1. The top horizontal surface, the left surface and the bottom surface at the adiabatic portion are:

$$u = v = 0, \quad \frac{\partial \chi^*}{\partial n} = 0, \quad \frac{\partial \Theta}{\partial n} = 0, \tag{10.17}$$

2. The left vertical surface and horizontal bottom surface at the heated portion are:

$$u = v = 0, \quad \frac{\partial \chi^*}{\partial n} = -\frac{D_T^*}{D_D^*}\frac{1}{N_{RT}}\frac{1}{1+\delta\Theta}\frac{\partial \Theta}{\partial n}, \quad \Theta = 1, \tag{10.18}$$

3. The right surface at the vertical movement:

$$u = v = 1, \quad \frac{\partial \chi^*}{\partial n} = -\frac{D_T^*}{D_D^*}\frac{1}{N_{RT}}\frac{1}{1+\delta\Theta}\frac{\partial \Theta}{\partial n}, \quad \Theta = 0, \tag{10.19}$$

4. On the interface surface $\Theta_{nf} = \Theta_w$:

$$u = v = 0, \quad \frac{\partial \chi^*}{\partial n} = -\frac{D_T^*}{D_D^*}\frac{1}{N_{RT}}\frac{1}{1+\delta\theta}\frac{\partial \Theta}{\partial n}, \quad \frac{\partial \Theta}{\partial n} = k_r\frac{\partial \Theta_w}{\partial n} \tag{10.20}$$

The thermal conductivity ratio is defined as $k_r = k_w/k_{nf}$.

10.2.2 Finite element method simulation

The weighted residual method of Galerkin finite element is utilized to solve the dimensionless governing equations introduced in Eqs. (10.11)–(10.14) with the recommended boundary conditions in dimensionless form from Eqs. (10.17) to (10.20). Navier–Stokes equations for the velocity vector $\bar{u}$ in 2D Cartesian coordinates (x, y) can be introduced as [18]:

$$\begin{aligned} u\frac{\partial u}{\partial x} + v\frac{\partial u}{\partial y} = & -\frac{\partial p}{\partial x} + \frac{\rho_f}{\rho_{nf}}\frac{\mu_{nf}}{\mu_f}\frac{1}{R_e}\left(\frac{\partial^2 u}{\partial x^2} + \frac{\partial^2 u}{\partial y^2}\right) \\ & + \frac{\rho_f}{\rho_{nf}}\frac{\sigma_{nf}}{\sigma_f}H_a^2(v\sin\gamma\cos\gamma - u\sin^2\gamma), \end{aligned} \tag{10.21}$$

$$\begin{aligned} u\frac{\partial v}{\partial x} + v\frac{\partial v}{\partial y} = & -\frac{\partial p}{\partial y} + \frac{\rho_f}{\rho_{nf}}\frac{\mu_{nf}}{\mu_f}\frac{1}{R_e}\left(\frac{\partial^2 v}{\partial x^2} + \frac{\partial^2 v}{\partial y^2}\right) + \frac{(\rho\beta)_f}{\rho_{nf}\beta_f}R_i\Theta \\ & + \frac{\rho_f}{\rho_{nf}}\frac{\sigma_{nf}}{\sigma_f}H_a^2(u\sin\gamma\cos\gamma - v\cos^2\gamma), \end{aligned} \tag{10.22}$$

Removing the pressure p by utilizing penalty parameter finite element technique ζ obtains:

$$p = -\zeta\left(\frac{\partial u}{\partial x} + \frac{\partial v}{\partial y}\right), \tag{10.23}$$

The Navier–Stokes equations in (x, y) directions are presented as:

$$\begin{aligned} u\frac{\partial u}{\partial x} + v\frac{\partial u}{\partial y} = & \frac{\partial \zeta}{\partial x}\left(\frac{\partial u}{\partial x} + \frac{\partial v}{\partial y}\right) + \frac{\rho_f}{\rho_{nf}}\frac{\mu_{nf}}{\mu_f}\frac{1}{R_e}\left(\frac{\partial^2 u}{\partial x^2} + \frac{\partial^2 u}{\partial y^2}\right) \\ & + \frac{\rho_f}{\rho_{nf}}\frac{\sigma_{nf}}{\sigma_f}H_a^2(v\sin\gamma\cos\gamma - u\sin^2\gamma), \end{aligned} \tag{10.24}$$

$$
\begin{aligned}
u\frac{\partial v}{\partial x}+v\frac{\partial v}{\partial y} &= \frac{\partial \zeta}{\partial y}\left(\frac{\partial u}{\partial x}+\frac{\partial v}{\partial y}\right)+\frac{\rho_f}{\rho_{nf}}\frac{\mu_{nf}}{\mu_f}\frac{1}{R_e}\left(\frac{\partial^2 v}{\partial x^2}+\frac{\partial^2 v}{\partial y^2}\right)+\frac{(\rho\beta)_f}{\rho_{nf}\,\beta_f}R_i\Theta \\
&+\frac{\rho_f}{\rho_{nf}}\frac{\sigma_{nf}}{\sigma_f}H_a^2\left(u\sin\gamma\cos\gamma-v\cos^2\gamma\right),
\end{aligned}
\tag{10.25}
$$

The weak structures corresponding to nondimensional Navier–Stokes Eqs. (10.24–10.25) over typical discretized triangular small elements (see Fig. 10.2) are given by:

$$
\begin{aligned}
\int_\Omega\left(\mathrm{E}_i u^k\frac{\partial u^k}{\partial x}+v^k\frac{\partial u^k}{\partial y}\right)dx\,dy &= \zeta\int_\Omega\frac{\partial \mathrm{E}_i}{\partial x}\left(\frac{\partial u^k}{\partial x}+\frac{\partial v^k}{\partial y}\right)dx\,dy \\
&+\frac{\rho_f}{\rho_{nf}}\frac{\mu_{nf}}{\mu_f}\frac{1}{R_e}\int_\Omega \mathrm{E}_i\left(\frac{\partial^2 u^k}{\partial x^2}+\frac{\partial^2 u^k}{\partial y^2}\right)dx\,dy \\
&+\frac{\rho_f}{\rho_{nf}}\frac{\sigma_{nf}}{\sigma_f}H_a^2\left(\mathrm{E}_i v^k\sin\gamma\cos\gamma-\mathrm{E}_i u^k\sin^2\gamma\right),
\end{aligned}
\tag{10.26}
$$

$$
\begin{aligned}
\int_\Omega\left(\mathrm{E}_i v^k\frac{\partial u^k}{\partial x}+v^k\frac{\partial u^k}{\partial y}\right)dx\,dy &= \lambda\int_\Omega\frac{\partial \mathrm{E}_i}{\partial y}\left(\frac{\partial u^k}{\partial x}+\frac{\partial v^k}{\partial y}\right)dx\,dy \\
&+\frac{\rho_f}{\rho_{nf}}\frac{\mu_{nf}}{\mu_f}\frac{1}{R_e}\int_\Omega \mathrm{E}_i\left(\frac{\partial^2 v^k}{\partial x^2}+\frac{\partial^2 v^k}{\partial y^2}\right)dx\,dy \\
&+\frac{(\rho\beta)_f}{\rho_{nf}\beta_f}R_i\int_\Omega \mathrm{E}_i\Theta^k dx\,dy \\
&+\frac{\rho_f}{\rho_{nf}}\frac{\sigma_{nf}}{\sigma_f}H_a^2\left(\mathrm{E}_i u^k\sin\gamma\cos\gamma-\mathrm{E}_i v^k\cos^2\gamma\right),
\end{aligned}
\tag{10.27}
$$

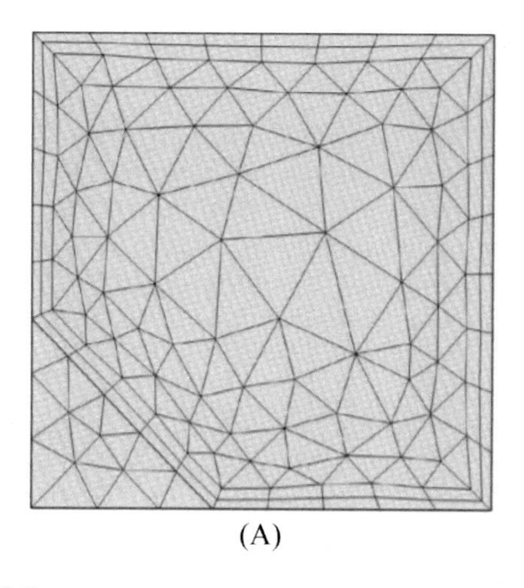
(A)

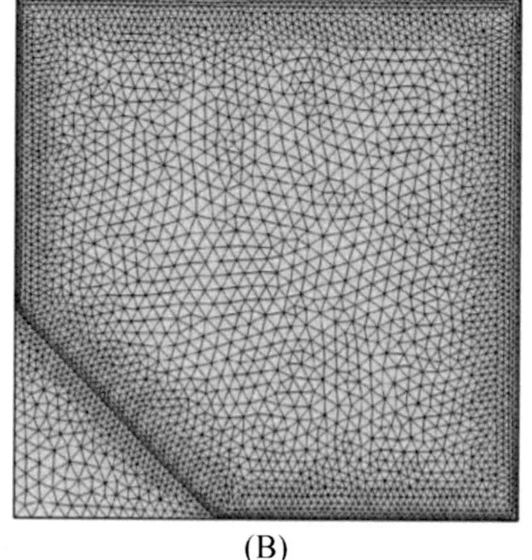
(B)

FIGURE 10.2

distributions of grid-points for grid size of (A) 222 and (B) 6702 elements [18].

The actualizing estimation interpolation functions for the velocity dispersion and temperature distribution can be introduced as:

$$u \approx \sum_{i=1}^{n} u_i \mathrm{E}_i(x, y),\ v \approx \sum_{i=1}^{n} v_i \mathrm{E}_i(x, y),\ \Theta \approx \sum_{i=1}^{n} \Theta_i \mathrm{E}_i(x, y), \tag{10.28}$$

The equations of weighted residual according to Galerkin finite element technique are:

$$\begin{aligned} Rs(1)_j = & \sum_{i=1}^{n} u_i \int_{\Omega} \left[\left(\sum_{i=1}^{n} u_i\, \mathrm{E}_i(x, y) \right) \frac{\partial \mathrm{E}_i}{\partial x} + \left(\sum_{i=1}^{n} v_i\, \mathrm{E}_i(x, y) \right) \frac{\partial \mathrm{E}_i}{\partial y} \right] \mathrm{E}_i dx\, dy \\ & + \lambda \left[\sum_{i=1}^{n} u_i \int_{\Omega} \frac{\partial \mathrm{E}_i}{\partial x} \frac{\partial \mathrm{E}_i}{\partial x} dx\, dy + \sum_{i=1}^{n} v_i \int_{\Omega} \frac{\partial \mathrm{E}_i}{\partial x} \frac{\partial \mathrm{E}_i}{\partial y} dx\, dy \right] \\ & - \frac{\rho_f}{\rho_{nf}} \frac{\mu_{nf}}{\mu_f} \frac{1}{R_e} \sum_{i=1}^{n} u_i \int_{\Omega} \left[\frac{\partial \mathrm{E}_i}{\partial x} \frac{\partial \mathrm{E}_i}{\partial x} + \frac{\partial \mathrm{E}_i}{\partial y} \frac{\partial \mathrm{E}_i}{\partial y} \right] dx\, dy \\ & + \frac{\rho_f}{\rho_{nf}} \frac{\sigma_{nf}}{\sigma_f} H_a^2 \left[\left(\sum_{i=1}^{n} v_i\, \mathrm{E}_i(x, y) \right) \sin\gamma \cos\gamma - \left(\sum_{i=1}^{n} u_i\, \mathrm{E}_i(x, y) \right) \sin^2\gamma \right], \end{aligned} \tag{10.29}$$

$$\begin{aligned} Rs(2)_j = & \sum_{i=1}^{n} v_i \int_{\Omega} \left[\left(\sum_{i=1}^{n} u_i\, \mathrm{E}_i(x, y) \right) \frac{\partial \mathrm{E}_i}{\partial x} + \left(\sum_{i=1}^{n} v_i\, \mathrm{E}_i(x, y) \right) \frac{\partial \mathrm{E}_i}{\partial y} \right] \mathrm{E}_i dxdy \\ & + \lambda \left[\sum_{i=1}^{n} u_i \int_{\Omega} \frac{\partial \mathrm{E}_i}{\partial x} \frac{\partial \mathrm{E}_i}{\partial x} dx\, dy + \sum_{i=1}^{n} v_i \int_{\Omega} \frac{\partial \mathrm{E}_i}{\partial x} \frac{\partial \mathrm{E}_i}{\partial y} dx\, dy \right] \\ & + \frac{\rho_f}{\rho_{nf}} \frac{\mu_{nf}}{\mu_f} \frac{1}{R_e} \sum_{i=1}^{n} v_i \int_{\Omega} \left[\frac{\partial \mathrm{E}_i}{\partial x} \frac{\partial \mathrm{E}_i}{\partial x} + \frac{\partial \mathrm{E}_i}{\partial y} \frac{\partial \mathrm{E}_i}{\partial y} \right] dx\, dymm \\ & + \frac{(\rho\, \beta)_f}{\rho_{nf}\, \beta_f} R_i \int_{\Omega} \mathrm{E}_i \left(\sum_{i=1}^{n} \Theta_i\, \mathrm{E}_i(x, y) \right) dxdy \\ & + \frac{\rho_f}{\rho_{nf}} \frac{\sigma_{nf}}{\sigma_f} H_a^2 \left[\left(\sum_{i=1}^{n} u_i\, \mathrm{E}_i(x, y) \right) \sin\gamma \cos\gamma - \left(\sum_{i=1}^{n} v_i\, \mathrm{E}_i(x, y) \right) \cos^2\gamma \right], \end{aligned} \tag{10.30}$$

where the superscript k denotes the inexact list, addendums i, j, and m refer to the hub number, the leftover number, and cycle number, respectively. The iteration of the Newton–Raphson calculation is utilized for the nonlinear terms. The intermingling of the arrangement is assumed if the overall mistake for every one of the factors gives the subsequent combination standards:

$$\frac{Er^{q+1} - Er^{q}}{Er^{q+1}} \le 10^{-5} \tag{10.31}$$

Fig. 10.3 demonstrates the qualification of streamlines, isotherms, and nanoparticle dissemination by means of expanding the Reynolds number. The nanofluid velocity is smoothed out at $R_e = 10$, and is affected by the right wall movement. The phenomenal temperature slope speaks to the advancement of the interface nanoparticles and left divider to various zones of the opening. Since the convection is overpowered by conduction (at low Reynolds number), and due to the effects of temperature angle and thermophoresis impacts, the dislodging of nanoparticles happens more.

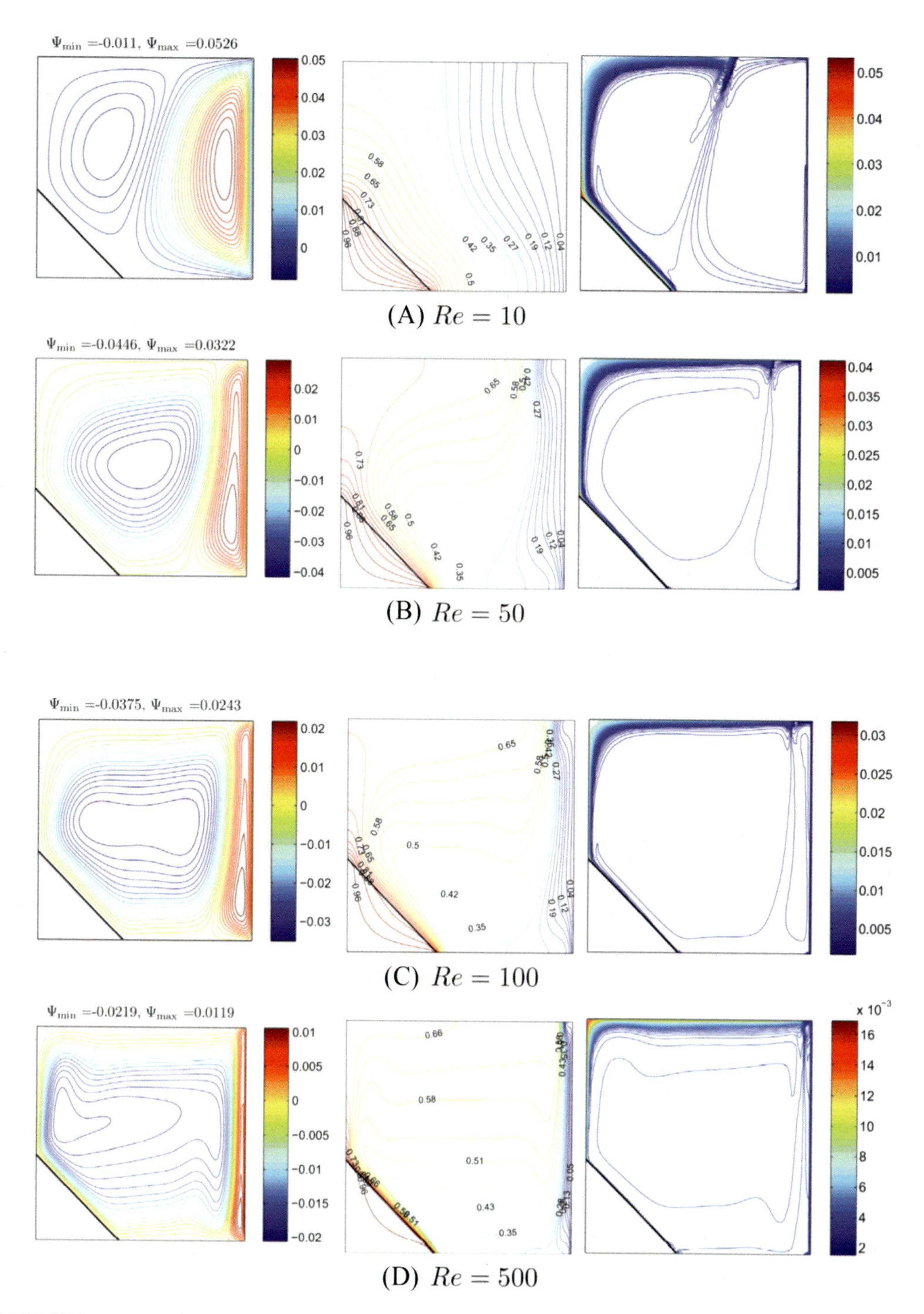

FIGURE 10.3

Variation of the streamlines (left), isotherms (middle), and nanoparticle distribution (right) evolution by Reynolds number (R_e) A) Re = 10, B) Re = 50, C) Re = 100 and D) Re = 500 [18].

10.3 Control volume finite element method

The control volume finite element or CVFEM technique consolidates fascinating qualities from both the finite volume and finite element strategies. The CVFEM was introduced by linear triangular finite elements, and linear quadrilateral components [20]. A few creators have improved CVFEM from in nanofluid applications. The characteristic and mixed convectional rectangular specialty piled up with water-based nanofluids for the entropy creation and transfer of heat was examined by Ogbana et al. [21]. The control volume based on finite element procedure was used to discretize the governing equations which demonstrate this case. The mesh was discretized dependent on the technique of finite element. The collected subcontrol volumes will be arising from the adjacent interfacing parts to shape of a control volume around the nodal focuses, where the troublesome space is discretized into direct straight quadrilateral limited components. The ferrofluid flow and natural convection analysis in a triangular enclosure is investigated mathematically by a procedure dependent on CVFEM by Dogonchi et al. [22]. In a similar manner the nanoparticles shape factor and nanofluid volume division versus heat transportation and stream fields have been researched. The governing partially differential equations for the ethylene glycol-based nanomaterial flow including electric power are settled by the assistance of CVFEM [23]; the radiation and different shape factors have likewise been fused in the model. CVFEM has been applied in the nanofluid flow field in designing and clinical branches [24–31]. The control volume finite component strategy numerical method for entropy creation and the transfer of heat for ordinary and mixed convection in a rectangular niche stacked up in nanofluids has been applied. This investigation intends to locate an ideal estimation for the volume fraction of nanoparticle, of which entropy creation is limited. The examination likewise considers the regular convection transfer of heat execution and entropy creation for various nanofluids in the nook.

10.3.1 Fundamental equations

The problem structure set up for the normal and mixed convection depiction appears in Fig. 10.4. A square contained zone with (H, L) length and width, respectively, stacked up with water-based nanofluids is shown in Fig. 10.4. Rectangular Cartesian coordinates x and y are applied in this physical problem. It is accepted that the nanoparticles and the applied fluid are at equilibrium in thermal distribution, and the chemical reaction is neglected. There is no heat transfer due to radiation or heat transfer of nanoparticle movement compared with with the working fluid. The fluid flow in this investigation is proposed to be a 2-D (two-dimensional) Newtonian constitutive model. The thermophysical characteristics of nanofluid are thought to be consistent. The variety of thickness in the buoyancy power term is resolved dependent on the estimate of the Boussinesq assumption.

The equations of continuity, Navier–Stokes, and the transfer of heat are as follows according to the above suppositions [32]:

$$\frac{\partial u}{\partial x} + \frac{\partial v}{\partial y} = 0, \tag{10.32}$$

$$\rho_{nf}\left(\frac{\partial u}{\partial t} + u\frac{\partial u}{\partial x} + v\frac{\partial u}{\partial y}\right) = -\frac{\partial p}{\partial x} + \mu_{nf}\left(\frac{\partial^2 u}{\partial x^2} + \frac{\partial^2 u}{\partial y^2}\right), \tag{10.33}$$

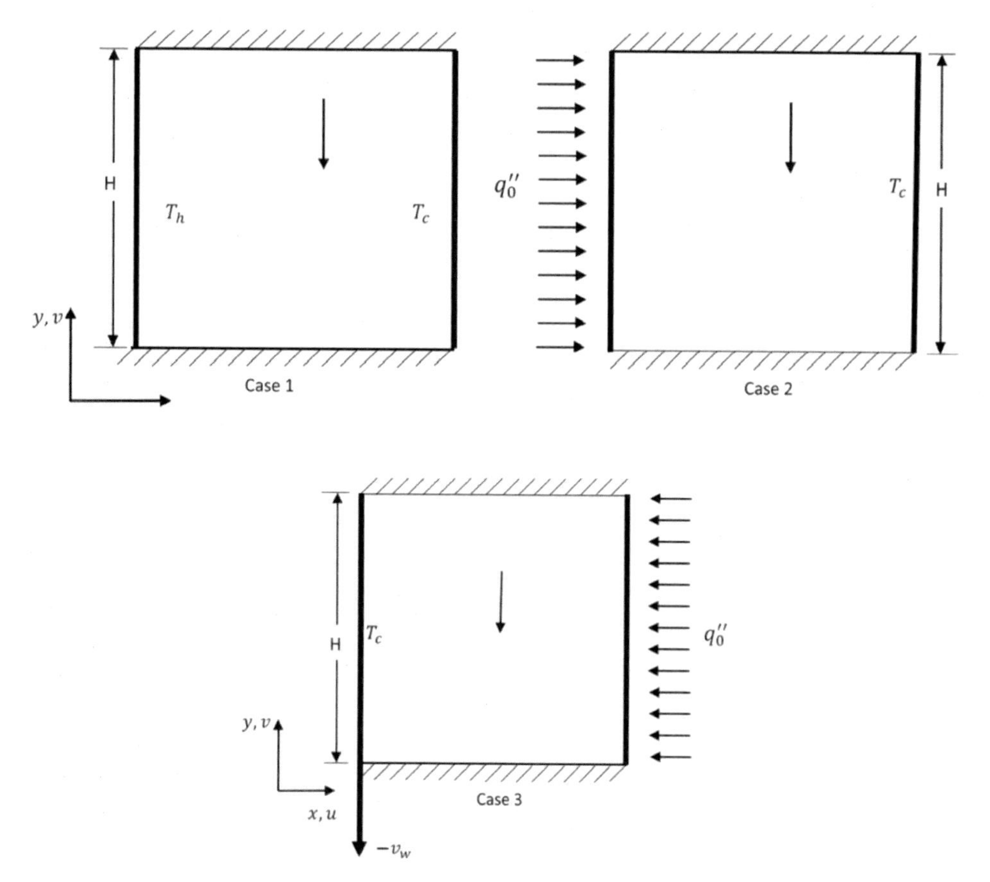

FIGURE 10.4

Schematic of the problem [32].

$$\rho_{nf}\left(\frac{\partial v}{\partial t}+u\frac{\partial v}{\partial x}+v\frac{\partial v}{\partial y}\right)=-\frac{\partial P}{\partial y}+\mu_{nf}\left(\frac{\partial^2 v}{\partial x^2}+\frac{\partial^2 v}{\partial y^2}\right)+(\rho\,\beta)_{nf}g(T-T_c), \tag{10.34}$$

$$(\rho\,C_p)_{nf}\left(\frac{\partial T}{\partial t}+u\frac{\partial T}{\partial x}+v\frac{\partial T}{\partial y}\right)=k_{nf}\left(\frac{\partial^2 T}{\partial x^2}+\frac{\partial^2 T}{\partial y^2}\right), \tag{10.35}$$

where the viable properties of nanofluids have been displayed in Eq. (10.16). The calculation of thermal conductivity and viable dynamic viscosity includes a more point by point definition dependent on exploratory relationships, for example the Maxwell–Garnett model [32–34]:

$$\mu_{nf}=\frac{\mu_{nf}}{(1-\varphi)^{2.5}},\ \frac{k_{nf}}{k_f}=\frac{k_p+2k_f-2(k_f-k_f)\varphi}{k_p+2k_f+2(k_f-k_f)\varphi}, \tag{10.36}$$

The thermophysical properties of nanoparticles and water are introduced in Table 10.1.

Table 10.1 Thermophysical properties of nanoparticles and base fluids.

Physical property	Al_2O_3	Cu	TiO_2	H_2O
Density, ρ (kg/m^3)	3970	8933	4250	993
Thermal conductivity, K (W/m K)	40	400	8.9538	0.613
Specific heat capacity, C_p (J/kg K)	765	385	686.5	4179
Thermal expansion coefficient, β (1/K)	8.5×10^{-6}	1.67×10^{-5}	9.0×10^{-6}	2.1×10^{-6}
Dynamic viscosity, μ (Pa.s)				8.9×10^{-4}

10.3.2 Modeling of numerical method

The differential equations which depict the governing problem (10.32)–(10.35) are discretized by a numerical technique which relies upon the control volume finite element technique. A spatial discretization and mesh rely upon the finite element strategy by means of quadrilateral straight finite components; at that point, the subcontrol volume encompassing the hub is neighbored by segments close to the control volume outline. Consider the area structure (r, w), which portrays the capacities of the shape and element properties, as shown in Fig. 10.3. The computational domain of this problem is discretized into straight finite quadrilateral components. The description of the control volume can be considered as subsurfaces and subcontrol volumes of four elements neighboring each other. Conservation equations are then fused over the time steps and limited control volumes secure the conservation equations due to discretization. Each element in the control volume is apportioned into four subcontrol volumes; these elements are associated with the individual part's center and the control volume. The control volumes are restricted by the subsurfaces, which correspond with the external limit of components. Neighborhoods are characterized by a proximal sorting system (r, w). The midpoint of a subsurface is described by the blend point (i p). The calculation variety and variable ϑ of general scalar, (for instance, profiles of velocity and temperature), at some arbitrary point inside the part can be portrayed viewing nodal values as follows:

$$x \approx \sum_{i=1}^{n} x_i M_i,\ y \approx \sum_{i=1}^{n} y_i M_i,\ \vartheta(r, w) \approx \sum_{i=1}^{n} M_i\,(r, w)_i, \tag{10.37}$$

Where Ω is near to nodes value, the shape functions $M_i(r, w)$ are characterized by:

$$M_1(r, w) = (1 + r)(1 + w)/4, \tag{10.38}$$

$$M_2(r, w) = (1 - r)(1 + w)/4, \tag{10.39}$$

$$M_3(r, w) = (1 - r)(1 - w)/4, \tag{10.40}$$

$$M_4(r, w) = (1 + r)(1 - w)/4, \tag{10.41}$$

The corresponding scalar derivatives are characterized by:

$$\frac{\partial \vartheta}{\partial x} = \sum_{i=1}^{4} \frac{\partial M_i}{\partial x} \vartheta_i, \tag{10.42}$$

$$\frac{\partial \vartheta}{\partial y} = \sum_{i=1}^{4} \frac{\partial M_i}{\partial y} \vartheta_i, \tag{10.43}$$

10.3.3 Discretization equation of general transport

The equalization of preservation over control volume is ensured by CVFEM. The conservation equation can be proposed as [34,35]:

$$\frac{\partial(\rho\vartheta)}{\partial t}+\nabla\cdot(\rho\vartheta\overline{u})-\nabla\cdot(\nabla\vartheta)=Z', \tag{10.44}$$

By utilizing the divergence theorem, and integrating Eq. (10.44), we have

$$\begin{aligned}&\int_t^{t+\Delta t}\int_{\mathrm{SCVi}}\frac{\partial(\rho\vartheta)}{\partial t}dV\,dt+\int_t^{t+\Delta t}\int_{\mathrm{SCVi}}(\rho\vartheta\overline{u})\,.\,ndS\,dt-\int_t^{t+\Delta t}\int_{\mathrm{SCVi}}(X\nabla\vartheta)\,.\,ndS\,dt\\&=\int_t^{t+\Delta t}\int_{\mathrm{SCVi}}Z'\,dV\,dt,\end{aligned} \tag{10.45}$$

where Si, SCVi, $\overline{u}$, and n are the surface zone, the subcontrol volume, the velocity field, and the normal unit vector to surface, respectively. To discretize the transient stockpiling term, a finite difference backward strategy can be utilized. An approximation for the transnational term is given by:

$$\int_t^{t+\Delta t}\int_{\mathrm{SCVi}}\frac{\partial(\rho\vartheta)}{\partial t}dV\,dt=J_i\left(\rho\Psi_i^{t+\Delta t}-\rho_i^t\right), \tag{10.46}$$

J_i is the determinant of Jacoby for some node i; and Ψ speaks to a nearby nodal esteem.

At given integrating point (i), the transnational term can be calculated by applying the finite difference backward method

$$\left.\frac{\partial\vartheta}{\partial t}\right|_{ip_i}=\frac{\vartheta_{ip_i}^{t+\Delta t}-\vartheta_{ip_i}^{t}}{\Delta t}, \tag{10.47}$$

Utilizing the strategy of upwind difference [32] to acquire estimations of the velocities at some integration point, the convectional term can be represented as:

$$\nabla\cdot(\rho\vartheta\overline{v})=\rho V\frac{\vartheta_{ip_i}-\vartheta_u}{A_c}, \tag{10.48}$$

where (L_c,ϑ_u) are the convectional length scale along the stream-wise flow and the means the upwind estimation respectively, and $U=\sqrt{u^2+v^2}$ speaks the fluid velocity magnitude. The bearing of the line divide between ϑ_{ip_i} and ϑ_u is demonstrated by an inclined upwinding methodology. The upstream worth, ϑ_u is constrained by addition of neighborhood hubs where there is a combination between the quadrant edge and the upwind course line, for instance, the upstream worth has such a relation with both local nodes 2 and local nodes 3;

$$\vartheta_u=g\vartheta_2+(1-g)\vartheta_3, \tag{10.49}$$

where g is the coefficient for the relating direct interjection for ϑ_u concerning Ψ_2 and Ψ_3. The operator of diffusion can be calculated by utilizing the central differencing strategy as:

$$\left.\nabla^2\omega\right|_{ip_i}=\frac{1}{L_d^2}\left(\sum_{i=1}^{4}M\Psi_i-\vartheta_{ip_i}\right), \tag{10.50}$$

The diffusion length scale s_d can be gotten as:

$$s_d^2=\left(\frac{2}{\Delta x^2}+\frac{2}{3\Delta y^2}\right)^{-1}, \tag{10.51}$$

In Eq. (10.45) the diffusion term is determined by base subsurfaces and on the left as:

$$\int_{t}^{t+\Delta t}\int_{\mathrm{SCVi}}(X\nabla\vartheta)\,.\,ndS\,dt = X\frac{\partial\vartheta}{\partial n}\Big|_{ip_i}\Delta t\Delta y_i - X\frac{\partial\vartheta}{\partial n}\Big|_{ip_i}\Delta t\Delta x_i, \tag{10.52}$$

The scalar variable and dispersion motion are associated subject to Fourier's law. Eq. (10.51) is surveyed using the estimation of the midpoint. The source term in Eq. (10.45) can be evaluated using the lumped surmise as:

$$\int_{t}^{t+\Delta t}\int_{\mathrm{SCVi}} Z^{\cdot} dV\,dt = J_i \cdot tZ^{\cdot}\Big|_{(1/2,1/2)}, \tag{10.53}$$

where the subscript, $(1/2, 1/2)$ addresses the close location at the point of convergence of SCVi: The source terms of the neighborhood can be constrained by direct substitution of the contrasting joining point factors using the Boussinesq estimation. At the point when all the combination point factors are subbed again into condition (10.45) at the joining point, the close-by altered lattices are applied to the point integration variable clearly with respect to nodal values alone. The grouping of all segments has been done after the subcontrol volume and formula of points integration which were calculated inside a component. At a certain hub Z, with four unlike enveloping segments, the control volume for hub Z is arranged by grouping the subcontrol volume responsibility from all four parts of the sharing hub Z. For example, consider the condition of vitality for the heat conduction term at close-by hub 3 in Fig. 10.5. The parity of vitality at hub 3 of subcontrol volume 1 can be imparted as:

$$\int_{V}\rho\,C_p\frac{\partial\Theta}{\partial t}dV = q_{2,1} + q_{4,1} + q_{01,1} + q_{02,1}, \tag{10.54}$$

where $q_{01,1}$ and $q_{02,1}$ speak to an outside element that contributes in a similar manner to the heat stream for the equilibrium of energy of subcontrol volume 1, where q indicates the transitional heat

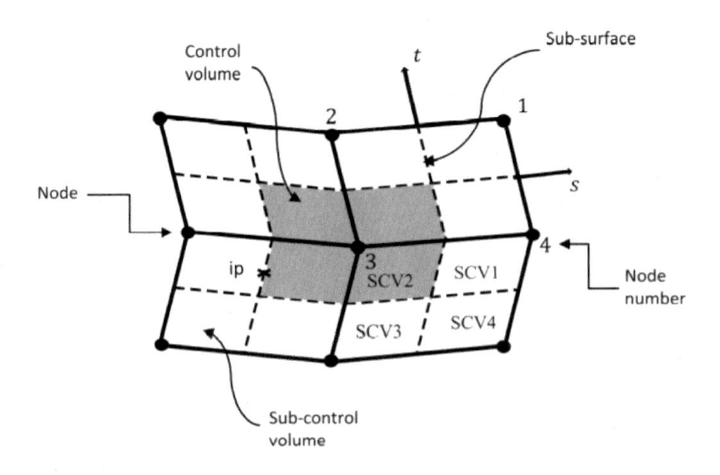

FIGURE 10.5

Schematic of FEM and control volume [34,35]. *FEM*, finite element method.

over the subsurface. The transitional heat inside a component, $q_{2,1}$; at subsurface 1 can be gotten as:

$$q_{2,1} = \sum_{i=1}^{4} \left\{ \int_0^1 \left(k \frac{\partial M_i}{\partial x} \frac{\partial y}{\partial t} - k \frac{\partial M_i}{\partial y} \frac{\partial x}{\partial t} \right)_{s=0} dt \right\} T_i, \tag{10.55}$$

The lumped capacitance estimation technique can be represented as [35];

$$\int_V \rho\, C_p \frac{\partial \Theta}{\partial t} dV = \rho\, C_p\, J_i \frac{\Theta_i^{t+\Delta t} - \Theta_i^t}{\Delta t}, \tag{10.56}$$

The superscripts Δt and t represent the of the old and new time steps.

10.4 Lattice Boltzmann method

The lattice Boltzmann method (LBM) is utilized to understand the velocity and temperature values utilizing the Bhatnagar–Gross–Krook (BGK) administrator. It is worth making reference that the cross section Boltzmann technique has some articulated points of interest that persuade the analysts to utilize this strategy, despite the fact that it is computationally costly with regard to the customary strategies, for example FV and FD. The benefits of LBM can be referenced as straight equation registering which makes this strategy more exact due to the local collision estimations. Additionally, this technique can deal with complex calculations and reenact the liquid velocity and heat transfer of incompressible lows precisely. In general. The significant preferences of LBM over other customary CFD techniques because of the way that the answer for the molecule conveyance capacities is express and simple for equal calculation and usage of limit conditions on complex basics [36]. The LBM uses two conveyance capacities, for the stream and temperature fields. It utilizes the demonstration of development of liquid particles to characterize naturally macroscopic parameters of the liquid stream. The fundamental type of LBM applies uniform Cartesian cells to the discrete area. Every cell of the matrix has a consistent number of dispersion capacities, which speak to the quantity of liquid. Particles develop in these isolated ways. The dissemination capacities are gotten by settling the lattice Boltzmann equation (LBE), which is an extraordinary type of the kinetic Boltzmann equation. The grid Boltzmann method can be worked on various cross sections, both cubic and three-sided, and with or without rest particles in the function of discrete distribution. A well-known system of categorizing the different technoques by lattice is the "DnQm" scheme. Where "Dn" refers to "n dimensions" and "Qm" denotes the "m speeds" [37]. The lattic Boltzmann strategy has been shown to be a compelling mathematical instrument for the assortment of complex liquid flow regimes which compared to ordinary techniques, contrasted and conventional computational liquid elements, LBM calculations are a lot simpler to be executed particularly in complex calculations and multicomponent streams. Fig. 10.6 shows the three scales engaged with framework investigation alongside their answering method [38]. In LBM, a quirk of issues decides the sort of grids to be utilized which tends to utilize DpQq documentation, as shown in Fig. 10.7 [39]. Most research papers on LBM for nanofluids with the transport of heat utilized D2Q9 type (lattice) for 2D issues since it has demonstrated to be exceptionally effective,

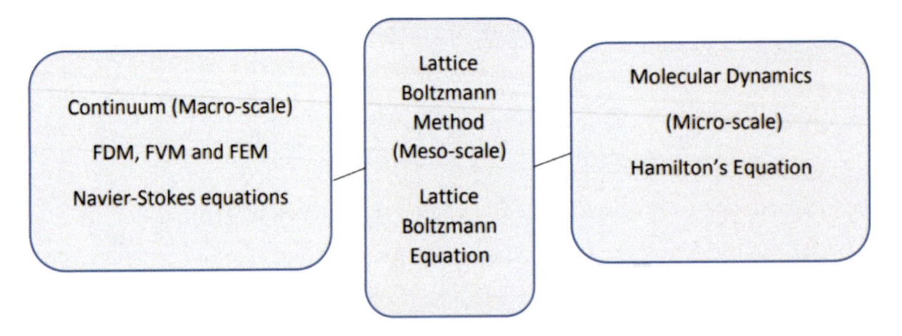

FIGURE 10.6

Scales at which systems are generally analyzed [38].

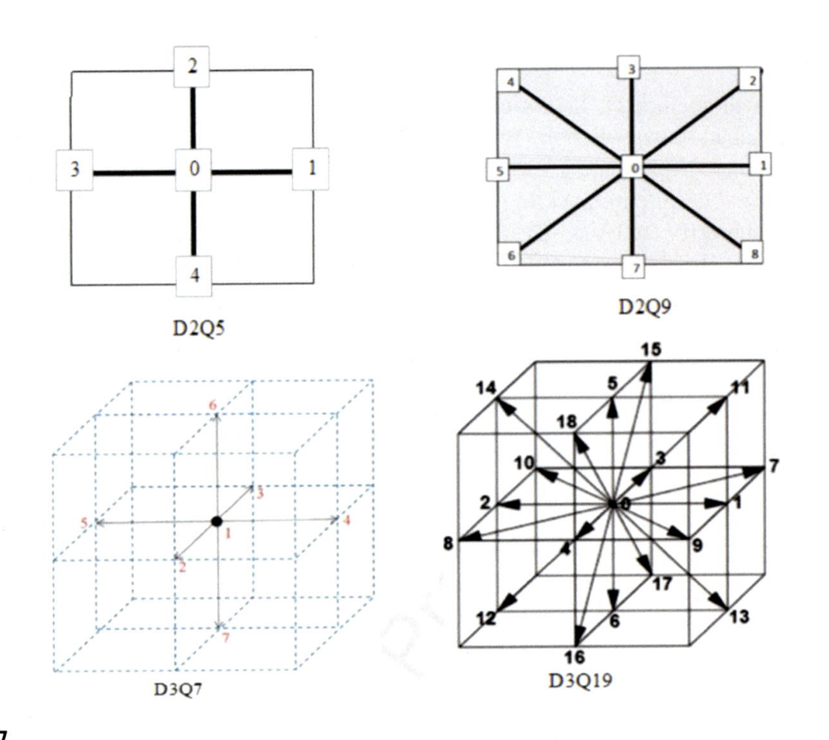

FIGURE 10.7

Schematic state of some basic lattices [39].

particularly for incompressible streams because of its capacity for flow velocity autonomy of pressure, Galilean invariance, and isotropy.

Rasul et al. [40] applied the LBM concerning fluid flow and vitality transport in a Γ-molded chamber which has a high temperature rectangular obstacles and is filled up with AL_2O_3/water nanofluid. The lattice Boltzmann reproduction of a microchannel heat sink loaded up with the nanofluid of alumina-water was demonstrated by Abolfazl [41]. The $Al_2O_3-H_2O$ natural convection for nanofuids and $(Al_2O_3/Cu)-H_2O$ composite nanofuids in a rectangular cavity was examined by Cong et al. [42], additionally they set up a two-stage LBM model. The impact of 50 nm Cu nanoparticles addition to the unadulterated water is examined by utilizing the LBM technique by Mehdi [43]; a two-component D2Q9 single relaxation-time model was utilized in his examination for both thermal and hydrodynamic equations. The lattice Boltzmann structure was utilized by Yuan [44] to reenact the hybrid nanofluid heat transfer and natural convection in a square walled-in area with a warming deterrent at high Rayleigh numbers.

10.4.1 The transport model of Lattice Boltzmann

Suppose that the framework can be clarified by appropriation function (a, b, t) where $\varphi(a, b, t)$ refers to the quantity of particles at time t situated among a and $a + da$ which have velocities among b and $b + dF$. In the last definition, F is defined as an external force which can be considered on a unit mass of molecule prompting a velocity change of that unit mass from b to $b + dF$ and furthermore, a positional change from a to $a + dt$ [45]. If the conveyance of particles before the power application is $\varphi(a, b, t)$, and distribution of the particles after the force is $\varphi(a + bdt, b + Fdt, t + dt)$, then by applying conservation law we have (the effect of particle collision is neglected)

$$\varphi(a + bdt, b + Fdt, t + dt)da\, db - \varphi(a, b, t)da\, db = 0, \tag{10.57}$$

The rate at which changes happen in the position of the function distribution when there is impact between particles is:

$$\varphi(a + bdt, b + Fdt, t + dt)da\, db - \varphi(a, b, t)da\, db = \psi(\varphi)da\, db\, dt, \tag{10.58}$$

where ψ is the collision operator. Some numerical goals decreased condition (10.58) to:

$$\frac{\partial \varphi_i}{\partial t} + c_i \cdot \Delta\varphi_i = \psi(\varphi) \tag{10.59}$$

It ought to be noted that condition (10.59) is without outer power. Eq. (10.59) is at mesoscale and it is utilized to figure plainly visible amounts, for example; liquid density ρ, liquid velocity $\leftarrow v$, and the energy e, respectively, according to [46]:

$$\rho(a, t) = \int \varepsilon\, \varphi(a, b, t)db, \tag{10.60}$$

$$\rho(a, t)\, v(a, t) = \int \varepsilon\, b\, \varphi(a, b, t)db, \tag{10.61}$$

$$\rho(a, t)\, e(a, t) = 0.5 \int \varepsilon\, b\, v_l^2\, \varphi(a, b, t)\, db, \tag{10.62}$$

where ε is the molecular mass and $v_l = a - v$ is the molecule velocity comparative with liquid velocity [38].

$$\psi(\varphi) = \frac{1}{\sigma}\left(\varphi_i - \varphi_i^{equl.}\right), \tag{10.63}$$

The discretized form (10.59) can be represented as:

$$\varphi_i(a + b_i\ \Delta t, t + \Delta t) - \varphi_i(a,t) = -\frac{\Delta t}{\sigma}\left[\varphi_i(a,t) - \varphi_i^{equl.}(a,t)\right], \tag{10.64}$$

$\varphi_i^{equl.}$ is the distribution function equilibrium state and its precision being issued subordinate for the most part given by [46];

$$\varphi_i^{equl.} = \pi_i S(a,t)\left[1 + \frac{b_i.v}{c_n^2} + \frac{1}{2}\left(\frac{b_i.v}{c_n^2}\right)^2 - \frac{1}{2}\frac{v^2}{c_n^2}\right], \tag{10.65}$$

where

$$b_i = \begin{cases} (0,0) & i = 0 \\ b\left(\cos\frac{(i-1)\pi}{4},\ \sin\frac{(i-1)\pi}{4}\right), & i = 1,3,5,7 \\ \sqrt{2b}\left(\cos\frac{(i-1)\pi}{4},\ \sin\frac{(i-1)\pi}{4}\right), & i = 2,4,6,8 \end{cases}$$

$$\pi_i = \begin{cases} \frac{4}{9} & i = 0 \\ \frac{1}{9} & i = 1,3,5,7 \\ \frac{1}{36} & i = 2,4,6,8 \end{cases}$$

π_i is weight factor fulfilling $\sum \pi_i = 1$, $c_n = \frac{1}{\sqrt{3}}$ is sound speed, $S(a,t)$ is the property to be resolved, and v is plainly visible velocity; all corresponding to the D2Q9 model. A presentation of an outside power term F, which might be a source or sink, identifies the well-known advancement of Eq. (10.66) utilized in LBM perceptible forecasts done by the second order molecule dissemination function [38]. Normally utilized in LBM nanofluid reproduction is the two-appropriation function strategy where one dissemination function computes flow characteristics like velocity and density, with the other conveyance function (Eq. 10.67) settling for temperature (heat move) or focus (mass exchange) circulation as given underneath [46]:

$$\varphi_i(a + b_i\Delta t, t + \Delta t) - \varphi_i(a,t) = -\frac{\Delta t}{\sigma_\varphi}\left[\varphi_i(a,t) - \varphi_i^{equl.}(a,t)\right] + F, \tag{10.66}$$

$$\cup_i(a + b_i\Delta t, t + \Delta t) - \cup_i(a,t) = -\frac{\Delta t}{\sigma_\cup}\left[\cup_i(a,t) - \cup_i^{equl.}(a,t)\right], \tag{10.67}$$

where σ_φ and $\sigma_\cup$ are unwinding boundaries for fluid flow and heat separately. Harmony dissemination capacities $\varphi_i^{equl.}$ and $\cup_i^{equl.}$ are extended through the Chapman–Enskong extension to fulfill naturally visible articulations. $\varphi_i^{equl.}$ and $\cup_i^{equl.}$ along with the moments of their functions distribution are communicated as [47–49]:

$$\varphi_i^{equl.} = \pi_i\rho\left[1 + 3\frac{b_i.v}{c_n^2} + 9\left(\frac{b_i.v}{c_n^2}\right)^2 - \frac{3}{2}\frac{v^2}{c_n^2}\right], \tag{10.68}$$

$$\cup_i^{equl.} = \pi_i T\left[1 + 3\frac{b_i.v}{c_n^2} + 9\left(\frac{b_i.v}{c_n^2}\right)^2 - \frac{3}{2}\frac{v^2}{c_n^2}\right], \tag{10.69}$$

$$\sum_i \varphi_i^{equl.} = \rho,\ \sum_i b_i\,\varphi_i^{equl.} = \rho\leftrightarrow v,\ \sum_i \cup_i^{equl.} = T, \tag{10.70}$$

10.4.2 Dynamic nanoparticle aggregation by lattice Boltzmann method

In the following example, LBM is used to explore the natural convection and nanoparticles migration or suspension in a rectangular enclosure, as shown in Fig. 10.8. The left and right sides of the enclosure have fixed constant temperatures T_H and T_L ($T_H > T_L$), respectively, while the other two sides are adiabatic [50].

The LBM-PBEs algorithm was examined as follows. For the cases of $Ra = 10^3 \sim 10^5$, $\Phi = 0.02$, and $AR = 1$, different grid numbers (16×16, 32×32, 64×64, 128×128, 256×256) are applied to calculate the average Nu, and the results are shown in Fig. 10.9 For grid numbers from 128×128 to 256×256, the average Nu becomes approximately constant. So, the grid number of $256 \times (256/AR)$ satisfies the grid independence and is used for all the modeling in this case.

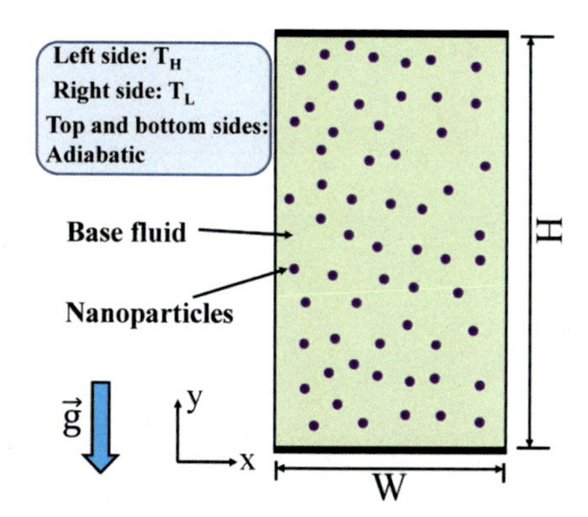

FIGURE 10.8

Geometry and boundary conditions of the investigated problem [50].

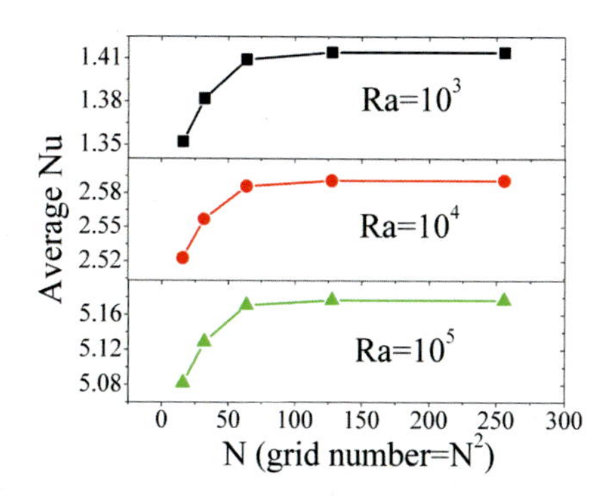

FIGURE 10.9

Average Nu over various grid numbers for $Ra = 10^3 \sim 10^5$, $\Phi = 0.02$, and $AR = 1$ [50].

Table 10.2 Comparison of the average Nu obtained from present model and previous ones when Ra = 10^5, $\Phi = 0.02$, and AR = 1.

Parameters	Present model LBM	Present model LBM-PBEs	Kahveci et al. [51]
Average Nu	5.214	5.177	5.173

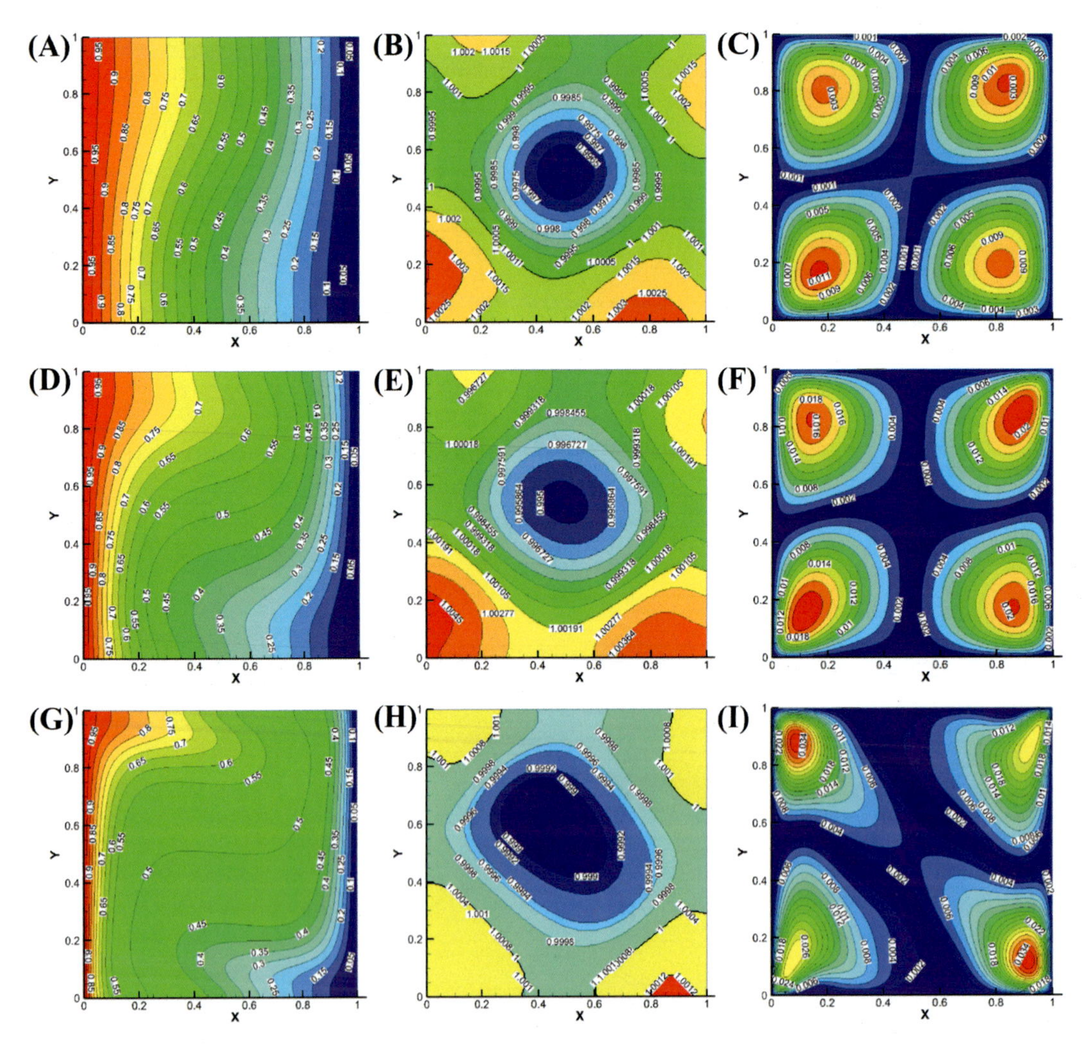

FIGURE 10.10

The temperature distribution (A), (D), and (G), solid volume fractions with average concentration 0.01 (the shown value corresponds to 100 times local concentration) (B), (E), and (H), and shear rates (C), (F), and (I), for Ra = 10^3, 10^4, and 10^5 and AR = 1. All data are with lattice unit [50].

Present results are also compared with previous theoretical studies as shown in Table 10.2. From this Table, it can be found that the average Nu of present models is very similar to that of Kahveci et al.'s [51] outcomes. It indicates that our algorithm is reliable and accurate.

Fig. 10.10 displays the distribution of various physical parameters, including temperature (A), (D), (G), concentration (B), (E), (H), and shear rate (C), (F), (I), of colloid suspension under natural convection in the cavity for $Ra = 10^3$, 10^4, and 10^5 and $AR = 1$. From Fig. 10.10A for $Ra = 10^3$, one can point out that the isotherms of colloid suspension in the square cavity are significantly curved with longitudinal asymmetries. The particle distribution in Fig. 10.10B confirms a concentric circles-like contour around the center of the cavity. By comparing with the streamlines of natural convection in square cavity at $Ra = 10^3$, it can be found that there are vortexes around the center of the cavity. Additionally, the concentration of particles around the center regime of the cavity is significantly less than that of the outside regime. It confirms that the nanoparticles in the colloid suspension are suffering centrifugal force due to the vortex which drives particles toward the outer region of the vortex [50–52].

10.5 Conclusion

In this chapter the heat and mass transfer modeling of nanofluids were investigated and the obtained governing equations were solved by the most efficient techniques such as the Finite Element Method (FEM), Control Volume Method (CVM), and Lattice Boltzmann Method (LBM). Based on the defined problem and boundary conditions, the challenges of each method are mentioned and possible solutions are introduced. It can be concluded that using the combined method such as Control Volume Finite Element Method (CVFEM) will be more suitable for these type of problems.

References

[1] U.S. Choi, J.A. Eastman, Enhancing Thermal Conductivity of Fluids with Nanoparticles, Argonne National Laboratory, Argonne, 1995.

[2] P. Sudarsana Reddy, A.J. Chamkha, Heat and mass transfer characteristics of nanofluid over horizontal circular cylinder, Ain. Shams. Eng. J. 9 (4) (2018) 707–716.

[3] M.H.A. Kamal, A. Ali, S. Shafie, N.A. Rawi, M.R. Ilias, g-Jitter effect on heat and mass transfer of 3D stagnation point nanofluid flow with heat generation, Ain. Shams. Eng. J. 11 (4) (2020) 1275–1294.

[4] S.-S. Ashrafmansouri, M.N. Esfahany, Mass transfer in nanofluids: a review, Int. J. Therm. Sci. 82 (2014) 84–99.

[5] X. Feng, D.W. Johnson, Mass transfer in SiO_2 nanofluids: a case against purported nanoparticle convection effects, Int. J. Heat Mass Transf. 55 (2012) 3447–3453.

[6] J.-Z. Jiang, S. Zhang, X.-L. Fu, L. Liu, B.-M. Sun, Review of gas–liquid mass transfer enhancement by nanoparticles from macro to microscopic, Heat Mass Transf. 55 (2019) 2061–2072.

[7] M. Sahar Goudarzi, A. Shekaramiz, E. Omidvar, A. Golab, Karimipour, A. Karimipour, Nanoparticles migration due to thermophoresis and Brownian motion and its impact on Ag-MgO/Water hybrid nanofluid natural convection, Powder Technol. 375 (20) (2020) 493–503.

[8] B.M. Tehrani, A. Rahbar-Kelishami, Influence of enhanced mass transfer induced by Brownian motion on supported nanoliquids membrane: experimental correlation and numerical modeling, Int. J. Heat Mass Transf. 148 (2020) 119034. Available from: https://doi.org/10.1016/j.ijheatmasstransfer.2019.119034.
[9] I.A. Badruddin, A. Khan, M.Y.I. Idris, N. Nik-Ghaali, N.J. Salman Ahmed, A. Al-Rashed, Simplified finite element algorithm to solve conjugate heat and mass transfer in porous medium, Int. J. Numer. Methods Heat Fluid Flow. 27 (11) (2017) 2481–2507.
[10] B. Vasu, A. Dubey, O.A. Bég, Finite element analysis of non-Newtonian magnetohemodynamic flow conveying nanoparticles through a stenosed coronary artery, Heat Transf. Asian Res. 49 (1) (2020) 33–66.
[11] A. Marjani, A. Taghvaie, N.A.S. Taleghani, S. Shirazian, Mass transfer modeling CO_2 absorption using nanofluids in porous polymeric membranes, J. Mol. Liq. 318 (2020) 114115.
[12] H.T. Kadhim, F.A. Jabbar, A. Rona, Cu-Al_2O_3 hybrid nanofluid natural convection in an inclined enclosure with wavy walls partially layered by porous medium, Int. J. Mech. Sci. 186 (2020) 105889.
[13] F.H. Ali, H.K. Hamzah, K. Egab, M. Arıcı, A. Shahsavar, Non-Newtonian nanofluid natural convection in a U-shaped cavity under magnetic field, Int. J. Mech. Sci. 186 (2020) 105887.
[14] M. Izadi, M.A. Sheremet, S.A.M. Mehryan, Natural convection of a hybrid nanofluid affected by an inclined periodic magnetic field within a porous medium, Chin. J. Phys. 65 (2020) 447–458.
[15] Y.-dong Tang, T. Jin, R.C.C. Flesch, Effect of mass transfer and diffusion of nanofluid on the thermal ablation of malignant cells during magnetic hyperthermia, Appl. Math. Model. 83 (2020) 122–135.
[16] M. Ghalambaz, M. Sabour, S. Sazgar, I. Pop, R. Trâmbiţaş, Insight into the dynamics of ferrohydrodynamic (FHD) and magnetohydrodynamic (MHD) nanofluids inside a hexagonal cavity in the presence of a non-uniform magnetic field, J. Magnetism Magnetic Mater. 497 (2020) 16602.
[17] E.D. Aboud, H.K. Rashid, H.M. Jassim, S.Y. Ahmed, S.O.W. Khafaji, H.K. Hamzah, et al., MHD effect on mixed convection of annulus circular enclosure filled with non-Newtonian nanofluid, Heliyon 6 (2020) e03773.
[18] A.I. Alsabery, T. Armaghani, A.J. Chamkha, I. Hashim, Two-phase nanofluid model and magnetic field effects on mixed convection in a lid-driven cavity containing heated triangular wall, Alex. Eng. J. 59 (1) (2020) 129–148.
[19] A.I. Alsabery, M.A. Ismael, A.J. Chamkha, I. Hashim, Mixed convection of Al_2O_3-water nanofluid in a double lid-driven square cavity with a solid inner insert using Buongiorno's twophase model, Int. J. Heat Mass Transf. 119 (2018) 939–961.
[20] M. Sheikholeslami, ISBN13: 9780128141526 Application of Control Volume Based Finite Element Method (CVFEM) for Nanofluid Flow and Heat Transfer, Elsevier Science Publishing Co Inc, 2018.
[21] P.U. Ogbana, G.F. Naterer, Control volume finite element method for entropy generation minimization in mixed convection of nanofluids, Numer. Heat Trans. Part. B: Fund. 75 (6) (2019) 363–382.
[22] A.S. Dogonchia, Z. Asghar, M. Waqas, CVFEM simulation for Fe_3O_4-H_2O nanofluid in an annulus between two triangular enclosures subjected to magnetic field and thermal radiation, Int. Commun. Heat Mass Transf. 112 (2020) 104449.
[23] T.K. Nguyen, F.A. Soomro, J.A. Ali, R.U. Haq, M. Sheikholeslami, A. Shafee, Heat transfer of ethylene glycol-Fe_3O_4 nanofluid enclosed by curved porous cavity including electric field, Phys. A 550 (2020) 123945.
[24] T.D. Manh, N.D. Nam, K. Jacob, A. Hajizadeh, H. Babazadeh, M. Mahjou, et al., Simulation of heat transfer in 2D porous tank in appearance of magnetic nanofluid, Phys. A 550 (2020) 123937.
[25] A.S. Dogonchi, M. Waqas, S.M. Seyyedi, M. Hashemi-Tilehnoee, D.D. Ganji, A modified Fourier approach for analysis of nanofluid heat generation within a semi-circular enclosure subjected to MFD viscosity, Int. Commun. Heat Mass Transf. 111 (2020) 104430.
[26] T.D. Manh, N.D. Nam, G.K. Abdulrahman, M.H. Khan, I. Tlili, A. Shafee, et al., Investigation of hybrid nanofluid migration within a porous closed domain, Phys. A: Stat. Mech. Appl. 551 (2020) 123960.

[27] T.K. Nguyen, M. Usman, M. Sheikholeslami, R.U. Haq, A. Shafee, A.K. Jilani, et al., Numerical analysis of MHD flow and nanoparticle migration within a permeable space containing non-equilibrium model, Phys. A 537 (2020) 122459.
[28] T. Zahra Abdelmalek, A.S. Tayebi, A.J. Dogonchi, D.D. Chamkha, Ganji, and Iskander. Role of various configurations of a wavy circular heater on convective heat transfer within an enclosure filled with nanofluid, Int. Commun. Heat Mass Transf. 113 (2020) 104525.
[29] D. Seyyed Masoud Seyyedi, On the entropy generation for a porous enclosure subject to a magnetic field: different orientations of cardioid geometry, Int. Commun. Heat Mass Transf. 116 (2020) 104712.
[30] S.A. Shehzad, M. Sheikholeslami, T. Ambreen, A. Shafee, Convective MHD flow of hybrid-nanofluid within an elliptic porous enclosure, Phys. Lett. A 384 (28) (2020) 126727.
[31] B.J. Gireesha, M. Umeshaiah, B.C. Prasannakumara, N.S. Shashikumar, M. Archana, Impact of nonlinear thermal radiation on magnetohydrodynamic three dimensional boundary layer flow of Jeffrey nanofluid over a nonlinearly permeable stretching sheet, Phys. A 549 (2020) 124051.
[32] P.U. Ogban, G.F. Naterera, Control volume finite element method for entropy generation minimization in mixed convection of nanofluids, Numer. Heat Transf. Part. B: Fund. 75 (6) (2019) 363–382.
[33] F. Garoosi, Presenting two new empirical models for calculating the effective dynamic viscosity and thermal conductivity of nanofluids, Powder Technol. 366 (2020) 788–820.
[34] C. Liu, Y. Qiaoa, B. Lv, T. Zhang, Z. Rao, Glycerol based binary solvent: thermal properties study and its application in nanofluids, Int. Commun. Heat Mass Transf. 112 (2020) 104491.
[35] G.F. Naterer, Advanced Heat Transfer, 2nd (ed.), CRC Press, Boca Raton, 2018.
[36] S. Wanga, C. Nana, J. Qiaoa, D. Huanga, N. Nabipour, D. Ross, Free convection and entropy generation in a nanofluid-filled star-ellipse annulus using lattice Boltzmann method supported by immersed boundary method, Int. J. Mech. Sci. 176 (2020) 105526.
[37] W. Cao. Investigation of the applicability of the lattice Boltzmann method to free-surface hydrodynamic problems in marine engineering. Fluids mechanics [physics.class-ph]. École centrale de Nantes (2019). HAL
[38] O. Aliu, H. Sakidin, J. Foroozesh, N. Yahya, Lattice Boltzmann application to nanofluids dynamics—a review, J. Mol. Liq. 300 (2020) 112284.
[39] H. Boyu, L. Shihua, G. Dongyan, C. Wei wei, L. Xinjun, Lattice Boltzmann simulation of double diffusive natural convection of nanofluids in an enclosure with heat conducting partitions and sinusoidal boundary conditions, Int. J. Mech. Sci. 161–162 (2019) 105003.
[40] R. Mohebbi, M. Izadi, H. Sajjadi, A.A. Delouei, M.A. Sheremet, Examining of nanofluid natural convection heat transfer in a Γ -shaped enclosure including a rectangular hot obstacle using the lattice Boltzmann method, Phys. A 526 (2019) 120831.
[41] Abolfazl Fattahi, LBM simulation of thermo-hydrodynamic and irreversibility characteristics of a nanofluid in microchannel heat sink under affecting a magnetic field, Energy Sources, Part A: Recovery, Utilization, and Environmental Effects, Taylor & Francis Group, LLC, 2020.
[42] C. Qi, J. Tang, G. Wang, Natural convection of composite nanofuids based on a two-phase lattice Boltzmann model, J. Therm. Anal. Calorim. 141 (2020) 277–287.
[43] M.H. Abadshapoori, LBM investigation of a Cu-water nanofluid over various configurations of pipes in the mixed convection flow, Heat Transf. (2020) 1–17.
[44] Y. Ma, Z. Yang, Simplified and highly stable thermal Lattice Boltzmann method simulation of hybrid nanofluid thermal convection at high Rayleigh numbers.
[45] A. Mohamad, The Boltzmann equation, Lattice Boltzmann Method, Springer, 2019, pp. 25–39.
[46] A.A. Mohamad, Lattice Boltzmann Method: Fundamentals and Engineering Applications with Computer Codes, Springer Science & Business Media, 2011.
[47] D. Zhang, K. Papadikis, S. Gu, Application of a high density ratio lattice Boltzmann model for the droplet impingement on flat and spherical surfaces, Int. J. Therm. Sci. 84 (2014) 75–85.

[48] M. Ikeda, P. Rao, L.J.C. Schaefer, A thermal multicomponent lattice Boltzmann model, Comput. Fluids 101 (2014) 250–262.
[49] K. Yaji, T. Yamada, M. Yoshino, T. Matsumoto, K. Izui, S. Nishiwaki, Topology optimization using the lattice Boltzmann method incorporating level set boundary expressions, J. Comput. Phys. 274 (2014) 158–181.
[50] S. Dongxing, M. Hatami, J. Zhou, D. Jing, Dynamic nanoparticle aggregation for a flowing colloidal suspension with nonuniform temperature field studied by a coupled LBM and PBE method, Ind. Eng. Chem. Res. 56 (38) (2017) 10886–10899.
[51] K. Kahveci, Buoyancy driven heat transfer of nanofluids in a tilted enclosure, J. Heat Trans. 132 (2010) 062501.
[52] M. Hatami, D. Jing, Nanofluids: mathematical, numerical, and experimental analysis, Elsevier, 2020. ISBN: 0081029330, 9780081029336.

Further reading

J. Forner-Escrig, R. Mondragon, L. Hernandez, R. Palma, Non-inear finite element modelling of light-to-heat energy conversion applied to solar nanofluids, Int. J. Mech. Sci. 188 (15) (2020) 105952.

CHAPTER

CFD simulation of nanofluids flow dynamics including mass transfer 11

Mohammad Hatami[1,2], Jiandong Zhou[2] and Dengwei Jing[2]

[1]*Department of Mechanical Engineering, Esfarayen University of Technology, Esfarayen, Iran* [2]*International Research Center for Renewable Energy, Xi'an Jiaotong University, Xi'an, P.R. China*

11.1 Heat and mass transfer in nanofluids

11.1.1 Thermal conductivity

Nanofluids' thermal conductivity has gained the attention of scientists and engineers, due to its large value in theoretical and practical fields. As Wu et al. [1] and Paul et al. [2] have reported, there are usually the following methods for measuring thermal conductivity: the transient hot-wire method [3–5], steady-state parallel plate [6], and 3-ω method [7]. Among these methods, the transient hot-wire technique is used most widely. For the hot method, it is based on a constant heat generation source and scattering the heat into an unlimited area. Nagasaka and Nagashima [8] introduced a novel and corrected version of the hot-wire cell and electrical system through coating the hot wire with an epoxy adhesive. Their technique had outstanding heat conduction and electrical insulation. Kostic et al. [9] also presented the details of a modified transient hot-wire technique as a fast and exact way for this problems.

According to the literature [10–16], there are four factors have the most influence on the thermal conductivity of nanofluids: temperature [17,18], concentration of solid particles [19,20], size of nanoparticles [21,22], and the shape of nanoparticles [23]. Several works focus on how these parameters affect the nanofluids' thermal conductivity through experimental methods. Besides the experimental investigation, various theoretical models have been proposed to guess the nanofluids thermal conductivity [10,24–27].

Many types of nanoparticles are available for use in heat and mass transfer applications of nanofluids [28–31]. Based on the literature, metal oxide nanoparticles can be introduced as the most applicable cases, which are added to different base fluids [32–37]. Al_2O_3, CuO, SiO_2, and TiO_2 are examples of these types of nanoparticles which more frequently are used with base liquids, such as water and ethylene glycol.

11.1.2 Nanoparticles concentration

To find the mass transfer of nanofluids, the nanoparticle concentration must be considered. Many scientists have examined the thermal conductivity enhancement by the volumetric loading of nanoparticles in suspension. Most of the researchers reported that by increasing the nanoparticle

Nanofluids and Mass Transfer. DOI: https://doi.org/10.1016/B978-0-12-823996-4.00011-2

concentration, the thermal conductivity is also increased. The increase of thermal conductivity in some reports was nonlinear with respect to particle concentration [38–41], as shown in Fig. 11.1. Hong et al. [38] examined the effect of Cu and Fe nanoparticles on thermal conductivity, the results showed that thermal conductivity of a Fe nanofluid was improved nonlinearly up to 18% when the nanoparticles volume fraction was amplified up to 0.55 vol.%. Also, Fe nanofluids showed a more effective thermal transport property than Cu nanofluids. Wen et al. [39] considered the carbon nanotube (CNT) concentration effect on the effective thermal conductivity by experimental analysis. They observed that the effective thermal conductivity is an increased nonlinear function of CNT concentration at very low concentrations. The ethylene glycol-based ZnO nanofluids also have been studied by Li et al. [40]. They reported that ZnO thermal conductivity depends strongly

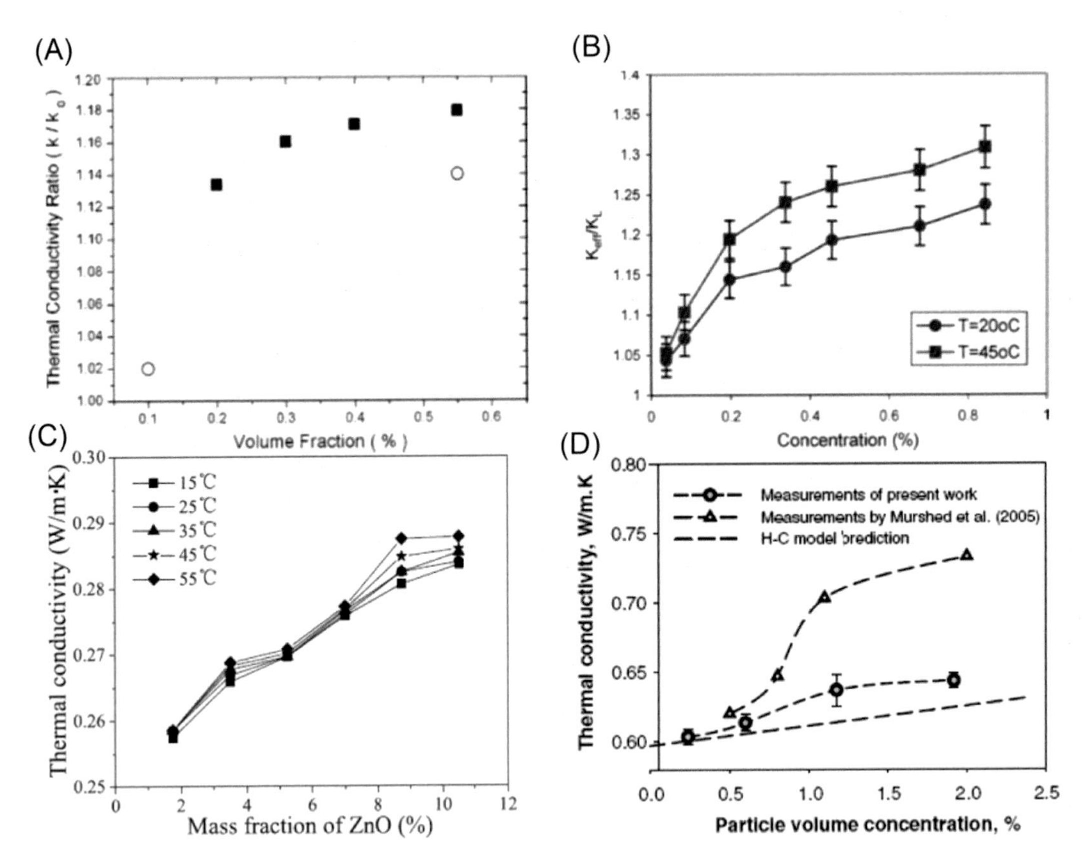

FIGURE 11.1

Variation of thermal conductivity of nanofluids with particle concentration. (A) Thermal conductivity of Fe nanofluids for different Fe nanoparticles concentration [38]. (B) Effect of CNT concentration on effective thermal conductivity of nanofluids at two temperature [39]. (C) Thermal conductivity distribution of ZnO-EG as a function of mass fraction. (D) Effect of particle concentration on the thermal conductivity of TiO_2 nanofluids. *CNT*, Carbon nanotube.

on particle concentration and increases nonlinearly with the concentrations of 1.75–10.5 wt.%. Another study also studied the thermal conductivity of aqueous suspensions of TiO_2 nanoparticles [41] and concluded that by increasing the nanoparticle concentrations, the measured effective thermal conductivity of nanofluids increases nonlinearly.

However, others found a linear relationship between the thermal conductivity of nanofluids and the particle concentrations [21,42–44]. Kim et al. [42] considered the thermal conductivity of several types of nanoparticles, that is ZnO, TiO_2, and Al_2O_3. They reported that the thermal conductivity of the nanofluids is linearly proportional to the particle concentration in the range of the measurement. Zhang et al. [43] have measured the effective thermal conductivity of five types of nanofluids, which are Au/toluene, CuO/water, TiO_2/water, Al_2O_3/water, and CNT/water nanofluids, by using the short-hot-wire method. The results showed that the effective thermal conductivity increases with increasing particle concentration, and the dependence is linear, and other works also showed the linear relationship function [45–59] (Fig. 11.2).

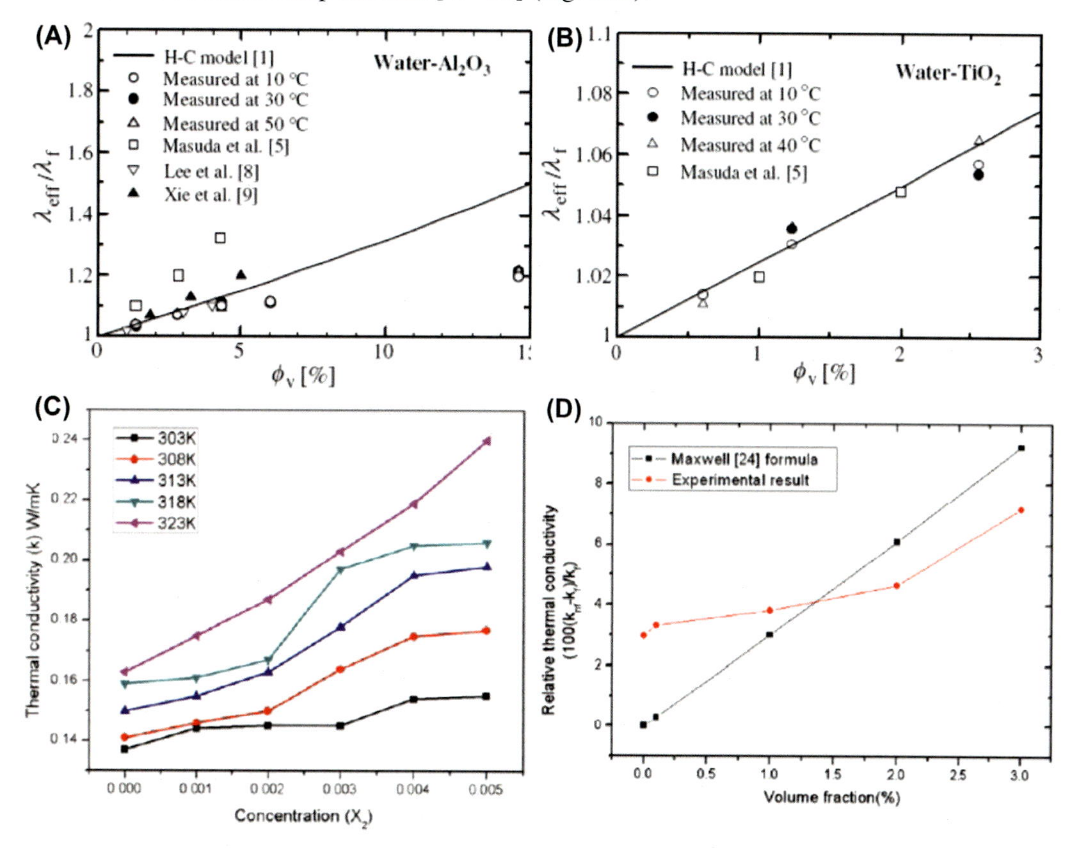

FIGURE 11.2

(A) Thermal conductivity of Al_2O_3/water nanofluids [43]. (B) Thermal conductivity of TiO_2/water nanofluids [43]. (C) Thermal conductivity versus concentration of MWCNT in the nanofluids [45]. (D) Thermal conductivity–volume fraction curve of SiC/DIW nanofluids [52]. *MWCNT,* Multiwalled carbon nanotube.

11.1.3 Nanoparticle size

One of the important parameters which has a significant effect on thermal conductivity enhancement is particle size. Actually, particle sizes help researchers to find difference nanofluids and micron-suspensions. The size of nanoparticles not only influences the suspension stability, but also affects the thermal conductivity of nanofluids. Many researchers tried to study the effect on enhancing the thermal conductivity of nanofluids during recent decades. Lee et al. [37] measured the thermal conductivities of oxide nanofluids by a transient hot-wire technique. They found that the effective thermal conductivity of nanofluids will be improved by decreasing the particle size. And then they compared this result with the reported data presented by Masuda et al. [60]. Wang et al. [61] also have compared the data of Masuda and Lee's with their own work, and the same results were obtained. Xuan and Li [62] found that larger particle size of Cu will lead to a decrease in the enhancement from 44% to 12%. So they concluded that to achieve the same thermal conductivity improvement for 100 nm particles compared to 35 nm ones, an increment of the nanoparticle concentration from 0.052 to 7.5 vol.% is required. Also, the larger particle size gives a lower thermal conductivity of nanofluids in the same context. Many researchers in the literature confirmed this reverse function between particle size and thermal conductivity enhancement [21,41,63–71] (Fig. 11.3).

However, this relation is not unalterable. The results of Refs. [72,73] showed that big cylindrical shaped multiwalled CNT (MWCNT) nanoparticles have a better enhanced performance with regard to thermal conductivity than small-sized spherical shaped ones when added to the same fluid. Moreover, in order to enhance the heat transfer, Pak and Cho [74] suggested choosing bigger particles, based on their results. Hwang et al. [73] and Beck et al. [75] also found similar conclusions. Hwang et al. [73] explored the thermal conductivity of six different kinds of nanofluids. They reported that CuO (33 nm) had better enhanced thermal conductivity than SiO_2 (12 nm) at the same concentration and base fluid. Beck et al. [75] considered seven types of nanofluids with different Al_2O_3 nanoparticles diameter from 8 to 282 nm and calculated the thermal conductivity enhancement in ethylene glycol or water as the base fluid. Their results showed that the thermal conductivity improvements in tested nanofluids reduces when the nanoparticle sizes reduces to under 50 nm. They mentioned that this reduction in the thermal conductivity of nanofluid is due to improved phonon scattering influence (Fig. 11.4).

11.1.4 Nanoparticle shape

The nanoparticles' shape also effects the nanofluids thermal conductivity. Xie et al. [72] considered the effects of the morphologies (size and shape) of the added solid nanoparticles on the enhancement of the thermal conductivity of the nanoparticle suspension for the first time. Their results showed that the larger cylindrical nanoparticles had better enhancement in thermal conductivity than smaller spherical nanoparticles when added to the same base fluid, which confirms the shape effect on thermal treatments. Similar results were also obtained by other works [76–80]. Moreover, if the nanoparticle has a large length-to-diameter ratio, the nanofluids would have high thermal conductivity enhancement [43,81]. In the same context, Choi et al. [82] measured the effective thermal conductivity and reported that MWCNTs showed an unnormal and nonlinear increase in thermal conductivity, while it was predicted to be linear. Also, they reported lower improvements for the fullerenes—carbon molecules in the shapes of a hollow sphere, tube, ellipsoid, etc. [83]. Putnam

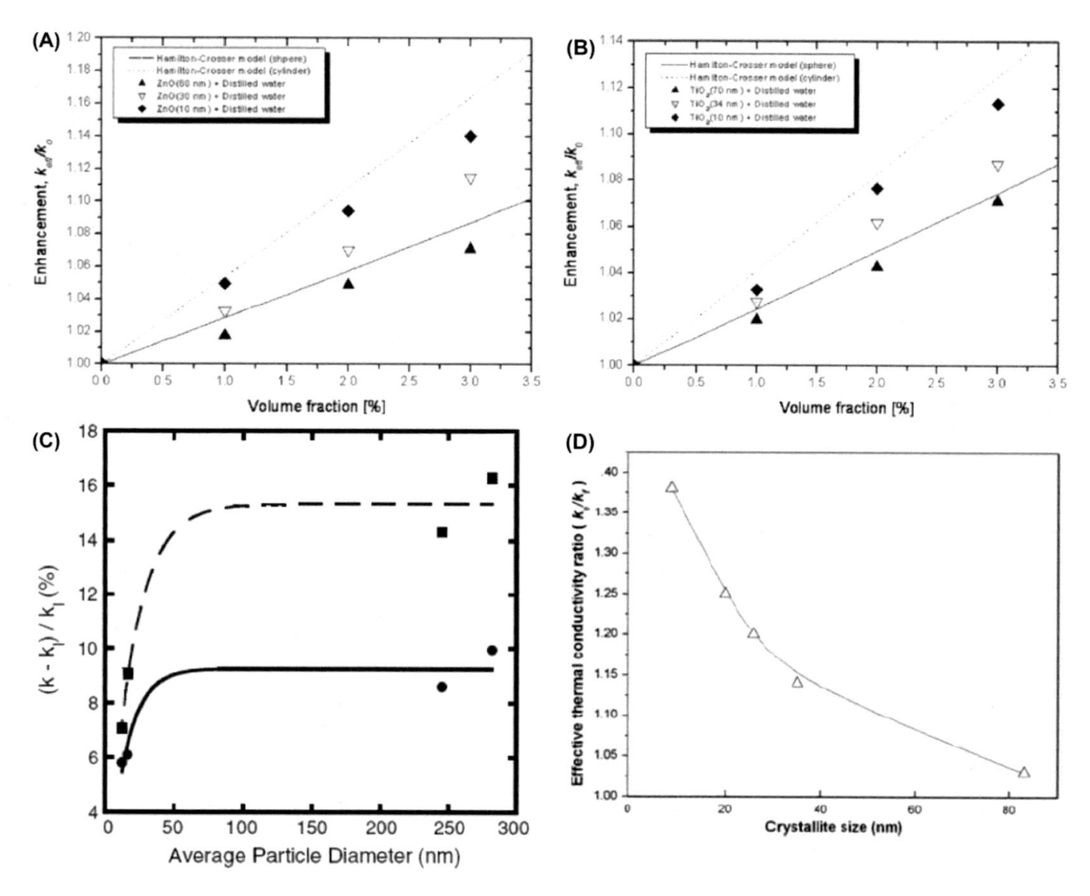

FIGURE 11.3

Thermal conductivity ratio: (A) ZnO/water and (B) TiO_2/water [63]. (C) Thermal conductivity enhancement for 2% (v/v) and 3% (v/v) alumina particles dispersed in ethylene glycol at 298K [65]. (D) Effect of crystallite size thermal conductivity ratio of ethylene glycol-based ($Al_{70}Cu_{30}$, 0.5 vol.%) nanofluid [68].

et al. [84] compared the C60–C70 fullerenes dispersed in toluene to the Au dispersed nanoparticles to find the thermal conductivity improvement effect. Their results concluded that fullerenes had lower improvements than Au at concentrations $<<1$ vol.%. Although, for higher volumetric loading, it showed the opposite result. Furthermore, Hwang et al. [72] by comparing the fullerenes with MWCNT in mineral nanooil observed lower enhancement of conductivity even at higher concentrations.

11.1.5 Nanoparticle thermal conductivity and base fluid

The nanoparticles which have higher thermal conductivity were recommended for use for enhancing heat transfer performance by Pak and Cho [74]. This finding was confirmed later by other

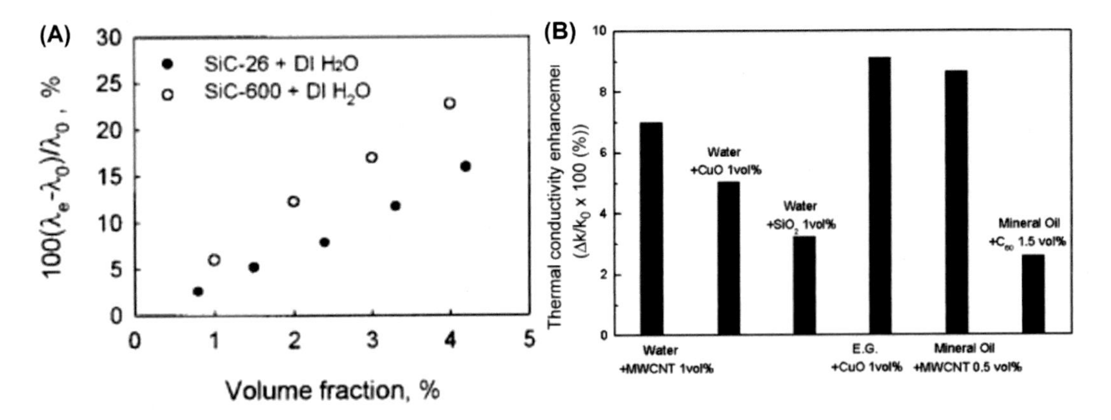

FIGURE 11.4

(A) Thermal conductivity ratios of suspensions containing different solid particles [72]. (B). CuO (33 nm) enhanced thermal conductivity higher than SiO_2 (12 nm) [73].

results [36,37,43,73,85]. In a similar condition, after comparing studies [86] and [87], the results showed that for a constant base fluid, MgO has the higher thermal conductivity than Al_2O_3, despite the larger MgO particle size (40 nm) compared to (13 nm) for Al_2O_3 nanoparticles. This outcome shows that particle thermal conductivity has more influence than size of nanoparticles. Eastman et al. [88] found that nanofluids including metallic nanoparticles can reach a large increment compared to either base fluid or oxide nanoparticles additives. In addition, compared to conventional heat transfer fluids (HTFs), ceramics encapsulating phase change nanoparticles additions have a better thermal conductivity enhancement [59]. Furthermore, they reported that those particles improved the heat transfer and thermal storage properties of HTFs. However, Yoo et al. [65] got the opposite conclusion; they reported that the thermal conductivity of suspended nanoparticles is not the main factor affecting the thermal conductivity of nanofluids.

For the same nanoparticle, the different base fluid also led to different thermal conductivity of nanofluids, regardless of nanoparticles shape [63]. This conclusion was also obtained in investigations [37,72,73,79,80], and observed greater thermal conductivity enhancement for ethylene glycol (EG) nanofluids compared to water-based nanofluids. However, the contrast conclusion was also reported by [76,89]. They found that at the same volume fraction, using EG as the base fluid led to achieving higher enhancements compared to the enhancement in the case of using EO. Chopkar et al. [69] also found that water-based nanofluids had lower thermal conductivity improvements compared to EG-based nanofluids at constant nanoparticles volume fraction. With the development of nanofluids, a new research style appeared. It uses mixed base fluids instead of a conventional single base fluid. Beck et al. [71] studied the thermal conductivity of nanofluids with a mixed base fluid of EG–water. The results indicated that for the novel base fluid, thermal conductivity indicated better improvements than scattering in traditional base fluids. As for the mixed base fluid, the mixing ratio is an important factor to affect the thermal conductivity of nanofluids. Abdolbaqi et al. [90] confirmed this fact and found that the thermal conductivity improvements of Al_2O_3 nanofluids with 40:60 wt.% BG–water mixing ratio was 24% at 2 vol.% at 80°C temperature. But this improvement decreased to 13% when changing the base fluid mixing ratio to 60:40 wt.% BG–water.

Many researchers have focused on finding the temperature effect on the thermal conductivity improvements. For instance, Patel et al. [91] reported that the nanofluids thermal conductivity improvements were directly related to temperature. After that, many researches also reported the similar conclusion [66,68,69,92–95]. Even the relationship between the thermal conductivity and the temperature is nonlinear [57,72]. However, high temperature will also lead to destabilization of the nanofluids.

11.2 CFD modeling

Although a lot of experimental work has been done on the heat and mass transfer of nanofluids, most researchers only compare the computational fluid dynamic (CFD) modeling cases with former numerical works or experimental cases. This may be an appropriate CFD method to verify the prediction of thermal and hydraulic characteristics of nanofluid flow. In this section, we discuss different CFD methods and try to report some numerical studies about nanofluid heat and mass transfer. Also we try to compare numerical outcomes with experimental data to see which method can give more accurate outcomes.

11.2.1 Single-phase approach

Actually the mixing of nanoparticles and base solution produces a multiphase fluid, but due to the presence of ultrafine nanoparticles, nanoparticles can certainly flow in the fluid, so a nanofluid can nearly be regarded as a traditional single-phase fluid (homogeneous). Furthermore, in the single-phase model, both the fluid and solid (particle) phases are considered to be in thermal equilibrium and transfer with a unit velocity [76]. For this single-phase approach with steady-state assumption, the governing equations are as follows [96]:

Mass conservation:

$$\nabla\cdot\left(\rho_{nf}\vec{V}\right) = 0 \tag{11.1}$$

Conservation of momentum:

$$\nabla\cdot\left(\rho_{nf}\vec{V}\vec{V}\right) = -\nabla P + \mu_{nf}\nabla^2\vec{V} - \rho_{nf}(\overline{vv})(\rho\beta)_{nf}(T - T_0)g \tag{11.2}$$

On the right-hand side of Eq. (11.2) the third and fourth terms refer to the turbulent flow and natural convection effects, respectively. It is clear that for a laminar and forced convection flow, described terms can be obtained as zero. Conservation of energy:

$$\nabla\cdot\left(\left(\rho C_p\right)_{nf}\vec{V}T\right) = \nabla\cdot\left(k_{nf}\nabla T - \left(\rho C_p\right)_{nf}\overline{vt}\right) \tag{11.3}$$

where $\left(\rho C_p\right)_{nf}\overline{vt}$ refers to turbulent flow regime effect in the energy conservation equation.

As far as the authors know and based on the literature, just limited papers have been published on the numerical mass transfer of nanofluids. Most of the papers include heat transfer and through the velocity and momentum analysis, the mass transfer is also discussed. But in a few papers the equation of mass transfer or concentration function is solved separately; this is presented in this chapter shortly.

A numerical validation of experimental data on laminar forced convection in a double tube and shell and tube heat exchangers filled by nanofluids is offered by Akhtari et al. [97]. They used the commercial ANSYS-FLUENT 12.1software based on the finite volume method (FVM) to find the effect of cold and hot flow rate, flow temperature, and nanoparticle concentration on the heat and mass transfer analysis. The unstructured quadrilateral element is used to discretize the computational domain for the two-dimensional arrangement of the double tube heat exchanger, and the quadrilateral mixed element is used for the three-dimensional configuration of the shell and tube heat exchanger. Results showed that the average comparative errors between the predicted results and the experimental data are 11.2% and 15.6%, correspondingly. It is also found that the total coefficient of heat transfer for shell tube and double tube heat exchanger rises with the increase of cold and hot volume flow rate, particle concentration, and inlet temperature of nanofluid.

Izadi et al. [98] investigated the 2D laminar forced convection of Al_2O_3—water nanofluids in the annulus. FVM was applied to discretize the single-phase governing equation and the first-order upwind method was used for convection and diffusion terms, also semiimplicit method was applied for velocity pressure coupling [99]. The nonuniform grid near the nozzle and wall is fine enough to solve the problem of velocity and temperature gradient. The predicted friction coefficient was in acceptable adjustment with the experimental outcomes. The results show that the nanoparticles volume fraction had a great influence on the temperature distribution, but had a small effect on the dimensionless axial velocity distribution.

Moraveji et al. [100] studied the convective heat transfer effect of nanofluid flow in the developing region of a constant heat flux tube using CFD techniques. The experimental results were obtained when the Reynolds number was 500—2500, the concentration of nanoparticles was 0%—6%, and the diameter of nanoparticles was 45 and 150 nm. The maximum error between the experimental outcomes [101] and the predicted heat transfer coefficient was 10%. The heat transfer coefficient improves with the nanoparticle concentration and Reynolds number rises, and decreases with the increase of axial position and diameter of the nanoparticles.

Hussein et al. [102] explored the effects of different cross sections of flat, elliptical, and circular tubes on the turbulent forced convection nanofluids flow, numerically. In their study, the rectangular element was used to grid the wall surface, while the triangular element was used for the inner space grids. Also, the standard k-ε model [103] was adopted. The computational results of the CFD model are compared with the experimental data in references [74,104]. When the volume fraction of TiO_2 was 4%, the compatibility coefficient of TiO_2/water nanofluids was 4% and 6%, correspondingly. Additionally, the friction coefficient for the circular tube was greater than that of other types. The effects of different types of nanofluids on fluid flow and heat transfer features of triangular MCH were investigated by Muhammad [105]. The FVM is used to solve the 3D, laminar, steady, and heat transfer governing equations, where the hexagonal element was used for numerical simulation. Compared with the experimental outcomes of Chein and Chuang [106], the numerical results of temperature difference (ΔT) between the outlet and inlet of MCHS and the velocity and pressure drop (ΔP) of MCHS are very consistent. Diamond—water and silver—water nanofluids were more suitable for heat transfer for the water-based nanofluid with different types of nanoparticles, such as Al_2O_3, Ag, CuO, diamond, SiO_2, and TiO_2.

Rashmi et al. [107] described a complete study on heat transfer improvements using CNTs as nanoparticles in base water and stabilized by gum Arabic. Experiments and simulations were approved for a laminar countercurrent heat exchanger. The single-phase numerical simulation of two-dimensional heat exchanger is carried out by using fluent V6.3 (Fig. 11.5). Similarly, Meng and Li [108] stated the natural convection of aqueous alumina nanofluids. They considered a 3D horizontal cylinder and developed a completely open source 3D CFD program. Taking into account the original open bubble solver "buoyancy bubble solver" and the newly developed temperature-dependent solver "buoyancy bubble solver," compared with the corresponding experimental results, the early solver is more suitable for 4% Al_2O_3/water, and the latter is more suitable for 2% Al_2O_3/water nanofluids. Rakhsha et al. [109] also applied the new open bubble solver to study the steady turbulent forced convection development of CuO–water nanofluids in the side helical tubes when the wall temperature was constant. Numerical outcomes show that the convective heat transfer and pressure drop increase by 6%–7% and 9%–10%, correspondingly, when CuO/water nanofluid is applied to replace pure water. The experimental results show that the heat transfer coefficient and pressure drop increase by 16%–17% and 14%–16%, correspondingly for some tube shapes and Re numbers. So, the consistency between the numerical and experimental outcomes of nanofluids is recognized due to the existence of centrifugal force, which may lead to a higher particle concentration near the outer wall, thus increasing the thermal conductivity and viscosity of the space, and affecting the accuracy of the numerical "uniform model," which assumes that the thermophysical properties of the whole region are constant. Kumar et al. [110], in a CFD analysis by FLUENT

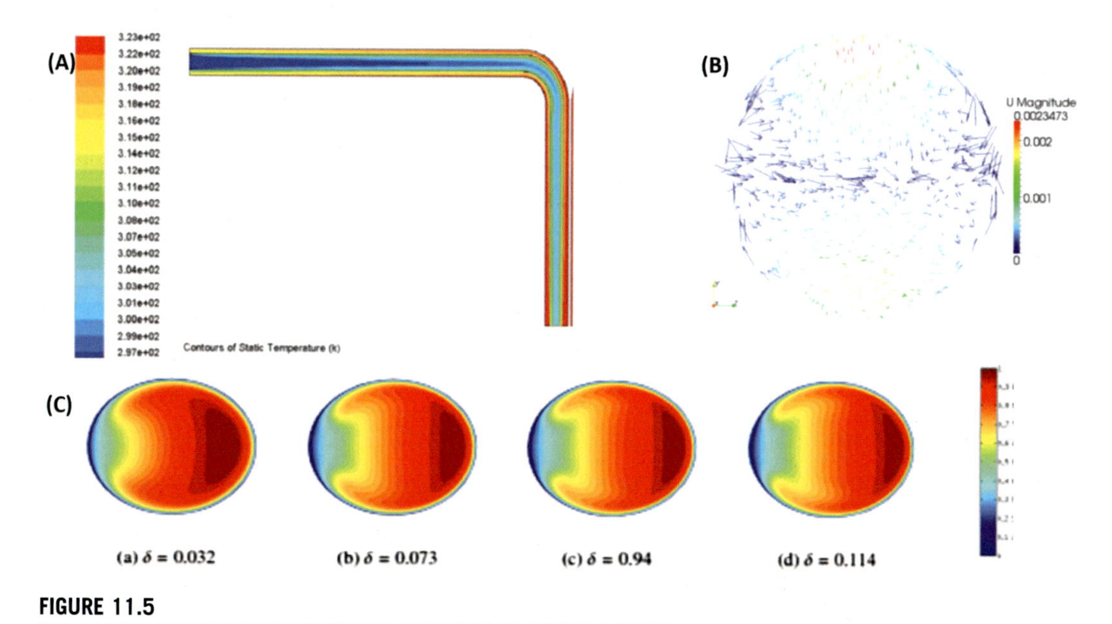

FIGURE 11.5

(A) Temperature contour of 0.5 vol.% nanofluid [96]. (B) Numerical study of convection in a horizontal cylinder filled with water-based alumina nanofluid [107]. (C) Effect of the curvature ratio on fully developed nondimensional temperature profile at $Re = 13.000$.

software for a spiral tube heat exchanger, reported that the average comparative errors of pressure drop and Nu results were 9.5% and 8.5%, respectively.

Many scientists have demonstrated the effectiveness of the numerical calculation by comparing the results with the traditional correlations for pressure drop and heat transfer of nanofluids. These correlations are improved based on the experimental data of soft theory and are suitable for finite geometries such as smooth tubes and closed bodies.

The forced convection of nanofluids with water as base fluid and Al_2O_3 as nanoparticles in a two-dimensional ribbed channel was numerically studied by Manca et al. [111]. In the range of Reynolds number (20,000–60,000), the flow state is turbulent with volume fraction (0%–4%) and particle size of 38 nm. Supposing a single-phase model, Menter's Shear Stress Transport or SST k-ω model [112] is used to solve the continuous, momentum, and energy turbulent and heat transfer governing equations by using the FLUENT program. In the process of mesh generation, the structural grid is considered, and the $Y^+ = 1$ grid is used for adjacent walls. The simulation results of Pak and Cho [74] and Maiga et al. [113] have a good correlation with the given Nusselt number for smooth channels. The results show that the average Nusselt number increases with the increase of particle volume concentration and Reynolds number, but the pump power needs to be increased.

Rostamani studied the hydrodynamics and thermal behavior of suspensions of Al_2O_3, CuO, and TiO_2 nanoparticles with different concentrations (0%–6%) in a 2D pipe under constant heat flux boundary conditions. All thermophysical properties of nanofluids are considered to be temperature dependent. The time averaged incompressible Navier–Stokes equations and energy equations are discretized by the control volume method. The Reynolds stress is modeled by K-ε turbulence model. Their predictions for the Nusselt numbers of CuO 6% and TiO_2 3% are very close to those of Pak and Cho [74], Maiga et al. [113], and Gnielinski [114].

It can be concluded from the literature review that the traditional single-phase model has been applied to study the convection heat transfer of several nanofluids. Actually this model is widely used due to its ease and high efficiency in computational results. But, in this method choosing the best and suitable correlation for the thermal properties is the main challenge to have the closest results to experimental outcomes. It cannot be understandable from the literature to find these thermal correlations to be suitable, so in this method researchers must focus on selecting appropriate effective properties and considering the chaotic motion of ultrafine particles. Due to these facts, the numerical calculations of the model may not be consistent with the experimental results completely.

11.2.2 Two-phase approach

Nanofluids are essentially two-phase fluids, so it can be expected that they may have some characteristics of solid–liquid mixtures. Therefore the classical two-phase flow theory is applied to nanofluids. For the two-phase method, two separate phases for fluid and nanoparticles must be considered which have different temperatures and velocities. Also, in this approach the theory of zero slip velocity between fluid and particles is no more usable [63]. The reason of this assumption is related to gravity, Brownian motion and diffusion, fluid and solid particles friction, sedimentation and dispersion. Since the motion between solid and fluid molecules is considered in the two-phase flow method, practical results can be obtained. Although the model can well understand the role of liquid phase and solid phase in the heat transfer procedure, it takes a long time to calculate and needs a high-precision computer. The two-stage method can be divided into two types: Eulerian–Eulerian and Lagrange–Eulerian models.

11.2.2.1 Eulerian–Eulerian model

The Eulerian–Eulerian model has been recommended for two-phase fluids containing a large number of nanoparticles as solid phases. Therefore for modeling purposes, it is clear that this is an appropriate model for nanofluids containing a large number of nanoparticles, even at low volume fraction. Three main Eulerian–Eulerian models are available in the literature, namely volume of fluid (VOF), mixture, and Eulerian model.

11.2.2.2 Mixture model

The mixture model is a model that we often use based on the mixture theory. The popularity of it is due to the following advantages:

- The theory is simple and the application is easy.
- Shorten the CPU use time and consequently less run time.
- Using a straightforward scheme for introducing the turbulence model to mixture one.
- For a wide range of two-phase flow, it has high accuracy.

This model has the following assumptions:

- The interaction between the two phases is strong, and the flow characteristics of the particles are good.
- All phases share one pressure.
- Both separate phases have related and special velocity vectors and in each region of the mixture a certain fraction (concentration) of phases is available.
- Spherical particles with uniform diameter sizes are considered as the secondary dispersed phases during the computational studies.
- The concentrations of nanoparticles of the secondary dispersed phases will be solved in the governing equations.

Conservation of momentum:

$$\nabla\cdot\left(\rho_m \vec{V}_m\right) = 0 \tag{11.4}$$

Conservation of momentum:

$$\nabla\cdot\left(\rho_m \vec{V}_m \vec{V}_m\right) = -\nabla P_m + \mu_m \nabla^2 \vec{V} - \nabla\cdot\left(\sum_{k=1}^{n} \varphi_k \rho_k \overline{V_k V_k}\right) + \nabla\cdot\left(\sum_{k=1}^{n} \varphi_k \rho_k \vec{V}_{dr,\&\#x000A0;k} \vec{V}_{dr,\&\#x000A0;k}\right) - \rho_m \beta_m (T - T_0) g \tag{11.5}$$

where $\vec{V}_{dr,k}$ refers to the drift velocity for the kth phase.

Volume fraction:

$$\nabla\cdot\left(\varphi_p \rho_p \vec{V}_m\right) = -\nabla\cdot\left(\varphi_p \rho_p \vec{V}_{dr,\&\#x000A0;p}\right) \tag{11.6}$$

Conservation of energy:

$$\nabla\cdot\left(\sum_{k=1}^{n}\varphi_k \vec{V}_k\left(\rho_k H_k + P\right)\right) = \nabla\cdot\left(k\nabla T - C_{P\rho_m}\overline{vt}\right) \tag{11.7}$$

$$\vec{V}_{dr,\&\#x000A0;k} = \vec{V}_k - \vec{V}_m \tag{11.8}$$

The slip velocity (relative velocity) can be introduced as the velocity of secondary phase (p) relative to the velocity of the primary phase (f):

$$\vec{V}_{pf} = \vec{V}_p - \vec{V}_f \tag{11.9}$$

The drift velocity is linked to the relative velocity as:

$$\vec{V}_{dr,\&\#x000A0;p} = \vec{V}_{pf} - \sum_{k=1}^{n}\frac{\varphi_k \rho_k}{\rho_m}\vec{V}_{fk} \tag{11.10}$$

In this model, we need to know that the mixed momentum conservation differential equation only needs to solve one starting velocity component. Additionally, the velocity of the dispersed phase is achieved from the algebraic equilibrium equations. This results in a significant reduction in computational effort. In addition, it should be noted that the main phase affects the second phase through resistance and turbulence, and the second phase affects the main phase through the reduction of average momentum and turbulence. To apply the mixing theory to the modeling of nanofluids, it does not depend on the governing equations of each phase, but solves the continuity, momentum, and fluid energy equations of the whole mixture.

It can be seen from the literature that Behzadmehr et al. [115] was the first researcher to study the turbulent heat transfer in a circular tube containing 1% copper/water nanofluid by using the mixture method. They examined their work with the experimental outcomes of Xuan and Li [116], and found that the mixture model was more accurate than the single-phase model. Mirasoumi and Behzadmehr [117] also used the same theory to analyze the laminar mixed convection of dilute phase Al_2O_3/water nanofluids in a horizontal tube under constant heat flux boundary conditions. The results show that with the increase of the volume fraction of nanoparticles, the secondary flow is enhanced, and the Nusselt number increases significantly, while the surface friction coefficient in the fully developed region changes little. With the help of this two-phase flow method, it is also found that due to the significant influence of viscous force, the concentration of particles near the wall is higher. In the previous two studies the discrete mesh is uniform in the circumferential direction and inhomogeneous in the other two directions. Saberiel [118] studied the laminar forced convection heat transfer of Al_2O_3/zirconia–water nanofluids in vertical heating tubes using single-phase and two-phase mixing models. Compared with the experimental data, the results show that the mixture model has a good predictive ability on the convective heat transfer coefficient of alumina/water and zirconia/water nanofluids, with an average relative error of 8% and 5%, respectively, while that of the single-phase model is 13% and 8%.

Moghadassi et al. [119] applied both single- and two-phase models for a nanofluid flow in a horizontal circular tube and found that the two-phase model with 4.76% deviation from experimental results was more accurate than the single-phase with 8.48% deviation. Also, Naphon and Nakharintr [120] made another comparison between models in a mini channel heat sink and found 3.74% and 1.66% deviation from experimental data for the single-phase and two-phase models, respectively. So, most of the recent researchers such as Corcione et al. [121] used the two-phase model for their modeling due to its accuracy and lower deviations from experimental results.

11.2.2.3 Eulerian model

Assumptions:

- The pressure must be jointed for both phases.
- Nanoparticles phase assumed to be a continuum of the fluid phase.
- For both primary and secondary phases, separate continuity, momentum, and energy equations were used.
- To calculate the volume of solid and fluid phases, an integrating of the volume fractions throughout the domain is required. The sum of all volume fractions must be unity.
- The differences of velocity and temperature values between the liquid and solid phases are explained numerically using the FVM.

Conservation of mass:

$$\nabla\cdot\left(\varphi_q\rho_q\vec{V}_q\right)=0 \tag{11.11}$$

where $\vec{V}_q=\int_V \varphi_q dV$, $\sum_{k=1}^{n}\varphi_q=1$, and q shows the phase.

Conservation of momentum (qth phase):

$$\nabla\cdot\left(\rho_q p_q\vec{V}_q\vec{V}_q\right)=-\varphi_q\nabla P+\varphi_q\mu_q\nabla^{2}\vec{V}+\varphi_q\nabla\cdot\left(\sum_{k=1}^{n}\varphi_k\rho_k\overline{V_k V_k}\right)+\varphi_q\rho_q\vec{g}+\sum_{k=1}^{n}\vec{R}_{pq}+\left(\vec{F}_{\text{collision}}+\vec{F}_{lift,\&\#x000A0;q}+\vec{F}_{vm,\&\#x000A0;q}\right) \tag{11.12}$$

Conservation of energy:

$$\nabla\cdot\left(\varphi_q\rho_q\vec{V}_q H_q\right)=-\nabla\cdot\left(k_q\nabla T_q\right)-\tau_q\text{UNDEFINEDratio;}\,\nabla\vec{V}_q+\sum_{p=1}^{n}\vec{Q}_{pq} \tag{11.13}$$

where $\vec{Q}_{pq}=h\vec{}(V_p-\vec{V}_q)$ is the heat exchange coefficient.

The laminar forced convection heat transfer of Cu–water nanofluids in an isothermally heated microchannel was numerically considered by Kalteh et al. [122]. The Eulerian two-fluid model was used to model the flow of nanofluids, and the governing equations of the two phases were solved by the finite element method (FEM). Like the first study, the relative temperature and velocity of liquid phase as well as nanoparticle phase were studied in detail. They observed that the relative temperature and velocity between the phases were very slight and insignificant, and the distribution of nanoparticles concentration was uniform. That's why they came to the conclusion that the assumption of nanofluids as homogeneous solutions is reasonable. In addition, when the Reynolds number and the volume concentration of nanoparticles are high and the particle size is small, the degree of heat transfer enhancement is greater, and the pressure drop increases slightly. Their results showed that the two-phase results were in better adjustment with the experimental outcomes, and the maximum error was 7.42%, while that of the homogeneous (single-phase) model is 12.61% [123].

Fani et al. [124] carried out CFD analysis to study the effect of nanoparticle size on the thermal properties and pressure drop of CuO–water nanofluids in trapezoidal MCH. The laminar, incompressible, and steady flows are numerically simulated by using the Eulerian two-phase flow numerical method. The FEM was used to solve the 3D flow and heat transfer equations. Their numerical studies show that the two-phase model produces larger heat transfer than the uniform model, increasing by about 15% in the entrance region of MCHS. In addition, with the increase of the size of nanoparticles, the pressure drop increases and the heat transfer decreases. The results show that the influence and contribution of the base solution on the thermal properties is greater than that of the nanoparticles. Kalteh and coworkers [122] extended the previous study to investigate the effects of different nanofluid types on heat transfer and pressure drop. The results show that the particle size is 100 nm. Compared with glycol and oil-based nanofluids, water-based nanofluids have the lowest pressure drop and the highest heat transfer coefficient. In addition, the dimensionless pressure drop of all nanoparticles is almost the same, the heat transfer coefficient of diamond/water is the highest, and that of SiO_2/water nanofluid is the lowest.

The single-phase model was compared with the two-phase model (mixed model and Eulerian model) to understand which model predicted the experimental results more correctly [125]. They found the turbulent convection heat transfer of Al_2O_3−water and TiO_2−water nanofluids in annular channels, and reported that the two-phase model was closer to the experimental results than the single-phase model. The expected value of the mixed model is slightly higher than that of the Eulerian model. In general, compared with the homogeneous model, the most important advantage of the Eulerian model is that nanofluids do not need an effective thermophysical model [122]. However, as declared in Refs. [125,126], it may not be as accurate as the mixture model.

11.2.2.4 Volume of fluid model

Assumptions:

- A continuity equation for the secondary phases must be solved to track the volume fraction of all phases in whole area.
- Velocity components shared by all phases are obtained through a single set of momentum equation solution.
- The sum of all volume fraction for the phases must be unity. So, the primary phase volume fraction value can be obtained.
- Physical properties must be obtained through a weighted-average of each phases. This is according to their volume fraction for each control volume.
- Shared temperature must be obtained through a particular energy equation.

Mass conversion for VOF model is defined as:

$$\nabla\cdot\left(\varphi_{\mathrm{q}}\vec{V}_{q\rho_{\mathrm{q}}}\right)=0 \tag{11.14}$$

where $\sum_{q=1}^{n}\varphi_{\mathrm{q}}=1$ and all characteristics can be obtained like $\mathrm{N}=\sum_{q=1}^{n}\varphi_{\mathrm{q}}N_q$

A few researchers have used the VOF method for nanofluids modeling. The single-phase and two-phase models in a horizontal pipe with uniform heat flux were compared and analyzed by Akbari et al. [127]. Three different two-phase flow models (VOF, mixed, Eulerian) were compared to simulate laminar mixed convection nanofluid flow. The outcomes showed that the hydrodynamic fields predicted by single-phase and two-phase models are almost the same, but the thermal fields are quite different. However, the convective heat transfer coefficient of the two-phase model is nearer to the experimental results than that of the single-phase model. Since the results of the three two-phase models are relatively consistent, the author prefers to use VOF (cheaper) model. A similar numerical study was carried out by the same research group [128], but in the case of flow in the tube. The realizable k-ε model with enhanced wall function was used in this analysis. The results show that the predicted results of the three two-phase models are basically consistent, but they are far from the experimental data and the numerical results of the single-phase model. However, the hydrodynamic fields of single-phase and two-phase models are almost equal. For the two-phase model, the results deviate greatly with the increase of particle volume fraction. However, in a different way, Moraveji and Ardehali [129] reported that CFD modeling based on the two-phase method is closer to the experimental data than the single-phase form. In addition, some studies [115,119,120,125,126,130] also confirmed this observation. It is obvious that the results of recent numerical calculation are in contradiction with the experimental results, and the query of which model is more correct, has not been solved.

11.2.2.5 Lagrangian–Eulerian model

Conservation of mass:

$$\nabla\cdot(\rho\vec{V}) = 0 \tag{11.15}$$

Conservation of momentum:

$$\nabla\cdot(\rho\vec{V}\vec{V}) = -\nabla P + \mu\nabla^2\vec{V} + S_m \tag{11.16}$$

Conservation of energy:

$$\nabla\cdot\left(\rho C_p\vec{V}T\right) = \nabla\cdot(k\nabla T) + S_e \tag{11.17}$$

Source/sink terms S_m and S_e refer to the integrated properties of momentum and energy exchange for base fluid when the particles are transferring through an element of the Eulerian phase of the base fluid with δV volume and they are equal to zero for the single-phase model.

The momentum transfer between the particles and base fluid is calculated as:

$$S_m = \frac{1}{\delta V}\sum_{np}\vec{F} \tag{11.18}$$

The equation of motion for the particles in the Lagrangian frame of reference can be calculated by:

$$m_p\frac{d\vec{V}_p}{dt} = \vec{F}_D + \vec{F}_L + \vec{F}_g + \vec{F}_b + \vec{F}_{br} \tag{11.19}$$

The right-hand side terms indicate the acting forces on the nanoparticle and refer to drag, Saffman's lift, gravity, buoyancy, and Brownian forces, in sequence.

The energy source term would be the heat transfer between the phases as:

$$S_e = \frac{1}{\delta V}\sum_{np} Nu_p\pi d_p k_p\left(T - T_p\right) \tag{11.20}$$

where Nu_p is obtained from the Ranz and Marshall correlation:

$$Nu_p = 2 + 0.6Re_p^{0.5}Pr^{1/3} \tag{11.21}$$

Bianco et al. [131] and Keshavarz Moraveji and Esmaeili [132] conducted single-phase and two-phase CFD simulation of alumina nanofluid in laminar forced convection flow through a circular tube considering the constant heat flux density. In both works, the Lagrange–Eulerian model is used. By comparing the average heat transfer coefficients of single-phase and two-phase models, the maximum variance is 11%. Compared with the available correlation given by Maiga et al. [113] and Heris et al. [133], their results are satisfactory. The results show that the temperature-dependent model has higher heat transfer coefficient and Nusselt number due to the minimum difference between wall temperature and body temperature. He et al. [134] investigated the TiO_2 nanofluids convective heat transfer through a straight tube under laminar flow numerically. Two different models are used in the analysis: model A—single-phase flow model; model B—Lagrange trajectory method. In the first mock exam, the interaction between the nanoparticle and the interaction between the substrate and the nanoparticles is considered in the mode B. The heat transfer coefficient of the model is slightly higher than that of the model A. The results indicate that the heat transfer of nanofluids is significantly enhanced, especially in the inlet region, which is in good agreement with the experimental outcomes. In another study by Mirzaei et al. [135] the laminar flow heat transfer of nanofluids in microchannels at constant wall temperature was studied by Lagrange–Eulerian method. The results based on the two-phase model show that the heat transfer coefficient is slightly improved compared with the homogeneous single-phase nanofluid technology. In addition, it is found that not only the thermal conductivity of copper/water and alumina/water nanofluids are affected by their thermal hydraulic behaviors, but also the thermal physical properties of other nanofluids should be considered. However, the distribution of particle volume fraction is uniform in most areas except near the wall, which has great influence on the heat transfer and flow features. The turbulent convective flow of water and water–silver nanofluids in a spiral tube was studied experimentally and numerically by Bahrmand et al. [136]. Based on the Lagrange–Eulerian two-phase flow method and RNG k-ε turbulence model considering four-way coupling collision, the numerical simulation was carried out by using ANSYS CFX commercial software. The two-stage method is much more precise than the uniform model. The turbulent kinetic energy and axial velocity of nanoparticles have no significant change. At the same time, it is observed that the heat transfer is improved more effectively by using the base liquid in the spiral tube than by using the nanofluid in the straight tube.

Mahdavi et al. [137] investigated the hybrid model of the Eulerian method and the discrete phase model (DPM) of the Lagrange method are used to guess the hydrodynamic and thermal treatments of nanofluids passing inside the vertical tubes in laminar flow. Three kinds of nanoparticles with different sizes, namely alumina, zirconia, and silica, were studied. The mixing model shows that the velocity and temperature curves of the two phases are the same, while the DPM model shows the ability to capture the slip velocity and temperature difference between the particles and the liquid. Behroyan et al. [138] reported the prediction results of five CFD simulations containing two single-phase models (Newton and non-Newtonian) and three two-phase models (Euler, mixing, and Lagrange) were comprehensively compared to study the turbulent forced flow of copper/water nanofluids in pipes with constant heat flux density on the pipe wall. They compared the numerical results with the experimental results in the literature, and recommended Newton single-phase and Lagrange–Eulerian models as reliable models.

The above literature review shows that the research of Lagrange–Eulerian two-phase flow method in turbulent heat transfer of nanofluid flow is not enough. Therefore some studies are needed to define the ability of the Lagrange–Eulerian model to simulate the characteristics of nanofluids in the turbulent region.

11.2.3 Other CFD approaches

11.2.3.1 Lattice Boltzmann method

Although traditional CFD methods are used, the lattice Boltzmann method (LBM) or thermal LBM have gained some attention for simulating the flow and heat transfer treatments of nanofluids in recent years. LBM uses the Boltzmann equation to model the flow instead of solving the Navier–Stokes equation. In LBM simulation, two different forms were more used: the typical D2Q9 (two-dimensional and nine-speed) square and D3Q19 (three-dimensional and 19-speed) cubic lattice structure [139]. Additionally, this method also has some well-known benefits, such as simple and efficient application of parallel coding, unified algorithm of multiphase flow, and easy processing of complex geometry [140]. Yuan and Yao [141] first proposed LBM to simulate the flow and energy transfer process in nanofluids, taking into account the forces on the flow, such as gravity, buoyancy, and interaction between brown and nanoparticles. LBM has been widely used in natural convection and mixed convection [142–147], and to a lesser extent for forced convection [148–151]. The results of these studies are consistent with the existing numerical outcomes based on the traditional CFD method. However, as far as we know, researchers using LBM did not bring any code validation based on the nanofluid work carried out by experiments to evaluate the accuracy of LBM for nanofluids. Therefore it is necessary to do further research in this area to see to what extent LBM provides reliable results for nanofluid simulation.

11.2.3.2 Finite element method

COMSOL Multiphysics can be introduced as a powerful open source software based on finite element analysis for solution of problems including single- and two-phase nanofluid flow modeling. In this software which has many modules for the multiphysics heat transfer in fluids (ht) and laminar flow (spf), modules can be used for single-phase modeling. Also, P2-P1 Lagrange elements and the Galerkin least-squares method are applied to reach the stability of problem in this type of common problems. In FEM the convergence criteria also must be defined, as well as other numerical techniques, in which the estimated error is placed as $\left|M^{m+1} - M^{m}\right| \leq 10^{-5}$, where M shows the common dependent variable and m refers to the iterations number [152].

11.2.3.3 Example of CFD nanofluid mass transfer

As shown in Fig. 11.6, a simple micromixer is designed with cross-shaped blades.

Here the two-phase governing equations presented by Hatami et al. [153] for the solution of current problem are considered. The natural convection analysis is performed for the case of fixed propeller to find also the mass transfer of nanoparticles as nanoparticle concentration around the blades. The outer wall of the micromixer (or crust) is assumed to be at constant T_c as the cold temperature, while the micromixer blades were kept at T_h as the high temperature. To avoid the singularity solution or zero answer the values, it is assumed that the nanoparticles concentration on the walls is a nonzero value as shown in Fig. 11.6. Furthermore, other required boundary conditions are presented in Fig. 11.6. Finally, for the steady-state condition the dimensional governing equation can be obtained as [153]:

$$\frac{\partial u}{\partial x} + \frac{\partial v}{\partial y} = 0 \tag{11.22}$$

$$\rho_{nf}\left[u\frac{\partial u}{\partial x} + v\frac{\partial u}{\partial y}\right] = -\frac{\partial p}{\partial x} + \mu_{nf}\left(\frac{\partial^2 u}{\partial x^2} + \frac{\partial^2 u}{\partial y^2}\right) \tag{11.23}$$

$$\rho_{nf}\left[u\frac{\partial v}{\partial x} + v\frac{\partial v}{\partial y}\right] = -\frac{\partial p}{\partial y} + \mu_{nf}\left(\frac{\partial^2 v}{\partial x^2} + \frac{\partial^2 v}{\partial y^2}\right) + (\rho\beta)_{nf}g(T - T_c) + (\rho\beta^*)_{nf}g(C - C_c) \tag{11.24}$$

$$u\frac{\partial T}{\partial x} + v\frac{\partial T}{\partial y} = \alpha_{nf}\left(\frac{\partial^2 T}{\partial x^2} + \frac{\partial^2 T}{\partial y^2}\right) + \frac{D_B}{C}\left(\frac{\partial C}{\partial x}\frac{\partial T}{\partial x} + \frac{\partial C}{\partial y}\frac{\partial T}{\partial y}\right) + \frac{D_T}{T}\left[\left(\frac{\partial T}{\partial x}\right)^2 + \left(\frac{\partial T}{\partial y}\right)^2\right] \tag{11.25}$$

$$u\frac{\partial C}{\partial x} + v\frac{\partial C}{\partial y} = D_B\left(\frac{\partial^2 C}{\partial x^2} + \frac{\partial^2 C}{\partial y^2}\right) + \frac{CD_T}{T}\left(\frac{\partial^2 T}{\partial x^2} + \frac{\partial^2 T}{\partial y^2}\right) + \frac{D_T}{T}\left(\frac{\partial C}{\partial x}\frac{\partial T}{\partial x} + \frac{\partial C}{\partial y}\frac{\partial T}{\partial y}\right) \tag{11.26}$$

where C, T, D_T, and D_B refer to dimensional concentration, temperature, thermophoretic diffusion, and Brownian diffusion coefficients of nanofluid, respectively. D_T and D_B were defined in [152]. By introducing the following nondimensional parameters:

$$\begin{gathered} X = \frac{x}{L};\quad Y = \frac{y}{L};\quad U = \frac{uL}{\alpha_{bf}};\quad V = \frac{vL}{\alpha_{bf}};\quad P = \frac{pL^2}{\rho_{bf}\alpha_{bf}^2} \\ \theta = \frac{T - T_c}{\Delta T},\quad \Phi = \frac{C - C_c}{\Delta C},\quad [\Delta T = T_h - T_c,\ \Delta C = C_h - C_c] \end{gathered} \tag{11.27}$$

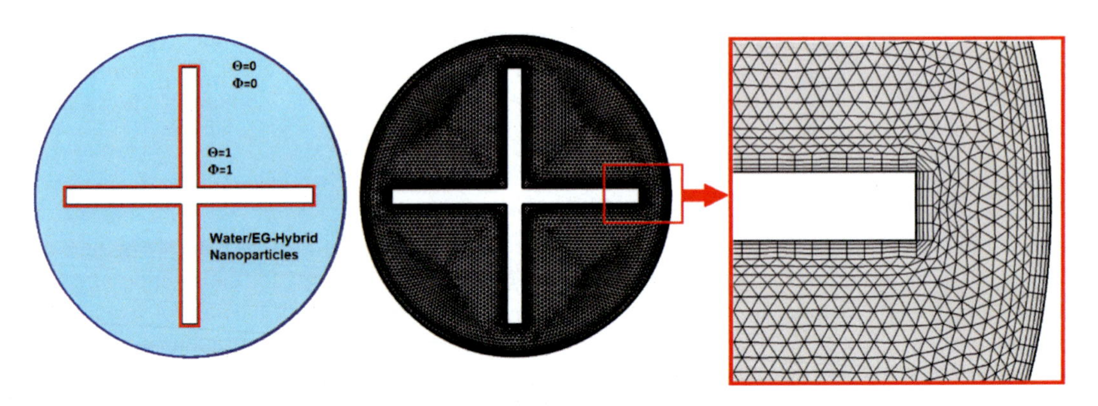

FIGURE 11.6

Cross section of micromixer and generated mesh.

The two-dimensional natural convection flow around the micromixer propellers using conservation of momentum, mass, energy, and concentration (as second phase) can be introduced as the following dimensionless shape:

$$\frac{\partial U}{\partial X}+\frac{\partial V}{\partial Y}=0 \tag{11.28}$$

$$U\frac{\partial U}{\partial X}+V\frac{\partial U}{\partial Y}=-\frac{\rho_{bf}}{\rho_{nf}}\frac{\partial P}{\partial X}+\frac{Pr\,\mu_{nf}}{\nu_{bf}\,\rho_{nf}}\left(\frac{\partial^2 U}{\partial X^2}+\frac{\partial^2 U}{\partial Y^2}\right) \tag{11.29}$$

$$U\frac{\partial V}{\partial X}+V\frac{\partial V}{\partial Y}=-\frac{\rho_{bf}}{\rho_{nf}}\frac{\partial P}{\partial Y}+\mathrm{Pr}\frac{\mu_{nf}}{\rho_{nf}\nu_{bf}}\left(\frac{\partial^2 V}{\partial X^2}+\frac{\partial^2 V}{\partial Y^2}\right)+\frac{(\rho\beta)_{nf}}{\rho_{nf}\beta_{bf}}Ra_T Pr\theta+Ra_C Pr\Phi \tag{11.30}$$

$$U\frac{\partial \theta}{\partial X}+V\frac{\partial \theta}{\partial Y}=\frac{\alpha_{nf}}{\alpha_{bf}}\left(\frac{\partial^2 \theta}{\partial X^2}+\frac{\partial^2 \theta}{\partial Y^2}\right)+\frac{1}{Le}\left(\frac{\partial \Phi}{\partial X}\frac{\partial \theta}{\partial X}+\frac{\partial \Phi}{\partial Y}\frac{\partial \theta}{\partial Y}\right)+\frac{Pr\,N_{\mathrm{TBT}}}{Sc}\left(\left(\frac{\partial \theta}{\partial X}\right)^2+\left(\frac{\partial \theta}{\partial Y}\right)^2\right) \tag{11.31}$$

$$U\frac{\partial \Phi}{\partial X}+V\frac{\partial \Phi}{\partial Y}=\frac{\mathrm{Pr}}{Sc}\left(\frac{\partial^2 \Phi}{\partial X^2}+\frac{\partial^2 \Phi}{\partial Y^2}\right)+\frac{Pr}{Sc}\left[N_{\mathrm{TBTC}}\left(\frac{\partial^2 \theta}{\partial X^2}+\frac{\partial^2 \theta}{\partial Y^2}\right)+N_{\mathrm{TBT}}\left(\frac{\partial \Phi}{\partial X}\frac{\partial \theta}{\partial X}+\frac{\partial \Phi}{\partial Y}\frac{\partial \theta}{\partial Y}\right)\right] \tag{11.32}$$

where the dimensionless parameters Ra_T and Ra_C represent the local thermal Rayleigh and solute Rayleigh numbers, respectively. Pr and Le indicate the Prandtl and modified Lewis numbers, respectively, and N_{TBTC}, N_{TBT}, and Sc numbers denote the dynamic diffusion parameter, dynamic thermodiffusion parameter, and Schmidt numbers, correspondingly. These parameters can be obtained by:

$$\begin{aligned} &Ra_C=\frac{L^3 g\Delta C\,(\rho\beta^*)_{nf}}{\upsilon_{bf}\alpha_{bf}\quad\rho_{nf}},\quad Ra_T=\frac{L^3 g\Delta T\beta_{bf}}{\upsilon_{bf}\alpha_{bf}},\quad Pr=\frac{\upsilon_{bf}}{\alpha_{bf}}\\ &Le=\frac{k_{bf}C_c}{(\rho c_p)_{bf}D_B\Delta C},\quad N_{\mathrm{TBTC}}=\frac{D_T\,\Delta T\,C_c}{D_B\,\Delta C\,T_c},\quad N_{\mathrm{TBT}}=\frac{D_T\,\Delta T}{D_B\;T_c},\quad Sc=\frac{\mu_{bf}}{\rho_{bf}D_B} \end{aligned} \tag{11.33}$$

where thermal properties of two different described hybrid nanofluids are introduced based on the available equations in the literature. Next Eqs. (11.34)–(11.37) are introduced for the hybrid TiO_2–CuO nanoparticles added to water as base fluid:

$$\rho_{hnf}=\rho_f(1-\varphi_2)\left[(1-\varphi_1)+\varphi_1\left(\frac{\rho_{s1}}{\rho_f}\right)\right]+\varphi_2\rho_{s2} \tag{11.34}$$

$$(\rho C_p)_{hnf}=(\rho C_p)_f(1-\varphi_2)\left[(1-\varphi_1)+\varphi_1\frac{(\rho c_p)_{s1}}{(\rho c_p)_f}\right]+\varphi_2(\rho c_p)_{s2} \tag{11.35}$$

$$\mu_{hnf}=\frac{\mu_f}{(1-\varphi_1)^{2.5}(1-\varphi_2)^{2.5}} \tag{11.36}$$

$$\frac{k_{hnf}}{k_{bf}}=\frac{k_{s2}+(s-1)k_{bf}-(s-1)\varphi_2(k_{bf}-k_{s2})}{k_{s2}+(s-1)k_{bf}+\varphi_2(k_{bf}-k_{s2})}\quad\text{where}\quad\frac{k_{bf}}{k_f}=\frac{k_{s1}+(s-1)k_f-(s-1)\varphi_1(k_f-k_{s1})}{k_{s1}+(s-1)k_f+\varphi_1(k_f-k_{s1})} \tag{11.37}$$

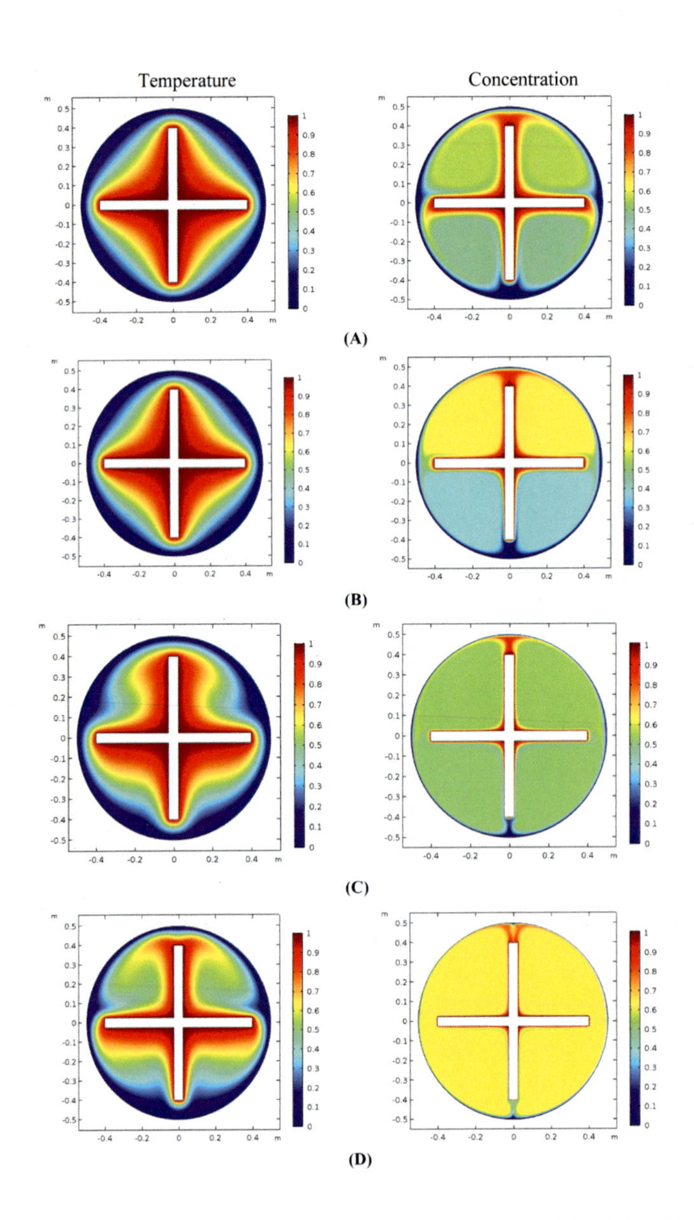

FIGURE 11.7

Temperature and nanoparticles concentration for hybrid TiO_2–CuO–water when $Ra_C = 10$, $\varphi = 0.02$: (A) $Ra_T = 1e3$, (B) $Ra_T = 1e4$, (C) $Ra_T = 1e5$, and (D) $Ra_T = 1e6$.

And Eqs. (11.38)–(11.43) have been used for ethylene glycol-based nanofluid with the hybrid (MoS_2–SiO_2) nanoparticles:

$$\frac{\rho_{hnf}}{\rho_f} = (1 - \varphi_2)\left[(1 - \varphi_1) + \varphi_1\left(\frac{\rho_{s1}}{\rho_f}\right)\right] + \varphi_2\frac{\rho_{s2}}{\rho_f} \tag{11.38}$$

$$\frac{(\rho C_p)_{hnf}}{(\rho c_p)_f} = (1-\varphi_2)\left[(1-\varphi_1)+\varphi_1\frac{(\rho c_p)_{s1}}{(\rho c_p)_f}\right]+\varphi_2\frac{(\rho c_p)_{s2}}{(\rho c_p)_f} \tag{11.39}$$

$$\mu_{hnf} = \frac{\mu_f}{(1-\varphi_1)^{2.5}(1-\varphi_2)^{2.5}} \tag{11.40}$$

$$\frac{k_{hnf}}{k_{bf}} = \frac{k_{s2}+2k_{bf}-2\varphi_2(k_{bf}-k_{s2})}{k_{s2}+2k_{bf}+\varphi_2(k_{bf}-k_{s2})} \text{ where } \frac{k_{bf}}{k_f} = \frac{k_{s1}+2k_f-2\varphi_1(k_f-k_{s1})}{k_{s1}+2k_f+\varphi_1(k_f-k_{s1})} \tag{11.41}$$

$$\frac{(\rho\beta)_{hnf}}{(\rho\beta)_f} = (1-\varphi_2)\left[(1-\varphi_1)+\varphi_1\frac{(\rho\beta)_{s1}}{(\rho\beta)_f}\right]+\varphi_2\frac{(\rho\beta)_{s2}}{(\rho\beta)_f} \tag{11.42}$$

$$\frac{\sigma_{hnf}}{\sigma_{bf}} = \frac{\sigma_{s2}+2\sigma_{bf}-2\varphi_2(\sigma_{bf}-\sigma_{s2})}{\sigma_{s2}+2\sigma_{bf}+\varphi_2(\sigma_{bf}-\sigma_{s2})} \quad \text{where} \quad \frac{\sigma_{bf}}{\sigma_f} = \frac{\sigma_{s1}+2\sigma_f-2\varphi_1(\sigma_f-\sigma_{s1})}{\sigma_{s1}+2\sigma_f+\varphi_1(\sigma_f-\sigma_{s1})} \tag{11.43}$$

Fig. 11.7 shows the results of nanoparticles concentrations and mass transfer around the blades as well as the temperature profiles using FEM. More details about the numerical methods for nanofluid modeling can be found in Ref. [154].

11.3 Conclusion

In this chapter the CFD simulation of nanofluids through single-phase and two-phase modeling were presented and three two-phase models, namely Eulerian–Eulerian model, mixture model, and Eulerian model, were introduced as efficient models. Also, in this chapter the VOF, LBM, and FEM were introduced to solve the nanofluid mass transfer problems using numerical codes or commercial software.

References

[1] D. Wu, H. Zhu, L. Wang, L. Liu, Critical issues in nanofluids preparation, characterization and thermal conductivity, Curr. Nanosci. 5 (2009) 103–112.

[2] G. Paul, M. Chopkar, I. Manna, P.K. Das, Techniques for measuring the thermal conductivity of nanofluids: a review, Renew. Sustain. Energy Rev. 14 (2010) 1913–1924.

[3] M. Kostic, K.C. Simham, Computerized, transient hot-wire thermal conductivity (HWTC) apparatus for nanofluid, in: L. Xi (Ed.), Proceedings of the Sixth WSEAS International Conference on Heat and Mass Transfer (HMT'09), WSEAS Press, 2009, pp. 71–78.

[4] P.L. Woodfield, J. Fukai, M. Fujii, Y. Takata, K. Shinzato, A two-dimensional analytical solution for the transient short-hot-wire method, Int. J. Thermophys. 29 (2008) 1278–1298. Available from: https://doi.org/10.1007/s10765-008-0469-y.

[5] S.W. Hong, Y.T. Kang, C. Kleinstreuer, J. Koo, Impact analysis of natural convection on thermal conductivity measurements of nanofluids using the transient hot-wire method, Int. J. Heat Mass Transf. 54 (2011) 3448–3456.

[6] N. Shalkevich, W. Escher, T. Büurgi, B. Michel, S. Lynda, D. Poulikakos, On the Thermal conductivity of gold nanoparticle colloids, Langmuir 26 (2) (2010) 663–670.

[7] Z.L. Wang, D.W. Tang, S. Liu, X.H. Zheng, N. Araki, Thermal-conductivity and thermal-diffusivity measurements of nanofluids by 3u method and mechanism analysis of heat transport, Int. J. Thermophys. 28 (2007) 1255–1268.

[8] Y. Nagasaka, A. Nagashima, Absolute measurement of the thermal conductivity of electrically conducting liquids by the transient hot-wire method, J. Phys. Sci. Instrum. 14 (1981) 1435–1440.

[9] P. Vadasz, Rendering the transient hot wire experimental method for thermal conductivity estimation to two-phase systems-theoretical leading order results, J. Heat Transf. 132 (2010) 081601.

[10] M.H. Ahmadi, M.A. Nazari, R. Ghasempour, H. Madah, M.B. Shafii, M.A. Ahmadi, Thermal conductivity ratio prediction of Al_2O_3/water nanofluid by applying connectionist methods, Colloids Surf. A Physicochem. Eng. Asp. (2018). Available from: https://doi.org/10.1016/j.colsurfa.2018.01.030.

[11] A. Alirezaie, M.H. Hajmohammad, M.R. Hassani Ahangar, M. Hemmat Esfe, Price performance evaluation of thermal conductivity enhancement of nanofluids with different particle sizes, Appl. Therm. Eng. 128 (2018) 373–380. Available from: https://doi.org/10.1016/J.APPLTHERMALENG.2017.08.143.

[12] I.O. Alade, T.A. Oyehan, I.K. Popoola, S.O. Olatunji, A. Bagudu, Modeling thermal conductivity enhancement of metal and metallic oxide nanofluids using support vector regression, Adv. Powder Technol. 29 (2018) 157–167. Available from: https://doi.org/10.1016/J.APT.2017.10.023.

[13] F. Jabbari, A. Rajabpour, S. Saedodin, Thermal conductivity and viscosity of nanofluids: a review of recent molecular dynamics studies, Chem. Eng. Sci. 174 (2017) 67–81. Available from: https://doi.org/10.1016/J.CES.2017.08.034.

[14] M.A. Ariana, B. Vaferi, G. Karimi, Prediction of thermal conductivity of alumina water-based nanofluids by artificial neural networks, Powder Technol. 278 (2015) 1–10. Available from: https://doi.org/10.1016/j.powtec.2015.03.005.

[15] A. Shahsavar, M.R. Salimpour, M. Saghafian, M.B. Shafii, An experimental study on the effect of ultrasonication on thermal conductivity of ferrofluid loaded with carbon nanotubes, Thermochim. Acta 617 (2015) 102–110. Available from: https://doi.org/10.1016/J.TCA.2015.08.025.

[16] B.T. Sone, A. Diallo, X.G. Fuku, A. Gurib-Fakim, M. Maaza, Biosynthesized CuO nanoplatelets: physical properties & enhanced thermal conductivity nanofluidics, Arab. J. Chem. (2017). Available from: https://doi.org/10.1016/J.ARABJC.2017.03.004.

[17] M. Afrand, M. Hemmat Esfe, E. Abedini, H. Teimouri, Predicting the effects of magnesium oxide nanoparticles and temperature on the thermal conductivity of water using artificial neural network and experimental data, Phys. E Low. Dimens. Syst. Nanostruct. 87 (2017) 242–247. Available from: https://doi.org/10.1016/j.physe.2016.10.020.

[18] M. Hemmat Esfe, M.H. Hajmohammad, Thermal conductivity and viscosity optimization of nanodiamond-Co_3O_4/EG (40:60) aqueous nanofluid using NSGA-II coupled with RSM, J. Mol. Liq. 238 (2017) 545–552. Available from: https://doi.org/10.1016/J.MOLLIQ.2017.04.056.

[19] N.N. Esfahani, D. Toghraie, M. Afrand, A new correlation for predicting the thermal conductivity of ZnO–Ag (50%–50%)/water hybrid nanofluid: an experimental study, Powder Technol. 323 (2018) 367–373. Available from: https://doi.org/10.1016/j.powtec.2017.10.025.

[20] S. Akilu, A.T. Baheta, K.V. Sharma, Experimental measurements of thermal conductivity and viscosity of ethylene glycol-based hybrid nanofluid with TiO_2-CuO/C inclusions, J. Mol. Liq. 246 (2017) 396–405. Available from: https://doi.org/10.1016/J.MOLLIQ.2017.09.017.

[21] C.H. Li, G.P. Peterson, The effect of particle size on the effective thermal conductivity of Al_2O_3-water nanofluids, J. Appl. Phys. 101 (2007) 044312. Available from: https://doi.org/10.1063/1.2436472.

[22] Y. Ren, H. Xie, A. Cai, Effective thermal conductivity of nanofluids containing spherical nanoparticles, J. Phys. D. Appl. Phys. 38 (2005) 3958–3961. Available from: https://doi.org/10.1088/0022-3727/38/21/019.

[23] R.M. Sarviya, V. Fuskele, Review on thermal conductivity of nanofluids, Mater. Today Proc. 4 (2017) 4022–4031. Available from: https://doi.org/10.1016/J.MATPR.2017.02.304.

[24] M. Hemmat Esfe, M. Afrand, W.-M. Yan, M. Akbari, Applicability of artificial neural network and non-linear regression to predict thermal conductivity modeling of Al_2O_3−water nanofluids using experimental data, Int. Commun. Heat Mass Transf. 66 (2015) 246−249. Available from: https://doi.org/10.1016/j.icheatmasstransfer.2015.06.002.

[25] M. Hemmat Esfe, S. Saedodin, N. Sina, M. Afrand, S. Rostami, Designing an artificial neural network to predict thermal conductivity and dynamic viscosity of ferromagnetic nanofluid, Int. Commun. Heat Mass Transf. 68 (2015) 50−57. Available from: https://doi.org/10.1016/j.icheatmasstransfer.2015.06.013.

[26] M. Hemmat Esfe, M.R.H. Ahangar, D. Toghraie, M.H. Hajmohammad, H. Rostamian, H. Tourang, Designing artificial neural network on thermal conductivity of Al_2O_3−water−EG (60−40%) nanofluid using experimental data, J. Therm. Anal. Calorim. 126 (2016) 837−843. Available from: https://doi.org/10.1007/s10973-016-5469-8.

[27] D. Dasgupta, K. Mondal, Thermal circuits based model for predicting the thermal conductivity of nanofluids, Int. J. Heat Mass Transf. 113 (2017) 806−818. Available from: https://doi.org/10.1016/J.IJHEATMASSTRANSFER.2017.05.132.

[28] W. Guo, G. Li, Y. Zheng, C. Dong, Measurement of the thermal conductivity of SiO_2 nanofluids with an optimized transient hot wire method, Thermochim. Acta (2018). Available from: https://doi.org/10.1016/j.tca.2018.01.008.

[29] N. Kumar, S.S. Sonawane, Experimental study of thermal conductivity and convective heat transfer enhancement using CuO and TiO_2 nanoparticles, Int. Commun. Heat Mass Transf. 76 (2016) 98−107. Available from: https://doi.org/10.1016/J.ICHEATMASSTRANSFER.2016.04.028.

[30] B. Wei, C. Zou, X. Li, Experimental investigation on stability and thermal conductivity of diathermic oil based TiO_2 nanofluids, Int. J. Heat Mass Transf. 104 (2017) 537−543. Available from: https://doi.org/10.1016/J.IJHEATMASSTRANSFER.2016.08.078.

[31] M. Vakili, M. Karami, S. Delfani, S. Khosrojerdi, K. Kalhor, Experimental investigation and modeling of thermal conductivity of CuO−water/EG nanofluid by FFBPANN and multiple regressions, J. Therm. Anal. Calorim. 129 (2017) 629−637. Available from: https://doi.org/10.1007/s10973-017-6217-4.

[32] S.S. Sanukrishna, M.J. Prakash, Experimental studies on thermal and rheological behaviour of TiO_2-PAG nanolubricant for refrigeration system, Int. J. Refrig. 86 (2018) 356−372. Available from: https://doi.org/10.1016/J.IJREFRIG.2017.11.014.

[33] R.K. Singh, A.K. Sharma, A.R. Dixit, A. Mandal, A.K. Tiwari, Experimental investigation of thermal conductivity and specific heat of nanoparticles mixed cutting fluids, Mater. Today Proc. 4 (2017) 8587−8596. Available from: https://doi.org/10.1016/J.MATPR.2017.07.206.

[34] D. Cabaleiro, J. Nimo, M.J. Pastoriza-Gallego, M.M. Piñeiro, J.L. Legido, L. Lugo, Thermal conductivity of dry anatase and rutile nano-powders and ethylene and propylene glycol-based TiO_2 nanofluids, J. Chem. Thermodyn. 83 (2015) 67−76. Available from: https://doi.org/10.1016/J.JCT.2014.12.001.

[35] M. Chandrasekar, S. Suresh, A. Chandra Bose, Experimental investigations and theoretical determination of thermal conductivity and viscosity of Al_2O_3/water nanofluid, Exp. Therm. Fluid Sci. 34 (2010) 210−216. Available from: https://doi.org/10.1016/J.EXPTHERMFLUSCI.2009.10.022.

[36] S.K. Das, N. Putra, P. Thiesen, W. Roetzel, Temperature dependence of thermal conductivity enhancement for nanofluids, J. Heat Transf. 125 (2003) 567. Available from: https://doi.org/10.1115/1.1571080.

[37] S. Lee, S.U.-S. Choi, S. Li, J.A. Eastman, Measuring thermal conductivity of fluids containing oxide nanoparticles, J. Heat Transf. 121 (1999) 280−289. Available from: https://doi.org/10.1115/1.2825978.

[38] T.T.-K. Hong, H.H.-S. Yang, C.J. Choi, Study of the enhanced thermal conductivity of Fe nanofluids, J. Appl. Phys. 97 (2005) 064311. Available from: https://doi.org/10.1063/1.1861145.

[39] D. Wen, Y. Ding, Effective thermal conductivity of aqueous suspensions of carbon nanotubes (carbon nanotube nanofluids), J. Thermophys. Heat Transf. 18 (2004) 481−485. Available from: https://doi.org/10.2514/1.9934.

[40] H. Li, L. Wang, Y. He, et al., Experimental investigation of thermal conductivity and viscosity of ethylene glycol based ZnO nanofluids, Appl. Therm. Eng. 88 (2014) 363–368. Available from: https://doi.org/10.1016/j.applthermaleng.2014.10.071.

[41] Y. He, Y. Jin, H. Chen, Y. Ding, D. Cang, H. Lu, Heat transfer and flow behaviour of aqueous suspensions of TiO_2 nanoparticles (nanofluids) flowing upward through a vertical pipe, Int. J. Heat Mass Transf. 50 (2007) 2272–2281. Available from: https://doi.org/10.1016/j.ijheatmasstransfer.2006.10.024.

[42] P.B. Maheshwary, C.C. Handa, K.R. Nemade, A comprehensive study of effect of concentration, particle size and particle shape on thermal conductivity of titania/water based nanofluid, Appl. Therm. Eng. 119 (2017) 79–88.

[43] X. Zhang, H. Gu, M. Fujii, Effective thermal conductivity and thermal diffusivity of nanofluids containing spherical and cylindrical nanoparticles, Exp. Therm. Fluid Sci. 31 (2007) 593–599.

[44] R. Agarwal, K. Verma, N.K. Agrawal, et al., Sensitivity of thermal conductivity for Al_2O_3 nanofluids, Exp. Therm. Fluid Sci. 80 (2017) 19–26.

[45] M. Premalatha, A.K.S. Jeevaraj, Investigation on the thermal conductivity and rheological properties of multiwalled carbon nanotube (MWCNT)—dowtherm a nanofluids, Part. Sci. Technol. 34 (2) (2015). 150731121429007.

[46] J. Philip, P. Shima, B. Raj, Evidence for enhanced thermal conduction through percolating structures in nanofluids, Nanotechnology 19 (2008) 305706.

[47] W. Duangthongsuk, S. Wongwises, Measurement of temperature-dependent thermal conductivity and viscosity of TiO_2-water nanofluids, Exp. Therm. Fluid Sci. 33 (2009) 706–714. Available from: https://doi.org/10.1016/j.expthermflusci.2009.01.005.

[48] A. Turgut, I. Tavman, M. Chirtoc, H.P. Schuchmann, C. Sauter, S. Tavman, Thermal conductivity and viscosity measurements of water-based TiO_2 nanofluids, Int. J. Thermophys. 30 (2009) 1213–1226. Available from: https://doi.org/10.1007/s10765-009-0594-2.

[49] D. Zhu, X. Li, N. Wang, X. Wang, J. Gao, H. Li, Dispersion behavior and thermal conductivity characteristics of Al_2O_3-H_2O nanofluids, Curr. Appl. Phys. 9 (2009) 131–139. Available from: https://doi.org/10.1016/j.cap.2007.12.008.

[50] F.M. Ali, W.M.M. Yunus, Study of the effect of volume fraction concentration and particle materials on thermal conductivity and thermal diffusivity of nanofluids, Japanese J. Appl. Phys. 50 (2011) 085201.

[51] W. Yu, H. Xie, L. Chen, Y. Li, Enhancement of thermal conductivity of kerosene-based Fe_3O_4 nanofluids prepared via phase-transfer method, Colloids Surf. A Physicochem. Eng. Asp. 355 (2010) 109–113. Available from: https://doi.org/10.1016/j.colsurfa.2009.11.044.

[52] S.W. Lee, S.D. Park, S. Kang, I.C. Bang, J.H. Kim, Investigation of viscosity and thermal conductivity of SiC nanofluids for heat transfer applications, Int. J. Heat Mass Transf. 54 (2011) 433–438. Available from: https://doi.org/10.1016/j.ijheatmasstransfer.2010.09.026.

[53] S. Harish, K. Ishikawa, E. Einarsson, S. Aikawa, T. Inoue, P. Zhao, et al., Temperature dependent thermal conductivity increase of aqueous nanofluid with single walled carbon nanotube inclusion, Mater. Express 2 (2012) 213–223. Available from: https://doi.org/10.1166/mex.2012.1074.

[54] S. Harish, K. Ishikawa, E. Einarsson, S. Aikawa, S. Chiashi, J. Shiomi, et al., Enhanced thermal conductivity of ethylene glycol with single-walled carbon nanotube inclusions, Int. J. Heat Mass Transf. 55 (2012) 3885–3890. Available from: https://doi.org/10.1016/j.ijheatmasstransfer.2012.03.001.

[55] R.S. Khedkar, S.S. Sonawane, K.L. Wasewar, Influence of CuO nanoparticles in enhancing the thermal conductivity of water and monoethylene glycol based nanofluids, Int. Commun. Heat Mass Transf. 39 (2012) 665–669. Available from: https://doi.org/10.1016/j.icheatmasstransfer.2012.03.012.

[56] M. Fakoor Pakdaman, M.A. Akhavan-Behabadi, P. Razi, An experimental investigation on thermophysical properties and overall performance of MWCNT/heat transfer oil nanofluid flow inside vertical helically coiled tubes, Exp. Therm. Fluid Sci. 40 (2012) 103–111. Available from: https://doi.org/10.1016/j.expthermflusci.2012.02.005.

[57] L.S. Sundar, M.H. Farooky, S.N. Sarada, M.K. Singh, Experimental thermal conductivity of ethylene glycol and water mixture based low volume concentration of Al_2O_3 and CuO nanofluids, Int. Commun. Heat Mass Transf. 41 (2013) 41–46. Available from: https://doi.org/10.1016/j.icheatmasstransfer.2012.11.004.
[58] S.A. Angayarkanni, J. Philip, Effect of nanoparticles aggregation on thermal and electrical conductivities of nanofluids, J. Nanofluids 3 (2014) 17–25.
[59] S. Cingarapu, D. Singh, E.V. Timofeeva, M.R. Moravek, Nanofluids with encapsulated tin nanoparticles for advanced heat transfer and thermal energy storage, Int. J. Energy Res. 38 (2014) 51–59. Available from: https://doi.org/10.1002/er.3041.
[60] H. Masuda, A. Ebata, K. Teramae, N. Hishinuma, Alteration of thermal conductivity and viscosity of liquid by dispersing ultra fine particles, Netsu Bussei 7 (1993) 227–233.
[61] X. Wang, X. Xu, S.U.S. Choi, Thermal conductivity of nanoparticle—fluid mixture, J. Thermophys. Heat Transf. 13 (1999) 474–480. Available from: https://doi.org/10.2514/2.6486.
[62] Y. Xuan, Q. Li, Heat transfer enhancement of nanofluids, Int. J. Heat Fluid Flow. 21 (2000) 58–64.
[63] S.H. Kim, S.R. Choi, D. Kim, Thermal conductivity of metal-oxide nanofluids: particle size dependence and effect of laser irradiation, J. Heat Transf. 129 (2007) 298–307. Available from: https://doi.org/10.1115/1.2427071.
[64] D.-H. Yoo, K.S. Hong, H.-S. Yang, Study of thermal conductivity of nanofluids for the application of heat transfer fluids, Thermochim. Acta 455 (2007) 66–69.
[65] M.P. Beck, Y. Yuan, P. Warrier, et al., The effect of particle size on the thermal conductivity of alumina nanofluids, J. Nanopart. Res. 11 (5) (2008) 1129–1136.
[66] H.A. Mintsa, G. Roy, C.T. Nguyen, D. Doucet, New temperature dependent thermal conductivity data for water-based nanofluids, Int. J. Therm. Sci. 48 (2009) 363–371. Available from: https://doi.org/10.1016/j.ijthermalsci.2008.03.009.
[67] R.S. Prasher, P. Bhattacharya, P.E. Phelan, Effect of aggregation kinetics on the thermal conductivity of nanoscale colloidal solutions (nanofluids), Nanoletters 6 (2006) 1529–1534.
[68] M. Chopkar, P.K. Das, I. Manna, Synthesis and characterization of nanofluid for advanced heat transfer applications, Scr. Mater. 55 (2006) 549–552. Available from: https://doi.org/10.1016/j.scriptamat.2006.05.030.
[69] M. Chopkar, S. Sudarshan, P.K. Das, I. Manna, Effect of particle size on thermal conductivity of nanofluid, Metall. Mater. Trans. A 39 (2008) 1535–1542. Available from: https://doi.org/10.1007/s11661-007-9444-7.
[70] M.P. Beck, Y. Yuan, P. Warrier, A.S. Teja, The thermal conductivity of alumina nanofluids in water, ethylene glycol, and ethylene glycol + water mixtures, J. Nanopart. Res. 12 (2010) 1469–1477. Available from: https://doi.org/10.1007/s11051-009-9716-9.
[71] M. Xing, J. Yu, R. Wang, Experimental investigation and modelling on the thermal conductivity of CNTs based nanofluids, Int. J. Therm. Sci. 104 (2016) 404–411. Available from: https://doi.org/10.1016/j.ijthermalsci.2016.01.024.
[72] H. Xie, J. Wang, T. Xi, Y. Liu, Thermal conductivity of suspensions containing nanosized SiC particles, Int. J. Thermophys. 23 (2002) 571–580.
[73] Y. Hwang, J.K. Lee, C.H. Lee, Y. Jung, S.I. Cheong, C.G. Lee, et al., Stability and thermal conductivity characteristics of nanofluids, Thermochim. Acta (2007) 455. Available from: https://doi.org/10.1016/j.tca.2006.11.036.
[74] B.C. Pak, Y.I.Y. Cho, Hydrodynamic and heat transfer study of dispersed fluids with submicron metallic oxide particles, Exp. Heat Transf. 11 (1998) 151–170. Available from: https://doi.org/10.1080/08916159808946559.
[75] M.P. Beck, Y. Yuan, P. Warrier, A.S. Teja, The effect of particle size on the thermal conductivity of alumina nanofluids, J. Nanopart. Res. 11 (2009) 1129–1136.
[76] M.-S. Liu, M.C.-C. Lin, I.-T. Huang, C.-C. Wang, Enhancement of thermal conductivity with carbon nanotube for nanofluids, Int. Commun. Heat Mass Transf. 32 (2005) 1202–1210.

[77] S.M.S. Murshed, K.C. Leong, C. Yang, Enhanced thermal conductivity of TiO_2 water based nanofluids, Int. J. Therm. Sci. 44 (2005) 367–373.
[78] Y. Ding, H. Alias, D. Wen, R.A. Williams, Heat transfer of aqueous suspensions of carbon nanotubes (CNT nanofluids), Int. J. Heat Mass Transf. 49 (2006) 240–250.
[79] H. Xie, H. Lee, W. Youn, M. Choi, Nanofluids containing multiwalled carbon nanotubes and their enhanced thermal conductivities, J. Appl. Phys. 94 (2003) 4967–4971. Available from: https://doi.org/10.1063/1.1613374.
[80] Y.J. Hwang, Y.C. Ahn, H.S. Shin, C.G. Lee, G.T. Kim, H.S. Park, et al., Investigation on characteristics of thermal conductivity enhancement of nanofluids, Curr. Appl. Phys. 6 (2006) 1068–1071. Available from: https://doi.org/10.1016/j.cap.2005.07.021.
[81] M.J. Assael, I.N. Metaxa, J. Arvanitidis, D. Christofilos, C. Lioutas, Thermal conductivity enhancement in aqueous suspensions of carbon multi-walled and double-walled nanotubes in the presence of two different dispersants, Int. J. Thermophys. 26 (2005) 647–664. Available from: https://doi.org/10.1007/s10765-005-5569-3.
[82] S.U.S. Choi, Z.G. Zhang, W. Yu, F.E. Lockwood, E.A. Grulke, Anomalous thermal conductivity enhancement in nanotube suspensions, Appl. Phys. Lett. 79 (2001) 2252–2254. Available from: https://doi.org/10.1063/1.1408272.
[83] B.I. Kharisov, O.V. Kharissova, U. Ortiz-Mendez, CRC Concise Encyclopedia of Nanotechnology, CRC Press, 2016.
[84] S.A. Putnam, D.G. Cahill, P.V. Braun, Z. Ge, R.G. Shimmin, Thermal conductivity of nanoparticle suspensions, J. Appl. Phys. 99 (2006) 084308. Available from: https://doi.org/10.1063/1.2189933.
[85] S.M.S. Murshed, K.C. Leong, C. Yang, Investigations of thermal conductivity and viscosity of nanofluids, Int. J. Therm. Sci. 47 (2008) 560–568.
[86] N.A. Usri, W.H. Azmi, R. Mamat, K.A. Hamid, G. Najafi, Thermal conductivity enhancement of Al_2O_3 nanofluid in ethylene glycol and water mixture, Energy Procedia 79 (2015) 397–402. Available from: https://doi.org/10.1016/j.egypro.2015.11.509.
[87] M. Hemmat Esfe, M. Afrand, A. Karimipour, W.-M. Yan, N. Sina, An experimental study on thermal conductivity of MgO nanoparticles suspended in a binary mixture of water and ethylene glycol, Int. Commun. Heat Mass Transf. 67 (2015) 173–175. Available from: https://doi.org/10.1016/j.icheatmasstransfer.2015.07.009.
[88] J.A. Eastman, S.U.S. Choi, S. Li, W. Yu, L.J. Thompson, Anomalously increased effective thermal conductivities of ethylene glycol-based nanofluids containing copper nanoparticles, Appl. Phys. Lett. 78 (2001) 718–720. Available from: https://doi.org/10.1063/1.1341218.
[89] R. Agarwal, K. Verma, N.K. Agrawal, R.K. Duchaniya, R. Singh, Synthesis, characterization, thermal conductivity and sensitivity of CuO nanofluids, Appl. Therm. Eng. 102 (2016) 1024–1036. Available from: https://doi.org/10.1016/j.applthermaleng.2016.04.051.
[90] H.E. Patel, S.K. Das, T. Sundararajan, A. Sreekumaran Nair, B. George, T. Pradeep, Thermal conductivities of naked and monolayer protected metal nanoparticle based nanofluids: manifestation of anomalous enhancement and chemical effects, Appl. Phys. Lett. 83 (2003) 2931–2933. Available from: https://doi.org/10.1063/1.1602578.
[91] C.H. Chon, K.D. Kihm, S.P. Lee, S.U.S. Choi, Empirical correlation finding the role of temperature and particle size for nanofluid (Al_2O_3) thermal conductivity enhancement, Appl. Phys. Lett. 87 (2005) 153107. Available from: https://doi.org/10.1063/1.2093936.
[92] M.K. Abdolbaqi, W.H. Azmi, R. Mamat, K.V. Sharma, G. Najafi, Experimental investigation of thermal conductivity and electrical conductivity of bioglycol—water mixture based Al_2O_3 nanofluid, Appl. Therm. Eng. 102 (2016) 932–941. Available from: https://doi.org/10.1016/j.applthermaleng.2016.03.074.
[93] S. Aberoumand, A. Jafarimoghaddam, M. Moravej, H. Aberoumand, K. Javaherdeh, Experimental study on the rheological behavior of silver-heat transfer oil nanofluid and suggesting two empirical based

correlations for thermal conductivity and viscosity of oil based nanofluids, Appl. Therm. Eng. (2016). Available from: https://doi.org/10.1016/j.applthermaleng.2016.01.148.

[94] R.S. Khedkar, N. Shrivastava, S.S. Sonawane, K.L. Wasewar, Experimental investigations and theoretical determination of thermal conductivity and viscosity of TiO_2−ethylene glycol nanofluid, Int. Commun. Heat Mass Transf. 73 (2016) 54−61. Available from: https://doi.org/10.1016/j.icheatmasstransfer.2016.02.004.

[95] Fluent Incorporated, Fluent 6.2 User Manual, 2006.

[96] L. Godson, B. Raja, D. Mohan Lal, S. Wongwises, Enhancement of heat transfer using nanofluids—an overview, Renew. Sustain. Energy Rev. 14 (2010) 629−641.

[97] M. Akhtari, M. Haghshenasfard, M. Talaie, Numerical and experimental investigation of heat transfer of α-Al_2O_3/water nanofluid in double pipe and shell and tube heat exchangers, Numer. Heat Transf. Part A: Appl. 63 (2013) 941−958.

[98] M. Izadi, A. Behzadmehr, D. Jalali-Vahida, Numerical study of developing laminar forced convection of a nanofluid in an annulus, Int. J. Therm. Sci. 48 (2009) 2119−2129.

[99] J.D. Anderson, Computational Fluid Dynamics, McGraw-Hill, New York, 1995.

[100] M.K. Moraveji, M. Darabi, S.M.H. Haddad, R. Davarnejad, Modeling of convective heat transfer of a nanofluid in the developing region of tube flow with computational fluid dynamics, Int. Commun. Heat Mass Transf. 38 (2011) 1291−1295.

[101] K.B. Anoop, T. Sundararajan, S.K. Das, Effect of particle size on the convective heat transfer in nanofluid in the developing region, Int. J. Heat Mass Transf. 52 (2009) 2189−2195.

[102] A.M. Hussein, K. Sharma, R. Bakar, K. Kadirgama, The effect of crosssectional area of tube on friction factor and heat transfer nanofluid turbulent flow, Int. Commun. Heat Mass Transf. 47 (2013) 49−55.

[103] B.E. Launder, D.B. Spaulding, Mathematical Models of Turbulence, Academic Press, New York, 1972.

[104] W. Duangthongsuk, S. Wongwises, An experimental study on the heat transfer performance and pressure drop of TiO_2-water nanofluids flowing under a turbulent flow regime, Int. J. Heat Mass Transf. 53 (2010) 334−344.

[105] H. Mohammed, P. Gunnasegaran, N. Shuaib, The impact of various nanofluid types on triangular microchannels heat sink cooling performance, Int. Commun. Heat Mass Transf. 38 (2011) 767−773.

[106] R. Chein, J. Chuang, Experimental microchannel heat sink performance studies using nanofluids, Int. J. Therm. Sci. 46 (2007) 57−66.

[107] W. Rashmi, M. Khalid, A.F. Ismail, R. Saidur, A. Rashid, Experimental and numerical investigation of heat transferr in CNT nanofluids, J. Exp. Nanosci. 10 (2015) 545−563.

[108] X. Meng, Y. Li, Numerical study of natural convection in a horizontal cylinder filled with water-based alumina nanofluid, Nanoscale Res. Lett. 10 (2015) 1−10.

[109] M. Rakhsha, F. Akbaridoust, A. Abbassi, M. Safar-Avval, Experimental and numerical investigation of turbulent forced convection flow of nano-fluid in helical coiled tube sat constant surface temperature, Powder Technol. (2015).

[110] P.M. Kumar, K. Palanisamy, J. Kumar, R. Tamilarasan, S. Sendhilnathan, CFD analysis of heat transfer and pressure drop in helically coiled heat exchangers using Al_2O_3/water nanofluid, J. Mech. Sci. Technol. 29 (2015) 697−705.

[111] O. Manca, S. Nardini, D. Ricci, A numerical study of nanofluid forced convection in ribbed channels, Appl. Therm. Eng. 37 (2012) 280−292.

[112] F.R. Menter, Two-equation eddy-viscosity turbulence models for engineering applications, AIAAJ 32 (1994) 1598−1605.

[113] S.E.B. Maiga, S.J. Palm, C.T. Nguyen, G. Roy, N. Galanis, Heat transfer enhancement by using nanofluids in forced convection flows, Int. J. Heat Fluid Flow. 26 (2005) 530−546.

[114] A. Bejan, Heat Transfer, John Wiley & Sons Inc., New York, 1993.

[115] A. Behzadmehr, M. Saffar-Avval, N. Galanis, Prediction of turbulent forced convection of a nanofluid in a tube with uniform heat flux using a two phase approach, Int. J. Heat Fluid Flow. 28 (2007) 211–219.

[116] Y. Xuan, Q. Li, Investigation on convective heat transfer and flow features of nanofluids, J. Heat Transf. 125 (2003) 151–155.

[117] S. Mirmasoumi, A. Behzadmehr, Numerical study of laminar mixed convection of a nanofluid in a horizontal tube using two-phase mixture model, Appl. Therm. Eng. 28 (2008) 717–727.

[118] M. Saberi, M. Kalbasi, A. lipourzade, Numerical study of forced convective heat transfer of nanofluids inside a vertical tube, Int. J. Therm. Technol. 3 (2013) 10–15.

[119] A. Moghadassi, E. Ghomi, F. Parvizian, A numerical study of water based Al_2O_3 and Al_2O_3–Cu hybrid nanofluid effect on forced convective heat transfer, Int. J. Therm. Sci. 92 (2015) 50–57.

[120] P. Naphon, L. Nakharintr, Numerical investigation of laminar heat transfer of nanofluid-cooled mini-rectangular fin heat sinks, J. Eng. Phys. Thermophys. 88 (2015) 666–675.

[121] M. Corcione, M. Cianfrini, A. Quintino, Enhanced natural convection heat transfer of nanofluids in enclosures with two adjacent walls heated and the two opposite walls cooled, Int. J. Heat Mass Transf. 8 (2015) 902–913.

[122] M. Kalteh, A. Abbassi, M. Saffar-Avval, J. Harting, Eulerian–Eulerian two-phase numerical simulation of nanofluid laminar forced convection in a micro-channel, Int. J. Heat Fluid Flow. 32 (2011) 107–116.

[123] M. Kalteh, A. Abbassi, M. Saffar-Avval, A. Frijns, A. Darhuber, J. Harting, Experimental and numerical investigation of nanofluid forced convection inside a wide microchannel heatsink, Appl. Therm. Eng. 36 (2012) 260–268.

[124] B. Fani, A. Abbassi, M. Kalteh, Effect of nanoparticles size on thermal performance of nanofluid in a trapezoidal microchannel-heat-sink, Int. Commun. Heat Mass Transf. 45 (2013) 155–161.

[125] A. Beheshti, M.K. Moraveji, M. Hejazian, Comparative numerical study of nanofluid heat transfer through an annular channel, Numer. Heat Transf. Part. A: Appl. 67 (2015) 100–117.

[126] R. Lotfi, Y. Saboohi, A. Rashidi, Numerical study of forced convective heat transfer of nanofluids: comparison of different approaches, Int. Commun. Heat Mass Transf. 37 (2010) 74–78.

[127] M. Akbari, N. Galanis, A. Behzadmehr, Comparative analysis of single and two-phase models for CFD studies of nanofluid heat transfer, Int. J. Therm. Sci. 50 (2011) 1343–1354.

[128] M. Akbari, N. Galanis, A. Behzadmehr, Comparative assessment of single and two-phase models for numerical studies of nanofluid turbulent forced convection, Int. J. Heat. Fluid Flow. 37 (2012) 136–146.

[129] M.K. Moraveji, R.M. Ardehali, CFD modeling (comparing single and two-phase approaches) on thermal performance of Al_2O_3/water nanofluid in mini-channel heat sink, Int. Commun. Heat Mass Transf. 44 (2013) 157–164.

[130] M. Hejazian, M.K. Moraveji, A comparative analysis of single and two-phase models of turbulent convective heat transfer in a tube for TiO_2 nanofluid with CFD, Numer. Heat Transf. Part A: Appl. 63 (2013) 795–806.

[131] V. Bianco, F. Chiacchio, O. Manca, S. Nardini, Numerical investigation of nanofluids forced convection in circular tubes, Appl. Therm. Eng. 29 (2009) 3632–3642.

[132] M. Keshavarz Moraveji, E. Esmaeili, Comparison between single-phase and two-phases CFD modeling of laminar forced convection flow of nanofluids in a circular tube under constant heat flux, Int. Commun. Heat Mass Transf. 39 (2012) 1297–1302.

[133] S. Zeinali Heris, M. Nasr Esfahany, S.G. Etemad, Experimental investigation of convective heat transfer of Al_2O_3/water nanofluid in circular tube, Int. J. Heat Fluid Flow. 28 (2007) 203–210.

[134] Y. He, Y. Men, Y. Zhao, H. Lu, Y. Ding, Numerical investigation into the convective heat transfer of TiO_2 nanofluids flowing through a straight tube under the laminar flow conditions, Appl. Therm. Eng. 29 (2009) 1965–1972.

[135] M. Mirzaei, M. Saffar-Avval, H. Naderan, Heat transfer investigation of laminar developing flow of nanofluids in a microchannel based on Eulerian–Lagrangian approach, Can. J. Chem. Eng. (2014).

[136] H. Bahremand, A. Abbassi, M. affar-Avval, Experimental and numerical investigation of turbulent nanofluid flow in helically coiled tubes under constant wall heat flux using Eulerian–Lagrangian approach, Powder Technol. 269 (2015) 93–100.

[137] M. Mahdavi, M. Sharifpur, J. Meyer, CFD modelling of heat transfer and pressure drops of nanofluids through vertical tubes in laminar flow by Lagrangian and Eulerian approaches, Int. J. Heat Mass Transf. 88 (2015) 803–813.

[138] I. Behroyan, et al., Turbulent forced convection of Cu–water nanofluid: CFD model comparison, Int. Commun. Heat Mass Transf. (2015). Available from: https://doi.org/10.1016/j.icheatmasstransfer.2015.07.014.

[139] J. Zhang, Lattice Boltzmann method for microfluidics: models and applications, Microfluid. Nanofluidics 10 (2011) 1–28.

[140] H.K. Jeong, H.S. Yoon, M.Y. Ha, M. Tsutahara, An immersed boundary-thermal lattice Boltzmann method using an equilibrium internal energy density approach for the simulation of flows with heat transfer, J. Comput. Phys. 229 (2010) 2526–2543.

[141] Y. Xuan, Z. Yao, Lattice Boltzmann model for nanofluids, Heat Mass Transf. 41 (2005) 199–205.

[142] F.-H. Lai, Y.-T. Yang, Lattice Boltzmann simulation of natural convection heat transfer of Al_2O_3/water nanofluids in a square enclosure, Int. J. Therm. Sci. 50 (2011) 1930–1941.

[143] E. Fattahi, M. Farhadi, K. Sedighi, H. Nemati, Lattice Boltzmann simulation of natural convection heat transfer in nanofluids, Int. J. Therm. Sci. 52 (2012) 137–144.

[144] M. Sheikholeslami, M. Gorji-Bandpy, S.M. Seyyedi, D.D. Ganji, H.B. Rokni, S. Soleimani, Application of LBM in imulation of natural convection in a nanofluid filled square cavity with curve boundaries, Powder Technol. 247 (2013) 87–94.

[145] H. Nemati, M. Farhadi, K. Sedighi, E. Fattahi, A.A.R. Darzi, Lattice Boltzmann simulation of nanofluid in lid-driven cavity, Int. Commun. Heat Mass Transf. 37 (2010) 1528–1534.

[146] M. Nabavitabatabayi, E. Shirani, M.H. Rahimian, Investigation of heat transfer enhancement in an enclosure filled with nanofluids using multiple relaxation time lattice Boltzmann modeling, Int. J. Commun. Heat Mass Transf. 38 (2011) 128–138.

[147] G.R. Kefayati, S.F. Hosseinizadeh, M. Gorji, H. Sajjadi, Lattice Boltzmann simulation of natural convection in tall enclosures using water/SiO_2 nanofluid, Int. Commun. Heat Mass Transf. 38 (2011) 798–805.

[148] Y. Xuan, K. Yu, Q. Li, Investigation on flow and heat transfer of nanofluids by the thermal Lattice Boltzmann model, Prog. Comput. Fluid Dyn. Int. J. 5 (2005) 13–19.

[149] A. Zarghami, S. Ubertini, S.F. Succi, Lattice Boltzmann modeling of thermal transport in nanofluids, Comput. Fluids 77 (2013) 56–65.

[150] Y.-T. Yang, F.-H. Lai, Numerical study of flow and heat transfer characteristics of alumina-water nanofluids in a microchannel using the lattice Boltzmann method, Int. Commun. Heat Mass Transf. 38 (2011) 607–614.

[151] M. Mohammadi Pirouz, M. Farhadi, K. Sedighi, H. Nemati, E. Fattahi, Lattice Boltzmann simulation of conjugate heat transferrin a rectangular channel with wall-mounted obstacles, Sci. Iran. 18 (2011) 213–221.

[152] G. Li, M. Aktas, Y. Bayazitoglu, A review on the discrete Boltzmann model for nanofluid heat transfer in enclosures and channels, Numer. Heat Transf. Part B Fund. 67 (6) (2015) 463–488.

[153] M. Hatami, M.J. Uddin, M. Hu, D. Jing, M. Javed, Two-phase natural convection analysis and hybrid nanoparticle migration around micromixer blades, Heat Transf. 49 (5) (2020) 3044–3065.

[154] S.M. Vanaki, P. Ganesan, H.A. Mohammed, Numerical study of convective heat transfer of nanofluids: a review, Renew. Sustain. Energy Rev. 54 (2016) 1212–1239.

CHAPTER

Mass transfer enhancement in liquid–liquid extraction process by nanofluids

12

Morteze Esfandyari[1] and Ali Hafizi[2]

[1]*Department of Chemical Engineering, University of Bojnord, Bojnord, Iran* [2]*Department of Chemical Engineering, Shiraz University, Shiraz, Iran*

12.1 Introduction

Liquid–liquid extraction, which is also known as extraction by solvent, is an indirect and sometimes direct extraction process in which the solution's components are separated by contacting with another nonsolution liquid [1]. The basic principal of separation in the liquid–liquid extraction is the difference in the chemical properties and this process is usually used when other processes like distillation, evaporation, and crystallization are not applicable [2]. Liquid–liquid extraction is one of the most important separation processes in chemical engineering which is used extensively in variant industries such as separation of metals and hydrometallurgy process, pharmacology, environment, oil, food industry, and nuclear fuels. Various liquid–liquid extraction systems are used in industry in order to perform a two-step mixing. One of the most important units that use this process is packed columns. These columns are usually used for continuous extraction phases, when the balance is only needed in a few steps. They are basically similar to spray columns, but the packages which exist in a winding path increase the scattered phase, which results in higher efficiency of the columns. In addition, the packs can increase the mass transfer by decomposing the drops of scattered phase and decrease the mixing [1,3].

The liquid–liquid extraction operation is very important in industries such as refineries, petrochemicals, purification of metals, pharmaceutical and food industries, and any improvement in the amount and intensity of mass transfer and the purity of the product can play an important role in increasing the efficiency of the mentioned industries. In addition, extraction by liquid requires using several devices such as spraying columns, rotating disk, filled columns, and pulse. Among these devices the spraying columns have many advantages such as simplicity, low cost of design, and construction and maintenance. However low efficiency of the spraying columns has limited their industrial applications.

Mass transfer effects the size of drops by changing two parameters: break and the cohesion drops. Due to mass transfer from the continuous phase to the scattered phase, an area will be created between two drops with low concentration; this concentration is lower than the concentration of media around the drop. The cross phase tension between two drops is relatively higher than the tension between phases around the drops which causes the Marangoni phenomenon. This

Nanofluids and Mass Transfer. DOI: https://doi.org/10.1016/B978-0-12-823996-4.00015-X

phenomenon runs the liquid from around the drops to the area between them and in conclusion the drops take distance and the coalition decreases between them. In other words, when the mass transfer direction is from continuous phase to a scattered phase the Marangoni phenomenon causes the instability of the system and increases the amount of the scattered phase by reinforcing the drop breaking mechanism [3].

Generally, the average diameter of drops, their size distribution, and the scattered phase maintenance are the key parameters in design and optimization of the extraction columns; as a result, the size of particles has a major impact on mass transfer. Particles with sizes between 1 and 100 nanometers are called nanoparticles, and they have a considerable effect on the mass transfer of extraction by the solvent processes due to their effect on decomposition and interlocking of the drops.

The drop properties affect all the related parameters which have an effect on hydrodynamic and mass transfer in the liquid–liquid extraction column. The average decrease of a parameter is important as it effects the residence of scattered phase, the rising rate of the drops, the capacity, and maximum volume of the extraction process. The average diameter (Sauter) of drops are measured as below:

$$\bar{d} = \frac{\sum n_i d_i^3}{\sum n_i d_i^2} \tag{12.1}$$

where n_i is the number of drops and d_i is diameter of the drops.

Nanofluid is defined as a fluid with the particles in the size range of 1–100 nanometers and hovering continuously in the base fluid. Addition of nanoparticles affects the thermophysical features of fluid. Nowadays extensive research has been done extensively in the nanofluid area. The studies have shown that using nanofluids in the base fluid causes the increase of heat transfer and mass transfer coefficients. The increase of these coefficients causes an increase of yield, and a decrease in the usage of devices and also the construction costs. Nanofluids are known as perfect inductors for heat and mass transfer when the nanoparticles are well separated in the base fluid. Based on extensive research it has been determined that continuous distribution of nanoparticles in a nanofluid increases the thermal conductivity coefficient of the fluid, forces the thermal convection coefficient, and also combines convection abnormally, which also increases heat transfer rate [4,5].

A nanofluid is composed of a base liquid and a nanofactor with variable size of metal or nonmetal nanoparticles, such as Cu, Al_2O_3, SiO_2, TiO_2, and Fe_3O_4. [6]. The main difference between nanofluids and regular suspensions is caused by the ultrafine size of suspended particles [7]. Actually, many factors, such as viscosity, temperature, volumetric fraction, size, shape, and clustering of particles, can affect the resulting mass transfer in nanofluids and most of the research has focused on volumetric fraction, thermal conductivity, and the Brownian movement of nanoparticles [1,8]. Many of the effective forces from a macro point of view become insignificant in terms of their efficiency by shrinking the size of particles and increasing their surface. Properties such as high special surface, movement and velocity of small particles, and high activity result in the choice of nanoparticles as a proper option for making effective suspensions in heat transfer. Some researchers have observed high mass transfer rates by using nanoparticles in bubble mass transfer systems [7]. The mechanism of mass transfer systems in the presence of nanoparticles has not been well determined due to the lack of experimental data. More research is needed in this field to determine the effective factors [8].

Table 12.1 Relationships of thermophysical properties of nanofluids.

Property	Relation	Parameter definition	References
Viscosity	$\mu_{nf} = \rho_p\left(kt_e - \frac{c}{t_e}\right)$	t_e: efflux time; ρ_p : density; k and c are the viscometer constants determined by calibration	[9]
	$\mu_{nf} = \mu_f + \frac{\rho_p V_B d_p^2}{72c\delta}$	ρ_p : density; d_p: denotes the particle diameter; δ: indicates the distance between the nanoparticles; C and V_B are the two functions of temperature; μ_f: viscosity of fluid	[2]
	$\frac{\mu_{nf}}{\mu_f} = \left(1+\frac{\phi}{100}\right)^{11.3}\left(1+\frac{T_{nf}}{70}\right)^{-0.038}\left(1+\frac{d_p}{170}\right)^{-0.061}$	φ: volume fraction of nanoparticle; T_{nf} : temperature of nanofluid; d_p : diameter of nanoparticle;	[2]
	$\mu_{nf} = \mu_f\left(\frac{1}{1-34.87\left(\frac{d_p}{d_f}\right)^{-0.3}\phi^{1.03}}\right)$	d_f : is the equivalent diameter of the base fluid molecule	[10]
Surface tension	$\rho_{nf} = \rho_{bf}(1-\phi) + \rho_p\phi$	ρ_p: density of nanoparticle; φ: volume fraction of nanoparticle; ρ_{bf} : density of base fluid	[9]
Tension	$\gamma = \frac{V\Delta\rho g}{2\pi rf\left(\frac{r}{\sqrt[3]{V}}\right)}$	r: outer radius of the capillary; g: the acceleration of gravity; $\Delta\rho$: difference in density between two phases; V: the volume of the drop that falls from the capillary into the organic phase	[2]

Based on the recent research about the effect of nanoparticles on liquid–liquid extraction and to analyze the efficiency of this process using nanofluid, it is necessary to determine thermophysical properties of the fluid, such as viscosity, density, and surface tension. These properties could be calculated by the equations tabulated in Table 12.1.

12.2 Mass transfer in nanofluids

12.2.1 Molecular penetration in nanofluids

Penetration coefficient is an indicator of the penetration power of one material in another and its amount depends on the intended compound and the conditions of media around it, such as temperature, pressure, and concentration. The rise of the penetration coefficient increases the mass transfer coefficient which is very important in the mass transfer process. Krishnamurthy et al. [11] are the pioneers in mass penetration in nanofluids. They have analyzed the effect of color penetration in water–aluminum oxide. They have shown that the penetration coefficient in nanofluids is higher than in distilled water (base fluid) and the maximum penetration coefficient can be achieved in a concentration of 0.5 vol.% of nanoparticles. The amount of penetration of color in pure water is equal to $5.2 \times 10^{-10} m^2 \cdot s^{-1}$ while it is equal to $1.1 \times 10^{-8} m^2 \cdot s^{-1}$ in nanofluid with 0.5 wt.%, which shows the increase of mass transfer coefficient in nanofluids. At low concentrations Brownian movement creates chaos fields in nanofluids, which can be the reason for the increase of penetration, but with increasing concentration of nanoparticles, they will stick together and get heavier which decreases the Brownian movement of nanoparticles and thus as a result the penetration coefficient will be reduced.

Kang et al. [12] analyzed the falling liquid film with two nanoparticles. They used H_2O/LiBr solution, metal nanoparticles, and carbon nanotubes in different concentrations. They finally came to a conclusion that by the increase of nanoparticles concentration and intensity of mass flowrate, steam absorption intensity will rise.

Based on the Olle experiments using a nanofluid including 1% of Fe_2O_3 nanoparticles covered by oleic acid, the mass transfer coefficient decreases 1.6 times while k_{la} saw an about sixfold rise [13].

Another similar experiment was done by Fang et al. [14]. They observed that the addition of copper nanoparticles to water increased the color penetration. They used different parts of florescent rhodamine B in Cu–water nanofluids at various temperatures and observed that mass transfer in water with 0.5% copper nanoparticles increased 10.71 times more than distilled water. Feng et al. [15] started analyzing the effect of penetration of sodium chloride in silica–water nanofluid and they observed that the addition of nanoparticles did not affect the penetration of sodium chloride. Most of the research performed on analyzing penetration of nanofluids has shown that there is an optimal concentration beyond which the penetration coefficient falls.

12.2.2 Calculation of mass transfer coefficient

General volumetric mass transfer coefficient is one of the important parameters in the design and choice of the optimal conditions for liquid–liquid extraction devices. It is a function of the general mass transfer coefficient, stability of scattered phase, and Sauter average diameter of drops. By writing mass balance:

$$\dot{m}_1 - \dot{m}_2 = \frac{dm}{dt} \tag{12.2}$$

$$-N_{Ar} S_r M_A = \frac{d(n_A M_A)}{dt} = \frac{V_A dC_A}{dt} \tag{12.3}$$

$$K_d \left(C_A - C_A^*\right) \times 4\pi r^2 = \frac{4}{3}\pi r^3 \frac{dC_A}{dt} \tag{12.4}$$

$$K_d \int_0^t dt = -\frac{r}{3} \int_{C_{A0}}^{C_A} \frac{dC_A}{\left(C_A - C_A^*\right)} \tag{12.5}$$

$$K_d = -\frac{d}{6t} \ln\left(\frac{C_A - C_A^*}{C_{A0} - C_A^*}\right) = -\frac{d}{6t} \ln\left(1 - \frac{C_{A0} - C_A}{C_{A0} - C_A^*}\right) \tag{12.6}$$

After integration and sorting Eqs. 12.1–12.6 the general mass transfer coefficient equation is as follows:

$$K_d = -\frac{d}{6t}\ln(1 - E) \tag{12.7}$$

where t is the average rising time of drops, d is the average diameter of drops, and E is the yield of extraction which is defined as below:

$$E = \frac{C_{A0} - C_A}{C_{A0} - C_A^*} \tag{12.8}$$

where C_A^* and C_{A0} are the balanced and resolvable concentration in drops before contacting the continuous phase, respectively. Eq. 12.9 is used for calculation of raising time of drops:

$$t = \frac{L\varepsilon\varphi A}{Q_d} \tag{12.9}$$

where L is the active height of column, ε is the porosity filler coefficient, A is cross section surface area of the column, φ is the stability of scattered phase, and Q_d is flow of scattered phase. The drops' special surface area is calculated by Eq. 12.10:

$$a = \frac{6\varphi\varepsilon}{d} \tag{12.10}$$

Eq. 12.11 is resulted by multiplying the drops' specific surface area and the general mass transfer coefficient. The general volumetric mass transfer coefficient is as follows:

$$K_d \cdot a = -\frac{\varphi\varepsilon}{t}\ln(1 - E) \tag{12.11}$$

12.3 Mass transfer of liquid–liquid extraction

Fig. 12.1 shows a liquid–liquid extraction process in which the extraction column can have trays or be packed. Additionally, the dispersed phase holder tank is usually placed higher than the Earth's surface to provide enough pressure for the transfer of the scattered phase. The solenoid valve's duty is to control the flow of scattered phase into the column and the nozzle is used as a spreader at the end of the column to measure the diameter of the drops by photographing which is the most common way for this process.

The liquid–liquid extraction process can be divided into two categories, including packed and column liquid–liquid extraction. And again, each of these processes can be divided into two categories: with an external field and without an external field, as shown in Fig. 12.2.

Using nanofluids has attracted the attention of many researchers from various fields due to achieving more functionality and better efficiency. Generally, mass transfer by nanoparticles concentration does not show the same trend of heat transfer and based on existing nanoparticles in scattered or continuous phase, positive or negative results have been reported. Undesired effects have been observed by the addition of nanoparticles to a continuous phase [16]. While there is an

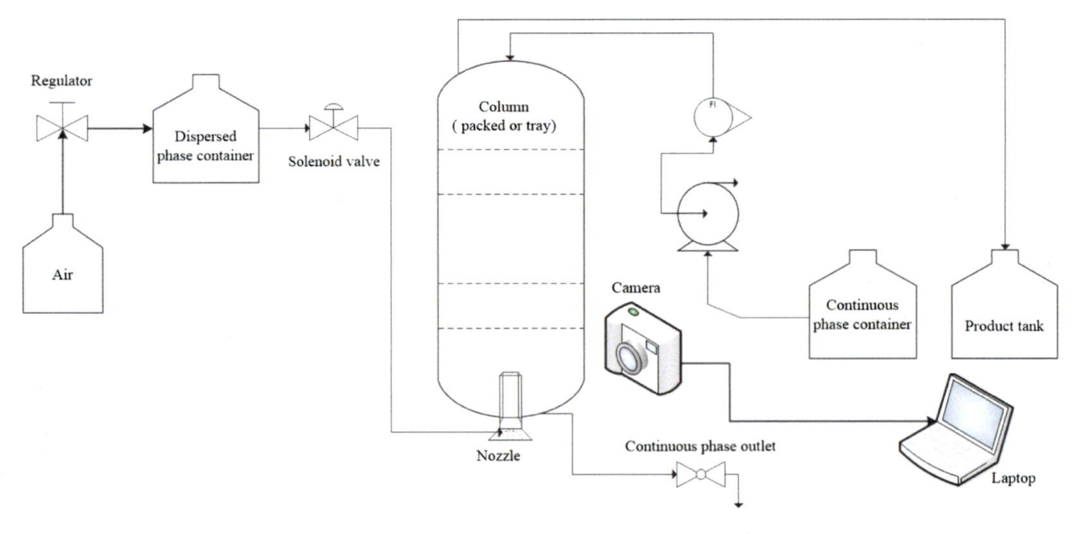

FIGURE 12.1

Schematic of a experimental apparatus for liquid–liquid extraction system.

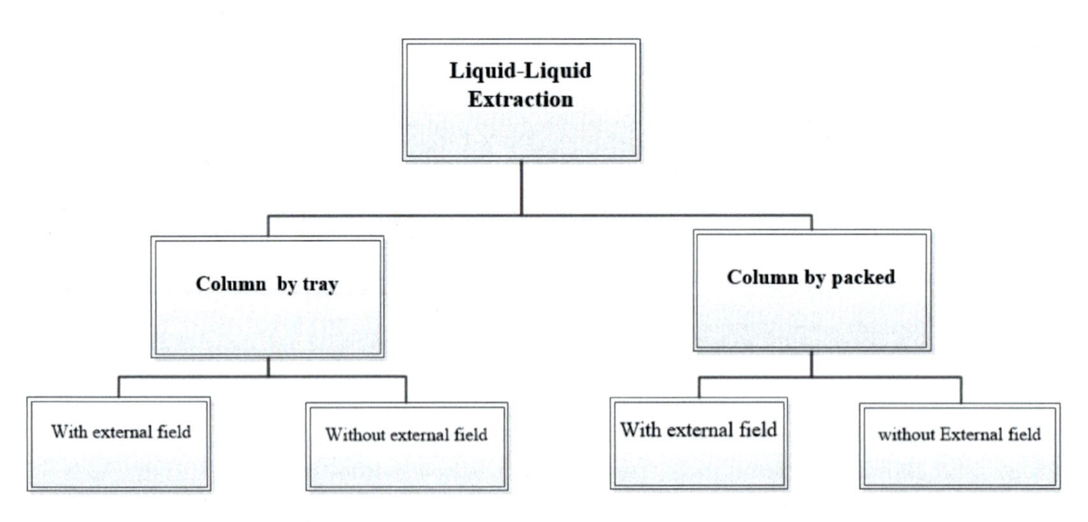

FIGURE 12.2

Schematic diagram of the liquid–liquid extraction.

increase of nanoparticles in particle transfer in continuous phases [7]. It is worth noting that most of the mass transfer increase in the presence of nanoparticles was detected in gas liquid systems. Research about gas liquid mass transfer with nanoliquid have been done by Ashrafmansoury and Nasr in Isfahan in 2014 [17].

In liquid–liquid extraction systems in experimental and industrial scales, water is one of the interacting components. The pH of water can change due to different resources and being exposed to air and carbon dioxide, where the increase of pH can result in a decrease of surface tension which in turn can decrease the amount of downfall. Moreover, it is reported that mass transfer will decrease by increasing pH. This can be related to absorbing hydroxyl ions which decreases the surface mobilization.

Temperature is another parameter in mass transfer which affects the physical features of the liquids and improves the performance of the unit. However, due to normal function at the room temperature, this parameter has been undermined. Researchers have shown that the general mass transfer coefficient increases with temperature and smaller drops are more apt in this matter.

The considerable increase of overall mass transfer of liquid–liquid transfer equipped with ultrasound transducer shows that release of ultrasound waves can improve mass transfer. This can be related to the provision of turbulence and stimulation at a small scale. Finally, the addition of chemicals causes the increase or decrease of mass transfer. The effect of addition of nanoparticles on the mass transfer function of liquid–liquid will be analyzed in the following. The properties of nanofluids have drawn the attention of different fields of transportation. The incredible features of nanoparticles in terms of increasing heat transfer encourages researchers to examine their function in mass transfer. In numerous studies it has been mentioned that the Brownian movement of nanoparticles is the main factor responsible for upgrading mass transfer. Brownian movement is known as the random movement of suspended nanoparticles in a basic liquid; particle movement makes a microconvection that causes mass or heat transfer. Where, Table 12.2 shows the effect of nanofluid on the performance of liquid–liquid extraction.

12.4 Nanoparticles in liquid–liquid extraction

Different nanoparticles based on metal oxides, such as SiO_2, ZnO, MgO, and Fe_3O_4, and nanostructures, such as carbon nanotubes, have been investigated in the liquid–liquid extraction process. Different properties of these particles, such as particle size distribution, mean particles size, hydrophobicity, and hydrophilicity, cause different effects on extraction performance. Some of the nanoparticles investigated recently are listed in Table 12.3.

12.4.1 SiO_2 nanoparticles

The nanoparticles of SiO_2 are widely applied in nanofluids due to their high stability, low price, easy synthesis procedure, monodispersed spheres, and the ability to attach different functional groups to them [25]. For such reasons, the SiO_2 nanoparticles are commonly used in liquid–liquid extraction. Different studies have investigated this nanoparticle in a single drop liquid–liquid extraction columns

The effect of SiO_2 nanoparticles with hydrophilic nature was examined in a single drop extraction by Goodarzi and Nasr Esfahany [16]. They added the nanoparticles to the water as a continuous phase in the ternary chemical system of toluene–acetic acid–water. The obtained results

Table 12.2 A list of studies about the influence of nanofluids on liquid–liquid extraction.

System	Characterization of system	Nanofluid		Chemical system	Results	References
		Type	Concentration or volume fraction			
Irregular packed liquid–liquid extraction columns	The stainless steel Raschig ring random packings Column total height: 1.6 m Internal diameter: 0.052 m External diameter: 0.055 m Packed height: 1.2 m	Spherical SiO_2 nanoparticles (10, 30, 80 nm)	0, 0.01, 0.05, and 0.1 vol.%	Water–acetic acid–toluene	Enhancement in mass transfer coefficient up to 42% at 0.05% concentration of nanoparticles	[3]
Irregular packed liquid–liquid extraction column	The stainless steel Raschig Ring random packings Column total height: 1.6 m Internal diameter: 0.052 m External diameter: 0.055 m Packed height:1.2 m	Spherical SiO_2, TiO_2, and ZrO_2 nanoparticles (10 nm)	0, 0.01, 0.05, and 0.1 vol.%	Toluene–acetic acid–water	The improvements of 35%, 245%, and 207% were attained for 0.05 vol.% of SiO_2, TiO_2, and ZrO_2 nanofluids	[18]
Pulsed liquid–liquid extraction	Contactor (0.114 m diameter and 0.51 m height) diameter of the nozzle: 0.0003 m	Modified magnetite (Fe_3O_4) and alumina (Al_2O_3)	0.0005, 0.001, 0.002, 0.003, 0.004, and 0.005 wt.%.	Toluene–acetic acid–water	The rate of mass transfer was increased 157% and 121% by 0.002 wt.% of magnetite and alumina nanoparticles	[7]
Stagnant and stirred liquid–liquid extraction	Extraction column, with a length of 0.5 m and width of 0.1 m, an electric motor (180 W) with rotor speed 330 and 450 rpm Temperature: 20 ± 0.5°C	Magnetite nanoparticles with a size of 17.1 ± 3.8 nm	0.001, 0.002, 0.003, 0.004, and 0.005 wt.%	Toluene–acetic acid–water	Improvement of 103.1% in general dispersed-phase mass transfer coefficient at the optimum concentration of 0.001 wt.%	[9]
Spray liquid–liquid extraction column	Column with an inner diameter of 0.056 m and a total height of 1.33 m temperature: 25°C	SiO_2	0.0005, 0.001, 0.005, and 0.01 vol.%	Toluene–acetic acid–water	Maximum enhancement of 47.4% and 107.5% in overall mass-transfer coefficient	[19]
Pulsed liquid–liquid extraction column	Fixed mass flow rates of dispersed (Q_d) and continuous (Q_c) phases (with ratio $Q_c/Q_d = 1.2$)	SiO_2	0.01, 0.05, and 0.1 vol.%	Kerosene–acetic acid–water	The addition of 0.1 vol.% nanoparticles advances the mass transfer performance by up to 60%	[20]
Liquid–liquid extraction	Column (0.05 m inside diameter and 0.5 m height) Syringe nozzle: 0.001 m Syringe volume: 50 mL Concentration inlet (succinic acid): 10 $g.L^{-1}$	ZnO, carbon nanotubes (CNT), and TiO_2 nanoparticles	0.025, 0.05, and 0.1 wt.%	n-Butanol–succinic acid–water	The obtained results revealed that the effect of ZnO nanoparticles on mass transfer enhancement is more than that of CNT and TiO_2 nanoparticles	[21]

Table 12.3 Recently experimental investigations of nanoparticles in liquid–liquid extraction columns.

Nanofluid	Particle concentration	Remark	References
Fe_3O_4–toluene–acetic acid	0.0005–0.005 wt.%	• The optimum mass transfer enhancement was achieved at a nanoparticle concentration of 0.002 wt.% • The rate of mass transfer is increased up to an average of 72% (the smallest drop about 157% and the largest drop 46%) • As the nanofluid concentration increases, the mass transfer coefficient first increases and then decreases	[7]
Al_2O_3–toluene–acetic acid	0.0005–0.005 wt.%	• The optimum mass-transfer enhancement was achieved at a nanoparticle concentration of 0.002 wt.% (maximum enhancements in the rate of mass transfer of 121%) • The rate of mass transfer rises to an average of about 75% (121% for the smallest and 55% for the largest) • Increase the mass transfer coefficient by 75% • As the nanofluid concentration increases, the mass transfer coefficient first increases and then decreases	[7]
Fe_3O_4–toluene–acetic acid	0.001–0.005 wt.%	• The optimum mass-transfer enhancement was achieved at a nanoparticle concentration of 0.001 wt.% (maximum enhancements in the rate of mass transfer of 103.1%) • As the nanofluid concentration increases, the mass transfer coefficient first increases and then decreases	[9]
SiO_2–toluene–acid acetic	0–0.01 vol.%	• The mass transfer coefficient of nanofluid depends on the concentration • The maximum increase in mass transfer coefficient was 35% for 0.05 vol.% of nanofluids • The effects of the presence of hydrophilic nanoparticles are greater than the Brownian motion of nanoparticles on the mass transfer coefficient	[18]
TiO_2–toluene–acid acetic	0–0.01 vol.%	• The mass transfer coefficient depends on the nanofluid concentration • The maximum increase in mass transfer coefficient was 245% for 0.05 vol.% of nanofluids • The effects of the presence of hydrophilic nanoparticles are greater than the Brownian motion of nanoparticles on the mass transfer coefficient	[18]
ZrO_2–toluene–acid acetic	0–0.01 vol.%	• The mass transfer coefficient depends on the nanofluid concentration	[18]

(*Continued*)

Table 12.3 Recently experimental investigations of nanoparticles in liquid–liquid extraction columns. ***Continued***

Nanofluid	Particle concentration	Remark	References
SiO_2–toluene–acid acetic	0–0.1 vol.%	• The maximum increase in mass transfer coefficient was 207% for 0.05 vol.% of nanofluids • The effects of the presence of hydrophilic nanoparticles are greater than the Brownian motion of nanoparticles on the mass transfer coefficient • The increase in mass transfer in nanofluids with smaller particles is more significant • The mass transfer coefficient in nanofluids is larger compared to the dispersed phase without particles • The maximum enhancement in the rate of mass transfer 42% was achieved for 0.05 vol.% of nanofluids	[3]
SiO_2–kerosene	0.01–0.1 vol.%	• In the presence of nanoparticles, the mass transfer coefficient increases between 4% and 60% • In the presence of nanofluids, the static and stable dispersed phase rates increased by 23%–398% and 23%–257%, respectively	[22]
SiO_2–kerosene	0.05 vol.%	• The presence of nanoparticles increases the dynamics by up to 70% • The static holdup increases in the presence of nanoparticles along the column	[23]
SiO_2–toluene	0.0005–0.01 vol.%	• Silica nanoparticles do not have a significant effect on hydrodynamic parameters • The maximum enhancement in the rate of mass transfer of 47% was achieved for 0.001 vol.% of nanofluids • The maximum enhancement of extraction efficiency 26% was achieved for 0.001 vol.% of nanofluids	[24]
SiO_2– toluene	0.0005–0.01 vol.%	• The maximum increase of the overall mass transfer coefficient in the direction of mass transfer from dispersed to continuous phase is 47.4% and 107.5%, respectively • No significant variation in holdup observed with increasing the concentration of nanoparticles	[19]

revealed the highest drop of 22% in mass transfer coefficient. It revealed that SiO_2 nanoparticles do not affect the hydrodynamic parameters of the liquid–liquid extraction process meaningfully.

Saien and Hasani [26] investigated SiO_2 nanoparticles with three particle size ranges of 11–14, 20–30, and 60–70 nm and different concentrations on the mass transfer of circulating drops of the toluene–acetic acid–water extraction system. Results revealed that increasing the size and concentration of silica nanoparticles increased the viscosity progressively with constant interfacial tension.

Thus, smaller particles are better due to their lower viscosity and higher intensity Brownian motion. The mean mass transfer rate enhancement of 51.8% resulted from the addition of SiO_2 nanoparticles with the concentration of about 0.003 wt.%.

Besides, according to the structure of such columns and the interactions of neighboring droplets, the attained results are not applicable to other extraction contactors. Thus some investigations on pulsed, spray, packed, and rotary disk columns have been performed during recent years. For this purpose, Bahmanyar et al. [20,22] examined the effect of the mass transfer of SiO_2–kerosene nanofluid drops on the efficiency of a pulsed extraction contactor. They found that static and dynamic dispersed phase holdups significantly increased with the nanofluids concentration. The highest improvement in mass transfer of 60% was achieved with 0.1% of SiO_2 concentration. In addition, the static and dynamic holdups related to dispersed phase were improved considerably with the nanofluids concentration.

Nematbakhsh and Rahbar [3] studied the effects of silica-based nanofluid on mass transfer improvement in an irregularly packed extraction contactor in an acetic acid, water, and toluene system. The SiO_2 nanoparticles with a mean size of 10, 30, or 80 nm and different concentrations were dispersed in toluene–acetic acid. The results revealed that the addition of silica nanoparticles enhanced the mass transfer coefficient. About 42% improvement in mass transfer coefficient was achieved using 0.05% of 10 nm SiO_2 nanoparticles.

12.4.2 ZnO, ZrO_2, and TiO_2 nanoparticles

Mirzazadeh et al. [21] investigated the effect of ZnO and TiO_2 nanoparticles on the liquid–liquid extraction of n-butanol–succinic acid–water in dropping and jetting fluid models. They found that the addition of the as-mentioned nanoparticles generally has a progressive influence on mass transfer rate especially in laminar flow. The obtained results revealed that ZnO nanoparticles improved the mass transfer rate more than TiO_2 nanoparticles. Additionally, Hatami et al. [27] investigated the effect of titanium dioxide nanoparticle on the mass transfer rate of liquid–liquid extraction of toluene and acetic acid. They showed that the superhydrophobic TiO_2 nanoparticles had a positive effect on mass transfer rate until an optimum weight percentage. Rahbar et al. [28] investigated the liquid–liquid extraction of n-butyl acetate with acetic acid and ZrO_2 nanoparticles in a spray extraction column. The obtained results showed that at the optimum concentration of this nanoparticle the mass transfer coefficient is at the maximum, while the addition of SiO_2 nanoparticles revealed a negative effect on this parameter. In addition, they proposed a correlation for the calculation of the Sherwood number in the presence of nanoparticles with good precision.

12.4.3 Al_2O_3 nanoparticles

Al_2O_3 nanoparticles have been studied to investigate their effect on the extraction performance and rate. For instance, liquid–liquid extraction of toluene–acetic acid–water was investigated using a nanofluid comprising Al_2O_3 [7]. The synthesized nanoparticles were functionalized with fatty acids in order to make them hydrophobic particles and to improve the dispersion in organic phases. The addition of 0.002 wt.% alumina nanoparticles improved the rate of mass transfer up to 121%.

12.4.4 MgO nanoparticles

MgO nanoparticles are not widely used in liquid–liquid extraction during the last years due to high price and higher solubility in acidic solvents. Sepehri et al. [29] investigated the effect of MgO nanoparticles on liquid–liquid extraction of acetic acid-water and organic phases (i.e., n-butanol, n-butyl acetate, toluene, and chloroform). Experimental results show that the MgO as a hydrophilic nanoparticle reduced the mass transfer coefficient. However, the MgO nanoparticles revealed better performance than the SiO_2 nanoparticles in the same test conditions.

12.4.5 Fe_3O_4 nanoparticles

According to the previously investigated nanoparticles, most nanoparticles improve the mass transfer performance in liquid–liquid extraction contactors. The application of magnetic nanoparticles could be helpful for a further increase in mass transfer performance along with easy separation and handling. The presence of magnetic nanoparticles in both organic and aqueous environments delivers stable colloidal suspensions. In addition, the magnetic field can help to improve lateral mixing in fluid flows [2,30].

The effect of magnetic nanoparticles on liquid–liquid extraction in a single drop extraction contactor was investigated by Saien et al. [31]. They investigated the effect of oscillating magnetic fields and magnetic nanoparticle concentration on mass transfer of the as-mentioned system. The attained results of improving the mass transfer rate were interpreted as Brownian motion microconvection in a magnetic field in the presence of magnetic nanoparticles. This phenomenon could be related to increasing the rate of collision affecting the heat and mass transfer efficiency of nanofluids. They mentioned that the Lorenz forces can present the lower mass transfer resistance with better transport of solute at higher magnetic field strengths. Saad et al. [32] showed that dispersive liquid–liquid microextraction of chloramphenicol in water improved with Fe_3O_4 nanoparticles, revealing that in the presence of decanoic acid, the extraction rate and efficiency are improved herein. They proposed this system as a simple, effective, suitable, and promising extraction method for the separation of chloramphenicol.

Vahedi et al. [33] studied the effect of Fe_3O_4 nanoparticles with a mean size of about 29 nm for the liquid–liquid extraction of toluene–acetic acid–water in a single drop extractor. The attained results revealed that the nanoparticle concentration has an optimum value for increasing the mass transfer rate. The mass transfer rate was improved by increasing the magnetic field strength due to Brownian motion. Subuhi et al. [34] studied the liquid–liquid microextraction of bisphenol and water using magnetic nanoparticles. This process was performed by the formation of rhamnolipid biosurfactant droplet that improved the mass transfer rate and extraction. Subsequently, the process was completed by the recovery of the extraction solvent with magnetic nanoparticles.

12.5 Conclusions and future outlooks

The present studies indicate the effect of nanofluids on improving the mass transfer coefficient. In this chapter, the recent studies on the application of nanofluids in improving the mass transfer coefficient of liquid–liquid extraction columns have been briefly studied. Although the mechanism of action for nanoparticles in increasing the mass transfer coefficient is still somewhat obscure, mainly most studies revealed that Brownian motion of nanoparticles affects the extraction efficiency.

As indicated in previous sections, nanotechnology has attained wide use in different industrial applications. The effect of nanoparticles on liquid–liquid extraction is one of the research areas that recently has attracted attention. Different experimental research has been accomplished to investigate the effect of various nanoparticles on mass transfer efficiency and hydrodynamic behavior of extraction systems.

Brownian diffusion, diffusiophoresis, liquid layering on the nanoparticle–liquid interface, and induced microconvection are the most important mechanisms proposed for the investigation nanofluids in extraction [2]. Some studies revealed that the performance of mass transfer in liquid–liquid extraction might be reduced using nanoparticles. The hydrophilic nature of some nanoparticles might reveal undesirable effects on liquid–liquid extraction.

The separation of nanoparticles from effluent streams is one of the most important challenges of nanoparticles in extraction processes. Among the researches in this field, some have investigated the handling and separation of nanoparticles in the extraction process, which seems to be the most important issue. The application of nanomaterials in the industrial scale increases the operating costs due to the need for sonication, stabilization, posttreatment, and make-up injection, generally. In addition, the maintenance process costs might be increased due to scaling and corrosive effects of some nanofluidic systems. The sedimentation and agglomeration of nanoparticles along with clogging inside the extraction contactor are the other difficulties of nanoparticles in liquid–liquid extraction systems. All these are some of the most important limiting parameters for the widespread application of nanoparticles in extraction processes. Consequently, the future industrial application of nanoparticles is based on the overall economic investigation of improved systems.

References

[1] J.D. Seader, E.J. Henley, D.K. Roper, Separation Process Principles, Wiley, New York, 1998.

[2] P. Amani, M. Amani, G. Ahmadi, O. Mahian, S. Wongwises, A critical review on the use of nanoparticles in liquid–liquid extraction, Chem. Eng. Sci. 183 (2018) 148–176.

[3] G. Nematbakhsh, A. Rahbar-Kelishami, The effect of size and concentration of nanoparticles on the mass transfer coefficients in irregular packed liquid–liquid extraction columns, Chem. Eng. Commun. 202 (2015) 1493–1501.

[4] C. Hanson, Recent Advances in Liquid-Liquid Extraction, Elsevier, 2013.

[5] J. Kim, Y.T. Kang, C.K. Choi, Soret and Dufour effects on convective instabilities in binary nanofluids for absorption application, Int. J. Refrig. 30 (2007) 323–328.

[6] G. Polidori, S. Fohanno, C.T. Nguyen, A note on heat transfer modelling of Newtonian nanofluids in laminar free convection, Int. J. Therm. Sci. 46 (2007) 739–744.

[7] J. Saien, H. Bamdadi, Mass transfer from nanofluid single drops in liquid–liquid extraction process, Ind. Eng. Chem. Res. 51 (2012) 5157–5166.

[8] S.K. Das, S.U.S. Choi, H.E. Patel, Heat transfer in nanofluids—a review, Heat. Transf. Eng. 27 (2006) 3–19.

[9] A. Hatami, Z. Azizi, D. Bastani, Investigation of the impact of synthesized hydrophobic magnetite nanoparticles on mass transfer and hydrodynamics of stagnant and stirred liquid–liquid extraction systems, Chem. Eng. Res. Des. 147 (2019) 305–318.

[10] M. Corcione, Empirical correlating equations for predicting the effective thermal conductivity and dynamic viscosity of nanofluids, Energy Convers. Manag. 52 (2011) 789–793.

[11] S. Krishnamurthy, P. Bhattacharya, P.E. Phelan, R.S. Prasher, Enhanced mass transport in nanofluids, Nano Lett. 6 (2006) 419–423.

[12] Y.T. Kang, H.J. Kim, K. Il Lee, Heat and mass transfer enhancement of binary nanofluids for H_2O/LiBr falling film absorption process, Int. J. Refrig. 31 (2008) 850–856.

[13] B. Olle, S. Bucak, T.C. Holmes, L. Bromberg, T.A. Hatton, D.I.C. Wang, Enhancement of oxygen mass transfer using functionalized magnetic nanoparticles, Ind. Eng. Chem. Res. 45 (2006) 4355–4363.

[14] X. Fang, Y. Xuan, Q. Li, Experimental investigation on enhanced mass transfer in nanofluids, Appl. Phys. Lett. 95 (2009) 203108.

[15] X. Feng, D.W. Johnson, Mass transfer in SiO_2 nanofluids: a case against purported nanoparticle convection effects, Int. J. Heat Mass Transf. 55 (2012) 3447–3453.

[16] H.H. Goodarzi, M.N. Esfahany, Experimental investigation of the effects of the hydrophilic silica nanoparticles on mass transfer and hydrodynamics of single drop extraction, Sep. Purif. Technol. 170 (2016) 130–137.

[17] S.-S. Ashrafmansouri, M.N. Esfahany, Mass transfer in nanofluids: a review, Int. J. Therm. Sci. 82 (2014) 84–99.

[18] A. Vesal, A. Rahbar-Kelishami, T. Mohammadi, Investigation of mass transfer coefficients in irregular packed liquid–liquid extraction columns in the presence of various nanoparticles, J. Part. Sci. Technol. 3 (2017) 113–120.

[19] S. Ashrafmansouri, M.N. Esfahany, Mass transfer into/from nanofluid drops in a spray liquid–liquid extraction column, AIChE J. 62 (2016) 852–860.

[20] A. Bahmanyar, N. Khoobi, M.M.A. Moharrer, H. Bahmanyar, Mass transfer from nanofluid drops in a pulsed liquid–liquid extraction column, Chem. Eng. Res. Des. 92 (2014) 2313–2323.

[21] A. Mirzazadeh Ghanadi, A. Heydari Nasab, D. Bastani, A.A. Seife Kordi, The effect of nanoparticles on the mass transfer in liquid–liquid extraction, Chem. Eng. Commun. 202 (2015) 600–605.

[22] A. Bahmanyar, N. Khoobi, M.R. Mozdianfard, H. Bahmanyar, The influence of nanoparticles on hydrodynamic characteristics and mass transfer performance in a pulsed liquid–liquid extraction column, Chem. Eng. Process. Process Intensif. 50 (2011) 1198–1206.

[23] M.A.G. Roozbahani, M.S. Najafabadi, K.N.H. Abadi, H. Bahmanyar, Simultaneous investigation of the effect of nanoparticles and mass transfer direction on static and dynamic holdup in pulsed-sieve liquid–liquid extraction columns, Chem. Eng. Commun. 202 (2015) 1468–1477.

[24] S.-S. Ashrafmansouri, M.N. Esfahany, The influence of silica nanoparticles on hydrodynamics and mass transfer in spray liquid–liquid extraction column, Sep. Purif. Technol. 151 (2015) 74–81.

[25] B. Sunden, F.S. Nanagedani, Prediction of mass transfer coefficient of the continuous phase in a structured packed extraction column in the presence of SiO_2 nanoparticles, Front. Heat Mass Transf. 14 (2020).

[26] J. Saien, R. Hasani, Hydrodynamics and mass transfer characteristics of circulating single drops with effect of different size nanoparticles, Sep. Purif. Technol. 175 (2017) 298–304.

[27] A. Hatami, D. Bastani, F. Najafi, Investigation the effect of super hydrophobic titania nanoparticles on the mass transfer performance of single drop liquid-liquid extraction process, Sep. Purif. Technol. 176 (2017) 107–119.

[28] A. Rahbar-Kelishami, S.N. Ashrafizadeh, M. Rahnamaee, The effect of type and concentration of nanoparticles on the mass transfer coefficients: experimental and Sherwood number correlating, Sep. Sci. Technol. 50 (2015) 1776–1784.

[29] M.S. Sepehri Sadeghian, D. Abooali, A. Rahbar-Kelishami, Impacts of SiO_2 and MgO nanoparticles on mass transfer performance of batch liquid–liquid extraction, Inorg. Nano-Metal Chem. 47 (2017) 677–680.

[30] M. Amani, M. Ameri, A. Kasaeian, The experimental study of convection heat transfer characteristics and pressure drop of magnetite nanofluid in a porous metal foam tube, Transp. Porous Media. 116 (2017) 959–974.

[31] J. Saien, H. Bamdadi, S. Daliri, Liquid–liquid extraction intensification with magnetite nanofluid single drops under oscillating magnetic field, J. Ind. Eng. Chem. 21 (2015) 1152–1159.

[32] S.M. Saad, N.A. Aling, M. Miskam, M. Saaid, N.N. Mohamad Zain, S. Kamaruzaman, et al., Magnetic nanoparticles assisted dispersive liquid–liquid microextraction of chloramphenicol in water samples, R. Soc. Open. Sci. 7 (2020) 200143.

[33] A. Vahedi, A.M. Dehkordi, F. Fadaei, Mass-transfer enhancement in single drop extraction in the presence of magnetic nanoparticles and magnetic field, AIChE J. 62 (2016) 4466–4479.

[34] N.E.A.M. Subuhi, S.M. Saad, N.N.M. Zain, V. Lim, M. Miskam, S. Kamaruzaman, et al., An efficient biosorption-based dispersive liquid–liquid microextraction with extractant removal by magnetic nanoparticles for quantification of bisphenol A in water samples by gas chromatography-mass spectrometry detection, J. Sep. Sci. 43 (16) (2020) 3294–3303.

PART

Applications of nanofluids as mass transfer enhancers

3

CHAPTER

Increasing mass transfer in absorption and regeneration processes via nanofluids

13

Meisam Ansarpour[1] and Masoud Mofarahi[1,2]

[1]*Department of Chemical Engineering, Faculty of Petroleum, Gas and Petrochemical Engineering, Persian Gulf University, Bushehr, Iran* [2]*Department of Chemical and Biomolecular Engineering, Yonsei University, Seoul, Republic of Korea*

13.1 Introduction

Nanofluids are colloids composed of nanoscale particles in a base fluid. Recently, many researchers have paid attention to nanoparticles because of their high potential in the heat and mass transfer processes. Nanoparticles are extensively employed in industries such as the ceramic, paint, coating, and food industries [1,2]. Fig. 13.1 shows the number of papers containing the "nanofluid" keyword in the last decade from the "Web of Science" source [3]. Choi and Eastman [4] were the first to utilize nanofluid and reported a significant increase in the thermal conductivity of nanofluid in comparison with the base fluid [4].

The mass transfer mechanisms can be classified into two major sections, that is convective mass transfer and mass diffusion. The Brownian motion of the particles and, consequently, the microconvection are mentioned as the effective parameters on the augmentation of both sections [5,6]. The movements of these small particles cause momentum transfer, change in velocity, and subsequently, microconvection, which would lead to the enhancement in heat and mass transfer processes [6,7]. The other mechanisms that affect the mass transfer enhancement due to using nanofluids (as depicted in Fig. 13.2) include bubble breaking, grazing effect, and hydrodynamic effect. These mechanisms are explained in the following:

- Through the bubble breaking mechanism, the larger bubbles break into smaller ones at the gas–liquid interface. When the number of bubbles increases, a larger interfacial area will be formed, which would enhance the mass transfer rate from the gas to the liquid phase [9,10].
- Through the grazing effect, solutes in the gas phase are adsorbed by the nanofluid and then transported to the liquid phase, quickly [11,12].
- The hydrodynamic effect states that the nanoparticles cover the bubble surface and prevent their coalescence, which decreases the specific interfacial area. As a consequence, smaller bubbles are formed [13].

Nanofluids and Mass Transfer. DOI: https://doi.org/10.1016/B978-0-12-823996-4.00014-8

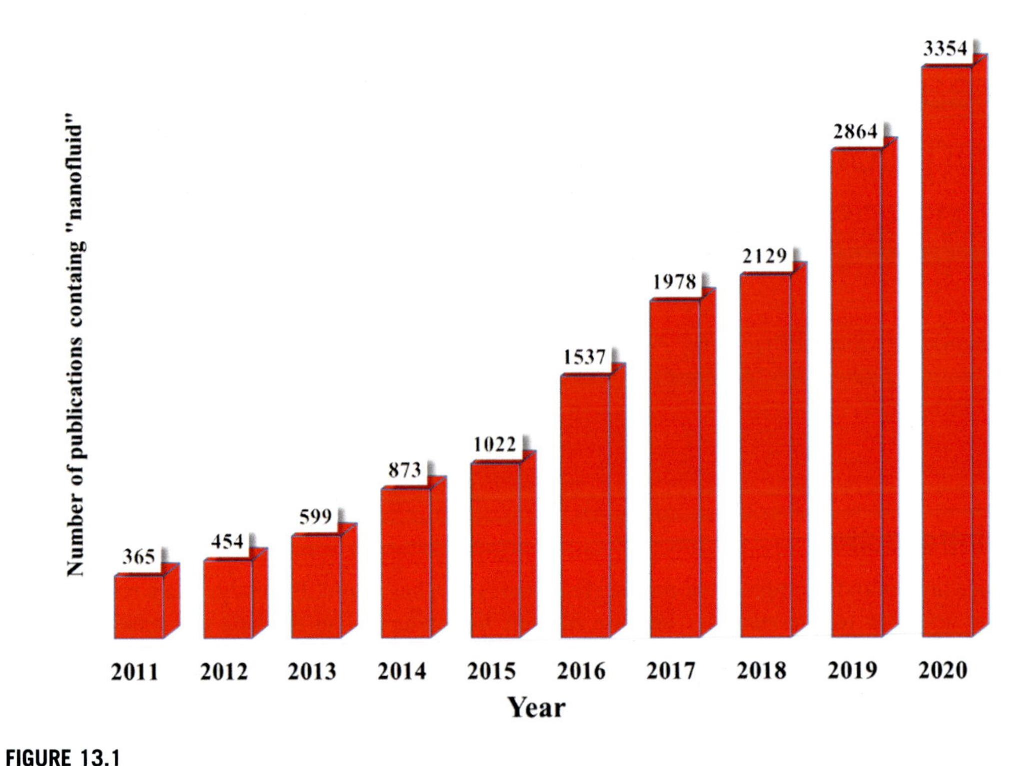

FIGURE 13.1

The number of publications containing “nanofluid,” 2010–2020 [3].

Despite the great features of nanofluids, there are negative mechanisms that decrease the nanofluid efficiency as follows:

- Aggregation is beneficial in the thermal conductivity enhancement of nanofluids. Besides, nanoparticle clusters do not increase mass transfer because of their solid phase (nanoparticles) [14,15].
- Increasing the tortuosity of the nanoparticle path decreases the mass diffusivity [16].
- Increasing the elasticity of nanofluids reduces the convective mass transfer [17–20].
- Lowering of liquid diffusion coefficient related to its viscosity increment, reduces the convective mass transfer [21].

Previous researchers reported an optimum concentration for adding nanofluids in each suspension. Hence, at concentrations above this optimal value, nanoparticles will become too dense and may cause negative effects [22–24]. Furthermore, extra quantities of nanoparticles during the absorption process can prevent the gas phase from contacting nanofluid, which results in reduced diffusion [25].

Generally, the most used processes to enhance mass transfer utilizing nanofluids are membrane contactor, falling film absorption, bubble absorption, tray column absorption, and nanofluids’

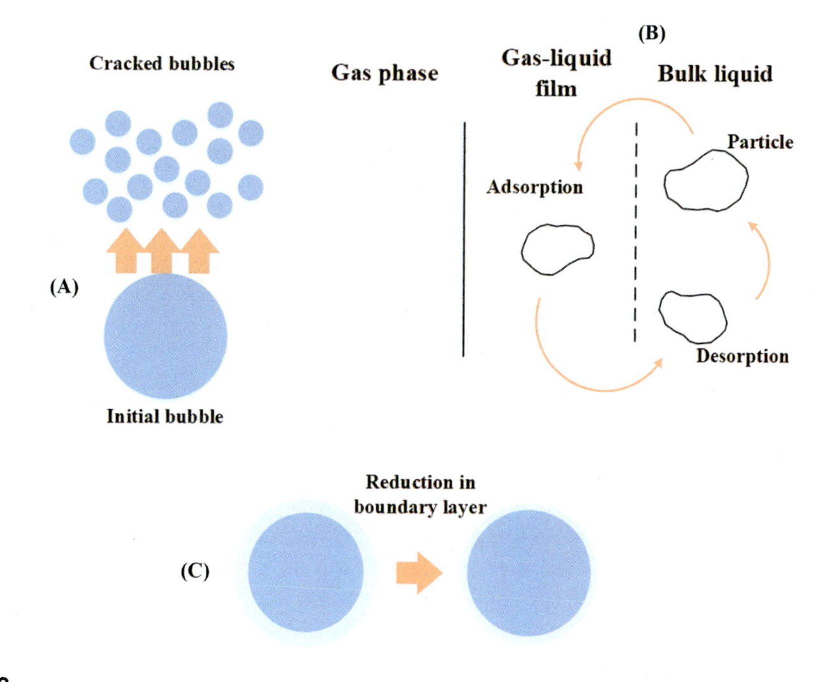

FIGURE 13.2

Absorption mechanisms in nanofluids: (A) bubble breaking; (B) grazing effect; (C) hydrodynamic effect [8].

effect on the regeneration process. All the processes mentioned above are described in the following sections.

13.2 Mass transfer enhancement in the absorption processes via nanofluids

In this section, mass transfer enhancement in various processes and the influence of using nanofluid on their performances are studied. Before that, some of major factors with regard to the use of nanofluids on the mass transfer phenomena are explained. Although many studies reported an increment in mass transfer by using nanofluids, some researchers observed no enhancement or even a reduction in mass transfer [16,24–27]. They concluded several factors are responsible: the diffusion coefficient, base fluid viscosity, pH of the liquid, surface tension, and intermolecular forces existing between nanoparticles for nanofluids are less than for the base fluid (water, etc.) [25,28]. Some of these factors are summarized in the following:

- Brownian motion: the Brownian motion of the particles causes microconvection and improves the mass transfer by creating a velocity gradient in the boundary layer [29].

- Nanofluids concentration: there is an optimum concentration for nanofluids according to gas solubility, and at the higher concentrations, the nanoparticles avoid the gas molecules to contact the liquid phase [22].
- pH: by adding more nanoparticles to the base fluid, the pH of the liquid reduces because of bonding with the hydroxyl group, and particle separation. Consequently, the pH changes are related to the absorption process [30].

The most common processes are described in the following.

13.2.1 Membrane contactor process

One of the most used processes in mass transfer phenomena like CO_2 absorption is the membrane contactor process. Qi and Cussler are the pioneer researchers who conducted experiments on gas absorption processes utilizing membrane contactors [31]. In a hollow fiber membrane contactor (HFMC) system, gas molecules diffuse through the shell side to the gas–liquid interface (across the membrane's pores). Consequently, gas molecules are absorbed by the absorbent. Hence, the membrane structure plays a significant role according to the gas diffusivity through the fiber pores. In other words, gas diffusivity must be large enough to prevent pores from getting filled with the liquid phase. It means that gas-filled pores have lower resistance compared to the liquid-filled pores [32–34]. The membrane fouling is known as a significant challenge in membrane contactor systems. After nanoparticles are treated by acid, they become hydrophilic, while the fibers (polypropylene) are hydrophobic. Due to these characteristics, the nanoparticles do not tend to make contact with fibers [35]. In a membrane contactor process, different factors should be considered such as liquid temperature, type of nanoparticles, liquid and gas flow rates, the concentration of nanofluids, and CO_2 inlet concentration.

The absorption process for the membrane contactor process is schematically described in Fig. 13.3. As can be seen from the figure, the gas enters the shell side. Nanofluid will enter through the fiber side after passing the plate heat exchanger to fix the fluid inlet temperature. Then, nanofluid exits from the hollow fiber and enters the heating section to facilitate CO_2 dispersing. Consequently, hot fluid enters the channel, then goes to a reservoir with a mixer to disperse the absorbed CO_2. The blue and red lines refer to the cold and hot fluids, respectively. Besides, the outlet gas from HFMC can be analyzed using gas chromatography (GC).

The main mass transfer resistance in both physical and some of the chemical absorption processes exist in the liquid side of the membrane [37,38]. The low driving force between liquid and gas phases arises from the cocurrent flow and low diffusivity of CO_2. Besides, the pseudo-wetting property of the fibers results in the formation of a film at the interface, which decreases the rate of the mass transfer [39].

The previous studies [40–42] reported that CO_2 solubility in water has a significant role in the absorption process through membrane contactors. The CO_2 solubility decreases significantly by increasing liquid temperature, which affects the absorption process [43]. When the liquid temperature rises, the water evaporation rate increases, which would be then condensed in the membrane's pores. According to the grazing effect, due to the adsorbing of gas molecules by the nanoparticles, gas concentration near the interface is reduced, which results in the absorption rate increase. The nanoparticles that exist near the interface make an effective layer due to breaking the diffusion

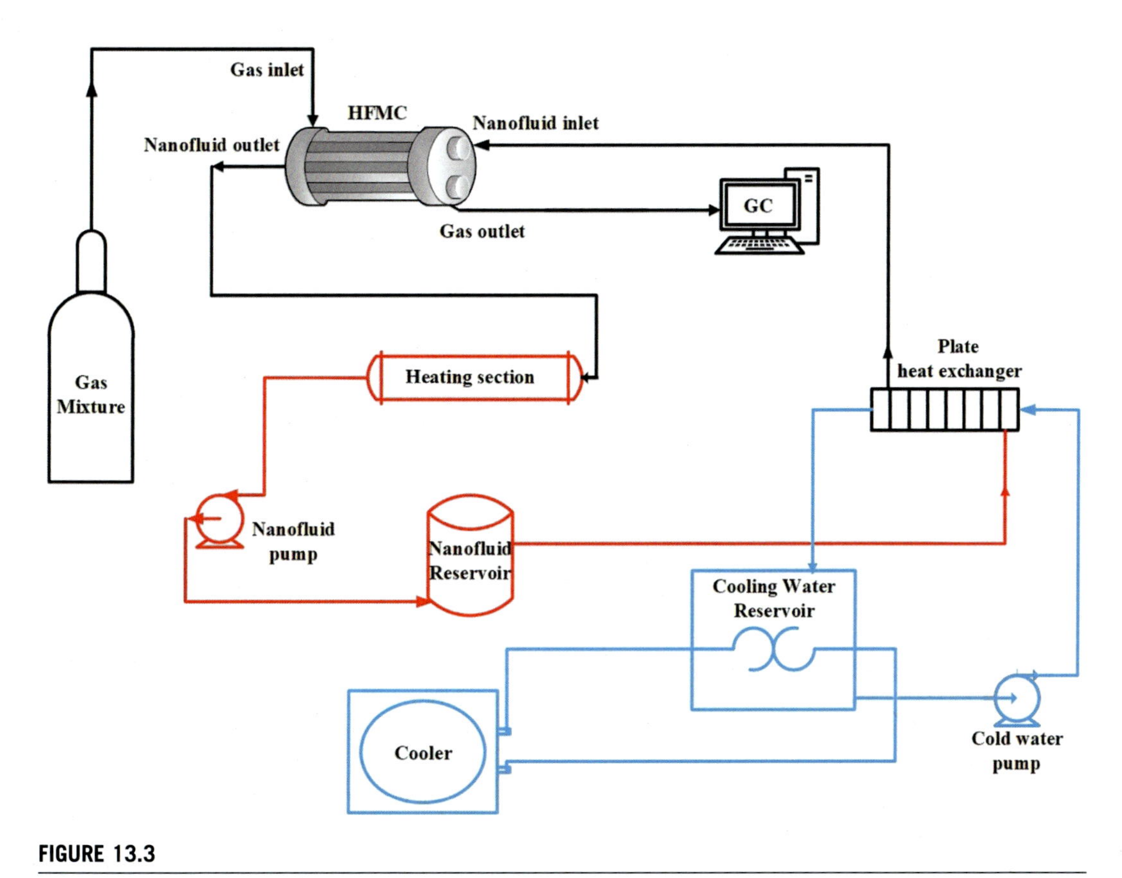

FIGURE 13.3

Schematic diagram of the hollow fiber membrane contactor system. [36].

boundary layer, namely gas–liquid boundary layer mixing. It also improves the gas diffusion due to the enhancement in the flow turbulence resulting in mass transfer coefficient enhancement [44].

Golkhar et al. utilized SiO_2 and carbon nanotube (CNT) nanofluids to investigate the CO_2 absorption in an HFMC. Their results revealed that the CO_2 removal efficiency decreased by increasing CO_2 inlet concentration. Moreover, increasing the liquid flow rate and decreasing liquid temperature resulted in increased CO_2 removal efficiency. Other research [41,45–47] also confirmed that increasing liquid flow rate would enhance CO_2 absorption efficiency and reduce the boundary layer thickness, which decreases the liquid mass transfer resistance. Another factor investigated by Golkhar et al. was the increased gas flow rate, which reduces CO_2 removal efficiency [35]. Peyravi et al. studied the CO_2 removal utilizing various nanofluids in the membrane contactor and introduced nanoparticles hydrodynamic diameter and the nanofluid stability as the crucial factors in nanoparticle selection [48]. Saidi et al. studied the CO_2 absorption in an HFMC in the presence of CNT and SiO_2 nanoparticles in the distilled water and (diethylaminobenzaldehyde) DEAB/water as the base fluids by using a 2D mathematical model. They reported that the CO_2 absorption was enhanced as a result of using nanofluids compared to the base fluids. They claimed an increased amine concentration enhances the mass transfer process because of the active molecules of the absorbent increase. Saidi et al. also reported that increasing the liquid flow rate and reducing the gas flow rate causes an increase in the CO_2 absorption because of the reduced CO_2 residence time in the membrane. Another investigated factor is the liquid temperature that showed different nanofluids results with water and DEAB/water base fluids. They concluded that by raising the nanofluid temperature, CO_2 physical absorption reduced (water-based nanofluids) CO_2. On the other hand, increasing nanofluid temperature causes an increase in the chemical absorption (water-based nanofluids) because of the increase in diffusion coefficients and the reaction rates [49].

Some of numerical results obtained by previous researchers for mass transfer in a HFMC are listed in Table 13.1.

In this subsection, the advantages and disadvantages, effective factors, and setup of membrane contactor process were described. Finally, some of the literature and their results were presented briefly in Table 13.1.

13.2.2 Falling film absorption process

Another interested process for mass transfer phenomena by researchers is the falling film process. In this process, a thin film of liquid falls along the interior walls of vertical tubes. The hydrodynamics of the liquid–vapor contact has a significant function in the falling film setup, since it controls the mass transfer rates and assigns the column fluid dynamics limits.

In a falling film process, gas absorption occurs when the molecules penetrate or diffuse into a falling film [50]. Mass transfer happens in the gas–liquid interface of a thin falling film. The knowledge of the heat and mass transfer rates of the equipment and processes is necessary for a reliable design. However, in wavy falling films, heat and mass transfer processes are mostly influenced by their unsteady hydrodynamic properties such as the film thickness, velocity distribution, and the changes in wall shear stress [51]. In the falling film process, a gravity-driven liquid flows

Table 13.1 Experimental research on the mass transfer enhancement in hollow fiber membrane contactor processes [36].

Nanoparticle	Base fluid	Concentration (wt.%)	Maximum CO_2 removal absorption	References
Al_2O_3 (40 nm)	Water	0.20	98.74%	[36]
Al_2O_3 (20 nm)	Water	0.20	95.40%	[36]
SiO_2	Water	0.20	93.16%	[36]
TiO_2 (25 nm)	Water	0.20	78.11%	[36]
TiO_2 (10–15 nm)	Water	0.20	76.25%	[36]
SiO_2	Water	0.50	37.7%	[35]
CNT	Water	0.50	50.0%	[35]
Fe_3O_4	Water	0.10	57.1%	[48]
CNT	Water	0.10	56.3%	[48]
SiO_2	Water	0.10	49.4%	[48]
Al_2O_3	Water	0.10	38.4%	[48]

on the outside of the tubes and at the same time, exchanges heat with a fluid that flows on the inside of the tubes. Furthermore, flow configurations have several benefits like high heat and mass transfer rates and low parasitic losses [52].

Suresh and Bhalerao investigated the effect of utilizing nanoferrofluid on the mass transfer efficiency under the oscillating magnetic field in a wetted wall column. They reported that when employing a 50 Hz magnetic field, a 50% enhancement of the mass transfer coefficient was achieved. This is because the magnetic field forces particles to participate in the diffusion film by mixing at the wall [53]. In a further study, Komati and Suresh investigated the CO_2 absorption efficiency in a wetted wall column for the CO_2/methyl diethanolamine (MDEA) system by utilizing nanoferrofluid/MDEA. They observed a 92.8% enhancement in the mass transfer coefficient by increasing Fe_3O_4 concentration in the ferrofluid up to 0.39 vol.%. They also reported that utilizing a periodic oscillating magnetic field made no significant enhancement in the absorption process. They claimed that the magnetic nanoparticles' size should be smaller than 15 nm to perform as single-domain magnetic nanoparticles [54].

Kang et al. studied the heat transfer and vapor absorption rates in a falling film with Fe and CNT nanoparticles (0.01–0.10 wt.%) in H_2O/LiBr as a base fluid. They observed enhancement in the vapor absorption rate by increasing nanofluids concentration. The maximum enhancement was 2.48 times higher than the base fluid in the case of 0.10 wt.% of CNT nanofluids [55]. In a comparative study, Yang et al. studied the influence of Al_2O_3, Fe_2O_3, and $ZnFe_2O_4$ nanoparticles in the ammonia/water base fluid in a falling film system. Their results demonstrated that by increasing nanofluids concentration ($\leq$ optimum concentration) the mass transfer rate increased. Further increase in the concentration reduced the mass transfer rate. Particularly, the optimal concentrations for Fe_2O_3, $ZnFe_2O_4$, and Al_2O_3 nanoparticles were 0.2, 0.1, and 0.2 wt.%, respectively. In an initial ammonia concentration of 15 wt.%, the corresponding absorption enhancements were 70%, 50%, and 30%, respectively. They stated that the results were attributable to various factors, including reduction in nanofluids viscosity, the grazing effect of nanoparticles, and the increase in microconvection. [56]. The vapor absorption rate in a falling film system was studied by Kim et al. utilizing binary nanofluids of SiO_2 nanoparticles and H_2O/LiBr base fluid. They observed that maximum enhancement in the mass transfer was 18% for a 0.005 vol.% nanofluid concentration. They also observed that the addition of 2-ethyl-1-hexanol as a surfactant reduced the enhancement of mass transfer [57]. Wen et al. proposed a novel nanofluid by adding multiwalled CNTs (MWCNTs) nanoparticles to LiCl/water solution as a base fluid for the dehumidification system. Polyvinyl pyrrolidone (PVP) was also added to the fluid as a surfactant to increase stability. They observed that dehumidification enhancements were 25.9% and 26.1% for the nanofluid and LiCl/H_2O-PVP solutions, respectively. This was due to the decrease in contact angle that reduced from 58.5° to 28.0° for LiCl and from 28.0° to 26.5° for another solution. The schematic diagram of the experimental setup is illustrated in Fig. 13.4 [58].

An increase in air flow rate results in an increment in dehumidification rates for the air flow rates lower than 0.06 kg/s. However, for higher flow rates, the dehumidification rate shows the inverse behavior. The dehumidification rate increases by adding PVP. However, adding nanoparticles to the LiCl/H_2O-PVP does not considerably affect the dehumidification rate. They also concluded that the liquid flow rate does not affect the dehumidification performance, which decreases with increasing liquid temperature. The dehumidification rate reduction can be attributed to the increase in solution temperature when the partial water vapor pressure increases [58].

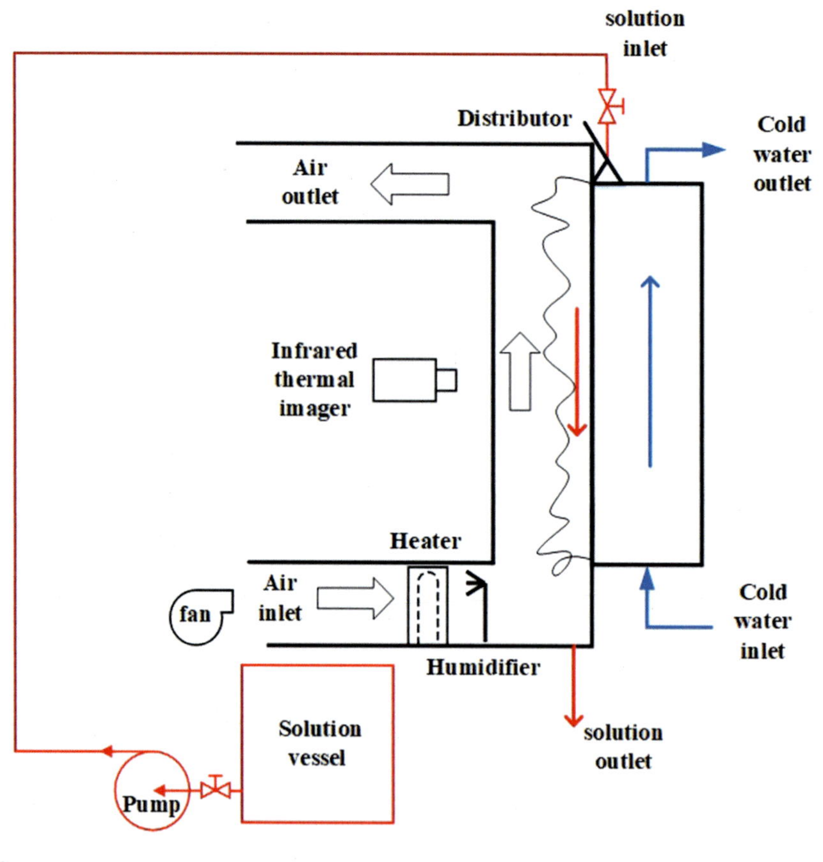

FIGURE 13.4

Schematic diagram of the falling film system [58].

Wang et al. studied the falling film absorption of LiBr solution in the presence of nanoparticles employing a mathematical model based on energy and mass conservations. They confirmed that utilizing nanofluids results in an enhancement in water vapor absorption rate. Moreover, by increasing the concentration of the nanofluid, the absorption efficiency increased. The mass transfer coefficient enhancement was reported to be 1.41 and 1.28 times larger than the base fluid for 0.10 and 0.05 wt.% nanofluids, respectively. They concluded that the major factors that affect their study were the Brownian motion of nanoparticles and the microconvection of the nanofluids. The Brownian motion makes the nanoparticles move and collide. Then, due to the nanoparticles' motion, a disturbance field is created, which causes the microconvention [59].

A summary of the previous research about the falling film absorption process is shown in Table 13.2.

In this subsection, the falling film process and its experimental conditions and setup were studied. Finally, studies and their results were reported in Table 13.2.

Table 13.2 Experimental research on the mass transfer enhancement in the falling film process [60].

Nanoparticle	Base fluid	Concentration (vol.%)	Enhancement ratio in mass transfer	References
Fe_3O_4	CO_2/ MDEA	0.020–0.390	1.928	[54]
Fe and CNT	LiBr/H_2O solution	0.01–0.1 wt.%	2.480 (0.1 wt.% nanofluid) 1.990 for (0.1 wt.% nanofluid)	[55]
Fe_3O_4	Pure water	0.05–1.00	48.00 (1.0 vol.% nanofluid)	[61]
$ZnFe_2O_4$, Fe_2O_3, and Al_2O_3	NH_3/H_2O solution	0.1–0.3 wt.%	1.700, 1.500, and 1.300 (0.2 wt.% Fe_2O_3, 0.1 wt.% $ZnFe_2O_4$, and 0.2 wt.% Al_2O_3)	[56]
SiO_2	LiBr/H_2O solution	0.001–0.010	1.180 (0.005 vol.% nanofluid)	[57]

13.2.3 Bubble absorption process

Another investigated process with high efficiency for mass transfer is the bubble column process. Bubble columns are an increasingly used application in the chemical processes. In this process equipment, long liquid residence time and liquid hold-up are very useful. The other advantages of using bubble absorption columns are their high liquid and vapor mass transfer coefficients, large absorption area, and low investment cost. But high pressure drop and reverse liquid flow count as bubble columns' disadvantages. The absorber columns can work in a cocurrent flow and countercurrent flow modes, continuously. Forecasting bubble behavior during the absorption is very important. In fact, bubble situations can be different. One situation is that the refrigerant is absorbed by sorbent when the bubble stream is nearly unrestricted, and another one is the bubble absorption happens in a tube when the bubble flow is influenced by the wall of tube [62]. Due to bubble columns' simple construction, it is easier to control the liquid stream's residence time and the heat and mass transfer coefficient will be higher than the other processes [63,64].

In the bubble absorption process, there are some evident differences in bubble size between the cases with and without the nanoparticles. For example, Kim et al. studied the bubble behavior via shadowgraph and observed the bubble size in the pure base fluid rises to 7.50 mm, while it increased to 5.83 mm after adding the Cu nanoparticle. Moreover, the nanofluids' bubble size is smaller than that of pure fluid. The bubbles are spherical and hemispherical in the cases with and without the nanoparticles, respectively. Moreover, the bubble residence time in the nanofluid is lower than the one in the base liquid. Hence, the bubble absorption enhances by adding nanoparticles to the binary mixture [65]. Fig. 13.5 sketches a schematic of the bubble type absorber when CO_2 is injected through the absorber's bottom to determine the absorption rate. The output gas flow from the input gas flow is subtracting [10].

The absorption mechanism of the bubble absorption process is addressed in Fig. 13.6. The stability of the nanofluids is very important due to nanofluids creating extra energy for the solution, which increases the total area of the bubbles. It means that this stability, besides the flowing gas bubbles, cracks the bubbles into smaller ones. Hence, the mass transfer area is enhanced.

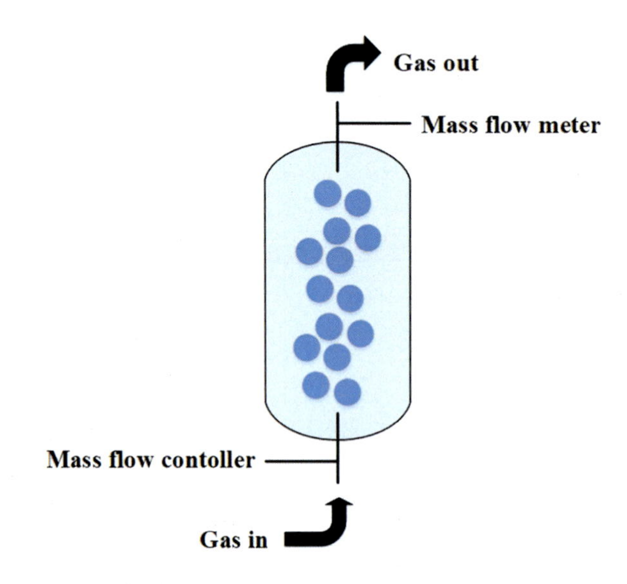

FIGURE 13.5

Schematic diagram of the bubble type absorber [10].

FIGURE 13.6

Absorption mechanism in the bubble absorber [10].

The smaller bubble sizes improve the absorption process and also the inner bubble pressure by the Laplace–Young equation (Eq. 13.1). Moreover, the gas solubility increases by reducing the gas bubbles, according to the Kelvin equation (Eq. 13.2) [66].

$$\Delta P = P_{in}\text{-}P_{out} = \frac{2\gamma}{R} \tag{13.1}$$

where γ is the surface tension, R is the sphere radius, and P refers to the pressure.

$$\mathrm{R}T\ln\frac{P}{P_0} = \frac{2\gamma V}{r} \tag{13.2}$$

In Eq. 13.2, R is the universal gas constant, T denotes the temperature, r is the radius of the droplet, and V is the liquid volume. Fig. 13.7 shows the schematic diagram of the bubble absorption system [29].

Kim et al. studied the effect of utilizing Al_2O_3, CuO, and Cu nanoparticles in concentrations lower than 0.10 wt.% in a compact NH_3/H_2O bubble absorption system. They reported that by increasing nanofluids concentration, the absorption rate increased. The maximum enhancement absorption ratio was 3.21 compared to the base fluid for 18.7 and 0.10 wt.% concentrations of ammonia and Cu nanoparticles in the solution, respectively. They concluded that the grazing effect was more effective than the type of nanoparticles. They also observed that the nanofluid bubble size was smaller than the solution without nanoparticles. And also, the bubble residence time in nanofluid was higher than that of the binary mixture base fluid [65]. Kang et al. investigated the impact of CNT nanoparticles in a bubble absorber on ammonia absorption in binary nanofluids and reported 20% enhancement due to utilizing 0.001 wt.% CNT nanoparticle [67]. Kim et al. investigated the effect of SiO_2 nanoparticles (30, 70, and 120 nm) for CO_2 removal in a bubble absorption

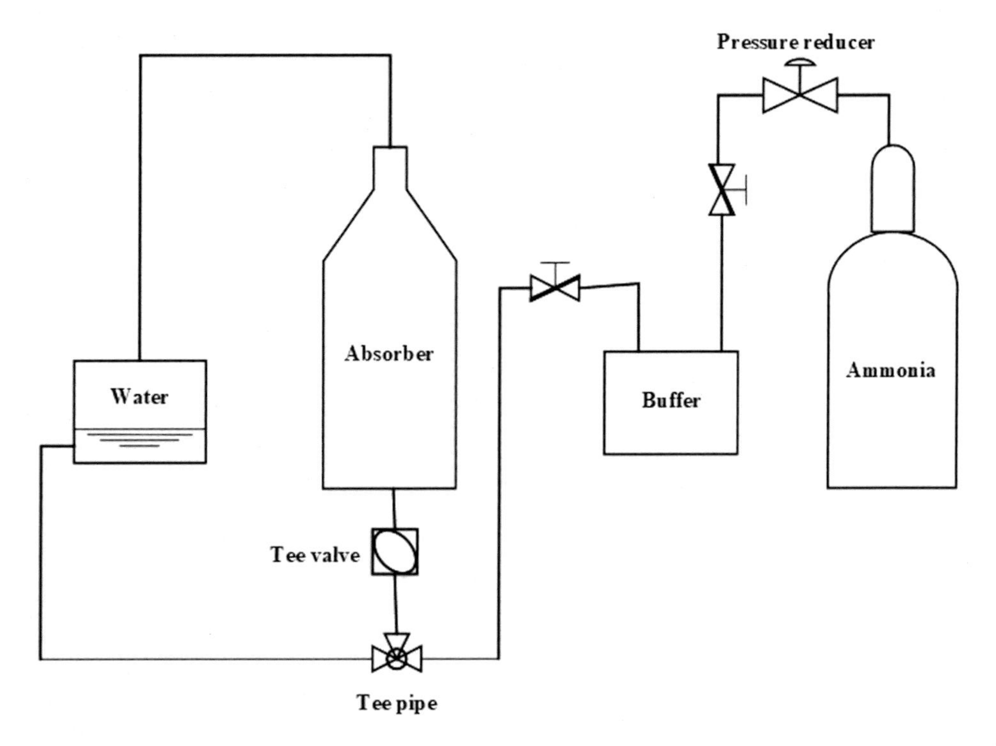

FIGURE 13.7

Schematic diagram of the bubble absorption system [29].

system with a concentrations range of 0.010–0.040 wt.%. The best enhancement was observed during the first minute of the process, which was 76% for the nanofluid with 0.021 wt.% concentration. They also concluded that the size of SiO_2 nanoparticles did not significantly affect absorption efficiency. They believed that the same type of nanoparticle provides an equal amount of energy to the base fluid. The gas bubbles in nanofluids are cracked to the same size and showed the same absorption enhancement even though the nanoparticles' sizes are different [55].

There are similar works in the literature that investigate the effect of adding nanoparticles to the base fluid. For example, Liu et al. studied the ammonia absorption process using FeO nanoparticles [68]. Ma et al. utilized the bubble absorption process to investigate the heat and mass transfer processes in a bubble absorption system for NH_3/H_2O employing CNTs–ammonia binary nanofluid [29]. To improve the NH_3/H_2O bubble absorption efficiency, Wu et al. utilized Ag binary nanofluid [69]. All this research observed enhancement in the absorption process after adding nanoparticles to the base liquids. In another work, SiO_2 and alumina/methanol nanofluids were employed by Lee et al. to study the CO_2 absorption and an increment in mass transfer due to adding nanoparticles was reported. They stated that this enhancement was due to the bubble's breakage and the observed decrease in the velocity disturbance field. Moreover, they concluded that the nanoparticle aggregation and $-$OH bond group of nanoparticles significantly affected CO_2 removal efficiency. The enhancement of absorption caused due to using nanofluids will reduce with increasing nanoparticle concentration higher than an optimum concentration [30].

To combine bubble absorption with a magnetic field, Wu et al. studied the Fe_3O_4 nanoparticles behavior in the ammonia/water bubble type absorber. They observed that this combination enhanced ammonia/water bubble absorption. They also reported 0.1 wt.% as an optimum nanoferrofluid concentration and found that absorption performance increased by increasing nanofluid concentrations. The maximum enhancement ratio (1.0812) was observed at the 20.0 wt.% initial ammonia concentration, 0.10 wt.% Fe_3O_4 nanofluid and the external magnetic field intensity of 280 mT. They concluded that the observed results could be due to the different factors such as increase in heat transfer, the disturbance of the absorption liquid and increase in bubble disruption (due to extrusion), distortion and elongation of bubbles through the direction of the magnetic field lines, surface tension decreasing and, consequently, creating the Marangoni convection effect which causes turbulence at the gas–liquid interface [70]. Pineda et al. observed a 40% enhancement in CO_2 removal utilizing 0.01 vol.% Al_2O_3/methanol nanofluids in a bubble absorption system. The enhancement contributed to the residence time and travel distance of CO_2 bubbles [71].

A bubble absorption process was employed by Darvanjooghi et al. [72] to study the influences of the nanoparticle size on the CO_2 removal in the water-based SiO_2 nanofluid. They reported that the CO_2 absorption was increased by increasing nanoparticle size. They also observed that by increasing the nanoparticle size, the renewal surface rate increased; however, diffusivity and liquid–film thickness reduced. Subsequently, the increase in mass transfer coefficients was also observed.

Pashaei et al. investigated the CO_2 absorption in a continuously stirred bubble column utilizing TiO_2, ZrO_2, and ZnO nanoparticles dispersed in the diethanolamine (DEA) solution with a concentration range of 0.01–0.10 wt.%. The effects of several factors such as solid loading, CO_2 partial pressure, type of the nanoparticle, and stirring speed were considered. They observed that the presence of nanoparticles resulted in a 43.8% increment in the CO_2 absorption. They concluded that increasing nanoparticle concentration up to an optimal value, increases the removal efficiency. For concentrations beyond that, nanoparticles could be an obstacle for the CO_2 absorption and reduce the CO_2 loading [73]. Table 13.3 shows a summary of experimental studies on the bubble type absorption process in the presence of nanoparticles.

The applied binary nanofluid may affect the heat and mass transfer rate of the bubble absorption process in four possible ways:

- Brownian motion: due to the nanoparticles' Brownian motion, CNTs develop microconvection in aqueous ammonia and, consequently, an increase in ammonia mass diffusion in the binary nanofluid would be observed.
- Grazing effect: Kars et al. [11] and Alper et al. [12] introduced the grazing effect caused by CNTs in the binary nanofluid. According to Kars et al., first, by increasing the concentration of the nanoparticles, the influence of the grazing effect on absorption increases and then remains constant [11].

Table 13.3 Experimental research on the mass transfer enhancement in the bubble type process [60].

Nanoparticle	Base fluid	Concentration (wt.%)	Absorption enhancement ratio	References
Al_2O_3, CuO, and Cu	NH_3/H_2O solution	0.010–0.10	3.21 (0.1 wt.% nanofluid)	[65]
Al_2O_3, CuO, and Cu	NH_3/H_2O solution	0.010–0.10	5.32 (0.1 wt.% Cu nanofluid)	[74]
CNT	NH_3 solution	0.10–0.30	1.162 (0.23 wt.% nanofluid)	[75]
SiO_2	Pure water and K_2CO_3/piperazine solution	0.01–0.04	1.24 (water-based 0.021 wt.% nanofluid) and 1.12 (0.021 wt.% nanofluid into the K_2CO_3/piperazine solution)	[55]
MWCNT	NH_3/H_2O solution	0.050–0.50	1.2 (0.2 wt.% nanofluid)	[29]
CNT and Al_2O_3	NH_3/H_2O solution	0.010–0.080 vol.%	1.16 (0.02 vol.% CNT nanofluid) and 1.18 (0.02 vol.% Al_2O_3 nanofluid)	[76]
Al_2O_3 and SiO_2	Methanol	0.0050–0.50 vol.%	1.056 (0.01 vol.% SiO_2 nanofluid) and 1.045 (0.01 vol.% Al_2O_3 nanofluid)	[30]
Al_2O_3	Methanol	0.0050–0.10 vol.%	1.083 (0.01 vol.% of nanofluid)	[77]
Ag	NH_3/H_2O solution	0.0050–0.020	1.55 (0.02 wt.% nanofluid)	[78]
Fe_3O_4	NH_3/H_2O solution	0.0050–0.20	1.0812 (0.1 wt.% nanofluid)	[70]
Al_2O_3	NaCl aqueous solution	0.0050–0.10 vol.%	1.125 (0.01 vol.% nanofluid)	[22]
CNT	$NH_3/LiNO_3$ solution	0.010, 0.020	Up to 1.64 and 1.48 (enhancement in mass flux)	[79]

- Thermal conductivity: CNTs in the binary nanofluid can affect both the thermal conductivity of aqueous ammonia [80] and the heat transfer in the absorption process.
- Gas hold-up: in the presence of nanofluids, the gas hold-up is higher than that of pure water at the same experimental conditions. When the gas hold-up increases, the gas–liquid interface area and, subsequently, the ammonia vapor absorption rate are enhanced [81].

In this section, the bubble absorption process and its advantages, disadvantages, and experimental setup were studied. Finally, some of the literature results were listed in Table 13.2.

13.2.4 Tray column absorption process

The tray column absorption process has not been well noted by researchers and there are few studies that have focused on this process enhancement using nanofluids. This may be due to the high cost of the tray column absorber experimental setup. There is a dynamic flow in a tray column absorber and the mass transfer enhancement is highly affected by the forces induced by the liquid and vapor movements.

Pineda et al. investigated the CO_2 removal in a tray column absorber in the presence of SiO_2 and Al_2O_3 nanoparticles dispersed in the methanol. They utilized an acrylic sieve tray column with 12 plates. The flat perforated plates were applied to prevent the liquid flowing downward due to the vapor velocity. A countercurrent flow pattern was applied. Mass flowmeters and thermocouples were located at the inlet and outlet of the column to measure the absorption rate and flow temperature. The results revealed that the optimum concentration of nanoparticles was 0.05 vol.%. They reported that the absorption rate was increased to 9.4% for Al_2O_3 and 9.7% for SiO_2 in the optimum nanofluid concentration, and then it was decreased. This could be because after the critical concentration, the nanoparticles become too dense in the fluid and therefore the self-diffusion coefficient is reduced, which results in the reduction of gas absorption [9]. The schematic diagram of the tray column absorber for the methanol/CO_2 system is indicated in Fig. 13.8 [9]. There are sieve trays through the column to prevent the liquid from descending through the holes by the vapor flow. A desiccant unit is placed to remove the absorbent from gas.

The main reason for the mass transfer enhancement in these kinds of tray column absorbers is not clear yet. However, the bubble breaking model (as shown in Fig. 13.6) explains that due to the suspending of the nanoparticles in the base fluid, the bubble surfaces are covered, the fluid's movement becomes more dynamic, and the nanoparticles collide at the gas–liquid interface. Therefore the coalescence of bubbles is prevented and smaller bubbles would be created. By increasing the number of bubbles, the interfacial area is enlarged, which will improve the mass transfer rate between the liquid and gas phases [13]. Another theory claimed that when nanoparticles collide, local turbulence is encouraged to form and the gas–liquid boundary layer is refreshed by liquid phase mixing [82]. Research by Krishnamurthy et al. [7] stated that the mass transfer enhancement is not related to the nanoparticles' Brownian motion. However, the velocity disturbance field in the fluid created by the motion of the nanoparticles affects the mass transfer process.

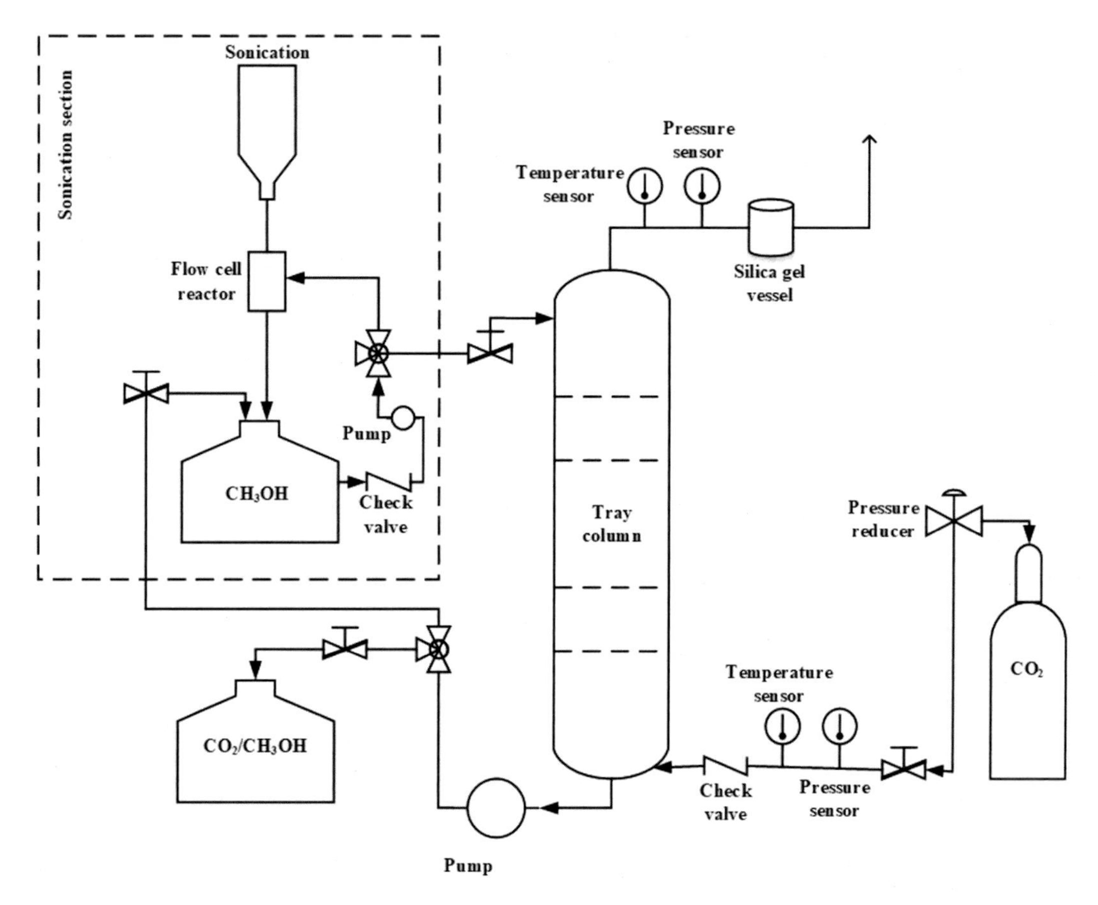

FIGURE 13.8

Schematic diagram of the tray column absorber [9].

13.3 Mass transfer enhancement in the regeneration process via nanofluids

Based on Henry's law of solubility, both absorption and regeneration processes require high energy consumption. To overcome this issue, nanofluids have been applied by researchers. The regeneration process by nanofluids is highly affected by heat transfer coefficients. The first mechanism is the activation energy effect model [8] (Fig. 13.9A). Activation energy is attained by either the heaters or the generated steam during the regeneration process. When nanoparticles receive this energy, they will move more actively. These dynamic motions force nanoparticles to collide. The dissolved gas in the absorbent will discharge more easily due to the collisions, and the regeneration performance would be enhanced. The second model

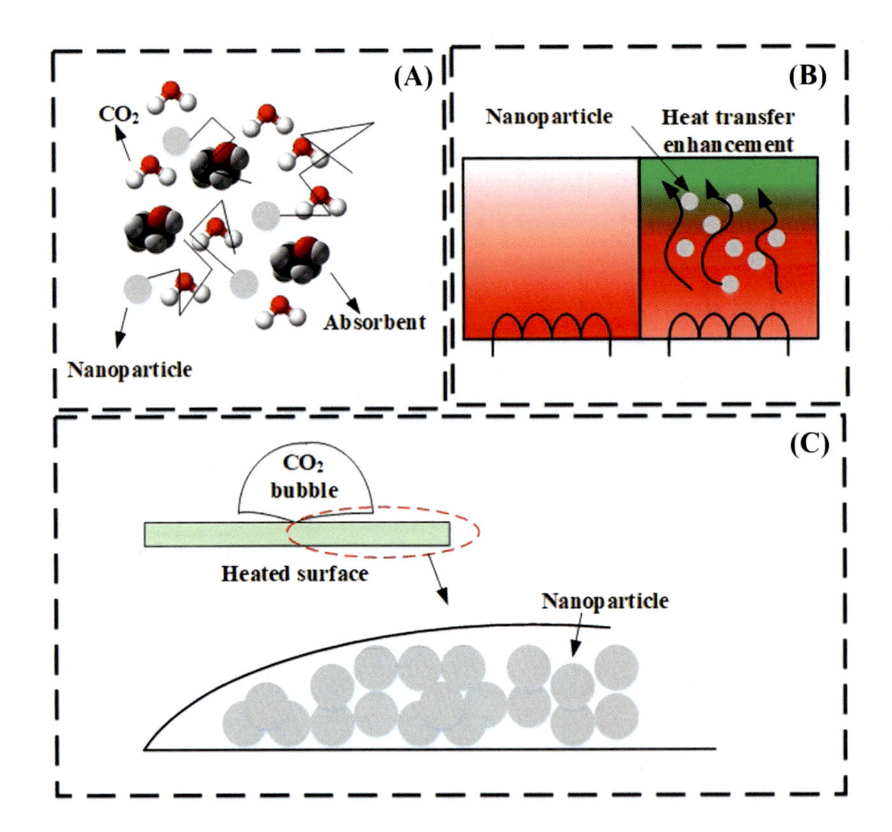

FIGURE 13.9

Mechanisms of regeneration enhancement in nanosorbents: (A) activation effect; (B) thermal effect; (C) surface effect [83].

focuses on the enhanced heat transfer in nanofluids, called the thermal effect model (Fig. 13.9B). Previous studies [84–86] have stated that adding nanoparticles to the system enhances thermal conductivity. Improved thermal activity will increase the operating temperature, which reduces the solubility of CO_2. Furthermore, the amount of energy consumed during the regeneration process decreases [85,87]. The third mechanism, which is the surface effect, states that the CO_2 regeneration process in the presence of physical absorbents has similar properties to boiling processes (Fig. 13.9C) [88]. During the nanoparticles' dispersion in the solution, they deposit on the boiling surface due to either natural convection or gravity. The boiling process occurs at a saturation point, where the phase change happens. Meanwhile, the CO_2 bubble regeneration occurs when the operating temperature is higher than the absorption temperature. Consequently, the bubble generation conditions are achieved more easily than the boiling process conditions. Therefore the CO_2 bubble generation can be easily affected by the nanoparticles.

A promising technology in the air-conditioning industry is a liquid desiccant cooling system (LDCS), known for energy-saving properties and precise humidity and temperature control [89]. The regenerator is the main tool in this system. Green energies, such as solar energy and wastewater heat, are employed in regeneration processes [90]. Surface modification is a newly proposed method that enhances the mass and heat transfer during regeneration. This method often changes the flow pattern on the absorber or regenerator, and provides a greater contact area for the mass and heat transfer. The suggested surface configurations include constant curvature surface (CCS) [91], surface treatment tubes [92,93], film-inverting structures [94,95], and plate-fin structures [96].

Galhorta stated that when CO_2 reacts with the Al_2O_3 surface, an extra amount of adsorbed bicarbonate and carbonate species are formed. In addition to this, the utilized nanoparticles (i.e., SiO_2 and Al_2O_3) have different characteristics. CO_2 is strongly absorbed on Al_2O_3 due to its high surface potential related to change in pH. Hence, the CO_2 desorption from Al_2O_3 surface would be difficult in the presence of SiO_2; the pressure of the system increased by the shift in concentration up to 0.01 vol.% and then decreased [97]. However, the final operating pressure during the regeneration process decreased by increasing the concentration of Al_2O_3 [8]. In a further study, Lee et al. [8] investigated the visualization of CO_2 bubble regeneration for SiO_2 and Al_2O_3 nanofluids. They observed that the bubble generation and absorption for Al_2O_3 nanofluid were more facile than the pure water and SiO_2 nanofluid. They also concluded that the cluster size affects the process significantly compared to the size of nanoparticles. It means, by floating nanoparticles clusters in the liquid, these floating particles reached bubble generation points. Therefore the regeneration efficiency increased due to the increase in CO_2 discharging through the increased regeneration sites [98].

Furthermore, the nanofluids improve both the regeneration and mass transfer processes. Improving these phenomena influences the economic aspects such as system miniaturization and energy-saving by advanced absorbents. The regeneration rate increased up to 16% and 22% after the first and second cycles for the Al_2O_3/methanol and five cycles for the SiO_2/methanol nanofluids, respectively [8]. Wen et al. studied the regeneration performance of LiCl falling film using PVP (as a surfactant) and MWNTs, experimentally and numerically. By adding PVP and applying mechanical methods, a steady nanofluid with MWNTs was formed. The regeneration rates of the LiCl/H_2O-PVP solution and nanofluid are 24.9% and 24.7%, which are higher than the LiCl/H_2O solution. This kind of behavior occurs due to when the contact angle decreases, there will be an increment in the mass transfer area and a reduction in falling film thickness [99]. Fig. 13.10 illustrates the schematic diagram of the LDCS [99].

Wen et al. [99] investigated the impacts of solution properties (temperature and flow rate), as well as the air properties (temperature, flow rate, and humidity), on the regeneration performance rate. They demonstrated that:

- An increase in the solution temperature will increase the mass transfer driving force, and the regeneration rate will increase.
- The solution flow rate hardly impacts the mass transfer coefficient. The regeneration rates for the given solutions will be ordered as: nanofluid > LiCl/H_2O-PVP > LiCl.
- Even when the air temperature increases to 9°C, the regeneration rate remains stable for all three solutions [99].

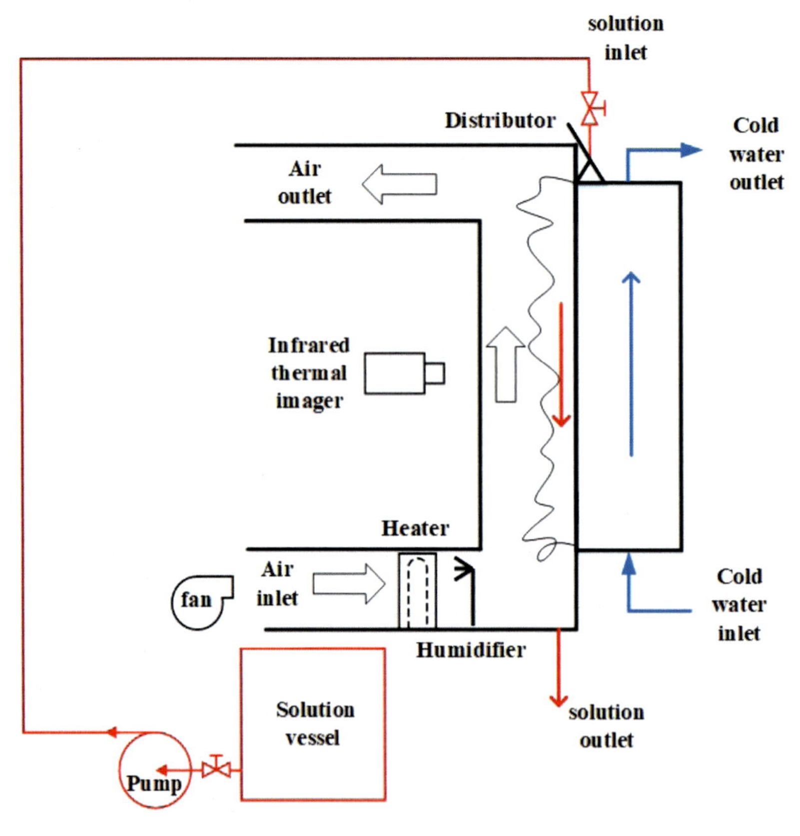

FIGURE 13.10

The schematic diagram of the LDCS [99].

13.4 Conclusion

This chapter has detailed the effect of using nanofluids on the mass transfer (including membrane contactor, falling film absorption, bubble absorption, and tray column absorption) and regeneration processes. The dominant mechanisms related to using nanofluids are described. Also, for each process, experimental setup, maximum mass transfer enhancement, various types of nanoparticle, and experiment conditions were studied individually. Moreover, the benefits and disadvantages have been investigated. Some of the published papers on the nanofluid effect on mass transfer have been reviewed. In the membrane contactor absorption process, utilizing nanofluids showed a positive effect on the mass transfer. Regarding the experimental setup, the type of nanoparticle, experiment conditions, and the concentration of nanofluid, up to 98% enhancements were reported. For the bubble absorption process, the results indicated a significant positive effect due to using adding nanoparticles to the base fluid. The enhancement ratio was reported up to 5.32 as a result of using 0.10 wt.% Cu nanofluid. Besides, in a falling film absorption process most of the researchers reported obvious enhancement due to the use of nanofluids in their experiments. For instance, the enhancement ratio was reported to be up to 5.32 as a result of using 0.10 wt.% nanoparticles. Also, at the high nanofluid concentration (1.0 vol.%), the absorption enhancement ratio was reported to be 48 times higher than base fluid by using Fe_3O_4 nanofluid. Finally, the mechanisms of the regeneration enhancement due to using nanofluids have been studied. For regeneration processes, nanofluids showed positive behavior on the retention efficiency, too. Previously researchers had claimed that using nanofluid in their studies increased the regeneration rate and also decreased the final operating pressure of the system. One of the most important factors when choosing the mass transfer process is nanofluid recyclability, which has not been sufficiently studied. So, it is necessary for further works to investigate nanofluid recyclability after use.

Abbreviations

CCS Constant curvature surface
CNT Carbon nanotube
DEA Diethanolamine
DEAB Diethylaminobenzaldehyde
DI Deionized
GC Gas chromatography
HFMC Hollow fiber membrane contactor
LDCS Liquid desiccant cooling system
LiBr Lithium bromide
LiCl Lithium chloride
MDEA Methyl diethanolamine
MWCNT Multiwalled carbon nanotube
PVP Polyvinyl pyrrolidone

References

[1] B.F. Gibbs, S. Kermasha, I. Alli, C.N. Mulligan, Encapsulation in the food industry: a review, Int. J. Food Sci. Nutr. 50 (3) (1999) 213–224.
[2] J.A. Lewis, Colloidal processing of ceramics, J. Am. Ceram. Soc. 83 (10) (2000) 2341–2359.
[3] Web of Science. (2020); Available from: <apps.webofknowledge.com>.
[4] S.U. Choi, J.A. Eastman, Enhancing Thermal Conductivity of Fluids with Nanoparticles, Argonne National Lab., Argonne, 1995.
[5] H. Mohammed, et al., Convective heat transfer and fluid flow study over a step using nanofluids: a review, Renew. Sustain. Energy Rev. 15 (6) (2011) 2921–2939.
[6] J. Saien, H. Bamdadi, Mass transfer from nanofluid single drops in liquid–liquid extraction process, Ind. Eng. Chem. Res. 51 (14) (2012) 5157–5166.
[7] S. Krishnamurthy, et al., Enhanced mass transport in nanofluids, Nano Lett. 6 (3) (2006) 419–423.
[8] J.S. Lee, J.W. Lee, Y.T. Kang, CO_2 absorption/regeneration enhancement in DI water with suspended nanoparticles for energy conversion application, Appl. Energy 143 (2015) 119–129.
[9] I.T. Pineda, et al., CO_2 absorption enhancement by methanol-based Al_2O_3 and SiO_2 nanofluids in a tray column absorber, Int. J. Refrig. 35 (5) (2012) 1402–1409.

[10] W.-G. Kim, et al., Synthesis of silica nanofluid and application to CO_2 absorption, Sep. Sci. Technol. 43 (11–12) (2008) 3036–3055.

[11] R. Kars, R. Best, A. Drinkenburg, The sorption of propane in slurries of active carbon in water, Chem. Eng. J. 17 (2) (1979) 201–210.

[12] E. Alper, B. Wichtendahl, W.-D. Deckwer, Gas absorption mechanism in catalytic slurry reactors, Chem. Eng. Sci. 35 (1) (1980) 217–222.

[13] V. Linek, M. Kordač, M. Soni, Mechanism of gas absorption enhancement in presence of fine solid particles in mechanically agitated gas–liquid dispersion, Eff. Mol. Diffusivity. Chem. Eng. Sci. 63 (21) (2008) 5120–5128.

[14] J. Veilleux, S. Coulombe, A total internal reflection fluorescence microscopy study of mass diffusion enhancement in water-based alumina nanofluids, J. Appl. Phys. 108 (10) (2010) 104316.

[15] J. Veilleux, S. Coulombe, A dispersion model of enhanced mass diffusion in nanofluids, Chem. Eng. Sci. 66 (11) (2011) 2377–2384.

[16] C. Gerardi, et al., Nuclear magnetic resonance-based study of ordered layering on the surface of alumina nanoparticles in water, Appl. Phys. Lett. 95 (25) (2009) 253104.

[17] S.-W. Park, B.-S. Choi, J.-W. Lee, Effect of elasticity of aqueous colloidal silica solution on chemical absorption of carbon dioxide with 2-amino-2-methyl-1-propanol, Korea Aust. Rheol. J. 18 (3) (2006) 133–141.

[18] S.W. Park, B.S. Choi, J.W. Lee, Chemical absorption of carbon dioxide into aqueous colloidal silica solution with diethanolamine, Sep. Sci. Technol. 41 (14) (2006) 3265–3278.

[19] S.-W. Park, et al., Chemical absorption of carbon dioxide into aqueous colloidal silica solution containing monoethanolamine, J. Ind. Eng. Chem. 13 (1) (2007) 133–142.

[20] S.-W. Park, et al., Absorption of carbon dioxide into aqueous colloidal silica solution with diisopropanolamine, J. Ind. Eng. Chem. 14 (2) (2008) 166–174.

[21] M. Zhou, W.F. Cai, C.J. Xu, A new way of enhancing transport process–the hybrid process accompanied by ultrafine particles, Korean J. Chem. Eng. 20 (2) (2003) 347–353.

[22] J.W. Lee, Y.T. Kang, CO_2 absorption enhancement by Al_2O_3 nanoparticles in NaCl aqueous solution, Energy 53 (2013) 206–211.

[23] J. Tinge, A. Drinkenburg, Absorption of gases into activated carbon—water slurries in a stirred cell, Chem. Eng. Sci. 47 (6) (1992) 1337–1345.

[24] X. Feng, D.W. Johnson, Mass transfer in SiO_2 nanofluids: a case against purported nanoparticle convection effects, Int. J. Heat Mass Transf. 55 (13) (2012) 3447–3453.

[25] A. Turanov, Y.V. Tolmachev, Heat-and mass-transport in aqueous silica nanofluids, Heat Mass Transf. 45 (12) (2009) 1583–1588.

[26] S. Ozturk, Y.A. Hassan, V.M. Ugaz, Interfacial complexation explains anomalous diffusion in nanofluids, Nano Lett. 10 (2) (2010) 665–671.

[27] V. Subba-Rao, P.M. Hoffmann, A. Mukhopadhyay, Tracer diffusion in nanofluids measured by fluorescence correlation spectroscopy, J. Nanopart. Res. 13 (12) (2011) 6313–6319.

[28] A. Azari, M. Derakhshandeh, An experimental comparison of convective heat transfer and friction factor of Al_2O_3 nanofluids in a tube with and without butterfly tube inserts, J. Taiwan. Inst. Chem. Eng. 52 (2015) 31–39.

[29] X. Ma, et al., Enhancement of bubble absorption process using a CNTs-ammonia binary nanofluid, Int. Commun. Heat Mass Transf. 36 (7) (2009) 657–660.

[30] J.W. Lee, et al., CO_2 bubble absorption enhancement in methanol-based nanofluids, Int. J. Refrig. 34 (8) (2011) 1727–1733.

[31] Z. Qi, E. Cussler, Microporous hollow fibers for gas absorption: I. Mass transfer in the liquid, J. Membr. Sci. 23 (3) (1985) 321–332.

[32] S. Boributh, et al., Effect of membrane module arrangement of gas–liquid membrane contacting process on CO_2 absorption performance: a modeling study, J. Membr. Sci. 372 (1) (2011) 75–86.
[33] R. Wang, et al., Influence of membrane wetting on CO_2 capture in microporous hollow fiber membrane contactors, Sep. Purif. Technol. 46 (1) (2005) 33–40.
[34] M.H. Al-Marzouqi, et al., Modeling of CO_2 absorption in membrane contactors, Sep. Purif. Technol. 59 (3) (2008) 286–293.
[35] A. Golkhar, P. Keshavarz, D. Mowla, Investigation of CO_2 removal by silica and CNT nanofluids in microporous hollow fiber membrane contactors, J. Membr. Sci. 433 (2013) 17–24.
[36] H. Mohammaddoost, et al., Experimental investigation of CO_2 removal from N_2 by metal oxide nanofluids in a hollow fiber membrane contactor, Int. J. Greenh. Gas. Control. 69 (2018) 60–71.
[37] M.C. Yang, E. Cussler, Designing hollow-fiber contactors, AIChE J. 32 (11) (1986) 1910–1916.
[38] A. Mansourizadeh, Experimental study of CO_2 absorption/stripping via PVDF hollow fiber membrane contactor, Chem. Eng. Res. Des. 90 (4) (2012) 555–562.
[39] M.H. El-Naas, et al., Evaluation of the removal of CO_2 using membrane contactors: membrane wettability, J. Membr. Sci. 350 (1–2) (2010) 410–416.
[40] P. Keshavarz, K.J. Fathi, E.S. Ayat, Mass transfer analysis and simulation of a hollow fiber gas-liquid membrane contactor, Iranian Journal of Science and Technology Transactions B- Engineering 32 (2008) 585–599.
[41] A. Mansourizadeh, A.F. Ismail, T. Matsuura, Effect of operating conditions on the physical and chemical CO_2 absorption through the PVDF hollow fiber membrane contactor, J. Membr. Sci. 353 (1) (2010) 192–200.
[42] H.A. Rangwala, Absorption of carbon dioxide into aqueous solutions using hollow fiber membrane contactors, J. Membr. Sci. 112 (2) (1996) 229–240.
[43] H.-J. Song, et al., Solubilities of carbon dioxide in aqueous solutions of sodium glycinate, Fluid Phase Equilibria 246 (1–2) (2006) 1–5.
[44] J. Kluytmans, et al., Mass transfer in sparged and stirred reactors: influence of carbon particles and electrolyte, Chem. Eng. Sci. 58 (20) (2003) 4719–4728.
[45] A. Haghtalab, M. Mohammadi, Z. Fakhroueian, Absorption and solubility measurement of CO_2 in water-based ZnO and SiO_2 nanofluids, Fluid Phase Equilibria 392 (2015) 33–42.
[46] R. Wang, D. Li, D. Liang, Modeling of CO_2 capture by three typical amine solutions in hollow fiber membrane contactors, Chem. Eng. Processing: Process. Intensif. 43 (7) (2004) 849–856.
[47] S.-H. Yeon, et al., Determination of mass transfer rates in PVDF and PTFE hollow fiber membranes for CO_2 absorption, Sep. Sci. Technol. 38 (2) (2003) 271–293.
[48] A. Peyravi, P. Keshavarz, D. Mowla, Experimental investigation on the absorption enhancement of CO_2 by various nanofluids in hollow fiber membrane contactors, Energy Fuels 29 (12) (2015) 8135–8142.
[49] M. Saidi, CO_2 absorption intensification using novel DEAB amine-based nanofluids of CNT and SiO_2 in membrane contactor, Chem. Eng. Processing: Process. Intensif. 149 (2020) 107848.
[50] M.A. Aroon, M.A. Khansary, S. Shirazian, Revisiting 'penetration depth'in falling film mass transfer, Chem. Eng. Res. Des. 155 (2020) 18–21.
[51] Z. Xu, B. Khoo, N. Wijeysundera, Mass transfer across the falling film: simulations and experiments, Chem. Eng. Sci. 63 (9) (2008) 2559–2575.
[52] A.K. Nagavarapu, S. Garimella, Experimentally validated models for falling-film absorption around microchannel tube banks: heat and mass transfer, Int. J. Heat Mass Transf. 139 (2019) 303–316.
[53] A. Suresh, S. Bhalerao, Rate intensification of mass transfer process using ferrofluids, Indian. J. Pure Appl. Phys. 40 (3) (2002) 172–184.

[54] S. Komati, A.K. Suresh, CO_2 absorption into amine solutions: a novel strategy for intensification based on the addition of ferrofluids, J. Chem. Technol. Biotechnol. 83 (8) (2008) 1094–1100.
[55] Y.T. Kang, H.J. Kim, K.I. Lee, Heat and mass transfer enhancement of binary nanofluids for H_2O/LiBr falling film absorption process, Int. J. Refrig. 31 (5) (2008) 850–856.
[56] L. Yang, et al., Experimental study on enhancement of ammonia–water falling film absorption by adding nano-particles, Int. J. Refrig. 34 (3) (2011) 640–647.
[57] H. Kim, J. Jeong, Y.T. Kang, Heat and mass transfer enhancement for falling film absorption process by SiO_2 binary nanofluids, Int. J. Refrig. 35 (3) (2012) 645–651.
[58] T. Wen, L. Lu, H. Zhong, Investigation on the dehumidification performance of LiCl/H_2O-MWNTs nanofluid in a falling film dehumidifier, Build. Environ. 139 (2018) 8–16.
[59] G. Wang, et al., Investigation on mass transfer characteristics of the falling film absorption of LiBr aqueous solution added with nanoparticles, Int. J. Refrig. 89 (2018) 149–158.
[60] S.-S. Ashrafmansouri, M.N. Esfahany, Mass transfer in nanofluids: a review, Int. J. Therm. Sci. 82 (2014) 84–99.
[61] S. Komati, A.K. Suresh, Anomalous enhancement of interphase transport rates by nanoparticles: effect of magnetic iron oxide on gas – liquid mass transfer, Ind. Eng. Chem. Res. 49 (1) (2010) 390–405.
[62] R. Zarzycki, A. Chacuk, Bubble columns, in: R. Zarzycki, A. Chacuk (Eds.), Absorption, Pergamon, 1993, pp. 601–619.
[63] C. Fleischer, S. Becker, G. Eigenberger, Detailed modeling of the chemisorption of CO_2 into NaOH in a bubble column, Chem. Eng. Sci. 51 (10) (1996) 1715–1724.
[64] I. Sreedhar, et al., Carbon capture by absorption–path covered and ahead, Renew. Sustain. Energy Rev. 76 (2017) 1080–1107.
[65] J.-K. Kim, J.Y. Jung, Y.T. Kang, The effect of nano-particles on the bubble absorption performance in a binary nanofluid, Int. J. Refrig. 29 (1) (2006) 22–29.
[66] A.W. Adamson, A.P. Gast, Physical Chemistry of Surfaces, Vol. 150., Interscience Publishers, New York., 1967.
[67] Y.T. Kang, B.-C. Kim, Absorption heat transfer enhancement in binary nanofluids, in: The 22nd International Congress of Refrigeration. No. 07-B2 (2007) p. 371.
[68] H. Liu, et al., Experimental study on enhancing ammonia bubble absorption by FeO nanofluid, Chem. Ind. Eng. Prog. 28 (7) (2009) 1138–1141.
[69] W. Wu, et al., Enhancement on NH_3/H_2O bubble absorption in binary nanofluids by mono nano Ag, CIESC J. 61 (5) (2010) 1112–1117.
[70] W.-D. Wu, et al., Nanoferrofluid addition enhances ammonia/water bubble absorption in an external magnetic field, Energy Build. 57 (2013) 268–277.
[71] I.T. Pineda, D. Kim, Y.T. Kang, Mass transfer analysis for CO_2 bubble absorption in methanol/Al_2O_3 nanoabsorbents, Int. J. Heat Mass Transf. 114 (2017) 1295–1303.
[72] M.H.K. Darvanjooghi, M.N. Esfahany, S.H. Esmaeili-Faraj, Investigation of the effects of nanoparticle size on CO_2 absorption by silica-water nanofluid, Sep. Purif. Technol. 195 (2018) 208–215.
[73] H. Pashaei, A. Ghaemi, CO_2 absorption into aqueous diethanolamine solution with nano heavy metal oxide particles using stirrer bubble column: hydrodynamics and mass transfer, J. Environ. Chem. Eng. (2020) 104110.
[74] J.-K. Kim, J.Y. Jung, Y.T. Kang, Absorption performance enhancement by nano-particles and chemical surfactants in binary nanofluids, Int. J. Refrig. 30 (1) (2007) 50–57.
[75] X. Ma, et al., Heat and mass transfer enhancement of the bubble absorption for a binary nanofluid, J. Mech. Sci. Technol. 21 (11) (2007) 1813.

[76] J.K. Lee, et al., The effects of nanoparticles on absorption heat and mass transfer performance in NH_3/H_2O binary nanofluids, Int. J. Refrig. 33 (2) (2010) 269–275.

[77] J.-Y. Jung, J.W. Lee, Y.T. Kang, CO_2 absorption characteristics of nanoparticle suspensions in methanol, J. Mech. Sci. Technol. 26 (8) (2012) 2285–2290.

[78] C. Pang, et al., Mass transfer enhancement by binary nanofluids (NH_3/H_2O + Ag nanoparticles) for bubble absorption process, Int. J. Refrig. 35 (8) (2012) 2240–2247.

[79] C. Amaris, M. Bourouis, M. Vallès, Passive intensification of the ammonia absorption process with NH_3/$LiNO_3$ using carbon nanotubes and advanced surfaces in a tubular bubble absorber, Energy 68 (2014) 519–528.

[80] X. Ma, et al., Experimental study on the thermal physical properties of a CNTs-ammonia binary nanofluid, in: International Conference on Micro/Nanoscale Heat Transfer, (2008).

[81] L.-S. Fan, et al., Bubbles in nanofluids, Ind. Eng. Chem. Res. 46 (12) (2007) 4341–4346.

[82] K.C. Ruthiya, et al., Influence of particles and electrolyte on gas hold-up and mass transfer in a slurry bubble column, Int. J. Chem. React. Eng. 4 (1) (2006).

[83] J.W. Lee, et al., Combined CO_2 absorption/regeneration performance enhancement by using nanoabsorbents, Appl. Energy 178 (2016) 164–176.

[84] J. Fan, L. Wang, Review of heat conduction in nanofluids, J. Heat Transf. 133 (4) (2011).

[85] C. Pang, et al., Thermal conductivity measurement of methanol-based nanofluids with Al_2O_3 and SiO_2 nanoparticles, Int. J. Heat Mass Transf. 55 (21–22) (2012) 5597–5602.

[86] P. Keblinski, et al., Mechanisms of heat flow in suspensions of nano-sized particles (nanofluids), Int. J. Heat Mass Transf. 45 (4) (2002) 855–863.

[87] A. Vatani, P.L. Woodfield, D.V. Dao, A survey of practical equations for prediction of effective thermal conductivity of spherical-particle nanofluids, J. Mol. Liq. 211 (2015) 712–733.

[88] S.M. Kwark, et al., Pool boiling characteristics of low concentration nanofluids, Int. J. Heat Mass Transf. 53 (5–6) (2010) 972–981.

[89] A.H. Abdel-Salam, C.J. Simonson, State-of-the-art in liquid desiccant air conditioning equipment and systems, Renew. Sustain. Energy Rev. 58 (2016) 1152–1183.

[90] Y. Yin, J. Qian, X. Zhang, Recent advancements in liquid desiccant dehumidification technology, Renew. Sustain. Energy Rev. 31 (2014) 38–52.

[91] N. Isshiki, K. Ogawa, Enhancement of heat and mass transfer by CCS tubes, in: Ab-Sorption 96, Proceedings of the International Ab-Sorption Heat Pump Conference, Montreal, Canada, (1996).

[92] C.W. Park, et al., Experimental correlation of falling film absorption heat transfer on micro-scale hatched tubes, Int. J. Refrig. 26 (7) (2003) 758–763.

[93] J.-I. Yoon, O.-K. Kwon, C.-G. Moon, Experimental investigation of heat and mass transfer in absorber with enhanced tubes, KSME Int. J. 13 (9) (1999) 640–646.

[94] X.-Y. Cui, et al., Investigation of plate falling film absorber with film-inverting configuration, J. Heat Transf. 131 (7) (2009).

[95] M.R. Islam, N. Wijeysundera, J. Ho, Performance study of a falling-film absorber with a film-inverting configuration, Int. J. Refrig. 26 (8) (2003) 909–917.

[96] M. Mortazavi, et al., Absorption characteristics of falling film LiBr (lithium bromide) solution over a finned structure, Energy 87 (2015) 270–278.

[97] P. Galhotra, Carbon dioxide adsorption on nanomaterials, PhD Dissertation, University of Iowa. DOI: 10.17077/etd.detbvuv9, (2010).

[98] S.B. White, Enhancement of boiling surfaces using nanofluid particle deposition, PhD Dissertation, University of Michigan. (2010).

[99] T. Wen, et al., Experimental and numerical study on the regeneration performance of LiCl solution with surfactant and nanoparticles, Int. J. Heat Mass Transf. 127 (2018) 154–164.

Further reading

E.S. Kim, J.-Y. Jung, Y.T. Kang, The effect of surface area on pool boiling heat transfer coefficient and CHF of Al_2O_3/water nanofluids, J. Mech. Sci. Technol. 27 (10) (2013) 3177–3182.

S.H. Kim, et al., Study of leidenfrost mechanism in droplet impacting on hydrophilic and hydrophobic surfaces, Int. J. Air-Conditioning Refrig. 21 (04) (2013) 1350028.

R.N. Wenzel, Surface roughness and contact angle, J. Phys. Chem. 53 (9) (1949) 1466–1467.

CHAPTER

Mass transfer basics and models of membranes containing nanofluids 14

Colin A. Scholes

Department of Chemical Engineering, The University of Melbourne, Melbourne, VIC, Australia

14.1 Introduction

Membranes are conceptually straightforward, a semipermeable material that separates two domains while allowing the passage of selective chemical species, whereas other chemical species experience a barrier. Membrane technology has been commercialized in a wide range of applications, with the most successful being water desalination; nano-, ultra-, and microfiltration as well as natural gas sweetening [1]. Membrane processes are primarily described by the size of the chemical compounds they selectively separate as well as the corresponding pore size present within the membrane material (Fig. 14.1).

Solvent extraction and gas–solvent absorption applications are increasing trialing membrane module configurations to improve their performance. Solvent extraction, also known as liquid–liquid extraction or partitioning, is used to separate and concentrate compounds based on their relative solubilities in two immiscible solvents, usually water and a nonpolar solvent. Gas–solvent absorption is the removal of gas and vapor molecules from the gas phase into a selective solvent, which may be through physical or chemical absorption mechanisms [2,3]. Utilizing a membrane configuration to separate the two phases in these processes enables the membrane to act as the mass transfer area and achieve higher separation performance outcomes compared to conventional packed solvent columns. Specifically, these applications take advantage of the membrane module's tight packing density to achieve much greater mass transfer area per unit volume and hence process intensification [4]. The rigid control over the flow of the phases also ensures issues such as channeling and foaming do not occur within the membrane modules. One of the most widely studied examples of gas–solvent absorption is for acidic gas removal, specifically carbon dioxide capture, and commonly known in this application as membrane contactors. This technology involves the transfer of CO_2 from industrial flue gas through a membrane where the CO_2 is chemically absorbed into a solvent on the permeate side. The membrane is ideally porous, to enable the rapid transfer of CO_2 from the gas phase to the solvent phase, if the pores remain gas-filled. Importantly, the technology uses the highly selective nature of the solvent to capture CO_2, while benefiting from the membrane module's control of the phases' flows. In addition, the flexibility and modular nature of membrane contactors enable the capture process to be undertaken in any orientation as well as accommodated in limited spaces and locations.

Nanofluids and Mass Transfer. DOI: https://doi.org/10.1016/B978-0-12-823996-4.00017-3

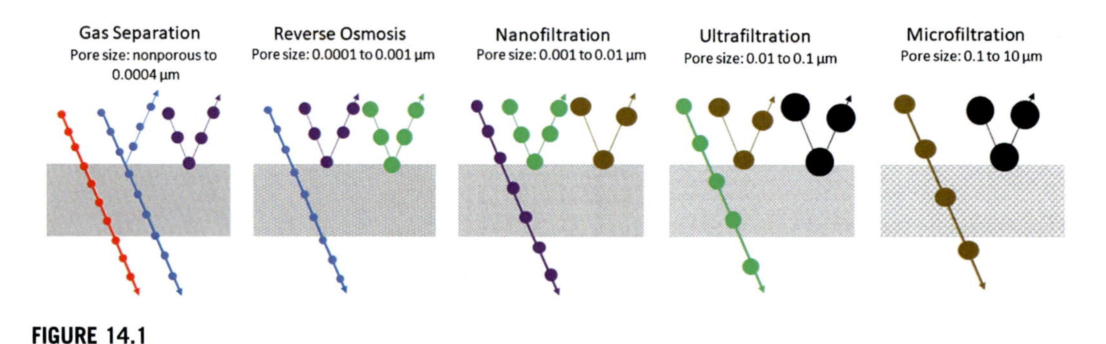

FIGURE 14.1

Membrane separation processes, based on the pore sizes present, which correspond to the molecular/particle size separation.

Membrane gas–solvent contactors conventionally use traditional solvents for CO_2 capture, such as ethanolamines, predominately monoethanolamine, as well as glycinate and potassium carbonate. While these solvents are well characterized and demonstrated to have improved performance in a membrane contactor configuration, there are several drawbacks that limit their deployment, such as wettability and surface tension [5], that have meant alternative solvents have been studied. Contactors have also been applied for sulfur dioxide capture from flue gas and water dehydration from flue gas.

Nanofluids are one such example, which have demonstrated potential for enhanced CO_2 capture [6–9]. Nanofluids are solvents that contain nanometer-sized particles (nanoparticles) as a colloidal suspension. The nanoparticles can be a range of metals, oxides, carbon nanotubes, or other materials, that remain suspended in the solvent during the absorption process. The importance of nanoparticles is that they confer novel properties on the base solvent that result in improved mass (and heat) transfer [10]. This is achieved through the additional mass transfer mechanisms the nanoparticles add to the process, derived from their interactions with the surrounding solvent as well as the phase interfaces present [11].

This chapter therefore details the theory of mass transfer in membrane contactor systems, focusing on the improvements to mass transfer imposed by nanofluids; demonstrating how membrane contactor systems can be successfully modeled to account for the additional enhancement conferred by nanofluids, as well as presenting fluid dynamic theory that accurately models experimental data. This presentation of mass transfer theory and models will therefore enable the optimization of mass transfer in nanofluid-based membrane systems.

14.2 Mass transfer enhancement effects

The mass transfer through a membrane gas–solvent contactor process is the product of four individual mass transfer stages [12,13]: mass transfer through the gas boundary layer to the membrane, mass transfer through the pores of the membrane, mass transfer across the gas–solvent interface, and finally mass transfer through the solvent boundary layer into the bulk solvent; these are

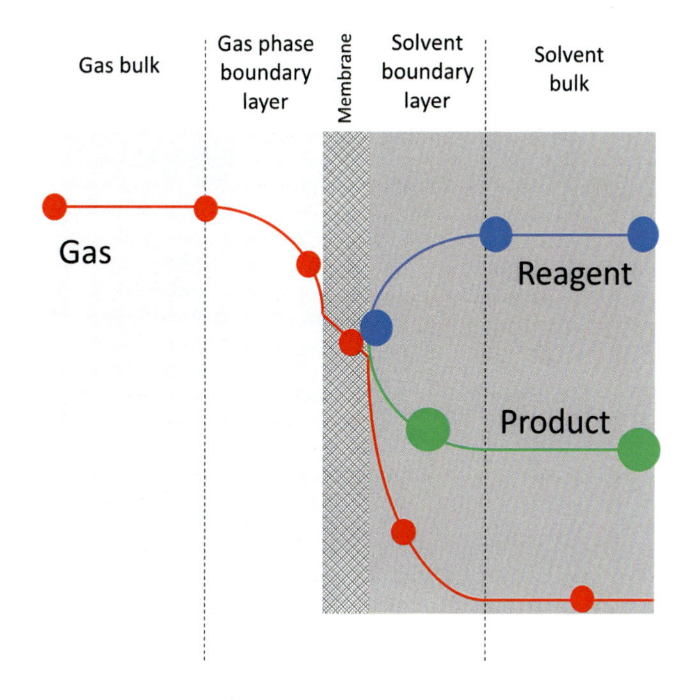

FIGURE 14.2

Mass transfer stages in a membrane gas–solvent contactor, based on a porous membrane, where the gas undergoes reaction in the solvent phase.

depicted in Fig. 14.2. These stages act as resistance to mass transfer. For chemical solvents, those that react with the captured gas, such as ethanolamines with CO_2, the mass transfer in the solvent boundary layer is enhanced by the reaction; with this enhancement dependent on the rate of reaction relative to the diffusivity of CO_2 in the solvent phase. This is modeled through the empirical enhancement factor in mass transfer theory.

The presence of nanofluids within the solvent phase also leads to additional enhancement in mass transfer, through their influence on the transport and diffusion of gases in the solvent phase. Two mechanisms are accepted to be the cause.

14.2.1 Grazing effect

The grazing effect, also sometime called the shuttle effect, is due to the concentration of nanoparticles at the gas–liquid interface, because of their general hydrophobicity, creating a gas–liquid–solid three-phase system [14,15]. The nanoparticles surface presents a domain to which the gas particles in the solvent boundary layer readily adsorb. This results in a decrease in the gas concentration in the solvent boundary layer, increasing the driving force for gas absorption. In addition, the gas-adsorbed nanoparticles diffuse away from the interface into the bulk solvent, where the gas molecules will

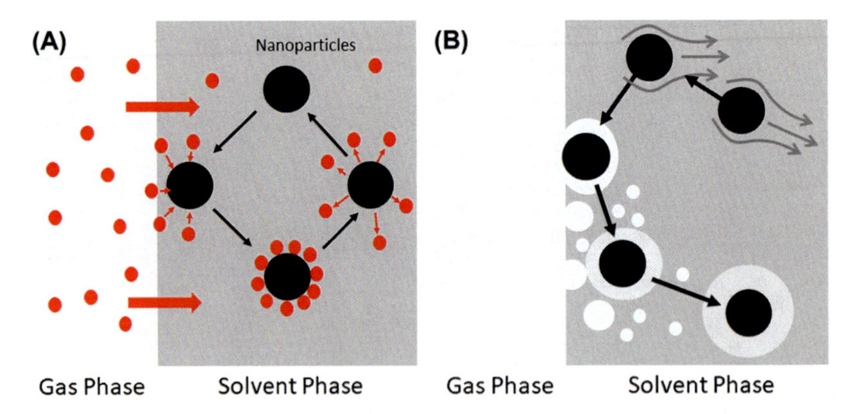

FIGURE 14.3

(A) Grazing and (B) hydrodynamic interfacial effects of nanoparticles that improve mass transfer.

desorb. This creates another mechanism for transferring gas molecules from the boundary layer into the bulk solvent and hence enhancing mass transfer. The grazing effect is demonstrated in Fig. 14.3.

14.2.2 Hydrodynamic effect in the gas–solvent boundary layer

The presence of nanoparticles in the solvent boundary layer also interrupts and breaks the boundary layer because of collisions between the interface and the nanoparticles [14,15]. This leads to a thinner boundary layer being formed as well as creating more turbulence in the layer, enhancing gas diffusion and mixing with the solvent phase. The impact on the hydrodynamics of the solvent boundary layer by nanoparticles is provided in Fig. 14.3.

In conventional solvent absorption processes, an additional enhancement mechanism around the inhibition of bubble coalescence is believed to be present. This directly correlates bubbles surface area with mass transfer area. However, this process has a very limited role in membrane gas-contactor systems, because the pores in the membrane act as the mass transfer area between the two phases and bubbling of gas in the solvent phase is strictly avoided.

14.3 Mass transfer theory

The overall objective for gas absorption is to maximize the gas flux through the membrane contactor into the lean solvent. The gas flux can be described as:

$$N_{Absorption} = K_{OV,L} LM(x^* - x) \tag{14.1}$$

where K_{OV} is the overall mass transfer coefficient (defined here by the solvent (L) phase) and LM is the log mean difference of the mass transfer driving force along the membrane module. The driving force for mass transfer is the difference between the gas mole fraction in the solvent bulk phase

(x) and the equilibrium mole fraction of gas in the solvent (x^*). This is dependent on the partial pressure in the gas entering and exiting the membrane module.

The overall mass transfer coefficient is the diffusion rate constant that measures the ability of a system to enable mass transfer; and importantly provides comparison between different membrane materials and solvent systems. The overall mass transfer coefficient is a function of mass transfer through the gas boundary layer, mass transfer through the membrane pores, and mass transfer through the solvent boundary layer. These stages act as a resistance to mass transfer and are usually expressed as a series [12,16]:

$$\frac{1}{K_{OV}} = \frac{1}{k_H k_g} + \frac{1}{k_H . k_{m,p}} + \frac{1}{E k_l} \quad (14.2)$$

where k_g is the physical mass transfer coefficient in the gas side boundary layer, k_H is the dimensionless Henry's Law constant, $k_{m,p}$ is the membrane pores, and k_l is the solvent side boundary layer. As there is a change in the phases, k_H is required to account for the conversion of the gas phase to the solvent phase. E is the enhancement factor due to any chemical reaction between gas and solvent (ignoring nanofluid effects).

Most membrane gas–solvent contactor systems are based on hollow-fiber bundle modules, and hence the geometry of the fiber can be incorporated into the overall mass transfer resistance equation [17]:

$$\frac{1}{K_{OV} d_i} = \frac{1}{k_H \cdot k_g d_O} + \frac{1}{k_H \cdot k_{m,p} d_{In}} + \frac{1}{E k_l d_i} \quad (14.3)$$

d_O, d_i, and d_{In} are the outer, inner, and logarithmic mean diameters of the membrane fiber, respectively.

The mass transfer coefficient for the membrane pores can be modeled by [18]:

$$k_{m,p} = \frac{D_g \varepsilon}{\delta_p \tau} \quad (14.4)$$

where D_g is the diffusivity of targeted gas species in the gas phase, ε is the porosity of the porous layer, δ_p is the thickness of the porous layer, and τ is the tortuosity of the porous layer. The tortuosity of membranes is often related to the porosity of the layer [19]:

$$\tau = \frac{(2-\varepsilon)^2}{\varepsilon} \quad (14.5)$$

This model assumes gas-filled pores, however generally there is partial wetting of the porous layer by the solvent and therefore the mass transfer coefficient for a wetted porous support layer can be modeled by:

$$k_{m,p} = \frac{m E D_L \varepsilon}{\delta_p \tau} \quad (14.6)$$

where D_L is the diffusivity of gas in the solvent phase and m the partition coefficient for the solvent (defined as the ratio at equilibrium of the solvent to gas concentration). For many membrane contactor systems, the pores are partially wetted and therefore the mass transfer coefficient can be defined in terms of the proportion of the pore space that is wetted (w) [20]:

$$\frac{1}{k_{m,p}} = \frac{\tau\delta_p}{\varepsilon}\left(\frac{1-w}{D_g} - \frac{mw}{D_L E}\right) \tag{14.7}$$

The overall mass transfer coefficient is particularly sensitive to partial pore wetting by the solvent, given that chemical diffusivity in liquids is often three orders of magnitude lower than that of the gas phase.

The mass transfer in the gas and solvent boundary layers are determined from mass transfer correlations that have been developed from theoretical and empirical methods. For the lumen side of the membrane hollow-fiber, the mass transfer coefficient can be modeled by the dimensionless Sherwood (Sh), Reynolds (Re), and Schmidt (Sc) numbers based on the Graetz–Leveque correlations [21]:

$$Sh_G = \frac{k_g d_i}{D_G} = 1.62\left(ScRe\left(\frac{d_i}{l}\right)\right)^{1/3} \quad Gz > 6 \tag{14.8}$$

$$Sh_G = \frac{k_g d_i}{D_g} = 0.5.ScRe\left(\frac{d_i}{l}\right) Gz < 6 \tag{14.9}$$

where l is the length of the membrane fiber.

Generally, the gas phase is on the lumen side of the membrane, and the solvent on the shell side. The mass transfer on the shell side boundary layer is dependent on the configuration of the membrane contactor, particularly packing density (φ) and the ratio of hydraulic diameter (d_h) [22]:

$$Sh_L = \frac{k_s d_h}{D_L} = A \cdot f(\varphi)\left(\frac{d_h}{l}\right)^{\alpha} Re^{\beta} Sc^{\gamma} \tag{14.10}$$

where A is a constant of proportionality. Generally, the exponent on the Schmidt number is ⅓ and the hydraulic diameter is determined by:

$$d_h = \frac{d_{cin}^2 - nd_o^2}{d_{cin} + nd_o} \tag{14.11}$$

where d_{cin} is the inner diameter of the contactor and n is the number of membrane fibers.

The shell side flow regime strongly influences the mass transfer coefficient. Parallel flow is along the membrane fibers and corresponds to countercurrent, while cross flow is tangential across the membrane fibers, usually established through a baffle arrangement within the module. The definition of Reynolds number is different for both flow regimes, and for parallel flow it is defined as [21]:

$$Re = \frac{4\rho Q}{(\pi\eta(d_{cin} + n \cdot d_o))} \tag{14.12}$$

while for cross flow it is defined as:

$$Re = \frac{2\rho Q(d_{cin} + d_{co})(ln(d_{cin}/d_{co}))}{\pi\eta n d_o l} \tag{14.13}$$

where ρ is density, η the viscosity, Q volumetric flowrate, n number of fibers in the module, d_{co} the outer diameter of the central tube for cross flow, and l the length of the fiber. For the same flowrate, cross flow configuration generally achieves greater mass transfer through the shell side boundary layer than parallel flow due to the difference in the boundary layer formed on the membrane fiber surface.

For parallel flow on the shell side of membrane contactors, several correlations have been established in the literature. Those proposed by Yang and Cussler are the most adopted by the literature, in part because of the wide range of flow conditions covered [21,23]:

$$Sh_s = \frac{kd_e}{D_L} = 1.25\left[\frac{d_e}{l}Re\right]^{0.93}[Sc]^{1/3} Re < 2100 \tag{14.14}$$

$$Sh_s = 0.116\left[Re^{2/3} - 125\right]\left[1 + \left(\frac{d_e}{l}\right)^{2/3}\right] 2100 < Re < 10,000 \tag{14.15}$$

$$Sh_s = 0.023Re^{0.8}Sc^{2/3} Re > 10,000 \tag{14.16}$$

Similarly, several correlations have been proposed for cross flow on the shell side of membrane contactors, along with the conditions over which the correlations are valid [22]. There is more deviation in these correlations and their various usage, given the different methods used to generate cross flow conditions.

14.3.1 Nanofluids mass transfer

For nanoparticles in the solvent phase, the Brownian motion of the particles is an important characteristic of the system. The hydrodynamics are influenced by the movement of these particles, which results in an increased velocity profile around the particle creating microconvection. This correspondingly increases the diffusion coefficient of the absorbing gas compounds. This increase in diffusion within the nanofluid can be described through the diffusion coefficient within the solvent and the diffusion addition due to the generated microconvection (D_{mc}), through the following expression:

$$D_{nf} = D_L + D_{mc} \tag{14.17}$$

The nanoparticles hydrodynamics are linked with their Brownian motion, with the self-diffusion coefficient of Brownian particles expressed by Stokes–Einstein diffusivity:

$$D^* = \frac{k_B T}{3d_p \pi \eta} \tag{14.18}$$

where k_B is the Boltzmann constant, T the temperature, and d_p is the diameter of nanoparticle. The corresponding root mean square velocity of the nanoparticles is therefore:

$$v_p = \sqrt{\frac{3k_B T}{m_p}} \tag{14.19}$$

where m_p is the mass of the nanoparticle. This velocity is the contribution of the nanoparticle's Brownian motion to the microconvection generated in the fluid. This then enables the nanoparticle Reynolds number to be described [24,25]:

$$Re_{nf} = \frac{1}{\nu}\sqrt{\frac{18k_B T}{\pi \rho d_p}} \tag{14.20}$$

where ν is the kinematic viscosity of the solvent with the nanoparticles present. For most nanofluids the $Re < 0.1$. Consequently, the Schmidt number of the nanofluid can also be defined:

$$Sc_{nf} = \frac{\nu}{D^*} \tag{14.21}$$

This enables the diffusion coefficient of a gas within the nanofluid to be described through correlations based on the configuration [26]:

$$D_{nf} = D_L\left(1 + A\phi^{\alpha} {Re_{nf}}^{\beta} {Sc_{nf}}^{\gamma}\right) \tag{14.22}$$

where A is a constant of proportionality and ϕ the volume fraction of the nanoparticles. Several correlations have been proposed in the literature and provided in Table 14.1.

The inclusion of diffusion coefficient for nanofluids within the membrane contactor mass transfer correlations is through the Sherwood number, as such a modified correlation based on equation 14.10 can be used:

$$Sh_L = \frac{k_s d_h}{D_L\left(1 + A\phi^{\alpha} {Re_{nf}}^{\beta} {Sc_{nf}}^{\gamma}\right)} = A \cdot f(\varphi)\left(\frac{d_h}{l}\right)^{\alpha} Re^{\beta} Sc^{\gamma} \tag{14.23}$$

The grazing effect is modeled differently, as the migration of gas molecules adsorbed to the nanoparticles is significantly slower than the diffusion coefficient of the gas molecules in the solvent. Rather, the grazing effect impacts the concentration of gas within the solvent phase and the coverage of nanoparticles at the gas–solvent interface can be expressed by the Langmuir isotherm [28]:

$$\alpha = \alpha_{max} \frac{k_s m_s}{1 + k_s m_s} \tag{14.24}$$

where α_{max} is the maximum nanoparticle coverage at the interface, k_s is the Langmuir affinity constant, and M_s is the nanoparticle loading in the solvent. Correspondingly, the transport of the loaded nanoparticles to the bulk solvent phase acts as an additional mechanism for mass transfer, which is observed in an additional enhancement factor (E_{gz}) for the system. This additional mechanism is accounted for in the modified Hatta modules for nanofluid systems [29]:

$$Ha = \sqrt{\frac{k_a D^*}{k_l^2}} \tag{14.25}$$

Table 14.1 Diffusion coefficient correlations for nanofluids in membrane contactor systems.

Correlation	References
$D_{nf} = D_L(1 + 640[\![Re_n f]\!]^{1.7}[\![Sc_{nf}]\!]^{1/3}(\phi^{1/3}/(1-\phi^{1/3})))$	Nagy, Feczko, and Koroknai [24]
$D_{nf} = D_L(1 + 1650\phi^{0.203}[\![Re_{nf}]\!]^{0.039}[\![Sc_{nf}]\!]^{(-1.064)})$	Rezakazemi et al. [27]

where k_a is the rate constant for adsorption to the nanoparticle. This enables a general expression for the enhancement factor due to the grazing effect to be established [30]:

$$E_{gz} = \sqrt{\frac{1 + \phi(m_{np} - 1) + (1 - \phi)Ha^2}{1 + Ha^2}} \tag{14.26}$$

where m_{np} is the partition coefficient for the gas molecules adsorbed to the nanoparticle relative to the solution. This is defined by the Henry's law for the solvent, surface area of the nanoparticles per unit volume (Γ), the volume fraction of nanoparticles in the solvent, as well as the gas adsorption coverage of the nanoparticles:

$$m_{np} = \frac{\alpha.\Gamma.\phi}{k_H} \tag{14.27}$$

If the adsorption reaction is sufficiently slow relative to mass transfer, then the enhancement factor can be approximated by [31]:

$$E_{gz} = \sqrt{1 + \phi(m_{np} - 1)} \tag{14.28}$$

The enhancement factor due to the grazing effect for different partition coefficients can be seen in Fig. 14.4, for a model system. For many chemical solvents undertaking acidic gas absorption,

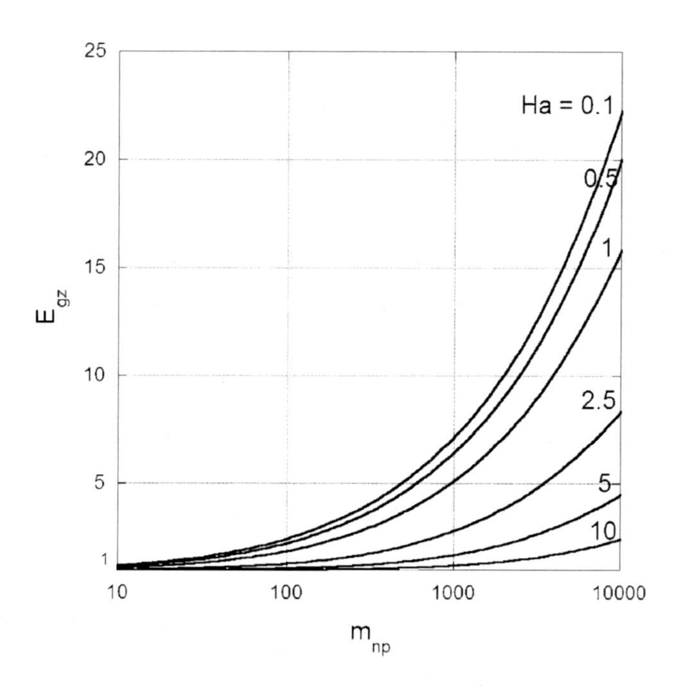

FIGURE 14.4

Enhancement factor (E_{gz}) as a function of nanoparticle partition coefficient (m_{np}) and Hatta modules for a nanofluid of volume fraction 0.05.

the enhancement factor for the chemical reaction is generally an order of magnitude greater than that observed for the grazing effect.

The inclusion of grazing effect enhancement factor for membrane contactors is within the observed increase in mass transfer through the solvent boundary layer, in conjunction with the enhancement factor imposed from chemical reaction present. This results in a modification to the overall mass transfer coefficient definition of the membrane contactor system, based on equation 14.3:

$$\frac{1}{K_{OV}d_i} = \frac{1}{k_H \cdot k_g d_0} + \frac{1}{k_H \cdot k_{m,p} d_{In}} + \frac{1}{(E_{gz} + E) k_l d_i} \tag{14.29}$$

14.3.2 Fluid dynamic models

A range of fluid dynamic models have been proposed in the literature to describe nanofluid behavior [26], including in membrane contactor systems. These models approximate the nanoparticles as spheres of a designated radius, which is incorporated in computational fluid dynamics analysis. These models assume a porous hollow-fiber membrane of specific characteristics provided in Fig. 14.5. The continuity, momentum, and mass transfer equations for the respective lumen, membrane, and shell sides can be readily defined [32]:

For the lumen side:

$$\eta \left[\frac{1}{r} \frac{\partial}{\partial r} \left(r \frac{\partial v_z}{\partial r} \right) \right] - \frac{dP}{dx} = 0 \tag{14.30}$$

with the boundary conditions:

$$\text{at} \quad r = 0, \frac{\partial v_z}{\partial r} = 0 \tag{14.31}$$

$$\text{at} \quad r = 0, \frac{\partial v_z}{\partial r} = 0 \tag{14.32}$$

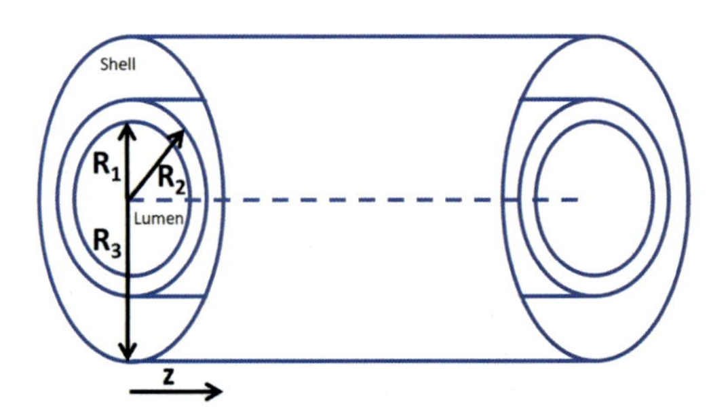

FIGURE 14.5

Definitions of a membrane fiber within the module shell for CFD modeling.

The lumen velocity profile is described by:

$$v_{z-lumen} = 2v_{lumen}\left[1 - \left(\frac{r}{R_1}\right)^2\right] \tag{14.33}$$

where v_{lumen} is the average velocity through the lumen.

The continuity equation for gas A in the lumen side is derived from Fick's law of diffusion:

$$D_{A-lumen}\left[\frac{\partial^2 C_{A-lumen}}{\partial r^2} + \frac{1}{r}\frac{\partial C_{A-lumen}}{\partial r} + \frac{\partial^2 C_{A-lumen}}{\partial z^2}\right] = v_{z-lumen}\frac{\partial C_{A-lumen}}{\partial z} \tag{14.34}$$

where $C_{A-lumen}$ is the concentration of gas A in the lumen and $D_{A-lumen}$ is the diffusion coefficient of gas A in the lumen. The continuity equation can be solved with the following boundary conditions:

$$\text{at } z = 0, C_{A-lumen} = C_{A,0} \tag{14.35}$$

$$\text{at } z = L, N_{A-lumen} = v_{z-lumen}C_{A-lumen} \tag{14.36}$$

$$\text{at } r = 0, \frac{\partial C_{A-lumen}}{\partial r} = 0 \tag{14.37}$$

$$\text{at } r = R_1, C_{A-lumen} = C_{A-memb} \tag{14.38}$$

where $N_{A-lumen}$ is the flux.

For the porous membrane, there is no velocity of components and therefore the continuity equation can be expressed as:

$$D_{A-memb}\left[\frac{\partial^2 C_{A-memb}}{\partial r^2} + \frac{1}{r}\frac{\partial C_{A-memb}}{\partial r} + \frac{\partial^2 C_{A-memb}}{\partial z^2}\right] = 0 \tag{14.39}$$

with the boundary conditions:

$$\text{at } z = 0, \frac{\partial C_{A-memb}}{\partial z} = 0 \tag{14.40}$$

$$\text{at } z = L, \frac{\partial C_{A-memb}}{\partial z} = 0 \tag{14.41}$$

$$\text{at } r = R_1, C_{A-memb} = C_{A-lumen} \tag{14.42}$$

$$\text{at } r = R_2, C_{A-memb} = k_H.C_{A-shell} \tag{14.43}$$

which assumes that the phase of the pores of the membrane are the same phase as the lumen.

The shell side is generally more complex, given the possible flow regimes and fiber packing present. However, a similar momentum conservation equation can be used:

$$\eta\left[\frac{1}{r}\frac{\partial}{\partial r}\left(r\frac{\partial v_z}{\partial r}\right)\right] - \frac{dP}{dx} = 0 \tag{14.44}$$

with the following boundary conditions:

$$\text{at } r = R_2, v_z = 0 \tag{14.45}$$

$$\text{at } r = R_3, \frac{v_z}{\partial r} = 0 \tag{14.46}$$

where the velocity distribution in the module shell is defined by:

$$v_{z-shell} = 2v_{shell}\left[1-\left(\frac{R_2}{R_3}\right)^2\right]\left[\frac{\left(\frac{r}{R_3}\right)^2-\left(\frac{R_2}{R_3}\right)^2+2ln\left(\frac{R_2}{r}\right)}{\left(\frac{R_2}{R_3}\right)^4-4\left(\frac{R_2}{R_3}\right)^2+4\left(\frac{R_2}{R_3}\right)+3}\right] \tag{14.47}$$

where v_{shell} is the average phase flow through the module shell.

The continuity equation for gas A in the module shell side is:

$$D_{A-shell}\left[\frac{\partial^2 C_{A-shell}}{\partial r^2}+\frac{1}{r}\frac{\partial C_{A-shell}}{\partial r}+\frac{\partial^2 C_{A-shell}}{\partial z^2}\right] = v_{z-shell}\frac{\partial C_{A-shell}}{\partial z} \tag{14.48}$$

with the following boundary conditions on the shell side:

$$\text{at } z=0, N_{A-shell} = v_{z-shell}C_{A-shell} \tag{14.49}$$

$$\text{at } z=L, C_{A-shell} = C_{A,0} \tag{14.50}$$

$$\text{at } r=R_2, C_{A-shell} = C_{A-memb} \tag{14.51}$$

$$\text{at } r=R_3, \frac{\partial C_{A-shell}}{\partial r} = 0 \tag{14.52}$$

The enhancement effects imposed by the nanofluid are incorporated into the relevant side of the module (lumen or shell, but not the membrane pores which are gas-filled), with the hydrodynamics effect influencing the diffusion coefficient of the gas in the solvent phase side. The grazing effect requires the addition of an extra term to the respective continuity equation: [33]

$$\frac{K_{OV}\Gamma}{1-\phi}(C_A - m_s) \tag{14.53}$$

This term is added with the concentration gradient in the z-direction term, on the right-side of the equation.

14.4 Theory verification

Mass transfer theory through membrane contactors, specifically for gas–solvent contactor systems, has been widely verified in the literature [5,12,16]. There is very good alignment between theoretical expectations and experimental outcomes. Where deviations do exist, this is almost exclusively due to the theoretical simulations not accounting for partially wetted pores, and hence overemphasizing mass transfer through the pores, which are assumed to be completely gas-filled [5,12]. Membrane contactor mass transfer theory has been used extensively to simulate large-scale gas absorption processes, with important outcomes for the technology [4,34,35].

There has been less theory [27,36] verification of nanofluids within membrane contactor systems, in part because these solvents have not been studied to the same degree as conventional solvents. Those studies that have been undertaken utilizing computational fluid dynamics with membrane contactor mass transfer theory have demonstrated good agreement with experimental data [32,33]. Three main nanofluids have been reported in the literature to verify the theory; those based on spherical SiO_2 nanoparticles, tubular carbon nanotubes, and spherical (α and γ)-Al_2O_3

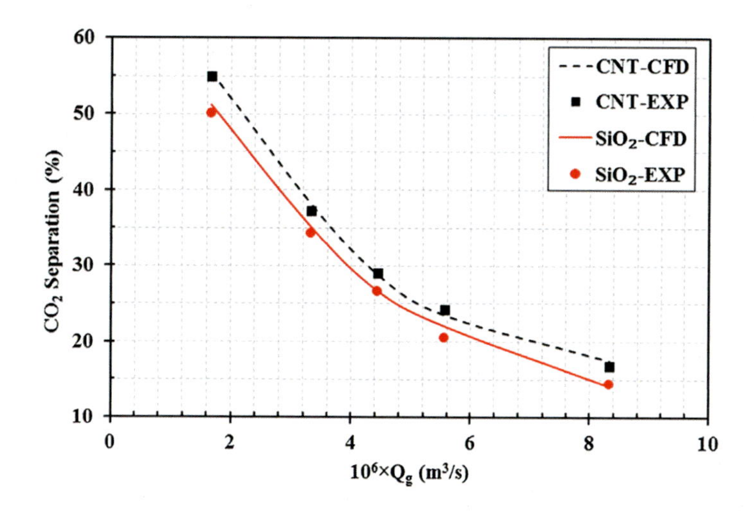

FIGURE 14.6

Comparison of CO_2 separation efficiency as a function of flowrate for nanofluids containing SiO_2 and carbon nanotubes between experimental data and CFD modeling.

Reproduced with permission from M. Rezakazemi, M. Darabi, E. Soroush, M. Mesbah, CO_2 absorption enhancement by water-based nanofluids of CNT and SiO_2 using hollow-fiber membrane contactor. Sep. Purif. Technol. 210 (2019) 920–926, doi:10.1016/j.seppur.2018.09.005.

nanoparticles. These studies have verified that nanofluids consisting of SiO_2 and Al_2O_3 particles achieve a CO_2 absorption increase of 16% and 20%, respectively, up to a weight fraction of 0.05%, while for carbon nanotubes absorption increases of 34% have been observed. This improved performance for carbon nanotubes is due to the higher adsorption capacity of the nanotubes, and their hydrophobicity means they strongly associate with the gas–solvent interface. Correspondingly, the modeling of these nanofluid membrane systems has demonstrated theory deviations of 9.5% or less. This is demonstrated in Fig. 14.6, for both SiO_2 and carbon nanotube systems [27]. However, to date a drawback of these theoretical studies with experimental data has been to position the nanofluid within the lumen side of the membrane. Conventional solvents are known to be more effective on the shell side of the module, to minimize pressure drop and pore wetting [37,38], and this will also be the case for nanofluids.

14.5 Conclusions

In summary, membrane contactor configurations, especially for gas absorption are efficient technology that can readily accommodate nanofluids for enhanced performance. The membrane geometry and the separation of the two phases presents unique opportunities for improved mass transfer through nanofluids. Here, the theory of mass transfer through membrane contactors is presented, in terms of both mass transfer through the membrane as well as concentration profiles within the respected domains (shell and lumen) of the membrane module, to enable computational fluid dynamic simulations. The presence of nanoparticles creates microconvection that increases diffusion coefficient for the transferring species in the nanofluid phase, which is modeled through an extension of the diffusion coefficient based on Brownian motion, influenced by the Reynolds and Schmidt numbers. The grazing effect, which delivers adsorbed gases to the bulk solvent, is modeled through an enhancement to mass transfer in the solvent boundary layer. This is dependent on the amount of adsorbed gas on the nanoparticles, their volume fraction within the nanofluid, as well as the diffusion coefficient of the nanoparticles. The enhancement effect produced by the grazing effect is minor compared to fast chemical reactions, but not negligible. Importantly, these mass transfer models have been verified against experimental data and demonstrated the robustness of their approaches.

The expectation is that membrane contactor technology incorporating nanofluids to enhance mass transfer will increasingly be adapted and implemented for a range of separation processes.

References

[1] W.S.W. Ho, K. Sirkar, Membrane Handbook, Springer, Boston, 1992, pp. 3–15.

[2] O. Falk-Pedersen, M.S. Gronvold, P. Nokleby, F. Bjerve, H.F. Svendsen, CO_2 capture with nembrane contactors, Int. J. Green Energy (2005) 157–165. Available from: https://doi.org/10.1081/GE-200058965.

[3] K. Simons, K. Nijmeijer, M. Wessling, Gas–liquid membrane contactors for CO_2 removal, J. Membr. Sci. 340 (1–2) (2009) 214–220. Available from: https://doi.org/10.1016/j.memsci.2009.05.035.

[4] E. Favre, H.F. Svendsen, Membrane contactors for intensified post-combustion carbon dioxide capture by gas–liquid absorption processes, J. Membr. Sci. 407–408 (2012) 1–7. Available from: https://doi.org/10.1016/j.memsci.2012.03.019.

[5] J.A. Franco, S.E. Kentish, J.M. Perera, G.W. Stevens, Poly(tetrafluoroethylene) sputtered polypropylene membranes for carbon dioxide separation in membrane gas absorption, Ind. Eng. Chem. Res. 50 (7) (2011) 4011–4020. Available from: https://doi.org/10.1021/ie102019u.

[6] J. Jiang, B. Zhao, Y. Zhuo, S. Wang, Experimental study of CO_2 absorption in aqueous MEA and MDEA solutions enhanced by nanoparticles, Int. J. Greenhouse Gas Control. 29 (2014) 135–141. Available from: https://doi.org/10.1016/j.ijggc.2014.08.004.

[7] I.P. Koronaki, M.T. Nitsas, C.A. Vallianos, Enhancement of carbon dioxide absorption using carbon nanotubes—a numerical approach, Appl. Therm. Eng. 99 (2016) 1246–1253. Available from: https://doi.org/10.1016/j.applthermaleng.2016.02.030.

[8] J.W. Lee, J.Y. Jung, S.G. Lee, Y.T. Kang, CO_2 bubble absorption enhancement in methanol-based nanofluids, Int. J. Refrig. 34 (8) (2011) 1727–1733. Available from: https://doi.org/10.1016/j.ijrefrig.2011.08.002.

[9] I.T. Pineda, C.K. Choi, Y.T. Kang, CO_2 gas absorption by CH_3OH based nanofluids in an annular contactor at low rotational speeds, Int. J. Greenh. Gas. Control. 23 (2014) 105–112. Available from: https://doi.org/10.1016/j.ijggc.2014.02.008.

[10] B.C. Pak, Y.I. Cho, Hydrodynamic and heat transfer study of dispersed fluids with submicron metallic oxide particles, Exp. Heat Transf. 11 (2) (1998) 151–170. Available from: https://doi.org/10.1080/08916159808946559.

[11] S. Krishnamurthy, P. Bhattacharya, P.E. Phelan, R.S. Prasher, Enhanced mass transport in nanofluids, Nano Lett. 6 (3) (2006) 419–423. Available from: https://doi.org/10.1021/nl0522532.

[12] J. Franco, D. DeMontigny, S. Kentish, J. Perera, G. Stevens, A study of the mass transfer of CO_2 through different membrane materials in the membrane gas absorption process, Sep. Sci. Technol. 43 (2) (2008) 225–244. Available from: https://doi.org/10.1080/01496390701791554.

[13] C.A. Scholes, S.E. Kentish, G.W. Stevens, D. deMontigny, Comparison of thin film composite and microporous membrane contactors for CO_2 absorption into monoethanolamine, Int. J. Greenhouse Gas Control. 42 (2015) 66–74. Available from: https://doi.org/10.1016/j.ijggc.2015.07.032.

[14] J.H. Kim, C.W. Jung, Y.T. Kang, Mass transfer enhancement during CO_2 absorption process in methanol/Al_2O_3 nanofluids, Int. J. Heat Mass Transf. 76 (2014) 484–491. Available from: https://doi.org/10.1016/j.ijheatmasstransfer.2014.04.057.

[15] Z. Zhang, J. Cai, F. Chen, H. Li, W. Zhang, W. Qi, Progress in enhancement of CO_2 absorption by nanofluids: a mini review of mechanisms and current status, Renew. Energy 118 (2018) 527–535. Available from: https://doi.org/10.1016/j.renene.2017.11.031.

[16] D. Demontigny, P. Tontiwachwuthikul, A. Chakma, Comparing the absorption performance of packed columns and membrane contactors, Ind. Eng. Chem. Res. 44 (15) (2005) 5726–5732. Available from: https://doi.org/10.1021/ie040264k.

[17] S. Khaisri, D. deMontigny, P. Tontiwachwuthikul, R. Jiraratananon, CO_2 stripping from monoethanolamine using a membrane contactor, J. Membr. Sci. 376 (1–2) (2011) 110–118. Available from: https://doi.org/10.1016/j.memsci.2011.04.005.

[18] S. Khaisri, D. deMontigny, P. Tontiwachwuthikul, R. Jiraratananon, A mathematical model for gas absorption membrane contactors that studies the effect of partially wetted membranes, J. Membr. Sci. 347 (1–2) (2010) 228–239. Available from: https://doi.org/10.1016/j.memsci.2009.10.028.

[19] M.E. Davis, Numerical methods and modeling for chemical engineers, Wiley, New York, (1984).

[20] M. Simioni, S.E. Kentish, G.W. Stevens, Membrane stripping: desorption of carbon dioxide from alkali solvents, J. Membr. Sci. 378 (1–2) (2011) 18–27. Available from: https://doi.org/10.1016/j.memsci.2010.12.046.

[21] M. Yang, E.L. Cussler, Designing hollow-fiber contactors, AIChE J. 32 (11) (1986) 1910–1916. Available from: https://doi.org/10.1002/aic.690321117.

[22] S. Shen, S.E. Kentish, G.W. Stevens, Shell-side mass-transfer performance in hollow-fiber membrane contactors, Solvent Extraction Ion. Exch. 28 (6) (2010) 817–844. Available from: https://doi.org/10.1080/07366299.2010.515176.

[23] E.L. Cussler, Diffusion mass transfer in fluid systems, Cambridge University Press, Cambridge, (1984).

[24] E. Nagy, T. Feczkó, B. Koroknai, Enhancement of oxygen mass transfer rate in the presence of nanosized particles, Chem. Eng. Sci. 62 (24) (2007) 7391–7398. Available from: https://doi.org/10.1016/j.ces.2007.08.064.

[25] R. Prasher, P. Bhattacharya, P.E. Phelan, Thermal conductivity of nanoscale colloidal solutions (nanofluids), Phys. Rev. Lett. 94 (2) (2005). Available from: https://doi.org/10.1103/PhysRevLett.94.025901.

[26] X. Yimin, Conception for enhanced mass transport in binary nanofluids, Heat Mass Transf. (2009) 277–279. Available from: https://doi.org/10.1007/s00231-009-0564-z.

[27] M. Rezakazemi, M. Darabi, E. Soroush, M. Mesbah, CO_2 absorption enhancement by water-based nanofluids of CNT and SiO_2 using hollow-fiber membrane contactor, Sep. Purif. Technol. 210 (2019) 920–926. Available from: https://doi.org/10.1016/j.seppur.2018.09.005.

[28] H. Vinke, P.J. Hamersma, J.M.H. Fortuin, Enhancement of the gas-absorption rate in agitated slurry reactors by gas-adsorbing particles adhering to gas bubbles, Chem. Eng. Sci. 48 (12) (1993) 2197–2210. Available from: https://doi.org/10.1016/0009-2509(93)80237-K.

[29] E. Dumont, H. Delmas, Mass transfer enhancement of gas absorption in oil-in-water systems: a review, Chem. Eng. Processing: Process. Intensif. 42 (6) (2003) 419–438. Available from: https://doi.org/10.1016/S0255-2701(02)00067-3.

[30] A.H.G. Cents, D.W.F. Brilman, G.F. Versteeg, Gas absorption in an agitated gas–liquid–liquid system, Chem. Eng. Sci. 56 (3) (2001) 1075–1083. Available from: https://doi.org/10.1016/S0009-2509(00)00324-9.

[31] W.J. Bruining, G.E.H. Joosten, A.A.C.M. Beenackers, H. Hofman, Enhancement of gas–liquid mass transfer by a dispersed second liquid phae, Chem. Eng. Sci. 41 (7) (1986) 1873–1877. Available from: https://doi.org/10.1016/0009-2509(86)87066-X.

[32] M. Ansaripour, M. Haghshenasfard, A. Moheb, Experimental and numerical investigation of CO_2 absorption using nanofluids in a hollow-fiber membrane contactor, Chem. Eng. Technol. 41 (2) (2018) 367–378. Available from: https://doi.org/10.1002/ceat.201700182.

[33] N. Hajilary, M. Rezakazemi, CFD modeling of CO_2 capture by water-based nanofluids using hollow fiber membrane contactor, Int. J. Greenhouse Gas Control. 77 (2018) 88–95. Available from: https://doi.org/10.1016/j.ijggc.2018.08.002.

[34] C.A. Scholes, Membrane contactors modelled for process intensification post combustion solvent regeneration, Int. J. Greenhouse Gas Control. 87 (2019) 203–210. Available from: https://doi.org/10.1016/j.ijggc.2019.05.025.

[35] E. Chabanon, C. Bouallou, J.C. Remigy, E. Lasseuguette, Y. Medina, E. Favre, et al., Study of an innovative gas–liquid contactor for CO_2 absorption, Energy Proc, 4, Elsevier Ltd., 2011, pp. 1769–1776. https://doi.org/10.1016/j.egypro.2011.02.052.

[36] M. Darabi, M. Rahimi, A. Molaei Dehkordi, Gas absorption enhancement in hollow fiber membrane contactors using nanofluids: modeling and simulation, Chem. Eng. Processing: Process. Intensif. 119 (2017) 7–15. Available from: https://doi.org/10.1016/j.cep.2017.05.007.

[37] C.A. Scholes, S.E. Kentish, A. Qader, Membrane gas-solvent contactor pilot plant trials for post-combustion CO_2 capture, Sep. Purif. Technol. 237 (2020). Available from: https://doi.org/10.1016/j.seppur.2019.116470.

[38] C.A. Scholes, A. Qader, G.W. Stevens, S.E. Kentish, Membrane gas-solvent contactor pilot plant trials of CO_2 absorption from flue gas, Sep. Sci. Technol. (Phila.) 49 (16) (2014) 2449–2458. Available from: https://doi.org/10.1080/01496395.2014.937499.

CHAPTER

Applications of membranes with nanofluids and challenges on industrialization

15

Colin A. Scholes

Department of Chemical Engineering, The University of Melbourne, Melbourne, VIC, Australia

15.1 Introduction to nanofluids and membranes

Membrane technology has significant potential to replace a range of conventional separation processes in numerous industries over the coming decades. This is because membrane processes have several advantages that make strong arguments for industry implementation [1]. These include the simplicity of the technology, a semipermeable material that allows selective compounds to transverse, while other compounds experience a barrier; membranes require no moving parts and have straightforward flowsheets; the modular nature of membranes means the technology can be readily linearly scaled to meet variable demand and arranged in any orientation. The greatest advantage of the technology is the incredibly high mass transfer area per unit volume achieved within membrane modules, which ensures highly efficient separation compared to alternative technologies [2].

Membrane separation is characterized by the pore sizes in the membrane. For example, water desalination through reverse osmosis is based on Angstrom-sized pores. Nano-, ultra-, and microfiltration are based on approximate pore sizes of 0.001, 0.01, and 0.1 μm, respectively; nanofiltration is able to selectively separate water and monovalent ions, ultrafiltration is also able to separate multivalent ions, and the larger pores of microfiltration are able to separate chemical compounds and small particles, see Fig. 15.1. Importantly, filtration applications are now dominated by membranes, with future challenges focused on ensuring process-resistant membrane materials, minimizing membrane fouling, and widening the range of operating conditions over which membranes function [1].

Pore diameters comparable or greater than microfiltration result in significant contact between the two domains on either side of the membrane and processes that use these types of membranes are associated with phase separation applications, such as liquid–liquid extraction and gas–solvent absorption [3]. In these applications, the two phases on either side of the membrane are immiscible and so the membrane acts as the mass transfer area through the pores. This approach is known as membrane contactors, where the membrane ensures a high contact area between the two phases, while at the same time preventing direct mixing.

Membrane contactors are now widely trialed for acidic gas absorption and liquid–liquid extraction in wastewater, biological, and mining applications [2,4]. In each of these applications, the solvent phase(s) present are of paramount importance to the process. As the solvent phase provides the selectivity to the process; be it to chemically react with CO_2 or an organic phase that enables

Nanofluids and Mass Transfer. DOI: https://doi.org/10.1016/B978-0-12-823996-4.00018-5

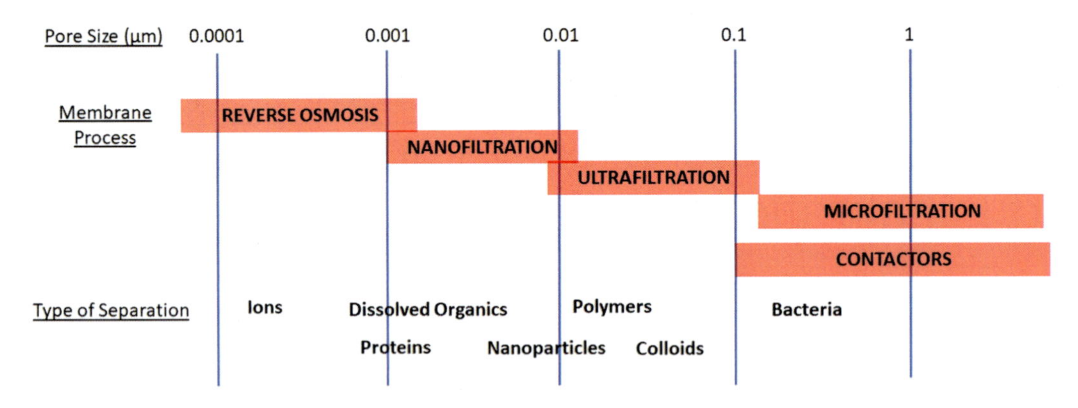

FIGURE 15.1

Size selection of membrane processes based on pore dimensions.

multivalent metal ions transfer. Therefore the types of solvent used are of importance and increasingly novel solvents are being trialed to enhance mass transfer. A good example of a novel solvent system in these membrane applications are nanofluids, which are solvents that consist of nanoparticles in a stable colloidal suspension [5]. The nanoparticles can be a range of materials, such as metals, oxides, carbon nanotubes, and other compounds, that are nanometer-sized. The presence of these nanoparticles improves mass and heat transfer properties on the solvent [6]. For mass transfer this is through the hydrodynamic interactions the nanoparticles have within the interfacial region, creating microconvection currents, disrupting the interface, and decreasing the solvent boundary layer [7]. In addition, the surface area of the nanoparticles can act as adsorption sites for compounds, such as gases, which create an additional mechanism for transporting through the solvent phase. This phenomenon is known as the grazing or shuttling effect [7]. Importantly, the use of nanofluids in membrane processes provides the opportunity to further improve membrane technology and achieve even greater separation performance.

This chapter details the applications of membrane technology where nanofluids are or can be used; specifically, liquid–liquid extraction, gas absorption, as well as nano- and ultrafiltration. This chapter describes the current state of the respective membrane applications, including industrial demonstration of the technology, as well as challenges that have so far limited implementation. The overall outcome is to highlight the advantages of membrane technology and provide the strategy needed to achieve full industry acceptance.

15.2 Nanofluid characteristics

A wide range of nanofluids have been reported in the literature, that usually consist of an aqueous solvent phase with various nanoparticle suspensions. Examples include oxides (Fe_3O_4, SiO_2, and Al_2O_3), metals (Au and Ag), carbon (nanotubes and fullerenes), as well as clays and other inorganic materials [8]. Each nanoparticle imposes unique properties onto the solvent phase, dependent on the type, size, shape, volume fraction, surface chemistry, and corresponding solvent conditions (organic, aqueous, pH, ionic concentration, two phases). The following characteristics of the nanoparticles are important, and should be well defined for any system:

- average particle size and diameter
- morphology (e.g., tubular, spherical)
- density
- specific surface area
- absorbed amount of the target solute and Langmuir affinity constant

The nanoparticle size and morphology influence the Brownian motion of the particle, which is associated with the translational diffusion of nanoparticles through the membrane system and around interfaces [7]. Brownian motion also influences the surrounding solvent and the establishment of microconvection currents which contribute to diffusion of chemical compounds into the solvent. These microconvection currents are influenced by the morphology of the nanoparticles, displacing solvent molecules within the Brownian motion wake.

The surface of nanoparticles is crucial to their functionality in membrane processes, especially at the phases' interfaces [9]. The free surface energy and hydrophobicity of the nanoparticles are key parameters that dictate their affinity for the respective solvent phases. For example, the hydrophobicity of carbon nanotubes means that they will be more concentrated at the interfaces for aqueous systems. The surface charge of nanoparticles is also important, mainly for aqueous phases but also in certain organic phases, as this influences the distribution of nearby ions in the solvent as well as the interaction with the interface. This is mainly represented in the electrical double layer that exists around the nanoparticles [10,11]. This usually consists of two layers, the inner region associated with adsorbed ions and the outer diffuse region in which ions are distributed according to the influence of electrical forces and random motion. This creates a charged shell around the nanoparticles that increases the region over which the nanoparticle has influence, as well as the intermolecular interactions the nanoparticles will have with the system. The intermolecular forces are important for the grazing effect in transporting compounds through the membrane interfacial region. Optimizing and maximizing the nanoparticle surface characteristics to be selective for specific chemical compounds enhances this process. These interactions are generally characterized by an adsorption model, where the system is dominated by physical intermolecular forces. For example, the selective adsorption of acidic gases, such as carbon dioxide, to metal oxide nanoparticles in gas–solvent absorption [12]. The importance of nanoparticle surface properties is illustrated in Fig. 15.2.

Nanofluids characteristics are readily reported in the literature and in this monograph, and this information is not repeated here. However, these details are critical to evaluate the improvement in fluid properties, based on theory and empirical definitions, and the characteristics provide the separation performance enhancements that benefit the subsequent membrane applications.

15.3 Membrane contactors

Membrane contactors are the most common membrane configuration that utilize nanofluids to optimize separation performance. These consist of porous membranes that enable the contact between

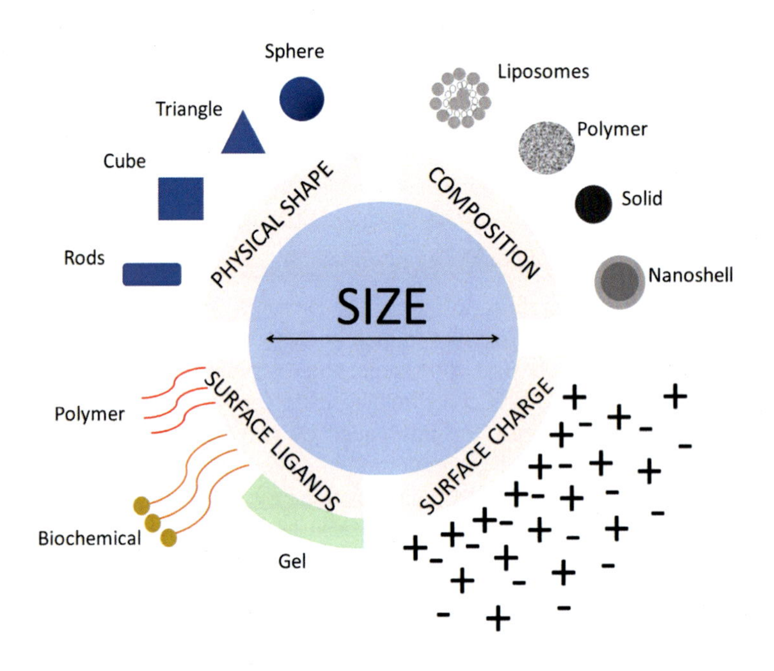

FIGURE 15.2

Nanoparticle properties' importance in separation processes.

two phases. The porous membranes are usually a fiber configuration, which are bunded into a hollow fiber module design [2–4], reminiscent of a shell and tube heat exchanger. Critically, mass transfer occurs by diffusion across the interface, just as in conventional contacting processes. As such, the enhancements provided by nanofluids in conventional processes are also present in membrane contactors. The membrane contactor usually provides no selectivity to the separation, though for nonporous membranes there is the potential for selectivity [13]. The manner in which the membrane fibers are bundled within the module is important, as packing density and flow regimes influence mass transfer. It is paramount that channeling between the membrane fibers be avoided, which can occur at very high or very low packing densities. Also it is important to ensure that solvent flow adequately interacts with the complete membrane fibers area, so that no stagnated dead volumes exist. Similarly, countercurrent flow regimes are generally not as efficient as cross flow regimes, and so baffle arrangements are often located within the module shell to create cross flow conditions and disrupt channeling effects [14,15].

There are several advantages that make membrane contactors very attractive as alternative separation technologies. These include:

- The mass transfer area, which is the membrane area, remains unaffected by changing phases' flow rates, because the two fluid flows are independent, and therefore issues such as flooding and entrainment are nonexisting.
- The interfacial area is well-known and constant, which enables more accurate simulations of the separation performance of the system.

- The separation of the two fluids means that phase density difference to direct fluid flow is not required. Therefore membrane contactors can be operated in any orientation and use fluids of identical density.
- The modular design of membranes enables the process to be operated over a wide range of capacities, simply by adding or subtracting the number of modules. This also means that membrane contactors can be accommodated in locations where space is at a premium or difficult to access.
- The contactor design enables separation efficiencies to overcome equilibrium limitations, through the selective extraction of one or more compounds into the respective phase in different modules, and the possibility of recycling streams.

Membrane contactors are commercially supplied for a range of applications, most notably for oxygenation and ozonation, bioremediation and wastewater treatment, as well as gas separation.

15.4 Membrane applications with nanofluids

Three membrane processes have application for nanofluids; which are liquid–liquid extraction and gas–solvent absorption for the separation of selective compounds, as well as ultrafiltration/nanofiltration, for the purification and concentration of nanoparticles within solvents. In the first two examples, the driving force for separation is the chemical potential gradient across the membrane, which is usually described as a concentration or activity difference. The filtration processes are based on a pressure driving force to achieve separation.

15.4.1 Liquid–liquid extraction

Liquid–liquid extraction is the separation of one or more components from one solvent into another immiscible solvent phase. Liquid–liquid extraction is most commonly found in mineral processing, hydrocarbon separation, and chemical purification. Membrane contactors undertaking liquid–liquid extraction are well-known in chemical processing, especially biological compound extraction, where mass transfer is usually from the aqueous to the organic phase. Hence, diffusion is first through the aqueous boundary layer, then through the aqueous–organic interface, followed by diffusion through the organic-phase filled membrane pores and finally through the organic phase boundary layer. A schematic of membrane contactors undertaking liquid–liquid extraction is provided in Fig. 15.3.

There are few articles on membrane contactors utilizing nanofluids for liquid–liquid extraction, but conventional liquid–liquid extraction has been enhanced with the use of nanofluids. For example, the addition of nanoparticles based on ZnO, TiO_2, and carbon nanotubes in the aqueous phase have been reported to double mass transfer rate of succinic acid from the organic phase [16]. Similarly, ZnO nanoparticles in ionic liquids have been demonstrated to be viable for the extraction of mercury and fungicides from water [17–19]. Organic-based nanofluids consisting of magnetite, alumina, and silica nanoparticles have all been demonstrated to improve mass transfer from the aqueous phase for several systems [20,21].

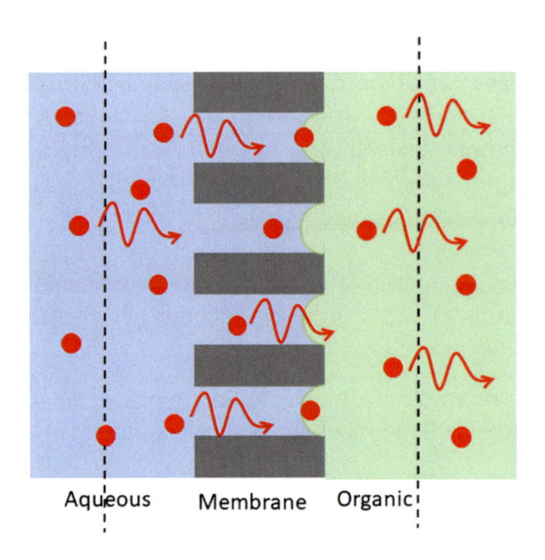

FIGURE 15.3

Schematic of membrane contactor for liquid–liquid extraction.

To maximize mass transfer efficiency several strategies are available, focused around encouraging mixing in the respective solvents boundary layers and facilitating the transport of the selective compounds across the interface. Nanofluids have potential in this application, given their ability to generate microconvection within the fluid's boundary, increasing diffusivity of components. However, a limitation of this strategy is that generally the nanofluids are present in the receiving solvent phase, and hence hydrodynamic enhancement only occurs in the organic phase boundary layer. Stabilization of nanoparticles in the organic phase requires functionalization of the surface with hydrophobic moieties, such as fatty acids [20]. For those systems where the rate-limiting mass transfer stage is the aqueous boundary layer, from which the solute is transferring from, the presence of nanoparticles has only a minor impact on separation efficiency. Similarly, the grazing effect of nanoparticles has only a minor role in solvent–solvent extraction, given the affinity for the organic phase is of paramount importance to ensure solute partition.

15.4.2 Gas–solvent absorption

Gas–solvent absorption is the separation of a gas or vapor compound into a solvent phase; which can either be through physical or chemical means. Gas–solvent absorption is most commonly found for acidic gas removal from natural gas [22]. The gas or vapor must have a strong affinity for the solvent phase; through physical absorption based on favorable intermolecular bonding, or through chemical absorption, where the gas is reacted into a soluble species. Physical absorption is often used for high pressure applications, where Henry's law absorption is favored, while lower pressure applications are more amenable to chemical absorption, where the reaction can enhance mass transfer [23].

Membrane contactors for this application can be porous, with direct contact between the gas and solvent phases, but increasingly nonporous membranes are being used to overcome issues with transmembrane pressure control and bubbling [13]. The mass transfer process is through the gas boundary layer, then diffusion through the pores of the membrane, which are gas filled, and then finally diffusion through the solvent boundary layer, and a schematic of the process provided in Fig. 15.4. The major resistance to mass transfer in this process is the solvent boundary layer [24], as the diffusion coefficient of gases is generally two orders of magnitude greater than the diffusion coefficient of liquids. Hence, to maximize mass transfer efficiency it is necessary to create additional mixing within the solvent boundary layer, which is achievable through nanofluids. The hydrodynamic effect of nanoparticles near the gas–solvent interface will lead to additional mixing from microconvection and disruption to the interface because of nanoparticle partial breakthrough into the gas phase, which results in thinning of the boundary layer. In addition, the grazing effect can be a significant mechanism for the delivery of adsorbed gas molecules into the bulk phase.

Several membrane contactor studies have been reported in the literature for CO_2 absorption into water-based nanofluids, with examples of nanoparticles used including SiO_2, ZnO, TiO_2, Fe_3O_4, and Al_2O_3 as well as forms of carbon nanotubes. The membrane contactors used have been porous polypropylene based, with pore sizes ranging from 100 to 200 nm in diameter. All the reported nanofluids have an increased absorption efficiency, with ZnO-based nanoparticles being the highest at 130% [25], with most other nanoparticles having performance increases of 25%–60% [26–28]. Interestingly, the magnitude of absorption enhancement is linked with solvent flow rate, with a decrease in efficiency improvement as the solvent flow rate increases [26]. This strongly correlates to the nanoparticles' hydrodynamic influence on the solvent boundary layer, which is more established and thicker at lower solvent flow rates. Importantly, the loading of nanoparticles in the nanofluid has a clear impact on performance, but the temptation to load the solvent with excess

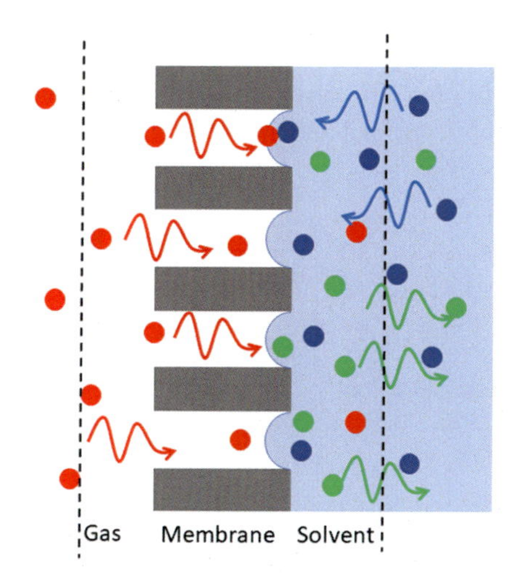

FIGURE 15.4

Schematic of membrane contactor for gas–solvent absorption, with a chemical solvent reacting with the target gas.

nanoparticles results in issues with nanofluid stability and viscosity, which limits its practicability [25]. The hydrodynamic details of the nanoparticles are demonstrated to be a key parameter in nanofluids success for CO_2 absorption through membrane contactors [27].

15.4.3 Ultrafiltration/nanofiltration

Ultrafiltration [29,30] and nanofiltration are the use of porous membranes to concentrate or remove suspended particles, including nanoparticles, from solution. Both filtration technologies are emerging as integral purification solutions to many industrial processes, such as vegetable oil processing, removal of heat-stable salts from chemical solvents, water softening, and wastewater purification [31,32]. Both ultrafiltration and nanofiltration utilize a pressure drop across the membrane as the driving force to separate both charged and uncharged components, based on their size. A schematic of how nanofiltration operates is provided in Fig. 15.5. Typical nanofiltration processes are characterized by the molecular weight cutoff of compounds that will permeate the membrane, while the ultrafiltration process is more often characterized by their size restriction [1]. Importantly, both filtration processes have the potential to prevent most macromolecules, including nanoparticles, from permeating through the membrane. For nanofluids, the application of ultrafiltration and nanofiltration is focused both on preparing the fluid, especially concentrating the nanoparticles in solution, as well as waste disposal of the nanoparticles at end of the nanofluid life.

Concentrating of nanofluids through filtration processes have been reported in the literature, mainly associated with generating the nanofluids and therefore little information has been provided about the details of the process. The second application of processing nanofluids for waste disposal is focused on removing the nanoparticles from the solvent to prevent nanoparticles being released into the environment. The use of nanofiltration and ultrafiltration for particle removal from waste solvents has been reported in the literature [33], but not for nanofluids, mainly because nanofluids have yet to be used on sufficient scale to warrant such waste processing. Nanoparticles incorporated into membranes used for nanofiltration processes have been reported in the literature to improve solvent cleanup, mainly in the lubricant and oil industries [34]. Importantly, it is anticipated that

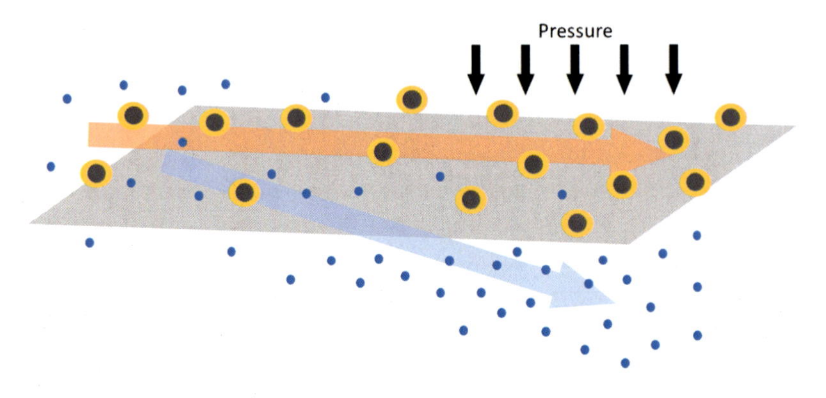

FIGURE 15.5

Schematic of ultrafiltration of nanofluids concentrating the nanoparticles.

nanofluid waste disposal will increasingly be a concern in the coming decade, given the recognized environmental impacts of nanoparticles [35,36].

15.5 Membrane process industrial demonstrations

Membrane contactor technology has been demonstrated on the pilot scale for several industrial processes; the most commonly reported one is carbon dioxide separation from industrial gases, although liquid–liquid extraction plants have also been undertaken. An example of three membrane contactor pilot plants for carbon capture have been undertaken by the author, focused on using conventional and novel solvents for the selective removal of CO_2 from both syngas and flue gas through various membrane contactor types. The Mulgrave project was a small-scale pilot plant focused on removing CO_2 from unshifted syngas, the product of coal gasification [37]; the H3 project was a larger-scale pilot plant focused on the recovery of CO_2 from flue gas, produced from a lignite based power station [38]; and the Vales Point project was a large-scale pilot plant recovering and purifying CO_2 from flue gas generated by a black coal power station [39]. The Vales Point project was a demonstration of a combined process, where membrane contactors were used to recovery CO_2 from the flue gas as well as used to regenerate the solvent producing the purified CO_2 project [40,41]. Photos of the pilot plants are provided in Fig. 15.6, demonstrating the respective scale of each project. These pilot plants utilized a range of membrane contactors, including porous polypropylene and polytetrafluoroethylene, as well as nonporous polyethylene and polydimethylsiloxane; with polydimethylsiloxane demonstrating the best performance if the active layer was ultrathin. While none of the pilot plant projects utilized nanofluids as the capture solvent, the solvents that were utilized consisted of conventional 30 wt.% monoethanolamine, 30 wt.%

FIGURE 15.6

Photos of the (A) Mulgrave, (B) H3, and (C) Vales Point project membrane contactor pilot plants.

potassium carbonate and a commercial solvent based on potassium glycinate. It was determined that both the monoethanolamine and glycinate solvents provided the best performance, due to the fast reaction kinetics with CO_2, which significantly enhanced the mass transfer process within the solvent boundary layer compared to potassium carbonate-based solvent. Hence, solvents that reduce the mass transfer resistance of the boundary layer, such as nanofluids, have a strong potential for CO_2 capture. The development of membrane contactors for this application has proceeded to the stage that the technology can now be implemented on an industrial scale, with a clear demonstration that consistent CO_2 absorption can be obtained for extend periods of time.

15.6 Challenges for membranes processes

There remain several challenges to the industrialization of membrane processes involving nanofluids, particular for membrane contactor technology.

A major limitation for membrane technology in liquid–liquid extraction is the potential for interface shift in the membrane pores. This changes the resistance to mass transfer in the process, which generally results in a loss of separation efficiency. Therefore interface stabilization between the two fluids is of paramount importance, which is influenced by operational conditions and the properties of the membrane and both solvents. For most liquid–liquid extraction processes, hydrophobic polypropylene-based membrane fibers are generally used [14]. As such the organic phase will spontaneously wet the membrane pores, and therefore a slight overpressure in the aqueous phase is required to avoid a breakthrough of the organic phase out of the pores and into the aqueous solvent. A similar phenomenon occurs in gas–solvent absorption, where the gas-filled pores become wetted with the solvent over time [24]. This results in significant reduction in mass transfer efficiency, as the diffusion through the solvent phase within the pores is lower than the gas phase. Both phenomena are associated with the surface tension (σ) between the two phases, the membrane pore dimensions (d), contact angle of the phase on the membrane pore walls (θ), and the transmembrane pressure. The critical parameter is the breakthrough pressure for the membrane pores, which arises from the capillary pressure and is defined as [42]:

$$\Delta P = \frac{4\sigma\cos\theta}{d}$$

Hence, for a membrane contactor process, it is important to keep the transmembrane pressure below the breakthrough pressure to ensure the correct phase remains within the membrane pores. Nanofluids generally have a higher surface tension than the base solvent, because the presence of nanoparticles increased the free energy of the interface [43]. However, the corresponding change to the wetting angle of the nanofluid on the membrane surface is unknown. This may result in a decrease in the breakthrough pressure for the membrane pores, meaning a higher operating transmembrane pressure is required compared to the base solvent, to avoid significant pore wetting. This poses a significant problem for the implementation of membrane contactor technology in industry, as the pilot plant trials have demonstrated the challenges of transmembrane control and the inability to rapidly control pressure fluctuations in the system [38].

Another issue that has not been studied in detail in the literature is the interaction of nanoparticles with the membrane surface and pore walls, especially adsorption. There is evidence in the literature of nanoparticles aggregating within membrane pores [44,45], which is due to the more favorable intermolecular interactions between the hydrophobic membrane polymer and many nanoparticles. Over time this adsorption of nanoparticles is likely to foul the membrane surface and for membrane contactor systems constrict and block pores. Cleaning procedures will need to be developed to remove the accumulated nanoparticles, probably comparable to those currently implemented to clean nano- and ultrafiltration membranes used in biological applications [46].

Finally, membrane stability over the long term remains a considerable issue for many of the processes where novel solvents are present. In liquid–liquid extraction the increasing use of ionic liquids and their potential to plasticize and swell the membrane material over the long term has not been fully characterized. Similarly, for gas–solvent absorption there is documented evidence that a range of CO_2-specific solvents degrade polymer materials; a good example of this is ethanolamine-based solvents deforming polyvinylidene membranes. This is also the case for ultra- and nanofiltration, with solvent resistant membrane materials being the focus of considerable research to cover some of the limitations of current commercial filtration membranes.

15.7 Conclusions

In summary, nanofluids will have the largest impact on membrane contactor processes, such as liquid–liquid extraction and gas absorption, because of the increased mass transfer generated in the solvent boundary layer by the presence of nanoparticles. Another important application will be nano- and ultrafiltration processes for processing the nanofluids, especially during waste disposal. While no current membrane process incorporating nanofluids has been demonstrated at any industrial scale, membrane processes that have been proven using conventional solvents highlight the effectiveness of the technology. However, there are several questions that still need to be addressed regarding the interaction of nanofluids with membrane technology. These issues are surmountable, and membrane technology has a positive future as a viable separation technology for a wide range of applications and systems. In the near future, nanofluids and membrane technology will be applied to pharmaceutical processing and purification, environmental treatment of industrial wastes, as well as energy applications.

References

[1] W.S.W. Ho, K. Sirkar, Membrane Handbook, 1992, Boston, Springer, pp 3–15.

[2] E. Drioli, E. Curcio, G. Di Profio, State of the art and recent progresses in membrane contactors, Chem. Eng. Res. Des. 83 (3 A) (2005) 223–233. Available from: https://doi.org/10.1205/cherd.04203.

[3] M. Stanojevic, B. Lazarevic, D. Radic, Review of membrane contactors designs and applications of different modules in industry, FME Trans. 31 (2003) 91–98.

[4] A. Gabelman, S.T. Hwang, Hollow fiber membrane contactors, J. Membr. Sci. 159 (1–2) (1999) 61–106. Available from: https://doi.org/10.1016/S0376-7388(99)00040-X.

[5] J. Jiang, B. Zhao, Y. Zhuo, S. Wang, Experimental study of CO_2 absorption in aqueous MEA and MDEA solutions enhanced by nanoparticles, Int. J. Greenh. Gas. Control. 29 (2014) 135–141. Available from: https://doi.org/10.1016/j.ijggc.2014.08.004.

[6] B.C. Pak, Y.I. Cho, Hydrodynamic and heat transfer study of dispersed fluids with submicron metallic oxide particles, Exp. Heat Transf. 11 (2) (1998) 151–170. Available from: https://doi.org/10.1080/08916159808946559.

[7] J.H. Kim, C.W. Jung, Y.T. Kang, Mass transfer enhancement during CO_2 absorption process in methanol/Al_2O_3 nanofluids, Int. J. Heat Mass Transf. 76 (2014) 484–491. Available from: https://doi.org/10.1016/j.ijheatmasstransfer.2014.04.057.

[8] Z. Zhang, J. Cai, F. Chen, H. Li, W. Zhang, W. Qi, Progress in enhancement of CO_2 absorption by nanofluids: a mini review of mechanisms and current status, Renew. Energy 118 (2018) 527–535. Available from: https://doi.org/10.1016/j.renene.2017.11.031.

[9] Z.U. Rehman, N. Ghasem, M. Al-Marzouqi, N. Abdullatif, Enhancement of carbon dioxide absorption using nanofluids in hollow fiber membrane contactor, Chin. J. Chem. Eng. (2020). Available from: https://doi.org/10.1016/j.cjche.2019.07.001.

[10] J.F. Hicks, D.T. Miles, R.W. Murray, Quantized double-layer charging of highly monodisperse metal nanoparticles, J. Am. Chem. Soc. 124 (44) (2002) 13322–13328. Available from: https://doi.org/10.1021/ja027724q.

[11] H. Zhao, H.H. Bau, The polarization of a nanoparticle surrounded by a thick electric double layer, J. Colloid Interface Sci. 333 (2) (2009) 663–671. Available from: https://doi.org/10.1016/j.jcis.2009.01.056.

[12] J.W. Lee, J.Y. Jung, S.G. Lee, Y.T. Kang, CO_2 bubble absorption enhancement in methanol-based nanofluids, Int. J. Refrig. 34 (8) (2011) 1727–1733. Available from: https://doi.org/10.1016/j.ijrefrig.2011.08.002.

[13] C.A. Scholes, S.E. Kentish, G.W. Stevens, D. deMontigny, Comparison of thin film composite and microporous membrane contactors for CO_2 absorption into monoethanolamine, Int. J. Greenh. Gas. Control. 42 (2015) 66–74. Available from: https://doi.org/10.1016/j.ijggc.2015.07.032.

[14] S. Shen, S.E. Kentish, G.W. Stevens, Shell-side mass-transfer performance in hollow-fiber membrane contactors, Solvent Extraction Ion. Exch. 28 (6) (2010) 817–844. Available from: https://doi.org/10.1080/07366299.2010.515176.

[15] S. Shen, S.E. Kentish, G.W. Stevens, Effects of operational conditions on the removal of phenols from wastewater by a hollow-fiber membrane contactor, Sep. Purif. Technol. 95 (2012) 80–88. Available from: https://doi.org/10.1016/j.seppur.2012.04.023.

[16] A.M. Ghanadi, A.H. Nasab, D. Bastani, A.A.S. Kordi, The effect of nanoparticles on the mass transfer in liquid–liquid extraction, Chem. Eng. Comm. 202 (2014) 600–605.

[17] M. Amde, J.F. Liu, Z.Q. Tan, D. Bekana, Ionic liquid-based zinc oxide nanofluid for vortex assisted liquid liquid microextraction of inorganic mercury in environmental waters prior to cold vapor atomic fluorescence spectroscopic detection, Talanta 149 (2016) 341–346. Available from: https://doi.org/10.1016/j.talanta.2015.12.004.

[18] M. Amde, Z.Q. Tan, R. Liu, J.F. Liu, Nanofluid of zinc oxide nanoparticles in ionic liquid for single drop liquid microextraction of fungicides in environmental waters prior to high performance liquid chromatographic analysis, J. Chromatogr. A 1395 (2015) 7–15. Available from: https://doi.org/10.1016/j.chroma.2015.03.049.

[19] A. Bahmanyar, N. Khoobi, M.R. Mozdianfard, H. Bahmanyar, The influence of nanoparticles on hydrodynamic characteristics and mass transfer performance in a pulsed liquid–liquid extraction column, Chem. Eng. Processing: Process. Intensif. 50 (11–12) (2011) 1198–1206. Available from: https://doi.org/10.1016/j.cep.2011.08.008.

[20] J. Saien, H. Bamdadi, Mass transfer from nanofluid single drops in liquid–liquid extraction process, Ind. Eng. Chem. Res. 51 (14) (2012) 5157–5166. Available from: https://doi.org/10.1021/ie300291k.

[21] A. Bahmanyar, N. Khoobi, M.M.A. Moharrer, H. Bahmanyar, Mass transfer from nanofluid drops in a pulsed liquid–liquid extraction column, Chem. Eng. Res. Des. 92 (11) (2014) 2313–2323. Available from: https://doi.org/10.1016/j.cherd.2014.01.024.

[22] I. Sreedhar, T. Nahar, A. Venugopal, B. Srinivas, Carbon capture by absorption – path covered and ahead, Renew. Sustain. Energy Rev. 76 (2017) 1080–1107. Available from: https://doi.org/10.1016/j.rser.2017.03.109.
[23] A.L. Kohl, R. Nielsen, Gas Purification, 5th ed., 1997, Houston, Gulf Professional Publishing.
[24] J. Franco, D. DeMontigny, S. Kentish, J. Perera, G. Stevens, A study of the mass transfer of CO_2 through different membrane materials in the membrane gas absorption process, Sep. Sci. Technol. 43 (2) (2008) 225–244. Available from: https://doi.org/10.1080/01496390701791554.
[25] P. Zare, P. Keshavarz, D. Mowla, Membrane absorption coupling process for CO_2 capture: application of water-based ZnO, TiO_2, and multi-walled carbon nanotube nanofluids, Energy Fuels 33 (2) (2019) 1392–1403. Available from: https://doi.org/10.1021/acs.energyfuels.8b03972.
[26] A. Golkhar, P. Keshavarz, D. Mowla, Investigation of CO_2 removal by silica and CNT nanofluids in microporous hollow fiber membrane contactors, J. Membr. Sci. 433 (2013) 17–24. Available from: https://doi.org/10.1016/j.memsci.2013.01.022.
[27] A. Peyravi, P. Keshavarz, D. Mowla, Experimental investigation on the absorption enhancement of CO_2 by various nanofluids in hollow fiber membrane contactors, Energy Fuels 29 (12) (2015) 8135–8142. Available from: https://doi.org/10.1021/acs.energyfuels.5b01956.
[28] I.T. Pineda, C.K. Choi, Y.T. Kang, CO_2 gas absorption by CH_3OH based nanofluids in an annular contactor at low rotational speeds, Int. J. Greenh. Gas. Control. 23 (2014) 105–112. Available from: https://doi.org/10.1016/j.ijggc.2014.02.008.
[29] S.C. Low, C. Liping, L.S. Hee, Water softening using a generic low cost nano-filtration membrane, Desalination 221 (1–3) (2008) 168–173. Available from: https://doi.org/10.1016/j.desal.2007.04.064.
[30] R. Chalatip, R. Chawalit, R. Nopawan, Removal of haloacetic acids by nanofiltration, J. Environ. Sci. 21 (1) (2009) 96–100. Available from: https://doi.org/10.1016/S1001-0742(09)60017-6.
[31] J.M. Gozálvez-Zafrilla, D. Sanz-Escribano, J. Lora-García, M.C. León Hidalgo, Nanofiltration of secondary effluent for wastewater reuse in the textile industry, Desalination (2008) 272–279. Available from: https://doi.org/10.1016/j.desal.2007.01.173.
[32] M. Mänttäri, K. Viitikko, M. Nyström, Nanofiltration of biologically treated effluents from the pulp and paper industry, J. Membr. Sci. 272 (1–2) (2006) 152–160. Available from: https://doi.org/10.1016/j.memsci.2005.07.031.
[33] B. Van der Bruggena, M. Mänttäri, M. Nyström, Drawbacks of applying nanofiltration and how to avoid them: a review, Sep. Purif. Technol. (2008) 251–263. Available from: https://doi.org/10.1016/j.seppur.2008.05.010.
[34] A.S. Mohruni, E. Yuliwati, S. Sharif, A.F. Ismail, Membrane technology for treating of waste nanofluids coolant: a review, in: AIP Conference Proceedings, vol. 1885, American Institute of Physics Inc., 2017. https://doi.org/10.1063/1.5002284.
[35] D.M. Aruguete, M.F. Hochella, Bacteria–nanoparticle interactions and their environmental implications, Environ. Chem. 7 (1) (2010) 3–9. Available from: https://doi.org/10.1071/EN09115.
[36] D.B. Warheit, C.M. Sayes, K.L. Reed, K.A. Swain, Health effects related to nanoparticle exposures: environmental, health and safety considerations for assessing hazards and risks, Pharmacol. Ther. 120 (1) (2008) 35–42. Available from: https://doi.org/10.1016/j.pharmthera.2008.07.001.
[37] C.A. Scholes, M. Simioni, A. Qader, G.W. Stevens, S.E. Kentish, Membrane gas-solvent contactor trials of CO_2 absorption from syngas, Chem. Eng. J. 195–196 (2012) 188–197. Available from: https://doi.org/10.1016/j.cej.2012.04.034.
[38] C.A. Scholes, A. Qader, G.W. Stevens, S.E. Kentish, Membrane gas-solvent contactor pilot plant trials of CO_2 absorption from flue gas, Sep. Sci. Technol. (Phila.) 49 (16) (2014) 2449–2458. Available from: https://doi.org/10.1080/01496395.2014.937499.
[39] C.A. Scholes, S.E. Kentish, A. Qader, Membrane gas-solvent contactor pilot plant trials for post-combustion CO_2 capture, Sep. Purif. Technol. 237 (2020). Available from: https://doi.org/10.1016/j.seppur.2019.116470.

[40] C.A. Scholes, S.E. Kentish, G.W. Stevens, D. deMontigny, Asymmetric composite PDMS membrane contactors for desorption of CO_2 from monoethanolamine, Int. J. Greenh. Gas. Control. 55 (2016) 195–201. Available from: https://doi.org/10.1016/j.ijggc.2016.10.008.
[41] C.A. Scholes, S.E. Kentish, G.W. Stevens, J. Jin, D. DeMontigny, Thin-film composite membrane contactors for desorption of CO_2 from monoethanolamine at elevated temperatures, Sep. Purif. Technol. 156 (2015) 841–847. Available from: https://doi.org/10.1016/j.seppur.2015.11.010.
[42] S.F. Shen, K.H. Smith, S. Cook, S.E. Kentish, J.M. Perera, T. Bowser, et al., Phenol recovery with tributyl phosphate in a hollow fiber membrane contactor: experimental and model analysis, Sep. Purif. Technol. 69 (1) (2009) 48–56. Available from: https://doi.org/10.1016/j.seppur.2009.06.024.
[43] S. Tanvir, L. Qiao, Surface tension of nanofluid-type fuels containing suspended nanomaterials, Nanoscale Res. Lett. 7 (2012). Available from: https://doi.org/10.1186/1556-276X-7-226.
[44] H. Rabiee, V. Vatanpour, M.H.D.A. Farahani, H. Zarrabi, Improvement in flux and antifouling properties of PVC ultrafiltration membranes by incorporation of zinc oxide (ZnO) nanoparticles, Sep. Purif. Technol. 156 (2015) 299–310. Available from: https://doi.org/10.1016/j.seppur.2015.10.015.
[45] A. Sotto, A. Boromand, R. Zhang, P. Luis, J.M. Arsuaga, J. Kim, et al., Effect of nanoparticle aggregation at low concentrations of TiO_2 on the hydrophilicity, morphology, and fouling resistance of PES–TiO_2 membranes, J. Colloid Interface Sci. 363 (2) (2011) 540–550. Available from: https://doi.org/10.1016/j.jcis.2011.07.089.
[46] X. Shi, G. Tal, N.P. Hankins, V. Gitis, Fouling and cleaning of ultrafiltration membranes: a review, J. Water Process. Eng. 1 (2014) 121–138. Available from: https://doi.org/10.1016/j.jwpe.2014.04.003.

CHAPTER

16 Enhanced carbon dioxide capture by membrane contactors in presence of nanofluids

Adolfo Iulianelli[1] and Kamran Ghasemzadeh[2]

[1]*Institute on Membrane Technology of the Italian National research Council (CNR-ITM), Rende, Italy*

[2]*Urmia University of Technology, Urmia, Iran*

16.1 Introduction

Currently, the research and the development of nanomaterials represent an important engineering sector. Several studies have focused on the utilization of nanoparticles in nanofluids at various volume concentrations and sizes, and they are particularly used in heat transfer processes [1]. Another emerging field of nanofluids technology application is related to the enhancement of the gas–liquid mass transfer and the absorption process [2]. Nanofluids are constituted of a colloidal mixture containing nanosized particles, which allows gas–liquid mass transfer for the depletion of the energy consumption, and the reduction of the equipment size and, consequently, the industrial manufacturing costs, thus favoring an intensification of the whole process, which is in good agreement with the principles of the Process Intensification Strategy (PIS) [3].

In this regard, membrane engineering results in being a viable paradigm to pursuing the PIS, as largely demonstrated by the membrane technologies applied in different industrial areas [4]. Membrane engineering involves the design, management, and control of the various membrane operations, which are selected according to the specific requests of an industrial process. It is currently applied in a variety of industrial fields such as water desalination [5] and wastewater treatments [6], agro-food and biorefinery [7], the petrochemical industry, and gas separation [8]. Nevertheless, the adoption of membrane engineering results in a consolidated strategy only in some industrial sectors, such as gas separation and water treatment, meanwhile emerging membrane operations are receiving growing attention for their indubitable advantages over the conventional systems. Among them, membrane contactors are ever more proposed in gas separation processes because they are contact devices favoring high efficiency in the separation of species due to their peculiar characteristics of large surface area per volume unit [9].

In the last decade, nanofluids were fruitfully adopted to improve the solute gas rate absorption in membrane contactors [10,11]. In particular, carbon nanotubes (CNT), Al_2O_3, Fe_3O_4, and SiO_2 were used as nanoparticle materials in a base fluid (distilled or deionized water, amine solutions) [12].

Considering that, nowadays, carbon dioxide is one of the main causes of global warming and a relevant environmental issue, the depletion of its emissions in the atmosphere represents not only a huge challenge but also a top priority, especially for the industrialized countries [13]. Today,

Nanofluids and Mass Transfer. DOI: https://doi.org/10.1016/B978-0-12-823996-4.00016-1

Table 16.1 CO_2 separation: comparison among conventional techniques, such as absorption, adsorption, and cryogenic, and membrane separation technology.

Sr. no.	Parameter	Absorption	Adsorption	Membrane	Cryogenic
1	Operating flexibility	Moderate	Moderate	High ($CO_2 > 20\%$) Low ($CO_2 < 20\%$)	Low
2	Response to variations	Rapid (5–15 min)	–	Instantaneous	Slow
3	Startup after variations	1 h	–	Extremely short (10 min)	8–24 h
4	Turndown	Down to 30%		Down to 10%	Down to 50%
5	Reliability	Moderate	Moderate	100%	Limited
6	Control requirement	High	High	Low	High
7	Ease of expansion	Moderate	Moderate	Very high	Very low
8	Energy requirement	4–6 MJ/kgCO_2	2–3 MJ/kgCO_2	0.5–6 MJ/kgCO_2	6–10 MJ/kgCO_2
9	CO_2 recovery	90%–98%	80%–95%	80%–90%	>95%

Modified from M.K. Mondal, H.K. Balsora, P. Varshney, Progress and trends in CO_2 capture/separation technologies: a review, Energy 46 (2012) 431–441 [14]. Copyright Elsevier (2020).

industrial carbon dioxide is separated/captured through the following processes: absorption in solvents or adsorption on solid sorbents, and cryogenic technique. The most usual carbon dioxide separation process in industry is represented by the removal of carbon dioxide from flue gas, which is carried out by absorption in solvents [14]. Membrane engineering shows various benefits over the traditional carbon dioxide capture technologies, as summarized in Table 16.1. The former highlights the advantages and drawbacks of membrane gas separation technologies over the traditional ones and how membrane engineering is particularly attractive for its easy reliability, stability, and removal efficiency.

Considering that carbon dioxide membrane gas separation is a consolidated technology at industrial scale, this chapter proposes the nanofluids enhanced membrane contactors to be an advanced and alternative technology for carbon dioxide removal from gaseous mixtures, highlighting how they represent a further option to the membrane gas separation technology and a step forward with respect to the traditional membrane contactors technology.

16.2 Membrane contactors technology

The utilization of hollow fiber membranes for gas separation was first proposed by Dow Chemical, constituting a recognized step forward in membrane engineering applied to gas separation [15]. Compared to other membrane typologies, hollow fiber membranes possess a larger membrane area per volume unit, showing high flexibility and self-mechanical support, easy operation, and facile reparation [16]. The aforementioned benefits induced several scientists to adopt hollow fiber membranes for membrane contactor applications. The former act as an interface, able to favor the compounds transport between gas and liquid phases, without the phases dispersion and the direct mixing [17]. Membrane contactors may be hence used as contact devices for the gas separation process, which is efficiently performed, even though they do not provide any selectivity because no mass transfer takes place selectively. In membrane contactors, the membrane function involves the contact between gas and liquid phases without mixing. The separation characteristics of a membrane contactor are hence due to a different solubility of the species present in the liquid phase. Therefore most of the membrane materials adopted in membrane contactors are porous in order to emphasize the mass transfer. Membrane contactors exercised at high pressure are necessarily based on asymmetric membrane solutions, which are constituted of a dense and thin layer deposited on a porous support [18].

The adoption of porous hydrophobic materials for fabricating membranes for membrane contactors should not allow the penetration of the liquid phase into the pores, being the former filled with gas in their whole volume.

Most of the membrane materials in the market used in membrane contactors applications are polymer based, whereas only a few applications deal with inorganic membranes. In more detail, a general subdivision of membranes may distinguish them into asymmetric and symmetric, or porous and dense (nonporous). Asymmetric membranes are based on the same polymer, whereas different polymers may be adopted in the case of composite membranes. In the case of the utilization of chemically stable polypropylene (PP) or polytetrafluoroethylene (PTFE) for membranes fabrication for use in membrane contactor devices, melting extrusion combined with stretching for the pores formation is the most adopted technique. On the other hand, polysulfone (PSF), polyetherimide (PEI), or polyvinylidene fluoride (PVDF) in fibers may be manufactured through different temperature-induced or nonsolvent-induced phase separation techniques [15–17].

Currently, hollow fiber membrane contactors are quite effective in the removal of acid gases such as carbon dioxide, sulfur dioxide, and hydrogen sulfide. In the natural gas sweetening and biogas treatment, membrane contactors were successfully utilized for hydrogen sulfide removal. In the case of natural gas sweetening, hollow fiber membranes may be based on PTFE, PVDF, or PSF as perfluorinated polymers [19–21], while they may be based on nonporous fibers from polydimethylsiloxane (PDMS) if used for biogas treatment [22,23]. In the field of carbon dioxide removal, the research on the membrane contactors adoption was not only focused on the development and utilization of new membrane materials but also on the performance of carbon dioxide-selective absorption liquids, particularly compatible with the fibers [24]. For instance, nanostructured membrane materials are suitable for applications as membrane contactors for high-pressure carbon dioxide removal from natural gas. Polytrimethylsilylpropyne (PTMSP) and polymethyl pentene (PMP) modified with nanoparticle fillers represent new polymers for membrane contactors [18].

In the market, the leader for membrane contactors manufacturing is Celgard LLC (United States), which fabricates and commercializes hollow fiber membrane contactor modules (Liqui-Cel), housing porous polypropylene membranes under fiber mat form. Liqui-Cel modules adopt a particular hollow fibers membranes interconnected by means of a polymer thread [25]. A large pilot plant developed by Kvaerner for carbon dioxide removal from natural gas via membrane contactors technology was built in Aberdeen (Scotland) and there was another pilot-scale study at the Statoil Gas Terminal at Kårstø (Norway) for flue gas treatment [26,27].

Membrane contactors technology was also investigated for the recovery of ammonia via chemical stripping. The former is unfavored at low pH because NH_4^+ tends to be greatly unstable. Indeed, a pH equal to 12, or higher than 9, requires a solution in which the ammonia plays the role of the dominant species. Industrial plants by Aliachem have been installed in Pardubice (Czech Republic), where ammonia is removed by membrane contactor modules and collected under solution (27 wt.% of ammonia), guaranteeing ammonia emission reduction in the atmosphere of about 99.9% [28].

16.3 Carbon dioxide separation by membrane contactors

The growing consumption of fossil fuels is mainly responsible for the large emission of carbon dioxide, which, combined with other polluting compounds, causes the sun's heat trapping and consequent global warming [29]. At industrial level, the exploitation of natural gas, which is seen as a cleaner source than other derived fossil fuels, involves the purification of methane to remove undesirable by-products, such as carbon dioxide and hydrogen sulfide, which could be responsible for the corrosion of pipelines and devices due to their acidic nature. Absorption processes are nowadays mostly adopted for the acidic gases removal from natural gas and flue gas [14]. Nevertheless, the gas–liquid absorption process does not show only benefits, but also drawbacks such as channeling, liquid overflow, frothing, etc. [30]. Hollow fiber membrane contactors are quite efficient carbon dioxide removal systems, which were largely studied in the last decade [22,24,31–35]. Membrane contactor modules are often constituted of bunches of hollow fiber membranes, which are surrounded by a shell. Generally, the liquid solvent flows through a distribution tube internal to the membrane bundle, whereas the gaseous stream flows from the casing side (Fig. 16.1), but it is possible to find the reverse solution. As stated above, the large membrane surface area per volume unit of a membrane contactor represents the most important advantage compared to traditional contacting columns [26,37].

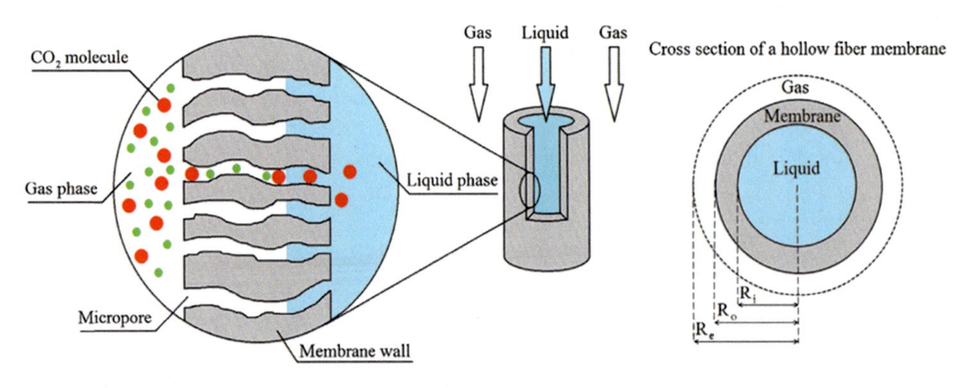

FIGURE 16.1

Schemes of a hollow fiber membrane contactor module for carbon dioxide removal.

Reprinted with permission from M. Mehdipour, P. Keshavarz, A. Seraji, S. Masoumi, Performance analysis of ammonia solution for CO_2 capture using microporous membrane contactors, Int. J. Greenh. Gas. Contr. 31 (2014) 16–24 [36]. Copyright Elsevier (2020).

16.3.1 Carbon dioxide absorption in presence of solid nanoparticles

The carbon dioxide absorption from gaseous mixtures may be positively influenced by the presence of nanofluids (CNT, SiO_2, Al_2O_3, and Fe_3O_4) in water and amine solutions. Rahmatmand et al. demonstrated that, at lower concentrations, Fe_3O_4 and CNT nanoparticles may allow superior carbon dioxide absorption; on the contrary, at higher concentrations Al_2O_3 and SiO_2 nanoparticles perform better [12]. Esmaeili et al. found that, in a bubble column, synthetic SiO_2 nanofluid permitted the simultaneous H_2S and CO_2 absorption better than the pristine fluid due to the hydrogen bonds between gas molecules and SiO_2 groups [38]. Furthermore, Al_2O_3 or SiO_2 nanoparticles were used in a bubble type absorber, observing improved carbon dioxide removal using methanol solvent [39]. Regarding the utilization of aqueous nanofluids in membrane modules to remove carbon dioxide from CO_2/air gas mixtures, it was demonstrated that CNT nanofluids may be a better choice than SiO_2 in terms of carbon dioxide separation performance [40–45]. In detail, at higher carbon dioxide fractions in the gaseous mixture, the CNT nanofluids removal fraction increased, whereas when using SiO_2 nanofluids it decreased by increasing the carbon dioxide concentration owing to the saturation. On the other hands, mass transfer was improved when using larger size SiO_2 particles [46]. Last but not least, the utilization of Al_2O_3 nanofluids in hollow fiber membrane contactors was also explored for removing carbon dioxide from molecular nitrogen, demonstrating that an aqueous concentration of Al_2O_3 nanofluids equal to 0.2 wt.% allowed the maximum carbon dioxide removal rate [47].

16.3.2 Nanofluids hollow fiber membranes: modeling studies

Enhanced nanofluids hollow fiber membranes were studied also from a theoretical point of view in order to study in depth the mass transfer adopting these particular systems. Hence, in addition to experimental studies, numerous models have been developed to investigate hollow fiber membrane systems. In previous works about, Amrei et al. [47] modeled the carbon dioxide removal utilizing an aqueous monoethanolamine solvent as a function of carbon dioxide concentration across the hollow fiber membrane module. This theoretical study took into account the gas flows inside the hollow fiber membranes exercised under non- or partially wetted conditions. Other studies studied the enhancement of the mass transfer analyzing the absorption rate in presence of nanosized droplets, combining experimental and modeled results. Carbon dioxide absorption in the presence of nanofluids (e.g., nano-Al_2O_3 and CNTs) was studied by Sumin et al. [44] using a stirred thermostatic reactor. They considered in the theoretical model the enhancement of absorption due to the nanoparticles. Furthermore, Koronaki et al. theoretically studied the carbon dioxide absorption improvement due to the presence of CNTs in a batch vessel [45], observing an enhancement of absorption over time up to achieving an equilibrium condition, commonly reached by batch systems. A 2D mathematical model was implemented by Darabi et al. [48] to simulate the carbon dioxide absorption in the presence of CNT nanofluid in a hollow fiber membrane contactor under nonwetted conditions, basing their work on the assumption that the fluid flows in the tube side under highly dilute conditions. In this regard, the effects due to microconvection were not considered, improving the absorbed solute diffusion.

16.3.3 Modeling of carbon dioxide removal in a hollow fiber membrane contactor

In the following, a theoretical approach to model the enhancement of the carbon dioxide absorption in the presence of aqueous nanofluids in a hollow fiber membrane contactor (membrane under partial wetting state) is proposed. The stagnant liquid film surrounding the solid nanoparticles represents the main cause of the flow resistance. The membrane contactor may be subdivided into three parts: tube, membrane, and shell parts. A further subdivision may describe the membrane contactor into five phases: dense and solid-free phases in the tube side, wetted and dry phases in the membrane, and shell-side gas-phase segment. The effect due to the nanoparticle motion is assumed acting in the dense phase of the tube side. On the tube side the nanofluid flow is modeled as a solid-free zone and dense phase. In the case of the carbon dioxide-based mixture flowing outside the hollow fiber membranes, the nanofluids flow inside the hollow fiber membrane tubes (lumen) in a countercurrent configuration, Fig. 16.2.

If the carbon dioxide-based mixture flows from the top side of the membrane contactor module under vertical configuration (at $z = L$), the nanofluids flow in the hollow fibers lumen from the bottom side (at $z = 0$). The carbon dioxide passes through the membrane pores in order to reach the nanofluid for diffusion from the gaseous mixture. Thus the solid nanoparticles adsorb the carbon dioxide that is dissolved in the nanofluid.

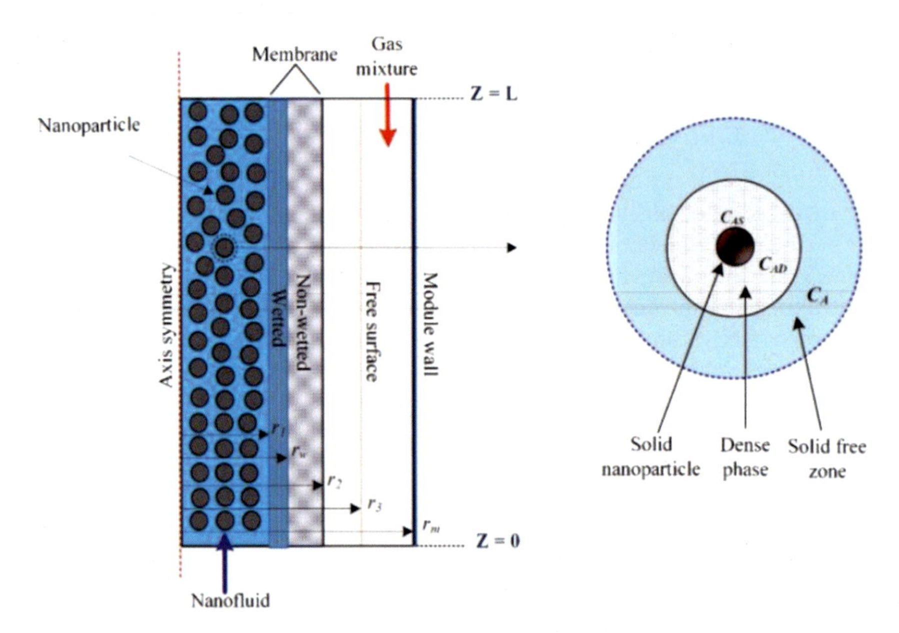

FIGURE 16.2

Scheme of the hollow fiber membrane contactor showing solid nanoparticles inside and the phases subdivision.

Reprinted from N. Ghasem, Modeling and simulation of CO_2 absorption enhancement in hollow-fiber membrane contactors using CNT–water-based nanofluids, J. Membr. Sci. Res. 5 (2019) 295–302 [10]. Copyright Elsevier (2020).

The polymeric membrane morphology regulates not only the rate of carbon dioxide removal, but also the carbon dioxide distribution factor, operating conditions, typology of absorbent, aqueous solvent, nanoparticle concentrations, and liquid and gas flow rates. The main mechanism about the material balance of a membrane contactor to be taken into account in presence of nanofluids are the Brownian motion and the grazing effect. On one hand, the nanoparticle Brownian motion leads to a velocity increase around the nanoparticle, with a resulting effect of microconvection and an enhanced mass diffusion, which is responsible for the modification of the diffusion coefficient [44]. The gas adsorption due to the nanoparticles presence at the gas–liquid interface is described by the grazing effect [48].

In the case where the nanofluid velocity is fully developed in the tube side, the nanoparticles' influence and consequent effects are neglected due to their low concentration. In the shell side, the velocity of the gas in the shell is regulated by the Happel's free surface model. The main assumptions that it is realistic to take into consideration are:

1. Steady-state and isothermal conditions; indeed, the fluid flow is kept at constant temperature, without any reaction development.
2. Homogeneous nanoparticles and uniform distribution at the interface layer.
3. Incompressible and Newtonian fluid flow, very low nanoparticle concentration in the water-based nanofluids, with consequent negligible effect of the nanoparticles presence in the liquid phase.
4. Gas–liquid equilibrium regulated by Henry's law.

It is possible to develop mathematical equations describing the system characteristics: for the tube side (solid and liquid phase), shell side, and wetted and nonwetted zones of the membrane. As a consequence, the continuity equations related to the carbon dioxide adsorption may be described according to the aforementioned subdivisions.

In the tube side (according to Fig. 16.2, $0 < r < r_1$):

Constituted of nanoparticles in a base fluid, the nanofluid flowing in the lumen of the hollow fiber membranes are able to remove the carbon dioxide contained in the gaseous stream via absorption due to the base fluid's (e.g., water) effect and via adsorption due to the nanoparticles' effect. The carbon dioxide concentration in the free solid nanoparticles solution (C_A) may be represented as:

$$\frac{D}{R^2}\left[\frac{\partial^2 C_A}{\partial \xi^2} + \frac{1}{\xi}\frac{\partial C_A}{\partial \xi}\right] + \frac{D}{L^2}\frac{\partial^2 C_A}{\partial \zeta^2} + \frac{u_{Z_t}}{L}\frac{\partial C_A}{\partial \zeta} = 0 \quad (16.1)$$

where the carbon dioxide concentration in the dense phase (C_{AD}) may be calculated through the equation:

$$\frac{D_n}{R^2}\left[\frac{\partial^2 C_{AD}}{\partial \xi^2} + \frac{1}{\xi}\frac{\partial C_{AD}}{\partial \xi}\right] + \frac{D_n}{L^2}\frac{\partial^2 C_{AD}}{\partial \zeta^2} + \frac{u_{Z_t}}{L}\frac{\partial C_{AD}}{\partial \zeta} = R_d \quad (16.2)$$

The dimensionless parameters inserted in Eqs. (16.1) and (16.2) are $\zeta = z/L$ and $\zeta = r/r_3$, where D is the carbon dioxide diffusion coefficient in the solid free volume, while D_n represents the diffusion coefficient of CO_2 in the dense solid phase. The former around the nanoparticles may be expressed as:

$$D_n = D(1 + 640Re^{1.7}Sc^{\frac{1}{3}}\phi) \quad (16.3)$$

ϕ represents the solid volume fraction, whereas Re is the Reynolds number of the nanoparticles under Brownian motion:

$$Re = \left(\frac{18kT\rho}{\pi d_p \rho_p \mu^2}\right)^{0.5} \tag{16.4}$$

k represents the Boltzmann's constant ($1.38 \times 10^{-23} J/K$), T the temperature (K), d_P and ρ_P the particle diameter and density, respectively, ρ the liquid density, μ the liquid viscosity. and Sc represents the Schmidt number:

$$Sc = \frac{\mu}{\rho D} \tag{16.5}$$

R_d is the adsorption rate:

$$R_d = k_p a_p (C_{AD} - C_{AS}) \tag{16.6}$$

k_P represents the solid–liquid mass transfer coefficient (m/s), C_{AS} the solute concentration at the particles interface (mol/m^3), C_{AD} the solute concentration in the suspension (mol/m^3), and a_P is the solid–liquid interfacial area (m^2/m^3). The adsorbed carbon dioxide amount on the solid per unit mass of particles may be represented by the coefficient q (mol/kg):

$$\phi \rho_p \frac{v_{zt}}{L} \frac{\partial q}{\partial \xi} = k_p a_p (C_{AD} - C_{AS}) \tag{16.7}$$

The adsorption of the carbon dioxide solute on the particles may be described as:

$$q = q_m \frac{k_d C_{AS}}{1 + k_d C_{AS}} \tag{16.8}$$

q_m represents the highest adsorbed gas solute quantity, while k_d (m^3/mol) is the solute adsorption coefficient. In the tube side, v_{zt} represents the velocity distribution, assumed to agree with the Newtonian laminar flow.

$$v_{zt} = \frac{2Q_L}{\pi r_1{}^2 n_t} \left(1 - \left(\frac{r}{r_1}\right)^2\right) \tag{16.9}$$

The boundary conditions are:

Solvent inlet side: $z = 0$	$C_A = C_{AD} = 0$	(fresh solvent)
Solvent exit side: $z = L$	$\frac{\partial C_A}{\partial \xi} = \frac{\partial C_{AD}}{\partial \xi} = 0$	(convective flux)
Tube center: $r = 0$	$\frac{\partial C_A}{\partial \xi} = \frac{\partial C_{AD}}{\partial \xi} = 0$	(axial symmetry)
Inner radius: $r = r_1$	$C_A = C_{wm}$	(solubility of CO_2 in solvent)

The steady-state material balance related to the carbon dioxide transport in the hollow fiber membrane contactor skin layer ($r_1 \leq r \leq r_2$), inside the wetted portion of the membrane ($r_1 \leq r \leq r_w$) may be represented by Eq. (16.10). No convection term is present due to the unique presence of diffusion, which takes place in the wetted membrane:

$$\frac{D_{mw}}{R^2}\left[\frac{\partial^2 C_{wm}}{\partial \xi^2} + \frac{1}{\xi}\frac{\partial C_{wm}}{\partial \xi}\right] + \frac{D_{mw}}{L^2}\frac{\partial^2 C_{wm}}{\partial \zeta} = 0 \tag{16.10}$$

C_{wm} represents the carbon dioxide fraction in the wetted membrane section, and $D_{mw} = D_t\ \varepsilon/\tau$ the carbon dioxide diffusivity in the wetted membrane section. In this zone, the boundary conditions are:

Tube-wetted-membrane interface:	$r = r_1$	$C_{At} = C_{wm}$
Wet-dry membrance interface:	$r = r_2$	$C_{wm} = C^{m}_{Am}$
Membrane inlet end:	$z = 0$	$\frac{\partial C_{wm}}{\partial z} = 0$
Membrane exit side	$z = L$	$\frac{\partial C_{wm}}{\partial z} = 0$

Inside the nonwetted portion of the membrane ($r_w \leq r \leq r_2$), the steady-state material balance related to the carbon dioxide transport does not take into account the convection term, because only diffusion takes place in the membrane:

$$\frac{D_m}{R^2}\left[\frac{\partial^2 C_m}{\partial \xi^2} + \frac{1}{\xi}\frac{\partial C_m}{\partial \xi}\right] + \frac{D_m}{L^2}\frac{\partial^2 C_m}{\partial \zeta^2} = 0 \tag{16.11}$$

In this zone, the boundary conditions are:

Interface of membrane-tube:	$r = r_w$	$C_m = C_{wm}/m$
Membrane-shell interface:	$r = r_2$	$C_A = C_{Ag}$
Dry membrane inlet end:	$z = 0$	$\frac{\partial C_m}{\partial z} = 0$
Dry membrane exit side	$z = L$	$\frac{\partial C_m}{\partial z} = 0$

$D_{mw} = D_t\ \varepsilon/\tau$ represents the carbon dioxide diffusivity in the nonwetted membrane section. The material balance of the gas solute in the shell side at steady state (C_{Ag}) may be represented by:

$$\frac{D_g}{R^2}\left[\frac{\partial^2 C_{Ag}}{\partial \xi^2} + \frac{1}{\xi}\frac{\partial C_{Ag}}{\partial \xi}\right] + \frac{D_g}{L^2}\frac{\partial^2 C_{Ag}}{\partial \zeta^2} + \frac{v_{zs}}{L}\frac{\partial C_{Ag}}{\partial \zeta} = 0 \tag{16.12}$$

In this zone, the boundary conditions are:

Gas inlet side:	$z = L$	$C_{Ag} = C_{A0}$	(inlet concentration)
Gas exit side:	$z = 0$	$\frac{\partial C_{Ag}}{\partial \xi} = 0$	(convective flux)
Free surface:	$r = r_3$	$\frac{\partial C_{Ag}}{\partial \xi} = 0$	(symmetry)
Shell membrane interface:	$r = r_2$	$C_{Ag} = C_{Am}$	

The axial velocity in the shell side may be expressed in correlation to the Happel's free surface model:

$$v_{zs} = \frac{2Q_g}{\pi\left(\frac{3r_3^4}{4} + \frac{r_2^4}{4} - r_2^2 r_3^2 - r_3^4 \mathrm{In}\left(\frac{r_3}{r_2}\right)nt\right.}\left[r^2 - r_2^2 - 2r_3^2 \mathrm{In}\left(\frac{r}{r_2}\right)\right] \tag{16.13}$$

Ghasem [10] compared theoretical data given by the simulations using COMSOL Multiphysics 5.4 software, based on the mass transfer theoretical approach described above, finding good agreement between experimental and simulation data, as reported in Fig. 16.3.

Ghasem [10] also theoretically demonstrated that, for a fixed nanoparticle concentration, both the inlet gas flow rate and the nanoparticle volume fraction did not influence the enhancement in

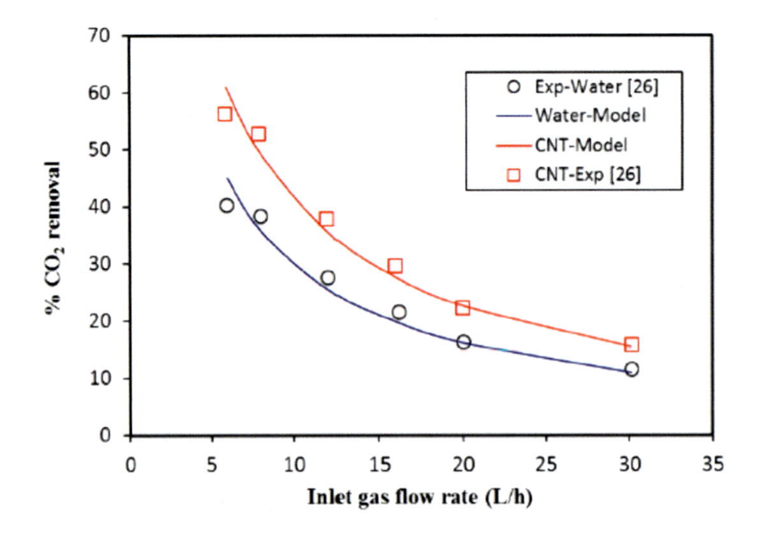

FIGURE 16.3

Comparison between experimental and theoretical data of carbon dioxide removal via hollow fiber membrane contactor.

Reprinted from N. Ghasem, Modeling and simulation of CO_2 absorption enhancement in hollow-fiber membrane contactors using CNT–water-based nanofluids, J. Membr. Sci. Res. 5 (2019) 295–302 [10].

carbon dioxide removal. Indeed, due to the saturation of the nanoparticles, a flow rate increase does not correspond to a relevant improvement. Similarly, by increasing the nanoparticle concentration, the carbon dioxide removal should theoretically increase. The higher the absorbed carbon dioxide, the higher the carbon dioxide removal. This can be due to an enhancement of the mass-transfer driving force. On the contrary, the influence of the inlet carbon dioxide mole fraction is low on its removal rate.

16.4 Conclusion and future outlooks

Packed bed or tray columns are currently being adopted for carrying out absorption processes. In particular, they are useful for acidic gases removal (i.e., carbon dioxide) from natural gas and flue gas. Nevertheless, the great benefits related to the adoption of conventional gas–liquid absorption columns are counterbalanced by a number of drawbacks, such as frothing, entrainment, channeling, and liquid overflow. In this regard, gas–liquid hollow fiber membrane contactors are an interesting and efficient technological option to remove carbon dioxide from gaseous mixtures. Nanofluids constitute a step forward for membrane contactors development, particularly because they are able to increase the solute absorption rate in membrane contactors.

In this chapter, a theoretical approach was followed to describe the mass transfer occurring inside a nanofluids featured hollow fiber membrane contactor used for carbon dioxide removal, analyzing how, for a fixed nanoparticle concentration, both the inlet gas flow rate and the nanoparticle volume fraction do not affect the enhancement in carbon dioxide removal. This is because the saturation of the nanoparticles do not allow the high removal of carbon dioxide.

Gas–liquid hollow fiber membrane contactors have attracted the attention of several researchers due to their high interfacial area per unit volume compared to conventional absorption processes from the viewpoint of the carbon dioxide separation from gaseous mixtures. The benefits due to high absorption performance of alkanolamines are, on the contrary, the main cause of membrane degradation if used as membrane contactors, corrosion issues if used in industrial gas absorbers, and high-energy consumption during solvent regeneration and circulation. As a consequence, researchers are deeply involved in looking for better absorbents and solid nanoparticles dispersed in water (nanofluids) could be a great choice for an environment-friendly substitute absorbent in membrane contactor applications, and these constitute the most interesting option for the future of carbon dioxide removal via membrane technology.

References

[1] R. Dharmalingam, K.K. Sivagnanaprabhu, B. Senthil Kumar, R. Thirumalai, Nano materials and nanofluids: an innovative technology study for new paradigms for technology enhancement, Procedia Eng. 97 (2014) 1434–1441.

[2] L. Godson, B. Raja, D. Mohan Lal, S. Wongwises, Enhancement of heat transfer using nanofluids—an overview, Ren. Sustain. Ener. Rev. 14 (2010) 629–641.

[3] A. Górak, A. Stankiewicz (Eds.), Research Agenda for Process Intensification: Towards a Sustainable World of 2050. Creative Energy—Energy Transition, 2011. Retrieved from: http://3me.tudelft.nl/fileadmin/Faculteit/3mE/Actueel/Nieuws/2011/docs/DSD_Research_Agenda.pdf.

[4] E. Drioli, A. Stankiewicz, F. Macedonio, Membrane engineering in process intensification—an overview, J. Membr. Sci. 380 (2011) 1–8.

[5] A. Ali, R.A. Tufa, F. Macedonio, E. Curcio, E. Drioli, Membrane technology in renewable-energy-driven desalination, Renew. Sustain. Ener. Rev. 81 (2018) 1–21.

[6] E. Drioli, A. Ali, Y.M. Lee, M. Al-Beirutty, F. Macedonio, Membrane operations for produced water treatment, Desalin. Water Treatm. 57 (2016) 14317–14335.

[7] R.W. Field, E. Bekassy-Molnar, F. Lipnizki, G. Vatai, Engineering aspects of membrane separation and application in food processingISBN 9781420083637Available from: https://doi.org/10.4324/9781315374901Contemporary Food Engineering Series, CRC Press, 2017pp. 1–390.

[8] A. Iulianelli, E. Drioli, Membrane engineering: latest advancements in gas separation and pre-treatment processes, petrochemical industry and refinery, and future perspectives in emerging applications, Fuel Proc. Techn. 206 (2020) 106464–106497.

[9] E. Drioli, A. Criscuoli, E. Curcio, ISBN 9780444522030 Membrane Contactors: Fundamentals, Applications and Potentialities, Elsevier, 2011pp. 1–516.

[10] N. Ghasem, Modeling and simulation of CO_2 absorption enhancement in hollow-fiber membrane contactors using CNT–water-based nanofluids, J. Membr. Sci. Res. 5 (2019) 295–302.

[11] P. Luis, T.V. Gerven, B.V. der Bruggen, Recent developments in membrane-based technologies for CO_2 capture, Prog. Energy Combust. Sci. 38 (2012) 419–448.

[12] B. Rahmatmand, P. Keshavarz, S. Ayatollahi, Study of absorption enhancement of CO_2 by SiO_2, Al_2O_3, CNT, and Fe_3O_4 nanoparticles in water and amine solutions, J. Chem. Eng. Data 61 (2016) 1378–1387.

[13] Y. Yin, K. Bowman, A.A. Bloom, J. Worden, Detection of fossil fuel emission trends in the presence of natural carbon cycle variability, Environ. Res. Lett. 14 (2019) 084050.

[14] M.K. Mondal, H.K. Balsora, P. Varshney, Progress and trends in CO_2 capture/separation technologies: a review, Energy 46 (2012) 431–441.
[15] G. Bakeri, S. Naeimifard, T. Matsuura, A.F. Ismail, A porous polyethersulfone hollow fiber membrane in a gas humidification process, RSC Adv. 5 (2015) 14448–14457.
[16] K. Kneifel, S. Nowak, W. Albrecht, R. Hilke, R. Just, K.V. Peinemann, Hollow fiber membrane contactor for air humidity control: modules and membranes, J. Membr. Sci. 276 (2006) 241–251.
[17] T.O. Leikness, M.J. Semmens, Vacuum degassing using microporous hollow fiber membranes, Sep. Purif. Technol. 22–23 (2000) 287–294.
[18] G.A. Dibrov, V.V. Volkov, V.P. Vasilevsky, A.A. Shutova, S.D. Bazhenov, V.S. Khotimsky, et al., Robust high-permeance PTMSP composite membranes for CO_2 membrane gas desorption at elevated temperatures and pressures, J. Membr. Sci. 470 (2014) 439–450.
[19] M. Hedayat, M. Soltanieh, S.A. Mousavi, Simultaneous separation of H_2S and CO_2 from natural gas by hollow fiber membrane contactor using mixture of alkanolamines, J. Membr. Sci. 377 (2011) 191–197.
[20] M.H. Al-Marzouqi, S.A. Marzouk, N. Abdullatif, High pressure removal of acid gases using hollow fiber membrane contactors: further characterization and long-term operational stability, J. Nat. Gas. Sci. Eng. 37 (2017) 192–198.
[21] R. Faiz, K. Li, M. Al-Marzouqi, H_2S absorption at high pressure using hollow fibre membrane contactors, Chem. Eng. Proc. Proc. Intens. 83 (2014) 33–42.
[22] P. Jin, C. Huang, Y. Shen, X. Zhan, X. Hu, L. Wang, et al., Simultaneous separation of H_2S and CO_2 from biogas by gas–liquid membrane contactor using single and mixed absorbents, Energy Fuels 31 (2017) 11117–11126.
[23] E. Tilahun, A. Bayrakdar, E. Sahinkaya, B. Çalli, Performance of polydimethylsiloxane membrane contactor process for selective hydrogen sulfide removal from biogas, Waste Managm 61 (2017) 250–257.
[24] S.D. Bazhenov, E.S. Lyubimova, Gas–liquid membrane contactors for carbon dioxide capture from gaseous streams, Petrol. Chem. 56 (2016) 889–914.
[25] Liqui-Cel® Membrane Contactors by 3M. https://multimedia.3m.com/mws/media/14124630/3m-liqui-cel-exf-6x28-series-membrane-contactorlc-1042-pdf.pdf (accessed March 2020).
[26] A. Mansourizadeh, A.F. Ismail, Hollow fiber gas–liquid membrane contactors for acid gas capture: a review, J. Hazard. Mater. 171 (2009) 38–53.
[27] Kvaerner Membrane Contactor Technology. https://www.kvaerner.com/ (accessed March 2020).
[28] R. Klaassen, P.H.M. Feron, A.E. Jansen, Membrane contactors in industrial applications, Chem. Eng. Res. Des. 83 (2005) 234–246.
[29] J.D. Figueroa, T. Fout, S. Plasynsky, H. McIlvried, R.D. Srivastava, Advances in CO_2 capture technology—the U.S department of energy's carbon sequestration program, Int. J. Greenh. Gas. Contr. 2 (2008) 9–20.
[30] E. Favre, Carbon dioxide recovery from post-combustion processes: can gas permeation membranes compete with absorption? J. Membr. Sci. 294 (2007) 50–59.
[31] Z.A. Tarsa, S.A.A. Hedayat, M. Rahbari-Sisakht, Fabrication and characterization of polyetherimide hollow fiber membrane contactor for carbon dioxide stripping from Monoethanolamine Solution, J. Membr. Sci. Res. 1 (2015) 118–123.
[32] N.M. Ghasem, M. Al-Marzouqi, Modeling and experimental study of carbon dioxide Absorption in a flat sheet membrane contactor, J. Membr. Sci. Res. 3 (2017) 57–63.
[33] N.M. Ghasem, M. Al-Marzouqi, L.P. Zhu, Preparation and properties of polyether sulfone hollow fiber membranes with o-xylene as additive used in membrane contactors for CO_2 absorption, Sep. Purif. Technol. 12 (2012) 1–10.
[34] N.M. Ghasem, M.H. Al-Marzouqi, A. Duaidar, Effect of quenching temperature on the performance of polyvinylidene fluoride microporous hollow fiber membranes fabricated via thermally induced phase

separation technique on the removal of CO_2 from CO_2−gas mixture, Int. J. Greenh. Gas. Control. 5 (2011) 1550−1558.

[35] A. Mansourizadeh, A.F. Ismail, A developed asymmetric PVDF hollow fiber membrane structure for CO_2 absorption, Int. J. Greenh. Gas. Control. 5 (2011) 374−380.

[36] M. Mehdipour, P. Keshavarz, A. Seraji, S. Masoumi, Performance analysis of ammonia solution for CO_2 capture using microporous membrane contactors, Int. J. Greenh. Gas. Contr. 31 (2014) 16−24.

[37] S. Alex, F. Biasotto, P.R. Rout, P. Bhunia, Membrane contactors: an overview of their applications, Technical Rep. (2017) 1−13. Available from: https://www.researchgate.net/publication/314137389.

[38] S.H. Esmaeili-Faraj, M. Nasr Esfahany, Absorption of hydrogen sulfide and carbon dioxide in water based nanofluids, Ind. Eng. Chem. Res. 55 (2016) 4682−4690.

[39] I.T. Pineda, J.W. Lee, I.K. Jung, T. Yong, CO_2 absorption enhancement by methanol-based Al_2O_3 and SiO_2 nanofluids in a tray column absorber, Int. J. Refrig. 35 (2013) 1402−1409.

[40] A. Golkhar, P. Keshavarz, D. Mowla, Investigation of CO_2 removal by silica and CNT nanofluids in microporous hollow fiber membrane contactors, J. Membr. Sci. 433 (2013) 17−24.

[41] M. Darabi, M. Rahimi, A.M. Dehkordi, Gas absorption enhancement in hollow fiber membrane contactor using nanofluids: Modeling and Simulation, Chem. Eng. Proc. Process Intensif. 119 (2017) 7−15.

[42] M. Rezakazemi, M. Darabi, E. Soroush, M. Mesbah, CO_2 absorption enhancement by water-based nanofluids of CNT and SiO_2 using hollow-fiber membrane contactor, Sep. Purif. Technol. 210 (2019) 920−926.

[43] A. Peyravi, P. Keshavarz, D. Mowla, Experimental investigation on the absorption enhancement of CO_2 by various nanofluids in hollow fiber membrane contactors, Energy Fuels 29 (2015) 8135−8142.

[44] L. Sumin, X. Min, S. Yan, D. Xiangjun, Experimental and theoretical studies of CO_2 absorption enhancement by nano-Al_2O_3 and carbon nanotube particles, Chin. J. Chem. Eng. 21 (2013) 983−990.

[45] I.P. Koronaki, M.T. Nitsas, Ch.A. Vallianos, Enhancement of carbon dioxide absorption using carbon nanotubes—a numerical approach, Appl. Therm. Eng. 99 (2016) 1246−1253.

[46] M. Hossein, K. Darvanjooghi, M. Esfahany, S. Esmaeili-Faraj, Investigation of the effects of nanoparticle size on CO_2 absorption by silica-water nanofluid, Sep. Purif. Technol. 195 (2018) 208−215.

[47] S.M.H.H. Amrei, S. Memardoost, A.M. Dehkordi, Comprehensive modeling and CFD simulation of absorption of CO_2 and H_2S by MEA solution in hollow fiber membrane reactors, AIChE J. 60 (2014) 657−672.

[48] M. Darabi, M. Rahimi, A.M. Dehkordi, Gas absorption enhancement in hollow fiber membrane contactor using nanofluids: modeling and Simulation, Chem. Eng. Process. Process Intens. 119 (2017) 7−15.

CHAPTER

17 Mass transfer improvement in hydrate formation processes by nanofluids

Fateme Etebari, Yasaman Enjavi, Mohammad Amin Sedghamiz and Mohammad Reza Rahimpour

Department of Chemical Engineering, Shiraz University, Shiraz, Iran

17.1 Introduction

In 1810 the first established natural gas hydrate, the crystalline component of chlorine and water, was identified in Humphrey Davy's investigation. However, at temperatures beyond the freezing point of water, he found the ice-like solid to be built and made of much more than water. Hydrate analysis was also carried out in 1823 by Michael Faraday, and the formulation of chlorine hydrate was determined and recorded quantitatively [1,2]. Wroblewski documented a carbon dioxide hydrate in 1882. Cailletet in 1878 reported acetylene hydrate and was the first to find that the development of these crystal structures resulted from a drastic pressure drop. In a study conducted by Woehler, the behavior of sulfide hydrate hydrogen was recorded in 1840. For more than 40 years, Villard and de Forcrand focused on this compound category. Methane, ethane, acetylene, and ethylene hydrates were documented by Villard [2]. Villard and de Forcrand, who conducted experiments in the late 19th and early 20th centuries, identified the first hydrates containing hydrocarbons, and calculated the balanced temperatures of 15 various compounds, mostly natural gas organisms, ambient pressure, hydrogen, methane, and propane hydrates. The key concern of these primary hydrate-related experiments was estimating the sum of hydrate-crystalline molecules (hydration percentage) by the guest molecule. A hydrate arrangement, containing six water molecules by guest components, was stated by Villard. Schroeder further suggested that early studies on the hydrate structure could be restricted to 15 compounds solely [1]. The hydrogen sulfide product and soluble alcohol were described as blended hydrogen sulfide and alcohol by De Forcrands [2]. He also noticed the significant "sulfhydrierten" group of hydrates where hydrogen sulfide could be unified in the hydrogen phase with a significant number of aliphatic sequences of halogen derivatives. Cailletet and Bordet identified the double hydrate of carbon dioxide and phosphine in 1882. In 1897 De Forcrand and Sully Thomas discovered a double hydrate of acetylene and carbon tetrachloride. Double hydrate of acetylene, ethylene, sulfur dioxide, carbon dioxide, and several compounds, including ethylene, bromide, methyl iodide, methyl bromide, etc., methylene iodide, were also recorded. Related molecules such as carbon, hydrate of ammonia, and ether also were mentioned by Hempel and Seidel. Crystalline hydrogen coupled with water is produced by methyl mercaptan [2]. Fowler et al. found that certain ammonium salts in a quaternary structure can produce hydrates at atmospheric temperature and pressure in 1940. Then Jeffrey identified the constructions

Nanofluids and Mass Transfer. DOI: https://doi.org/10.1016/B978-0-12-823996-4.00013-6

of this hydrate using X-ray crystallography and discovered that both water molecules and anions were essential components of the cages. Semiclathrate hydrate is the name given to organic hydrates [3].

In the early 19th century, attempts were mainly focused on discovering various forms of hydrates and the hydrate-developing environments. While hydrate-forming criteria were examined during the 19th century by investigators of a broad spectrum of substances, the hydrates' industrial value was not shown until the 20th century [1].

17.1.1 Gas hydrates formation

The ice-like crystalline solids that consist of water (host), and gas molecules (guest), which occur at some temperatures and pressures, constitute natural gas hydrates. The gas molecules, including CH_4, CO_2, N_2, and H_2, are found in these hydrates' hydrogen bonding cavities [3–6]. The hydrate formation requires four conditions: hydrate and gas molecules, elevated pressure, and low temperature. Efficient variables of hydrate formulation separate into conditions for the activity, surfactant usages, or nanoparticles. Pressure, temperature, water present in the device, reactor blending velocity, and reactor capacity are the key operating considerations [7,8]. Nonstoichiometric ice-like incorporation is produced at low temperatures and high pressures is gas hydrates or clathrate hydrates [9]. Exothermal production of hydrate is essential to heat and mass flow [7,10]. Natural gas hydrate is also an ice-like clathrate, consisting of hydrogen bonds and light compounds such as methane that occupy the holes via Van der Waals forces [11]. Since gas hydrates often produce >85 mol.% water, their characteristics are known as variants of ice [12]. Gas hydrates, for example, are 20 times greater than ice in terms of their mechanical power [13,14]. The most significant distinction is that ice is a pure ingredient, whereas hydrates will not be shaped without correctly sized guests. It is classified as ice Ih (hexagonal ice) with molecular water as the most typical solid shape. Each hydrogen molecule in ice (solid lines) is bound to four other molecules in basically tetrahedral edges [6,15]. Cage-like vacancies with comparatively big pores containing hydrogen bonds are created to capture the guest compounds [7]. Since the host water molecules have powerful hydrogen bonding, there is no chemical contact except Van der Waals between the guest and host components [16]. Gas hydrates are similar to ice but are often established at much higher than the ice point temperature [17].

Based on the thermodynamic situations and the gaseous molecules' size, hydrates may possess various crystallography constructions. Both forms of cavities are found in these arrangements. In these cavities, the SI is the most popular one, and various frameworks can be created by capturing different compounds. The cavity structure I (sI), cubic structure II (SII), and hexagonal structure H (sH) vary in size and shape; these are the three major crystallographic models for gas hydrates. The three types of hydrate structure are shown in Fig. 17.1. The gas hydrate composition is defined largely by the size of the guest gas particles embedded through cavities. Composition I consists of guest compounds, including methane, ethane, carbon dioxide, and hydrogen sulfide, with diameters around 4.2–6 Å. As single guests, nitrogen and small molecules with hydrogen ($d < 4.2$ Å) generate structure II. Larger guest shapes lead to structure II (6 Å $< d <$ 7 Å) compounds, including propane and iso-butane. Nevertheless, tiny gases like Xe and Ar may form hydrate I and II constructs. Still, larger molecules (typically 7 Å $< d <$ 9 Å) such as isopentane or hexane(2,2-dimethyl butane) can form structure H when accompanied by smaller molecules such as methane, hydrogen sulfide,

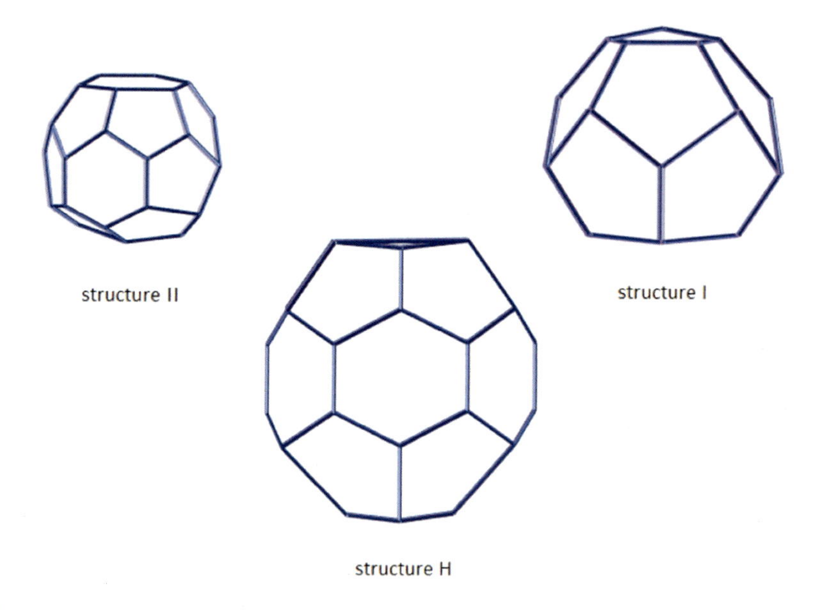

FIGURE 17.1

Types of gas hydrate crystal structures: cubic structure I (sI), cubic structure II (SII), and hexagonal structure H(sH).

or nitrogen [6,13,17–20]. The ratio of the number of water molecules to entrapped gas molecules, the hydrate number, depends on the gas composition and pressure [17].

17.1.2 The gas hydrate formation process

Hydrate formation involves two steps, known as nucleation and growth. Hydrate nucleation involves the growth and dispersion of tiny water and gas clusters to reach a critical scale to grow continuously. The nucleation stage is a quantitatively hard to detect phenomenon including tens to thousands of compounds. Experimental data suggest that nucleation is statistically possible and not deterministic [21]. The measurement of the hydrate's growth level allows some encouragement for the forming of the hydrate following the fortuitous existence of hydrate crystal nucleation. Regarding the crystal growth rate following nucleation, there is only a small range of reliable evidence. The hydrate growth remains essential for many nucleation criteria, including a shift in equilibrium, surface region, stimulation, water background, and gas structure. Even so, mass and heat exchange are essential throughout the growth period. The mass transportation of gas into the hydrate surface is crucial. The process may be overwhelmed due to the high relative gas content of the hydrate (even 15 mol.% gas), which is at least two orders of magnitude greater when compared with methane solubilization. However, the exothermic heat formed by hydrate can affect growth. Considering the changes at the molecular level, three factors may be considered: (1) crystal growth kinetics at the surface of hydrate; (2) transfer of mass of the components to the surface of the growing crystal; and (3) the thermal radiation due to hydrate formation induced by the exothermic energy of the surface of the rising crystal.

17.1.3 Application of gas hydrates

Notwithstanding their harmful impact on the petroleum and gas production sector, clathrate hydrates have gained much consideration during the past few years. Several positive applications of clathrate hydrates have also been published in the literature over the years, including storage and transport of natural gas [22–24], carbon dioxide accumulation and isolation [25,26], gas mixture segregation [1,27], desalination of water [28,29], and cold storage [30].

17.1.3.1 Natural gas storage and transportation

The main nonrenewable energy resource, with massive worldwide supplies, is natural gas (NG) and CO_2 pollution can be mitigated by consuming natural gas, rather than coal and oil. In 2018 the consumption of natural gas worldwide was approximately 3.86 billion ^{m3}, while in China, the annual level of NG utilization expanded by 16.56%. The use of natural gas is expected to rise up to 2040. The request for production of powerful storage and transport technology is rising due to the high use of natural gas [9]. It is also claimed that nearly 70% of the overall gas supply is either too far from a pipeline or too limited for liquefaction. It has been proposed that distribution of gas streamed in a hydrated type is financially sustainable as facilities and machinery investments are reduced. The high hydrated gas storage capability makes this possible [7,30–32]. Significant amounts of methane can be contained by solidified natural gases (SNG) in clathrate hydrate shape, therefore 1 m^3 gas can maintain 160–180 m^3 of gas on normal temperature and pressure (STP) situations [9,30,33].

17.1.3.2 CO_2 and greenhouse gases capture and isolation

The ignition of fossil gas in dwellings and trucks, and the burning of fire load gas in refineries, petrochemicals, and thermal plants are the major sources of carbon dioxide. CO_2 is one of the most important contaminants due to the formation of a greenhouse effect that induces heating and a steady rise in the Earth's temperature. Greenhouse gas isolation and preservation is the most critical ecological concern of the current time. The problem includes the development of modern facilities and the adoption of new approaches. Many innovations are being established, and hydrate forming is a useful approach of trapping CO_2 [12]. CO_2 hydrate collection is a special less power-intensive process. The application is based on the fast entanglement of CO_2 by the formulation of hydrate in mild formation settings, and the discharge of the trapped gas by the disintegration of hydrate [1,16,34,35].

There are certain benefits of the hydrate crystallization techniques. Water, which offers inexpensive and greener chemicals applications, is a core component of CO_2 hydrate formation since water is very prevalent and a life-living potential chemical. It is important to remember that the use of gas hydrate developers will decrease the requirements of hydrate formation energy. The carbon dioxide collection should be quarantined following isolation [30,36]. Such greenhouse gases, including hydrogen sulfide (H_2S), sulfur hexafluoride (SF_6), and 1,1,1,2-tetrafluoroethane (R-134a), have been examined in addition to carbon dioxide and methane. Since a high hydrogen level of sulfide in gas streams raises the risk of solid sulfur deposition in the forming, well boiling, and processing factories, particularly at high temperatures and pressures, the isolation of hydrogen sulfide from gas flows is an essential activity for the petroleum industry [1]. There are other different applications of gas hydrate, for example, water desalination, cool storage technology, and in the future energy resource [28–30].

17.2 Hydrate inhibition versus the hydrate promotion

In order to avoid or minimize the intensity of hydrate production in gas pipes, the key tasks of gas hydrate creation are mainly focused upon thermodynamic and kinetic inhibitors. In addition, inhibitors which have thermodynamics or kinetic properties can be used to suppress hydrate formation within gas transmission pipelines. Alcohol, glycol, and inorganic salts are generally the most popular thermodynamic hydrate inhibitors that are able efficiently to alter hydrate-forming conditions' pressures and reduce temperatures through the pressure–temperature equilibrium curve. In addition, several ionic liquids have been used for this intention previously [1,37–39]. The level of reduction in the aqueous process relies on the alcohol type and quantity and/or electrolyte [17].

In a series that begins with methanol and finishes with butanol, alcohols form hydrogen bonds between their group of hydroxyls and water; the hydrocarbon end of alcohols exerts a similar clustering effect on the hydrate constructors' water molecules. Therefore two effects are identified from the alcohols competing with the soluble polar molecules in shaping clusters on water: the major effect is the binding of the hydrogen group with the water molecules' hydroxyl group, while the minor effect is that the alcohol's terminal hydrocarbon, which directly competes with hydrate for guest and host molecules in water, organizes as solvent clusters [40].

Slow level of hydrate formation, limited storage space, extreme situations of buildup, and environmental catastrophic boosters are key challenges for the deployment of technologies focused on hydrate formation [35]. As previously stated, a big concern with the industrial application of clathrate hydrates is the sluggish kinetics of methane hydrate formation, implying prolonged initiation times and lower growth levels [41]. Researchers use various approaches, including the effective surface materials and nanoparticles, in this respect to accelerate the amount of hydrate formation [16].

The hydrate production and gas accumulation ability of hydrates has been improved with anionic, cationic, and nonionic surfactants. Nanofluids recently demonstrate an improved potential for heat and mass transport, as a suspension of nanoscaled fragments of basic fluids, to improve the performance of gas absorption compared with their basic fluids [37]. One of the big obstacles for its industrial use is the sluggish pace of hydrate formation. Consequently, a lot of experiments have recently been conducted to improve hydrate formation output by utilizing additives including surfactants and nanoparticles. These additives, called kinetic boosters, typically improve the hydrate formation frequency by raising mass and energy transmission [10].

There have been a numerous studies aimed at growing the hydrate formation frequency by hydrate boosters, namely surfactants supplementation, shaking, and boiling [30]. In spite of the lack of efficacy and environmental considerations, nanoparticles and amino acids are a source of good SDS-like kinetics for hydrate boosters [30].

17.3 Nanofluid in hydrate formation/inhibition process

The nanofluid idea originated in the 1990s [42]. Colloids in nanoparticles have unique physical characteristics that make them valuable for different applications, including painting, insulation, ceramics, drug delivery, and the food industry [15].

Certain materials are used in producing nanoparticles, which include composite materials including nanoparticle core–polymer shell composites; these materials include nitride ceramics (SiN, AlN), oxide ceramics (CuO, Al_2O_3), metals (Ag, Fe, Al, Cu, Au), semiconductors (TiO_2, SiO_2), carbide ceramics (TiC, SiC), and single-, double-, or multiwall carbon nanotubes (SWCNT, DWCNT, MWCNT). Thanks to its strong thermal conductance, abundance, relatively low cost, and environment-friendly nature, water is used as a conventional host liquid [43].

Fluids in this type include limited concentrations of solid particles (100 nm and smaller in size typically less than 10%), and a high defined surface among particles and liquids [44].

An innovation strategy to enhance the formulation of gas hydrate kinetics by heat and mass exchange is the insertion of nanoparticles to the aqueous phase. Nanoparticles synthesis, particularly attrition, pyrolysis, and hydrothermal, has been developed using various methodologies. The suspensions of the nanoparticles in the base fluid have demonstrated an improvement in the capacity of exchange of heat and mass to strengthen the dissolution of gas and heat in contrast with their base fluids [12].

17.3.1 Improvement of mass exchange during hydrate formation by nanofluids

The heat and mass transport in the water/gas interfaces can be improved by any agent. This will increase the pace of hydrate formation and the quantity of trapped gas completely. Thus these challenges can be mitigated by facilitating heat and mass transmission processes [7]. In this phase heat transfer is very crucial because of the exothermic nature of the gas hydrate development. During the nucleation stage, the heat produced will destroy nucleons and thus reduce the growth of crystalline materials. In order to preserve the process pace, the heat produced in the process should be removed. For this reason, the usage of high thermal conductivity fluids, creating an improvement in heat flow from the reactor to the environment, is an effective method. Since most nanoparticles, including the metal and nonmetal particles, turn water into nanofluids with large heat transmission coefficients [45], the use of nanofluids rather than basic fluids is extremely successful in speeding up the hydrate formation. Because of their nanometric form and scale, nanostructures may also improve the field of heat exchange. Thanks to their very small dimensions, nanostructures may provide more active positions to nucleate the hydrate-forming phase. Thermal characteristics, form, porosity, scale, thickness, and resilience of specific fluids are the most important promoter criteria for nanoparticles [7,46]. The concentration is more important than the particle scale [45].

The increased heat exchange is based upon the appropriate conductance of the nanocopper, whereas the increased mass transmission may be attributed to the existence of the broadly defined surface region, which will not only establish nucleation areas, but will also accumulate huge quantities of gas/liquid [47]. In addition, to improving mass transmission, increasing gas intake and improving the solubility of gases within nanofluids, the existence of nanoparticles in water can contribute to improved gas hydrate storage ability. Via the exacerbation of mass transport, the existence of nanoparticles contributes to higher levels of hydrate development. The transmission process of gas into aqueous phases is improved by the increased surface area and coefficient of mass. The usage rate of gas is thus improved.

The mass exchange process is improved by alleviating the surface tension because of the high surface area of nanoparticles [45,48,49].

Li et al. initially introduced water-based nanofluids in 2006 to speed up hydrate generation [47]. This article provides an important approach for heat and mass exchange during the improvement of hydrate formation and separation through ultrafine particles of HFC134a (CH_2FCF_3). The laboratory findings imply that introducing nanocopper improves the heat and mass transmission of the HFC134a hydrate production. Through a rise of the nanocopper mass portion below the critical point of dissociation, the formation steps will be reduced. The dissociation pressure of basic HFC134 hydrate at a particular temperature was greatly increased in the nanofluid. The crucial point of separation often moves from the previous point (283.15K, 414.86 kPa) to the latter (282.65K, 401.35 kPa). As the mass fraction of the nanocopper transition increases, the dissociation curve does not move by lowering the interfacial tension, regardless of its large surface region.

Ghozatloo et al. used G-Hummers to enhance the process of hydrate formation. This nanostructure forms a nanofluid when added to water and improves the coefficient of heat transfer. Also, it is highly hydrophilic when solved in water. Thanks to their nanometric size and shape, nanostructures may improve mass transfer area. Due to their very tiny size, nanostructures can provide a higher number of active sites, which may be used in nucleation in the process of hydrate formation. Ghozatloo et al. calculated the amount of dissolved gas in the first 33.4 min to be 0.023 mol, indicating a 23-fold higher gas dissolution in the nanofluid than pure water. Yet, they found that the dissolution time decreases by 41.4%, which may be attributed to the increased mass transfer area in nanoparticles' presence [7]. The unique microstructure characteristics of graphene oxide, including functional groups and a wide region of special region, can contribute to a high performance of heat and mass exchange, a broad dissolution of gases, and rapid levels of nucleation and development. This research explored the development of CO_2 hydrate with and without nanoparticles containing graphene oxide. Graphene oxide was seen to decrease the acceleration periods by 53%–74.3%, and to raise the gas intake by 5.1%–15.9% through various device pressures. The optimal concentration was calculated as 50 ppm based on the findings. The nanoparticles were expected to increase thermal conductivity not only dramatically but also to strengthen heat and mass exchange. Nano-Al_2O_3 has demonstrated heat-conductivity higher than the base fluid lacking nanoparticles [50].

In a study in 2012, SiO_2, and Al_2O_3 nanoparticle suspensions were produced in methanol (nanofluid) and investigated for CO_2 absorption application in a tray column absorber, which is an acrylic tray column consisting of 12 plates. As a sieve tray-type column with flat-perforated plates, the velocity of vapor prevents the liquid from passing down through holes, and the countercurrent flow brings methanol liquid and CO_2 gas into contact. Two mass flowmeters are inserted in the test section to measure the rate of absorption. Based on the results, the maximum improved absorption rates were 9.7% and 9.4% for SiO_2 and Al_2O_3 particles (in comparison with pure methanol), respectively. Also, it was revealed that SiO_2 nanoparticles were better candidates compared with Al_2O_3 nanoparticles. Moreover, 0.05 vol.% of nanoparticles was proven to provide the optimum condition in terms of CO_2 absorption improvement in these experimental conditions [51]. More experiments have shown that CO_2 solubility has been significantly improved with Al_2O_3 nanofluids-containing aqueous NaCl solution. Ultrasonic therapy is used to prepare the nanofluids which demonstrate the highest durability. It is observed that the nanofluid-induced CO_2 absorption is up to 8.3% greater than the purified methanol [52].

CO_2 solvability measurements in distilled water and two 0.1 wt.% nanofluids prove that the volume of absorbed CO_2 is higher in the nanoparticles' presence. It also revealed ZnO nanofluids to be more efficient compared with SiO_2 nanofluids in all laboratory settings. Besides, CO_2 solubility was investigated at 5°C in 0.05, 0.1, 0.5, and 1 wt.% ZnO nanofluid with the pressure ranging from 1 to 22 bar, which proved that CO_2 absorption was improved as the ZnO mass loading increased in all measurements. The process of carbon absorption is explored comprehensively in the nanofluids, and pure water findings are opposed to the solubility of gases [53]. SiO_2 nanoparticles may also be used to facilitate the development of CH_4 hydrate which decreases the desorption level of CH_4, shortens the induction period and massively improves the pace of growth [54]. In 2014 Choi et al. studied the new absorbents to form CO_2 hydrate at atmospheric pressure, and to evaluate the effects of surfactants and additives on the formation rate and the induction time of CO_2 hydrate. THF (tetrahydrofuran) is used as a surfactant and SDS (sodium dodecyl sulfate) and nanoparticles such as Al_2O_3 are used as the additives. It is found that the maximum CO_2 hydrate formation rate is enhanced up to 3.74 times by adding 0.6 wt.% of SDS and 0.2 wt.% of Al_2O_3 nanoparticles compared to the formation rate without the surfactants. Finally, it is concluded that THF 10 wt.% and SDS 0.6 wt.% with Al_2O_3 0.2 wt.% is the optimum condition for CO_2 hydrate formation rate enhancement [55].

Compared to the pure water system, the impact of several Al_2O_3/ZnO nanoparticle concentrations in the aqueous solutions in the kinetics of $CH_4 + C_2H_6 + C_3H_8$ hydrate formation was investigated. Also, gas consumption was measured to be approximately 121% when using two different types of nanoparticles. Even the storage potential was enhanced, mostly thanks to blended nanoparticles [56].

The impact of nanofluids of carbon during the production of methane hydrate was investigated by Park et al. Throughout the presence of carbon nanofluids, the rate of methane absorbed during the development of methane hydrate increased by 4.5 relative to distilled water [57]. When nanoparticles are applied to the base fluid, they improve the fluid nucleation and conduction. Nevertheless, owing to elevated surface energy, these nanoparticles appear to clot with the creation of an unstable nanofluid. This is why many scientists have used surfactants to stabilize the nanoparticles in the basic fluid [10].

Moreover, thanks to high motion and high surface area/volume ratio of the CTAB surfactant stabilized Fe_3O_4 nanofluid reduced the resistance against mass transfer at the solution–hydrate and the gas–solution interfaces. Nanoparticles are regularly arranged toward the magnetic field, thus reducing movement resistance against CO_2 in the solution and facilitating hydrate formation [16].

Fe_3O_4 surface-specific nanoparticles and Brownian movements decrease the mass exchange tolerance at the interface between the gas and liquid. The existence of surfactants often raises the mass movement frequency from the gas to the liquid phase, which enhances the intake of gas. Extremely conductive Fe_3O_4 nanoparticles accelerate heat transmission of the fundamental fluid. The involvement of nanoparticles in the base fluid raises the nucleation positions and variability and, due to Brownian movement and highly specialized surface region, it increases the mass transport. Fe_3O_4 nanoparticles improve the mass exchange and carbon dioxide solvability throughout the aqueous environment by declining the gas–liquid interfacial tension, raising the interaction zone across phases and moving gas molecules from the liquid–gas interface to the liquid. This also enhances the pace of nucleation and hydrate development [10]. Furthermore, the SiO_2 nanoparticles were used for the enhancement of the formation of methane hydrate. The results indicated the system's steady gas intake for numerous freezing–thawing cycles to improve gas storage stability. Accordingly, the former studies showed that the nanoparticles could enhance the hydrate formation kinetics in addition to gas storage enhancement in the processes of hydrate formation [58].

The enhanced solubility in the presence of ZnO nanoparticles may impact the conditions of the hydrate formation of equilibrium by changing water activity in the liquid phase. Conversely, as a metal oxide, ZnO nanoparticles enhanced the heat and mass transfer in a way that hydrate formation kinetic was intensified. As observed, the CO_2 consumption rate was enhanced using ZnO nanofluid instead of pure water with an initial pressure of 2.2 and 2.6 MPa, so that nanoparticle presence resulted in hydrate growth rate improvement due to mass transfer intensification [37].

The impact of suspension of polymers of nanoparticles on stabilization and retention capability of methane gas hydrate was studied by Ganji et al. in 2013. Any studied additives were able to efficiently improve the stability of the formulated hydrate. When these additions were present the hydrate produced was higher than the pure water content [59].

Zhou et al. have been studying the impact of graphite nanoparticles on the CO_2 hydrate development [60]. The findings demonstrated a favorable impact of the nanographite particles on CO_2 hydrate generation. Compared to distilled water in the presence of graphite nanoparticles, the induction period of CO_2 hydrate dropped by 80.8% and the maximal CO_2 intake rose by 12.8%. In addition, the hydrate response in the presence of nanoparticles increased by 98.8% around 400 min. Also, note that in the systems with both CTAB and Cu nanoparticles' presence, higher values are observed for the mean apparent rate constant. In other words, the combination of enhancement of mass transfer in CTAB presence and the higher number of stable nuclei at the growth stage onset in the presence of Cu nanoparticles better improve the growth rate of methane hydrate formation [41].

Mass transfer characteristics are not affected by the small amount of nanoparticles in the aqueous solution. Film resistance is reduced by the movement and motion of CuO nanoparticles at the water/gas interface. As a result, mass transfer improves. The gas transfer rate into the aqueous phase increases by the elevated mass transfer coefficient and surface area. Consequently, it leads to a higher rate of gas consumption [49].

The development and separation mechanism of HFC134a (CH_2FCF_3) hydrate is enhanced by introducing nanoparticles of CuO to the refrigerant water mixture during heat and mass exchange [61]. The MWCNTs were used to enhance the methane hydrate levels and gas consumption by Park et al. [62]. Their findings revealed that the quantity of gas absorbed was around 300% greater than that of water, and that the duration of development of the hydrate dropped at about 0.004% of the MWCNT at a low temperature.

Arjang et al. [63] assessed the methane hydrate generation activation period and the methane gas level with silver nanoparticles. Their findings indicated that the methane hydrate induction period was decreased by 85% and 73.9%. On the other hand, methane hydrate production's methane consumption rate improved by 7.4% and 33.7% with an initial pressure of 5.7 and 4.7 MPa, respectively. Mohammadi et al. [64] evaluated SDS's effect and the synthesized silver nanoparticles on CO_2 production and preservation rate. Based on this analysis, silver nanoparticles and SDS were significantly impacted. Nonetheless, the SDS–silver blend enhanced the storage potential of CO_2 by 93.9%. Abdi-Khanghah et al. assessed the ZnO nanofluid's effect on the generation of CH_4 hydrate through conducting experiments at 276.65K and 274.65K with the primary pressures of 6.0

and 5.0 MPa. They stated CH_4 hydrate generation level was improved using ZnO nanofluids [65]. Another study provides an evaluation into the usage of nanofluids with graphene oxide (GO) to boost CH_4 isolation from low-concentration coalbed methane (LCCBM) through the creation of gas hydrate. Throughout various experiments on diverse levels of graphene oxide nanoparticles and function stresses, the impact of GO nanofluids on the filtration of LCCBM was examined. Both the development and usage of hydrate were reported to be higher by introducing nanoparticles with graphene oxide to the liquid phase. In addition, the performance of CH_4 isolation relative to GON-free systems has been improved considerably. Upon two-stage hydrate isolation of LCCBM, the concentration of CH_4 was increased from 30 to 76 mol.%. GO nanofluids however have an immense capacity for enhanced kinetics and purification performance of LCCBM [66].

Inside a hollow condenser of fiber membrane, a two-dimensional (2D) template for the elimination of CO_2 from a gas mixture was created. The silica and carbon nanotubes (CNT) have been used as absorbing agents. Nanofluids have also been employed. Calculating fluid mechanics technique addressed the governing equations (CFD). The findings were correlated with the observational findings and the reliability of the evolved system of mass transfer was verified by good conventions. The findings revealed that the addition of nanofluids to water improves the elimination of CO_2. The grazing influence, Brownian movement, and spread coefficient of growth are responsible for this change. The performance of CO_2 elimination has also been shown to rise as the fluid flow rate has improved. CO_2 absorption has also been revealed to increase the levels of nanofluid. CNT's nanofluid efficiency is much greater than that of nanosilica [67].

In another investigation the development of the method of conversion of gas to hydrate utilized NGO-based nanofluids for the CO_2 conservation and sequestration. Various experiments have been performed under various operating conditions using a novel GTH reactor to explore the prerequisite operating conditions to maximize the CO_2 recovery consuming minimum energy. The mentioned technique is promising to develop high-performance and low-cost hydrate-based CO_2 recovery plants consuming low power [68].

Hydrates are deemed an effective process constrained by low levels of development for the transportation and preservation of natural gas. While using MRI technology in this research, the kinetic of SiO_2 nanofluids and porous medium on hydrate generation were explored. The findings demonstrated that hydrates, close to the wall, developed, and expanded into pores containing appropriate water. A substantial decrease in induction phase and an improvement in gas consumption while hydrate forming were observed throughout the concentrations variance analysis. This research offers empirical foundations for natural gas preservation and transport via incorporating microporous medium and nanofluids [69].

Carbon nanotubes (CNTs) have been documented for hydrate production as an important kinetic booster. The extreme accumulation of hydrophobic nanotubes will, however, diminish their effectiveness. This study was conducted by a process for either the sonication or the milling of the ball through carbon nanotubes covered by $\pi-\pi$-conjugated molecules of responsive red 195 (RR 195). These biocompatible CNTs were incredibly soluble and stable in water. Using water-soluble coated CNTs (RR 195@CNTs) in the hydrate reaction system reduced the formation duration to the shortest observed time (i.e., 203 ± 20 min) while improving gas storage capacity to 140 ± 2 v/v compared with 40 ± 8 v/v in the case of using deionized water. The scattered CNTs, which were packed via RR 195, not only improved the mass transport but also expanded regions, all of which facilitated the speedy formulation of hydrate [70].

Graphite nanofluids for the implementation of solidified gas storage regarding gas hydrates were developed and used in this experiment. In contrast with liquid and SDS solutions, it was observed that the hydrate nucleation and growth level of CH_4 in graphite nanofluids were increased. Although the ability for storage of gas was not substantially improved, induction and average time for hydrate generation were shortened compared with fluid water. On the other hand, the gas consumption obtained in graphite nanofluids was larger than that achieved in other nanofluids structures including the ZnO, CuO, and Fe_3O_4 nanofluids. In this research, it was discovered that the mass and heat distributions for CH_4 hydrate formation were facilitated by the use of graphite nanofluids. The development of CH_4 hydrate was decreased as the experiments continued because a considerable amount of CH_4 hydrate collected on the boundary of the gas/liquid. Under such circumstances, gas propagation and mass transport were delayed [9].

Nanoparticles present in the base fluid enhance the nucleation positions, framework diversity, and mass exchange due to Brownian movement and highly specific regions. The induction cycle for the Fe_3O_4 nanoparticle suspension was shortened by these conditions relative to both solutions with surfactant. Fe_3O_4 nanoparticles improve mass transmission and carbon dioxide solubility through decreasing gas–liquid interfacial tension in the aqueous settings. The interaction layer among phases and diffusion of gas molecules from the gas–liquid interface into the liquid also improves. This also accelerates the amount of nucleation and hydrate development. The film tolerance seen between fluid and gas phases is reduced by the addition of surfactants. Consequently, liquid gas solubility and mass transport rates have improved [10].

The process to describe this improvement in mass transfers is still theoretical. Some hypotheses, like the hydrodynamic motion, suggest that particles may raise the interfacial region by coating the surface of the bubbles and keeping the bubbles from coagulation. The particles will also clash, creating local agitation and refreshing the gas/liquid boundary layer by injecting the layer into bulk fluid. Nevertheless, there is also no general method to maximize mass transmission. It is concluded that the Brownian movement of nanoparticles is not specifically responsible for increased mass transport, yet primarily because of the velocity disruption of the fluid produced by nanoparticles displacement. For bubble models, several mass transfer mechanisms are established, including bubble columns and stirred tanks. In a complex flow device, like the tray column absorber, however, the powers of liquid and vapor diffusion are considered to be more dominating in the strengthening of the mass exchange [51].

The advantages of nanoparticle application in hydrate promotion/inhibition procedure are so attractive for the researchers to investigate different types of nanofluids in the experimental/simulation procedures. However, a comparative table of nanofluid applications in hydrate applications are presented in Table 17.1.

17.4 Conclusions and future trends

There has been a dramatic increase in gas hydrate research over the last decade. Accelerating the hydrate formation rate and increasing gas storage capacity are the main contributing factors for commercially feasible technologies. Growing attention has been recently paid to nanofluids because of their potential for augmenting transfer processes, that is, heat and mass transfer. Various studies in recent years have shown that the use of nanofluids, such as Al_2O_3, ZnO, graphene oxide, CNTs, and many other nanomaterials, increases the mass transfer coefficient in the formation of hydrates.

Table 17.1 Nanofluid application in a hydrate process.

Authors	Type of nanomaterial	Effect of nanomaterial	References
Li et al. (2006)	HFC134a (CH_2FCF_3)	Improves the heat and mass transmission.	[47]
Ghozatloo et al. (2015)	G-Hummers	Improves the coefficient of heat and mass transfer.	[7]
Zhu et al. (2009)	Nanoparticles containing graphene oxide	Raises the gas intake, increases thermal conductivity.	[50]
Torres Pineda et al. (2012)	SiO_2 and Al_2O_3 nanoparticle suspensions	Improves absorption rate compared with the pure methanol, SiO_2 nanoparticle is a better candidate than Al_2O_3 nanoparticle.	[51]
Jung et al. (2012)	Al_2O_3 nanofluids-containing aqueous NaCl solution	Absorption of CO_2 is up to 8.3% greater than purified methanol.	[52]
Haghtalab et al. (2015)	ZnO nanofluid	Improves absorption of CO_2, ZnO nanofluids to be more efficient compared with SiO_2 nanofluids.	[53]
Prasad et al. (2012)	SiO_2	Decreases the desorption level of CH_4, shortens the induction period, improves the pace of growth.	[54]
Choi et al. (2014)	0.2 wt.% Al_2O_3 and 0.6 wt.% SDS nanoparticles with THF (tetrahydrofuran) as surfactant	The maximum CO_2 hydrate formation rate is enhanced up to 3.74 times by adding 0.6 wt.% of SDS and 0.2 wt.% of Al_2O_3 nanoparticles.	[55]
Mohammadi et al. (2016)	SDS and ZnO nanoparticle	Storage potential enhanced, gas consumption was measured to be approximately 121%.	[37]
Park et al. (2012)	Carbon nanofluids	The rate of methane absorbed during the development of methane hydrate increased by 4.5 relative to distilled water.	[57]
Firoozabadi et al. (2018)	CTAB surfactant stabilized Fe_3O_4 nanofluid	Reduces the resistance against mass transfer at the solution–hydrate and the gas–solution interfaces.	[16]
Rajabi Firoozabadi et al. (2020)	Fe_3O_4 nanofluids with an SDS	Accelerates heat transmission, raises the nucleation positions, increases the mass transport, improve the mass exchange and carbon dioxide solvability, enhances the pace of nucleation and hydrate development.	[10]
Chari et al. (2013)	SiO_2 nanoparticles	Improves gas storage stability, enhances the hydrate formation kinetics.	[58]
Ganji et al. (2013)	Polymers of nanoparticles	Efficiently improves the stability of the formulated hydrate.	[59]
Zhou et al. (2014)	Graphite nanoparticles	The induction period of CO_2 hydrate dropped by 80.8%, the maximal CO_2 intake rose by 12.8%, the hydrate response in the presence of nanoparticles increased by 98.8% around 400 min.	[60]

Table 17.1 Nanofluid application in a hydrate process. ***Continued***

Authors	Type of nanomaterial	Effect of nanomaterial	References
Pahlavanzadeh et al. (2016)	CTAB and Cu nanoparticles	Enhancement of mass transfer, improves the growth rate of methane hydrate formation.	[41]
Najibi et al. (2015)	CuO nanoparticles	Film resistance is reduced, mass transfer improves, rate of gas consumption improves.	[49]
Kakati et al. (2016)	SDS, Al_2O_3, ZnO	The rate of change registered was almost 150% relative to pure water, improves heat and mass transmission.	[56]
Xuan et al. (2009)	CuO	Heat and mass exchange.	[61]
Park et al. (2010)	MWCNTs	The quantity of gas absorbed was around 300% greater than that of water.	[62]
Arjang et al. (2013)	Silver nanoparticle	The methane hydrate induction period decreased, methane consumption rate improved.	[63]
Mohammadi et al. (2014)	SDS and silver nanoparticles	Enhances the storage potential of CO_2 by 93.9%.	[64]
Abdi-Khanghah et al. (2018)	ZnO nanofluid	CH_4 hydrate generation level was improved.	[65]
Li et al. (2020)	Graphene oxide nanoparticles (GON)	Performance of CH_4 isolation relative to GON-free systems has been improved.	[66]
Hajilary et al. (2018)	Nanosilica and carbon nanotubes (CNTs)	Mass transfer was verified by good conventions. CO_2 absorption increased, CNT's nanofluid efficiency is much greater than that of nanosilica.	[67]
Zarenezhad et al. (2016)	Nanographene oxide nonconductives	High mass transfer rate, imposes a high CO_2 gas consumption rate.	[68]
Cheng et al. (2020)	SiO_2 nanofluids	Decrease in induction phase, improvement in gas consumption.	[69]
Song et al. (2019)	Carbon nanotubes (CNTs)	Reduces the formation duration, improving gas storage capacity, improves the mass transport.	[70]
Lu et al. (2020)	Graphite nanofluids	The ability for storage of gas was not substantially improved, induction and average time for hydrate generation were shortened, mass and heat distributions for CH_4 hydrate, formation were facilitated, the gas consumption obtained in graphite nanofluids was larger.	[9]

Given that hydrate has been recognized as a source of energy storage and a suitable method for moving and transporting gas in recent years, it is predicted to have a wider application in the not too distant future. What is clear, however, is that more research is needed on the effect of nanofluids on improving mass transfer in the formation of gas hydrates.

Abbreviations

CNT	Carbon nanotube
DWCNT	Double-walled carbon nanotubes
GO	Graphene oxide
LCCBM	Coalbed methane
MWCNT	Multiwalled carbon nanotubes
NG	Natural gas
(R-134a)	1,1,1,2-Tetrafluoroethane
RR 195	Responsive red 195
SNG	Solidified natural gases
SWCNT	Single wall carbon nanotubes

References

[1] A. Eslamimanesh, A.H. Mohammadi, D. Richon, P. Naidoo, D. Ramjugernath, Application of gas hydrate formation in separation processes: a review of experimental studies, J. Chem. Thermodyn. 46 (2012) 62–71.

[2] E.G. Hammerschmidt, Formation of gas hydrates in natural gas transmission lines, Ind. Eng. Chem. 26 (8) (1934) 851–855.

[3] D.L. Li, S.M. Sheng, Y. Zhang, D.Q. Liang, X.P. Wu, Effects of multiwalled carbon nanotubes on CH_4 hydrate in the presence of tetra-n-butyl ammonium bromide, RSC Adv. 8 (18) (2018) 10089–10096.

[4] H. Sarlak, A. Azimi, S. Mostafa, T. Ghomshe, M. Mirzaei, Thermodynamic study of CO_2 hydrate formation in the presence of SDS and graphene oxide nanoparticles, Adv. Environ. Technol. 4 (2018) (2019) 233–240.

[5] P. Englezos, J.D. Lee, Gas hydrates: a cleaner source of energy and opportunity for innovative technologies, Korean J. Chem. Eng. 22 (5) (2005) 671–681.

[6] M. Similarities, Structures and 2 Molecular Similarities to Ice, 2008.

[7] A. Ghozatloo, M. Hosseini, M. Shariaty-Niassar, Improvement and enhancement of natural gas hydrate formation process by Hummers' graphene, J. Nat. Gas. Sci. Eng. 27 (2015) 1229–1233.

[8] J. Verrett, D. Posteraro, P. Servio, Surfactant effects on methane solubility and mole fraction during hydrate growth, Chem. Eng. Sci. 84 (2012) 80–84.

[9] Y.Y. Lu, B. Bin Ge, D.L. Zhong, Investigation of using graphite nanofluids to promote methane hydrate formation: application to solidified natural gas storage, Energy 199 (2020).

[10] S. Rajabi Firoozabadi, M. Bonyadi, A comparative study on the effects of Fe_3O_4 nanofluid, SDS and CTAB aqueous solutions on the CO_2 hydrate formation, J. Mol. Liq. 300 (2020).

[11] J. Sloan, Clathrate hydrates: the other common solid water phase, Ind. Eng. Chem. Res. 39 (9) (2000) 3123–3129.

[12] S. Said, et al., A study on the influence of nanofluids on gas hydrate formation kinetics and their potential: application to the CO_2 capture process, J. Nat. Gas. Sci. Eng. 32 (2016) 95–108.

[13] M. Tariq, D. Rooney, E. Othman, S. Aparicio, M. Atilhan, M. Khraisheh, Gas hydrate inhibition: a review of the role of ionic liquids, Ind. Eng. Chem. Res. 53 (46) (2014) 17855–17868.

[14] L.A. Stern, S.H. Kirby, W.B. Durham, Peculiarities of methane clathrate hydrate formation and solid-state deformation, including possible superheating of water ice, Sci. (80-.) 273 (5283) (1996) 1843–1848.

[15] D.K. Lonsdale, P. R. S. L. A, The structure of ice, Proc. R. Soc. London. Ser. A. Math. Phys. Sci. 247 (1251) (1958) 424–434.

[16] S.R. Firoozabadi, M. Bonyadi, A. Lashanizadegan, Experimental investigation of Fe_3O_4 nanoparticles effect on the carbon dioxide hydrate formation in the presence of magnetic field, J. Nat. Gas. Sci. Eng. 59 (2018) 374–386.

[17] K. Nasrifar, M. Moshfeghian, A model for prediction of gas hydrate formation conditions in aqueous solutions containing electrolytes and/or alcohol, J. Chem. Thermodyn. 33 (9) (2001) 999–1014.

[18] T.C.W. Mak, Hexamethylenetetramine hexahydrate: a new type of clathrate hydrate, 2799 (1965) (2010) 0–7.

[19] Polyhedral clathrate hydrates. XI. Structure of tetramethylammonium hydroxide pentahydrate, 2338 (1966).

[20] A. Eling, Keynote article, 1994.

[21] H.D. Processes, Formation and 3 Hydrate Dissociation Processes, 2008.

[22] A.A. Khokhar, E.D. Sloan, J.S. Gudmundsson, Natural gas storage properties of structure H hydrate, Ann. N. Y. Acad. Sci. 912 (2000) 950–957.

[23] W. Wang, C.L. Bray, D.J. Adams, A.I. Cooper, Methane storage in dry water gas hydrates, J. Am. Chem. Soc. 130 (35) (2008) 11608–11609.

[24] Z.G. Sun, R. Wang, R. Ma, K. Guo, S. Fan, Natural gas storage in hydrates with the presence of promoters, Energy Convers. Manag. 44 (17) (2003) 2733–2742.

[25] M.C. Dichicco, S. Laurita, M. Paternoster, G. Rizzo, R. Sinisi, G. Mongelli, Serpentinite carbonation for CO_2 sequestration in the southern apennines: preliminary study, Energy Procedia 76 (2015) 477–486.

[26] H. Yang, et al., Progress in carbon dioxide separation and capture: a review, J. Environ. Sci. 20 (1) (2008) 14–27.

[27] D. Aaron, C. Tsouris, Separation of CO_2 from flue gas: a review, Sep. Sci. Technol. 40 (1–3) (2005) 321–348.

[28] H. Fakharian, H. Ganji, A. Naderifar, Saline produced water treatment using gas hydrates, J. Environ. Chem. Eng. 5 (5) (2017) 4269–4273.

[29] J. Javanmardi, M. Moshfeghian, Energy consumption and economic evaluation of water desalination by hydrate phenomenon, Appl. Therm. Eng. 23 (7) (2003) 845–857.

[30] B. Lal, O. Nashed, Chemical Additives for Gas Hydrates, 2020.

[31] H.P. Veluswamy, S. Kumar, R. Kumar, P. Rangsunvigit, P. Linga, Enhanced clathrate hydrate formation kinetics at near ambient temperatures and moderate pressures: application to natural gas storage, Fuel 182 (2016) 907–919.

[32] H.P. Veluswamy, et al., Rapid methane hydrate formation to develop a cost effective large scale energy storage system, Chem. Eng. J. 290 (2016) 161–173.

[33] X. Wang, A.J. Schultz, Y. Halpern, Kinetics of methane hydrate formation from polycrystalline deuterated ice, J. Phys. Chem. A 106 (32) (2002) 7304–7309.

[34] D. Sadeq, S. Iglauer, M. Lebedev, C. Smith, A. Barifcani, Experimental determination of hydrate phase equilibrium for different gas mixtures containing methane, carbon dioxide and nitrogen with motor current measurements, J. Nat. Gas. Sci. Eng. 38 (2017) 59–73.

[35] S. Yan, W. Dai, S. Wang, Y. Rao, S. Zhou, Graphene oxide: an effective promoter for CO_2 hydrate formation, Energies 11 (7) (2018) 1–13.

[36] V.T. John, K.D. Papadopoulos, G.D. Holder, A generalized model for predicting equilibrium conditions for gas hydrates, AIChE J. 31 (2) (1985) 252–259.

[37] M. Mohammadi, A. Haghtalab, Z. Fakhroueian, Experimental study and thermodynamic modeling of CO_2 gas hydrate formation in presence of zinc oxide nanoparticles, J. Chem. Thermodyn. 96 (2016) 24–33.

[38] M. Zare, A. Haghtalab, A.N. Ahmadi, K. Nazari, Experiment and thermodynamic modeling of methane hydrate equilibria in the presence of aqueous imidazolium-based ionic liquid solutions using electrolyte cubic square well equation of state, Fluid Phase Equilib. 341 (2013) 61–69.

[39] M. Królikowski, M. Królikowska, H. Hashemi, P. Naidoo, D. Ramjugernath, U. Domańska, Experimental study of carbon dioxide gas hydrate formation in the presence of zwitterionic compounds, J. Chem. Thermodyn. 137 (2019) 94–100.

[40] F. However, Techniques for 4 Estimation Phase Equilibria of Natural Gas Hydrates, 2008.

[41] H. Pahlavanzadeh, S. Rezaei, M. Khanlarkhani, M. Manteghian, A.H. Mohammadi, Kinetic study of methane hydrate formation in the presence of copper nanoparticles and CTAB, J. Nat. Gas. Sci. Eng. 34 (2016) 803–810.

[42] X. Fang, Y. Xuan, Q. Li, Experimental investigation on enhanced mass transfer in nanofluids, Appl. Phys. Lett. 95 (20) (2009) 7–10.

[43] S.S. Ashrafmansouri, M. Nasr Esfahany, Mass transfer in nanofluids: a review, Int. J. Therm. Sci. 82 (1) (2014) 84–99.

[44] C. Pang, J.W. Lee, Y.T. Kang, Review on combined heat and mass transfer characteristics in nanofluids, Int. J. Therm. Sci. 87 (2015) 49–67.

[45] O. Nashed, B. Partoon, B. Lal, K.M. Sabil, A.M. Shariff, Review the impact of nanoparticles on the thermodynamics and kinetics of gas hydrate formation, J. Nat. Gas. Sci. Eng. 55 (2018) 452–465.

[46] P.S.R. Prasad, V.D. Chari, Preservation of methane gas in the form of hydrates: use of mixed hydrates, J. Nat. Gas. Sci. Eng. 25 (2015) 10–14.

[47] J. Li, D. Liang, K. Guo, R. Wang, S. Fan, Formation and dissociation of HFC134a gas hydrate in nano-copper suspension, Energy Convers. Manag. 47 (2) (2006) 201–210.

[48] Y. Zhong, R.E. Rogers, Surfactant effects on gas hydrate formation, Chem. Eng. Sci. 55 (19) (2000) 4175–4187.

[49] H. Najibi, M. Mirzaee Shayegan, H. Heidary, Experimental investigation of methane hydrate formation in the presence of copper oxide nanoparticles and SDS, J. Nat. Gas. Sci. Eng. 23 (2015) 315–323.

[50] D. Zhu, X. Li, N. Wang, X. Wang, J. Gao, H. Li, Dispersion behavior and thermal conductivity characteristics of Al_2O_3-H_2O nanofluids, Curr. Appl. Phys. 9 (1) (2009) 131–139.

[51] I. Torres Pineda, J.W. Lee, I. Jung, Y.T. Kang, CO_2 absorption enhancement by methanol-based Al_2O_3 and SiO_2 nanofluids in a tray column absorber, Int. J. Refrig. 35 (5) (2012) 1402–1409.

[52] J.Y. Jung, J.W. Lee, Y.T. Kang, CO_2 absorption characteristics of nanoparticle suspensions in methanol, J. Mech. Sci. Technol. 26 (8) (2012) 2285–2290.

[53] A. Haghtalab, M. Mohammadi, Z. Fakhroueian, Absorption and solubility measurement of CO_2 in water-based ZnO and SiO_2 nanofluids, Fluid Phase Equilib. 392 (2015) 33–42.

[54] P.S.R. Prasad, V.D. Chari, D.V.S.G.K. Sharma, S.R. Murthy, Effect of silica particles on the stability of methane hydrates, Fluid Phase Equilib. 318 (2012) 110–114.

[55] J.W. Choi, J.T. Chung, Y.T. Kang, CO_2 hydrate formation at atmospheric pressure using high efficiency absorbent and surfactants, Energy 78 (2014) 869–876.

[56] H. Kakati, A. Mandal, S. Laik, Promoting effect of Al_2O_3/ZnO-based nanofluids stabilized by SDS surfactant on $CH_4 + C_2H_6 + C_3H_8$ hydrate formation, J. Ind. Eng. Chem. 35 (2016) 357–368.

[57] S.S. Park, E.J. An, S.B. Lee, W. gee Chun, N.J. Kim, Characteristics of methane hydrate formation in carbon nanofluids, J. Ind. Eng. Chem. 18 (1) (2012) 443–448.

[58] V.D. Chari, D.V.S.G.K. Sharma, P.S.R. Prasad, S.R. Murthy, Methane hydrates formation and dissociation in nano silica suspension, J. Nat. Gas. Sci. Eng. 11 (2013) 7–11.
[59] H. Ganji, J. Aalaie, S.H. Boroojerdi, A.R. Rod, Effect of polymer nanocomposites on methane hydrate stability and storage capacity, J. Pet. Sci. Eng. 112 (2013) 32–35.
[60] S.D. Zhou, Y.S. Yu, M.M. Zhao, S.L. Wang, G.Z. Zhang, Effect of graphite nanoparticles on promoting CO_2 hydrate formation, Energy Fuels 28 (7) (2014) 4694–4698.
[61] Y. Xuan, Conception for enhanced mass transport in binary nanofluids, Heat Mass Transf.: Wärme- und Stoffuebertragung 46 (2) (2009) 277–279.
[62] S.S. Park, S.B. Lee, N.J. Kim, Effect of multi-walled carbon nanotubes on methane hydrate formation, J. Ind. Eng. Chem. 16 (4) (2010) 551–555.
[63] S. Arjang, M. Manteghian, A. Mohammadi, Effect of synthesized silver nanoparticles in promoting methane hydrate formation at 4.7MPa and 5.7MPa, Chem. Eng. Res. Des. 91 (6) (2013) 1050–1054.
[64] A. Mohammadi, M. Manteghian, A. Haghtalab, A.H. Mohammadi, M. Rahmati-Abkenar, Kinetic study of carbon dioxide hydrate formation in presence of silver nanoparticles and SDS, Chem. Eng. J. 237 (2014) 387–395.
[65] M. Abdi-Khanghah, M. Adelizadeh, Z. Naserzadeh, H. Barati, Methane hydrate formation in the presence of ZnO nanoparticle and SDS: application to transportation and storage, J. Nat. Gas. Sci. Eng. 54 (2018) 120–130.
[66] J.B. Li, D.L. Zhong, J. Yan, Improving gas hydrate-based CH_4 separation from low-concentration coalbed methane by graphene oxide nanofluids, J. Nat. Gas. Sci. Eng. 76 (2020) 103212.
[67] N. Hajilary, M. Rezakazemi, CFD modeling of CO_2 capture by water-based nanofluids using hollow fiber membrane contactor, Int. J. Greenh. Gas. Control. 77 (2018) 88–95.
[68] B. Zarenezhad, V. Montazeri, Nanofluid-assisted gas to hydrate (GTH) energy conversion for promoting CO_2 recovery and sequestration processes in the petroleum industry, Pet. Sci. Technol. 34 (1) (2016) 37–43.
[69] Z. Cheng, Y. Zhao, W. Liu, Y. Liu, L. Jiang, Y. Song, Kinetic analysis of nano-SiO_2 promoting methane hydrate formation in porous medium, J. Nat. Gas. Sci. Eng. 79 (2020) 103375.
[70] Y.M. Song, F. Wang, S.J. Luo, R.B. Guo, D. Xu, Methane hydrate formation improved by water-soluble carbon nanotubes via Π-Π conjugated molecules functionalization, Fuel 243 (189) (2019) 185–191.

CHAPTER

Mass transfer enhancement in solar stills by nanofluids

18

Ali Behrad Vakylabad

Department of Materials, Institute of Science and High Technology and Environmental Sciences, Graduate University of Advanced Technology, Kerman, Iran

18.1 Introduction

Today, cheap solar energy (4×10^{15} mW) is increasingly available, and its use with high efficiency through the development of new technologies, including nanotechnology, is advancing and developing rapidly [1,2]. Simply put, solar stills are devices for purifying or distilling water through cycles of evaporation and condensation in which heat absorbed from the sun is used to evaporate water and then the evaporated water is cooled and collected while the dissolved salts, suspended solids, and other dissolved impurities are left. This process has efficiencies of nearly 100%. The importance of this technology and its development lie in the growing water crisis in the world. According to U.N. estimates, about 3.6 billion people in the world are now facing this crisis, and this figure may reach 5.7 billion people by 2050 [3]. Although the performance of a solar still unit depends on many factors [4,5], including parameters related to the environment (solar radiation, ambient temperature, relative humidity, wind velocity, cloud and dust cover), design (evaporation area, water depth, condensing cover angle, thermal storage materials and additives, solar tracking and reflector, insulation), and operation (water feeding, position of stepped solar still, maintenance), in all these cases, nanofluids can significantly increase the efficiency and desalination productivity of solar stills [6]. The most common nanomaterials used in solar still−related nanofluids are Al_2O_3 [6,7], MnO_2 [8], SiC [9], SiO_2 [10], SnO_2[6], TiO_2 [11], ZnO [6], Fe_2O_3 [6], Cu_2O [12], CuO [13], Cu [10], carbon nanotubes [14], and carbon [15]. Among the various nanomaterials, alumina-based nanofluids are the most common type of nanofluid. The reason for this is the low production cost of this type of nanofluid along with its high technical efficiency, for example in terms of thermal conductivity (40 W/mK) [6].

The most important feature of nanomaterials is the possibility of manipulating their specific properties during their production. For example, nanomaterials with hydrophobic properties [16], magnetic, and superparamagnetic [17] properties and controlled morphology [18,19] (e.g., spherical [20], three-dimensional hierarchical flower-like [21]) can be obtained with different particle size distributions (PSD) [20,22,23], each of which can exhibit a wide range of specific properties

Nanofluids and Mass Transfer. DOI: https://doi.org/10.1016/B978-0-12-823996-4.00019-7

depending on their application [24,25]. Solar still units are a new field of application for nanofluids, many aspects of which are not yet known and require a great deal of research and development. The purpose of this chapter is to investigate the role of nanofluids in mass transfer and the productivity of the solar stills.

18.2 Components of a solar still unit

An environmentally sustainable solar treatment unit (solar still) distills and purifies salt water or water with soluble materials (brackish, contaminated, and sea water). This is done by using free solar energy to evaporate the water, and the water vapor is cooled and collected in later stages, the end result of which is water purification and delivery of pure water. This technology is used in areas where drinking water is not available. When exposed to sunlight, clean water can be obtained from salt water or other contaminated water or even from plants. The same method of solar treatment is inspired by the method of rainfall. Water vapor with very high mobility (compared to the main saline water molecules) from solar evaporation is collected by creating a condensation system. With this process in solar stills, various impurities, such as salts and heavy metals and even various organisms and microorganisms, are deposited. On the basis of the design instructions, configurations, and production capacity, the various solar treatment units can be divided into six main categories: roof-type solar still, water film cooling over glass cover of a still including evaporation effects, passive active solar still, multieffect diffusion-type solar still, tilted wick solar still, and solar still made up of tubes for sea water desalting [26,27]. In this regard, basic analysis of the transfer phenomena in the units leads to a comprehensive understanding of their design principles. As a result, a comprehensive theoretical guide is available to continuously modify and optimize their structure [27–31]. Today, solar desalination plants are developing rapidly, with 14,451 desalination plants operating at a total capacity of about 60 million cubic meters per day. In Iran the first phase of the solar desalination system superproject, which has a capacity of 200,000 cubic meters per day, was recently inaugurated to transfer desalinated water from the Persian Gulf. In the first part of this project, water is delivered from the desalination plant in Bandar Abbas and is pumped through the transmission line of steel pipes with a diameter of 1600 mm to the Gol-Gohar mine in Sirjan. In addition, to the advantages especially related to solar stills, such as convenience and cheapness in design and construction, economic studies show the dominance of this equipment over other conventional methods, so that the cost of producing each kilogram of water with this method is estimated to be about 0.014 USD [32].

The simple design of solar stills enables them to act simultaneously as an absorber for solar energy radiation and a converter to convert this energy into thermal energy. A wide variety of design features are considered for this equipment. The basis of evaluation of these various designs is to increase the production efficiency or productivity of this equipment (Fig. 18.1) [26,27]. Fig. 18.2 shows the different parts of a typical solar still unit. This figure shows the main part where nanofluids with different properties can be used. Nanofluids increase the production ability of a solar treatment unit typically by increasing the temperature of various parts through improving heat transfer (44%–74% improvement) [13,33,34]. These parts include the basin, the brine fluid, the clear glass of condensation, and the vapor temperature. However, the increase in heat transfer

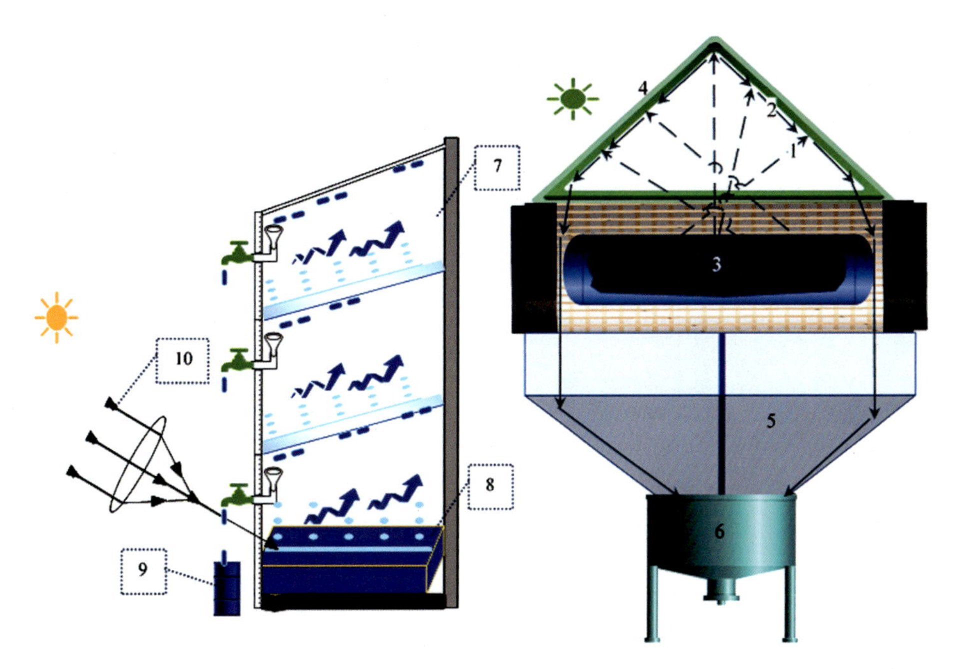

FIGURE 18.1

Different designs of solar water purifier units. (1) Evaporated water, (2) Condensed water, (3) Brackish water reservoir, (4) Transparent glass cover, (5) Condenser with cooling coils, (6) Storage tank for purified water, (7) Multistage solar still, (8) Brine water tank, (9) Distilled water collectors, (10) Glass magnifier lens (focused sun radiation).

and mass transfer are so intertwined that it may not be possible to separate them, or the role of heat transfer is so prominent that the role of mass transfer cannot be excluded from the heart of the productivity [35].

18.3 Effective parameters in solar stills

Parameters that affect the performance of solar desalination are solar radiation intensity, which is a function of the day of the year and latitude; wind speed; ambient temperature; glass cover temperature; solar still geometry; absorber plate surface; inlet water temperature; cap (clear glass) angle; water depth; glass thickness; percentage diffuse radiation extinction coefficient of the glass cover and reflectance of the still liner. Among these, the meteorological parameters, such as wind speed or solar radiation intensity (the first three factors), cannot be controlled, but the rest of the factors can be modified to an optimized state. Solar stills can be made in different shapes, such as pool, pipe, chimney, spherical, wick, and other designs. Each shape has its own advantages and disadvantages and may be modified to improve performance, such as the use of reflectors, phase change

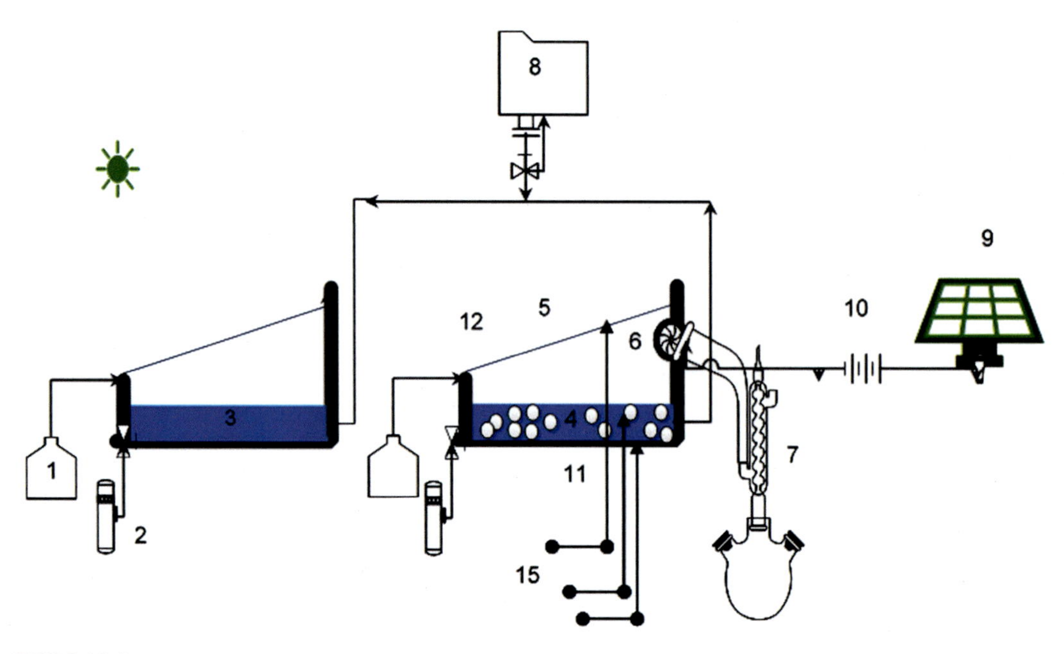

FIGURE 18.2

Simplified schematic representation of a solar water treatment unit. (1) Collector of condensate water; (2) Brine residue to waste; (3) Saline water; (4) Nanofluidized brine; (5) Clear glass cover; (6) Vacuum fan; (7) Condenser; (8) Storage tank of brackish water; (9) Solar cell for energy supply; (10) Rechargeable battery; (11) Basin; (12) Distillate trough; (13) Ordinary solar still;. (14) Modified solar still; (15) Heat sensors.

materials, and other methods [35,36]. There have been theoretical or empirical attempts to formulate the relationship between the effective parameters and productivity of the solar stills. For example, Eq. (18.1) shows the empirical relationship of different parameters and the performance of solar still equipment [5]:

$$P_d = -1.39 + 0.894H_d + 0.033t_a - 0.017V - 0.008\theta - 1.2\frac{\delta}{l} \tag{18.1}$$

where P_d = the productivity of the solar still (L/m^2/day), H_d = the power received from solar irradiation (kW·h/m^2), t_a = normal environment temperature (°C), V = velocity of wind (m/s), θ = angle of the clear glass inclination (deg), and $\frac{\delta}{l}$ = brine depth to frontal height of still ratio. Of course, more experimental and semiempirical relationships can be found for any particular situation. In most of these relationships the footprint of mass can be seen in different forms [26].

The use of assistive devices such as vacuum fans significantly demonstrates this interplay between mass transfer and heat transfer [26]. Although nanofluids increase heat retention and transfer to different parts, the use of a vacuum fan leads to a significant increase in mass transfer (water vapor particles) from the saline water-based fluid unit to the condensing unit, thereby increasing the system efficiency. Perhaps the increase in evaporation of the base fluid can simply be considered as a result of the increase in mass transfer by nanofluids, which itself is significantly dependent on the increase in heat transfer power of the nanofluids. This concept is formulated in Eqs. (18.2)–(18.4) [13]:

$$J_{eV}^{*} = \sqrt{\frac{m}{2\pi k_B}}\left\{\sigma_e^{*}\frac{P_{sL}}{\sqrt{T_L}} - \sigma_e^{*}\frac{P_V}{\sqrt{T_V}}\right\} \tag{18.2}$$

$$\sigma_e^{*} = \frac{P_{sL}}{\sqrt{T_L}}\exp\left\{(DOF+4)\left(1-\frac{T_V}{T_L}\right)\right\}\left(\frac{T_V}{T_L}\right)^{DOF+4} \tag{18.3}$$

$$\sigma_c^{*} = \sqrt{\frac{T_V}{T_L}}\exp\left\{-(DOF+4)\left(1-\frac{T_V}{T_L}\right)\right\}\left(\frac{T_V}{T_L}\right)^{DOF+4} \tag{18.4}$$

where J_{eV}^{*} = the evaporation rate, m = mass of a molecule, k_B = the Boltzmann constant, P_{sL} = the saturated vapor pressure of the liquid at the vapor-liquid interface, P_V = real vapor pressure at the vapor-liquid interface, T_L = the temperature of the liquid at the vapor-liquid interface, T_V = temperature of the vapor at the vapor-liquid interface, σ_e^{*} = evaporation coefficient, σ_c^{*} = condensation coefficient, and DOF = degrees of freedom for the vibrational frequency ($3n-6$ for nonlinear molecules, and $3n-5$ for linear molecules, where n = the number of atoms in the molecule). Once again, given that the temperatures of the fluid and nanofluid entering the system are the same, on the basis of these formulas, it can be concluded that the only driving force for higher efficiency lies in the P_{sL} parameter. Increased saturated vapor pressure (equivalent to increased mass transfer) has been proven in several studies [37,38]. Also, it has been found that the saturated vapor pressure and consequently the evaporation rate increase significantly by adding nanoparticles (with different properties, such as abrasive nanoparticles) to water [38].

Various modifications and optimizations in the solar still system, in terms of both operation and design, can increase the efficiency of the system or its production capacity by more than 100%. For example, although the use of aluminum oxide nanoparticles has increased the production capacity of the system by 12.2% [39], the creation of a vacuum system with the same nanoparticles has increased the production efficiency of the unit by 116% [40]. Although solar still devices seem simple, to achieve high efficiency, they need to be constantly optimized with creative and scientific methods. However, at each stage, it is necessary to conduct economic feasibility studies to select the best process option from technical and economic points of view. For example, among the nanofluids containing TiO_2, CuO, and GO, nanofluids with CuO have both the highest production capacity and the lowest cost. Therefore nanofluids with CuO are technically and economically considered the best option [11].

The characteristics of nanoparticles as well as their dispersivity have a significant effect on the production efficiency of the system. Hematite nanoparticles show low efficiency in production, owing to poor dispersion and stability. Conversely, aluminum oxide nanoparticles show the best production efficiency among the studied nanoparticles (Al_2O_3, SnO_2, ZnO, SnO_2, Fe_2O_3) [6]. Zeta potential is a good criterion for predicting the stability of colloidal fluid, so a potential higher than ± 61 mV indicates the excellent stability of nanomaterial colloids [6]. Stability, thermal conductivity, and environmental conditions play a major role in the productivity of the solar still. The higher the thermal conductivity of the nanofluid, the greater the absorption of solar radiation. This phenomenon in turn leads to a greater temperature difference between the base fluid and the clear glass cover, which in turn means increased evaporation and increased mass transfer. However, it is important to note that the base fluid, dispersion agents, and nanoparticle type, as well as the appropriate pH adjustment to provide the optimal zeta potential for maximum dispersion, are the effective determinants in the process from the nanofluid viewpoint.

18.4 Productivity of solar stills

One of the practical approaches to assess the suitability of energy conversion processes and systems or energy distribution systems is to use exergy analysis. Conventional energy analysis methods do not provide the efficient tools required to accurately evaluate the performance of an energy conversion system. However, exergy analysis provides a more complete and accurate energy analysis. This type of analysis involves the use of the concepts of exergy, equilibrium, and efficiency to evaluate and improve energy in a system [41,42].

Thermodynamics in general can analyze the behavior, performance, and efficiency of any energy conversion system. However, classical analysis methods are based on the first law of thermodynamics and the principles of energy conservation. In this way, the input energy and output energy of a system are considered, and energy losses within the system are omitted. In contrast, the exergy analysis method is defined on the basis of the maximum amount of work generated by a stream or system when it reaches equilibrium relative to the reference environment. Therefore this type of exergy analysis is a measure of the usefulness or quality of energy [42].

Basically, according to the second law of thermodynamics, the solar stills can be analyzed with the general exergetic method as follows (Eqs. 18.5–18.9):

$$\sum \dot{Ex}_{in} - \sum \dot{Ex}_{out} = \sum \dot{Ex}_{des} \tag{18.5}$$

where

$\sum \dot{Ex}_{in}$ = the exergy input
$\sum \dot{Ex}_{out}$ = the exergy output
$\sum \dot{Ex}_{des}$ = exergy destruction

Exergy efficiency (η_{EX}) for solar treatment systems is simply obtained by dividing the output exergy by the input exergy, which is equivalent to the exergy of distilled water ($Ex_{evaporation}$) to the insulation exergy (Ex_{input}).

$$\eta_{EX} = \frac{Ex_{output}}{Ex_{input}} = \frac{Ex_{evaporation}}{Ex_{input}} \tag{18.6}$$

Considering the concepts of exergy, the output exergy per hour is as follows (Eq. 18.7):

$$\sum \dot{Ex}_{out} = \left(Ex_{evaporation}\right) = \frac{(\dot{m}_{evaporation} \times h_{fg})}{3600} \times \left(1 - \frac{T_a + 273}{T_w + 273}\right) \tag{18.7}$$

where T_a = ambient temperature, T_w = water temperature, h_{fg} = heat transfer coefficient related to the convective and radiative phenomena from the glass cover to ambient, ($Wm^{-2}/°C$), and $\dot{m}_{evaporation}$ = the amount of distilled water.

For a solar treatment unit, the radiation collected by this system is defined by the parameter (Ex_{input}). Therefore the input exergy to the solar still unit can be calculated as follows:

$$\sum Ex_{input} = \sum Ex_{sun\ radiation} \tag{18.8}$$

$$Ex_{sun\ radiation} = A_b \times I(t) \times \left[1 - \frac{4}{3} \times \left[\frac{T_a}{T_{sun}}\right] + \frac{1}{3} \times \left[\frac{T_a}{T_{sun}}\right]^4\right] \tag{18.9}$$

In Eq. (18.9) the temperature of the sun (T_{sun}) is considered to be equal to 6000 K, $I(t)$ = solar radiation, and A_b = effective area of base (basin) of the solar purification system,

The relationship between mass transfer law and exergy analysis as a measure of the proximity of a process to the ideal thermodynamic state has been proven. Therefore exergy improvement is directly related to the improvement of mass transfer, which means the thermodynamic relationship and mass transfer from exergy site. Although collecting these items in an explicit formula may require further and more detailed laboratory and theoretical research, any concrete improvement in the heat transfer and thermodynamic conditions of the system toward the ideal situation implicitly means improving mass transfer conditions, especially in heat-based systems such as solar stills [43].

18.5 On the mass transfer of solar stills

As was mentioned earlier in the chapter, no specific boundaries can be considered in the phenomena intertwined in the solar treatment system in terms of heat transfer, mass transfer, and design and implementation parameters. Therefore considering that heat transfer in solar stills has been extensively studied, here we can conclude by reviewing the parameters of how the mass transfer characteristics of the equipment improve as a result of improving these parameters. Here, the calculation of internal heat transfer is considered. By definition, heat transfer is intended to be exchanged through the liquid inside the base part of the solar treatment unit to the transparent inner clear glass cover. Heat transfer relations in the solar unit occur through radiation, evaporation, and convection [44,45].

18.5.1 Inferred mass transfer from heat transfer mechanism

The relationship between different transfer phenomena helps to provide a more and better explanation and interpretation of each of them. Although the heat transfer behavior by different mechanisms in a solar still is clearly detected, mass transfer could directly and indirectly follow this behavior [46–50].

18.5.1.1 Convection

This type of heat transfer may be indicated between the basin fluid and the inside surface of the clear glass cover, and calculated as follows [45]:

$$Q_{c,w-g} = h_{c,w-g}A(T_w - T_g) \tag{18.10}$$

where T_w = basin fluid temperature, T_g = transparent glass temperature, and $h_{c,w-g}$ = heat transfer coefficient.

The heat transfer coefficient is due to convection, which is a function of pressure and temperature in different areas inside the cell of the solar treatment unit (Eq. 18.11):

$$h_{c,w-g} = 0.884\left\{(T_w - T_g) + \frac{[P_w - P_g][T_w + 273]}{[268.9 \times 10^3 - P_w]}\right\}^{\frac{1}{3}} \tag{18.11}$$

where P_w = partial pressure of vapor phase at fluid temperature ($N \cdot m^{-2}$); P_g = partial pressure of vapor gas at temperature of the inner transparent glass ($N \cdot m^{-2}$).

These pressures themselves (p_w and p_g) are function of temperatures (T_w and T_g) in different parts of the system:

$$P_w = \exp\left\{25.317 - \frac{5144}{(T_w + 273)}\right\} \tag{18.12}$$

$$P_g = \exp\left\{25.317 - \frac{5144}{(T_g + 273)}\right\} \tag{18.13}$$

The parameters of Eqs. (18.12) and (18.13) have been previously defined.

A more precise relationship between nanofluids and increased production and evaporation can further highlight the role of mass transfer here. For example, the effect of pressure on volumetric mass transfer coefficient has been proven [51]. The following empirical equation (Eq. 18.14) based on the experiments shows a strong relationship between the partial pressure of the system and its volumetric mass transfer [51]:

$$k_L a = C\varepsilon_G^D\left(\rho Q^2 d_0^{-3} \gamma^{-1}\right)^E \left(\frac{P}{P_0}\right)^F \tag{18.14}$$

where a = vapor-fluid interfacial area per unit volume of the fluid ($m^2 \cdot m^{-3}$); k_L = mass transfer coefficient of fluid phase ($m \cdot s^{-1}$); the empirical constants obtained experimentally for water are $C = 7.33 \times 0^{-3}$; $D = 0.702$; $E = 0.122$; $F = 0.151$; ε_G^D = gas or vapor phase hold up; ρ = fluid density ($kg \cdot m^{-3}$); Q = volumetric flow rate of vapor phase under the still chamber ($m^3 \cdot s^{-1}$); d_0 = effective length of the fluid-vapor interface (mm); γ = surface tension of the fluid ($N \cdot m^{-1}$); P = partial pressure in the still chamber (Pa); and P_0 = standard environment pressure (1.013×10^5 Pa).

18.5.1.2 Radiation

There is radiant heat transfer for two surfaces that are at different temperatures. In the case of a solar still, these two surfaces are the base fluid and the transparent cover. Radiant heat transfer for these conditions in a solar still can be calculated as follows [45,52]:

$$Q_{r,w-g} = h_{r,w-g} A(T_w - T_g) \tag{18.15}$$

The radiation-related heat transfer coefficient for the base fluid and clear glass is estimated as follows:

$$h_{r,w-g} = \varepsilon_{\text{eff}}\sigma\left[(T_w + 273)^2 + (T_g + 273)^2\right](T_w + T_g + 546) \tag{18.16}$$

In Eq. (18.16) the ε_{eff} parameter can be calculated as follows (Eq. 18.17):

$$\varepsilon_{eff} = \left(\frac{1}{\varepsilon_w} + \frac{1}{\varepsilon_g} - 1\right)^{-1} \tag{18.17}$$

The two new parameters in Eqs. (18.16) and (18.17) are: ε_{eff} = effective emissivity of fluid surface (0.993 W/m^2K^4) and σ = the Stefan-Boltzmann constant (5.67×10^{-8} W/m^2K^4).

18.5.1.3 Evaporation

The heat transfer due to evaporation between the fluid and glass surfaces is obtained as follows [45]:

$$Q_{e.w-g} = h_{e,w-g}A(T_w - T_g) \tag{18.18}$$

The heat transfer coefficient due to evaporation for the two surfaces is determined as follows:

$$h_{e,w-g} = \frac{16.273 \times 10^{-3} \times h_{c,w-g}\left[P_w - P_g\right]}{[T_w - T_g]} \tag{18.19}$$

In Eq. (18.19) the pressure difference between the surfaces also plays a major role. According to the identification of heat transfer sources in the solar still system, the total heat transfer coefficient is the sum of the transfer coefficients due to the three mentioned sources (convection, radiation, and evaporation) (Eq. 18.20):

$$h_{t,w-g} = h_{c,w-g} + h_{r,w-g} + h_{e,w-g} \tag{18.20}$$

18.5.2 Perspectives of mass transfer

As can be deduced from the heat transfer equations in different ways, in all these mechanisms the trace of the vapor pressure difference in different parts of the solar still leads to an intensification of the desired mass transfer (i.e., greater and improved mobility of the vapor phase). Of course, one of the most important mechanisms for mass and heat transfer in the presence of nanofluids is the Brownian motion of nanoparticles, which itself leads to an increase in transfer phenomena in various environments, including the solar still system. However, there are several parameters associated with the used nanomaterials that determine their role in improving their performance in terms of transfer phenomena. These parameters have not yet been carefully considered. Some of them are PSD, morphology, volume percentage, and dispersibility of nanomaterials [53].

The viscosity and density adversely affect the mass transfer, which can also be seen in the following experimental equation [54,55]:

$$K_La = 2.26u_G^{0.62}u_L^{0.35}\mu_L^{-0.1}\left[\frac{(1-\alpha^2)}{\alpha^2}\right]^{0.4} \tag{18.21}$$

$$e = \rho_L g u_G + \left(\frac{2\pi^2\rho_L}{3}\right)\left[\frac{(1-\alpha^2)}{C_0^2\alpha^2 h}\right](AF)^3 \tag{18.22}$$

$$K_La = 0.15e^{0.31}u_G^{0.3}u_L^{0.35}\mu_L^{-0.13} \tag{18.23}$$

where K_La = mass transfer coefficient, u_G = superficial velocity of gas or vapor (m/s), u_L = superficial velocity of fluid (m/s), μ_L = viscosity of fluid (cP), ρ_L = density of fluid (kg/m^3), α = fractional free area, e = power distribution or propagation rate (W/m^3), g = gravitational acceleration (m/s^2), C_0 = coefficient related to los, h = geometrical parameter or spacing (m), A = geometrical parameter, and F = frequency (Hz).

As can be seen in these experimental relationships (Eqs. 18.21–18.23), the presence of nanoparticles can increase the viscosity (μ_L) and density (ρ_L) of the fluid and in turn add a certain complexity to the system. This is one of the reasons for the disagreement about the effect of nanofluids in improving mass transfer. According to these empirical equations, viscosity has an adverse impact. But density positively affect the mass transfer. Of course, with different expressions the improvement of mass transfer rate can be related to the changes in the concentration of the desired component by nanofluid in different areas inside the solar still. Eqs. (18.24)–(18.26) express the mass transfer rate as a function of the difference in concentration of the gas phase component [54]:

$$K_La = f(C_s, C_t, C_0) \tag{18.24}$$

$$ln\left(\frac{C_s - C_0}{C_s - C_t}\right) = K_Lat \tag{18.25}$$

$$\frac{ln\left(\frac{C_s - C_0}{C_s - C_t}\right)}{t} = K_La \tag{18.26}$$

To adapt these equations to the solar still system, C_s can be considered as the saturated vapor concentration at the base fluid surface, C_0 is the initial vapor phase concentration, and C_t represents the vapor phase concentration at time t on the transparent glass surface. By using the second law of thermodynamics, it has been proven that increasing the entropy due to the addition of an alumina-water nanofluid additive can improve the production capacity of a solar still unit by up to 25%. In this regard, the modeling shows an increase in the Nusselt number to 18% with an increase in the volume percentage of nanoalumina (Al_2O_3) from 0% to 5%. This increase in thermodynamic irregularity (entropy) in the base fluid and the inner surface of the transparent glass eventually leads to better mass transfer and increased unit production capacity [56].

Of course, as was mentioned earlier in the chapter, using a combination of operating parameters and nanofluids can increase the production capacity of a solar treatment unit by more than 100%. The composition of a nanoalumina-water nanofluid with an externally installed condenser on a solar still unit has increased production by 116% [39]. Also, by using a combination of copper oxide (Cu_2O) nanofluid in water (0.08% by volume) and thermoelectric cooling modules, the unit production capacity has increased up to 80%, and the exergy efficiency has increased up to more than 112% [12,57].

The calculations show that Brownian motion is the main dispersion factor of particles in nanofluids used in solar still. This diffusion of nanoparticles in the base fluid means an increase in solar vapor generation, which is associated with the intensification of convective mass transfer from brackish water base fluid to the distillate section. This view proves once again that there is a profound dependence and indivisible entanglement between the transmission phenomena (heat and mass). In some cases, the relationship between temperature and mass changes has even been formulated (Eq. 18.27). Engineered titanium nitride nanoparticles such as floating photothermal agents can increase this parameter Δm as a representative of mass transfer by creating a significant temperature difference in the liquid-air interface [30,31].

$$\frac{\pi D^2}{4} h_c \Delta T_{\text{surface}}(\text{t}) = -L\Delta m \tag{18.27}$$

where $\Delta T_{\text{surface}}(t)$ = temperature difference relative to time at the liquid-steam interface, h_c = the heat transfer coefficient due to convection, and Δm = vaporized weight of water. Depending on the materials, h_c ranges from 50 to 10,000 $\text{W} \cdot \text{m}^2 \cdot \text{K}^{-1}$. It is very important to know that the same h_c coefficient in steam vapor mode reaches the maximum value of this spectrum (10,000 $\text{W} \cdot \text{m}^2 \cdot \text{K}^{-1}$) while (natural) convection has the lowest value (500 $\text{W} \cdot \text{m}^2 \cdot \text{K}^{-1}$). In the nanoscopic world this phenomenon can be defined by transferring the mass of energetic vapor particles in differential elements and delivering them to other particles by spreading through Brownian motion [30,58].

As was mentioned above, there is an intertwined relationship between the transmission phenomena in a solar still, so there can be a direct relationship between the mass flow and thermal changes in the still [59]:

$$\dot{m_{ew}} = \frac{h_{ew}(T_w - T_g) \times 3600}{\varphi} \tag{18.28}$$

where $\dot{m_{ew}}$ = hourly yield per unit area (kg/m$^2 \cdot$h), h_{ew} = evaporative heat transfer coefficient from fluid to cover (W/m^2 °C), T_w = temperature of the fluid water (°C), T_g = temperature of the glass cover (°C), and φ = latent heat of vaporization (J/kg).

The combination of increased heat in the presence of nanoparticles according to Eq. (18.29) gives elevated power to the particle diffusion (D_p) [31,60]:

$$D_p = \frac{k_B T}{3\pi \mu_{nf} d_{np}} \tag{18.29}$$

However, the relationship between these complex parameters simultaneously with the addition of nanomaterials with different types, such as Janus nanoparticles, can be investigated by simulating the molecular, nanoscale, microscopic, and macroscopic levels with numerical modeling by using molecular dynamics and density functional theory (DFT), which is one of the future objectives of this scientific field [61].

Besides studying the basis for molecular-scale transmission phenomena, identifying the effective factors and their effect on the final productivity is essential. These factors can be divided into three categories:

1. The climatic factors, including solar radiation, wind speed, and ambient temperature with direct relationship but dust with inverse relationship (environment dust deposition on glass reduces radiation transmittance).
2. The design parameters, including single slope/double slope (better performance of single slope, owing to its high absorption of the sun's radiation) fluid depth in the basin (adverse effect on productivity); slope of the cover glass (dependent on the geographical location, the best inclination is the state solar radiation close to normal to the cover); type of solar still (designing auxiliary equipment to improve performance such as magnifying Fresnel lens to concentrate

radiation or vacuum fan for improved evaporation); hybrid solar still (increased efficiency with coupling the various types of the stills) [62–64]; stepped solar still (increased productivity through reduced thermal inertia of the water mass, small trays minimize the area of the basin) [65–67]; material selection (better performance of glass than transparent plastic materials); absorbent and radiation storage with better performance (phase change material such as paraffin wax and stearic acid as storage media show efficient performance) [68,69]; external and internal reflector (with the aim of increased sunlight irradiation) [70,71], sun-tracking system (getting more sun radiation means more productive cells) [72–74]; thickness of insulation (direct relationship with productivity) [75–77]; and gap distance between fluid surface and glass cover (the less gap, the more productivity).

3. The operational parameters comprising nanomaterials in base fluid (improving mass and heat transfer to result more productivity) [10,39,56], coloring of water fluid (significant increase in productivity of the base fluid with dye, owing to increased absorptivity of the solar irradiation) [78], fluid flow (directly relates the productivity to the flow because of enhanced heat transfer from the cover glass to the fluid) [79], salt concentration (negative effect on the still efficiency) [80], forced convection inside the solar still (considerable positive effect. like such as vacuum fan and condenser (Fig. 18.2) to promote mass and heat transfer), surfactant additives (may have various positive effects on productivity, especially along with nanofluid to improve dispersivity of nanomaterials) [81], thermophysical properties of the binary mixture of water vapor and dry air (considerably affects convective heat and mass transfer) [82,83].

18.6 Economic viewpoints

Of course, in all these issues, economics will have the first and last word. In addition, to the technical conditions, the use of solar water desalination must also satisfy the appropriate economic conditions, that is not only must it be high in terms of performance, but it must also be cost-effective [84–86]. The cost of water production depends on a number of factors, including maintenance costs, control systems, and energy consumption. The cost of a solar still is obtained from the following relation (Eq. 18.30):

$$C = \frac{10 \quad I\left(\overline{LA} + \overline{MR} + \overline{TI}\right) + 1000(K_C + s)}{A(Y_D + Y_R)} \tag{18.30}$$

where I = total investment, $\overline{LA}$ = annual targets and depreciation rate, $\overline{MR}$ = annual maintenance and repair work and materials, $\overline{TI}$ = annual tax and insurance costs, K_C = wages and labor costs, s = total fixed amount of saline water supply cost, Y_D = annual unit distilled water yield ($L \cdot m^{-2}$), Y_R = annual yield of unit of collected rainwater, and A = cross-sectional area of the solar desalination plant. Since other types and addition equipment for optimization, such as rotary solar, still requires additional costs for rotation and control of the collector direction, they should be considered (Eq. 18.31). Hence the expression (C_{ext}) must be added to Eq. (18.30):

$$C = \frac{10 \quad I\left(\overline{LA} + \overline{MR} + \overline{TI}\right) + 1000(K_C + s)}{A(Y_D + Y_R)} + C_{ext} \tag{18.31}$$

Economic calculations with different perspectives have estimated the cost of producing water with solar still to be between about $0.007–0.176/L. These estimates show that although solar stills are environmentally green, they have a long way to go to compete economically with classic desalination methods [42,87,88]. One of the solutions is the use of low-cost materials for making parts and cheap and available nanomaterials as nanofluids for the design and manufacture of these equipment. These structural adjustments (modified double-slope solar still) can make significant improvements in exergy efficiency up to above 146% and cost saving more than 30% compared to the conventional solar still [88]. These results encourage hopes for faster industrial development of this green technology [88].

18.7 Conclusion

According to the studies and related literature, the ultimate efficiency of solar still equipment is strongly dependent on operational and design parameters that are fully integrated. Therefore it is not possible and might not even be scientifically correct to separate a part of it as mass transfer due to nanofluids. The effect of nanofluids on increasing parameters such as vapor pressure at the vapor-liquid interface (P_{sL}) ultimately leads to a significant increase in evaporation rate, which means increased mass transfer from this perspective. Also, in terms of operation and design, optimized equipment of the system, such as the use of vacuum-generating fans, significantly improves this effect, that is increasing evaporation and mass transfer from this point of view. All in all, the temperature difference between two different phases of liquid-liquid or liquid-vapor in various parts is the main driving force corresponding to the difference in vapor pressure, which itself means mass transfer. Theoretical studies and the fit of laboratory experimental data to empirical equations show well that the penetration or diffusive resistance is the main controlling factor of mass transfer in solar still conditions. This controlling parameter can reduce through improving the partial pressure difference with nanofluids. On the other hand, the presence of nanomaterials with their specific properties, such as PSD, morphology, and the special characteristics of the nanomaterial in fluids, and their behavior is not fully understood, thereby introducing certain complexities into the system. One of these complexities can be the viscosity or density of nanofluids in the sense that on the one hand, increasing the volume of nanomaterials may lead to improved transfer phenomena, but this increase can be reversed by increasing the viscosity of the fluid. Therefore the optimal limit for the activity of nanomaterials in fluids must be determined. However, most research has focused on the significant effects of nanofluids on improving the efficiency and productivity of the solar stills. Along with all the important advantages, including green and environmentally friendly production that is hypothesized for solar stills, there are some disadvantages that should be at the subject of further research and development. Some of these shortcomings are the selection of low-cost but effective materials, lower productivity and efficiency, applicability in only small areas, very sensitive equipment, glass temperature, high capital cost, and shallow depth of brackish water under solar treatment. To develop research in this field, the effective parameters in the solar still equipment can be briefly divided into three main categories: climatic, design, and operation parameters. Climatic factors include solar radiation, ambient temperature, wind speed, dust categorized as climatic factors with uncontrollable nature. The next two categories are controllable. A wide range of design parameters are single slope/double slope, water depth in the basin, inclination of the cover, type of solar still, hybrid solar still, stepped solar still, the selection of the material, energy absorption and storing materials, external and internal reflector, sun-tracking system, insulation thickness, and gap distance. Other parameters are nanomaterials with various properties, coloring of fluid, flow regime of base fluid, salt concentration, forced convection inside solar still, binary mixture thermophysical properties, and surfactant additives. As these effective parameters become more sophisticated, the need to use intelligent simulation methods (artificial intelligence) using soft calculations is more than ever to better determine the future steps for making this technology economically more feasible. Understanding the scientific aspects of transfer phenomena (mass and heat) using a variety of nanomaterials in the base fluid to improve the productivity of solar still equipment requires numerical modeling using DFT, which will be included as part of future research objectives.

References

[1] O. Edenhofer, et al., Renewable Energy Sources and Climate Change Mitigation: Special Report of the Intergovernmental Panel on Climate Change, Cambridge University Press, 2011.
[2] A. Kongkanand, R. Martínez, P.V. Domínguez, et al., Single wall carbon nanotube scaffolds for photoelectrochemical solar cells. Capture and transport of photogenerated electrons, Nano lett. 7 (3) (2007) 676–680.
[3] A. Boretti, L. Rosa, Reassessing the projections of the world water development report, NPJ Clean Water 2 (1) (2019) 1–6.
[4] M.S.S. Abujazar, et al., The effects of design parameters on productivity performance of a solar still for seawater desalination: a review, Desalination 385 (2016) 178–193.
[5] A.S. Nafey, et al., Parameters affecting solar still productivity, Energy Convers. Manag. 41 (16) (2000) 1797–1809.
[6] T. Elango, A. Kannan, K.K. Murugavel, Performance study on single basin single slope solar still with different water nanofluids, Desalination 360 (2015) 45–51.
[7] S. Shanmugan, S. Palani, B. Janarthanan, Productivity enhancement of solar still by PCM and Nanoparticles miscellaneous basin absorbing materials, Desalination 433 (2018) 186–198.
[8] E.H. Bani-Hani, C. Borgford, K. Khanafer, Applications of porous materials and nanoparticles in improving solar desalination systems, J. Porous Media 19 (2016) 11.
[9] W. Chen, et al., Experimental investigation of SiC nanofluids for solar distillation system: stability, optical properties and thermal conductivity with saline water-based fluid, Int. J. Heat Mass Transf. 107 (2017) 264–270.
[10] O. Mahian, et al., Nanofluids effects on the evaporation rate in a solar still equipped with a heat exchanger, Nano Energy 36 (2017) 134–155.
[11] D.D.W. Rufuss, et al., Effects of nanoparticle-enhanced phase change material (NPCM) on solar still productivity, J. Clean. Prod. 192 (2018) 9–29.
[12] S. Nazari, H. Safarzadeh, M. Bahiraei, Performance improvement of a single slope solar still by employing thermoelectric cooling channel and copper oxide nanofluid: an experimental study, J. Clean. Prod. 208 (2019) 1041–1052.
[13] S. Sharshir, et al., The effects of flake graphite nanoparticles, phase change material, and film cooling on the solar still performance, Appl. Energy 191 (2017) 358–366.
[14] W. Chen, et al., Application of recoverable carbon nanotube nanofluids in solar desalination system: an experimental investigation, Desalination 451 (2019) 92–101.
[15] A. Madani, G. Zaki, Yield of solar stills with porous basins, Appl. Energy 52 (2–3) (1995) 273–281.

[16] M.M. Jalili, et al., Investigating the variations in properties of 2-pack polyurethane clear coat through separate incorporation of hydrophilic and hydrophobic nano-silica, Prog. Org. Coat. 59 (1) (2007) 81–87.
[17] E. Darezereshki, et al., Single-step synthesis of activated carbon/γ-Fe_2O_3 nano-composite at room temperature, Mater. Sci. Semicond. Process. 16 (1) (2013) 221–225.
[18] B. Wiley, Y. Sun, Y. Xia, Polyol synthesis of silver nanostructures: control of product morphology with Fe (II) or Fe (III) species, Langmuir 21 (18) (2005) 8077–8080.
[19] F. Iskandar, L. Gradon, K. Okuyama, Control of the morphology of nanostructured particles prepared by the spray drying of a nanoparticle sol, J. Colloid Interface Sci. 265 (2) (2003) 296–303.
[20] A.B. Vakylabad, et al., A procedure for processing of pregnant leach solution (PLS) produced from a chalcopyrite-ore bio-heap: CuO Nano-powder fabrication, Hydrometallurgy 163 (2016) 24–32.
[21] G. Tian, et al., 3D hierarchical flower-like TiO_2 nanostructure: morphology control and its photocatalytic property, CrystEngComm 13 (8) (2011) 2994–3000.
[22] E. Darezereshki, et al., Innovative impregnation process for production of γ-Fe_2O_3–activated carbon nanocomposite, Mater. Sci. Semicond. Process. 27 (2014) 56–62.
[23] E. Darezereshki, et al., Direct thermal decomposition synthesis and characterization of hematite (α-Fe_2O_3) nanoparticles, Mater. Sci. Semicond. Process. 15 (1) (2012) 91–97.
[24] K.J. Iversen, M.J. Spencer, Effect of ZnO nanostructure morphology on the sensing of H_2S gas, J. Phys. Chem. C 117 (49) (2013) 26106–26118.
[25] I.E. Stewart, M.J. Kim, B.J. Wiley, Effect of morphology on the electrical resistivity of silver nanostructure films, ACS Appl. Mater. Interfaces 9 (2) (2017) 1870–1876.
[26] A. Kaushal, Solar stills: a review, Renew. Sustain. Energy Rev. 14 (1) (2010) 446–453.
[27] V. Velmurugan, K. Srithar, Performance analysis of solar stills based on various factors affecting the productivity—a review, Renew. Sustain. Energy Rev. 15 (2) (2011) 1294–1304.
[28] P. Khalilmoghadam, A. Rajabi-Ghahnavieh, M.B. Shafii, A novel energy storage system for latent heat recovery in solar still using phase change material and pulsating heat pipe, Renewable Energy 163 (2021) 2115–2127.
[29] Z. Xu, et al., Ultrahigh-efficiency desalination via a thermally-localized multistage solar still, Energy Environ. Sci. 13 (3) (2020) 830–839.
[30] A.D., Phan, et al., Confinement effects on the solar thermal heating process of TiN nanoparticle solutions, Phys. Chem. Chem. Phys. 21 (36) (2019) 19915–19920.
[31] A. Iqbal, et al., Evaluation of the nanofluid-assisted desalination through solar stills in the last decade, J. Environ. Manag. 277 (2021) 111415.
[32] M. Gaur, G. Tiwari, Optimization of number of collectors for integrated PV/T hybrid active solar still, Appl. Energy 87 (5) (2010) 1763–1772.
[33] S. Sharshir, et al., Enhancing the solar still performance using nanofluids and glass cover cooling: experimental study, Appl. Therm. Eng. 113 (2017) 684–693.
[34] P.R. Prasad, et al., Energy efficient solar water still. I, nt. J. ChemTech Res. 3 (4) (2011) 1781–1787.
[35] V. Belessiotis, S. Kalogirou, E. Delyannis, Thermal Solar Desalination: Methods and Systems, Elsevier, 2016.
[36] K. Selvaraj, A. Natarajan, Factors influencing the performance and productivity of solar stills-a review, Desalination 435 (2018) 181–187.
[37] C.Y. Tso, C.Y. Chao, Study of enthalpy of evaporation, saturated vapor pressure and evaporation rate of aqueous nanofluids, Int. J. Heat Mass Transf. 84 (2015) 931–941.
[38] Z. Huang, et al., Hydrophobically modified nanoparticle suspensions to enhance water evaporation rate, Appl. Phys. Lett. 109 (16) (2016) 161602.
[39] A. Kabeel, Z. Omara, F. Essa, Enhancement of modified solar still integrated with external condenser using nanofluids: an experimental approach, Energy Convers. Manag. 78 (2014) 493–498.
[40] L. Sahota, G. Tiwari, Effect of Al_2O_3 nanoparticles on the performance of passive double slope solar still, Solar Energy 130 (2016) 260–272.

[41] A. Kumar, et al., A review on exergy analysis of solar parabolic collectors, Solar Energy 197 (2020) 411–432.
[42] S. Shoeibi, et al., Application of simultaneous thermoelectric cooling and heating to improve the performance of a solar still: an experimental study and exergy analysis, Applied Energy 263 (2020) 114581.
[43] S. Xia, L. Chen, F. Sun, Effects of mass transfer laws on finite time exergy, J. Energy Inst. 83 (4) (2010) 210–216.
[44] G. Tiwari, V. Dimri, A. Chel, Parametric study of an active and passive solar distillation system: energy and exergy analysis, Desalination 242 (1–3) (2009) 1–18.
[45] S.W. Sharshir, et al., Energy and exergy analysis of solar stills with micro/nano particles: a comparative study, Energy Convers. Manag. 177 (2018) 363–375.
[46] A. Ahsan, T. Fukuhara, Mass and heat transfer model of tubular solar still, Solar Energy 84 (7) (2010) 1147–1156.
[47] M. Phadatare, S. Verma, Influence of water depth on internal heat and mass transfer in a plastic solar still, Desalination 217 (1–3) (2007) 267–275.
[48] A.K. Tiwari, G. Tiwari, Effect of water depths on heat and mass transfer in a passive solar still: in summer climatic condition, Desalination 195 (1–3) (2006) 78–94.
[49] F.P. Incropera, et al., Fundamentals of Heat and Mass Transfer, Wiley, 2007.
[50] T.L. Bergman, et al., Fundamentals of Heat and Mass Transfer, Wiley, 2011.
[51] H. Kojima, J. Sawai, H. Suzuki, Effect of pressure on volumetric mass transfer coefficient and gas holdup in bubble column, Chem. Eng. Sci. 52 (21–22) (1997) 4111–4116.
[52] S.M. Parsa, et al., Experimental investigation at a summit above 13,000 ft on active solar still water purification powered by photovoltaic: a comparative study, Desalination 476 (2020) 114146.
[53] O. Bait, M. Si–Ameur, Enhanced heat and mass transfer in solar stills using nanofluids: a review, Solar Energy 170 (2018) 694–722.
[54] X. Feng, D.W. Johnson, Mass transfer in SiO_2 nanofluids: a case against purported nanoparticle convection effects, Int. J. Heat Mass Transf. 55 (13–14) (2012) 3447–3453.
[55] B. Chen, C. Liu, Pressure drop and mass transfer coefficients in concurrent screen plate bubble column, Chem. Eng. Commun. 100 (1) (1991) 113–134.
[56] S. Rashidi, et al., Volume of fluid model to simulate the nanofluid flow and entropy generation in a single slope solar still, Renewable Energy 115 (2018) 400–410.
[57] M. Bahiraei, et al., Using neural network optimized by imperialist competition method and genetic algorithm to predict water productivity of a nanofluid-based solar still equipped with thermoelectric modules, Powder Technol 366 (2020) 571–586.
[58] Y.A. Cengel, S. Klein, W. Beckman, Heat Transfer: A Practical Approach, Vol. 141, WBC McGraw-Hill, Boston, 1998.
[59] S. Kumar, G. Tiwari, H. Singh, Annual performance of an active solar distillation system, Desalination 127 (1) (2000) 79–88.
[60] Einstein, A., Investigations on the Theory of the Brownian Movement. 1956: Courier Corporation.
[61] A.B. Vakylabad, Treatment of highly concentrated formaldehyde effluent using adsorption and ultrasonic dissociation on mesoporous copper iodide (CuI) nano-powder, J. Environ. Manag. 285 (2021) 112085.
[62] K. Voropoulos, E. Mathioulakis, V. Belessiotis, Experimental investigation of a solar still coupled with solar collectors, Desalination 138 (1–3) (2001) 103–110.
[63] K. Voropoulos, E. Mathioulakis, V. Belessiotis, A hybrid solar desalination and water heating system, Desalination 164 (2) (2004) 189–195.
[64] Z. Omara, M.A. Eltawil, E.A. ElNashar, A new hybrid desalination system using wicks/solar still and evacuated solar water heater, Desalination 325 (2013) 56–64.
[65] V. Velmurugan, et al., Integrated performance of stepped and single basin solar stills with mini solar pond, Desalination 249 (3) (2009) 902–909.

[66] V. Velmurugan, et al., Performance analysis in stepped solar still for effluent desalination, Energy 34 (9) (2009) 1179–1186.
[67] A. Kabeel, et al., Theoretical and experimental parametric study of modified stepped solar still, Desalination 289 (2012) 12–20.
[68] M. Sakthivel, S. Shanmugasundaram, Effect of energy storage medium (black granite gravel) on the performance of a solar still, Int. J. Energy Res. 32 (1) (2008) 68–82.
[69] A. El-Sebaii, et al., Thermal performance of a single basin solar still with PCM as a storage medium, Appl. Energy 86 (7–8) (2009) 1187–1195.
[70] H. Tanaka, Y. Nakatake, Theoretical analysis of a basin type solar still with internal and external reflectors, Desalination 197 (1–3) (2006) 205–216.
[71] H. Tanaka, Experimental study of a basin type solar still with internal and external reflectors in winter, Desalination 249 (1) (2009) 130–134.
[72] S. Abdallah, O. Badran, Sun tracking system for productivity enhancement of solar still, Desalination 220 (1–3) (2008) 669–676.
[73] S.A. Kalogirou, Design and construction of a one-axis sun-tracking system, Solar Energy 57 (6) (1996) 465–469.
[74] A.-J.N. Khalifa, S.S. Al-Mutawalli, Effect of two-axis sun tracking on the performance of compound parabolic concentrators, Energy Convers. Manag. 39 (10) (1998) 1073–1079.
[75] A.J.N. Khalifa, A.M. Hamood, On the verification of the effect of water depth on the performance of basin type solar stills, Solar Energy 83 (8) (2009) 1312–1321.
[76] A. Al-Karaghouli, W. Alnaser, Performances of single and double basin solar-stills, Applied Energy 78 (3) (2004) 347–354.
[77] H. Al-Hinai, M. Al-Nassri, B. Jubran, Effect of climatic, design and operational parameters on the yield of a simple solar still, Energy Convers. Manag. 43 (13) (2002) 1639–1650.
[78] D. Dutt, et al., Performance of a double-basin solar still in the presence of dye, Appl. Energy 32 (3) (1989) 207–223.
[79] S. Aboul-Enein, A. El-Sebaii, E. El-Bialy, Investigation of a single-basin solar still with deep basins, Renew. Energ. 14 (1–4) (1998) 299–305.
[80] A. Nafey, M. Mohamad, M. Sharaf, Enhancement of solar water distillation process by surfactant additives, Desalination 220 (1–3) (2008) 514–523.
[81] M.K. Gnanadason, et al., Effect of nanofluids in a vacuum single basin solar still, Int. J. Adv. Eng. Res. Stud. 12 (2011) 171–177.
[82] P. Tsilingiris, The influence of binary mixture thermophysical properties in the analysis of heat and mass transfer processes in solar distillation systems, Solar Energy 81 (12) (2007) 1482–1491.
[83] A.F. Muftah, et al., Factors affecting basin type solar still productivity: a detailed review, Renew. Sustain. Energy Rev. 32 (2014) 430–447.
[84] H.E. Fath, et al., Thermal-economic analysis and comparison between pyramid-shaped and single-slope solar still configurations, Desalination 159 (1) (2003) 69–79.
[85] H. Panchal, et al., Economic and exergy investigation of triangular pyramid solar still integrated to inclined solar still with baffles, Int. J. Ambient Energy 40 (6) (2019) 571–576.
[86] R. Sathyamurthy, et al., A review of integrating solar collectors to solar still, Renew. Sustain. Energy Rev. 77 (2017) 1069–1097.
[87] F. Alshammari, M. Elashmawy, M.M. Ahmed, Cleaner production of freshwater using multi-effect tubular solar still, J. Clean. Prod. 281 (2021) 125301.
[88] K. Elmaadawy, et al., Performance improvement of double slope solar still via combinations of low cost materials integrated with glass cooling, Desalination 500 (2021) 114856.

CHAPTER

Application of nanofluids in drug delivery and disease treatment

19

Yasaman Enjavi, Mohammad Amin Sedghamiz and Mohammad Reza Rahimpour
Department of Chemical Engineering, Shiraz University, Shiraz, Iran

19.1 Introduction

To have a high quality of different characteristics like heat transfer, lubrication, drug delivery, and enhanced oil recovery, nanofluid—a suspension of nanoparticles having one dimension not more than 100 nm at least—which exhibits higher thermal, rheological, and wettability properties can be utilized [1].

Nanofluids can be categorized as one of the branches of nanotechnology. At the moment, there are a lot of research works that are being implemented to find out the advantages of nanofluids in biomedical applications. Heightening the effectiveness of delivery to a specific location is the primary goal of nanofluids in order to enhance the healing index of blends and diminish aggregation in body locations to prevent toxicity [2–4]. Since blocking micrometric blood vessels is not possible by nanofluids, this feature allows nanofluid to operate well for drug delivery [2].

19.2 Nanofluids

Nanofluids are known as a permanent suspension of particles with a size of less than 100 nm in a base fluid like water. To assess the parameters that influence the capability of drug delivery, several research works have been carried out to set up various arithmetical models [2,5,6].

19.2.1 Different varieties of nanofluids

First of all, it is worthy of note that a nanofluid is a fluid having nanoscale-distributed particles. There are many compounds of nanoparticles, including single elements (e.g., Cu, Fe, and Ag), single oxides (e.g., CuO, Cu_2O, Al_2O_3, and TiO_2), multielement oxides (e.g., $CuZnFe_4O_4$, $NiFe_2O_4$, and $ZnFe_2O_4$), alloys (e.g., Cu-Zn, Fe-Ni, and Ag-Cu), metal carbides (e.g., SiC, B4C, and ZrC), metal nitrides (e.g., SiN, TiN, and AlN), and carbon materials (e.g., graphite, carbon nanotubes, and diamond) blended in different base fluids, such as water, ethanol, EG, oil, and refrigerants; this is the way that nanofluids are formed. Single material and hybrid nanofluids are the two main groups of nanofluids [7–10].

Nanofluids and Mass Transfer. DOI: https://doi.org/10.1016/B978-0-12-823996-4.00012-4

19.2.1.1 Single material nanofluids

Choi [11] was the first researcher to propose single material nanofluids. Today, this type of nanofluid is the standard form of nanofluids employed. A single type of nanoparticle, as mentioned earlier, is utilized to create the suspension by means of various preparation approaches. Several researchers expressed that the single material nanofluids have a great quality practically, thanks to having promising thermophysical properties in comparison with their base fluid [12–16].

19.2.1.2 Hybrid nanofluid

To achieve this advanced type of nanofluids, mixing of different varieties of nanoparticles suspended in a base fluid is required [7].

19.2.2 Preparation of nanofluids

The preparation method has a significant influence on the stability of nanofluids [17]. Since two crucial characteristics of nanoparticles, that is, purity and particle size, are clarified by the preparation approach, it plays a key role in determining the nanofluids' effectiveness [2]. Two methods of single- and two-step are employed to produce nanofluids. In the former, simultaneous preparation and dispersion of nanomaterials in the liquid state occur. In the latter, the production of nanoparticles is the first step. At that point, to have a nanosuspension, the dispersion process is carried out in a base fluid via various mixing procedures [17].

19.2.2.1 Single-step method

As mentioned earlier, in this method, the production and dispersion of nanoparticles in the principal liquid are concurrent. In general, the preparation of nanofluids will be performed when a chemical reaction occurs. There are several procedures for nanofluid preparation; some of them are listed in the following: microwave irradiation procedure, vapor deposition (VP) approaches, thermal decomposition, grafting, submerged arc nanoparticle synthesis system (SANSS), and phase transfer procedures. It should be mentioned that the preparation method of some nanofluids is determined by the characteristics of nanoparticles and base fluid, interactions that occur between them, and the product application prerequisites. The difference between single- and two-step methods is the extremely stable dispersion of nanoparticles in the first method compared to the second one. Moreover, there is no substantial agglomeration in the single-step method-prepared nanofluids. To acquire improved stability, more extended dispersion times of nanoparticles in liquids are obligatory.

The single-step method has many shortcomings regardless of its superior stability. The applications of single-step prepared nanofluids are restricted to the low vapor pressure processes. The other difficult situation belonging to the single-step method is related to the control of the nanoparticles' size and structures. Furthermore, the complete conversion of reactants into products has never happened, and the presence of a high percentage of various impurities is possible. Another sticky situation in this method is its highly nonaffordable preparation process compared to the two-step method [17,18].

19.2.2.2 Two-step method

Recently, nanofluids have been broadly prepared by the two-step method. In this technique, physical or chemical synthesis-acquired dry nanopowders are dispersed in a liquid media to prepare nanofluids. Using this process, we can commercially obtain high purity and different controlled size nanoparticles on an industrial scale. Usually, the dispersion of nanoparticles in a base fluid is performed with the help of shear mixing. Another difference between single- and two-step methods is the agglomeration rate, which is higher in the second one than the first.

A variety of concentrations concerning nanofluids is obtainable in the two-step preparation process, whereas there are no changes in the concentration of nanofluid prepared by the single-step method. In this procedure, two phenomena of dilution and evaporation can commonly change nanoparticles' concentration in the host fluid. To assess the stability of nanosuspensions, a characterizing study is performed before and after the preparation process.

For the duration of different processes consisting of storage, transportation, and handling, dry powders' initial size probably changes up to microns owing to the governing attractive interactions between two particles. Numerous research has been carried out into the preparation process of various nanofluids utilizing a two-step method. The two-step method creates a condition in which the preparation process of nanofluids is exceptionally cost-effective. In this method, the preparation occurs at the bulk scale [1,17].

19.2.3 Nanofluid stability assessment methods

Because of van der Waals attraction forces, nanomaterials tend to become agglomerated, resulting in sedimentation; hence, nanofluid stability should be considered a significant factor. Using cationic Gemini surfactant, Li and colleagues enhanced nanofluids' stability, which was made of gold and silver nanoparticles dispersed in water [2,19].

To determine the stability of nanofluid, two parameters of pH and spacer length are crucial. Higher stability is accessible by 1,3-bis(cetyltrimethylammonium) propane dibromide when we have an extensive range of nanofluid pH, and also, if higher thermal stability is required, we can employ another dibromide, 1,8-bis(cetyltrimethylammonium) octane. Researchers realized that negatively charged (2,2,6,6-tetramethylpiperidin-1-yl) oxyl-oxidized cellulose nanofibers, which are not very heavy, are excellent stabilizers to be utilized in the thermal application of water-based nanofluids [2,20].

Since surfactants heighten the wettability of nanoparticles, one can state that nanofluids' stability is enhanced by using them. Besides influencing surfactants on stability, pH is a critical parameter as well; the higher the pH value, the greater the stability of nanofluids is expected. For instance, the monovalent copper oxide with a 9.5 pH in water is more stable than alumina whose pH is equal to 8.

There are different methods to determine the stability of nanoparticles. These methods are categorized into different types, for example, light-scattering procedures involved the SEM and TEM methods [2,21], Zeta potential measurements [1,17,22], centrifugation and sedimentation method [6,17,23,24]. In these methods, the stability criteria are defined, and in each method, the criteria should be satisfied to produce the stable nanomaterial.

Other methods are also applied to ensure the stability of the nanoparticles: electron microscopy with optical microscope and energy dispersive X-ray (EDX) analysis [17], spectral analysis with

Table 19.1 Categorization of nanofluid stabilization method.

Stabilization method	Physical		Ultrasonication
			Ball milling
			Magnetic and mechanical stirrer
			High-pressure homogenizer
	Chemical	Electrostatic stabilization	Ionic surfactant
			pH adjustment
			Covalent functionalization
		Steric stabilization	Polymer
			Nonionic surfactant
		Electrostatic stabilization	Ionic polymer
			Ionic liquid

ultraviolet–visible (UV–Vis) spectroscopy [25], and finally the 3 ω method to analyze the stability of nanofluids to investigate the agglomeration and sedimentation of particles [1,7].

19.2.4 Nanofluid stabilization procedure

To increase the stability of a nanofluid suspension, you can utilize a wide range of various stabilization procedures. Stabilization procedures are divided into two categories: mechanical and chemical, as shown in Table 19.1. These two types are categorized separately into different sections, which are presented in Table 19.1.

19.3 Nanofluid-based delivery system

In this intricate and growing world of medication, one of the important issues about drug development is related to drug delivery to its active site in remedially satisfactory amounts. By development of pharmacological knowledge, numerous novel drugs having small- to nanosize molecules, such as proteins and peptides, have been discovered. However, there are many uncertainties in achieving an illness-free circumstance as the final objective in the medical industry. Of these ambiguous situations, we can refer to several footraces concerning physicochemical and molecular complexities of the "free" drugs, unavailability, and underdosing of a majority of the biological/pathological targets. Hence, drug delivery systems (DDS) are utilized to manage these difficult conditions. These systems can be expressed by a formula or an instrument simplifying the giving of a medicine to the human body. This happens simultaneously as DDSs make better their pharmacokinetic and biodispersion profiles in addition to the effectiveness and security of the entire treatment. Targeting the drugs (and DDS) consists of the development of the system uniqueness with respect to the pharmacologically related target in the human body. Targeted drug delivery systems (TDDS) include giving medicine of the DDS to the patient, DDS delivery at the target (pathological) site, active ingredients liberation in/around the target, and preventing undefined poisonousness in regular cells. Paul Ehrlich was the first person to understand the notion of the targeted drugs being a magic bullet in the early 20th century. After that, many line of attacks have been technologically advanced to realize targeting [26–28].

TDDS is a system that can be generally comprehended as operating by targeted drug delivery systems as listed below:

- Simplify the healing ingredient in order to reach the action site from the administration site. It should be mentioned that the target can be various things like organ, tissue, cell, or particular cell organelles.
- Liberation of the healing payload in its dynamic state in/around the target site which presents operational healing levels at the action site.
- Keep the drug/gene from the damaging environmental aspects, for instance, pH, enzymes, and so on.
- Prevent toxicity or insanitary interactions of the drug/gene on undefined regular cells, also simplify the giving of medicine of inferior dosages in order to attain healing/analytical advantages.

Many investigations have been accomplished concerning targeted drug delivery (TDS) and have presented numerous possibilities of implementation of the aforementioned operates:

- Navigate targeting to the action site, for example, regional applications for skin illnesses.
- Utilizing exterior incentives, for example, ultrasound.
- Chemically modify the drug in order to perfectly improve its physicochemical characteristics for the delivery consisting of a prodrug attitude of attaching a promoiety to the drug.
- Various nanocarriers including liposomes, polymeric micelles, polymeric nanoparticles, and solid lipid nanoparticles are employed; they can be also functionalized additionally through attaching targeting ligands and antibodies.

A TDDS is supremely effective when it has the succeeding features:

- The drug–conjugate/drug carrier should reach the intended action site, which can be an organ, tissue, cell, or cell organelles, having negligible undefined accumulation.
- The effect of drug/gene on the intended action site is not deactivated or changed by chemically conjugating or physically encapsulating the drug/gene with the targeting ligands or carriers. TDDS can look after the drug from environmental issues like enzymatic breaking down until it reaches the target.
- In order to reach the intended action site, the ligand or carrier activity and operation should not be deactivated or changed by chemically conjugating or physically encapsulating the drug/gene with the targeting ligands or carriers.

19.4 Targeted drug delivery

Targeted drug delivery at the action site can be performed via direct procedures such as direct injection, catheter, gene-gun, and so on. Although these systems indicate direct delivery, intrusiveness is not patient serviceable and is costly to fulfill in several circumstances. Consequently, much work, including chemical, physical, and biological alterations with or without employing carriers, have been done in order to develop TDDS [26,29,30].

Several modifications have been accomplished to make targeting drug delivery better, that is, the investigation of structure activity interactions for developing the physicochemical characteristics. There is a problem with the drugs intended for brain delivery. This is the difficulty of their penetration through the blood–brain barrier (BBB); hence, they may perhaps be made to be lipophilic further in order to improve BBB penetration. This is the reason they should have a small size. The drug pharmacokinetics improves when prodrugs are made. Small-molecule drugs are chemically changed through attaching "promoeities" rendered pharmacologically inactive and are metabolically activated in the living organism into active drugs only after reaching their intended target. Drugs can be conjugated with antibodies, peptides, aptamers, folic acid, and so on to produce targeted prodrugs [26,31].

Instead, the drugs can be combined by different carriers or systems just at the nanoscale. These carriers and systems can be liposomes, polymeric micelles, polymeric nanoparticles, polymer–drug conjugates, nanogels, carbon nanotubes, and the like. To effectively deliver the drugs or genes, using nanosystems is strongly recommended. The nanosystems are utilized because the pharmacokinetic performance of the nanocarriers loaded with drugs is determined by them, not by the drugs or genes, and thus control of the whole operation by using additional targeting is worry-free. The diameter of most nanoparticles described in this chapter is less than 300 nm. There is a dependency on drugs/drug carrier systems. The two most important targeting modes are known as passive and active targeting [32].

19.4.1 Passive (physiology-based) targeting

Passive targeting exists in the human body. Receptors are naturally targeted by hormones, neurotransmitters, growth factors, and so on at their action sites, for instance, insulin and insulin receptors. Drugs also obey this notion. Passive targeting is defined as the aggregation of drugs/drug carrier systems at the intended action site via the action of physicochemical and physiological elements [33].

Passive targeting is utilized when the existence of leaky vasculature with outsized gaps in the epithelial layers of blood vessels is detected. This occurs in some cases of infected tissues in bowel disease, inflammatory rheumatoid arthritis, and malignant tumor tissues; here, it is conceivable to passively target the administered nanocarriers of proper sizes to extravasate into the target tissue. Even though the tumor tissue has restricted lymphatic drainage and the inflammatory tissues have operational lymphatic drainage, passive targeting can still exploit the inflammatory diseases. In liver diseases, because of large openings in the liver, the accumulation of nanocarriers is perceived so that this can be utilized to target the liver. This occurrence in which the nanocarriers gather into the diseased tissues owing to moveable fenestrations and/or lymphatic drainage is designated as enhanced permeability and retention (EPR) effect [34–37].

The reticuloendothelial system (RES), consisting of macrophages and mononuclear phagocytes, releases fundamentally influenced nanocarriers. To treat contagions influencing the RES (for instance, leishmaniasis and malaria), the aforementioned information can be employed for passive targeting of the macrophages as well as lymph nodes and spleen [38].

To accumulate high dosages of nanocarriers at the target sites, several revisions, such as the attachment of polyethylene glycol (PEG), are implemented in order that nanocarriers become long-circulating, thus preventing RES and allowing enough time for accumulation [39].

The existence of interior incentives like pH difference, redox systems, and so on in the diseased tissues aids in passive targeting. Incentive-sensitive drug targeting systems will be prompted by dint of such incentives to release the drug only at the target site and conserve the typical tissues. A lot of researchers have investigated various incentives-reactive systems [40–45].

19.4.2 Active targeting

Although passive targeting has presented noteworthy results, lots of research works have been focused on active targeting to give superior control on precise drug delivery. Suitable alterations and functionalization of the drugs or drug carriers give them attraction regarding particular receptors/markers on cells, tissues, or organs. Different factors, e.g., illness, intended target organ, and more targetable constituents on the target organs/cells, must be taken into account to better attach the ingredients to the target organ. Alterations on the drugs or drug carriers include using various ligands like peptides, antibodies, sugars, lectins, and so on. As a consequence, during the giving of a medicine to the human body, the targeting moieties will make it possible for the drug/drug carriers to proficiently reach only the intended action sites and prevent undefined gatherings and connected lateral effects [26].

19.4.3 Physical targeting

Variations of physical factors in the bioenvironment are the basis of physical targeting. These activating factors are divided into two categories of endogenous (temperature, pH, and redox condition) and/or exogenous (applied magnetic field, sound, and heat). The utilized nanocarriers come to be dynamic contributors in treatment. Also, they are denoted as incentives-receptive delivery systems. It should be noted that under definite pathological conditions, there is a substantial difference between the microbioenvironment of the target site and the standard physiological condition. These differences, such as acidosis-initiated low pH, low pH in a tumor, hyperthermia-initiated high temperature, and tumor redox condition, are exploited as a stimulus to accumulate and release drugs. The exceptional advantage of such a system is that although the drug is dispersed all over the human body, its healing activity is clarified only at the target site. Although different carriers, including liposomes, micelles, hydrogels, and dendrimers, are utilized, polymeric particles are a prevalent group of actors in this approach. Likewise, triggered delivery systems, which react to exogenous incentives, consist of nanocarriers having ferromagnetic features that can be effortlessly contrived under the impact of an outwardly practical magnetic field. Hence, targeting is obtainable. This type of targeting application is restricted because of requiring a high-strength magnetic field plus blood flow rate. Anyway, researchers observed the effective performance of theranostic systems. Correspondingly, sound-reacted vehicles have been utilized for drug/gene delivery and theranostic goals. It was clear that ultrasound concentration disturbs intravascular endothelial cells, which make tiny holes to convey the drug into the target tissue, as detected in the blood–brain barrier/blood–tumor barrier. The most important series of sound-supported drug delivery vehicles comprise microbubbles, micelles, liposomes, and perfluorocarbon nanoparticles, showing

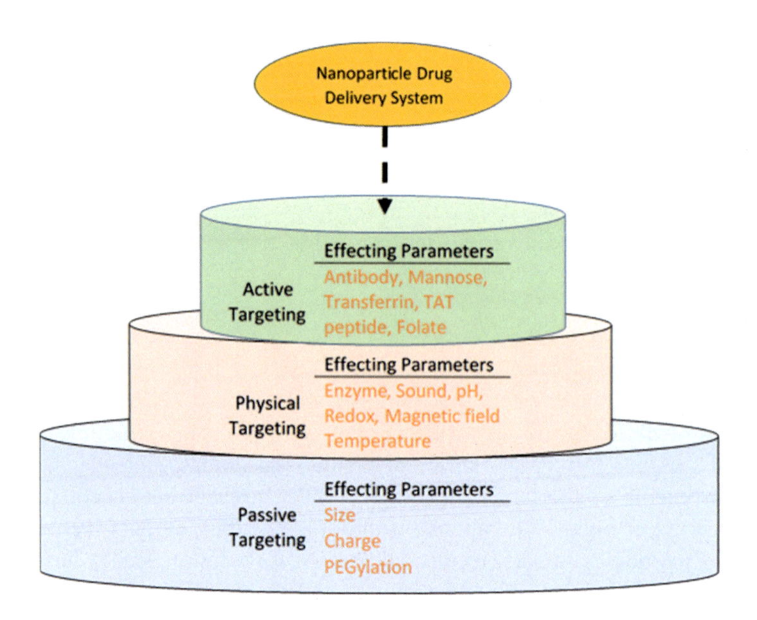

FIGURE 19.1

Different types of drug delivery systems.

remarkably improved thrombolysis and various deliveries, including transdermal delivery, delivery to solid tumors, and delivery through the blood–brain barrier. Fig. 19.1 shows the different systems for the targetting drug delivery [46].

19.5 Applications of the nanofluid-based delivery system

Although several applications of nanofluids and nanoparticles in the biomedical industry are confirmed, there are some lateral effects of the customary approaches to cancer healing. During the use of drugs or radiation, if keeping adjacent healthy tissue safe is critical, Fe-based nanoparticles should be employed as delivery vehicles. Utilizing magnets exterior to the human body allows the guiding of Fe-based nanoparticles to a tumor through the bloodstream. Nanofluids can be employed to keep surgery safe by making operative cooling near the surgical area. This heightens the survival chance of the patient and diminishes the threat of organ injury. Using magnetic nanoparticles in biofluids as delivery vehicles for drugs or radiation can present innovative procedures to treat cancer. It should be mentioned that absorbing further power by magnetic nanoparticles compared to microparticles at magnetic fields with AC leads to using magnetic ones because of the stamina of the human body. Due to a further tendency of nanoparticles for adhering to tumor cells compared to normal ones, AC magnetic field-stimulated magnetic nanoparticles are an excellent choice for cancer rehabilitation. The malfunction (concerning the repair process) and DNA damage induced by heat and radiation, in that order, result in a combined effect with respect to radiation and hyperthermia [47–49]. Different applications of nanomaterials are presented in Table 19.2.

Table 19.2 Different type of nanomaterials and applications [50].

Type of nanomaterial	Component	Application
Metallic compounds	Gold, silver, iron oxide, titanium oxide, titanium and ceramic	Pregnancy test, biosensors, drug delivery, photothermal, magnetic resonance imaging, vaccine delivery, cell separation, tissue regeneration, cancer, protein adsorption, implants
Inorganic compounds	Mesoporous compounds, for example, silica, hydroxyapatite	Drug and gene delivery, treatment of pathogen, repairing the bone defect
Carbon compounds	CNT, graphene, etc.	Drug delivery, tumors ablation
Polymeric compounds	Polymeric compounds, for example, starch, chitosan, poly amides	Drug therapy and delivery, sensing, tissue regeneration
Biological protein peptide-based compounds	Gelatin, protein from milk, legumin	Gene therapy and delivery, antimicrobial to infections, proteins delivery

19.5.1 Antibacterial activity of nanofluids

For various distinctive types of nanoparticles, the nanofluids comprising the nanoparticles with antibacterial activities or drug delivery features will reveal some pertinent characteristics. Organic antibacterial ingredients regularly have lower stability, specifically when temperatures or pressures are high. Therefore inorganic ingredients, for example, metal and metal oxides, have been of great interest in recent years since they are capable of surviving in severe process circumstances. By studying the antibacterial behavior of zinc oxide nanofluids, it is clear they have bacteriostatic activity against *Escherichia coli*. The antibacterial activity upsurges by increasing and decreasing nanoparticle concentration and particle size, respectively. At high concentrations of zinc oxide, several direct interactions between ZnO and bacteria membrane occur, which are approved by electrochemical measurements [51–53].

Zhang and coworkers investigated the inactivation productivity of a sonophotocatalytic process by ZnO nanofluids, which include different ultrasonic parameters like power density, frequency, and time. They reported a 20% increase in inactivation productivity through combining ultrasonic irradiation and the photocatalytic process in the presence of normal light. By comparing the inactivation productivity of photocatalytic, ultrasonic, and sonocatalytic processes with *E. coli*, it was realized that the inactivation productivity of sonophotocatalysis is the highest. Photocatalysis, sonocatalysis, and ultrasonic irradiation are ranked next, respectively [54].

Yadav et al. compared the antimicrobial activity of three nanofluids, which are based on trimetallic Au/Pt/Ag, monometallic Au, and bimetallic Au/Pt nanoparticles. A successive chemical reduction process-aided green microwave method was used to prepare trimetallic nanofluid. After that, it was studied with the help of several facilities consisting of UV–Vis spectroscopy, X-ray diffraction (XRD), scanning electron microscopy (SEM), and high-resolution transmission electron microscopy (HRTEM). Agar disc diffusion test was employed to analyze the antibacterial characteristics of metallic nanofluids against numerous microorganisms. Furthermore, minimum inhibitory concentration (MIC) values of metallic nanofluids were measured. By examining the effective antibacterial activity at a very low concentration of metal, it was realized that trimetallic nanofluid is a paramount factor in comparison with two other mentioned nanofluids. According to the acquired results, trimetallic nanofluids can be used in therapeutic studies as well as pharmaceutical industries [55].

A research work conducted by Umoren et al. scrutinized the antimicrobial behavior belonging to chitosan/silver nanofluids compared to Gram-positive *Bacillus licheniformis, Staphylococcus haemolyticus, Bacillus cereus*, and *Micrococcus luteus*, and Gram-negative *Pseudomonas aeruginosa, Pseudomonas citronellolis*, and *Escherichia coli* bacteria. The polydispersed nanoparticles in the chitosan matrix have great stability. Values of +46 and +56 mV were reported as the zeta potential belonging to the silver nanoparticles in DPLE- and HEL-mediated composites, respectively. Based on the FTIR analysis, the involvement of free carboxylates in the plant biomaterial in the stabilization process is sensible. Since HEL has a robust capability as a reducing agent compared to DPLE, its produced nanoparticles have a smaller size (about 8.0–36 nm) in contrast with DPLE ones (about 10–43 nm). Both DPLE- and HEL-mediated composites successfully hinder the evolution of the desired bacteria. However, the latter has a more powerful effect, which is due to the smaller size of nanoparticles. According to the abovementioned results, HEL extract can be employed in the green manufacture of prospective antimicrobial chitosan/silver nanofluids for biomedical and packing applications [56].

Khan et al. comprehensively synthesized Guar gum/gelatin/silver nanocomposite (GG/Gl/Ag-N-composite) through an uncomplicated in situ technique via maltose sugar reduction with a variable concentration of $AgNO_3$ solution. Tetracycline was used in a well diffusion procedure to analyze the antibacterial behavior of synthesized composites. The GG/Gl/Ag-N-composites were determined to show exceptional antibacterial characteristic against both Gram-positive *Staphylococcus aureus* and Gram-negative *Escherichia coli* and *Pseudomonas aeruginosa* [57].

19.5.2 Applications in cancer therapy

More than 100 diseases fall under the category of cancer. Cancer leads to irrepressibly growing undesirable cells, so it can damage body tissues. Graphene-based nanosensors can be used to examine and identify cancer from symptoms. The aforementioned biosensors are extremely susceptible and precise for the mission. Based on the literature, gold nanoparticles can be utilized to treat cancer. Moreover, another therapy for cancer is magnetite fluid hyperthermia. It is worthy of note that biocompatible magnetite nanoparticles are utilized to play the role of a heat mediator because they have fewer lateral effects and better productivity compared to the other nanoparticles. Besides, magnetic particles, which are based on ferrite, are employed to treat hyperthermia. Because of the high similarity of iron oxide ferrofluids and hemoglobin, they are rarely utilized. These ferrofluids adhere to other metals like zinc, nickel, and manganese. Using SWCNT as a thermal intensifier is recommended. Also, some biocompatible molecules, for example, iron oxide, are employed to be openly injected into the desired tissue [58–64].

New information applicable to breast cancer therapy has been recently obtained by scientists. This is related to the unique formulations of nanocomposites initiated from ZnO nanoparticles. Using two processes of sol–gel hydrothermal and fast quenching- and surface modification-involved coprecipitation, Abdolmohammadi and colleagues synthesized two forms of ZnO nanomaterials. The cytotoxic effects on the development of the breast cancer cell lines MCF-7 were assessed through 3-(4,5-dimethylthiazol-2-yl)-2,5-diphenyl tetrazolium bromide (MTT) testing.

By heightening the concentration of ZnO nanofluid at 48 and 72 h of therapy, they observed decreasing cell viability of the breast cancer cell line MCF-7. The IC_{50} value of MCF-7 cells after 72 h of treatment with the first product ZnO (a) and second one ZnO (c) were measured as 51.02 and 48.63 μg/mL, respectively ($P < .05$) [65].

In cancer therapy, injecting magnetic nanoparticles into the blood vessels together with the insertion of a magnet closer to the tumor is one of the efficacious techniques. The dynamics of these nanoparticles can occur under the activity of the peristaltic waves produced on the walls of the irregular narrowing channel. Examining the nanofluid flow under such activity may be extremely helpful to treat cancer tissues. In selected investigations, a recently defined peristaltic conveyance of Carreau nanofluids under the influence of a magnetic field in the narrowing asymmetric channel is systematically scrutinized. If the wavelength is long and the Reynolds number is low, then we will have precise terminologies for temperature and nanoparticle fraction fields, axial velocity, stream function, pressure gradient, and shear stress. In conclusion, different developing parameters on the physical quantities are of interest. Increasing the Hartmann number and thermophoresis parameter leads to enhancing pressure rise [66–68].

Gold nanoparticles (GNPs) are significant in biomedicine and are encouraging agents for therapy. These types of nanoparticles have been used in numerous applications as drug carriers, photovoltaic and contrast agents, and radio infections. Furthermore, because gold nanoparticles are capable of encapsulating huge quantities of healing molecules, they are chosen as operational drug transfer compounds and delivery drugs. Due to a lot of characteristics of gold nanoparticles, they are at the center of interest for employment in cancer treatment. They can extensively diffuse all over the body owing to their small size. The most influential point is that they can bind several proteins and drugs and can be aggressively directed to cancer cells to treat them, and destroy bacteria. Tumor selective photothermal treatment is a favorable option when the nanoparticle used is made of gold because the high atomic number of gold results in generating the heat which is utilized in the aforementioned therapy. Huang and Mostafa El-Sayed developed the optical features and applications of gold nanoparticles in identifying cancer from its symptoms and photothermal rehabilitation. GNPs have also been at the center of interest thanks to their exclusive and strong plasmon resonance in the observable range as well as their applications in biomedical. In summary, a few researchers have studied GNPs flow through two coaxial tubes. Hatami and Hamzehnezhad, together with coworkers, analyzed the third-grade non-Newtonian fluid-carrying GNPs in a porous and hollow vessel via various systematic approaches. Articles about the flow with nanoparticles and nanofluids include [60,69–85].

There is a dependency between biomedical applications and the characteristics of magnetic nanoparticles (MNPs); their characteristics are influenced by the type of applied MNPs, synthesis techniques, the interaction between particles, distribution of particle size, and particle size and morphology of NPs. Intrinsically, an appropriate synthesis process should be chosen to achieve particular performance (established upon definite biomedical applications). There are two categories—in vitro (outside the body) and in vivo (inside the body)—of general biomedical applications with MNPs. Diagnostic processes like separation/selection, magnetic relaxometry, and magnetic resonance imaging (MRI) are the applications of the first category.

The other diagnostic processes like nuclear magnetic resonance imaging and therapeutic applications (for instance, drug delivery and magnetic hyperthermia) are the applications of the second category [86].

Hyperthermia is a healing technique to treat cancer; the expression "hyperthermia" is initiated from two Greek words, "hyper" and "therme," that is, "rise" and "heat," for the reason that this situation is related to heightening body temperature. Busch and Coley perceived that a sarcoma vanished after a very high fever; this discovery showed the reaction of protected systems with respect to bacterial contamination. On the basis of this research work, high temperatures can kill cancer cells so that their growth can be stopped at a temperature range of 41°C–46°C or lower than 47°C for at least 20–60 min (differs in the collected works). Although this method has given rise to predictions and noteworthy developments, it may also result in opposition and disappointment on account of detrimental effects, including blisters, burns, and pain, which are quickly enhanced in healthy cells. Consequently, hyperthermia is locally employed, rather than exposing the whole body (WB) to high temperatures, to defeat confrontational lateral effects and heighten the productivity of the therapy [87–91].

Hyperthermia, due to rudimentary difficulties connected to local hyperthermia, has not been very operational to treat severe cancers. These problems can refer to heterogeneous temperature distribution in tumor mass and powerlessness to stop overheating at the deep-seated tumor area. For that reason, an innovative approach should be proposed to handle these dangerous concerns. Incidentally, researchers have developed nanotechnology providing a secure, stress-free, and operative therapy methodology. Using MNPs with high Ms shows that heat will be created to improve the hyperthermia productivity. In principle, MNPs can be injected locally or through the intravascular area inside the neighborhood of exterior ACMF. This practice results in concentration-created heat on the influenced cells, and the process is named magnetic hyperthermia or magnetic nanofluid hyperthermia therapy (MNFHT). In this procedure, magnetic fluids are utilized as stable colloidal suspensions of NPs in liquid media like water or hydrocarbon fluids [92,93].

19.6 Conclusion and future trends

Medical and biomedical applications of nanofluids are categorized among the wide variety of nanofluids applications. In this study, different procedures of nanofluid preparation, stability measurements, and applications in drug delivery systems were investigated. Particle size, shape, and requirement of the drug are strategic parameters of a well-organized healing effect. To attain this, requires the choice of the appropriate procedures of nanofluid preparation and nanoparticle encapsulation. Peristaltic flow is a significant mechanism to carry nanofluid to the target site. There is an opportunity to perform further comprehensive numeric and experimentation studies in this regard. Nanofluid stability is a crucial parameter if effective drug delivery is desired. There are lots of approaches to assess drug delivery. Nanofluids have a wide variety of applications in the therapeutic field (e.g., in PCR) to be utilized as antibacterial agents. Moreover, magnetite and gold nanoparticles can be employed as a heat mediator. However, based on the results and applications which are survived in different published articles, different suggestions are recommended. In the application of drug delivery, the main aim is the prevention of aggregation and acceptable stability. Different types of magnetic nanoparticles are used for cancer therapy in which the magnetic nanoparticles are able to heat the tumor and kill the cells. According to a small number of investigations on the applications of different nanofluids, finding new applications of nanofluids in biomedical and medical issues as well as increasing the in vivo and in vitro testing trials on the nanofluids agents are necessary.

References

[1] S. Chakraborty, P.K. Panigrahi, Stability of nanofluid: a review, Appl. Therm. Eng. (2020) 115259.

[2] P. Thakur, et al., Nanofluids-based delivery system, encapsulation of nanoparticles for stability to make stable nanofluids, Encapsulation Active Molecules Their Delivery System, Elsevier, Amsterdam, 2020, p. 141.

[3] D.K. Mishra, N. Balekar, P.K. Mishra, Nanoengineered strategies for siRNA delivery: from target assessment to cancer therapeutic efficacy, Drug. Delivery Transl. Res. 7 (2) (2017) 346–358.

[4] M. Talekar, T.-H. Tran, M. Amiji, Translational nano-medicines: targeted therapeutic delivery for cancer and inflammatory diseases, AAPS J. 17 (4) (2015) 813–827.

[5] D. Tripathi, O.A. Bég, A study on peristaltic flow of nanofluids: application in drug delivery systems, Int. J. Heat Mass Transf. 70 (2014) 61–70.

[6] F. Abbasi, T. Hayat, A. Alsaedi, Peristaltic transport of magneto-nanoparticles submerged in water: model for drug delivery system, Phys. E: Low-dimensional Syst. Nanostructures 68 (2015) 123–132.

[7] N. Ali, J.A. Teixeira, A. Addali, A review on nanofluids: fabrication, stability, and thermophysical properties, J. Nanomaterials (2018) 2018.

[8] M. Gupta, et al., A review on thermophysical properties of nanofluids and heat transfer applications, Renew. Sustain. Energy Rev. 74 (2017) 638–670.

[9] N.K. Gupta, A.K. Tiwari, S.K. Ghosh, Heat transfer mechanisms in heat pipes using nanofluids—a review, Exp. Therm. Fluid Sci. 90 (2018) 84–100.

[10] L.S. Sundar, et al., Hybrid nanofluids preparation, thermal properties, heat transfer and friction factor—a review, Renew. Sustain. Energy Rev. 68 (2017) 185–198.

[11] S.U. Choi, J.A. Eastman, *Enhancing thermal conductivity of fluids with nanoparticles*, No. ANL/MSD/CP-84938; CONF-951135-29. Argonne National Lab., IL (United States) (1995).

[12] M.M. Tawfik, Experimental studies of nanofluid thermal conductivity enhancement and applications: a review, Renew. Sustain. Energy Rev. 75 (2017) 1239–1253.

[13] M. Modak, S.S. Chougule, S.K. Sahu, An experimental investigation on heat transfer characteristics of hot surface by using CuO—water nanofluids in circular jet impingement cooling, J. Heat Transf. 140 (1) (2018).

[14] L. Yang, K. Du, A comprehensive review on heat transfer characteristics of TiO_2 nanofluids, Int. J. Heat Mass Transf. 108 (2017) 11–31.

[15] W. Azmi, et al., Potential of nanorefrigerant and nanolubricant on energy saving in refrigeration system—a review, Renew. Sustain. Energy Rev. 69 (2017) 415–428.

[16] K. Reddy, N.R. Kamnapure, S. Srivastava, Nanofluid and nanocomposite applications in solar energy conversion systems for performance enhancement: a review, Int. J. Low-Carbon Technol. 12 (1) (2017) 1–23.

[17] V.S. Korada, N.H.B. Hamid, Engineering Applications of Nanotechnology: From Energy to Drug Delivery, Springer, 2017.

[18] W. Yu, H. Xie, L.-H. Liu, A review on nanofluids: preparation, stability mechanisms, and applications, J. Nanomaterials 2012 (711) (2011) 128.

[19] D. Li, et al., Stability properties of water-based gold and silver nanofluids stabilized by cationic gemini surfactants, J. Taiwan. Inst. Chem. Eng. 97 (2019) 458–465.

[20] W.-K. Hwang, et al., Enhancement of nanofluid stability and critical heat flux in pool boiling with nanocellulose, Carbohydr. Polym. 213 (2019) 393–402.

[21] F. Farahmandghavi, M. Imani, F. Hajiesmaeelian, Silicone matrices loaded with levonorgestrel particles: impact of the particle size on drug release, J. Drug. Delivery Sci. Technol. 49 (2019) 132–142.

[22] S. Bhattacharjee, DLS and zeta potential—what they are and what they are not? J. Controlled Rel. 235 (2016) 337–351.

[23] S.U. Ilyas, R. Pendyala, N. Marneni, Settling characteristics of alumina nanoparticles in ethanol-water mixtures. in Applied Mechanics and Materials, Trans Tech Publ, 2013.
[24] P.C. Hiemenz, P.C. Hiemenz, Principles of Colloid and Surface Chemistry, vol. 188, M. Dekker, New York, 1986.
[25] B. LotfizadehDehkordi, A. Ghadimi, H.S. Metselaar, Box–Behnken experimental design for investigation of stability and thermal conductivity of TiO_2 nanofluids, J. Nanopart. Res. 15 (1) (2013) 1369.
[26] R.S.R. Murthy, Polymeric micelles in targeted drug delivery, Targeted Drug Delivery: Concepts and Design, Springer, 2015, pp. 501–541.
[27] K. Strebhardt, A. Ullrich, Paul Ehrlich's magic bullet concept: 100 years of progress, Nat. Rev. Cancer 8 (6) (2008) 473–480.
[28] R. Langer, Drugs on target, Science 293 (5527) (2001) 58–59.
[29] C. Kleinstreuer, et al., A new catheter for tumor targeting with radioactive microspheres in representative hepatic artery systems. Part I: impact of catheter presence on local blood flow and microsphere delivery, J. Biomech. Eng. 134 (5) (2012).
[30] A. Dunaevsky, The gene-gun approach for transfection and labeling of cells in brain slices, Neural Development, Springer, 2013, pp. 111–118.
[31] H.-K. Han, G.L. Amidon, Targeted prodrug design to optimize drug delivery, AAPS Pharmsci 2 (1) (2000) 48–58.
[32] V.P. Torchilin, Nanoparticulates as Drug Carriers, Imperial College Press, 2006.
[33] M.C. Garnett, Targeted drug conjugates: principles and progress, Adv. Drug. Delivery Rev. 53 (2) (2001) 171–216.
[34] B.J. Crielaard, et al., Drug targeting systems for inflammatory disease: one for all, all for one, J. Controlled Rel. 161 (2) (2012) 225–234.
[35] T. Stylianopoulos, EPR-effect: utilizing size-dependent nanoparticle delivery to solid tumors, Therapeutic Delivery 4 (4) (2013) 421–423.
[36] Y. Matsumura, H. Maeda, A new concept for macromolecular therapeutics in cancer chemotherapy: mechanism of tumoritropic accumulation of proteins and the antitumor agent smancs, Cancer Res. 46 (12 Part 1) (1986) 6387–6392.
[37] V. Torchilin, Tumor delivery of macromolecular drugs based on the EPR effect, Adv. Drug. Delivery Rev. 63 (3) (2011) 131–135.
[38] N.S. Santos-Magalhães, V.C.F. Mosqueira, Nanotechnology applied to the treatment of malaria, Adv. Drug. Delivery Rev. 62 (4-5) (2010) 560–575.
[39] V.P. Torchilin, Targeted pharmaceutical nanocarriers for cancer therapy and imaging, AAPS J. 9 (2) (2007) E128–E147.
[40] X. Zhang, Y. Lin, R.J. Gillies, Tumor pH and its measurement, J. Nucl. Med. 51 (8) (2010) 1167–1170.
[41] S. Singh, A.R. Khan, A.K. Gupta, Role of glutathione in cancer pathophysiology and therapeutic interventions, J. Exp. Ther. Oncol. 9 (4) (2012) 303–316.
[42] R.R. Sawant, et al., Polyethyleneimine-lipid conjugate-based pH-sensitive micellar carrier for gene delivery, Biomaterials 33 (15) (2012) 3942–3951.
[43] H. Wu, L. Zhu, V.P. Torchilin, pH-sensitive poly (histidine)-PEG/DSPE-PEG co-polymer micelles for cytosolic drug delivery, Biomaterials 34 (4) (2013) 1213–1222.
[44] J. Liu, et al., Redox-responsive polyphosphate nanosized assemblies: a smart drug delivery platform for cancer therapy, Biomacromolecules 12 (6) (2011) 2407–2415.
[45] Y.-J. Pan, et al., Redox/pH dual stimuli-responsive biodegradable nanohydrogels with varying responses to dithiothreitol and glutathione for controlled drug release, Biomaterials 33 (27) (2012) 6570–6579.
[46] R.K. Kesharwani, Two-Volume Set Drug Delivery Approaches and Nanosystems, CRC Press, 2020.

[47] R. Saidur, K. Leong, H.A. Mohammed, A review on applications and challenges of nanofluids, Renew. Sustain. Energy Rev. 15 (3) (2011) 1646–1668.

[48] V. Sridhara, et al., Nanofluids—a new promising fluid for cooling, Trans. Indian. Ceram. Soc. 68 (1) (2009) 1–17.

[49] W. Yu, et al., Review and Assessment of Nanofluid Technology for Transportation and Other Applications, Argonne National Lab. (ANL), Argonne, IL, 2007.

[50] Y. Yang, et al., Applications of nanotechnology for regenerative medicine; healing tissues at the nanoscale, Principles of Regenerative Medicine, Elsevier, 2019, pp. 485–504.

[51] D.K. Devendiran, V.A. Amirtham, A review on preparation, characterization, properties and applications of nanofluids, Renew. Sustain. Energy Rev. 60 (2016) 21–40.

[52] W. Yu, H. Xie, A review on nanofluids: preparation, stability mechanisms, and applications, J. Nanomaterials (2012) 2012.

[53] L. Zhang, et al., Investigation into the antibacterial behaviour of suspensions of ZnO nanoparticles (ZnO nanofluids), J. Nanopart. Res. 9 (3) (2007) 479–489.

[54] L. Zhang, et al., Sonophotocatalytic inactivation of E. coli using ZnO nanofluids and its mechanism, Ultrason. Sonochem. 34 (2017) 232–238.

[55] N. Yadav, et al., Trimetallic Au/Pt/Ag based nanofluid for enhanced antibacterial response, Mater. Chem. Phys. 218 (2018) 10–17.

[56] S.A. Umoren, et al., Preparation of silver/chitosan nanofluids using selected plant extracts: characterization and antimicrobial studies against gram-positive and gram-negative bacteria, Materials 13 (7) (2020) 1629.

[57] N. Khan, D. Kumar, P. Kumar, Silver nanoparticles embedded guar gum/gelatin nanocomposite: green synthesis, characterization and antibacterial activity, Colloid Interface Sci. Commun. 35 (2020) 100242.

[58] C. De Martel, et al., Global burden of cancers attributable to infections in 2008: a review and synthetic analysis, Lancet Oncol. 13 (6) (2012) 607–615.

[59] G. Eskiizmir, Y. Baskın, K. Yapıcı, Graphene-based nanomaterials in cancer treatment and diagnosis, in, Fullerens, Graphenes and Nanotubes, Elsevier, 2018, pp. 331–374.

[60] K.S. Mekheimer, et al., Peristaltic blood flow with gold nanoparticles as a third grade nanofluid in catheter: application of cancer therapy, Phys. Lett. A 382 (2–3) (2018) 85–93.

[61] P. Das, M. Colombo, D. Prosperi, Recent advances in magnetic fluid hyperthermia for cancer therapy, Colloids Surf. B: Biointerfaces 174 (2019) 42–55.

[62] S. Laurent, et al., Magnetic fluid hyperthermia: focus on superparamagnetic iron oxide nanoparticles, Adv. Colloid Interface Sci. 166 (1-2) (2011) 8–23.

[63] I. Sharifi, H. Shokrollahi, S. Amiri, Ferrite-based magnetic nanofluids used in hyperthermia applications, J. Magnetism Magnetic Mater. 324 (6) (2012) 903–915.

[64] A. Sohail, et al., A review on hyperthermia via nanoparticle-mediated therapy, Bull. du. Cancer 104 (5) (2017) 452–461.

[65] M.H. Abdolmohammadi, et al., Application of new ZnO nanoformulation and Ag/Fe/ZnO nanocomposites as water-based nanofluids to consider in vitro cytotoxic effects against MCF-7 breast cancer cells, Artif. Cell. Nanomed. B. 45 (8) (2017) 1769–1777.

[66] M. Kothandapani, J. Prakash, The peristaltic transport of Carreau nanofluids under effect of a magnetic field in a tapered asymmetric channel: application of the cancer therapy, J. Mech. Med. Biol. 15 (03) (2015) 1550030.

[67] N.S. Akbar, S. Nadeem, Z.H. Khan, Numerical simulation of peristaltic flow of a Carreau nanofluid in an asymmetric channel, Alex. Eng. J. 53 (1) (2014) 191–197.

[68] R. Ellahi, The effects of MHD and temperature dependent viscosity on the flow of non-Newtonian nanofluid in a pipe: analytical solutions, Appl. Math. Model. 37 (3) (2013) 1451–1467.

[69] X. Huang, M.A. El-Sayed, Gold nanoparticles: optical properties and implementations in cancer diagnosis and photothermal therapy, J. Adv. Res. 1 (1) (2010) 13–28.
[70] K.P. Kumar, W. Paul, C.P. Sharma, Green synthesis of gold nanoparticles with *Zingiber officinale* extract: characterization and blood compatibility, Process. Biochem. 46 (10) (2011) 2007–2013.
[71] M. Hatami, J. Hatami, D.D. Ganji, Computer simulation of MHD blood conveying gold nanoparticles as a third grade non-Newtonian nanofluid in a hollow porous vessel, Computer Methods Prog. Biomedicine 113 (2) (2014) 632–641.
[72] A. Hamzehnezhad, et al., Heat transfer and fluid flow of blood flow containing nanoparticles through porous blood vessels with magnetic field, Math. Biosci. 283 (2017) 38–47.
[73] T. Elnaqeeb, K.S. Mekheimer, F. Alghamdi, Cu-blood flow model through a catheterized mild stenotic artery with a thrombosis, Math. Biosci. 282 (2016) 135–146.
[74] K.S. Mekheimer, et al., Simultaneous effect of magnetic field and metallic nanoparticles on a micropolar fluid through an overlapping stenotic artery: blood flow model, Phys. Essays 29 (2) (2016) 272–283.
[75] K.S. Mekheimer, M.S. Mohamed, T. Elnaqeeb, Metallic nanoparticles influence on blood flow through a stenotic artery, Int. J. Pure Appl. Math. 107 (1) (2016) 201.
[76] M. Sheikholeslami, M. Shamlooei, Magnetic source influence on nanofluid flow in porous medium considering shape factor effect, Phys. Lett. A 381 (36) (2017) 3071–3078.
[77] M. Sheikholeslami, Numerical simulation of magnetic nanofluid natural convection in porous media, Phys. Lett. A 381 (5) (2017) 494–503.
[78] M. Sheikholeslami, S. Shehzad, CVFEM for influence of external magnetic source on Fe_3O_4-H_2O nanofluid behavior in a permeable cavity considering shape effect, Int. J. Heat Mass Transf. 115 (2017) 180–191.
[79] M. Sheikholeslami, H.B. Rokni, Simulation of nanofluid heat transfer in presence of magnetic field: a review, Int. J. Heat Mass Transf. 115 (2017) 1203–1233.
[80] M. Sheikholeslami, Lattice Boltzmann method simulation for MHD non-Darcy nanofluid free convection, Phys. B: Condens. Matter 516 (2017) 55–71.
[81] R. Ellahi, S. Rahman, S. Nadeem, Blood flow of Jeffrey fluid in a catherized tapered artery with the suspension of nanoparticles, Phys. Lett. A 378 (40) (2014) 2973–2980.
[82] M. Akbarzadeh, et al., A sensitivity analysis on thermal and pumping power for the flow of nanofluid inside a wavy channel, J. Mol. Liq. 220 (2016) 1–13.
[83] J. Esfahani, et al., Influences of wavy wall and nanoparticles on entropy generation over heat exchanger plat, Int. J. Heat Mass Transf. 109 (2017) 1162–1171.
[84] S. Nadeem, S. Ijaz, Theoretical analysis of metallic nanoparticles on blood flow through stenosed artery with permeable walls, Phys. Lett. A 379 (6) (2015) 542–554.
[85] T. Hayat, et al., Peristaltic transport of nanofluid in a compliant wall channel with convective conditions and thermal radiation, J. Mol. Liq. 220 (2016) 448–453.
[86] L.H. Reddy, et al., Magnetic nanoparticles: design and characterization, toxicity and biocompatibility, pharmaceutical and biomedical applications, Chem. Rev. 112 (11) (2012) 5818–5878.
[87] P. Gas, Essential Facts on the History of Hyperthermia and Their Connections with Electromedicine. arXiv preprint arXiv:1710.00652, 2017.
[88] W. Busch, Aus der Sitzung der medicinischen Section vom 13 November 1867. Berl Klin Wochenschr, 1868. 5, p. 137.
[89] W.B. Coley II, Contribution to the knowledge of sarcoma, Ann. Surg. 14 (3) (1891) 199.
[90] H. Chiriac, et al., In vitro cytotoxicity of Fe–Cr–Nb–B magnetic nanoparticles under high frequency electromagnetic field, J. Magnetism Magnetic Mater. 380 (2015) 13–19.

[91] A. Hervault, N.T.K. Thanh, Magnetic nanoparticle-based therapeutic agents for thermo-chemotherapy treatment of cancer, Nanoscale 6 (20) (2014) 11553–11573.
[92] Z. Hedayatnasab, F. Abnisa, W.M.A.W. Daud, Review on magnetic nanoparticles for magnetic nanofluid hyperthermia application, Mater. Des. 123 (2017) 174–196.
[93] N. Thorat, et al., Highly water-dispersible surface-functionalized LSMO nanoparticles for magnetic fluid hyperthermia application, N. J. Chem. 37 (9) (2013) 2733–2742.

CHAPTER

Environmental and industrialization challenges of nanofluids

20

Nazanin Abrishami Shirazi[1] and Mohammad Reza Rahimpour[2]

[1]School of Environment, College of Engineering, University of Tehran, Tehran, Iran [2]Department of Chemical Engineering, Shiraz University, Shiraz, Iran

20.1 Introduction

Environmental problems such as global warming and melting glaciers, widespread climate change, etc. are the result of human overuse of energy resources, especially fossil fuels [1,2]. Therefore one of the most important issues in modern life is energy supply and reduction of its waste, which leads to energy savings. To produce cleaner and more efficient energy sources and uses, nanotechnology has been considered in many industries and applications. Although many of these applications do not instantly affect energy transfer, each has the potential to reduce the need for electricity, distilled fuel oil, or natural gas, which is otherwise transmitted by the energy transmission system [3].

Recently, a new generation of fluids called "nanofluids" has emerged during dramatic advances in nanotechnology. Considering the benefits, researchers have used many nanofluids in their studies [4–6]. Nanofluids are partly a new fluids category that include a base fluid with nanosized particles suspended within them. The size of these nanoparticles is 1–100 nm and they are generally made of a metal or metal oxide. Nanoparticles enhance the conductivity and convection coefficients of the base fluids and allow heat to pass more effectively through the coolant [7]. This is while other methods that tried to increase the thermal conductivities of the common fluids encountered various problems. Therefore the development of nanoparticles led to a major breakthrough in this field [8–10]. Two common base fluids which host these nanosized particles are water and ethylene glycol. However, many new fluids have been recently proposed as base fluids, like amines, ionic liquids, etc. with various applications [11,12]. The major benefits of nanofluids are [7]:

- The higher specific surface area and, as a result, the greater rate of heat exchange between particles and liquids.
- High scattering fixity with predominant Brownian moving of particles.
- Less pumping power required as compared to pure liquid to gain the same heat transfer intensification.
- System miniaturization due to decreased particle blockage compared to common slurries.
- Adjustable properties, including thermal conductivity and surface wettability, by particle doping changing to suit several usages.

Nanofluids and Mass Transfer. DOI: https://doi.org/10.1016/B978-0-12-823996-4.00020-3

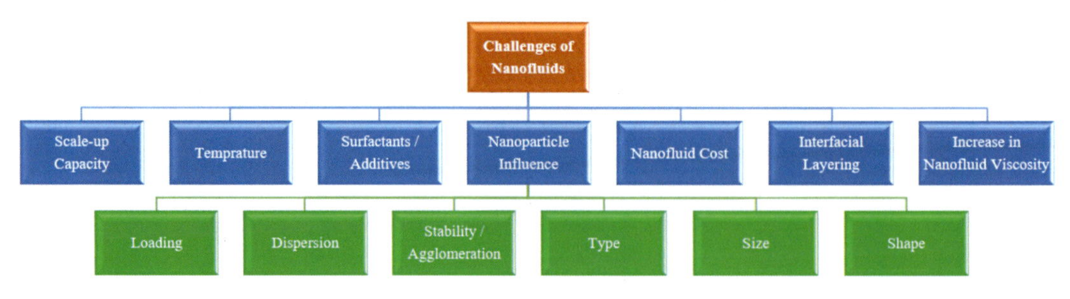

FIGURE 20.1

Nanofluids and nanoparticles and their challenges [23].

Much research has been conducted on the dominant heat transfer characteristics of nanofluids, especially their thermal conductivity and convective heat transfer. Based on research by Eastman et al. [13], Liu et al. [14], Huang et al. [15], Yu et al. [16], and Mintza et al. [17], the thermal conductivity of nanofluids is much higher than the conventional refrigerants. On the other hand, Zeinali Harris et al. [18], Kim et al. [19], Jung et al. [20], and Sharma et al. [21] have provided reports of increased convective heat transfer of these nanofluids and mentioned that utilizing them in industries such as in heat exchanging systems is hopeful with these features. However, various factors may hinder the development and applications of nanofluids, including long-term stability, increased pumping power and pressure slump, thermal efficiency of nanofluids in turbulent flow and fully developed area, lower specific heat, and higher production costs [22].

The aim of this chapter is to provide a detailed discussion on disadvantages of using nanofluids and major challenges facing both human life and industries to utilize these newly developed fluids instead of common commercial fluids. These challenges are summarized in Fig. 20.1.

20.2 The devastating consequences of nanotechnology

Despite the many benefits of nanoparticles and nanotechnology, these materials can also have potential hazards. Because nanoparticles can threaten the environment and human health, the potential risks of these new methods must be contemplated. Although nanotechnology increases the efficiency of products, these particle's sizes, which is one of their important attributes, can endanger the environment and health. Nanoparticles, which have a smaller size than plant pollen and common allergens, can produce allergies and hurt the immune system of humans and other living organisms. Some of these particles enter the lungs after respiration, and can damage the alveolar sacs. Macrophages try to dispatch and prevent them from penetrating to the bloodstream. However, macrophages are not able to detect particles smaller than 70 nm in diameter, so they are able to easily circulate in the blood. Like titanium dioxide and carbon black, which are broadly utilized in industrial processes and also lead to air pollution, nanoparticles cause inflammation and damage to the skin and accumulate in the lungs. Particles of titanium dioxide and zinc yield free radicals in the body, which change the DNA structure and cause tumors and cancer [23].

20.2.1 The nanoparticles effect on human health

Nowadays, the production of nanoparticles has extended and become more industrial which leads to more contact with them [24,25]. They have many applications, including electronics, food industry, cosmetics, medical equipment, and internal applications such as water purification, etc., and thus the more widespread use of these substances, the more human beings are affected by these particles, which causes short-term and long-term toxicity [26,27].

The nanoparticles characteristics such as surface energy, chemical structure, shape and morphology, size, and surface area can affect their function and biological effects. A special property can turn a nanoparticle into a harmless or hazardous particle. Very small nanoparticles are able to infiltrate the lymphatic system and bloodstream through different barriers in the body. They are able to penetrate various organs and tissues, disrupting biological structures and overshadowing their natural function [28]. Although pharmacists have not yet determined whether nanoparticles cause any diseases, these particles are widely used to decrease the side effects and toxicity of the drugs [29].

One of the positive aspects of these particles is their ability to cross the blood–brain barrier which allows them to be used to deliver drugs to the brain. The tiny size of nanoparticles also allows drugs to be transported to different cells and cell parts, including the core [30,31]. Nanoparticle carriers are also used to deliver drugs to the liver and spleen reticuloendothelial network more effectively [32]. Nanoparticles around 20 nm in size are able to enter the cell solely without the use of the endocytic systems [33]. High adhesion of nanoscale drugs enhances their efficiency. For instance, the binding of certain proteins like albumin to nanoparticles leads to the activation of immune reactions [34,35]. However, to use the nanotechnology potential in nanomedicine, the toxicity and safety of these materials must be considered. When nanomaterials are used in drug delivery systems, the therapeutic ratio (ratio between dose needed for clinical effect and doses due to side effects) is increased [36,37].

Gold nanoparticles are utilized in medical uses, such as imaging of molecules and genes, drug delivery, and new drugs production in the cancer remedy [38–40]. Cells can consume gold nanoparticles without causing cellular toxicity [41,42]. If one is exposed to nanoparticles for a long time, he/she is likely to experience serious injuries such as heart disease, respiratory and lung problems, and premature death [43]. Aspiration of these substances, ingestion, and direct contact endangers human health and, of course, poses a risk to the environment [44,45]. The smaller the particle size, the greater their penetration into the distal regions of the respiratory tract, resulting in increased sediment in the alveolar sacs.

Nanoparticles behave as free radicals after penetration into biological tissues and cause oxidative stress reactions. These reactions cause inflammation, which is followed by tissue destruction processes. Recently, autophagic dysfunction has been discovered as one of the mechanisms associated with nanoparticles in endothelial cell toxicity [46]. Cellular reactants such as inflammatory response, oxidative stress, signaling pathway, apoptosis, and genotoxic effects are supposedly related to the titanium nanoparticles' toxicity, but this has not yet been scientifically proven [47].

Many studies have proven the toxicity of Ag nanoparticles to humans. These particles enter the bloodstream through ingestion, inhalation, and intraperitoneal injection and poison the lungs, liver, brain, skin, and cardiovascular system. Silver nanoparticles are not degraded in any cytotoxic amounts. This leads to damage to DNA, cell membranes, mitochondria, and leads to oxidative stress reactions resulting in growth and structural abnormalities and apoptosis [48–50]. Gatwin and

Webster [51] cultured cells in an environment containing titanium and alumina. They proved that these nanoparticles significantly slow down cell growth. Another study revealed that very small gold nanoparticles could pierce the cell membrane and bind to DNA [52]. It is also proven that nanoparticles with a size of 14 nm are able to penetrate the cell membrane and get into the vacuoles inside the cell [53]. These particles, along with vacuoles, enter various cell parts and disrupt cellular normal functions, including proliferation, motility, and adhesion. Hsin et al. demonstrated that Ag nanoparticles activate reactive oxygen species and deplete cytochrome C to the cytosol and transport Bax protein to the mitochondria [54] which will lead to mitochondrial apoptosis [55].

As aforementioned, toxicological studies in humans and animals have demonstrated that nanoparticles can produce toxic effects on the immune system, relative oxidative stress disorders, lung and inflammatory diseases, and brain disorders. However, large amounts of these particles should enter the body to produce these effects [56]. Nevertheless, some studies have shown that nanoparticles made from nontoxic and relatively harmless degradable polymers can cause cell toxicity and immunotoxicity as well [57,58]. Lam et al. [59], Paddle-Ledinak et al. [60], and Poon and Burd [61] separately investigated the effect of wound dressings containing Ag nanoparticles on skin wounds. The results have proved that nanoparticles cause cell poisoning of keratinocytes. Besides, it has been shown that these nanoparticles are able to be released in human sweat [62].

Some studies have shown that the respiration of nanoparticles causes their deposition in the olfactory mucosa in the pharyngeal region. After settling, these nanoparticles travel through the olfactory nerves to the cerebellum and brain cortex [63,64]. These tiny particles are able to cross the blood–rain barrier. Neurotoxicity and concentration amine depletion in neurons and endocrine glands caused by these substances have detrimental effects on physiology and brain function [24,25,65].

20.2.2 The nanoparticles effect on the environment

Beside the effects on the human health, nanoparticles have devastating consequences on the environment and various organisms. Leakage of nanoparticles from laboratories to the environment as well as the disposal of laboratory wastes has caused pollution of their surroundings. Nanoparticles can pollute soil, water, and even the air. Bacteria may mutate rapidly in the presence of nanoparticles, and potentially threaten the environment and human health. Such particles, which are smaller than natural allergens and plant pollen, are able to attack both the human and animal immune systems. Since living organisms may not have strong defense mechanisms against engineered nanoparticles, they are more dangerous than natural particles [66–69]. In one study, researchers exposed zebrafish to Ag nanoparticles of various sizes, 10, 35, 600, and 1600 nm. Large amounts of nanoparticles smaller than 10 nm penetrated in the liver and intestinal tissues of fish [70]. Despite such harms, the use of nanoparticles in some areas might be beneficial for the environment. Researchers believe that energy consumption can be reduced with the aid of nanotechnology and it may help to control pollutants as well as make new environment-friendly materials [71].

20.3 Nanofluids utilization challenges

Although the use of nanofluids in a wide range of fields seems hopeful, there are also obstacles to the development of the use of these materials. There exist disagreement with the results obtained by various researchers and nanofluid suspensions suffer from poor stability. Besides, many of their properties, such as thermal conductivity and Brownian motion, change with temperature. The variation of such properties must be carefully checked for convective heat transfer in nanofluids. On the other hand, studies usually have been done with oxide particles concentrating where the resulting fluids have high viscosity and need high pumping power. It is worth mentioning that the energy transfer in nanofluids with low concentration of pure metal nanoparticles is 100 times better than oxides [72]. The use of nanofluids in heating pipes has increased the efficiency and significantly reduced the thermal resistance of the pipe. However, recent studies have shown the accumulation and deposition of particles in microchannel heat sinks [73].

Because synthetic nanofluids are innovative liquids that have been reported with good heat transfer and are in the study phase, there are many challenges to their use and synthesis that need to be explored and analyzed. There are many differences between experimental results and theoretical predictions of composite nanofluid species. Despite much research done on their behavior, their intricate mechanisms and rheological alterations of composite nanofluids heat transfer are still unclear to researchers. For the synthesis of different types of composite nanoparticles, such as Al_2O_3-MWCNT, CNT-Al_2O_3, Cu-TiO_2, CNT-Cu, CNT-Au, etc., it is important to select their proper production method. This is because each nanoparticle has unique attributes, and therefore must be considered behaviorally to be properly selected for the considered application [23].

The composite nanoparticles synthesis is very complex and the procedure is more complicated, time consuming, and more expensive than conventional nanofluid synthesis [74]. Sustainability is one of the most important challenges for conventional and composite nanofluids. The nanoparticles' stability has been well established by various methods for common nanofluids. But because two dissimilar materials are dispersed in a base liquid, the stability of hybrid nanofluids is more difficult due to the surface charges, which are different for each particle.

A nanofluid's stability depends on the surfactant type, the level of pH, and the type of nanocomposites suspended in the base liquid. According to some studies the thermal conductivity of common nanofluids decreases over time, but this may be different for composite nanofluids due to their stability [75].

Researches showed that several factors, including temperature, nanoparticle volume fraction, nanoparticle size, their shape, and the effect of Brownian motion, affect the thermal conductivity, viscosity, etc. of nanofluids. Nevertheless, there is no general law that can predict the various characteristics of nanofluids. Therefore various complex methods such as neural network have been used to forecast these characteristics. These networks have much less error compared to mathematical models and experimental data [76,77]. However, it might be better to use a general model to forecast the hybrid nanofluids characteristics, because of its simplicity.

20.3.1 Long-term stability of nanoparticles scattering

One of the basic needs of nanofluid usages is the long-term stability of nanoparticle dispersion. The problem is the provision of homogeneous suspensions, because of the very powerful van der Waals interactions in nanoparticles which always make bulk [15]. The nanofluids stability is directly related to the increase in thermal conductivity, so the better dispersion behavior, the greater the thermal conductivity of nanofluids [78]. In contrast, weak stability disrupts the system efficiency and changes desired results into unfortunate ones. When the particles make clusters, they begin to settle to the base of the working device, which increases the thermal resistance and raise the pumping power of the fluid by curbing the flow paths. The highest stability cycle has been reported in the studies of Sander et al. [79] and Hussein [80] as being equal to 60 days. Yarmand et al. [81] also reported a 60-day shelf life with GNP-Ag/water hybrid nanofluid. Although this is a great achievement, more effort is required to utilize hybrid nanofluids in practical applications.

A variety of physical or chemical methods are used to create stable nanofluids, including adding a surfactant, correcting the surface of suspended particles, or applying excessive force to suspended particle clusters. For the dispersion of hydrophobic particles in an aqueous solution, applying dispersing agents and surface-active agents is common [82].

According to studies by Eastman et al. [83], the thermal conductivity of ethylene glycol-based nanofluids containing 0.3% copper nanoparticles reduces over time. During their study, they measured the thermal conductivity of nanofluids twice: the first stage of measurement was within 2 days and the second was 2 months after production. The results showed that the thermal conductivity of fresh nanofluids is slightly higher than nanofluids that have been produced for up to 2 months. This is probably due to the decreased stability of the nanoparticle scattering over time. On the other hand, nanoparticles may tend to coalescence if stored for long periods of time. Comparison of the Al_2O_3 nanofluid stability over time has been performed by Lee and Mudawar [84]. Based on the results, nanofluids produced for 30 days show a slight concentration and accumulation gradient compared to fresh nanofluids. This means that the thermal performance of the nanofluid may be affected in the long term.

In order to prevent the cooling ducts from becoming blocked, the sedimentation of the nanofluid particles must be carefully monitored. Research results suggest that the addition of surfactant should be carefully controlled because excessive amounts of surfactant have a detrimental effect on viscosity, thermal properties, and chemical stability of the nanofluids [85]. The stability issue has led researchers to evaluate the pH values of several particles and their compositions, test multiple base liquids, use mixers, such as ultrasonicators and magnetic stirrers, and their effect on pumping power. The addition of surfactant causes the surface of the particles to be affected, resulting in a screening effect on the heat transfer performance of the nanoparticles. Surfactants may entail physical and/or chemical difficulties [86].

Akhgar and Tughraei [87] investigated the thermal conductivity of TiO_2/MWCNT composite nanofluids prepared in dual water/EG base liquid. The stability of TiO_2/water and MWCNT/water nanofluids were measured using imaging method (the simplest stability assessment method). The stability study on different pH values revealed that TiO_2 nanofluid has better stability after 48 hours at pH 9. Nonetheless, due to its hydrophobicity, MWCNT is not able to disperse in water at any pH, and after 48 hours, the particles in all samples were settled at the bottom of the flask. To solve this difficulty, they utilized the CTAB surfactant in this solution and, created a diffuse solution of MWCNT.

Hence, to select nanomaterials, stability must be attentively considered. Asadi et al. [88] used the zeta potential analysis method to determine the stability of $Mg(OH)_2$-MWCNT/engine oil. The authors found that the visual observation technique for dark colored nanofluids such as MWCNT, engine oil, etc. cannot be trusted and is not realistic. The results of most studies indicate that zeta potential values cause different distributions of nanofluid microparticles in various stability classes [89–91]:

- unsatisfactory stability (absolute zeta potential (AZP) < 30 mV)
- admissible stability ($30 \text{ mV} \leq \text{AZP} \leq 45$ mV)

- good stability (45 mV ≤ AZP ≤ 60 mV)
- superior stability (AZP > 60 mV)

Wei et al. [90] stated an amount of 52 mV for the SiC/TiO_2 oil composite nanofluid and found that the prepared solution remained durable after 10 days. In a research presented by Safi et al. [92] with $MWNT/TiO_2$ nanofluids, they discovered a zeta potential of −47/5 mV is suitable for the hybrid nanofluids studied within an admissible span.

There have been many studies on stability measurement methods, in which researchers provide discussions on various operating parameters [93–95]. According to these studies, several factors such as particle composition, ultrasound, stirring time, pH, base liquid, surfactant, particle concentration, and liquid temperature can affect the nanofluid stability. Researchers should focus more on the above factors to achieve better nanoparticles dispersion.

20.3.2 Increased pressure loss and pumping power

One of the important parameters in the nanofluids application in heat exchange instruments is the pressure generated during the flow through them. Nanoparticles lead to a notable enhancement in the pressure loss, resulting in higher needed pumping power when using the nanofluids. Experimental data show that pumping power can be enhanced by about 40% compared to water for a certain flow rates and as attended, more pressure loss is observed [96]. It should also be noted that the exact shape of the corrugation remarkably affects the friction factor, and even very tiny alterations in manufacturing can lead to remarkable variations in pressure loss [97].

There is a very close relation between pressure loss and cooling pump power. Few factors such as density and viscosity affect the pressure loss of the coolant. It seems that coolers with higher density and viscosity have rather more pressure loss. This has led to less use of nanofluids as coolants. The viscosities of Al_2O_3 nanofluids based on water and ZnO nanofluids based on ethylene glycol were investigated by Lee et al. [84] and Yu et al. [98]. According to the results, the nanofluids viscosity is higher than the base fluids and both properties are related to the volume fraction of nanoparticles [99]. Vaso et al. [100] studied the thermal design of a compact heat exchanger applying nanofluids. In this study, it was found that the pressure loss of 4% Al_2O_3 + H_2O nanofluids is almost twice that of the base fluid. The authors stated that there was a main increase in nanofluids pressure loss and pumping power in the plate heat exchanger

The nanofluid viscosity depends on the type of particles and their concentration. Studies have shown that the viscosity of nanofluids is generally greater than the base liquid [101]. The relation presented by Williams et al. [102] provides a fairly good prediction in this regard. Compared to the measurements presented by Pak and Chou [103], TiO_2 nanofluid shows a higher viscosity for 3% particle volume concentration.

Enhancing the viscosity of water nanoparticle suspensions is directly related to enhancing the particles concentration in the suspension [104]. In industrial heat exchangers, where a big volume of nanofluid is required and upside flow is needed, the replacement of ordinary liquids by nanofluids appears to be impossible. Panzali et al. [105] perceived a doubling of the scaled viscosity of the suspensions (e.g., nanofluids) compared to water. This significantly increases the pressure loss and thus the pumping power needed when applying nanofluids. According to the experimental data, the pumping power enhanced by approximately 40% compared to water for specific flow rates.

Besides, for a certain thermal function, the volumetric flow for water and nanofluid is actually the same, while the pumping power needed for nanofluid is twice as high as the corresponding amount for water due to the fluids higher kinematic viscosity [97]. In fact, the addition of nanoparticles might change the fluid flow characteristics [106]. Sundar et al. [79] produced the MWCNT-Fe_3O_4/water composite nanofluid by scattering MWCNT-Fe_3O_4 nanocomposite materials in distilled water for an experimental study of the pumping power. According to the results, the penalty for pumping power was 1.18 times relative to the base fluid with an increase in Nusselt number by 31.10% at the Reynolds number of 22,000. Similar results have also been reported and confirmed for a variety of nanofluids by CFD (Computational Fluid Dynamics) analysis using the three-dimensional tube pattern [107].

20.3.3 Thermal performance of nanofluids in turbulent flow and fully developed region

Recently, researchers have focused more on the convective heat transfer efficiency and nanofluids thermal conductivity. In recent years, discrepancies have been found in the results reported by researchers, so there is an issue that needs to be carefully considered regarding the thermal efficiency of nanofluids in upset flow. For example, it has been stated that despite 8% laminar flow recovery, no improvement in convective heat transfer has been seen for amorphous carbon nanofibers in turbulent flow [108]. Nevertheless, it was stated that the heat transfer coefficient of TiO_2−water nanofluids was greater than that of the base liquids [109]. This characteristic increases with rising Reynolds number and particle concentration from 0.2% to 2%. Albeit this study showed that a 26% increase could be observed for nanofluids with 1% of TiO_2 nanoparticles, it showed conflicting results at a volume break of 2.0%. Another study showed that the heat conduction coefficient of nanofluids in this circumstance is 14% lower than the base liquids.

To conclude, the thermal efficiency of nanofluids in a well-expanded area is one of their other disadvantages. The convective heat conduction coefficient of nanofluids in a small Reynolds number has the highest value at the pipe inlet, begins to reduce with axial distances, and finally becomes almost constant in fully developed areas [110].

20.3.4 Less specific heat

Previous research has shown that the specific heat of the nanofluid is less than that of the base fluid [87]. CuO/ethylene glycol nanofluids, SiO_2/ethylene glycol nanofluids, and Al_2O_3/ethylene glycol nanofluids show less specific heat than base fluids [99]. However, in order for an ideal refrigerant to be able to remove more heat, the refrigerant must have a higher specific heat.

20.3.5 Nanofluids production and usage price

One of the reasons that may prevent the use of nanofluids in industry is the high cost of their production. As mentioned before, nanofluids are produced via one-stage or two-stage methods. However, both methods need advanced instruments. Therefore the high cost of nanofluids is one of the disadvantages of their utilization [84]. One of the advantages of the single-stage method is the better stability of the nanofluid produced. However, this method suffers from the requirement for advanced instruments with expensive raw materials, while it produces nanofluids in small quantities. In the other technique of producing nanofluids, that is two-step technique, both the nanoparticles and the base fluid are supplied from the vendors and only the scattering procedure is handled by changing the particles concentration. With the two-step technique, nanofluids can be produced in high volumes for commercial applications particles synthesis is avoided [86].

On the other hand, composite nanofluids are more cost-effective than single nanofluids [111]. Xuan et al. [112] investigated the absorption of solar energy using TiO_2 and Ag single and composite nanofluids. According to this study, the composite nanofluid has a good efficiency and offers an amplified temperature relative to TiO_2 nanofluid. In addition, the temperature was the same as the temperature of Ag nanofluid. Therefore the price of TiO_2/Ag nanocomposite is further down. Hence, the authors recommend the use of nanocomposites instead of single nanoparticles.

In other studies, the use of SiO_2 nanoparticles in combination with other nanoparticles has been suggested by researchers due to their low cost and eligible chemical and physical properties [113–115]. Alirezaei et al. [116] concluded that nanofluids comprising metal oxides have greater PPF and are more economically viable than metal nanofluids. Wcis'lik [117] conducted an excellent study on the cost of nanofluids and presented a model for estimating the proposed cost based on the latest market rate. In his model, less work is required to minimize generation costs, while it is necessary to conduct an economic study before selecting the appropriate blend of nanoparticles for the intended application, along with examining the heat transfer properties. Work began on minimizing the cost of nanofluids production, and today these studies are existing for the interest of scientists [118–120].

20.4 Conclusions

Although nanofluids provide better heat and mass transfer performances, they face real challenges with regard to the environment, human health, and industries. Tiny particles can easily penetrate living organisms and cause damage. Producing such expensive particles may not be economic for many industries while nanofluids are not stable for a long period of time and consume much more pumping power. All these challenges make the commercial utilization of nanofluids almost impossible at the present time, but many efforts are being made to find proper solutions.

References

[1] A.A. Nassani, et al., Resource management for green growth: ensure environment sustainability agenda for mutual exclusive global gain, Environ. Prog. Sustain. Energy 38 (4) (2019) 13132.

[2] J.R. Vázquez-Canteli, et al., Fusing TensorFlow with building energy simulation for intelligent energy management in smart cities, Sustain. Cities Soc. 45 (2019) 243–257.

[3] D. Elcock, Potential Impacts of Nanotechnology on Energy Transmission Applications and Needs, Argonne National Lab.(ANL), Argonne, IL, 2007.

[4] A.A. Alnaqi, et al., Effects of magnetic field on the convective heat transfer rate and entropy generation of a nanofluid in an inclined square cavity equipped with a conductor fin: Considering the radiation effect, Int. J. Heat Mass Transf. 133 (2019) 256–267.

[5] A.A. Al-Rashed, et al., Entropy generation of boehmite alumina nanofluid flow through a minichannel heat exchanger considering nanoparticle shape effect, Phys. A: Stat. Mech. Appl. 521 (2019) 724–736.

[6] S. Aghakhani, et al., Effect of replacing nanofluid instead of water on heat transfer in a channel with extended surfaces under a magnetic field, Int. J. Numer. Methods Heat Fluid Flow. (2019).

[7] S. Choi, D. Singer, H. Wang, Developments and applications of non-Newtonian flows, Asme Fed. 66 (1995) 99–105.

[8] A. Prakash, et al., Investigation on Al_2O_3 nanoparticles for nanofluid applications-a review, IOP Conference Series: Materials Science and Engineering, IOP Publishing, 2018.

[9] S. Rashidi, et al., A concise review on the role of nanoparticles upon the productivity of solar desalination systems, J. Therm. Anal. Calorim. 135 (2) (2019) 1145–1159.

[10] M.U. Sajid, H.M. Ali, Recent advances in application of nanofluids in heat transfer devices: a critical review, Renew. Sustain. Energy Rev. 103 (2019) 556–592.

[11] N.J. Bridges, A.E. Visser, E.B. Fox, Potential of nanoparticle-enhanced ionic liquids (NEILs) as advanced heat-transfer fluids, Energy Fuels 25 (10) (2011) 4862–4864.

[12] Z. Zhang, et al., Progress in enhancement of CO_2 absorption by nanofluids: a mini review of mechanisms and current status, Renew. Energy 118 (2018) 527–535.

[13] J.A. Eastman, et al., Enhanced Thermal Conductivity Through the Development of Nanofluids, Argonne National Lab, IL, 1996.

[14] M.S. Liu, et al., Enhancement of thermal conductivity with CuO for nanofluids, Chem. Eng. Tech. 29 (1) (2006) 72–77.

[15] Y. Hwang, et al., Thermal conductivity and lubrication characteristics of nanofluids, Curr. Appl. Phys. 6 (2006) e67–e71.

[16] W. Yu, et al., Investigation of thermal conductivity and viscosity of ethylene glycol based ZnO nanofluid, Thermochim. Acta 491 (1–2) (2009) 92–96.

[17] H.A. Mintsa, et al., New temperature dependent thermal conductivity data for water-based nanofluids, Int. J. Therm. Sci. 48 (2) (2009) 363–371.

[18] S.Z. Heris, M.N. Esfahany, S.G. Etemad, Experimental investigation of convective heat transfer of Al_2O_3/water nanofluid in circular tube, Int. J. Heat Fluid Flow. 28 (2) (2007) 203–210.

[19] D. Kim, et al., Convective heat transfer characteristics of nanofluids under laminar and turbulent flow conditions, Curr. Appl. Phys. 9 (2) (2009) e119–e123.

[20] J.-Y. Jung, H.-S. Oh, H.-Y. Kwak. Forced convective heat transfer of nanofluids in microchannels, in: ASME International Mechanical Engineering Congress and Exposition, 2006.

[21] K. Sharma, L.S. Sundar, P. Sarma, Estimation of heat transfer coefficient and friction factor in the transition flow with low volume concentration of Al_2O_3 nanofluid flowing in a circular tube and with twisted tape insert, Int. Commun. Heat Mass Transf. 36 (5) (2009) 503–507.

[22] R. Saidur, K. Leong, H.A. Mohammed, A review on applications and challenges of nanofluids, Renew. Sustain. Energy Rev. 15 (3) (2011) 1646–1668.

[23] L. Yang, et al., An updated review on the properties, fabrication and application of hybrid-nanofluids along with their environmental effects, J. Clean. Prod. 257 (2020) 120408.

[24] S.M. Hussain, et al., The interaction of manganese nanoparticles with PC-12 cells induces dopamine depletion, Toxicol Sci. 92 (2) (2006) 456–463.

[25] K.M. Chow, et al., Retrospective review of neurotoxicity induced by cefepime and ceftazidime, Pharmacother: J. Hum. Pharmacol Drug. Ther. 23 (3) (2003) 369–373.
[26] N. Vigneshwaran, et al., Functional finishing of cotton fabrics using silver nanoparticles, J. Nanosci. Nanotechnol. 7 (6) (2007) 1893–1897.
[27] T.M. Tolaymat, et al., An evidence-based environmental perspective of manufactured silver nanoparticle in syntheses and applications: a systematic review and critical appraisal of peer-reviewed scientific papers, Sci. Total Environ. 408 (5) (2010) 999–1006.
[28] M.A. Zoroddu, et al., Toxicity of nanoparticles, Curr. Med. Chem. 21 (33) (2014) 3837–3853.
[29] W.H. De Jong, P.J. Borm, Drug delivery and nanoparticles: applications and hazards, Int. J. Nanomed. 3 (2) (2008) 133.
[30] J. Koziara, et al., The blood-brain barrier and brain drug delivery, J. Nanosci. Nanotechnol. 6 (9–10) (2006) 2712–2735.
[31] S.B. Tiwari, M.M. Amiji, A review of nanocarrier-based CNS delivery systems, Curr. Drug. Delivery 3 (2) (2006) 219–232.
[32] M. Gupta, A.K. Gupta, In vitro cytotoxicity studies of hydrogel pullulan nanoparticles prepared by AOT/N-hexane micellar system, J. Pharm. Pharm Sci. 7 (1) (2004) 38–46.
[33] M. Edetsberger, et al., Detection of nanometer-sized particles in living cells using modern fluorescence fluctuation methods, Biochem Biophys Res. Commun. 332 (1) (2005) 109–116.
[34] L. Nobs, et al., Poly (lactic acid) nanoparticles labeled with biologically active Neutravidin™ for active targeting, Eur. J. Pharmaceutics Biopharmaceutics 58 (3) (2004) 483–490.
[35] L. Prinzen, et al., Optical and magnetic resonance imaging of cell death and platelet activation using annexin A5-functionalized quantum dots, Nano Lett. 7 (1) (2007) 93–100.
[36] D.B. Buxton, et al., Recommendations of the national heart, lung, and blood institute nanotechnology working group, Circulation 108 (22) (2003) 2737–2742.
[37] M. Ferrari, Cancer nanotechnology: opportunities and challenges, Nat. Rev. Cancer 5 (3) (2005) 161–171.
[38] L.R. Hirsch, et al., Nanoshell-mediated near-infrared thermal therapy of tumors under magnetic resonance guidance, Proc. Natl Acad. Sci. 100 (23) (2003) 13549–13554.
[39] J.F. Hainfeld, D.N. Slatkin, H.M. Smilowitz, The use of gold nanoparticles to enhance radiotherapy in mice, Phys. Med. Biol. 49 (18) (2004) N309.
[40] C. Loo, et al., Nanoshell-enabled photonics-based imaging and therapy of cancer, Technol. Cancer Res. Treat. 3 (1) (2004) 33–40.
[41] E.E. Connor, et al., Gold nanoparticles are taken up by human cells but do not cause acute cytotoxicity, Small 1 (3) (2005) 325–327.
[42] D. Shenoy, et al., Surface functionalization of gold nanoparticles using hetero-bifunctional poly (ethylene glycol) spacer for intracellular tracking and delivery, Int. J. Nanomed. 1 (1) (2006) 51.
[43] D.Y. Pui, S.-C. Chen, Z. Zuo, *PM2*. 5 in China: measurements, sources, visibility and health effects, and mitigation, Particuology 13 (2014) 1–26.
[44] W.G. Kreyling, M. Semmler-Behnke, W. Möller, Health implications of nanoparticles, J. Nanopart. Res. 8 (5) (2006) 543–562.
[45] Y. Gao, T. Yang, J. Jin, Nanoparticle pollution and associated increasing potential risks on environment and human health: a case study of China, Environ. Sci. Pollut. Res. 22 (23) (2015) 19297–19306.
[46] Y. Cao, The toxicity of nanoparticles to human endothelial cells, Cellular and Molecular Toxicology of Nanoparticles, Springer, 2018, pp. 59–69.
[47] B. Song, et al., Contribution of oxidative stress to TiO_2 nanoparticle-induced toxicity, Environ. Toxicol. Pharmacol 48 (2016) 130–140.
[48] S. Arora, et al., Cellular responses induced by silver nanoparticles: in vitro studies, Toxicol. Lett. 179 (2) (2008) 93–100.

[49] S. Arora, et al., Interactions of silver nanoparticles with primary mouse fibroblasts and liver cells, Toxicol. Appl. Pharmacology 236 (3) (2009) 310–318.
[50] M. Ahamed, M.S. AlSalhi, M. Siddiqui, Silver nanoparticle applications and human health, Clin Chim. Acta 411 (23–24) (2010) 1841–1848.
[51] L.G. Gutwein, T.J. Webster, Increased viable osteoblast density in the presence of nanophase compared to conventional alumina and titania particles, Biomaterials 25 (18) (2004) 4175–4183.
[52] M. Tsoli, et al., Cellular uptake and toxicity of Au55 clusters, Small 1 (8-9) (2005) 841–844.
[53] N. Pernodet, et al., Adverse effects of citrate/gold nanoparticles on human dermal fibroblasts, Small 2 (6) (2006) 766–773.
[54] Y.-H. Hsin, et al., The apoptotic effect of nanosilver is mediated by a ROS-and JNK-dependent mechanism involving the mitochondrial pathway in NIH3T3 cells, Toxicol. Lett. 179 (3) (2008) 130–139.
[55] X. Deng, et al., Nanosized zinc oxide particles induce neural stem cell apoptosis, Nanotechnology 20 (11) (2009) 115101.
[56] R.D. Handy, B.J. Shaw, Toxic effects of nanoparticles and nanomaterials: implications for public health, risk assessment and the public perception of nanotechnology, Health, Risk Soc. 9 (2) (2007) 125–144.
[57] T. Xia, et al., Cationic polystyrene nanosphere toxicity depends on cell-specific endocytic and mitochondrial injury pathways, ACS Nano 2 (1) (2008) 85–96.
[58] M.A. Maurer-Jones, et al., Toxicity of therapeutic nanoparticles, 2009.
[59] P. Lam, et al., In vitro cytotoxicity testing of a nanocrystalline silver dressing (Acticoat) on cultured keratinocytes, Br. J. Biomed. Sci. 61 (3) (2004) 125–127.
[60] J.E. Paddle-Ledinek, Z. Nasa, H.J. Cleland, Effect of different wound dressings on cell viability and proliferation, Plastic Reconstr Surg. 117 (7S) (2006) 110S–118S.
[61] V.K. Poon, A. Burd, In vitro cytotoxity of silver: implication for clinical wound care, Burns 30 (2) (2004) 140–147.
[62] K. Kulthong, et al., Determination of silver nanoparticle release from antibacterial fabrics into artificial sweat, Part. Fibre Toxicol. 7 (1) (2010) 8.
[63] A. Elder, et al., Translocation of inhaled ultrafine manganese oxide particles to the central nervous system, Environ. Health Perspect. 114 (8) (2006) 1172–1178.
[64] G. Oberdörster, et al., Translocation of inhaled ultrafine particles to the brain, Inhalation Toxicol. 16 (6–7) (2004) 437–445.
[65] D. Screnci, M.J. McKeage, Platinum neurotoxicity: clinical profiles, experimental models and neuroprotective approaches, J. Inorg. Biochem. 77 (1–2) (1999) 105–110.
[66] A.P. Popov, et al., TiO_2 nanoparticles as an effective UV-B radiation skin-protective compound in sunscreens, J. Phys. D: Appl. Phys. 38 (15) (2005) 2564.
[67] B. Nowack, T.D. Bucheli, Occurrence, behavior and effects of nanoparticles in the environment, Environ. Pollut. 150 (1) (2007) 5–22.
[68] R.J. Griffitt, et al., Effects of particle composition and species on toxicity of metallic nanomaterials in aquatic organisms, Environ. Toxicol. Chem: An. Int. J. 27 (9) (2008) 1972–1978.
[69] H. Tjälve, J. Henriksson, Uptake of metals in the brain via olfactory pathways, Neurotoxicology 20 (2–3) (1999) 181–195.
[70] K.-T. Kim, et al., Silver nanoparticle toxicity in the embryonic zebrafish is governed by particle dispersion and ionic environment, Nanotechnology 24 (11) (2013) 115101.
[71] J.F. Sargent, Nanotechnology and Environmental Health and Safety: Issues for Consideration, DIANE Publishing, 2008.
[72] W. Lai, et al., A review of convective heat transfer with nanofluids for electronics packaging, Thermal and Thermomechanical Proceedings 10th Intersociety Conference on Phenomena in Electronics Systems, 2006, ITHERM 2006, 2006, IEEE.

[73] S. Peyghambarzadeh, et al., Performance of water based CuO and Al_2O_3 nanofluids in a Cu–Be alloy heat sink with rectangular microchannels, Energy Convers. Manag. 86 (2014) 28–38.
[74] G. Sharma, et al., Novel development of nanoparticles to bimetallic nanoparticles and their composites: a review, J. King Saud. Univ-Sci 31 (2) (2019) 257–269.
[75] Y.-j Hwang, et al., Stability and thermal conductivity characteristics of nanofluids, Thermochim. Acta 455 (1–2) (2007) 70–74.
[76] P.M. Kumar, B. Rajappa, A review on prediction of thermo physical properties of heat transfer nanofluids using intelligent techniques, Mater. Today: Proc. 21 (2020) 415–418.
[77] M. Ramezanizadeh, et al., A review on the applications of intelligence methods in predicting thermal conductivity of nanofluids, J. Therm. Anal. Calorim. 138 (1) (2019) 827–843.
[78] D. Wen, et al., Review of nanofluids for heat transfer applications, Particuology 7 (2) (2009) 141–150.
[79] L.S. Sundar, M.K. Singh, A.C. Sousa, Enhanced heat transfer and friction factor of MWCNT–Fe_3O_4/water hybrid nanofluids, Int. Commun. Heat Mass Transf. 52 (2014) 73–83.
[80] A.M. Hussein, Thermal performance and thermal properties of hybrid nanofluid laminar flow in a double pipe heat exchanger, Exp. Therm. Fluid Sci. 88 (2017) 37–45.
[81] H. Yarmand, et al., Graphene nanoplatelets–silver hybrid nanofluids for enhanced heat transfer, Energy Convers. Manag. 100 (2015) 419–428.
[82] I. Fratoddi, Hydrophobic and hydrophilic Au and Ag nanoparticles. Breakthroughs and perspectives, Nanomaterials 8 (1) (2018) 11.
[83] J.A. Eastman, et al., Anomalously increased effective thermal conductivities of ethylene glycol-based nanofluids containing copper nanoparticles, Appl. Phys. Lett. 78 (6) (2001) 718–720.
[84] J. Lee, I. Mudawar, Assessment of the effectiveness of nanofluids for single-phase and two-phase heat transfer in micro-channels, Int. J. Heat Mass Transf. 50 (3–4) (2007) 452–463.
[85] C. Choi, H. Yoo, J. Oh, Preparation and heat transfer properties of nanoparticle-in-transformer oil dispersions as advanced energy-efficient coolants, Curr. Appl. Phys. 8 (6) (2008) 710–712.
[86] H. Babar, H.M. Ali, Towards hybrid nanofluids: preparation, thermophysical properties, applications, and challenges, J. Mol. Liq. 281 (2019) 598–633.
[87] A. Akhgar, D. Toghraie, An experimental study on the stability and thermal conductivity of water-ethylene glycol/TiO_2-MWCNTs hybrid nanofluid: developing a new correlation, Powder Technol. 338 (2018) 806–818.
[88] A. Asadi, et al., An experimental and theoretical investigation on heat transfer capability of Mg (OH) 2/MWCNT-engine oil hybrid nano-lubricant adopted as a coolant and lubricant fluid, Appl. Therm. Eng. 129 (2018) 577–586.
[89] L. Vandsburger, Synthesis and Covalent Surface Modification of Carbon Nanotubes for Preparation of Stabilized Nanofluid Suspensions, McGill University, 2009.
[90] B. Wei, C. Zou, X. Li, Experimental investigation on stability and thermal conductivity of diathermic oil based TiO_2 nanofluids, Int. J. Heat Mass Transf. 104 (2017) 537–543.
[91] S. Aberoumand, A. Jafarimoghaddam, Experimental study on synthesis, stability, thermal conductivity and viscosity of Cu–engine oil nanofluid, J. Taiwan. Inst. Chem. Eng. 71 (2017) 315–322.
[92] M. Safi, et al., Preparation of MWNT/TiO_2 nanofluids and study of its thermal conductivity and stability, 2014.
[93] M.U. Sajid, H.M. Ali, Thermal conductivity of hybrid nanofluids: a critical review, Int. J. Heat Mass Transf. 126 (2018) 211–234.
[94] A. Ghadimi, H. Metselaar, B. Lotfizadehdehkordi, Nanofluid stability optimization based on UV-Vis spectrophotometer measurement, J. Eng. Sci. Technol. 10 (2015) 32–40.
[95] W. Yu, H. Xie, L.-H. Liu, A review on nanofluids: preparation, stability mechanisms, and applications, J. Nanomater 2012 (711) (2011) 128.

[96] A. Kanaris, A. Mouza, S. Paras, Optimal design of a plate heat exchanger with undulated surfaces, Int. J. Therm. Sci. 48 (6) (2009) 1184–1195.
[97] N.A.C. Sidik, et al., A review on preparation methods and challenges of nanofluids, Int. Commun. Heat Mass Transf. 54 (2014) 115–125.
[98] W. Yu, et al., Review and Assessment of Nanofluid Technology for Transportation and Other Applications, Argonne National Lab.(ANL), Argonne, IL, 2007.
[99] P.K. Namburu, et al., Numerical study of turbulent flow and heat transfer characteristics of nanofluids considering variable properties, Int. J. Therm. Sci. 48 (2) (2009) 290.
[100] V. Vasu, K. Rama Krishna, A. Kumar, Heat transfer with nanofluids for electronic cooling, Int. J. Mater. Product. Technol. 34 (1–2) (2009) 158–171.
[101] C.T. Nguyen, et al., Heat transfer enhancement using Al_2O_3–water nanofluid for an electronic liquid cooling system, Appl. Therm. Eng. 27 (8–9) (2007) 1501–1506.
[102] W. Williams, J. Buongiorno, L.-W. Hu, Experimental investigation of turbulent convective heat transfer and pressure loss of alumina/water and zirconia/water nanoparticle colloids (nanofluids) in horizontal tubes, J. Heat Transf. 130 (4) (2008).
[103] B.C. Pak, Y.I. Cho, Hydrodynamic and heat transfer study of dispersed fluids with submicron metallic oxide particles, Exp. Heat Transf. An. Int. J. 11 (2) (1998) 151–170.
[104] S. Wu, et al., Thermal energy storage behavior of Al_2O_3–H_2O nanofluids, Thermochim. Acta 483 (1–2) (2009) 73–77.
[105] M. Pantzali, A. Mouza, S. Paras, Investigating the efficacy of nanofluids as coolants in plate heat exchangers (PHE), Chem. Eng. Sci. 64 (14) (2009) 3290–3300.
[106] B. Takabi, H. Shokouhmand, Effects of Al 2 O 3–Cu/water hybrid nanofluid on heat transfer and flow characteristics in turbulent regime, Int. J. Mod. Phys. C. 26 (04) (2015) 1550047.
[107] A.A. Minea, Hybrid nanofluids based on Al_2O_3, TiO_2 and SiO_2: numerical evaluation of different approaches, Int. J. Heat Mass Transf. 104 (2017) 852–860.
[108] J. Kim, Y.T. Kang, C.K. Choi, Soret and Dufour effects on convective instabilities in binary nanofluids for absorption application, Int. J. Refrig. 30 (2) (2007) 323–328.
[109] W. Duangthongsuk, S. Wongwises, An experimental study on the heat transfer performance and pressure drop of TiO_2–water nanofluids flowing under a turbulent flow regime, Int. J. Heat Mass Transf. 53 (1–3) (2010) 334–344.
[110] G. Ding, et al., The migration characteristics of nanoparticles in the pool boiling process of nanorefrigerant and nanorefrigerant–oil mixture, Int. J. Refrig. 32 (1) (2009) 114–123.
[111] M.H. Esfe, S. Esfandeh, M. Rejvani, Modeling of thermal conductivity of MWCNT-SiO 2 (30: 70%)/EG hybrid nanofluid, sensitivity analyzing and cost performance for industrial applications, J. Therm. Anal. Calorim. 131 (2) (2018) 1437–1447.
[112] Y. Xuan, Q. Li, Investigation on convective heat transfer and flow features of nanofluids, J. Heat Transf. 125 (1) (2003) 151–155.
[113] M. Nabil, et al., An experimental study on the thermal conductivity and dynamic viscosity of TiO_2–SiO_2 nanofluids in water: ethylene glycol mixture, Int. Commun. Heat Mass Transf. 86 (2017) 181–189.
[114] K.A. Hamid, et al., Experimental investigation of nanoparticle mixture ratios on TiO_2–SiO_2 nanofluids heat transfer performance under turbulent flow, Int. J. Heat Mass Transf. 118 (2018) 617–627.
[115] F. Selimefendigil, H.F. Öztop, Analysis and predictive modeling of nanofluid-jet impingement cooling of an isothermal surface under the influence of a rotating cylinder, Int. J. Heat Mass Transf. 121 (2018) 233–245.
[116] A. Alirezaie, et al., Price-performance evaluation of thermal conductivity enhancement of nanofluids with different particle sizes, Appl. Therm. Eng. 128 (2018) 373–380.

[117] S. Wciślik, A simple economic and heat transfer analysis of the nanoparticles use, Chem. Pap. 71 (12) (2017) 2395–2401.

[118] M. Kheradmandfard, et al., Ultra-fast, highly efficient and green synthesis of bioactive forsterite nanopowder via microwave irradiation, Mater. Sci. Eng: C. 92 (2018) 236–244.

[119] N.C. Maji, H.P. Krishna, J. Chakraborty, Low-cost and high-throughput synthesis of copper nanopowder for nanofluid applications, Chem. Eng. J. 353 (2018) 34–45.

[120] A. Vodopyanov, et al., High rate production of nanopowders by the evaporation–condensation method using gyrotron radiation, in: EPJ Web of Conferences, 2017. EDP Sciences.

Index

Note: Page numbers followed by "*f*" and "*t*" refer to figures and tables, respectively.

A

B

C

D

E

O

U

V

W

X

Z

Printed in the United States
by Baker & Taylor Publisher Services

NANOFLUIDS AND MASS TRANSFER

Explores mass transfer processes in the presence of nanofluids, offers fundamental information of nanofluids and mass transfer modelling and industrialization of nanofluids

In recent decades, efficiency enhancement of refineries and chemical plants has become a focus of research and development groups. Use of nanofluids in absorption, regeneration, liquid-liquid extraction, and membrane processes can lead to mass transfer and heat-transfer enhancement in processes, which results in increased efficiency in all these processes. ***Nanofluids and Mass Transfer*** introduces the role of nanofluids in improving mass transfer phenomena and expressing their characteristics and properties.

The book also covers the theory and modelling procedures in detail and finally illustrates various applications of nanofluids in mass transfer enhancement in various processes such as absorption, regeneration, liquid-liquid extraction and membrane processes, and how nanofluids can increase mass transfer in processes.

Key Features

- Introduces specifications of nanofluids and mechanisms of mass transfer enhancement by nanofluids in various mass transfer processes
- Discusses mass transfer enhancement in various mass transfer processes such as, absorption, regeneration, liquid-liquid extraction and membrane processes
- Offers mass transfer modelling and flow behaviour of nanofluids
- Provides industrialization and scale up challenges of nanofluids utilization

Editors

Mohammad Reza Rahimpour, Ph.D., professor in Chemical Engineering at Shiraz University, Iran
Mohammad Amin Makarem, Ph.D., research associate, Shiraz University, Iran
Mohammad Reza Kiani research associate, University of Tehran, Iran
Mohammad Amin Sedghamiz, Ph.D., research associate, Shiraz University, Iran

Technology and Engineering /
Chemical and Biochemical

ELSEVIER elsevier.com/books-and-journals